aplia™ SUCCEED in your course with Aplia!

Take your understanding of biology to the next level with **Aplia for Biology**, an interactive online tool that complements the text. Aplia helps you learn and understand key concepts through focused assignments, exceptional text/art integration, and immediate feedback.

Active reinforcement for increased success: Exercises in Aplia can be repeated in alternate versions to help you learn the material. Aplia's immediate feedback helps you increase your confidence and understanding of complex processes.

Click on any one of the following three crosses to proceed. There is no right or wrong choice, but you will be responsible for analyzing the inheritance pattern of a particular trait based on observations of the parents you choose and the pups that they produce.

Option 1:
• German shepherd X basset hound

Option 2:
• **Saluki X basset hound**

Option 3:
• dachshund X Boston terrier

Saluki   **Basset hound**

The following photograph shows two offspring from your chosen cross. These offspring are representative of all offspring produced in this cross.

Based on these results, which of the following traits is recessive?
- ○ Short, bent legs of the bassett
- ○ Long, straight legs of the saluki

Interactive figures for better understanding: These dynamic figures help you focus on sequential processes one step at a time without losing the context of the entire process.

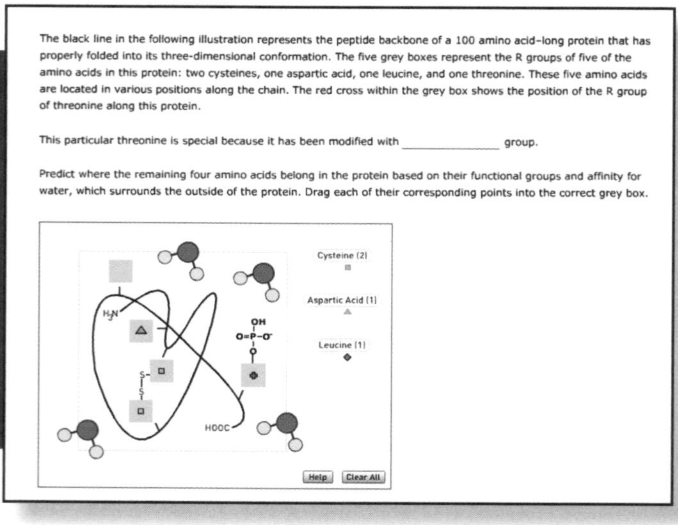

The black line in the following illustration represents the peptide backbone of a 100 amino acid–long protein that has properly folded into its three-dimensional conformation. The five grey boxes represent the R groups of five of the amino acids in this protein: two cysteines, one aspartic acid, one leucine, and one threonine. These five amino acids are located in various positions along the chain. The red cross within the grey box shows the position of the R group of threonine along this protein.

This particular threonine is special because it has been modified with _____ group.

Predict where the remaining four amino acids belong in the protein based on their functional groups and affinity for water, which surrounds the outside of the protein. Drag each of their corresponding points into the correct grey box.

Cysteine (2)

Aspartic Acid (1)

Leucine (1)

Help   Clear All

Get actively involved in the process of learning and doing biology

Interactive animations and videos within problems reinforce key concepts, tying your homework to the textbook and helping you visualize dynamic biological processes.

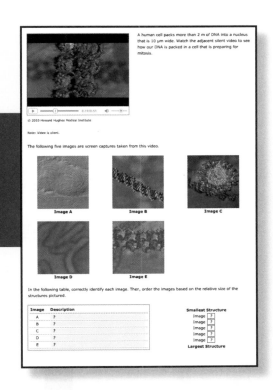

A human cell packs more than 2 m of DNA into a nucleus that is 10 μm wide. Watch the adjacent silent video to see how our DNA is packed in a cell that is preparing for mitosis.

© 2010 Howard Hughes Medical Institute

Note: Video is silent.

The following five images are screen captures taken from this video.

Image A Image B Image C

Image D Image E

In the following table, correctly identify each image. Then, order the images based on the relative size of the structures pictured.

Image	Description		Smallest Structure	
A	?		Image	?
B	?		Image	?
C	?		Image	?
D	?		Image	?
E	?		Image	?
			Largest Structure	

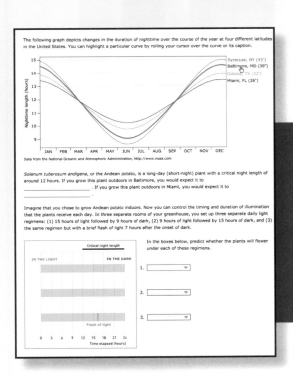

The following graph depicts changes in the duration of nighttime over the course of the year at four different latitudes in the United States. You can highlight a particular curve by rolling your cursor over the curve or its caption.

Syracuse, NY (43°)
Baltimore, MD (39°)
Odessa, TX (32°)
Miami, FL (26°)

Data from the National Oceanic and Atmospheric Administration, http://www.noaa.com

Solanum tuberosum andigena, or the Andean potato, is a long-day (short-night) plant with a critical night length of around 12 hours. If you grow this plant outdoors in Baltimore, you would expect it to _____. If you grow this plant outdoors in Miami, you would expect it to _____.

Imagine that you chose to grow Andean potato indoors. Now you can control the timing and duration of illumination that the plants receive each day. In three separate rooms of your greenhouse, you set up three separate daily light regimens: (1) 15 hours of light followed by 9 hours of dark, (2) 9 hours of light followed by 15 hours of dark, and (3) the same regimen but with a brief flash of light 7 hours after the onset of dark.

In the boxes below, predict whether the plants will flower under each of these regimens.

1.
2.
3.

Diverse and engaging problems include data analysis, experimental and observational approaches to research, real-world applications, and global problems to increase your critical thinking, analytic, and problem-solving skills.

BIOLOGY
THE DYNAMIC SCIENCE
VOLUME 1

BIOLOGY

THE DYNAMIC SCIENCE THIRD EDITION
VOLUME 1

Russell Hertz McMillan

BROOKS/COLE
CENGAGE Learning

Australia • Brazil • Japan • Korea • Mexico • Singapore • Spain • United Kingdom • United States

BROOKS/COLE
CENGAGE Learning

Biology: The Dynamic Science, **Third Edition, Volume 1**

Peter J. Russell, Paul E. Hertz, Beverly McMillan

Publisher: Yolanda Cossio

Developmental Editors: Shelley Parlante, Jake Warde, Suzannah Alexander

Assistant Editor: Alexis Glubka

Editorial Assistants: Lauren Crosby, Sean Cronin

Media Editor: Lauren Oliveira

Brand Manager: Nicole Hamm

Market Development Manager: Tom Ziolkowski

Content Project Manager: Hal Humphrey

Art Director: John Walker

Manufacturing Planner: Karen Hunt

Rights Acquisitions Specialist: Tom McDonough

Production Service: Dan Fitzgerald, Graphic World Inc.

Photo Researchers: Josh Garvin, Janice Yi, Q2A/Bill Smith

Text Researcher: Isabel Saraiva

Copy Editor: Graphic World Inc.

Illustrators: Dragonfly Media Group, Steve McEntee, Graphic World Inc.

Text and Cover Designer: Jeanne Calabrese

Cover Image: Klaus Nigge/National Geographic Stock

Compositor: Graphic World Inc.

About the Cover: The Philippine eagle is among the largest three eagles in the world. It is a rare species, little known and hardly ever photographed.

This eagle is endemic to the Philippines—a country whose population is ever-growing while its forest areas have come down to a mere 5 percent of what originally covered the whole country. As the Philippine eagle can survive only in the rainforest, the situation for this species is most dramatic.

Klaus Nigge has been to the Philippines three times, where he visited people who are working hard to rescue this eagle. Joint efforts enabled him to find a pair of this species in one of the last remaining forest areas of Mindanao, where he spent several weeks up in the treetops to record how the eagles raised their young.

For product information and technology assistance, contact us at **Cengage Learning Customer & Sales Support, 1-800-354-9706.**

For permission to use material from this text or product, submit all requests online at **www.cengage.com/permissions.** Further permissions questions can be e-mailed to **permissionrequest@cengage.com.**

Volume 1:

ISBN-13: 978-1-133-59204-4

ISBN-10: 1-133-59204-X

Brooks/Cole
20 Davis Drive
Belmont, CA 94002-3098
USA

Cengage Learning is a leading provider of customized learning solutions with office locations around the globe, including Singapore, the United Kingdom, Australia, Mexico, Brazil, and Japan. Locate your local office at **www.cengage.com/global.**

Cengage Learning products are represented in Canada by Nelson Education, Ltd.

To learn more about Brooks/Cole visit **www.cengage.com/brookscole**

Purchase any of our products at your local college store or at our preferred online store **www.CengageBrain.com.**

Printed in Canada
1 2 3 4 5 6 7 16 15 14 13 12

Brief Contents

Peter J. Russell received a B.Sc. in Biology from the University of Sussex, England, in 1968 and a Ph.D. in Genetics from Cornell University in 1972. He has been a member of the Biology faculty of Reed College since 1972 and is currently a professor of biology, emeritus. Peter taught a section of the introductory biology course, a genetics course, and a research literature course on molecular virology. In 1987 he received the Burlington Northern Faculty Achievement Award from Reed College in recognition of his excellence in teaching. Since 1986, he has been the author of a successful genetics textbook; current editions are *iGenetics: A Molecular Approach, iGenetics: A Mendelian Approach,* and *Essential iGenetics.* Peter's research was in the area of molecular genetics, with a specific interest in characterizing the role of host genes in the replication of the RNA genome of a pathogenic plant virus, and the expression of the genes of the virus; yeast was used as the model host. His research has been funded by agencies including the National Institutes of Health, the National Science Foundation, the American Cancer Society, the Department of Defense, the Medical Research Foundation of Oregon, and the Murdoch Foundation. He has published his research results in a variety of journals, including *Genetics, Journal of Bacteriology, Molecular and General Genetics, Nucleic Acids Research, Plasmid,* and *Molecular and Cellular Biology.* Peter has a long history of encouraging faculty research involving undergraduates, including cofounding the biology division of the Council on Undergraduate Research in 1985. He was Principal Investigator/Program Director of a National Science Foundation Award for the Integration of Research and Education (NSF–AIRE) to Reed College, 1998 to 2002.

Aaron Kinard

Paul E. Hertz was born and raised in New York City. He received a B.S. in Biology from Stanford University in 1972, an A.M. in Biology from Harvard University in 1973, and a Ph.D. in Biology from Harvard University in 1977. While completing field research for the doctorate, he served on the Biology faculty of the University of Puerto Rico at Rio Piedras. After spending two years as an Isaac Walton Killam Postdoctoral Fellow at Dalhousie University, Paul accepted a teaching position at Barnard College, where he has taught since 1979. He was named Ann Whitney Olin Professor of Biology in 2000, and he received The Barnard Award for Excellence in Teaching in 2007. In addition to serving on numerous college committees, Paul chaired Barnard's Biology Department for eight years and served as Acting Provost and Dean of the Faculty from 2011 to 2012. He is the founding Program Director of the Hughes Science Pipeline Project at Barnard, an undergraduate curriculum and research program that has been funded continuously by the Howard Hughes Medical Institute since 1992. The Pipeline Project includes the Intercollegiate Partnership, a program for local community college students that facilitates their transfer to four-year colleges and universities. He teaches one semester of the introductory sequence for Biology majors and pre-professional students, lecture and laboratory courses in vertebrate zoology and ecology, and a year-long seminar that introduces first-year students to scientific research. Paul is an animal physiological ecologist with a specific research interest in the thermal biology of lizards. He has conducted fieldwork in the West Indies since the mid-1970s, most recently focusing on the lizards of Cuba. His work has been funded by the NSF, and he has published his research in such prestigious journals as *The American Naturalist, Ecology, Nature, Oecologia,* and *Proceedings of the Royal Society.* In 2010, he and his colleagues at three other universities received funding from NSF for a project designed to detect the effects of global climate warming on the biology of *Anolis* lizards in Puerto Rico.

Courtesy of Beverly McMillan

Beverly McMillan has been a science writer for more than 25 years. She holds undergraduate and graduate degrees from the University of California, Berkeley, and is coauthor of a college text in human biology, now in its tenth edition. She has also written or coauthored numerous trade books on scientific subjects and has worked extensively in educational and commercial publishing, including eight years in editorial management positions in the college divisions of Random House and McGraw-Hill.

Welcome to the third edition of *Biology: The Dynamic Science.* The book's title reflects the speed with which our knowledge of biology is growing. Although biologists have made enormous progress in solving the riddles posed by the living world, every discovery raises new questions and provides new opportunities for further research. As in the prior two editions, we have encapsulated the dynamic nature of biology in the third edition by explaining biological concepts—and the data from which they are derived—in the historical context of each discovery and by describing what we know now and what new discoveries will be likely to advance the field in the future.

Building on a strong foundation . . .

The first two editions of this book provided students with the tools they need to learn fundamental biological concepts, processes, and facts. More important, they enabled students to think like scientists. Our approach encourages students to think about biological questions and hypotheses through clear examples of hypothesis development, observational and experimental tests of hypotheses, and the conclusions that scientists draw from their data. The many instructors and students who have used the book have generously provided valuable feedback about the elements that enhanced student learning. We have also received comments from expert reviewers. As a result of these inputs, every chapter has been revised and updated, and some units have been reorganized. In addition, the third edition includes new or modified illustrations and photos as well as some new features.

Emphasizing the big picture . . .

In this textbook, we have applied our collective experience as teachers, researchers, and writers to create a readable and understandable introduction that provides a foundation for students who choose to enroll in more advanced biology courses in the future. We provide straightforward explanations of fundamental concepts presented, where appropriate, from the evolutionary perspective that binds the biological sciences together. Recognizing that students in an introductory biology course face a potentially daunting amount of material, we strive to provide an appropriate balance between facts and concepts, taking great care to provide clear explanations while maintaining the narrative flow. In this way students not only see the big picture, but they understand how we achieved our present knowledge. Having watched our students struggle to navigate the many arcane details of college-level introductory biology, we constantly remind ourselves and each other to "include fewer facts, provide better explanations, and maintain the narrative flow," thereby enabling students to see the big picture. Clarity of presentation, thoughtful organization, a logical and seamless flow of topics within chapters, and carefully designed illustrations are key to our approach.

Focusing on research to help students engage the living world as scientists . . .

A primary goal of this book is to sharpen and sustain students' curiosity about biology, rather than dulling it with a mountain of disconnected facts. We can help students develop the mental habits of scientists and a fascination with the living world by conveying our passion for biological research. We want to excite students not only with *what biologists know* about the living world but also with *how they know it* and *what they still need to learn.* In doing so, we can encourage some students to accept the challenge and become biologists themselves, posing and answering important new questions through their own innovative research. For students who pursue other careers, we hope that they will leave their introductory—and perhaps only—biology course armed with intellectual skills that will enable them to evaluate future discoveries with a critical eye.

In this book, we introduce students to a biologist's "ways of knowing." Research biologists constantly integrate new observations, hypotheses, questions, experiments, and insights with existing knowledge and ideas. To help students engage the world as biologists do, we must not simply introduce them to the current state of knowledge. We must also foster an appreciation of the historical context within which those ideas developed, and identify the future directions that biological research is likely to take.

To achieve these goals, our explanations are rooted in the research that established the basic facts and principles of biology. Thus, a substantial proportion of each chapter focuses on studies that define the state of biological knowledge today. When describing research, we first identify the hypothesis or question that inspired the work and then relate it to the broader topic under discussion. Our research-oriented theme teaches students, through example, how to ask scientific questions and pose hypotheses, two key elements of the scientific process.

Because advances in science occur against a background of research, we also give students a feeling for how biologists of the past formulated basic knowledge in the field. By fostering an appreciation of such discoveries, given the information and theories available to scientists in their own time, we can help students understand the successes and limitations of what we consider cutting edge today. This historical perspective also encourages students to view biology as a dynamic intellectual enterprise, not just a collection of facts and generalities to be memorized.

We have endeavored to make the science of biology come alive by describing how biologists formulate hypotheses and evaluate them using hard-won data; how data sometimes tell only part of a story; and how the results of studies often end up posing more questions than they answer. Although students might prefer simply to learn the "right" answer to a question, they must be encouraged to embrace "the unknown," those gaps in knowledge that create opportunities for further research. An appreciation of what biologists do *not* yet know will draw more students into the field. And by defining *why* scientists do not understand interesting phenomena, we encourage students to think critically about possible solutions and to follow paths dictated by their own curiosity. We hope that this approach will encourage students to make biology a part of their daily lives by having informal discussions and debates about new scientific discoveries.

Presenting the story line of the research process . . .

In preparing this book, we developed several special features to help students broaden their understanding of the material presented and of the research process itself. A Visual Tour of these features and more begins on page xiii.

- The chapter openers, entitled *Why It Matters . . .* , are engaging, short vignettes designed to capture students' imaginations and whet their appetites for the topic that the chapter addresses. In many cases, this feature tells the story of how a researcher or researchers arrived at a key insight or how biological research solved a major societal problem, explained a fundamental process, or elucidated a phenomenon. The *Why It Matters . . .* also provides a brief summary of the contents of the chapter.
- To complement this historical or practical perspective, each chapter closes with a brief essay entitled *Unanswered Questions,* prepared by an expert or experts in the field. These essays identify important unresolved issues relating to the chapter topic and describe cutting-edge research that will advance our knowledge in the future.
- Each chapter includes a short, boxed essay entitled *Insights from the Molecular Revolution,* which describes how molecular tools allow scientists to answer questions that they could not have posed even 30 years ago. Most *Insights* focus on a single study and include sufficient detail for its content to stand alone.
- Many chapters are further supplemented with one or more short, boxed essays involving three different aspects of research. *Focus on Basic Research* essays describe how research has provided understanding of basic biological principles. *Focus on Applied Research* essays describe research designed to solve practical problems in the world, such as those relating to health or the environment. *Focus on Model Research Organisms* essays introduce model research organisms—such

as *Escherichia coli, Drosophila, Arabidopsis, Caenorhabditis,* the mouse, and *Anolis*—and explain why they are used as subjects for in-depth analysis.

Three types of specially designed *research figures* provide more detailed information about how biologists formulate and test specific hypotheses by gathering and interpreting data. The research figures are listed on the endpapers at the back of the book.

- *Experimental Research* figures describe specific studies in which researchers used both experimental and control treatments—either in the laboratory or in the field—to test hypotheses or answer research questions by manipulating the system they studied.
- *Observational Research* figures describe specific studies in which biologists have tested hypotheses by comparing systems under varying natural circumstances.
- *Research Method* figures provide examples of important techniques, such as the scientific method, cloning a gene, DNA microarray analysis, plant cell culture, producing monoclonal antibodies, radiometric dating, and cladistic analysis. Each *Research Method* figure leads a student through the purpose of the technique and protocol and describes how scientists interpret the data it generates.

Integrating effective, high-quality visuals into the narrative . . .

Today's students are accustomed to receiving ideas and information visually, making the illustrations and photographs in a textbook important. Our illustration program provides an exceptionally clear supplement to the narrative in a style that is consistent throughout the book. Graphs and anatomical drawings are annotated with interpretative explanations that lead students, step by step, through the major points they convey.

For the second edition, we undertook a rigorous review of all the art in the text. The publishing team identified the key elements of effective illustrations. In focus groups and surveys, instructors helped us identify the "Key Visual Learning Figures" covering concepts or processes that demand premier visual learning support. Each of these figures was critiqued by our Art Advisory Board to ensure its usability and accuracy. For the third edition, we again evaluated each illustration and photograph carefully and made appropriate changes to improve their use as teaching tools. New illustrations for the edition were created in the same style as existing ones.

For the third edition, important figures were developed as *Closer Look* figures; a Summary and a concluding *Think Like a Scientist* question are designed to enhance student learning. Many *Closer Look* figures involve key biological processes, such as meiosis, transcription, muscle contraction, the cohesion-tension mechanism of water transport in plants, ecological interactions between predators and prey, and the haplodiploidy genetic system in social insects.

Organizing chapters around important concepts . . .

As authors and college teachers, we understand how easily students can get lost within a chapter. When students request advice about how to read a chapter and learn the material in it, we usually suggest that, after reading each section, they pause and quiz themselves on the material they have just encountered. After completing all of the sections in a chapter, they should quiz themselves again, even more rigorously, on the individual sections and, most important, on how the concepts developed in the different sections fit together. Accordingly, we have adopted a structure for each chapter to help students review concepts as they learn them.

- The organization within chapters presents material in digestible sections, building on students' knowledge and understanding as they acquire it. Each major section covers one broad topic. Each subsection, titled with a declarative sentence that summarizes the main idea of its content, explores a narrower range of material.
- Whenever possible, we include the derivation of unfamiliar terms so that students will see connections between words that share etymological roots. Mastery of the technical language of biology will allow students to discuss ideas and processes precisely. At the same time, we have minimized the use of unnecessary jargon.
- *Study Break* questions follow every major section. These questions encourage students to pause at the end of a section and review what they have learned before going on to the next topic within the chapter. Short answers to these questions appear in an appendix.

Encouraging active learning, critical thinking, and self-assessment of learning outcomes . . .

The third edition of *Biology: The Dynamic Science* includes a new active learning feature, *Think Like a Scientist,* which is designed to help students think analytically and critically about research presented in the chapter. *Think Like a Scientist* questions appear at the ends of *Experimental Research* figures, *Observational Research* figures, *Closer Look* figures, *Insights from the Molecular Revolution* boxes, and *Unanswered Questions.*

The new edition also includes *Think Outside the Book,* an active learning feature introduced in the second edition. *Think Outside the Book* activities have been designed to encourage students to explore biology directly or through electronic resources. Students may engage in these activities either individually or in small groups.

Supplementary materials at the end of each chapter help students review the material they have learned, assess their understanding, and think analytically as they apply the principles developed in the chapter to novel situations. Many end-of-chapter questions also serve as good starting points for class discussions or out-of-class assignments.

- *Review Key Concepts* provides a summary of important ideas developed in the chapter, referencing specific figures and tables in the chapter. These *Reviews* are no substitute for reading the chapter, but students may use them as a valuable outline of the material, filling in the details on their own.
- *Understand & Apply* includes five types of end-of-chapter questions and problems that focus on the chapter's factual content while encouraging students to apply what they have learned: (1) *Test Your Knowledge* is a set of 10 questions (with answers in an appendix) that focus on factual material; (2) *Discuss the Concepts* involves open-ended questions that emphasize key ideas, the interpretation of data, and practical applications of the material; (3) *Design an Experiment* questions help students hone their critical thinking skills by asking them to test hypotheses that relate to the chapter's main topic; (4) *Interpret the Data* questions help students develop analytical and quantitative skills by asking them to interpret graphical or tabular results of experimental or observational research experiments for which the hypotheses and methods of analysis are presented; and (5) *Apply Evolutionary Thinking* asks students to answer a question in relation to the principles of evolutionary biology.

Helping students master key concepts throughout the course . . .

Teachers know that student effort is an important determinant of student success. Unfortunately, most teachers lack the time to develop novel learning tools for every concept—or even every chapter—in an introductory biology textbook. To help address this problem, we are pleased to offer **Aplia for Biology,** an automatically graded homework management system tailored to this edition. For students, Aplia provides a structure within which they can expand their efforts, master key concepts throughout the course, and increase their success. For faculty, Aplia can help transform teaching and raise productivity by requiring more—and more consistent—effort from students without increasing faculty workloads substantially. By providing students with continuous exposure to key concepts and their applications throughout the course, Aplia allows faculty to do what they do best—respond to questions, lead discussions, and challenge the students.

We hope you agree that we have developed a clear, fresh, and well-integrated introduction to biology as it is understood by researchers today. Just as important, we hope that our efforts will excite students about the research process and the biological discoveries it generates.

The enhancements we have made in the third edition of *Biology: The Dynamic Science* reflect our commitment to provide a text that introduces students to new developments in biology while fostering active learning and critical thinking. As a part of this effort, we have added *Closer Look* figures that integrate a major concept into a highlighted visual presentation. The key concept is stated briefly at the top, shown in detail through one or more illustrations, and summarized at the bottom. A *Think Like a Scientist* question invites students to apply the figure concept(s) to a related problem or issue. We have also incorporated *Think Like a Scientist* questions into *Insights from the Molecular Revolution* and *Unanswered Questions,* as well as into *Experimental Research* and *Observational Research* figures.

We have also made important changes in coverage to follow recent scientific advances. A new Chapter 19, Genomes and Proteomes, introduces methods of genomics and proteomics along with examples of new discoveries and insights. In addition, we now devote two chapters to plant diversity, discussing seedless plants in Chapter 28 and seed plants in Chapter 29. Finally, we've consolidated our treatment of animal behavior into a single Chapter 56 (Animal Behavior), which integrates various approaches to this subfield of biology. Beyond these major organizational changes, we have made numerous improvements to update and clarify scientific information and to engage students as interested readers and active learners, as well as responsive scientific thinkers. The following sections highlight some of the new content and organizational changes in this edition.

Unit One: Molecules and Cells

To make molecular and cellular processes easier to grasp, this unit incorporates explanatory material into many more illustrations. For example, in Chapter 3 (Biological Molecules: The Carbon Compounds of Life), Table 3.1 now presents more information on the roles of functional groups of organic molecules and a new Figure 3.3 clarifies the concept of stereoisomers. In Chapter 5 (The Cell: An Overview), we have combined the diagrams of animal and plant cells in Figure 5.9 and labeled functions of the organelles. In Chapter 8 (Harvesting Chemical Energy: Cellular Respiration), a new overview diagram of glycolysis (Figure 8.7) helps students understand basic concepts. Chapter 10 (Cell Division and Mitosis) features a new discussion and illustration of the tight pairing of chromatids (sister chromatid cohesion) during mitosis.

New references to molecular aspects of evolution have been integrated into Unit One chapters to emphasize evolution as the theme unifying the subfields of the biological sciences. For example, in Chapter 5 (The Cell: An Overview), the discussion of the mitochondrial matrix now highlights how equivalent structures in bacteria led scientists to propose and develop the endosymbiotic theory. In Chapter 6 (Membranes and Transport), we point out that the close similarity of bilayer membranes in all cells—prokaryotic and eukaryotic—is evidence that the basic structure of membranes evolved during the earliest stages of life on Earth, and has been conserved ever since. In Chapter 9 (Photosynthesis), a new section, *Evolution of Photosynthesis and Cellular Respiration,* summarizes the evolutionary development of these processes.

Unit Two: Genetics

Chapter 11 (Meiosis: The Cellular Basis of Sexual Reproduction) includes fuller descriptions of homologous chromosomes and sex chromosomes. Extensive revision and expansion of Chapter 16 (Regulation of Gene Expression) provides more thorough coverage of gene regulation and the operon model. We now introduce the role of DNA-binding proteins in prokaryotic as well as eukaryotic gene regulation and include more detail on the activation of regulatory molecules. We have also added detail on combinatorial gene regulation, along with a figure showing a specific example, and have added new information and an illustration of how growth factors and growth-inhibiting factors affect cell division. Chapter 17 (Bacterial and Viral Genetics) includes new information on how horizontal gene transfer contributes to genome evolution in prokaryotes, and evidence of its possible contribution to eukaryotic genome evolution.

A new Chapter 19 (Genomes and Proteomes) focuses on the methods of genomics and the information it generates. This chapter describes how genome sequences are determined and annotated, how genes in genomes are identified and characterized, and how studies have generated new information on the evolution of genes and of genomes. It also includes examples of how genomics has become a source of new discoveries in many fields, including human physiology and evolutionary biology.

Unit Three: Evolutionary Biology

In Chapter 21 (Microevolution: Genetic Changes within Populations), *Observational Research* Figure 21.11 now shows more clearly how opposing forces of directional selection produce stabilizing selection. Plant speciation by alloploidy and polyploidy is shown in parallel illustrations in one figure (22.16), allowing easy comparison. In Chapter 23 (Paleobiology and Macroevolution), Figure 23.15 clarifies our understanding of the rise and fall of plant lineages through evolutionary time. Chapter 24 (Systematics and Phylogenetics: Revealing the Tree of Life) includes a new example of how systematists construct phylogenetic trees with genetic distance data and includes a clarified discussion of statistical methods used to construct phylogenetic trees. Reworked phylogenetic trees throughout Chapter 24 are now fully consistent in presentation.

Unit Four: Biodiversity

In Chapter 26 (Prokaryotes: Bacteria and Archaea), a new *Insights from the Molecular Revolution* describes how changes in gene expression in the bacterium that causes gingivitis help govern its transition from a free-living state to a biofilm. A revised and expanded section discusses the five subgroups of proteobacteria. In Chapter 27 (Protists), we've added the nucleariids to the Opisthokont group, along with evidence that they may be more closely related to fungi than to animals.

Plant diversity is now covered in two chapters. Chapter 28 (Seedless Plants) describes trends in land plant evolution and the characteristics of bryophytes and seedless vascular plants, and Chapter 29 (Seed Plants) focuses on adaptations and distinguishing features of gymnosperms and flowering plants. Chapter 28 also includes a new *Unanswered Questions* essay, and Chapter 29 presents a new *Insights from the Molecular Revolution* feature on plant genome evolution. Chapter 30 (Fungi) presents an updated discussion of the evolution of multicellular animals and fungi from different opisthokont ancestors. Changes to Chapter 31 (Animal Phylogeny, Acoelomates, and Protostomes) include color-coding of anatomical illustrations of invertebrates to distinguish structures arising from endoderm, mesoderm, and ectoderm, and a new Table 31.1 providing a phylogenetic overview of the phyla presented in the chapter. In Chapter 32 (Deuterostomes: Vertebrates and Their Closest Relatives), a new *Insights from the Molecular Revolution* feature describes a study of the evolutionary gains and losses of genes that code for olfactory receptor proteins in various clades of mammals. Figure 32.38, which shows timelines for the species of hominins, has been updated with recently discovered fossils. The discussion of human evolution includes recent genomic studies of the relationship between Neanderthals and modern humans.

Unit Five: Plant Structure and Function

Chapter 33 (The Plant Body) features clearer illustrations of plant growth. Chapter 34 (Transport in Plants) has more focused discussions and illustrations of water movements in roots and the physiology of stomatal function. A new *Unanswered Questions* explores research in plant metabolomics. Chapter 36 (Reproduction and Development in Flowering Plants) includes refined diagrams of floral whorls and self-incompatibility, an updated *Insights from the Molecular Revolution* on trichome development, and an updated *Experimental Research* figure on studies of floral organ identity genes. Chapter 37 (Plant Signals and Responses to the Environment) begins with a new *Why It Matters* essay presenting the diverse adaptations of creosote bush *(Larrea tridentata)* to environmental challenges such as extended drought. The chapter also has been reorganized, with the discussion of signal transduction pathways and second messenger systems now included in the introduction to plant hormones. New art illustrates current thinking on different signal transduction mechanisms in plant cells.

Unit Six: Animal Structure and Function

In Chapter 43 (Muscles, Bones, and Body Movements), a new *Insights from the Molecular Revolution* presents experiments on exercise training in racehorses. We have updated and clarified the discussion of immunity in Chapter 45 (Defenses against Disease) and added new material on how microbial pathogens are detected and how pathogens may sometimes escape recognition by the immune system. In Chapter 47 (Animal Nutrition), we have added detail on absorption in the small intestine. Chapter 48 (Regulating the Internal Environment) has expanded coverage of mammalian kidney function and the role of countercurrent heat exchanges in maintaining body temperature. In Chapter 50 (Animal Development), we have revised and expanded Section 50.5, The Cellular Basis of Development, including new information on apoptosis during development and on molecular mechanisms of induction.

Unit Seven: Ecology and Behavior

In Chapter 51 (Ecology and the Biosphere), improved illustrations clarify the effects of latitudinal and seasonal variations in incoming solar radiation. In Chapter 52 (Population Ecology), we have updated Figures 52.22 and 52.23 on human population growth. In Chapter 53 (Population Interactions and Community Ecology), we have improved Figure 53.22 showing the food web. We have also added informative labels to Figure 53.25, which shows the effects of storms on corals. Chapter 55 (Biodiversity and Conservation Biology) features a new Figure 55.16 illustrating the species–area relationship. A unified Chapter 56 (Animal Behavior) concludes the text, integrating the discussions of genetic and experiential bases of animal behavior, the neurophysiological and endocrinological control of specific behaviors, and the ecology and evolution of several broad categories of animal behavior.

Develop a deep understanding of the core concepts in biology and build a strong foundation for future courses.

BIOLOGY
THE DYNAMIC SCIENCE THIRD EDITION
Russell Hertz McMillan

Welcome to **Biology: The Dynamic Science, Third Edition,** by Peter J. Russell, Paul E. Hertz, and Beverly McMillan. The authors convey their passion for biology as they guide you to an understanding of what scientists know about the living world, how they know it, and what they still need to learn. The pages that follow highlight a few of the many ways that they have made this book a great learning tool for you. You'll also find information about dynamic online resources, as well as print materials that will help you master key concepts and succeed in the course.

Aplia for Biology, an interactive online tool that complements the text and helps you learn and understand key concepts through focused assignments, an engaging variety of problem types, exceptional text/art integration, and immediate feedback.

A BIG PICTURE FOCUS

Straightforward explanations of fundamental concepts bind the biological sciences together and enable you to see the big picture. Easy-to-use learning tools point out the topics covered in each chapter, show why they are important, and help you learn the material.

22

Two closely related bird species, purple martins *(Progne subis)* and tree swallows *(Tachycineta bicolor)* perching together on a branch in Crane Creek, Ohio.

STUDY OUTLINE

22.1 What Is a Species?
22.2 Maintaining Reproductive Isolation
22.3 The Geography of Speciation
22.4 Genetic Mechanisms of Speciation

Speciation

FIGURE 22.1 Birds of paradise. A male Count Raggi's bird of paradise *(Paradisaea raggiana)* tries to attract the attention of a female (not pictured) with his showy plumage and conspicuous display. There are 43 known bird of paradise species, 35 of them found only on the island of New Guinea.

Why it matters . . . In 1927, nearly 100 years after Darwin boarded the *Beagle,* a young German naturalist named Ernst Mayr embarked on his own journey, to the highlands of New Guinea. He was searching for rare "birds of paradise" **(Figure 22.1).** These birds were known in Europe only through their ornate and colorful feathers, which were used to decorate ladies' hats. On his trek through the remote Arfak Mountains, Mayr identified 137 bird species (including many birds of paradise) based on differences in their size, plumage, color, and other external characteristics.

To Mayr's surprise, the native Papuans—who were untrained in the ways of Western science, but who hunted these birds for food and feathers—had their own names for 136 of the 137 species he had identified. The close match between the two lists confirmed Mayr's belief that the *species* is a fundamental level of organization in nature. Each species has a unique combination of genes underlying its distinctive appearance and habits. Thus, people who observe them closely—whether indigenous hunters or Western scientists—can often distinguish one species from another.

Mayr also discovered some remarkable patterns in the geographical distributions of the bird species in New Guinea. For example, each mountain range he explored was home to some species that lived nowhere else. Closely related species often lived on different mountaintops, separated by deep valleys of unsuitable habitat. In 1942, Mayr published the book *Systematics and the Origin of Species,* in which he described the role of geography in the evolution of new species; the book quickly became a cornerstone of the modern synthesis (which was outlined in Section 20.3).

480

◀ **Study Outline** provides an overview of main chapter topics and key concepts. Each section breaks the material into a manageable amount of information, so you can develop understanding as you acquire knowledge.

◀ **Why It Matters** sections at the beginning of each chapter capture the excitement of biology and help you understand why the topic is important and how the material that follows fits into the big picture.

▶ **Study Break** sections encourage you to pause and think about the material you have just encountered before moving to the next section.

STUDY BREAK 22.1

1. How do the morphological, biological, and phylogenetic species concepts differ?
2. What is clinal variation?

Your study of biology focuses not only on **what** scientists know about the living world but also **how** they know it. Use these unique features to learn how scientists ask scientific questions, pose hypotheses, and test them.

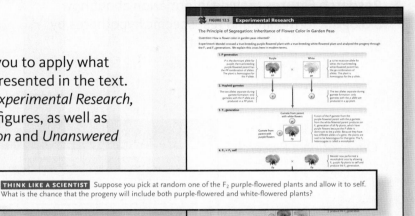

NEW!

"Think Like a Scientist" questions ask you to apply what you have learned beyond the material presented in the text. These questions are incorporated into *Experimental Research*, *Observational Research*, and *Closer Look* figures, as well as into *Insights from the Molecular Revolution* and *Unanswered Questions* boxes.

THINK LIKE A SCIENTIST Suppose you pick at random one of the F_2 purple-flowered plants and allow it to self. What is the chance that the progeny will include both purple-flowered and white-flowered plants?

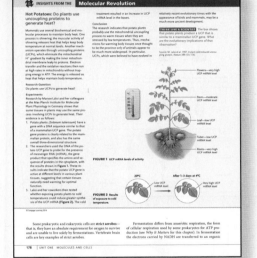

Insights from the Molecular Revolution essays highlight how molecular technologies allow researchers to answer questions that they could not even pose 20 or 30 years ago.

Unanswered Questions explore important unresolved issues identified by experts in the field and describe cutting-edge research that will advance our knowledge in the future.

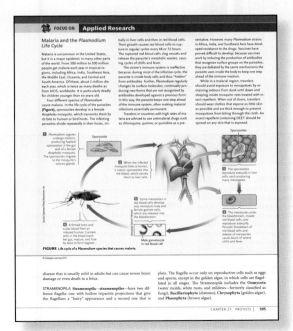

Focus on Research boxes present research topics in more depth.

Focus on Applied Research describes how scientific research has solved everyday problems.

Focus on Basic Research describes seminal research that provided insight into an important problem.

Focus on Model Research Organisms explains why researchers use certain organisms as research subjects.

ENGAGE LIKE A SCIENTIST

Be Active. Get involved in the process of learning and doing biology.

▶ **Research Figures** provide information about how biologists formulate and test specific hypotheses by gathering and interpreting data.

Research Method

Observational Research

Experimental Research

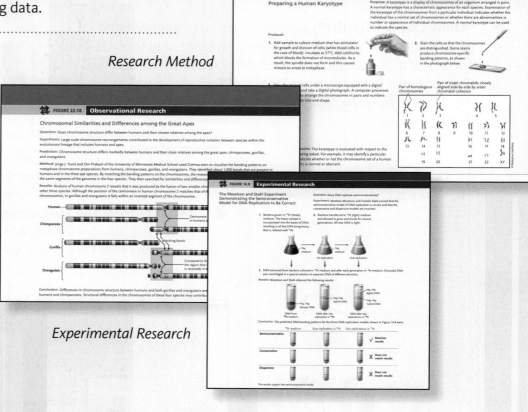

▶ **Interpret the Data** exercises, drawn from published biological research, help you build your skills in analyzing figures and reading graphs or tables.

Interpret the Data

Peter and Rosemary Grant of Princeton University have studied the ecology and evolution of finches on the Galápagos Islands since the early 1970s. They have shown that finches with large bills (as measured by bill depth; see **Figure**) can eat both small seeds and large seeds, but finches with small bills can only eat small seeds. In 1977, a severe drought on the island of Daphne Major reduced seed production by plants. After the birds consumed whatever small seeds they found, only large seeds were still available. The resulting food shortage killed a majority of the medium ground finches (*Geospiza fortis*) on Daphne Major; their population plummeted from 751 in 1976 to just 90 in 1978. The Grants' research also documented a change in the distributions of bill depths in the birds from 1976 to 1978, as illustrated in the graphs to the right. In light of what you now know about the relationship between bill size and food size for these birds, interpret the change illustrated in the graph. What type of natural selection does this example illustrate?

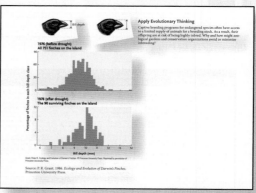

▶ **Think Outside the Book** activities help you think analytically and critically as you explore the biological world, either on your own or as part of a team.

THINK OUTSIDE THE BOOK

Earlier in the chapter we mentioned the fact that cloned animals may have many genes whose expression is abnormal compared to gene expression in a noncloned animal. Individually or collaboratively, outline the steps you would take experimentally to determine, on a genome-wide scale, if genes are abnormally expressed in a cloned mammal. Your answer should include how the experiment reveals both qualitative and quantitative differences in gene expression.

THINK OUTSIDE THE BOOK

Access the web page for the Tree of Life project at http://www.tolweb.org/tree/. Select a group of animals or plants that is of interest to you, and study the structure of its phylogenetic tree. How many major clades does it include? On the basis of what shared derived characters are those clades defined?

Spectacular illustrations—developed with great care—help you visualize biological processes, relationships, and structures.

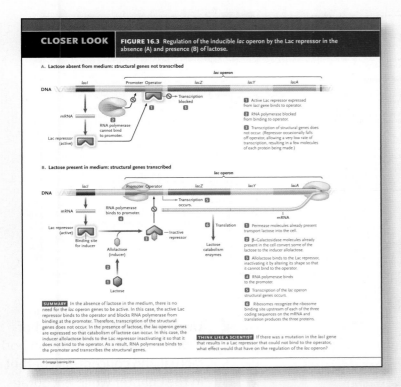

CLOSER LOOK | **FIGURE 16.3** Regulation of the inducible *lac* operon by the Lac repressor in the absence (A) and presence (B) of lactose.

NEW!

◀ **"Closer Look" figures** help you gain a better understanding of a major concept through a visual presentation, usually a detailed, multistep diagram. The figures end with a *Summary* and a *Think Like a Scientist* question.

▶ **Illustrations of complex biological processes** are annotated with *numbered step-by-step explanations* that lead you through all the major points. Orientation diagrams are inset on figures and help you identify the specific biological process being depicted and where the process takes place.

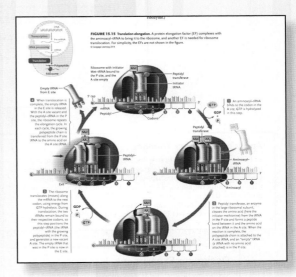

FIGURE 15.15 Translation elongation. A protein elongation factor (EF) complexes with the aminoacyl-tRNA to bring it to the ribosome, and another EF is needed for ribosome translocation. For simplicity, the EFs are not shown in the figure.

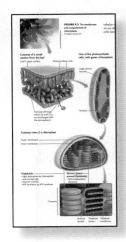

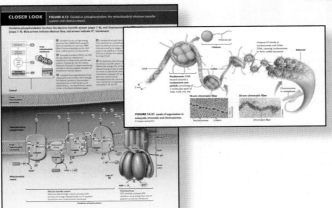

◀ **Macro-to-Micro views** help you visualize the levels of organization of biological structures and how systems function as a whole.

REVIEW

End-of-chapter material encourages you to review, assess your understanding, think analytically, and apply what you have learned to novel situations.

▶ **Review Key Concepts** provides an outline summary of important ideas developed in the chapter and references the chapter's figures and tables.

REVIEW KEY CONCEPTS

To access the course materials and companion resources for this text, please visit www.cengagebrain.com.

18.1 DNA Cloning

- Producing multiple copies of genes by cloning is a common first step for studying the structure and function of genes, or for manipulating genes. Cloning involves cutting genomic DNA and a cloning vector with the same restriction enzyme, joining the fragments to produce recombinant plasmids, and introducing those plasmids into a living cell such as a bacterium, where replication of the plasmid takes place (Figures 18.1–18.3).
- A clone containing a gene of interest may be identified among a population of clones by using DNA hybridization with a labeled nucleic acid probe (Figure 18.5).
- A genomic library is a collection of clones that contains a copy of every DNA sequence in the genome. A cDNA (complementary DNA) library is the entire collection of cloned cDNAs made from the mRNAs isolated from a cell. A cDNA library contains only sequences from the genes that are active in the cell when the mRNAs are isolated.

18.2 Applications of DNA Technologies

- Recombinant DNA and PCR techniques are used in DNA molecular testing for human genetic disease mutations. One approach exploits restriction site differences between normal and mutant alleles of a gene that create restriction fragment length polymorphisms (RFLPs) which are detectable by DNA hybridization with a labeled nucleic acid probe (Figures 18.8 and 18.9).
- Human DNA fingerprints are produced from a number of loci in the genome characterized by short, tandemly repeated sequences that vary in number in all individuals (except identical twins). To produce a DNA fingerprint, the PCR is used to amplify the region of genomic DNA for each locus, and the lengths of the PCR products indicate the alleles an individual has for the repeated sequences at each locus. DNA fingerprints are widely used to establish paternity, ancestry, or criminal guilt (Figure 18.10).
- Genetic engineering is the introduction of new genes or genetic information to alter the genetic makeup of humans, other animals, plants, and microorganisms such as bacteria and yeast. Genetic engineering primarily aims to correct hereditary defects, improve domestic animals and crop plants, and provide proteins for medicine, research, and other applications (Figures 18.11–18.13 and 18.15).
- Genetic engineering has enormous potential for research and applications in medicine, agriculture, and industry. Potential risks include unintended damage to living organisms or to the environment.

Animation: How Dolly was created

Animation: DNA fingerprinting

Animation: Transferring genes into plants

UNDERSTAND & APPLY

Test Your Knowledge

1. A complementary DNA library (cDNA) and a genomic library are similar in that both:
   a. use bacteria to make eukaryotic proteins.
   b. provide information on whether genes are active.
   c. contain all of the DNA of an organism cut into pieces.
   d. clone mRNA.
   e. depend on cloning in a living cell to produce multiple copies of the DNA of interest.

2. Why do the cDNA libraries produced from two different cell types in the human body often contain different cDNAs?
   a. Because different expression vectors must be used to insert cDNAs into different cell types.
   b. Because different cell types contain different numbers of chromosomes.
   c. Because different cell types contain different genomic DNA sequences.
   d. Because different genes are transcribed in different cell types.
   e. Because different cell types contain different restriction enzymes.

3. The point at which a restriction enzyme cuts DNA is determined by
   a. the sequence of nucleotides.
   b. the length of the DNA molecule.
   c. whether it is closer to the 5′ end or 3′ end of the DNA molecule.
   d. the number of copies of the DNA molecule in a bacterial cell.
   e. the location of a start codon in a gene.

4. Restriction endonucleases, ligases, plasmids, *E. coli*, electrophoretic gels, and a bacterial gene resistant to an antibiotic are all required for:
   a. dideoxyribonucleotide analysis.
   b. PCR.
   c. DNA cloning.
   d. DNA fingerprinting.
   e. DNA sequencing.

5. After a polymerase chain reaction (PCR), agarose gel electrophoresis is often used to:
   a. amplify the DNA.
   b. convert cDNA into genomic DNA.
   c. convert cDNA into messenger RNA.
   d. verify that the desired DNA sequence has been amplified.
   e. synthesize primer DNA molecules.

6. Restriction fragment length polymorphisms (RFLPs):
   a. are produced by reaction with restriction endonucleases and are detected by Southern blot analysis.
   b. are of the same length for mutant and normal β-globin alleles.
   c. determine the sequence of bases in a DNA fragment.
   d. have in their middle short fragments of DNA that are palindromic.
   e. are used as vectors.

7. DNA fingerprinting, which is often used in forensics, paternity testing, and for establishing ancestry:
   a. compares one stretch of the same DNA between two or more people.
   b. measures different lengths of DNA from many repeating noncoding regions.
   c. requires the largest DNA lengths to run for the greatest distance on a gel.
   d. requires amplification after the gels are run.
   e. can easily differentiate DNA between identical twins.

8. Which of the following is needed both in using bacteria to produce proteins and in genetic engineering of human cells?
   a. DNA fingerprinting based on microsatellite sequences
   b. insertion of a transgene into an expression vector
   c. restriction fragment length polymorphism (RFLP)
   d. screening of a cDNA library by DNA hybridization
   e. antibiotic resistance

9. Dolly, a sheep, was an example of reproductive (germ-line) cloning. Required to perform this process was:
   a. implantation of uterine cells from one strain into the mammary gland of another.
   b. fusion of the mammary cell from one strain with an enucleated egg of another strain.
   c. fusion of an egg from one strain with the egg of a different strain.
   d. fusion of an embryonic diploid cell with an adult haploid cell.
   e. fusion of two nucleated mammary cells from two different strains.

10. Which of the following is *not* true of somatic cell gene therapy?
    a. White blood cells can be used.
    b. Somatic cells are cultured, and the desired DNA is introduced into them.
    c. Cells with the introduced DNA are returned to the body.
    d. The technique is still very experimental.
    e. The inserted genes are passed to the offspring.

◀ **Understand & Apply** end-of-chapter questions focus on both factual and conceptual content in the chapter while encouraging you to apply what you have learned.

Apply Evolutionary Thinking

In PCR, researchers use a heat-stable form of DNA polymerase from microorganisms that are able to grow in extremely high temperatures. Given what you learned in Chapter 3 about protein folding, and in Chapter 4 about the effects of temperature on enzymes, would you predict that the amino acids of heat-stable DNA polymerase enzymes would have evolved so they can form stronger chemical attractions with each other, or weaker chemical attractions? Explain your answer.

◀ **Apply Evolutionary Thinking** asks you to interpret a relevant topic in relation to the principles of evolutionary thinking.

Design an Experiment

Suppose a biotechnology company has developed a GMO, a transgenic plant that expresses *Bt* toxin. The company sells its seeds to a farmer under the condition that the farmer may plant the seed, but not collect seed from the plants that grow and use it to produce crops in the subsequent season. The seeds are expensive, and the farmer buys seeds from the company only once. How could the company show experimentally that the farmer has violated the agreement and is using seeds collected from the first crop to grow the next crop?

◀ **Design an Experiment** challenges your understanding of the chapter and helps you deepen your understanding of the scientific method as you consider how to develop and test hypotheses about a situation that relates to a main chapter topic.

Discuss the Concepts

1. What should juries know to be able to interpret DNA evidence? Why might juries sometimes ignore DNA evidence?

2. A forensic scientist obtained a small DNA sample from a crime scene. In order to examine the sample, he increased its quantity by cycling the sample through the polymerase chain reaction. He estimated that there were 50,000 copies of the DNA in his original sample. Derive a simple formula and calculate the number of copies he will have after 15 cycles of the PCR.

◀ **Discuss the Concepts** enables you to participate in discussions on key questions to build your knowledge and learn from others.

Aplia for Biology
Get involved with biology content using Aplia!
Aplia's focused assignments and active learning opportunities (including randomized questions, exceptional text/art integration, and immediate feedback) get you involved with biology and help you think like a scientist. For more information, visit **www.aplia.com/biology.**

Interactive problems and figures help you visualize dynamic biological processes and integrate concepts, art, media, and homework practice.

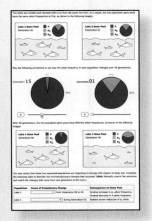

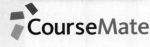

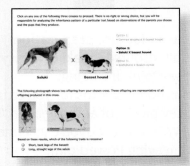

CourseMate
Make the grade with CourseMate.
The more you study, the better the results. Make the most of your study time by accessing everything you need to succeed in one place. **Biology CourseMate** includes an interactive eBook, which allows you to take notes, highlight, bookmark, search the text, and use in-context glossary definitions, as well as practice quizzes, animations, flashcards, videos, and more. Visit **www.cengage.com/coursemate** to learn more.

Quick Prep for Biology
Hit the ground running with *Quick Prep!*
Quick Prep for Biology covers the prerequisite math, chemistry, vocabulary, and study skills you need to succeed in introductory biology. Printed Access Card ISBN: 978-1-133-17690-9.

Study Guide
Study more efficiently and improve your performance on exams!
This invaluable guide includes key terms, labeling exercises, self-quizzes, review questions, and critical-thinking exercises to help you develop a better understanding of the concepts of the course. ISBN: 978-1-133-95464-4.

INSTRUCTOR RESOURCES

Save time with these innovative resources!

Aplia for Biology

Get your students engaged and motivated with Aplia for Biology.
Help your students learn key concepts via Aplia's focused assignments and active learning opportunities that include randomized, automatically graded questions, exceptional text/art integration, and immediate feedback. Aplia has a full course management system that can be used independently or in conjunction with other course management systems such as Blackboard and WebCT. Visit **www.aplia.com/biology**.

PowerLecture with JoinIn™

All the resources you need for faster, easier lecture preparation
PowerLecture's single DVD consolidates all resources, making it easy for you to create customized lectures with Microsoft® PowerPoint, including:
- **All art and photos from the book** plus bonus photos for you to use. All the art is available with removable labels, and key pieces of art have been "stepped" so they can be presented one segment at a time. With one click you can launch animations and videos related to the chapter, without leaving PowerPoint.
- **New 3-D animations** that help bring important concepts to life with topics such as DNA Translation, Mitosis, and Photosynthesis.
- **Brooks/Cole Video Library** (featuring BBC Motion Gallery Video Clips) that contains more than 40 high-quality videos you can use alongside the text.

ISBN: 978-1-133-95466-8.

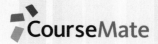

WebTutor™ on WebCT™ and Blackboard

Jump-start your course with customizable, rich, text-specific content!
Whether you want to Web-enable your class or put an entire course online, WebTutor delivers. WebTutor offers a wide array of resources, including media assets, quizzing, web links, exercises, flashcards, and more. WebTutor is available for both WebCT and Blackboard and in versions that include an eBook and/or Virtual Biology Labs.
Visit: **webtutor.cengage.com** to learn more.

CourseMate

Engaging. Affordable. Trackable.
Cengage Learning's Biology **CourseMate** brings course concepts to life with interactive learning, study, and exam preparation tools that support the textbook. Watch student comprehension soar as your class works with the book and the text-specific website. **CourseMate** goes beyond the book to deliver what you need.
Visit **www.cengage.com/login** for more information.

For a complete list of supplements available with Russell, Hertz, and McMillan's *Biology: The Dynamic Science,* **Third Edition**, visit **www.cengage.com/biology**.

Revising a text from edition to edition is an exciting and rewarding project, and the helpful assistance of many people enabled us to accomplish the task in a timely manner.

Yolanda Cossio provided the essential support and continual encouragement to bring the project to fruition.

Our Developmental Editors, Shelley Parlante and Jake Warde, served as pilots for the generation of this book. They provided very helpful guidance as the manuscript matured. They compiled, interpreted, and sometimes deconstructed reviewer comments; their analyses and insights have helped us tighten the narrative and maintain a steady course. Suzannah Alexander helped to organize our art development program and kept it on track; she also offered helpful suggestions on many chapters.

We are grateful to Alexis Glubka for coordinating the print supplements and our Editorial Assistant Lauren Crosby for managing all our reviewer information.

We offer many thanks to Lauren Oliveira and Shelley Ryan, who supervised our partnership with our technology authors and media advisory board. Their collective efforts allowed us to create a set of tools that support students in learning and instructors in teaching.

We thank the Aplia for Biology team, Qinzi Ji, Andy Marinkovich, and John Kyte for building a learning solution that is truly integrated with our text.

We appreciate the help of the production staff led by Hal Humphrey at Cengage and Dan Fitzgerald at Graphic World. We thank our Creative Director Rob Hugel and Art Director John Walker.

The outstanding art program is the result of the collaborative talent, hard work, and dedication of a select group of people. The meticulous styling and planning of the program are credited to Steve McEntee and to Dragonfly Media Group (DMG), led by Mike Demaray. The DMG group created hundreds of complex, vibrant art pieces. Steve's role was crucial in overseeing the development and consistency of the art program; he was the original designer for the *Experimental Research, Observational Research,* and *Research Methods* figures. We would like to thank Cecie Starr for the use of selected art pieces.

We also wish to acknowledge Tom Ziolkowski, our Marketing Manager, whose expertise ensured that all of you would know about this new book.

Peter Russell thanks Joel Benington of St. Bonaventure University for his extensive reviewing and contributions to the new Chapter 19, Genomes and Proteomes; Stephen Arch of Reed College for his expert input, valuable discussions, and advice during the revision of the Unit Six chapters on Animal Structure and Function; Arthur Glasfeld of Reed College for his expert input during the revision of the chemistry chapters;

and Daniel J. Fairbanks of Utah Valley University for providing additional material on evolution for my chapters. Paul Hertz thanks Hilary Callahan, John Glendinning, and Brian Morton of Barnard College for their generous advice on many phases of this project; Eric Dinerstein of the World Wildlife Fund for his contributions to the discussion of Conservation Biology; and Joel Benington for his expertise on genomic issues in systematics. Paul especially thanks Jamie Rauchman for extraordinary patience and endless support as this book was written (and rewritten, and rewritten again) as well as his thousands of past students at Barnard College, who have taught him at least as much as he has taught them. Beverly McMillan once again thanks John A. Musick, Acuff Professor Emeritus at the College of William and Mary—and an award-winning teacher and mentor to at least two generations of college students—for patient and thoughtful discussions about effective ways to present the often complex subject matter of biological science.

We would also like to thank our advisors and contributors:

Supplements Authors

David Asch, *Youngstown State University*
Carolyn Bunde, *Idaho State University*
Albia Duggar, *Miami Dade College*
Frederick B. Essig, *University of South Florida*
Brent Ewers, *University of Wyoming*
Anne Galbraith, *University of Wisconsin–LaCrosse*
Alan Hecht, *Hofstra University*
Kathleen Hecht, *Nassau Community College*
Qinzi Ji, *Instructional Curriculum Specialist*
William Kroll, *Loyola University Chicago–Lake Shore*
Todd Osmundson, *University of California, Berkeley*
Debra Pires, *University of California, Los Angeles*
Elena Pravosudova, *University of Reno, Nevada*
Jeff Roth-Vinson, *Cottage Grove High School*
Mark Sheridan, *North Dakota State University*
Gary Shin, *California State University, Long Beach*
Michael Silva, *El Paso Community College*
Michelle Taliaferro, *Auburn University, Montgomery*
Jeffrey Taylor, *State University of New York, Canton*
Catherine Anne Ueckert, *Northern Arizona University*
Jyoti Wagle, *Houston Community College, Central College*
Alexander Wait, *Missouri State University*

Media and Aplia for Biology Reviewers and Class Testers

Thomas Abbott, *University of Connecticut*
David Asch, *Youngstown State University*
John Bell, *Brigham Young University*
Anne Bergey, *Truman State University*
Gerald Bergtrom, *University of Wisconsin–Milwaukee*

Scott Bowling, *Auburn University*

Joi Braxton-Sanders, *Northwest Vista College*

Carolyn Bunde, *Idaho State University*

Jung H. Choi, *Georgia Institute of Technology*

Tim W. Christensen, *East Carolina University*

Patricia J. S. Colberg, *University of Wyoming*

Robin Cooper, *University of Kentucky*

Karen Curto, *University of Pittsburgh*

Joe Demasi, *Massachusetts College of Pharmacy and Health Science*

Nicholas Downey, *University of Wisconsin–LaCrosse*

Albia Dugger, *Miami-Dade College*

Natalie Dussourd, *Illinois State University*

Lisa Elfring, *University of Arizona*

Bert Ely, *University of South Carolina*

Kathleen Engelmann, *University of Bridgeport*

Helene Engler, *Science Writer*

Monika Espinasa, *State University of New York at Ulster*

Michael Ferrari, *University of Missouri–Kansas City*

David Fitch, *New York University*

Paul Fitzgerald, *Northern Virginia Community College*

Steven Francoeur, *Eastern Michigan University*

Daria Hekmat-Scafe, *Stanford University*

Jutta Heller, *Loyola University Chicago–Lake Shore*

Ed Himelblau, *California Polytechnic State University–San Luis Obispo*

Justin Hoffman, *McNeese State University*

Kelly Howe, *University of New Mexico*

Carrie Hughes, *San Jacinto College (Central Campus)*

Ashok Jain, *Albany State University*

Susan Jorstad, *University of Arizona*

Judy Kaufman, *Monroe Community College*

David Kiewlich, *Research Biologist*

Christopher Kirkhoff, *McNeese State University*

Richard Knapp, *University of Houston*

William Kroll, *Loyola University Chicago–Lake Shore*

Nathan Lents, *John Jay College*

Janet Loxterman, *Idaho State University*

Susan McRae, *East Carolina University*

Brad Mehrtens, *University of Illinois at Urbana-Champaign*

Jennifer Metzler, *Ball State University*

Bruce Mobarry, *University of Idaho*

Jennifer Moon, *The University of Texas at Austin*

Robert Osuna, *State University of New York at Albany*

Matt Palmer, *Columbia University*

Roger Persell, *Hunter College*

Michael Reagan, *College of Saint Benedict and Saint John's University*

Ann Rushing, *Baylor University*

Jeanne Serb, *Iowa State University*

Leah Sheridan, *University of Northern Colorado*

Mark Sheridan, *North Dakota State University*

Nancy N. Shontz, *Grand Valley State University*

Michael Silva, *El Paso Community College*

Julia Snyder, *Syracuse University*

Linda Stabler, *University of Central Oklahoma*

Mark Staves, *Grand Valley State University*

Eric Strauss, *University of Wisconsin–LaCrosse*

Mark Sturtevant, *Oakland University*

Mark Sugalski, *Southern Polytechnic State University*

David Tam, *University of North Texas*

Salvatore Tavormina, *Austin Community College*

Rebecca Thomas, *Montgomery College*

David H. Townson, *University of New Hampshire*

David Vleck, *Iowa State University*

Neal Voelz, *St. Cloud State University*

Camille Wagner, *San Jacinto College (Central Campus)*

Miryam Wahrman, *William Paterson University*

Alexander Wait, *Missouri State University*

Suzanne Wakim, *Butte Community College*

Johanna Weiss, *Northern Virginia Community College*

Lisa Williams, *Northern Virginia Community College*

Marilyn Yoder, *University of Missouri–Kansas City*

Martin Zahn, *Thomas Nelson Community College*

Reviewers and Contributors

Thomas D. Abbott, *University of Connecticut*

Lori Adams, *University of Iowa*

Heather Addy, *The University of Calgary*

Adrienne Alaie-Petrillo, *Hunter College–CUNY*

Richard Allison, *Michigan State University*

Terry Allison, *The University of Texas–Pan American*

Phil Allman, *Gulf Coast University*

Tracey M. Anderson, *University of Minnesota Morris*

Deborah Anderson, *Saint Norbert College*

Robert C. Anderson, *Idaho State University*

Andrew Andres, *University of Nevada, Las Vegas*

Steven M. Aquilani, *Delaware County Community College*

Stephen Arch, *Reed College*

Jonathan W. Armbruster, *Auburn University*

Peter Armstrong, *University of California, Davis*

John N. Aronson, *The University of Arizona*

Joe Arruda, *Pittsburgh State University*

Karl Aufderheide, *Texas A&M University*

Charles Baer, *University of Florida*

Gary I. Baird, *Brigham Young University*

Aimee Bakken, *University of Washington*

Marica Bakovic, *University of Guelph*

Mitchell F. Balish, *Miami University*

Michael Baranski, *Catawba College*

W. Brad Barbazuk, *University of Florida*

Michael Barbour, *University of California, Davis*

Timothy J. Baroni, *State University of New York at Cortland*

Edward M. Barrows, *Georgetown University*

Anton Baudoin, *Virginia Polytechnic Institute and State University*

Penelope H. Bauer, *Colorado State University*

Erwin A. Bautista, *University of California, Davis*

Kevin Beach, *The University of Tampa*

Mike Beach, *Southern Polytechnic State University*

Ruth Beattie, *University of Kentucky*

Robert Beckmann, *North Carolina State University*

Jane Beiswenger, *University of Wyoming*

Asim Bej, *University of Alabama at Birmingham*

Michael C. Bell, *Richland College*

Andrew Bendall, *University of Guelph*

Joel H. Benington, *St. Bonaventure University*

Anne Bergey, *Truman State University*

William L. Bischoff, *The University of Toledo*

Catherine Black, *Idaho State University*

Andrew Blaustein, *Oregon State University*

Anthony H. Bledsoe, *University of Pittsburgh*

Harriette Howard-Lee Block,
Prairie View A&M University

Dennis Bogyo, *Valdosta State University*

David Bohr, *University of Michigan*

Emily Boone, *University of Richmond*

Hessel Bouma III, *Calvin College*

Nancy Boury, *Iowa State University*

Scott Bowling, *Auburn University*

Robert S. Boyd, *Auburn University*

Laurie Bradley,
Hudson Valley Community College

William Bradshaw, *Brigham Young University*

J. D. Brammer, *North Dakota State University*

G. L. Brengelmann, *University of Washington*

Randy Brewton,
University of Tennessee–Knoxville

Bob Brick, *Blinn College–Bryan*

Mirjana Brockett,
Georgia Institute of Technology

William Bromer, *University of Saint Francis*

William Randy Brooks,
Florida Atlantic University–Boca Raton

Mark Browning, *Purdue University*

Gary Brusca, *Humboldt State University*

Alan H. Brush, *University of Connecticut*

Arthur L. Buikema, Jr., *Virginia Polytechnic
Institute and State University*

Carolyn Bunde, *Idaho State University*

E. Robert Burns,
University of Arkansas for Medical Sciences

Ruth Buskirk, *The University of Texas at Austin*

David Byres,
Florida Community College at Jacksonville

Christopher S. Campbell,
The University of Maine

Angelo Capparella, *Illinois State University*

Marcella D. Carabelli,
Broward Community College–North

Jeffrey Carmichael,
University of North Dakota

Bruce Carroll,
North Harris Montgomery Community College

Robert Carroll, *East Carolina University*

Patrick Carter, *Washington State University*

Christine Case, *Skyline College*

Domenic Castignetti,
Loyola University Chicago–Lake Shore

Peter Chen, *College of DuPage*

Jung H. Choi, *Georgia Institute of Technology*

Kent Christensen,
University of Michigan Medical School

James W. Clack, *Indiana University–
Purdue University Indianapolis*

John Cogan, *Ohio State University*

Patricia J. S. Colberg, *University of Wyoming*

Linda T. Collins,
University of Tennessee–Chattanooga

Lewis Coons, *University of Memphis*

Robin Cooper, *University of Kentucky*

Joe Cowles, *Virginia Polytechnic Institute
and State University*

George W. Cox, *San Diego State University*

David Crews, *The University of Texas at Austin*

Paul V. Cupp, Jr., *Eastern Kentucky University*

Karen Curto, *University of Pittsburgh*

Anne M. Cusic,
The University of Alabama at Birmingham

David Dalton, *Reed College*

Frank Damiani, *Monmouth University*

Melody Danley, *University of Kentucky*

Deborah Athas Dardis,
Southeastern Louisiana University

Rebekka Darner, *University of Florida*

Peter J. Davies, *Cornell University*

Fred Delcomyn,
University of Illinois at Urbana-Champaign

Jerome Dempsey,
University of Wisconsin–Madison

Philias Denette,
Delgado Community College–City Park

Nancy G. Dengler, *University of Toronto*

Jonathan J. Dennis, *University of Alberta*

Daniel DerVartanian, *University of Georgia*

Donald Deters, *Bowling Green State University*

Kathryn Dickson,
California State University, Fullerton

Eric Dinerstein, *World Wildlife Fund*

Kevin Dixon,
University of Illinois at Urbana-Champaign

Nick Downey,
University of Wisconsin–LaCrosse

Gordon Patrick Duffie,
Loyola University Chicago–Lake Shore

Charles Duggins, *University of South Carolina*

Carolyn S. Dunn,
University of North Carolina–Wilmington

Kathryn A. Durham,
Luzerne County Community College

Roland R. Dute, *Auburn University*

Melinda Dwinell,
Medical College of Wisconsin

Gerald Eck, *University of Washington*

Gordon Edlin, *University of Hawaii*

William Eickmeier, *Vanderbilt University*

Jamin Eisenbach, *Eastern Michigan University*

Ingeborg Eley,
Hudson Valley Community College

Paul R. Elliott, *Florida State University*

John A. Endler, *University of Exeter*

Kathleen Engelmann,
University of Bridgeport

Helene Engler, *Science Consultant and Lecturer*

Robert B. Erdman,
Florida Gulf Coast University

Jose Luis Ergemy, *Northwest Vista College*

Joseph Esdin,
University of California, Los Angeles

Frederick B. Essig, *University of South Florida*

Brent Ewers, *University of Wyoming*

Daniel J. Fairbanks, *Utah Valley University*

Piotr G. Fajer, *Florida State University*

Richard H. Falk, *University of California, Davis*

Ibrahim Farah, *Jackson State University*

Mark A. Farmer, *University of Georgia*

Jacqueline Fern, *Lane Community College*

Michael B. Ferrari,
University of Missouri–Kansas City

David H. A. Fitch, *New York University*

Daniel P. Fitzsimons,
University of Wisconsin–Madison

Daniel Flisser, *Camden County College*

R. G. Foster, *University of Virginia*

Austin W. Francis Jr.,
Armstrong Atlantic University

Dan Friderici, *Michigan State University*

J. W. Froehlich, *The University of New Mexico*

Anne M. Galbraith,
University of Wisconsin–LaCrosse

Paul Garcia,
Houston Community College–Southwest

E. Eileen Gardner,
William Paterson University

Umadevi Garimella,
University of Central Arkansas

David W. Garton,
Georgia Institute of Technology

John R. Geiser, *Western Michigan University*

Robert P. George, *University of Wyoming*

Stephen George, *Amherst College*

Tim Gerber, *University of Wisconsin–LaCrosse*

John Giannini, *St. Olaf College*

Joseph Glass, *Camden County College*

Florence Gleason,
University of Minnesota Twin Cities

Scott Gleeson, *University of Kentucky*

John Glendinning, *Barnard College*

Elizabeth Godrick, *Boston University*

Judith Goodenough,
University of Massachusetts Amherst

H. Maurice Goodman, *University of
Massachusetts Medical School*

Bruce Grant, *College of William and Mary*

Becky Green-Marroquin,
Los Angeles Valley College

Christopher Gregg,
Louisiana State University

Katharine B. Gregg,
West Virginia Wesleyan College

John Griffin, *College of William and Mary*

Erich Grotewold, *Ohio State University*

Samuel Hammer, *Boston University*

Aslam Hassan, *University of Illinois at Urbana-Champaign*

Albert Herrera,
University of Southern California

Wilford M. Hess, *Brigham Young University*

Martinez J. Hewlett,
The University of Arizona

R. James Hickey, *Miami University*

Christopher Higgins,
Tarleton State University

Phyllis C. Hirsch, *East Los Angeles College*

Carl Hoagstrom, *Ohio Northern University*

Stanton F. Hoegerman,
College of William and Mary

Kelly Hogan, *University of North Carolina*

Ronald W. Hoham, *Colgate University*

Jill A. Holliday, *University of Florida*

Margaret Hollyday, *Bryn Mawr College*

John E. Hoover, *Millersville University*

Howard Hosick, *Washington State University*

William Irby, *Georgia Southern University*

John Ivy, *Texas A&M University*

Alice Jacklet, *University at Albany, State University of New York*

John D. Jackson,
North Hennepin Community College

Jennifer Jeffery,
Wharton County Junior College

Eric Jellen, *Brigham Young University*

Rick Jellen, *Brigham Young University*

John Jenkin, *Blinn College–Bryan*

Dianne Jennings,
Virginia Commonwealth University

Leonard R. Johnson,
The University of Tennessee College of Medicine

Walter Judd, *University of Florida*

Prem S. Kahlon, *Tennessee State University*

Thomas C. Kane, *University of Cincinnati*

Peter Kareiva, *University of Washington*

Gordon I. Kaye, *Albany Medical College*

Greg Keller,
Eastern New Mexico University–Roswell

Stephen Kelso, *University of Illinois at Chicago*

Bryce Kendrick, *University of Waterloo*

Bretton Kent, *University of Maryland*

Jack L. Keyes, *Linfield College Portland Campus*

David Kiewlich,
Science Consultant and Research Biologist

Scott L. Kight, *Montclair State University*

John Kimball, *Tufts University*

Hillar Klandorf, *West Virginia University*

Michael Klymkowsky,
University of Colorado at Boulder

Loren Knapp, *University of South Carolina*

Richard Knapp, *University of Houston*

David Kooyman, *Brigham Young University*

Olga Ruiz Kopp, *Utah Valley State University*

Ana Koshy,
Houston Community College–Northwest

Donna Koslowsky, *Michigan State University*

Kari Beth Krieger,
University of Wisconsin–Green Bay

David T. Krohne, *Wabash College*

William Kroll,
Loyola University Chicago–Lake Shore

Josepha Kurdziel, *University of Michigan*

Allen Kurta, *Eastern Michigan University*

Howard Kutchai, *University of Virginia*

Paul K. Lago, *The University of Mississippi*

John Lammert, *Gustavus Adolphus College*

William L'Amoreaux,
College of Staten Island–CUNY

Brian Larkins, *The University of Arizona*

William E. Lassiter,
University of North Carolina–Chapel Hill

Shannon Lee,
California State University, Northridge

Lissa Leege, *Georgia Southern University*

Matthew Levy,
Case Western Reserve University

Harvey Liftin,
Broward Community College–Central

Tom Lonergan, *University of New Orleans*

Lynn Mahaffy, *University of Delaware*

Charly Mallery, *University of Miami*

Alan Mann, *University of Pennsylvania*

Paul Manos, *Duke University*

Kathleen Marrs, *Indiana University–Purdue University Indianapolis*

Robert Martinez, *Quinnipiac University*

Patricia Matthews,
Grand Valley State University

Joyce B. Maxwell,
California State University, Northridge

Jeffrey D. May, *Marshall University*

Geri Mayer, *Florida Atlantic University*

Jerry W. McClure, *Miami University*

Andrew G. McCubbin,
Washington State University

Mark McGinley, *Texas Tech University*

Jacqueline S. McLaughlin,
Penn State University–Lehigh Valley

F. M. Anne McNabb, *Virginia Polytechnic Institute and State University*

Mark Meade, *Jacksonville State University*

Bradley Mehrtens,
University of Illinois at Urbana-Champaign

Amee Mehta, *Seminole State University*

Michael Meighan,
University of California, Berkeley

Catherine Merovich, *West Virginia University*

Richard Merritt, *Houston Community College*

Jennifer Metzler, *Ball State University*

Ralph Meyer, *University of Cincinnati*

Melissa Michael,
University of Illinois at Urbana-Champaign

James E. "Jim" Mickle,
North Carolina State University

Hector C. Miranda, Jr.,
Texas Southern University

Jasleen Mishra,
Houston Community College–Southwest

Jeanne M. Mitchell, *Truman State University*

David Mohrman,
University of Minnesota Medical School Duluth

John M. Moore, *Taylor University*

Roderick M. Morgan,
Grand Valley State University

David Morton, *Frostburg State University*

Alexander Motten, *Duke University*

Alan Muchlinski,
California State University, Los Angeles

Michael Muller,
University of Illinois at Chicago

Richard Murphy, *University of Virginia*

Darrel L. Murray,
University of Illinois at Chicago

Allan Nelson, *Tarleton State University*

David H. Nelson, *University of South Alabama*

Jacalyn Newman, *University of Pittsburgh*

David O. Norris, *The University of Colorado*

Bette Nybakken, *Hartnell College*

Victoria Ochoa, *El Paso Community College, Rio Grande Campus*

Tom Oeltmann, *Vanderbilt University*

Bruce F. O'Hara, *University of Kentucky*

Diana Oliveras,
The University of Colorado at Boulder

Alexander E. Olvido,
Virginia State University

Todd W. Osmundson,
University of California, Berkeley

Robert Osuna,
State University of New York, Albany

Karen Otto, *The University of Tampa*

William W. Parson,
University of Washington School of Medicine

James F. Payne, *The University of Memphis*

Craig Peebles, *University of Pittsburgh*

Joe Pelliccia, *Bates College*

Kathryn Perez, *University of Wisconsin–LaCrosse*

Vinnie Peters, *Indiana University–Purdue University Fort Wayne*

Susan Petro, *Ramapo College of New Jersey*

Debra Pires, *University of California, Los Angeles*

Jarmila Pittermann, *University of California, Santa Cruz*

Thomas Pitzer, *Florida International University*

Roberta Pollock, *Occidental College*

Steve Vincent Pollock, *Louisiana State University*

Elena Pravosudova, *University of Nevada, Reno*

Jerry Purcell, *San Antonio College*

Jason M. Rauceo, *John Jay College of Criminal Justice*

Kim Raun, *Wharton County Junior College*

Michael Reagan, *College of Saint Benedict and Saint John's University*

Tara Reed, *University of Wisconsin–Green Bay*

Melissa Murray Reedy, *University of Illinois at Urbana-Champaign*

Lynn Robbins, *Missouri State University*

Carolyn Roberson, *Roane State Community College*

Laurel Roberts, *University of Pittsburgh*

George R. Robinson, *State University of New York, Albany*

Kenneth Robinson, *Purdue University*

Frank A. Romano, *Jacksonville State University*

Michael R. Rose, *University of California, Irvine*

Michael S. Rosenzweig, *Virginia Polytechnic Institute and State University*

Linda S. Ross, *Ohio University*

Ann Rushing, *Baylor University*

Scott D. Russell, *University of Oklahoma*

Linda Sabatino, *Suffolk Community College*

Tyson Sacco, *Cornell University*

Peter Sakaris, *Southern Polytechnic State University*

Frank B. Salisbury, *Utah State University*

Mark F. Sanders, *University of California, Davis*

Stephen G. Saupe, *College of Saint Benedict and Saint John's University*

Andrew Scala, *Dutchess Community College*

John Schiefelbein, *University of Michigan*

Deemah Schirf, *The University of Texas at San Antonio*

Kathryn J. Schneider, *Hudson Valley Community College*

Jurgen Schnermann, *University of Michigan Medical School*

Thomas W. Schoener, *University of California, Davis*

Brian Shea, *Northwestern University*

Mark Sheridan, *North Dakota State University*

Dennis Shevlin, *The College of New Jersey*

Rebecca F. Shipe, *University of California, Los Angeles*

Nancy N. Shontz, *Grand Valley State University*

Richard Showman, *University of South Carolina*

Jennifer L. Siemantel, *Cedar Valley College*

Michael Silva, *El Paso Community College*

Bill Simcik, *Lone Star College–Tomball*

Robert Simons, *University of California, Los Angeles*

Roger Sloboda, *Dartmouth College*

Jerry W. Smith, *St. Petersburg College*

Nancy Solomon, *Miami University*

Christine C. Spencer, *Georgia Institute of Technology*

Bruce Stallsmith, *The University of Alabama in Huntsville*

Richard Stalter, *College of St. Benedict and St. John's University*

Sonja Stampfler, *Kellogg Community College*

Karl Sternberg, *Western New England College*

Pat Steubing, *University of Nevada, Las Vegas*

Karen Steudel, *University of Wisconsin–Madison*

Tom Stidham, *Texas A&M University*

Richard D. Storey, *The Colorado College*

Tara Stoulig, *Southeastern Louisiana University*

Brian Stout, *Northwest Vista College*

Gregory W. Stunz, *Texas A&M University*

Mark T. Sugalski, *Southern Polytechnic State University*

Michael A. Sulzinski, *The University of Scranton*

Marshall Sundberg, *Emporia State University*

David Tam, *University of North Texas*

David Tauck, *Santa Clara University*

Salvatore Tavormina, *Austin Community College*

Jeffrey Taylor, *Slippery Rock University of Pennsylvania*

Franklyn Te, *Miami Dade College*

Roger E. Thibault, *Bowling Green State University*

Ken Thomas, *Northern Essex Community College*

Megan Thomas, *University of Nevada, Las Vegas*

Patrick Thorpe, *Grand Valley State University*

Ian Tizard, *Texas A&M University*

Terry M. Trier, *Grand Valley State University*

Robert Turner, *Western Oregon University*

Joe Vanable, *Purdue University*

William Velhagen, *New York University*

Linda H. Vick, *North Park University*

J. Robert Waaland, *University of Washington*

Alexander Wait, *Missouri State University*

Douglas Walker, *Wharton County Junior College*

James Bruce Walsh, *The University of Arizona*

Fred Wasserman, *Boston University*

R. Douglas Watson, *The University of Alabama at Birmingham*

Chad M. Wayne, *University of Houston*

Cindy Wedig, *The University of Texas–Pan American*

Michael N. Weintraub, *The University of Toledo*

Edward Weiss, *Christopher Newport University*

Mark Weiss, *Wayne State University*

Adrian M. Wenner, *University of California, Santa Barbara*

Sue Simon Westendorf, *Ohio University*

Ward Wheeler, *American Museum of Natural History, Division of Invertebrate Zoology*

Adrienne Williams, *University of California, Irvine*

Elizabeth Willott, *The University of Arizona*

Mary Wise, *Northern Virginia Community College*

Charles R. Wyttenbach, *The University of Kansas*

Robert Yost, *Indiana University–Purdue University Indianapolis*

Yunde Zhao, *University of California, San Diego*

Heping Zhou, *Seton Hall University*

Xinsheng Zhu, *University of Wisconsin–Madison*

Adrienne Zihlman, *University of California, Santa Cruz*

Unanswered Questions Contributors

Chapter 2
Li Li, *Pennsylvania State University, University Park*

Chapter 3
Michael S. Brown and Joseph L. Goldstein,
University of Texas Southwestern Medical School

Chapter 4
Ulrich Müller, *University of California, San Diego*

Chapter 5
Matthew Welch, *University of California, Berkeley*

Chapter 6
Peter Agre, *Johns Hopkins Malaria Research Institute*

Chapter 7
Jeffrey Blaustein, *University of Massachusetts Amherst*

Chapter 8
Gail A. Breen, *University of Texas at Dallas*

Chapter 9
David Kramer, *Washington State University*

Chapter 10
Raymond Deshaies, *California Institute of Technology*

Chapter 11
Monica Colaiácovo, *Harvard Medical School*

Chapter 12
Nicholas Katsanis, *Duke University*

Chapter 13
Michelle Le Beau and Angela Stoddart,
The University of Chicago

Chapter 14
Janis Shampay, *Reed College*

Chapter 15
Harry Noller, *University of California, Santa Cruz*

Chapter 16
Mark A. Kay, *Stanford University School of Medicine*

Chapter 17
Gerald Baron, *Rocky Mountain Laboratories*

Chapter 18
John F. Engelhardt and Tom Lynch, *University of Iowa*

Chapter 19
Larisa H. Cavallari,
University of Illinois at Chicago College of Pharmacy

Chapter 20
Douglas J. Futuyma, *Stony Brook University*

Chapter 21
Mohamed Noor, *Duke University*

Chapter 22
Jerry Coyne, *University of Chicago*

Chapter 23
Elena M. Kramer, *Harvard University*

Chapter 24
Richard Glor, *University of Rochester*

Chapter 25
Andrew Pohorille,
National Aeronautics and Space Administration (NASA)

Chapter 26
Stephen D. Bell and Rachel Y. Samson, *Oxford University*

Chapter 27
Geoff McFadden, *University of Melbourne*

Contents

BIOLOGY

THE DYNAMIC SCIENCE
VOLUME 1

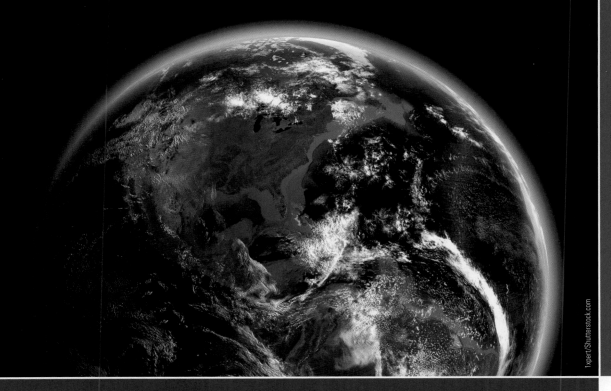

Earth, a planet teeming with life, is seen here in a satellite photograph.

1xpert/Shutterstock.com

Introduction to Biological Concepts and Research

Why it matters . . . Life abounds in almost every nook and cranny on our planet Earth. A lion creeps across an African plain, ready to spring at a zebra. The leaves of a sunflower in Kansas turn slowly through the day, keeping their surfaces fully exposed to rays of sunlight. Fungi and bacteria in the soil of a Canadian forest obtain nutrients by decomposing dead organisms. A child plays in a park in Madrid, laughing happily as his dog chases a tennis ball. In one room of a nearby hospital, a mother hears the first cry of her newborn baby; in another room, an elderly man sighs away his last breath. All over the world, countless organisms are born, live, and die every moment of every day. How did life originate, how does it persist, and how is it changing? Biology, the science of life, provides scientific answers to these questions.

What *is* life? Offhandedly, you might say that although you cannot define it, you know it when you see it. The question has no simple answer, because life has been unfolding for billions of years, ever since nonliving materials assembled into the first organized, living cells. Clearly, any list of criteria for the living state only hints at the meaning of "life." Deeper scientific insight requires a wide-ranging examination of the characteristics of life, which is what this book is all about.

Over the next semester or two, you will encounter examples of how organisms are constructed, how they function, where they live, and what they do. The examples provide evidence in support of concepts that will greatly enhance your appreciation and understanding of the living world, including its fundamental unity and striking diversity. This chapter provides a brief overview of these basic concepts. It also describes some of the ways in which biologists conduct research, the process by which they observe nature, formulate explanations of their observations, and test their ideas.

FIGURE 1.1 Living organisms and inanimate objects. Living organisms, such as this lizard (*Iguana iguana*), have characteristics that are fundamentally different from those of inanimate objects, like the rock on which it is sitting.

Image copyright Mishella, 2010. Used under license from Shutterstock.com

1.1 What Is Life? Characteristics of Living Organisms

Picture a lizard on a rock, slowly turning its head to follow the movements of another lizard nearby **(Figure 1.1).** You know that the lizard is alive and that the rock is not. At the atomic and molecular levels, however, the differences between them blur. Lizards, rocks, and all other matter are composed of atoms and molecules, which behave according to the same physical laws. Nevertheless, living organisms share a set of characteristics that collectively set them apart from nonliving matter.

The differences between a lizard and a rock depend not only on the kinds of atoms and molecules present, but also on their organization and their interactions. Individual organisms are at the middle of a hierarchy that ranges from the atoms and molecules within their bodies to the assemblages of organisms that occupy Earth's environments. Within every individual, certain biological molecules contain instructions for building other molecules, which, in turn, are assembled into complex structures. Living organisms must gather energy and materials from their surroundings to build new biological molecules, grow in size, maintain and repair their parts, and produce offspring. They must also respond to environmental changes by altering their chemistry and activity in ways that allow them to survive. Finally, the structure and function of living organisms often change from one generation to the next.

Life on Earth Exists at Several Levels of Organization, Each with Its Own Emergent Properties

The organization of life extends through several levels of a hierarchy **(Figure 1.2).** Complex biological molecules exist at the lowest level of organization, but by themselves, these molecules are not alive. The properties of life do not appear until they are arranged into cells. A **cell** is an organized chemical system that includes many spe-

Biosphere

All regions of Earth's crust, waters, and atmosphere that sustain life

1xpert/Shutterstock.com

Ecosystem

Group of communities interacting with their shared physical environment

Jamie & Judy Wild/Danita Delimont

Community

Populations of all species that occupy the same area

Ron Sefton/Bruce Coleman USA/Photoshot

Population

Group of individuals of the same kind (that is, the same species) that occupy the same area

Jamie & Judy Wild/Danita Delimont

Multicellular organism

Individual consisting of interdependent cells

Edward Snow/Bruce Coleman/Photoshot

FIGURE 1.2 The hierarchy of life. Each level in the hierarchy of life exhibits emergent properties that do not exist at lower levels. The middle four photos depict a rocky intertidal zone on the coast of Washington State.
© Cengage Learning 2014

Cell

Smallest unit with the capacity to live and reproduce, independently or as part of a multicellular organism

cialized molecules surrounded by a membrane. A cell is the lowest level of biological organization that can survive and reproduce—as long as it has access to a usable energy source, the necessary raw materials, and appropriate environmental conditions. However, a cell is alive only as long as it is organized as a cell; if broken into its component parts, a cell is no longer alive even if the parts themselves are unchanged. Characteristics that depend on the level of organization of matter, but do not exist at lower levels of organization, are called **emergent properties.** Life is thus an emergent property of the organization of matter into cells.

Many single cells, such as bacteria and protozoans, exist as **unicellular organisms.** By contrast, plants and animals are **multicellular organisms.** Their cells live in tightly coordinated groups and are so interdependent that they cannot survive on their own. For example, human cells cannot live by themselves in nature because they must be bathed in body fluids and supported by the activities of other cells. Like individual cells, multicellular organisms have emergent properties that their individual components lack; for example, humans can learn biology.

The next, more inclusive level of organization is the **population,** a group of organisms of the same kind that live together in the same place. The humans who occupy the island of Tahiti and a group of sea urchins living together on the coast of Washington State are examples of populations. Like multicellular organisms, populations have emergent properties that do not exist at lower levels of organization. For example, a population has characteristics such as its birth or death rate—that is, the number of individual organisms who are born or die over a period of time—that do not exist for single cells or individual organisms.

Working our way up the biological hierarchy, all the populations of different organisms that live in the same place form a **community.** The algae, snails, sea urchins, and other

organisms that live along the coast of Washington State, taken together, make up a community. The next higher level, the **ecosystem,** includes the community *and* the nonliving environmental factors with which it interacts. For example, a coastal ecosystem comprises a community of living organisms, as well as rocks, air, seawater, minerals, and sunlight. The highest level, the **biosphere,** encompasses all the ecosystems of Earth's waters, crust, and atmosphere. Communities, ecosystems, and the biosphere also have emergent properties. For example, communities can be described in terms of their *diversity*—the number and types of different populations they contain—and their *stability*—the degree to which the populations within the community remain the same through time.

Living Organisms Contain Chemical Instructions That Govern Their Structure and Function

The most fundamental and important molecule that distinguishes living organisms from nonliving matter is **deoxyribonucleic acid (DNA; Figure 1.3).** DNA is a large, double-stranded, helical molecule that contains instructions for assembling a living organism from simpler molecules. We recognize bacteria, trees, fishes, and humans as different because differences in their DNA produce differences in their appearance and function. (Some nonliving systems, notably certain viruses, also contain DNA, but biologists do not consider viruses to be alive because they cannot reproduce independently of the organisms they infect.)

DNA functions similarly in all living organisms. As you will discover in Chapters 14 and 15, the instructions in DNA are copied into molecules of a related substance, **ribonucleic acid (RNA),** which then directs the synthesis (production) of different protein molecules **(Figure 1.4). Proteins** carry out most of the activities of life, including the synthesis of all other biological

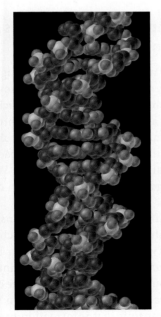

FIGURE 1.3 Deoxyribonucleic acid (DNA). A computer-generated model of DNA illustrates that it is made up of two strands twisted into a double helix.
© Cengage Learning 2014

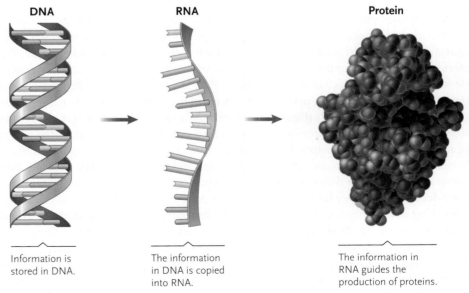

DNA	RNA	Protein
Information is stored in DNA.	The information in DNA is copied into RNA.	The information in RNA guides the production of proteins.

FIGURE 1.4 **The pathway of information flow in living organisms.** Information stored in DNA is copied into RNA, which then directs the construction of protein molecules. The protein shown here is lysozyme.
© Cengage Learning 2014

molecules. This pathway is preserved from generation to generation by the ability of DNA to copy itself so that offspring receive the same basic molecular instructions as their parents.

Living Organisms Engage in Metabolic Activities

Metabolism, described in Chapters 8 and 9, is another key property of living cells and organisms. **Metabolism** describes the ability of a cell or organism to extract energy from its surroundings and use that energy to maintain itself, grow, and reproduce. As a part of metabolism, cells carry out chemical reactions that assemble, alter, and disassemble molecules **(Figure 1.5).** For example, a growing sunflower plant carries out **photosynthesis,** in which the electromagnetic energy in sunlight is absorbed and converted into chemical energy. The cells of the plant store some chemical energy in sugar and starch molecules, and they use the rest to manufacture other biological molecules from simple raw materials obtained from the environment.

Sunflowers concentrate some of their energy reserves in seeds from which more sunflower plants may grow. The chemical energy stored in the seeds also supports other organisms, such as insects, birds, and humans, that eat them. Most organisms, including sunflower plants, tap stored chemical energy through another metabolic process, **cellular respiration.** In cellular respiration complex biological molecules are broken down with oxygen, releasing some of their energy content for cellular activities.

Energy Flows and Matter Cycles through Living Organisms

With few exceptions, energy from sunlight supports life on Earth. Plants and other photosynthetic organisms absorb energy from sunlight and convert it into chemical energy. They use this chemical energy to assemble complex molecules, such as sugars, from simple raw materials, such as water and carbon dioxide. As such, photosynthetic organisms are the **primary producers** of the food on which all other organisms rely **(Figure 1.6).** By contrast, animals are **consumers:** directly or indirectly, they feed on the complex molecules manufactured by plants. For example, zebras tap directly into the molecules of plants when they eat grass, and lions tap into it indirectly when they eat zebras. Certain bacteria and fungi are **decomposers:** they feed on the remains of dead organisms, breaking down complex biological molecules into simpler raw materials, which may then be recycled by the producers.

As you will see in Chapter 54, much of the energy that photosynthetic organisms trap from sunlight *flows* within and between populations, communities, and ecosystems. But because the transfer of energy from one organism to another is not 100% efficient, a portion of that energy is lost as heat. Although some animals can use this form of energy to maintain body tempera-

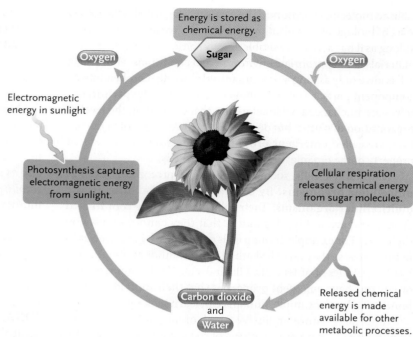

FIGURE 1.5 Metabolic activities. Photosynthesis converts the electromagnetic energy in sunlight into chemical energy, which is stored in sugars and starches built from carbon dioxide and water; oxygen is released as a by-product of the reaction. Cellular respiration uses oxygen to break down sugar molecules, releasing their chemical energy and making it available for other metabolic processes.
© Cengage Learning 2014

ture, it cannot sustain other life processes. By contrast, matter—nutrients such as carbon and nitrogen—*cycles* between living organisms and the nonliving components of the biosphere, to be used again and again (see Figure 1.6).

Living Organisms Compensate for Changes in the External Environment

All objects, whether living or nonliving, respond to changes in the environment; for example, a rock warms up on a sunny day and cools at night. But only living organisms have the capacity to detect environmental changes and *compensate* for them through controlled responses. Diverse and varied *receptors*—molecules or larger structures located on individual cells and body surfaces—can detect changes in external and internal conditions. When stimulated, the receptors trigger reactions that produce a compensating response.

For example, your internal body temperature remains reasonably constant, even though the environment in which you live is usually either cooler or warmer than you are. Your body compensates for these environmental variations and maintains its internal temperature at about 37° Celsius (C). When the environmental temperature drops significantly, receptors in your skin detect the change and transmit that information to your brain. Your brain may send a signal to your muscles, causing you to shiver, thereby releasing heat that keeps your body temperature from dropping below its optimal level. When the environmental temperature rises significantly, glands in your skin secrete sweat, which evaporates, cooling the skin and its underlying blood supply. The cooled blood circulates internally and keeps your body temperature from rising above 37°C. People also compensate behaviorally by dressing warmly on a cold winter day or jumping into a swimming pool in the heat of summer. Keeping your internal temperature within a narrow range is one example of **homeostasis**—a steady internal condition maintained by responses that compensate for changes in the external environment. As described in Units 5 and 6, all organisms have mechanisms that maintain homeostasis in relation to temperature, blood chemistry, and other important factors.

Living Organisms Reproduce and Many Undergo Development

Humans and all other organisms are part of an unbroken chain of life that began at least 3.5 billion years ago. This chain continues today through **reproduction,** the process by which parents produce offspring. Offspring generally resemble their parents because the parents pass copies of their DNA—with all the accompanying instructions for virtually every life process—to their offspring. The transmission of DNA (that is, genetic information) from one generation to the next is called **inheritance.** For example, the eggs produced by storks hatch into little storks, not into pelicans, because they inherited stork DNA, which is different from pelican DNA.

Multicellular organisms also undergo a process of **development,** a series of programmed changes encoded in DNA, through which a fertilized egg divides into many cells that ultimately are transformed into an adult, which is itself capable of reproduction. As an example, consider the development of a moth **(Figure 1.7).** This insect begins its life as a tiny egg that contains all the instructions necessary for its development into an adult moth. Following these instructions, the egg first hatches into a caterpillar, a larval form adapted for feeding and rapid growth. The caterpillar increases in size until internal chemical signals indicate that it is time to spin a cocoon and become a pupa. Inside its cocoon, the pupa undergoes profound developmental changes that remodel its body completely. Some cells die; others multiply and become organized in different patterns. When these transformations are complete, the adult moth emerges from the cocoon. It is equipped with structures and behaviors, quite different from those of the caterpillar, that enable it to reproduce.

KEY

→ Energy transfer

〰 Energy ultimately lost as heat

FIGURE 1.6 **Energy flow and nutrient recycling.** In most ecosystems, energy flows from the sun to producers to consumers to decomposers. On the African savanna, the sun provides energy to grasses (producers); zebras (primary consumers) then feed on the grasses before being eaten by lions (secondary consumers); fungi (decomposers) absorb nutrients and energy from the digestive wastes of animals and from the remains of dead animals and plants. All of the energy that enters an ecosystem is ultimately lost from the system as heat. Nutrients move through the same pathways, but they are conserved and recycled.

© Cengage Learning 2014

The sequential stages through which individuals develop, grow, maintain themselves, and reproduce are known collectively as the **life cycle** of an organism. The moth's life cycle includes egg, larva, pupa, and adult stages. Through reproduction, adult moths continue the cycle by producing the sperm and eggs that unite to form the fertilized egg, which starts the next generation.

Populations of Living Organisms Change from One Generation to the Next

Although offspring generally resemble their parents, individuals with unusual characteristics sometimes suddenly appear in a population. Moreover, the features that distinguish these oddballs are

FIGURE 1.7 Life cycle of an atlas moth (*Attacus atlas*).

A. Egg   **B. Larva**   **C. Pupa**   **D. Adult**

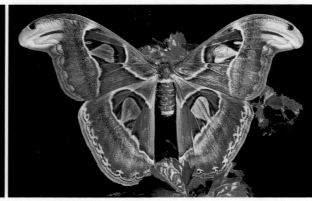

Photographs by Jack de Coningh/Animals Animals

often inherited by their offspring. Our awareness of the inheritance of unusual characteristics has had an enormous impact on human history because it has allowed plant and animal breeders to produce crops and domesticated animals with especially desirable characteristics.

Biologists have observed that similar changes also take place under natural conditions. In other words, populations of all organisms change from one generation to the next, because some individuals experience changes in their DNA and they pass those modified instructions along to their offspring. We introduce this fundamental process, **biological evolution,** in the next section. Although we explore biological evolution in great detail in Unit 3, every chapter in this book—indeed, every idea in biology—references our understanding that all biological systems are the products of evolutionary change.

STUDY BREAK 1.1

1. List the major levels in the hierarchy of life, and identify one emergent property of each level.
2. What do living organisms do with the energy they collect from the external environment?
3. What is a life cycle?

1.2 Biological Evolution

All research in biology—ranging from analyses of the precise structure of biological molecules to energy flow through the biosphere—is undertaken with the knowledge that biological evolution has shaped life on Earth. Our understanding of the evolutionary process reveals several truths about the living world: (1) all populations change through time, (2) all organisms are descended from a common ancestor that lived in the distant past, and (3) evolution has produced the spectacular diversity of life that we see around us. Evolution is the unifying theme that links all the subfields of the biological sciences, and it provides cohesion to our treatment of the many topics discussed in this book.

Darwin and Wallace Explained How Organisms Change through Time

How do evolutionary changes take place? One important mechanism was first explained in the mid-nineteenth century by two British naturalists, Charles Darwin and Alfred Russel Wallace. On

a five-year voyage around the world, Darwin observed many "strange and wondrous" organisms. He also found fossils of species that are now extinct (that is, all members of the species are dead). The extinct forms often resembled living species in some traits but differed in others. Darwin originally believed in special creation—the idea that living organisms were placed on Earth in their present numbers and kinds and have not changed since their creation. But he became convinced that species do not remain constant with the passage of time: instead, they change from one form to another over generations. Wallace came to the same conclusion through his observations of the great variety of plants and animals in the jungles of South America and Southeast Asia.

Darwin also studied the process of evolution through observations and experiments on domesticated animals. Pigeons were among his favorite experimental subjects. Domesticated pigeons exist in a variety of sizes, colors, and shapes, but all of them are descended from the wild rock dove **(Figure 1.8).** Darwin noted that pigeon breeders who wished to promote a certain characteristic, such as elaborately curled tail feathers, selected individuals with the most curl in their feathers as parents for the next generation. By permitting only these birds to mate, the breeders fostered the desired characteristic and gradually eliminated or reduced other traits. The same practice is still used today to increase the frequency of desirable traits in tomatoes, dogs, and other domesticated plants and animals. Darwin called this practice **artificial selection.** He termed the equivalent process that occurs in nature **natural selection.**

In 1858, Darwin and Wallace formally summarized their observations and conclusions explaining biological evolution. (1) Most organisms can produce numerous offspring, but environmental factors limit the number that actually survive and reproduce. (2) Heritable variations allow some individuals to compete more successfully for space, food, and mates. (3) These successful individuals somehow pass the favorable characteristics to their offspring. (4) As a result, the favorable traits become more common in the next generation, and less successful traits become less common. This process of natural selection results in evolutionary change. Today, evolutionary biologists recognize that natural selection is just one of several potent evolutionary processes, as described in Chapter 21.

Over many generations, the evolutionary changes in a population may become extensive enough to produce a population of organisms that is distinct from its ancestors. Nevertheless, parental and descendant species often share many characteristics, allowing researchers to understand their relationships and reconstruct their shared evolutionary history, as described below and in Chapter 24. Starting with the first organized cells, this aspect of evolutionary change has contributed to the diversity of life that exists today.

Darwin and Wallace described evolutionary change largely in terms of how natural selection changes the commonness or rarity of particular variations over time. Their intellectual achievement was remarkable for its time. Although Darwin and Wallace understood the central importance of variability among organisms to the process of evolution, they could not explain how new variations arose or how they were passed to the next generation.

Wild rock dove

FIGURE 1.8 Artificial selection. Using artificial selection, pigeon breeders have produced more than 300 varieties of domesticated pigeons from ancestral wild rock doves *(Columba livia)*.

Photographs courtesy Derrell Fowler, Tecumseh, Oklahoma

Mutations in DNA Are the Raw Materials That Allow Evolutionary Change

Today, we know that both the origin and the inheritance of new variations arise from the structure and variability of DNA, which is organized into functional units called **genes.** Each gene contains the code (that is, the instructions for building) for a protein molecule or one of its parts. Proteins are the molecules that establish the structures and perform important biological functions within organisms.

Variability among individuals—the raw material molded by evolutionary processes—arises ultimately through **mutations,** random changes in the structure, number, or arrangement of DNA molecules. Mutations in the DNA of reproductive cells (that is, sperm and eggs) may change the instructions for the development of offspring that the reproductive cells produce. Many mutations are of no particular value to individuals bearing them, and some turn out to be harmful. On rare occasions, however, a mutation is beneficial under the prevailing environmental conditions. Beneficial mutations increase the likelihood that individuals carrying the mutation will survive and reproduce. Thus, through the persistence and spread of beneficial mutations among individuals and their descendants, the genetic makeup of a population will change from one generation to the next.

Adaptations Enable Organisms to Survive and Reproduce in the Environments Where They Live

Favorable mutations may produce **adaptations,** characteristics that help an organism survive longer or reproduce more under a particular set of environmental conditions. To understand how organisms benefit from adaptations, consider an example from the recent literature on *cryptic coloration* (camouflage) in animals.

Many animals have skin, scales, feathers, or fur that matches the color and appearance of the background in their environment, enabling them to blend into their surroundings. Camouflage makes it harder for predators to identify and then catch them—an obvious advantage to survival. Animals that are not camouflaged are often just sitting ducks.

The rock pocket mouse *(Chaetodipus intermedius),* which lives in the deserts of the southwestern United States, is mostly nocturnal (that is, active at night). At most desert localities, the rocks are pale brown, and rock pocket mice have sandy-colored fur on their backs. However, at several sites, the rocks—remnants of lava flows from now-extinct volcanoes—are black; here, the rock pocket mice have black fur on their backs. Thus, like the sandy-colored mice in other areas, they are camouflaged in their habitats, the types of areas in which they live **(Figure 1.9A).** Camouflage appears to be important to these mice because owls, which locate prey using their exceptionally keen eyesight, frequently eat nocturnal desert mice.

Examples of cryptic coloration are well documented in scientific literature, and biologists generally interpret them as adaptations that reduce the likelihood of being captured by a predator. Michael W. Nachman, Hopi E. Hoekstra, and their colleagues at the University of Arizona explored the genetic and evolutionary basis for the color difference between rock pocket mice that live on light and dark backgrounds. In an article published in 2003, they reported the results of an analysis of mice sampled at six sites in southern Arizona and New Mexico. In two regions (Pinacate, AZ, and Armendaris, NM), both light and dark rocks were present, allowing the researchers to compare mice that lived on differently colored backgrounds. Two other sites had only light rocks and sandy-colored mice.

A. Camouflage in rock pocket mice (Chaetodipus intermedius)
Sandy-colored mice are well camouflaged on pale rocks, and black mice are well camouflaged on dark rocks (top); but mice with fur that does not match their backgrounds (bottom) are easy to see.

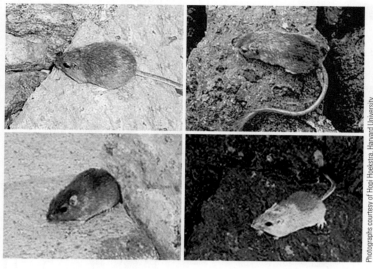

Photographs courtesy of Hopi Hoekstra, Harvard University

B. Distributions of rock pocket mice with light and dark fur
At sites in Arizona and New Mexico, mouse fur color closely matched the color of the rocks where they lived. The pie charts show the proportion of mice with sandy-colored or black fur, N = the number of mice sampled at each site. The bars beneath the pie charts indicate the rock color.

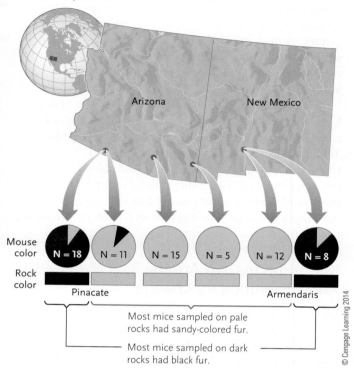

© Cengage Learning 2014

FIGURE 1.9 Adaptive coloration in rock pocket mice *(Chaetodipus intermedius)*.

Nachman and his colleagues found that nearly all of the mice they captured on dark rocks had dark fur and that nearly all of the mice they captured on light rocks had light fur **(Figure 1.9B)**. The researchers then studied the structure of *Mc1r*, a gene known to influence fur color in laboratory mice; random mutations in this gene can produce fur colors ranging from light to dark in any population of mice, regardless of the habitat it occupies. The 17 black mice from Pinacate all shared certain mutations in their *Mc1r* gene, which established four specific changes in the structure of the Mc1r protein. However, none of the 12 sandy-colored mice from Pinacate carried these mutations. The exact match between the presence of the mutations and the color of the mouse strongly suggests that these mutations in the *Mc1r* gene are responsible for the dark fur in the mice from Pinacate. These data on the distributions of light and dark mice coupled with analyses of their DNA suggest that the color difference is the product of specific mutations that were favored by natural selection. In other words, natural selection *conserved* random mutations that produced black fur in mice that live on black rocks.

Nachman's team then analyzed the *Mc1r* gene in the dark and light mice from Armendaris and in the light mice at two intermediate sites. Because the mice in these regions also closely matched the color of their environments, the researchers expected to find the *Mc1r* mutations in the dark mice but not in the light mice. However, none of the mice from Armendaris

shared any of the mutations that apparently contribute to the dark color of mice from Pinacate. Thus, mutations in some other gene or genes, which the researchers have not yet identified, must be responsible for the camouflaging black coloration of mice that live on black rocks in Armendaris.

The example of an adaptation provided by the rock pocket mice illustrates the observation that genetic differences often develop between populations. Sometimes these differences become so great that the organisms develop different appearances and adopt different ways of life. If they become different enough, biologists may regard them as distinct types, as described in Chapter 22. Over immense spans of time, evolutionary processes have produced many types of organisms, which constitute the diversity of life on Earth. In the next section, we survey this diversity and consider how it is studied.

STUDY BREAK 1.2

1. What is the difference between artificial selection and natural selection?
2. How do random changes in the structure of DNA affect the characteristics of organisms?
3. What is the usefulness of being camouflaged in natural environments?

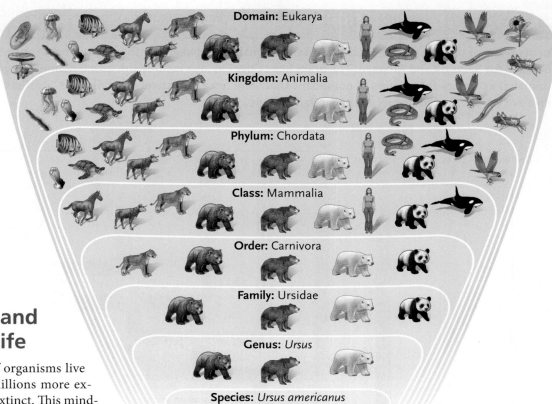

FIGURE 1.10 Traditional hierarchical classification. The classification of the American black bear *(Ursus americanus)* illustrates how each species fits into a nested hierarchy of ever-more inclusive categories. The following sentence can help you remember the order of categories in a classification, from *Domain* to *Species:* Diligent Kindly Professors Cannot Often Fail Good Students.
© Cengage Learning 2014

Domain: Eukarya

Kingdom: Animalia

Phylum: Chordata

Class: Mammalia

Order: Carnivora

Family: Ursidae

Genus: *Ursus*

Species: *Ursus americanus*

1.3 Biodiversity and the Tree of Life

Millions of different kinds of organisms live on Earth today, and many millions more existed in the past and became extinct. This mind-boggling biodiversity, the product of evolution, represents the many ways in which the common elements of life have combined to survive and reproduce. To make sense of the past and present diversity of life on Earth, biologists analyze the evolutionary relationships of these organisms and use classification systems to keep track of them. As described in Chapter 24, the task is daunting, and there is no clear consensus on the numbers and kinds of divisions and categories to use. Moreover, our understanding of evolutionary relationships is constantly changing as researchers develop new analytical techniques and learn more about extinct and living organisms.

Researchers Traditionally Defined Species and Grouped Them into Successively More Inclusive Hierarchical Categories

Biologists generally consider the species to be the most fundamental grouping in the diversity of life. As described in Chapter 22, a **species** is a group of populations in which the individuals are so similar in structure, biochemistry, and behavior that they can successfully interbreed. Biologists recognize a **genus** (plural, *genera*) as a group of similar species that share recent common ancestry. Species in the same genus usually also share many characteristics. For example, a group of closely related animals that have large bodies, four stocky legs, long snouts, shaggy hair, non-retractable claws, and short tails are classified together in the genus *Ursus,* commonly known as bears.

Each species is assigned a two-part **scientific name:** the first part identifies the genus to which it belongs, and the second part

designates a particular species within that genus. In the genus *Ursus,* for example, *Ursus americanus* is the scientific name of the American black bear; *Ursus maritimus,* the polar bear, and *Ursus arctos,* the brown bear, are two other species in the same genus. Scientific names are always written in italics, and only the genus name is capitalized. After its first mention in a discussion, the genus name is frequently abbreviated to its first letter, as in *U. americanus.*

In a traditional classification, biologists first identified species and then grouped them into successively more inclusive categories **(Figure 1.10):** related genera are placed in the same **family,** related families in the same **order,** and related orders in the same **class.** Related classes are grouped into a **phylum** (plural, *phyla*), and related phyla are assigned to a **kingdom.** In recent years, biologists have added the **domain** as the most inclusive group.

Today Biologists Identify the Trunks, Branches, and Twigs on the Tree of Life

For hundreds of years, biologists classified biodiversity within the hierarchical scheme described above, mostly using structural similarities and differences as clues to evolutionary relationships. With the development of new techniques late in the twentieth century, biologists began to use the precise structure of DNA and other biological molecules to trace the evolutionary pathways

An overview of the Tree of Life illustrates the relationships between the three domains. Branches and twigs are not included for Bacteria and Archaea. The branches of the Eukarya include three well-defined kingdoms (Plantae, Fungi, and Animalia) as well as five groups of organisms that were once collectively described as "protists" (marked with ✳); biologists have not yet clarified their evolutionary relationships.

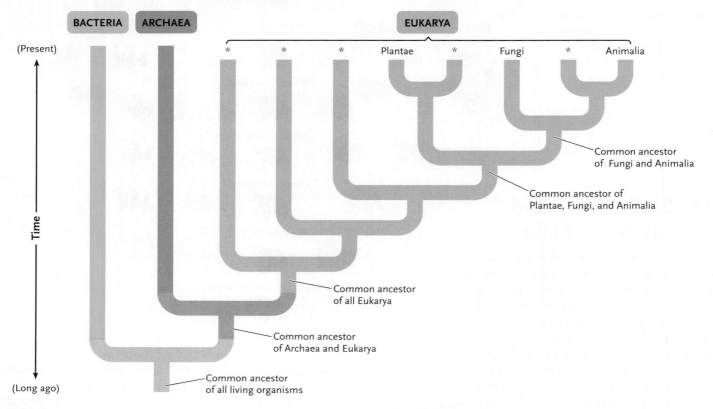

SUMMARY Phylogenetic trees contain more information than simple hierarchical classifications do because the trees illustrate which ancestors gave rise to which descendants, as well as when those evolutionary events occurred. Each fork between trunks, branches, and twigs on the phylogenetic tree represents an evolutionary event in which one ancestral species gave rise to two descendant species. Detailed phylogenetic trees illustrate how, over time, descendant species gave rise to their own descendants, producing the great diversity of life.

THINK LIKE A SCIENTIST Given the structure of the Tree of Life represented in this figure, do you think that animals are more closely related to fungi or to plants?

© Cengage Learning 2014

---

through which biodiversity evolved. This new approach allows the comparison of species as different as bread molds and humans because all living organisms share the same genetic code. It also provides so much data that biologists are now able to construct very detailed **phylogenetic trees** (*phylon* = race; *genetikos* = origin)—illustrations of the evolutionary pathways through which species and more inclusive groups appeared—for all organisms **(Figure 1.11)**. Phylogenetic trees, described further in Chapter 24, are like family genealogies spanning the many millions of years that evolution has been occurring. In the phylogenetic trees you will encounter in Units 3 and 4 of this book, time is usually represented vertically, with forks closer to the base of the tree representing evolutionary events in the distant past and those near the top representing more recent evolutionary events.

In many ways, information in a phylogenetic tree parallels the traditional hierarchical classification, because organisms on the same branch share the common ancestor that is represented at the base of their branch. If the base of a branch that includes two species is near the bottom of the tree, biologists would judge the species to be only distant relatives because their ancestries separated very long ago. By contrast, if the base of the branch containing two species is close to the top of the tree, we would describe the species as being close relatives. Major branches on the tree are therefore roughly equivalent to kingdoms and phyla;

A. *Escherichia coli,* **a prokaryote**

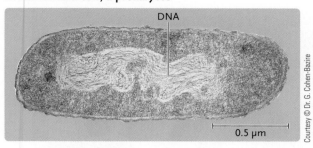

DNA

0.5 μm

Courtesy © Dr. G. Cohen-Bazire

B. *Paramecium aurelia,* **a eukaryote**

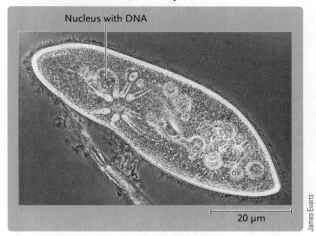

Nucleus with DNA

20 μm

James Evarts

FIGURE 1.12 Prokaryotic and eukaryotic cells. **(A)** *Escherichia coli,* a prokaryote, lacks the complex internal structures apparent in **(B)** *Paramecium aurelia,* a eukaryote.

progressively smaller branches represent classes, orders, families, and genera. The twigs represent species or the individual populations they comprise.

Since 1994, with substantial support from the National Science Foundation, biologists have collaborated on the Tree of Life web project to share and disseminate their discoveries about how all organisms on Earth are related. The "Tree of Life" has been reconstructed from data on the genetics, structure, metabolic processes, and behavior of living organisms, as well as data gathered from the fossils of extinct species. It is constantly updated and revised as scientists accumulate new data.

Three Domains and Several Kingdoms Form the Major Trunks and Branches on the Tree of Life

Biologists distinguish three domains—Bacteria, Archaea, and Eukarya—each of which is a group of organisms with characteristics that set it apart as a major trunk on the Tree of Life. Species in two of the three domains, Bacteria and Archaea, are described as **prokaryotes** (*pro* = before; *karyon* = nucleus). Their DNA is suspended inside the cell without being separated from other cellular components **(Figure 1.12A)**. By contrast, the domain Eukarya comprises organisms that are described as **eukaryotes** (*eu* =

typical) because their DNA is enclosed in a nucleus, a separate structure within the cells **(Figure 1.12B)**. The nucleus and other specialized internal compartments of eukaryotic cells are called **organelles** ("little organs").

THE DOMAIN BACTERIA The Domain Bacteria **(Figure 1.13A)** comprises unicellular organisms (bacteria) that are generally visible only under the microscope. These prokaryotes live as producers, consumers, or decomposers almost everywhere on Earth, utilizing metabolic processes that are the most varied of any group of organisms. They share with the archaeans a relatively simple cellular organization of internal structures and DNA, but bacteria have some unique structural molecules and mechanisms of photosynthesis.

THE DOMAIN ARCHAEA Similar to bacteria, species in the Domain Archaea (*arkhaios* = ancient) **(Figure 1.13B),** known as archaeans, are unicellular, microscopic organisms that live as producers or decomposers. Many archaeans inhabit extreme environments—hot springs, extremely salty ponds, or habitats with little or no oxygen—that other organisms cannot tolerate. They have some distinctive structural molecules and a primitive form of photosynthesis that is unique to their domain. Although archaeans are prokaryotic, they have some molecular and biochemical characteristics that are typical of eukaryotes, including features of DNA and RNA organization and processes of protein synthesis.

THE DOMAIN EUKARYA All the remaining organisms on Earth, including the familiar plants and animals, are members of the Domain Eukarya **(Figure 1.13C)**. Organisms with eukaryotic cell structure are currently described as "protists" or classified as members of one of three well-defined kingdoms: Plantae, Fungi, and Animalia.

The "Protists". The term "protists" describes a diverse set of single-celled and multicellular eukaryotic species. They do not constitute a kingdom, because they do not share a unique common ancestry (see asterisks in Figure 1.11). The most familiar "protists" are "protozoans," which are primarily unicellular, and algae, which range from single-celled, microscopic species to large, multicellular seaweeds. "Protozoans" are consumers and decomposers, but almost all algae are photosynthetic producers.

The Kingdom Plantae. Members of the Kingdom Plantae are multicellular organisms that, with few exceptions, carry out photosynthesis; they, therefore, function as producers in ecosystems. Except for the reproductive cells (pollen) and seeds of some species, plants do not move from place to place. The kingdom includes the familiar flowering plants, conifers, and mosses.

The Kingdom Fungi. The Kingdom Fungi includes a highly varied group of unicellular and multicellular species, among them the yeasts and molds. Most fungi live as decomposers by breaking down and then absorbing biological molecules from dead organisms. Fungi do not carry out photosynthesis.

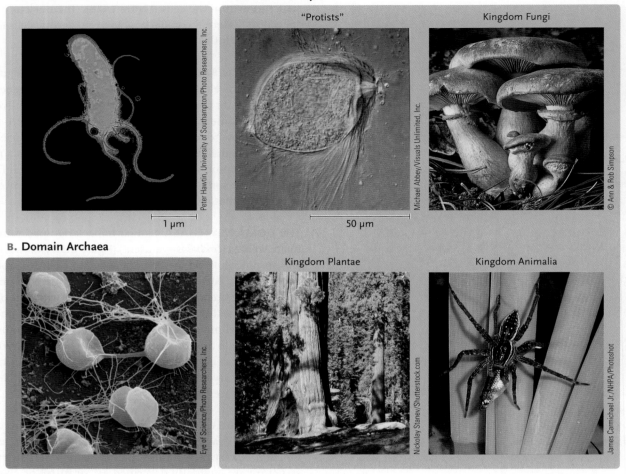

A. Domain Bacteria

C. Domain Eukarya

"Protists"

Kingdom Fungi

1 µm

50 µm

B. Domain Archaea

Kingdom Plantae

Kingdom Animalia

FIGURE 1.13 **Three domains of life. (A)** This member of the Domain Bacteria *(Helicobacter pylori)* causes ulcers in the digestive systems of humans. **(B)** These organisms from the Domain Archaea *(Pyrococcus furiosus)* live in hot ocean sediments near an active volcano. **(C)** The Domain Eukarya includes the "protists" and three kingdoms in this book. The "protists" are represented by a trichomonad *(Trichonympha* species) that lives in the gut of a termite. Coast redwoods *(Sequoia sempervirens)* are among the largest members of the Kingdom Plantae; the picture shows the large trunk of an older tree in the foreground. The Kingdom Fungi includes the big laughing mushroom *(Gymnopilus* species), which lives on the forest floor. Members of the Kingdom Animalia are consumers, as illustrated by the fishing spider *(Dolomedes* species), which is feasting on a minnow it has captured.

The Kingdom Animalia. Members of the Kingdom Animalia are multicellular organisms that live as consumers by ingesting "protists" and organisms from all three domains. The ability to move actively from one place to another during some stage of their life cycles is a distinguishing feature of animals. The kingdom encompasses a great range of organisms, including groups as varied as sponges, worms, insects, fishes, amphibians, reptiles, birds, and mammals.

Now that we have introduced the characteristics of living organisms, basic concepts of evolution, and biological diversity, we turn our attention to the ways in which biologists examine the living world to make new discoveries and gain new insights about life on Earth.

STUDY BREAK 1.3

1. What is a major difference between prokaryotic and eukaryotic organisms?
2. In which domain and kingdom are humans classified?

THINK OUTSIDE THE BOOK

Learn more about the Tree of Life web project by visiting this web site: http://tolweb.org/tree/.

1.4 Biological Research

The entire content of this book—every observation, experimental result, and generality—is the product of **biological research,** the collective effort of countless individuals who have worked to understand every aspect of the living world. This section describes how biologists working today pose questions and find answers to them.

Biologists Confront the Unknown by Conducting Basic and Applied Research

As you read this book, you may at first be uncomfortable discovering how many fundamental questions have not yet been answered. How and where did life begin? How exactly do genes govern the growth and development of an organism? What triggers the signs of aging? Scientists embrace these "unknowns" as opportunities to apply creative thinking to important problems. To show you how exciting it can be to venture into unknown territory, most chapters close with a discussion of Unanswered Questions. Although the concepts and facts that you will learn are profoundly interesting, you will discover that unanswered questions are even more exciting. In many cases, we do not even know *how* you and other scientists of your generation will answer these questions.

Research science is often broken down into two complementary activities—basic research and applied research—that constantly inform one another. Biologists who conduct **basic research** often seek explanations about natural phenomena to satisfy their own curiosity and to advance our collective knowledge. Sometimes, they may not have a specific practical goal in mind. For example, some biologists study how lizards control their body temperatures in different environments. At other times, basic research is inspired by specific practical concerns. For example, understanding how certain bacteria attack the cells of larger organisms might someday prove useful for the development of a new antibiotic (that is, a bacteria-killing agent). Many chapters in this book include a *Focus on Basic Research,* which describes particularly elegant or insightful basic research that advanced our knowledge.

Other scientists conduct **applied research,** with the goal of solving specific practical problems. For example, biomedical scientists conduct applied research to develop new drugs and to learn how illnesses spread from animals to humans or through human populations. Similarly, agricultural scientists try to develop varieties of important crop plants that are more productive and more pest-resistant than the varieties currently in use. Examples of applied research are presented throughout this book, some of them described in detail as a *Focus on Applied Research.*

The Scientific Method Helps Researchers Crystallize and Test Their Ideas

People have been adding to our knowledge of biology ever since our distant ancestors first thought about gathering food or hunting game. However, beginning about 500 years ago in Europe, inquisitive people began to understand that direct observation is the most reliable and productive way to study natural phenomena. By the nineteenth century, researchers were using the **scientific method,** an investigative approach to acquiring knowledge in which scientists make observations about the natural world, develop working explanations about what they observe, and then test those explanations by collecting more information.

Grade school teachers often describe the scientific method as a stepwise, linear procedure for observing and explaining the world around us **(Figure 1.14).** But because scientists usually think faster than they can work, they often undertake the different steps simultaneously. Application of the scientific method requires both curiosity and skepticism: successful scientists question the current state of our knowledge and challenge old concepts with new ideas and new observations. Scientists like to be shown *why* an idea is correct, rather than simply being told that it is: explanations of natural phenomena must be backed up by objective evidence rooted in observation and measurement.

Most important, scientists share their ideas and results through the publication of their work. Publications typically include careful descriptions of the methods employed and details of the results obtained so that other researchers can repeat and verify their findings at a later time.

Biologists Conduct Research by Collecting Observational and Experimental Data

Biologists generally use one of two complementary approaches—*descriptive science* and *experimental science*—or a combination of the two to advance our knowledge. In many cases, they collect **observational data,** basic information on biological structures or the details of biological processes. This approach, which is sometimes called descriptive science, provides information about systems that have not yet been well studied. For example, biologists are now collecting observational data about the precise chemical structure of the DNA in different species of organisms for the purpose of understanding their evolutionary relationships.

In other cases, researchers collect **experimental data,** information that describes the result of a careful manipulation of the system under study. This approach, which is known as experimental science, often answers questions about why or how systems work as they do. For example, a biologist who wonders whether a particular snail species influences the distribution of algae on a rocky shoreline might remove the snail from some enclosed patches of shoreline and examine whether the distribution of algae changes as a result. Similarly, a geneticist who wants to understand the role of a particular gene in the functioning of an organism might make mutations in the gene and examine the consequences.

Researchers Often Test Hypotheses with Controlled Experiments

Research on a previously unexplored system usually starts with basic observations (step 1 in Figure 1.14). Once the facts have been carefully observed and described, scientists may develop a

FIGURE 1.14 **Research Method**

The Scientific Method

Purpose: The scientific method is a method of inquiry that allows researchers to crystallize their thoughts about a topic and devise a formal way to test their ideas by making observations and collecting measurable data.

Protocol:

1. Make detailed observations about a phenomenon of interest.

Observations

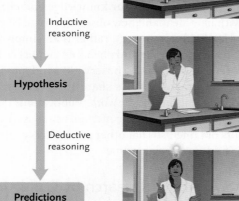

Inductive reasoning

2. Use inductive reasoning to create a testable hypothesis that provides a working explanation of the observations. Hypotheses may be expressed in words or in mathematical equations. Many scientists also formulate alternative hypotheses (that is, alternative explanations) at the same time.

Hypothesis

Deductive reasoning

3. Use deductive reasoning to make predictions about what you would observe if the hypothesis were applied to a novel situation.

Predictions

4. Design and conduct a controlled experiment (or new observational study) to test the predictions of the hypothesis. The experiment must be clearly defined so that it can be repeated in future studies. It must also lead to the collection of measurable data that other researchers can evaluate and reproduce if they choose to repeat the experiment themselves.

Experiments

Interpreting the Results: Compare the results of the experiment or new observations with those predicted by the hypothesis. Scientists often use formal statistical tests to determine whether the results match the predictions of the hypothesis.

If the results do not match the predictions, the hypothesis is refuted, and it must be rejected or revised.

If the statistical tests suggest that the prediction was correct, the hypothesis is confirmed—until new data refute it in the future.

hypothesis to explain them (step 2). In a 2008 report, the National Academy of Sciences defined a *hypothesis* as "a tentative explanation for an observation, phenomenon, or scientific problem that can be tested by further investigation." And whenever scientists create a hypothesis, they simultaneously define—either explicitly or implicitly—a **null hypothesis,** a statement of what they would see if the hypothesis being tested is not correct.

Many scientists structure their hypotheses with one crucial requirement: it must be *falsifiable* by experimentation or further observation. In other words, scientists must describe an idea in such a way that, if it is wrong, they will be able to demonstrate that it is wrong. The principle of falsifiability helps scientists define testable, focused hypotheses. Hypotheses that are testable and falsifiable fall within the realm of science, whereas those that cannot be falsified—although possibly valid and true—do not fall within the realm of science.

Hypotheses generally explain the relationship between **variables,** environmental factors that may differ among places or organismal characteristics that may differ among individuals. Thus, hypotheses yield testable **predictions** (step 3), statements about what the researcher expects to happen to one variable if another variable changes. Scientists then test their hypotheses and prediction with experimental or observational tests that generate relevant data (step 4). And if data from just one study refute a scientific hypothesis (that is, demonstrate that its predictions are incorrect), the scientist must modify the hypothesis and test it again or abandon it altogether. However, no amount of data can *prove* beyond a doubt that a hypothesis is correct; there may always be a contradictory example somewhere on Earth, and it is impossible to test every imaginable example. That is why scientists say that positive results *are consistent with, support,* or *confirm* a hypothesis.

To make these ideas more concrete, consider a simple example of hypothesis development and testing. Say that a friend gives you a plant that she grew on her windowsill. Under her loving care, the plant always flowered. You place the plant on your windowsill and water it regularly, but the plant never blooms. You know that your friend always gave fertilizer to the plant—your observation—and you wonder whether fertilizing the plant would make it flower. In other words, you create a hypothesis with a specific prediction: "This type of plant will flower if it receives fertilizer." This is a good scientific hypothesis because it is falsifiable. To test the hypothesis, you would simply give the plant fertilizer. If it blooms, the data—the fact that it flowers—confirm your hypothesis. If it does not bloom, the data force you to reject or revise your hypothesis.

One problem with this experiment is that the hypothesis does not address other possible reasons that the plant did not flower. Maybe it received too little water. Maybe it did not get enough sunlight. Maybe your windowsill was too cold. All of these explanations could be the basis of **alternative hypotheses,** which a conscientious scientist always considers when designing experiments. You could easily test any of these hypotheses by providing more water, more hours of sunlight, or warmer temperatures to the plant.

But even if you provide each of these necessities in turn, your efforts will not definitively confirm or refute your hypothesis unless you introduce a control treatment. The **control,** as it is often called, represents a null hypothesis; it tells us what we would see in the absence of the experimental manipulation. For example, your experiment would need to compare plants that received fertilizer (the experimental treatment) with plants grown without fertilizer (the control treatment). The presence or absence of fertilizer is the **experimental variable,** and in a controlled experiment, everything except the experimental variable—the flower pots, the soil, the amount of water, and exposure to sunlight—is exactly the same, or as close to exactly the same as possible. Thus, if your experiment is well controlled **(Figure 1.15),** any difference in flowering pattern observed between plants that receive the experimental treatment (fertilizer) and those that receive the control treatment (no fertilizer) can be attributed to the experimental variable. If the

plants that receive fertilizer did not flower more than the control plants, you would reject your initial hypothesis. The elements of a typical experimental approach, as well as our hypothetical experiment, are summarized in Figure 1.15. Figures that present observational and experimental research using this basic format are provided throughout this book.

Notice that in the preceding discussion we discussed plants (plural) that received fertilizer and plants that did not. Nearly all experiments in biology include **replicates,** multiple subjects that receive either the same experimental treatment or the same control treatment. Scientists use replicates in experiments because individuals typically vary in genetic makeup, size, health, or other characteristics—and because accidents may disrupt a few replicates. By exposing multiple subjects to both treatments, we can use a statistical test to compare the average result of the experimental treatment with the average result of the control treatment, giving us more confidence in the overall findings. Thus, in the fertilizer experiment we described, we might expose six or more individual plants to each treatment and compare the results obtained for the experimental group with those obtained for the control group. We would also try to ensure that the individuals included in the experiment were as similar as possible. For example, we might specify that they all must be the same age or size.

When Controlled Experiments Are Unfeasible, Researchers Employ Null Hypotheses to Evaluate Observational Data

In some fields of biology, especially ecology and evolution, the systems under study may be too large or complex for experimental manipulation. In such cases, biologists can use a null hypothesis to evaluate observational data. For example, Paul E. Hertz of Barnard College studies temperature regulation in lizards. As in many other animals, a lizard's body temperature can vary substantially as environmental temperatures change. Research on many lizard species has demonstrated that they often compensate for fluctuations in environmental temperature—that is, maintain thermal homeostasis—by perching in the sun to warm up or in the shade when they feel hot.

Hertz hypothesized that the crested anole, *Anolis cristatellus,* a lizard species in Puerto Rico, regulates its body temperature by perching in patches of sun when environmental temperatures are low. To test this hypothesis, Hertz needed to determine what he would see if lizards were *not* trying to control their body temperatures. In other words, he needed to know the predictions of a null hypothesis that states: "Lizards do not regulate their body temperature, and they select perching sites at random with respect to factors that influence body temperature" **(Figure 1.16).** Of course, it would be impossible to force a natural population of lizards to perch in places that define the null hypothesis. Instead, he and his students created a population of artificial lizards, copper models that served as lizard-sized, lizard-shaped, and lizard-colored thermometers. Each hollow copper model was equipped with a built-in temperature-sensing wire that can be plugged into an electronic thermometer. After constructing

FIGURE 1.15 **Experimental Research**

Hypothetical Experiment Illustrating the Use of Control Treatment and Replicates

Question: Your friend fertilizes a plant that she grows on her windowsill, and it flowers. After she gives you the plant, you put it on your windowsill, but you do not give it any fertilizer and it does not flower. Will giving the plant fertilizer induce it to flower?

Friend added fertilizer   You did not add fertilizer

Experiment: Establish six replicates of an experimental treatment (identical plants grown with fertilizer) and six replicates of a control treatment (identical plants grown without fertilizer).

Experimental Treatment   **Control Treatment**

Add fertilizer

No fertilizer

Possible Result 1: Neither experimental nor control plants flower.

Experimentals   **Controls**

Conclusion: Fertilizer alone does not cause the plants to flower. Consider alternative hypotheses and conduct additional experiments, each testing a different experimental treatment, such as the amount of water or sunlight the plant receives or the temperature to which it is exposed.

Possible Result 2: Plants in the experimental group flower, but plants in the control group do not.

Experimentals   **Controls**

Conclusion: The application of fertilizer induces flowering in this type of plant, confirming your original hypothesis. Pat yourself on the back and apply to graduate school in plant biology.

THINK LIKE A SCIENTIST Suppose you were interested in studying the effects of both fertilizer and water on the flowering of your plants. How would you design an experiment that addressed the two experimental variables at the same time?

the copper models, Hertz and his students verified that the models reached the same internal temperatures as live lizards under various laboratory conditions. They then traveled to Puerto Rico and hung 60 models at randomly selected positions in each of the habitats where this lizard species lives.

How did the copper models allow Hertz and his students to interpret their data? Because the researchers placed these inanimate objects at random positions in the lizards' habitats, the percentages of models observed in sun and in shade provided a measure of how sunny or shady a particular habitat was. In other words, the copper models established the null hypothesis about the percentage of lizards that would perch in sunlit spots just by chance. Similarly, the temperatures of the models provided a null hypothesis about what the temperatures of lizards would be if they perched at random in their habitats. Hertz and his students gathered data on the use of sunny perching places and temperatures from both the copper models and live lizards. By comparing the behavior and temperatures of live lizards with

FIGURE 1.16 | **Observational Research**

A Field Study Using a Null Hypothesis

Anolis cristatellus

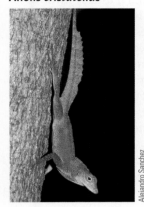

Alejandro Sanchez

Hypothesis: *Anolis cristatellus,* the crested anole, uses patches of sun and shade to regulate its body temperature.

Null Hypothesis: If these lizards do not regulate their body temperature, individuals would select perching sites at random with respect to environmental factors that influence body temperature.

Method: The researchers created a set of hollow, copper lizard models, each equipped with a temperature-sensing wire. At study sites where the lizards live in Puerto Rico, the researchers hung 60 models at random positions in trees. They observed how often live lizards and the randomly positioned copper models were perched in patches of sun or shade, and they measured the temperatures of live lizards and the copper models. Data from the randomly positioned copper models define the predictions of the null hypothesis.

Results: The researchers compared the frequency with which live lizards and the copper models perched in sun or shade as well as the temperatures of live lizards and the copper models. The data revealed that the behavior and temperatures of *A. cristatellus* were different from those of the randomly positioned models, therefore confirming the original hypothesis.

Copper *Anolis* model

Kevin de Queiroz, National Museum of Natural History, Smithsonian Institution

Percentage of models and lizards perched in sun or shade

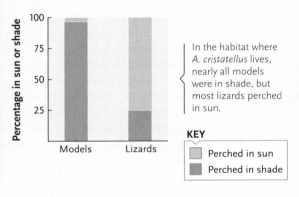

In the habitat where *A. cristatellus* lives, nearly all models were in shade, but most lizards perched in sun.

KEY
- Perched in sun
- Perched in shade

Temperatures of models and lizards

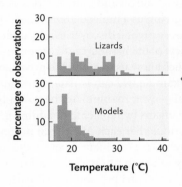

Body temperatures of *A. cristatellus* were significantly higher than those of the randomly placed models.

Conclusion: *A. cristatellus* uses patches of sun and shade to regulate its body temperature.

THINK LIKE A SCIENTIST Based on the data presented in the two graphs, how would you describe the variability in temperature among shaded sites in this environment?

Source: P.E. Hertz. 1992. Temperature regulation in Puerto Rican *Anolis* lizards: A field test using null hypotheses. *Ecology* 73:1405–1417.

© Cengage Learning 2014

the random "behavior" and random temperatures of the copper models, they demonstrated that *A. cristatellus* did, in fact, regulate its body temperature (see Figure 1.16).

Biologists Often Use Model Organisms to Study Fundamental Biological Processes

Certain species or groups of organisms have become favorite subjects for laboratory and field studies because their characteristics make them relatively easy subjects of research. In most cases, biologists began working with these **model organisms** because they have rapid development, short life cycles, and small adult size. Thus, researchers can rear and house large numbers of them in the laboratory. Also, as fuller portraits of their genetics and other aspects of their biology emerge, their appeal as research subjects grows because biologists have a better understanding of the biological context within which specific processes occur.

Because many forms of life share similar molecules, structures, and processes, research on these small and often simple organisms provides insights into biological processes that operate in larger and more complex organisms. For example, early analyses of inheritance in a fruit fly *(Drosophila melanogaster)* established our basic understanding of genetics in all eukaryotic organisms. Research in the mid-twentieth century with the bacterium *Escherichia coli* demonstrated the mechanisms that control whether the information in any particular gene is used to manufacture a protein molecule. In fact, the body of research with *E. coli* formed the foundation that now allows scientists to make and clone (that is, produce multiples copies of) DNA molecules. Similarly, research on a tiny mustard plant *(Arabidopsis thaliana)* is providing information about the genetic and molecular control of development in all plants, including important agricultural crops. Other model organisms facilitate research in ecology and evolution. For example, *Anolis cristatellus* is just 1 of more than 400 *Anolis* species. The geographic distribution of these species allows researchers to study general processes and interactions that affect the ecology and evolution of all forms of life. You will read about eight of the organisms most frequently used in research in *Focus on Model Organisms* boxes distributed throughout this book.

Molecular Techniques Have Revolutionized Biological Research

In 1941, George Beadle and Edward Tatum used a simple bread mold *(Neurospora crassa)* as a model organism to demonstrate that genes provide the instructions for constructing certain proteins. Their work represents the beginning of the molecular revolution. In 1953, James Watson and Francis Crick determined the structure of DNA, giving us a molecular vision of what a gene is. In the years since those pivotal discoveries, our understanding of the molecular aspects of life has increased exponentially because many new techniques allow us to study life processes at the molecular level. For example, we can isolate individual genes and study them in detail—even manipulate them—in the test tube. We can modify organisms by replacing or adding genes. We can explore the interactions that each individual protein in the cell has with other proteins. We can identify and characterize each of the genes in an organism to unravel its evolutionary relationships. The list of experimental possibilities is nearly endless.

This molecular revolution has made it possible to answer questions about biological systems that we could not even ask just a few years ago. For example: What specific DNA changes are responsible for genetic diseases? How is development controlled at the molecular level? What genes do humans and chimpanzees share? In particular, the continuing analysis of the structure of DNA in many organisms is fueling a new intensity of scientific enquiry focused on the role of whole genomes (all of the DNA of an organism) in directing biological processes. To give you a sense of the exciting impact of molecular research on all areas of biology, most chapters in this book include a box that focuses on *Insights from the Molecular Revolution.*

Advances in molecular biology have also revolutionized applied research. DNA "fingerprinting" allows forensic scientists to identify individuals who left molecular traces at crime scenes. **Biotechnology,** the manipulation of living organisms to produce useful products, has also revolutionized the pharmaceutical industry. For example, insulin—a protein used to treat the metabolic disorder diabetes—is now routinely produced by bacteria into which the gene coding for this protein has been inserted. Current research on gene therapy and the cloning of stem cells also promises great medical advances in the future.

Scientific Theories Are Grand Ideas That Have Withstood the Test of Time

When a hypothesis stands up to repeated experimental tests, it is gradually accepted as an accurate explanation of natural events. This acceptance may take many years, and it usually involves repeated experimental confirmations. When many different tests have consistently confirmed a hypothesis that addresses many broad questions, it may become regarded as a **scientific theory.** In a 2008 report, the National Academy of Sciences defined a *scientific theory* as "a plausible or scientifically acceptable, well-substantiated explanation of some aspect of the natural world … that applies in a variety of circumstances to explain a specific set of phenomena and predict the characteristics of as yet unobserved phenomena."

Most scientific theories are supported by exhaustive experimentation; thus, scientists usually regard them as established truths that are unlikely to be contradicted by future research. Note that this use of the word *theory* is quite different from its informal meaning in everyday life. In common usage, the word *theory* most often labels an idea as either speculative or downright suspect, as in the expression "It's only a theory." But when scientists talk about theories, they do so with respect for ideas that have withstood the test of many experiments.

Because of the difference between the scientific and common usage of the word *theory,* many people fail to appreciate the extensive evidence that supports most scientific theories. For example, virtually every scientist accepts the theory of evolution as fully supported scientific truth: all species change with time, new species are formed, and older species eventually die off. Although evolutionary biologists debate the details of how evolutionary processes bring about these changes, very few scientists doubt that the theory of evolution is essentially correct. Moreover, *no scientist who has tried to cast doubt on the theory of evolution has ever devised or conducted a study that disproves any part of it.* Unfortunately, the confusion between the scientific and common usage of the word *theory* has led, in part, to endless public debate about supposed faults and inadequacies in the theory of evolution.

Curiosity and the Joy of Discovery Motivate Scientific Research

What drives scientists in their quest for knowledge? The motivations of scientists are as complex as those driving people toward any goal. Intense curiosity about ourselves, our fellow creatures, and the chemical and physical objects of the world and their interactions is a basic ingredient of scientific research. The discovery of information that no one knew before is as exciting to a scientist as finding buried treasure. There is also an element of play in science, a joy in the manipulation of scientific ideas and apparatus, and the chase toward a scientific goal. Biological research also has practical motivations—for example, to cure disease or improve agricultural productivity. In all of this research, one strict requirement of science is honesty—without honesty in the gathering and reporting of results, the work of science is meaningless. Dishonesty is actually rare in science, not least because repetition of experiments by others soon exposes any funny business.

Whatever the level of investigation or the motivation, the work of every scientist adds to the fund of knowledge about us and our world. For better or worse, the scientific method—that inquiring and skeptical approach—has provided knowledge and technology that have revolutionized the world and improved the quality of human life immeasurably. This book presents the fruits of the biologists' labors in the most important and fundamental areas of biological science—cell and molecular biology, genetics, evolution, systematics, physiology, developmental biology, ecology, and behavioral science.

STUDY BREAK 1.4

1. In your own words, explain the most important requirement of a scientific hypothesis.
2. What information did the copper lizard models provide in the study of temperature regulation described earlier?
3. Why do biologists often use model organisms in their research?
4. How would you respond to a nonscientist who told you that Darwin's ideas about evolution were "just a theory"?

THINK OUTSIDE THE BOOK

Learn more about the scientific method and the richness of the scientific enterprise by visiting this web site, maintained by researchers at the University of California at Berkeley: http://undsci.berkeley.edu/.

REVIEW KEY CONCEPTS

To access the course materials and companion resources for this text, please visit www.cengagebrain.com.

1.1 What Is Life? Characteristics of Living Organisms

- Living systems are organized in a hierarchy, each level having its own emergent properties (Figure 1.2): cells, the lowest level of organization that is alive, are organized into unicellular or multicellular organisms; populations are groups of organisms of the same kind that live together in the same area; an ecological community comprises all the populations living in an area, and ecosystems include communities that interact through their shared physical environment; the biosphere includes all of Earth's ecosystems.

- Living organisms have complex structures established by instructions coded in their DNA (Figure 1.3). The information in DNA is copied into RNA, which guides the production of protein molecules (Figure 1.4). Proteins carry out most of the activities of life.

- Living cells and organisms engage in metabolism, obtaining energy and using it to maintain themselves, grow, and reproduce. The two primary metabolic processes are photosynthesis and cellular respiration (Figure 1.5).

- Energy that flows through the hierarchy of life is eventually released as heat. By contrast, matter is recycled within the biosphere (Figure 1.6).

- Cells and organisms use receptors to detect environmental changes and trigger a compensating reaction that allows the organism to survive.

- Organisms reproduce, and their offspring develop into mature, reproductive adults (Figure 1.7).

- Populations of living organisms undergo evolutionary change as generations replace one another over time.

Animation: Insect development

Animation: One-way energy flow and materials cycling

Animation: Levels of Organization of Life

1.2 Biological Evolution

- The structure, function, and types of organisms in populations change with time. According to the theory of evolution by natural selection, certain characteristics allow some organisms to survive better and reproduce more than others in their population. If the instructions that produce those characteristics are coded in DNA, successful characteristics will become more common in later generations. As a result, the characteristics of the offspring generation will differ from those of the parent generation (Figure 1.8).

- The instructions for many characteristics are coded by segments of DNA called *genes,* which are passed from parents to offspring in reproduction.

- Mutations—changes in the structure, number, or arrangement of DNA molecules—create variability among individuals. Variability is the raw material of natural selection and other processes that cause biological evolution.

- Over many generations, the accumulation of favorable characteristics may produce adaptations, which enable individuals to survive longer or reproduce more (Figure 1.9).

- Over long spans of time, the accumulation of different adaptations and other genetic differences between populations has produced the diversity of life on Earth.

Animation: Adaptive coloration in Moths

1.3 Biodiversity and the Tree of Life

- Scientists classify organisms in a hierarchy of categories. The species is the most fundamental category, followed by genus, family, order, class, phylum, and kingdom as increasingly inclusive categories (Figure 1.10). Analyses of DNA structure now allow biologists to construct the Tree of Life, a model of the evolutionary relationships among all known living organisms (Figure 1.11).

- Most biologists recognize three domains—Bacteria, Archaea, and Eukarya—based on fundamental characteristics of cell structure and molecular analysis. The Bacteria and Archaea each include one kingdom; the Eukarya is divided into the "protists" and three kingdoms: Plantae, Fungi, and Animalia (Figures 1.12 and 1.13).

Animation: Life's diversity

1.4 Biological Research

- Biologists conduct basic research to advance our knowledge of living organisms and applied research to solve practical problems.

- The scientific method allows researchers to crystallize and test their ideas. Scientists develop hypotheses—working explanations about the relationships between variables. Scientific hypotheses must be falsifiable (Figure 1.14).

- Scientists may collect observational data, which describe particular organisms or the details of biological processes, or experimental data, which describe the results of an experimental manipulation.

- A well-designed experiment considers alternative hypotheses and includes control treatments and replicates (Figure 1.15). When experiments are unfeasible, biologists often use null hypotheses, explanations of what they would see if their hypothesis was wrong, to evaluate data (Figure 1.16).

- Model organisms, which are easy to maintain in the laboratory, have been the subject of much research.

- Molecular techniques allow detailed analysis of the DNA of many species and the manipulation of specific genes in the laboratory.

- A scientific theory is a set of broadly applicable hypotheses that have been supported by repeated tests under many conditions and in many different situations. The theory of evolution by natural selection is of central importance to biology because it explains how life evolved through natural processes.

Animation: Bacteriophage mice experiment

UNDERSTAND & APPLY

Test Your Knowledge

1. What is the lowest level of biological organization that biologists consider to be alive?
   a. a protein
   b. DNA
   c. a cell
   d. a multicellular organism
   e. a population of organisms

2. Which category falls immediately below "class" in the systematic hierarchy?
   a. species
   b. order
   c. family
   d. genus
   e. phylum

3. Which of the following represents the application of the "scientific method"?
   a. comparing one experimental subject to one control subject
   b. believing an explanation that is too complex to be tested
   c. using controlled experiments to test falsifiable hypotheses
   d. developing one testable hypothesis to explain a natural phenomenon
   e. observing a once-in-a-lifetime event under natural conditions

4. Houseflies develop through a series of programmed stages from egg, to larva, to pupa, to flying adult. This series of stages is called:
   a. artificial selection.
   b. respiration.
   c. homeostasis.
   d. a life cycle.
   e. metabolism.

5. Which structure allows living organisms to detect changes in the environment?
   a. a protein
   b. a receptor
   c. a gene
   d. RNA
   e. a nucleus

6. Which of the following is *not* a component of Darwin's theory as he understood it?
   a. Some individuals in a population survive longer than others.
   b. Some individuals in a population reproduce more than others.
   c. Heritable variations allow some individuals to compete more successfully for resources.
   d. Mutations in genes produce new variations in a population.
   e. Some new variations are passed to the next generation.

7. What role did the copper lizard models play in the field study of temperature regulation?
   a. They attracted live lizards to the study site.
   b. They measured the temperatures of live lizards.
   c. They established null hypotheses about basking behavior and temperatures.
   d. They scared predators away from the study site.
   e. They allowed researchers to practice taking lizard temperatures.

8. Which of the following questions best exemplifies basic research?
   a. How did life begin?
   b. How does alcohol intake affect aging?
   c. How fast does H1N1 flu spread among humans?
   d. How can we reduce hereditary problems in purebred dogs?
   e. How does the consumption of soft drinks promote obesity?

9. When researchers say that a scientific hypothesis must be falsifiable, they mean that:
   a. the hypothesis must be proved correct before it is accepted as truth.
   b. the hypothesis has already withstood many experimental tests.
   c. they have an idea about what will happen to one variable if another variable changes.
   d. appropriate data can prove without question that the hypothesis is correct.
   e. if the hypothesis is wrong, scientists must be able to demonstrate that it is wrong.

10. Which of the following characteristics would *not* qualify an animal as a model organism?
    a. It has rapid development.
    b. It has a small adult size.
    c. It has a rapid life cycle.
    d. It has unique genes and unusual cells.
    e. It is easy to raise in the laboratory.

Discuss the Concepts

1. Viruses are infectious agents that contain either DNA or RNA surrounded by a protein coat. They cannot reproduce on their own, but they can take over the cells of the organisms they infect and force those cells to produce more virus particles. Based on the characteristics of living organisms described in this chapter, should viruses be considered living organisms?

2. While walking in the woods, you discover a large rock covered with a gelatinous, sticky substance. What tests could you perform to determine whether the substance is inanimate, alive, or the product of a living organism?

3. Explain why control treatments are a necessary component of well-designed experiments.

Design an Experiment

Design an experiment to test the hypothesis that the color of farmed salmon is produced by pigments in their food.

Interpret the Data

While working in Puerto Rico, Paul E. Hertz and his students studied a second species of lizard, the yellow-chinned anole, *Anolis gundlachi*, using the procedures described in Figure 1.16. Their results for copper models and living lizards are in the **Figure**. Based on these data, do you think that *A. gundlachi* regulates its body temperature? Why or why not?

Apply Evolutionary Thinking

When a biologist first tested a new pesticide on a population of insects, she found that only 1% of the insects survived their exposure to the poison. She allowed the survivors to reproduce and discovered that 10% of the offspring survived exposure to the same concentration of pesticide. One generation later, 50% of the insects survived this experimental treatment. What is a likely explanation for the increasing survival rate of these insects over time?

Percentage of models and lizards perched in sun or shade

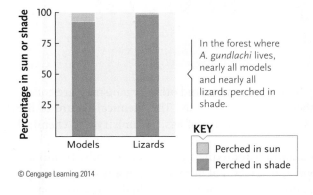

In the forest where *A. gundlachi* lives, nearly all models and nearly all lizards perched in shade.

KEY

☐ Perched in sun
■ Perched in shade

© Cengage Learning 2014

Temperatures of models and lizards

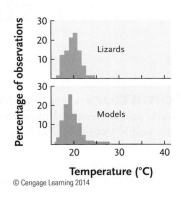

Body temperatures of *A. gundlachi* were not significantly different from those of the randomly placed models.

© Cengage Learning 2014

Anolis gundlachi

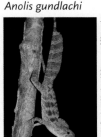

Kevin de Queiroz, National Museum of Natural History, Smithsonian Institution

2

© Markus Botzek/AgeFotostock

Life as we know it would be impossible without water, a small inorganic compound with unique properties.

Life, Chemistry, and Water

Why it matters . . . We—like all plants, animals, and other organisms—are collections of atoms and molecules linked together by chemical bonds. The chemical nature of life makes it impossible to understand biology without knowledge of basic chemistry and chemical interactions.

For example, the element selenium is a natural ingredient of rocks and soils. In minute amounts it is necessary for the normal growth and survival of humans and many other animals. However, in high concentrations selenium is toxic. For instance, in 1983, high concentrations of selenium killed or deformed thousands of waterfowl at the Kesterson National Wildlife Refuge in the San Joaquin Valley of California. The selenium had built up over decades as irrigation runoff washed selenium-containing chemicals from the soil into the water of the refuge. Since the problem was identified, engineers have diverted agricultural drainage water from the area, and the Kesterson refuge is now being restored.

Can selenium be prevented from accumulating in the environment? Norman Terry and his coworkers at the University of California, Berkeley found that some wetland plants could remove up to 90% of the selenium in wastewater from a gasoline refinery. The plants convert much of the selenium into a relatively nontoxic gas, methyl selenide, which can pass into the atmosphere without harming plants and animals.

To test further the ability of plants to remove selenium, Terry and his coworkers grew wetland plants in 10 experimental plots watered by runoff from agricultural irrigation **(Figure 2.1)**. The researchers measured how much selenium remained in the soil of the plots, how much was incorporated into plant tissues, and how much escaped into the air as a gas. Terry's results indicate that, before the runoff trickles through to local ponds, the plants in his plots reduce selenium to nontoxic levels, less than two parts per billion. Such applications of chemical and biological knowl-

FIGURE 2.1 Researcher Norman Terry in an experimental wetlands plot in Corcoran, California. Terry tested the ability of cattails, bulrushes, and marsh grasses to reduce selenium contamination in water draining from irrigated fields.

Living Organisms Are Composed of about 25 Key Elements

Four elements—carbon, hydrogen, oxygen, and nitrogen—make up more than 96% of the weight of living organisms. Seven other elements—calcium, phosphorus, potassium, sulfur, sodium, chlorine, and magnesium—contribute most of the remaining 4%. Several other elements occur in organisms in quantities so small (<0.01%) that they are known as **trace elements. Figure 2.2** compares the relative proportions of different elements in a human, a plant, Earth's crust, and seawater, and lists the most important trace elements in a human. The proportions of elements in living organisms, as represented by the human and the plant, differ markedly from those of Earth's crust and seawater; these differences reflect the highly ordered chemical structure of living organisms.

Trace elements are vital for normal biological functions. For example, iodine makes up only about 0.0004% of a human's weight. However, a lack of iodine in the human diet severely impairs the function of the thyroid gland, which produces hormones that regulate metabolism and growth (see Chapter 42). Symptoms of iodine deficiency include lethargy, apathy, and sensitivity to cold temperatures. Prolonged iodine deficiency causes a *goiter,* a condition in which the thyroid gland enlarges so much that the front of the neck swells significantly. Once a

edge to decontaminate polluted environments are known as **bioremediation.** Bioremediation could help safeguard our food supplies, our health, and the environment.

The reactions involving selenium show the importance of understanding and applying chemistry in biology. Many thousands of chemical reactions take place inside living organisms. Decades of research have taught us much about these reactions and have confirmed that the same laws of chemistry and physics govern both living and nonliving things. Therefore, we can apply with confidence information obtained from chemical experiments in the laboratory to the processes inside living organisms. An understanding of the relationship between the structure of chemical substances and their behavior is the first step in learning biology.

2.1 The Organization of Matter: Elements and Atoms

Selenium is an example of an **element**—a pure substance that cannot be broken down into simpler substances by ordinary chemical or physical techniques. All **matter** of the universe—anything that occupies space and has mass—is composed of elements and combinations of elements. Ninety-two different elements occur naturally on Earth, and more than fifteen artificial elements have been synthesized in the laboratory.

Seawater	
Oxygen	88.3
Hydrogen	11.0
Chlorine	1.9
Sodium	1.1
Magnesium	0.1
Sulfur	0.09
Potassium	0.04
Calcium	0.04
Carbon	0.003
Silicon	0.0029
Nitrogen	0.0015
Strontium	0.0008

Human	
Oxygen	65.0
Carbon	18.5
Hydrogen	9.5
Nitrogen	3.3
Calcium	2.0
Phosphorus	1.1
Potassium	0.35
Sulfur	0.25
Sodium	0.15
Chlorine	0.15
Magnesium	0.05
Iron	0.004
Iodine	0.0004

Pumpkin	
Oxygen	85.0
Hydrogen	10.7
Carbon	3.3
Potassium	0.34
Nitrogen	0.16
Phosphorus	0.05
Calcium	0.02
Magnesium	0.01
Iron	0.008
Sodium	0.001
Zinc	0.0002
Copper	0.0001

Earth's crust	
Oxygen	46.6
Silicon	27.7
Aluminum	8.1
Iron	5.0
Calcium	3.6
Sodium	2.8
Potassium	2.6
Magnesium	2.1
Other elements	1.5

FIGURE 2.2 The proportions by mass of different elements in seawater, the human body, a fruit, and Earth's crust. Trace elements in humans include boron, chromium, cobalt, copper, fluorine, iodine, iron, manganese, molybdenum, selenium, tin, vanadium, and zinc, as well as variable traces of other elements.

© Cengage Learning 2014

common condition, goiter has almost been eliminated by adding iodine to table salt, especially in regions where soils, and therefore crops grown in those soils, are iodine-deficient.

Elements Are Composed of Atoms, Which Combine to Form Molecules

Elements are composed of individual **atoms**—the smallest units that retain the chemical and physical properties of an element. Any given element has only one type of atom. Several million atoms arranged side by side would be needed to equal the width of the period at the end of this sentence.

Atoms are identified by a standard one- or two-letter symbol. For example, the element carbon is identified by the single letter *C*, which stands for both the carbon atom and the element, and iron is identified by the two-letter symbol *Fe* (*ferrum* = iron). **Table 2.1** lists the chemical symbols of these and other atoms common in living organisms.

Atoms, combined chemically in fixed numbers and ratios, form the **molecules** of living and nonliving matter. For example, the oxygen we breathe is a molecule formed from the chemical combination of two oxygen atoms, and a molecule of the carbon dioxide we exhale contains one carbon atom and two oxygen atoms. The name of a molecule is written in chemical shorthand as a **formula,** using the standard symbols for the elements and using subscripts to indicate the number of atoms of each element in the molecule. The subscript is omitted for atoms that occur only once in a molecule. For example, the formula for an oxygen molecule is written as O_2 (two oxygen atoms); for a carbon dioxide molecule, the formula is CO_2 (one carbon atom and two oxygen atoms).

Molecules whose component atoms are different (such as carbon dioxide) are called **compounds.** The chemical and physical properties of compounds are typically distinct from those of their atoms or elements. For example, we all know that water is a liquid at room temperature. We also know that water does not burn. However, the individual elements of water—hydrogen and oxygen—are gases at room temperature, and both are highly reactive.

2.2 Atomic Structure

Each element consists of one type of atom. However, all atoms share the same basic structure **(Figure 2.3).** Each atom consists of an **atomic nucleus,** surrounded by one or more smaller, fast-moving particles called **electrons.** Although the electrons occupy more than 99.99% of the space of an atom, the nucleus makes up more than 99.99% of its total mass.

The Atomic Nucleus Contains Protons and Neutrons

All atomic nuclei contain one or more positively charged particles called **protons.** The number of protons in the nucleus of each kind of atom is referred to as the **atomic number.** This number does not vary and thus specifically identifies the atom. The smallest atom, hydrogen, has a single proton in its nucleus, so its

TABLE 2.1	Atomic Number and Mass Number of the Most Common Elements in Living Organisms		
Element	Symbol	Atomic Number	Mass Number of the Most Common Form
Hydrogen	H	1	1
Carbon	C	6	12
Nitrogen	N	7	14
Oxygen	O	8	16
Sodium	Na	11	23
Magnesium	Mg	12	24
Phosphorus	P	15	31
Sulfur	S	16	32
Chlorine	Cl	17	35
Potassium	K	19	39
Calcium	Ca	20	40
Iron	Fe	26	56
Iodine	I	53	127

© Cengage Learning 2014

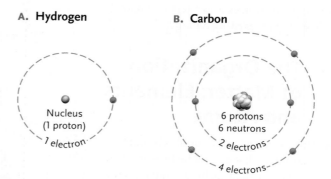

FIGURE 2.3 Atomic structure. The nucleus of an atom contains one or more protons and, except for the most common form of hydrogen, a similar number of neutrons. Fast-moving electrons, in numbers equal to the protons, surround the nucleus. The most common form of hydrogen, the simplest atom, has a single proton in its nucleus and a single electron. Carbon, a more complex atom, has a nucleus surrounded by electrons at two levels. The electrons in the outer level follow more complex pathways than shown here.
© Cengage Learning 2014

atomic number is 1. The heaviest naturally occurring atom, uranium, has 92 protons in its nucleus and therefore has an atomic number of 92. Carbon with six protons, nitrogen with seven protons, and oxygen with eight protons have atomic numbers of 6, 7, and 8, respectively (see Table 2.1).

With one exception, the nuclei of all atoms also contain uncharged particles called **neutrons.** Neutrons occur in variable numbers approximately equal to the number of protons. The lone exception is the most common form of hydrogen, which has just a single proton in its nucleus. There are two less common forms of hydrogen as well. One form, named deuterium, has one neutron and one proton in its nucleus. The other form, named tritium, has two neutrons and one proton.

Other atoms also have common and less common forms with different numbers of neutrons. For example, the most common form of the carbon atom has six protons and six neutrons in its nucleus, but about 1% of carbon atoms have six protons and seven neutrons in their nuclei and an even smaller percentage of carbon atoms have six protons and eight neutrons.

The distinct forms of the atoms of an element, all with the same number of protons but different numbers of neutrons, are called **isotopes (Figure 2.4).** The various isotopes of an atom differ in mass and other physical characteristics, but all have essentially the same chemical properties. Therefore, organisms can use any hydrogen or carbon isotope, for example, without a change in their chemical reactions.

A neutron and a proton have almost the same mass, about 1.66×10^{-24} grams (g). This mass is defined as a standard unit, the **dalton** (Da), named after John Dalton, a nineteenth-century English scientist who contributed to the development of atomic theory. Atoms are assigned a **mass number** based on the total number of protons and neutrons in the atomic nucleus (see Table 2.1). Electrons are ignored in determinations of atomic mass because

Isotopes of hydrogen

^{1}H	^{2}H (deuterium)	^{3}H (tritium)
1 proton	1 proton	1 proton
	1 neutron	2 neutrons
Atomic number = 1	Atomic number = 1	Atomic number = 1
Mass number = 1	Mass number = 2	Mass number = 3

Isotopes of carbon

^{12}C	^{13}C	^{14}C
6 protons	6 protons	6 protons
6 neutrons	7 neutrons	8 neutrons
Atomic number = 6	Atomic number = 6	Atomic number = 6
Mass number = 12	Mass number = 13	Mass number = 14

FIGURE 2.4 **The atomic nuclei of hydrogen and carbon isotopes.** Note that isotopes of an atom have the same atomic number but different mass numbers.

© Cengage Learning 2014

the mass of an electron, at only 1/1,800th of the mass of a proton or neutron, does not contribute significantly to the mass of an atom. Thus, the mass number of the hydrogen isotope with one proton in its nucleus is 1, and its mass is 1 Da. The mass number of the hydrogen isotope deuterium is 2, and the mass number of tritium is 3. These hydrogen mass numbers are written as ^{2}H and ^{3}H. The carbon isotope with six protons and six neutrons in its nucleus has a mass number of 12; the isotope with six protons and seven neutrons has a mass number of 13, and the isotope with six protons and eight neutrons has a mass number of 14 (see Figure 2.4). These carbon mass numbers are written as ^{12}C, ^{13}C, and ^{14}C, or carbon-12, carbon-13, and carbon-14, respectively. However, all carbon isotopes have the same atomic number of 6, because this number reflects only the number of protons in the nucleus.

What is the meaning of *mass* as compared to *weight*? **Mass** is the amount of matter in an object, whereas **weight** is a measure of the pull of gravity on an object. Mass is constant, but the weight of an object may vary because of differences in gravity. For example, a piece of lead that weighs 1 kilogram (kg) on Earth is weightless in an orbiting spacecraft, even though its mass is the same in both places. However, as long as an object is on Earth's surface, its mass and weight are equivalent. Thus, the weight of an object in the laboratory accurately reflects its mass.

The Nuclei of Some Atoms Are Unstable and Tend to Break Down to Form Simpler Atoms

The nuclei of some isotopes are unstable and break down, or *decay,* giving off particles of matter and energy that can be detected as **radioactivity.** The decay transforms the unstable, radioactive isotope—called a **radioisotope**—into an atom of another element. The decay continues at a steady, clocklike rate, with a constant proportion of the radioisotope breaking down at any instant. The rate of decay is not affected by chemical reactions or environmental conditions such as temperature or pressure. For example, the carbon isotope ^{14}C is unstable and undergoes radioactive decay in which one of its neutrons splits into a proton and an electron. The electron is ejected from the nucleus, but the proton is retained, giving a new total of seven protons and seven neutrons, which is characteristic of the most common form of nitrogen. Thus, the decay transforms the carbon atom into an atom of nitrogen.

Because unstable isotopes decay at a clocklike rate, they can be used to estimate the age of organic material, rocks, or fossils that contain them. These techniques have been vital in dating animal remains and tracing evolutionary lineages, as described in Chapter 23. Isotopes are also used in biological research as **tracers** to label molecules so that they can be tracked as they pass through biochemical reactions. Radioactive isotopes of carbon (^{14}C), phosphorus (^{32}P), and sulfur (^{35}S) can be traced easily by their radioactivity. A number of stable, nonradioactive isotopes, such as ^{15}N (called heavy nitrogen), can be detected by their mass differences and have also proved valuable as tracers in biological experiments. *Focus on Applied Research* describes some applications of radioisotopes in medicine.

Using Radioisotopes to Save Lives

Radioisotopes are widely used in medicine to diagnose and cure disease, to produce images of diseased body organs, and to trace the locations and routes followed by individual substances marked for identification by radioactivity. One example of their use is in the diagnosis of thyroid gland disease. The thyroid is the only structure in the body that absorbs iodine in quantity. The size and shape of the thyroid, which reflect its health, are measured by injecting a small amount of a radioactive iodine isotope into the patient's bloodstream. After the isotope is concentrated in the thyroid, the gland is then scanned by an apparatus that uses the radioactivity to produce an image of the gland on X-ray film (the energy of the radioisotope decay exposes the film). Examples of what the scans may show are presented in the **Figure.** Another application uses the fact that radioactive thallium is not taken up by regions of the heart muscle with poor circulation to detect coronary artery disease. Other isotopes are used to detect bone injuries and defects, including injured, arthritic, or abnormally growing segments of bone.

Treatment of disease with radioisotopes takes advantage of the fact that radioactivity in large doses can kill cells (radiation generates highly reactive chemical groups that break and disrupt biological molecules). Dangerously overactive thyroid glands are treated by giving patients a dose of radioactive iodine calculated to destroy just enough thyroid cells to reduce activity of the gland to normal levels. In radiation therapy, cancer cells are killed by bombarding them with radiation emitted by radium-226 or cobalt-60. As much as is possible, the radiation is focused on the tumor to avoid destroying nearby healthy tissues. In some forms of chemotherapy for cancer, patients are given radioactive substances at levels that kill cancer cells without also killing the patient.

Normal   Enlarged   Cancerous

FIGURE Scans of human thyroid glands after iodine-123 (^{123}I) was injected into the bloodstream. The radioactive iodine becomes concentrated in the thyroid gland.

© Cengage Learning 2014

The Electrons of an Atom Occupy Orbitals around the Nucleus

In an atom, the number of electrons surrounding the nucleus is equal to the number of protons in the nucleus. An electron carries a negative charge that is exactly equal and opposite to the positive charge of a proton. The equality of numbers of electrons and protons in an atom balances the positive and negative charges and makes the total structure of an atom electrically neutral.

An atom is often drawn with electrons following a pathway around the nucleus similar to planets orbiting a sun. The reality is different. The speed of electrons in motion around the nucleus approaches the speed of light. At any instant, an electron may be in any location with respect to its nucleus, from the immediate vicinity of the nucleus to practically infinite space. An electron moves so fast that it almost occupies all the locations at the same time. However, the electron passes through some locations much more frequently than others. The locations where an electron occurs most frequently around the atomic nucleus define a path called an orbital. An **orbital** is essentially the region of space where the electron "lives" most of the time. Although either one or two electrons may occupy an orbital, the most stable and balanced condition occurs when an orbital contains a pair of electrons.

Electrons are maintained in their orbitals by a combination of attraction to the positively charged nucleus and mutual repulsion because of their negative charge. The orbitals take different shapes depending on their distance from the nucleus and their degree of repulsion by electrons in other orbitals.

Under certain conditions, electrons may pass from one orbital to another within an atom, enter orbitals shared by two or more atoms, or pass completely from orbitals in one atom to orbitals in another. As discussed later in this chapter, the ability of electrons to move from one orbital to another underlies the chemical reactions that combine atoms into molecules.

Orbitals Occur in Discrete Layers around an Atomic Nucleus

Within an atom, electrons are found in regions of space called **energy levels,** or more simply, **shells.** Within each energy level, electrons are grouped into orbitals. The lowest energy level of an atom, the one nearest the nucleus, may be occupied by a maximum of two electrons in a single orbital **(Figure 2.5A).** This orbital, which has a spherical shape, is called the 1s orbital. (The "1" signifies that the orbital is in the energy level closest to the nucleus, and the "s" signifies the shape of the orbital, in this case, spherically symmetric around the nucleus.) Hydrogen has one electron in this orbital, and helium has two.

Atoms with atomic numbers between 3 (lithium) and 10 (neon) have two energy levels, with two electrons in the 1s orbital and one to eight electrons in orbitals at the next highest energy level. The electrons at this second energy level occupy one spherical orbital, called the 2s orbital **(Figure 2.5B),** and as many as three orbitals that are pushed into a dumbbell shape by repulsions between electrons, called 2p orbitals **(Figure 2.5C).** The orbitals for neon are shown in **Figure 2.5D.**

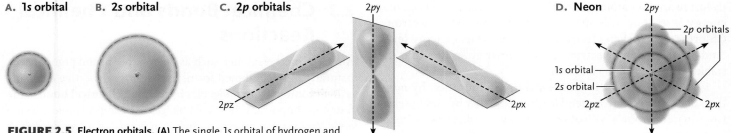

A. 1s orbital   **B. 2s orbital**   **C. 2p orbitals**   **D. Neon**

FIGURE 2.5 Electron orbitals. **(A)** The single 1s orbital of hydrogen and helium approximates a sphere centered on the nucleus. **(B)** The 2s orbital. **(C)** The 2p orbitals lie in the three planes x, y, and z, each at right angles to the others. **(D)** In atoms with two energy levels, such as neon, the lowest energy level is occupied by a single 1s orbital as in hydrogen and helium. The second, higher energy level is occupied by a maximum of four orbitals—a spherical 2s orbital and three dumbbell-shaped 2p orbitals.
© Cengage Learning 2014

Larger atoms have more energy levels. The third energy level, which may contain as many as 18 electrons in 9 orbitals, includes the atoms from sodium (11 electrons) to argon (18 electrons). (**Figure 2.6** shows the 18 elements that have electrons in the lowest three energy levels only.) The fourth energy level may contain as many as 32 electrons in 16 orbitals. In all cases, the total number of electrons in the orbitals is matched by the number of protons in the nucleus. However, whatever the size of an atom, the outermost energy level typically contains one to eight electrons occupying a maximum of four orbitals.

FIGURE 2.6 **The atoms with electrons distributed in one, two, or three energy levels.** The atomic number of each element (shown in boldface in each panel) is equivalent to the number of protons in its nucleus.
© Cengage Learning 2014

The Number of Electrons in the Outermost Energy Level of an Atom Determines Its Chemical Activity

The electrons in an atom's outermost energy level are known as **valence electrons** (*valentia* = power or capacity). Atoms in which the outermost energy level is not completely filled with electrons tend to be chemically reactive, whereas those with a completely filled outermost energy level are nonreactive, or inert. For example, hydrogen has a single, unpaired electron in its outermost and only energy level, and it is highly reactive; helium has two valence electrons filling its single orbital, and it is inert. For atoms with two or more energy levels, only those with unfilled outer energy levels are reactive. Those with eight electrons completely filling the four orbitals of the outer energy level, such as neon and argon, are stable and chemically unreactive (see Figure 2.6).

Atoms with outer energy levels that contain electrons near the stable numbers tend to gain or lose electrons to reach the stable configuration. For example, sodium has two electrons in its first energy level, eight in the second, and one in the third and outermost level (see Figure 2.6). The outermost electron is read-

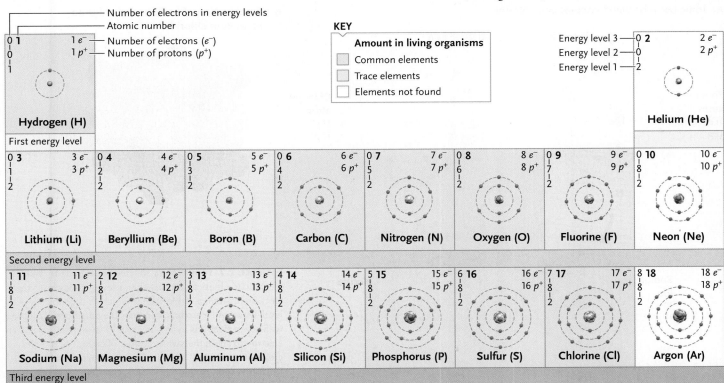

ily lost to another atom, giving the sodium atom a stable second energy level (now the outermost level) with eight electrons. Chlorine, with seven electrons in its outermost energy level, tends to take up an electron from another atom to attain the stable number of eight electrons.

Atoms that differ from the stable configuration by more than one or two electrons tend to attain stability by *sharing* electrons in joint orbitals with other atoms rather than by gaining or losing electrons completely. Among the atoms that form biological molecules, electron sharing is most characteristic of carbon, which has four electrons in its outer energy level and thus falls at the midpoint between the tendency to gain or lose electrons. Oxygen, with six electrons in its outer level, and nitrogen, with five electrons in its outer level, also share electrons readily. Hydrogen may either share or lose its single electron. The relative tendency to gain, share, or lose valence electrons underlies the chemical bonds and forces that hold the atoms of molecules together.

STUDY BREAK 2.2

1. **Where are protons, electrons, and neutrons found in an atom?**
2. **The isotopes carbon-11 and oxygen-15 do not occur in nature, but they can be made in the laboratory. Both are used in a medical imaging procedure called positron emission tomography. How many protons and neutrons are in carbon-11 and in oxygen-15?**
3. **What determines the chemical reactivity of an atom?**

2.3 Chemical Bonds and Chemical Reactions

Atoms of inert elements, such as helium, neon, and argon, occur naturally in uncombined forms, but atoms of reactive elements tend to combine into molecules by forming **chemical bonds.** The four most important chemical linkages in biological molecules are *ionic bonds, covalent bonds, hydrogen bonds,* and *van der Waals forces.* Chemical reactions occur when atoms or molecules interact to form new chemical bonds or break old ones.

Ionic Bonds Are Multidirectional and Vary in Strength

Ionic bonds result from electrical attractions between atoms that gain or lose valence electrons completely. A sodium atom (Na) readily loses a single electron to achieve a stable outer energy level, and chlorine (Cl) readily gains an electron:

$$\text{Na}^{\cdot} \ + \ \cdot \ddot{\text{C}}\ddot{\text{l}}: \ \rightarrow \ \text{Na}^{+} \ :\ddot{\text{C}}\ddot{\text{l}}:^{-}$$

(The dots in the preceding formula represent the electrons in the outermost energy level.) After the transfer, the sodium atom, now with 11 protons and 10 electrons, carries a single positive charge. The chlorine atom, now with 17 protons and 18 electrons, carries a single negative charge. In this charged condition, the sodium and chlorine atoms are called **ions** instead of atoms and are written as Na^{+} and Cl^{-} **(Figure 2.7).** A positively charged ion such as Na^{+} is called a **cation,** and a negatively charged ion such

A. **Ionic bond between sodium and chlorine**

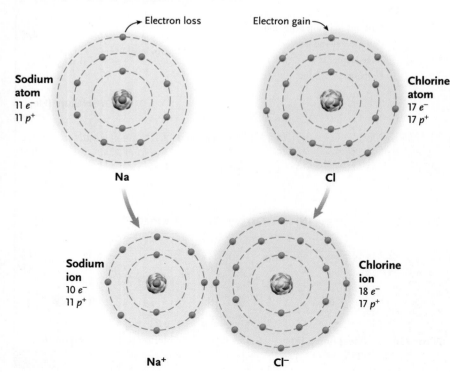

B. **Combination of sodium and chlorine in sodium chloride**

FIGURE 2.7 **Formation of an ionic bond.** Sodium, with one electron in its outermost energy level, readily loses that electron to attain a stable state in which its second energy level, with eight electrons, becomes the outer level. Chlorine, with seven electrons in its outer energy level, readily gains an electron to attain the stable number of eight. The transfer creates the ions Na^{+} and Cl^{-}. The combination of Na^{+} and Cl^{-} forms sodium chloride (NaCl), common table salt.
© Cengage Learning 2014

A. Shared orbitals of methane (CH₄)

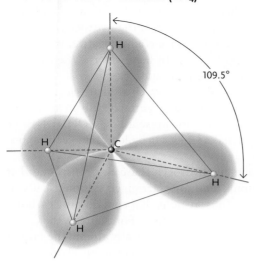

109.5°

B. Space-filling model of methane

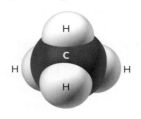

D. Cholesterol

Hydrogen

Carbon

C. A carbon "building block" used to make molecular models

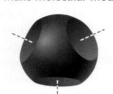

Oxygen

FIGURE 2.8 Covalent bonds shared by carbon. (A) The four covalent bonds of carbon in methane (CH₄) are shown as shared orbitals. The bonds extend outward from the carbon nucleus at angles of 109.5° from each other (dashed lines). The red lines connecting the hydrogen nuclei form a regular tetrahedron with four faces. **(B)** In the space-filling model of methane, the diameter of the sphere representing an atom shows the approximate limit of its electron orbitals. **(C)** A tetrahedral carbon "building block." One of the four faces of the block is not visible. **(D)** Carbon atoms assembled into rings and chains forming a complex molecule.
© Cengage Learning 2014

as Cl⁻ is called an **anion.** The difference in charge between cations and anions creates an attraction—the ionic bond—that holds the ions together in solid NaCl (sodium chloride).

Many other atoms that differ from stable outer energy levels by one electron, including hydrogen, can gain or lose electrons completely to form ions and ionic bonds. When a hydrogen atom loses its single electron to form a hydrogen ion (H⁺), it consists of only a proton and is often simply called a proton to reflect this fact. A number of atoms with outer energy levels that differ from the stable number by two or three electrons, particularly metallic atoms such as calcium (Ca²⁺), magnesium (Mg²⁺), and iron (Fe²⁺ or Fe³⁺), also lose their electrons readily to form cations and to join in ionic bonds with anions.

Ionic bonds are common among the forces that hold ions, atoms, and molecules together in living organisms because these bonds have three key features: (1) they exert an attractive force over greater distances than any other chemical bond; (2) their attractive force extends in all directions; and (3) they vary in strength depending on the presence of other charged substances. That is, in some systems, ionic bonds form in locations that exclude other charged substances, setting up strong and stable attractions that are not easily disturbed. For example, iron ions in the large biological molecule hemoglobin are stabilized by ionic bonds; these iron ions are key to the molecule's distinctive chemical properties. In other systems, particularly at molecular surfaces exposed to water molecules, ionic bonds are relatively weak, allowing ionic attractions to be established or broken quickly. For example, as part of their activity in speeding biological reactions, many enzymatic proteins bind and release molecules by forming and breaking relatively weak ionic bonds.

Covalent Bonds Are Formed by Electrons in Shared Orbitals

Covalent bonds form when atoms share a pair of valence electrons rather than gaining or losing them. For example, if two hydrogen atoms collide, the single electron of each atom may join in a new, combined two-electron orbital that surrounds both nuclei. The two electrons fill the orbital and, therefore, the hydrogen atoms tend to remain linked stably together in the form of molecular hydrogen, H₂. The linkage formed by the shared orbital is a covalent bond.

In molecular diagrams, a covalent bond is designated by a pair of dots or a single line that represents a pair of shared electrons. For example, in H₂, the covalent bond that holds the molecule together is represented as H:H or H—H.

Unlike ionic bonds, which extend their attractive force in all directions, the shared orbitals that form covalent bonds extend between atoms at discrete angles and directions, giving covalently bound molecules distinct, three-dimensional forms. For biological molecules such as proteins, which are held together primarily by covalent bonds, the three-dimensional form imparted by these bonds is critical to their functions.

An example of a molecule that contains covalent bonds is methane, CH₄, the main component of natural gas. Carbon, with four unpaired outer electrons, typically forms four covalent bonds to complete its outermost energy level, here with four atoms of hydrogen. The four covalent bonds are fixed at an angle of 109.5° from each other, forming a tetrahedron **(Figure 2.8A, B).** The tetrahedral arrangement of the bonds allows carbon "building blocks" **(Figure 2.8C)** to link to each other in both branched and unbranched chains and rings **(Figure 2.8D).** Such structures form the backbones of an almost unlimited variety of molecules. Carbon can also form double bonds, in which atoms share two pairs of electrons, and triple bonds, in which atoms share three pairs of electrons.

Oxygen, hydrogen, nitrogen, and sulfur also share electrons readily to form covalent linkages, and they commonly combine with carbon in biological molecules. In these linkages with carbon, oxygen typically forms two covalent bonds; hydrogen, one; nitrogen, three; and sulfur, two.

Unequal Electron Sharing Results in Polarity

Electronegativity is the measure of an atom's attraction for the electrons it shares in a chemical bond with another atom. The more electronegative an atom is, the more strongly it attracts shared electrons. Among atoms, electronegativity increases as the number of protons in the nucleus increases and as the distance of electrons from the nucleus increases.

Although all covalent bonds involve the sharing of valence electrons, they differ widely in the degree of sharing. Depending on the difference in electronegativity between the bonded atoms, covalent bonds are classified as **nonpolar covalent bonds** or **polar covalent bonds.** In a nonpolar covalent bond, electrons are shared equally, whereas in a polar covalent bond, they are shared unequally. When electron sharing is unequal, as in polar covalent bonds, the atom that attracts the electrons more strongly carries a partial negative charge, δ^- ("delta minus"), and the atom deprived of electrons carries a partial positive charge, δ^+ ("delta plus"). The atoms carrying partial charges may give the whole molecule partially positive and negative ends; in other words, the molecule is *polar*, hence the name given to the bond.

Nonpolar covalent bonds are characteristic of molecules that contain atoms of one element, such as elemental hydrogen (H_2) and oxygen (O_2), although there are some exceptions. Polar covalent bonds are characteristic of molecules that contain atoms of different elements.

For example, in water, an oxygen atom forms polar covalent bonds with two hydrogen atoms. Because the oxygen nucleus with its eight protons attracts electrons much more strongly than the hydrogen nuclei do, the bonds are strongly polar **(Figure 2.9).** In addition, the water molecule is asymmetric, with the oxygen atom located on one side and the hydrogen atoms on the other. This arrangement gives the entire molecule an unequal charge distribution, with the hydrogen end partially positive and the oxygen end partially negative, and makes water molecules strongly polar. In fact, water is the primary biological example of a polar molecule.

Oxygen, nitrogen, and sulfur, which all share electrons unequally with hydrogen, are located asymmetrically in many biological molecules. Therefore, the presence of —OH, —NH, or —SH groups tends to make regions in biological molecules containing them polar.

Although carbon and hydrogen share electrons somewhat unequally, these atoms tend to be arranged symmetrically in biological molecules. Thus, regions that contain only carbon–hydrogen chains are typically nonpolar. For example, the C—H bonds in methane are located symmetrically around the carbon atom (see Figure 2.8), so their partial charges cancel each other and the molecule as a whole is nonpolar.

Polar Molecules Tend to Associate with Each Other and Exclude Nonpolar Molecules

Polar molecules attract and align themselves with other polar molecules and with charged ions and molecules. These **polar associations** create environments that tend to exclude nonpolar molecules. When present in quantity, the excluded nonpolar molecules tend to clump together in arrangements called **nonpolar associations;** these nonpolar associations reduce the surface area exposed to the surrounding polar environment. Polar molecules that associate readily with water are identified as **hydrophilic** (*hydro* = water; *philic* = having an affinity or preference for). Nonpolar substances that are excluded by water and other polar molecules are identified as **hydrophobic** (*phobic* = having an aversion to or fear of).

Polar and nonpolar associations can be demonstrated with an apparatus as simple as a bottle containing water and vegetable oil. If the bottle has been placed at rest for some time, the nonpolar oil and polar water form separate layers, with the oil on top. If you shake the bottle, the oil becomes suspended as spherical droplets in the water; the harder you shake, the smaller the oil droplets become (the spherical form of the oil droplets exposes the least surface area per unit volume to the watery polar surroundings). If you place the bottle at rest, the oil and water quickly separate again into distinct polar and nonpolar layers.

Hydrogen Bonds Also Involve Unequal Electron Sharing

When hydrogen atoms are made partially positive by sharing electrons unequally with oxygen, nitrogen, or sulfur, they may be attracted to nearby oxygen, nitrogen, or sulfur atoms made partially negative by unequal electron sharing in a different covalent bond. This attractive force, the **hydrogen bond,** is illustrated by a dotted line in structural diagrams of molecules **(Figure 2.10A).**

FIGURE 2.9 Polarity in the water molecule, created by unequal electron sharing between the two hydrogen atoms and the oxygen atom and the asymmetric shape of the molecule. The unequal electron sharing gives the hydrogen end of the molecule a partial positive charge, δ^+ ("delta plus"), and the oxygen end of the molecule a partial negative charge, δ^- ("delta minus"). Regions of deepest color indicate the most frequent locations of the shared electrons. The orbitals occupied by the electrons are more complex than the spherical forms shown here.

© Cengage Learning 2014

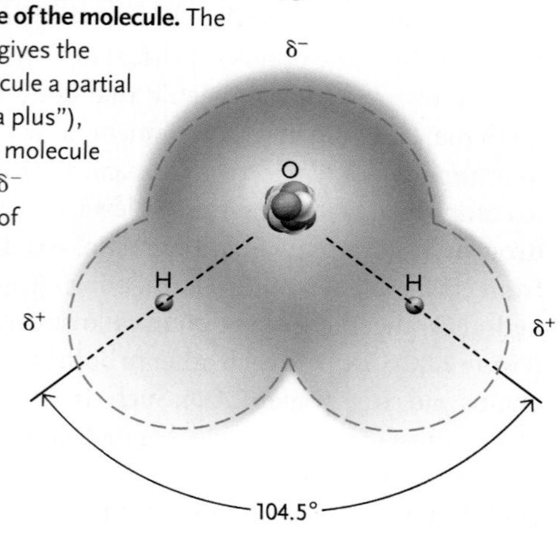

A hydrogen bond (dotted line) between the hydrogen of an —OH group and a nearby nitrogen atom, which also shares electrons unequally with another hydrogen. Regions of deepest blue indicate the most likely locations of electrons.

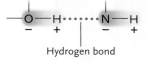

B. **Stabilizing effect of hydrogen bonds**

Multiple hydrogen bonds stabilize the backbone chain of a protein molecule into a spiral called the alpha helix. The spheres labeled *R* represent chemical groups of different kinds.

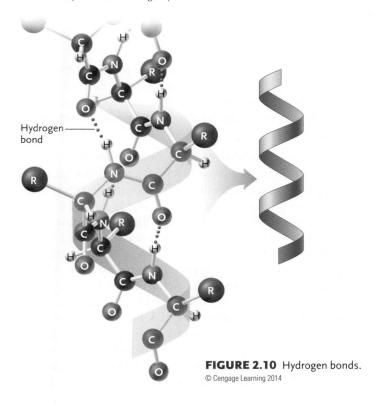

FIGURE 2.10 Hydrogen bonds.
© Cengage Learning 2014

Hydrogen bonds may be *intramolecular* (between atoms in the same molecule) or *intermolecular* (between atoms in different molecules).

Individual hydrogen bonds are weak compared with ionic and covalent bonds. However, large biological molecules may offer many opportunities for hydrogen bonding, both within and between molecules. When numerous, hydrogen bonds are collectively strong and lend stability to the three-dimensional structure of molecules such as proteins **(Figure 2.10B)**. Hydrogen bonds between water molecules are responsible for many of the properties that make water uniquely important to life (see Section 2.4 for a more detailed discussion).

The weak attractive force of hydrogen bonds makes them much easier to break than covalent and ionic bonds, particularly when elevated temperature increases the movements of molecules. Hydrogen bonds begin to break extensively as temperatures rise above 45°C and become practically nonexistent at 100°C. The disruption of hydrogen bonds by heat—for instance, the bonds in proteins—is one of the primary reasons most organisms cannot survive temperatures much greater than 45°C. Thermophilic (temperature-loving) organisms, which live at temperatures higher than 45°C, some at 120°C or more, have different molecules from those of organisms that live at lower temperatures. The proteins in thermophiles are stabilized at high temperatures by van der Waals forces and other noncovalent interactions.

Van der Waals Forces Are Weak Attractions over Very Short Distances

Van der Waals forces are even weaker than hydrogen bonds. These forces develop over very short distances between nonpolar molecules or regions of molecules when, through their constant motion, electrons accumulate by chance in one part of a molecule or another. This process leads to zones of positive and negative charge, making the molecule polar. If they are oriented in the right way, the polar parts of the molecules are attracted electrically to one another and cause the molecules to stick together briefly. Although an individual bond formed with van der Waals forces is weak and transient, the formation of many bonds of this type can stabilize the shape of a large molecule, such as a protein.

A striking example of the collective power of van der Waals forces concerns the ability of geckos, a group of tropical lizard species, to cling to and walk up vertical smooth surfaces **(Figure 2.11)**. The toes of the gecko are covered with millions of hairs, called *setae* (singular, *seta*). At the tip of each seta are hundreds of thousands of pads, called *spatulae*. Each pad forms a weak interaction—using van der Waals forces—with molecules on the smooth surface. Magnified by the huge number of pads involved, the attractive forces are 1,000 times greater than necessary for the gecko to hang on a vertical wall or even from a ceiling. To climb a wall, the animal rolls the setae onto the surface and then peels them off like a piece of tape. Understanding the gecko's remarkable ability to climb has led to the development of gecko tape, a superadhesive prototype tape capable of holding a 3-kg weight with a 1-cm^2 (centimeter squared) piece.

Bonds Form and Break in Chemical Reactions

In **chemical reactions,** atoms or molecules interact to form new chemical bonds or break old ones. As a result of bond formation or breakage, atoms are added to or removed from molecules, or the linkages of atoms in molecules are rearranged. When any of these alterations occur, molecules change from one type to another, usually with different chemical and physical properties. In biological systems, chemical reactions are accelerated by molecules called *enzymes* (which are discussed in more detail in Chapter 4).

The atoms or molecules entering a chemical reaction are called the **reactants,** and those leaving a reaction are the **products.** A chemical reaction is written with an arrow showing the direction of the reaction; reactants are placed to the left of

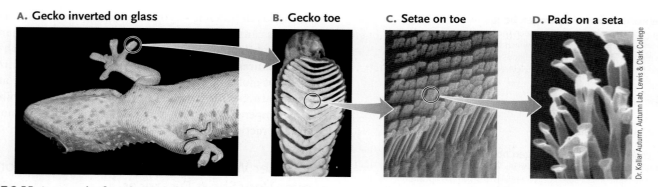

FIGURE 2.11 An example of van der Waals forces in biology. (A) A Tokay gecko *(Gekko gecko)* climbing while inverted on a glass plate. **(B)** Gecko toe. **(C)** Setae on a toe. Each seta is about 100 micrometers (μm; 0.004 inch) long. **(D)** Pads (spatulae) on an individual seta. Each pad is about 200 nanometers (nm; 0.000008 inch) wide—smaller than the wavelength of visible light.

the arrow, and products are placed to the right. Both reactants and products are usually written in chemical shorthand as formulas.

For example, the overall reaction of photosynthesis, in which carbon dioxide and water are combined to produce sugars and oxygen (see Chapter 9), is written as follows:

$$6\ CO_2 + 6\ H_2O \rightarrow C_6H_{12}O_6 + 6\ O_2$$

$$\underset{\text{carbon dioxide}}{} \quad \underset{\text{water}}{} \quad \underset{\text{a sugar}}{} \quad \underset{\text{molecular oxygen}}{}$$

The number in front of each formula indicates the number of molecules of that type among the reactants and products (the number 1 is not written). Notice that there are as many atoms of each element to the left of the arrow as there are to the right, even though the products are different from the reactants. This balance reflects the fact that in such reactions, atoms may be rearranged but not created or destroyed. Chemical reactions written in balanced form are known as **chemical equations.**

With the information about chemical bonds and reactions provided thus far, you are now ready to examine the effects of chemical structure and bonding, particularly hydrogen bonding, in the production of the unusual properties of water, the most important substance to life on Earth.

STUDY BREAK 2.3

1. How does an ionic bond form?
2. How does a covalent bond form?
3. What is electronegativity, and how does it relate to nonpolar covalent bonds and polar covalent bonds?
4. What is a chemical reaction?

2.4 Hydrogen Bonds and the Properties of Water

All living organisms contain water, and many kinds of organisms live directly in water. Even those that live in dry environments contain water in all their structures—different organisms range from 50% to more than 95% water by weight. The water inside

organisms is crucial for life: it is required for many important biochemical reactions and plays major roles in maintaining the shape and organization of cells and tissues. Key properties of water molecules that make them so important to life include the following:

- Hydrogen bonds between water molecules produce a *water lattice.* (A lattice is a cross-linked structure.) The lattice arrangement makes water denser than ice and gives water several other properties that make it a highly suitable medium for the molecules and reactions of life. For example, *water absorbs or releases relatively large amounts of energy as heat without undergoing extreme changes in temperature.* This property stabilizes both living organisms and their environments. Also, the water lattice has an unusually high internal *cohesion* (resistance of water molecules to separate). This property plays an important role, for example, in water transport from the roots to the leaves of plants (see Chapter 34). Further, water molecules at surfaces facing air are even more resistant to separation, producing the force called *surface tension,* which, for example, allows droplets of water to form and small insects and spiders to walk on water.
- The polarity of water molecules in the hydrogen-bond lattice contributes to the formation of distinct polar and nonpolar environments that are critical to the organization of cells.
- Water is a *solvent* for charged or polar molecules, meaning that it is a solution in which such molecules can dissolve. Water's solvent properties made life possible because it enabled the chemical reactions needed for life to evolve. The chemical reactions in all organisms take place in aqueous solutions.
- Water molecules separate into ions. Those ions are important for maintaining an environment within cells that is optimal for the chemical reactions that occur there.

A Lattice of Hydrogen Bonds Gives Water Several Unusual, Life-Sustaining Properties

Hydrogen bonds form readily between water molecules in both liquid water and ice. In liquid water, each water molecule establishes an average of 3.4 hydrogen bonds with its neighbors, forming an arrangement known as the **water lattice (Figure 2.12A)**. In liquid water, the hydrogen bonds that hold the lattice together constantly break and reform, allowing the water molecules to break loose from the lattice, slip past one another, and reform the lattice in new positions.

THE DIFFERING DENSITIES OF WATER AND ICE In ice, the water lattice is a rigid, crystalline structure in which each water molecule forms four hydrogen bonds with neighboring molecules **(Figure 2.12B)**. The rigid **ice lattice** spaces the water molecules farther apart than the water lattice. Because of this greater spacing, water has the unusual property of being about 10% less dense when solid than when liquid. (Almost all other substances are denser in solid form than in liquid form.) Hence, water filling a closed glass vessel will expand, breaking the vessel when the water freezes. At atmospheric pressure, water reaches its greatest density at a temperature of 4°C, while it is still a liquid.

Because it is less dense than liquid water, ice forms at the surface of a body of water and remains floating at the surface. The ice creates an insulating layer that helps keep the water below from freezing. If ice were denser than liquid water, a body of water would freeze from the bottom up, killing most aquatic plants and animals in it.

THE BOILING POINT AND TEMPERATURE-STABILIZING EFFECTS OF WATER The hydrogen-bond lattice of liquid water retards the escape of individual water molecules as the water is heated. As a result, relatively high temperatures and the addition of consider-able heat are required to break enough hydrogen bonds to make water boil. Its high boiling point maintains water as a liquid over the wide temperature range of 0°C to 100°C. Similar molecules that do not form an extended hydrogen-bond lattice, such as H_2S (hydrogen sulfide), have much lower boiling points and are gases rather than liquids at room temperature. Without its hydrogen-bond lattice, water would boil at −81°C. If this were the case, most of the water on Earth would be in gaseous form and life as described in this book would not have developed and evolved.

As a result of water's stabilizing hydrogen-bond lattice, it also has a relatively high **specific heat**—that is, the amount of energy as heat required to increase the temperature of a given quantity of water. As heat energy flows into water, much of it is absorbed in the breakage of hydrogen bonds. As a result, the temperature of water increases relatively slowly as heat energy is added. For example, a given amount of heat energy increases the temperature of water by only half as much as that of an equal quantity of ethyl alcohol. High specific heat allows water to absorb or release relatively large quantities of heat energy without undergoing extreme changes in temperature; this gives it a moderating and stabilizing effect on both living organisms and their environments.

The specific heat of water is measured in **calories.** This unit, used both in the sciences and in dieting, is the amount of heat energy required to raise 1 g of water by 1°C (technically, from 14.5°C to 15.5°C at one atmosphere of pressure). This amount of heat is known as a "small" calorie and is written with a small *c*. The unit most familiar to dieters, equal to 1,000 small calories, is

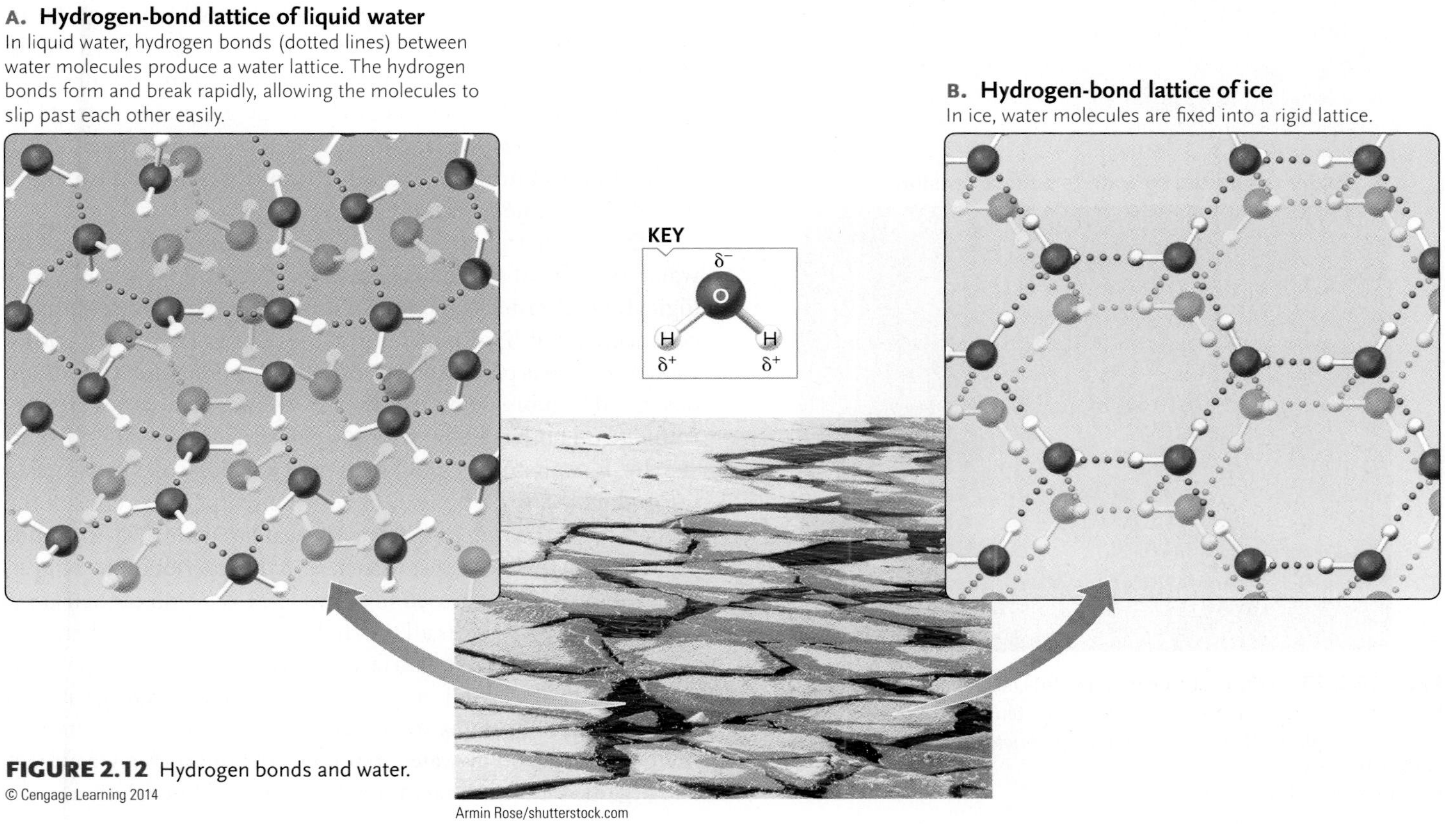

A. Hydrogen-bond lattice of liquid water
In liquid water, hydrogen bonds (dotted lines) between water molecules produce a water lattice. The hydrogen bonds form and break rapidly, allowing the molecules to slip past each other easily.

B. Hydrogen-bond lattice of ice
In ice, water molecules are fixed into a rigid lattice.

KEY

δ^-
O
H   H
δ^+ δ^+

FIGURE 2.12 Hydrogen bonds and water.
© Cengage Learning 2014

Armin Rose/shutterstock.com

written with a capital *C* as a **Calorie;** the same 1,000-calorie unit is known scientifically as a **kilocalorie (kcal).** A 250-Calorie candy bar therefore really contains 250,000 calories.

A large amount of heat, 586 calories per gram, must be added to give water molecules enough energy of motion to break loose from liquid water and form a gas. This required heat, known as the **heat of vaporization,** allows humans and many other organisms to cool off when hot. In humans, water is released onto the surface of the skin by more than 2.5 million sweat glands. The heat energy absorbed by the water in sweat as the sweat evaporates cools the skin and the underlying blood vessels. The heat loss helps keep body temperature from increasing when environmental temperatures are high. Plants use a similar cooling mechanism as water evaporates from their leaves (see Chapter 34).

COHESION AND SURFACE TENSION   The high resistance of water molecules to separation, provided by the hydrogen-bond lattice, is known as internal **cohesion.** For example, in land plants, cohesion holds water molecules in unbroken columns in microscopic conducting tubes that extend from the roots to the highest leaves.

A. Creation of surface tension by unbalanced hydrogen bonding

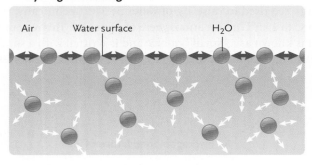

B. Spider supported by water's surface tension

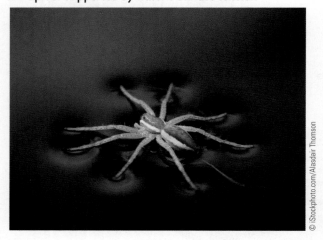

© iStockphoto.com/Alasdair Thomson

FIGURE 2.13 Surface tension in water. (A) Unbalanced hydrogen bonding places water molecules under lateral tension where a water surface faces the air. **(B)** A raft spider *(Dolomedes fimbriatus)* is supported by the surface tension of water.
© Cengage Learning 2014

As water evaporates from the leaves, water molecules in the columns, held together by cohesion, move upward through the tubes to replace the lost water. This movement raises water from roots to the tops of the tallest trees (see discussion in Chapter 34). Maintenance of the long columns of water in the tubes is aided by **adhesion,** in which molecules "stick" to the walls of the tubes by forming hydrogen bonds with charged and polar groups in molecules that form the walls of the tubes.

Water molecules at surfaces facing air can form hydrogen bonds with water molecules beside and below them but not on the sides that face the air. This unbalanced bonding produces a force that places the surface water molecules under tension, making them more resistant to separation than the underlying water molecules **(Figure 2.13A).** The force, called **surface tension,** is strong enough to allow small insects or arachnids such as raft spiders to walk on water **(Figure 2.13B).** Surface tension also causes water to form water droplets; the surface tension pulls the water in around itself to produce the smallest possible area, which is a spherical bead or droplet.

The Polarity of Water Molecules in the Hydrogen-Bond Lattice Contributes to Polar and Nonpolar Environments in and around Cells

The polarity of water molecules in the hydrogen-bond lattice gives water other properties that make it unique and ideal as a life-sustaining medium. In liquid water, the lattice resists invasion by other molecules unless the invading molecule also contains polar or charged regions that can form competing attractions with water molecules. If present, the competing attractions open the water lattice, creating a cavity into which the polar or charged molecule can move. By contrast, nonpolar molecules are unable to disturb the water lattice. The lattice thus excludes nonpolar substances, forcing them to form the nonpolar associations that expose the least surface area to the surrounding water—such as the spherical droplets of oil that form when oil and water are mixed together and shaken.

The distinct polar and nonpolar environments created by water are critical to the organization of cells. For example, biological membranes, which form boundaries around and inside cells, consist of lipid molecules with dual polarity: one end of each molecule is polar, and the other end is nonpolar. (Lipids are described in more detail in Chapter 3.) The membranes are surrounded on both sides by strongly polar water molecules. Exclusion by the water molecules forces the lipid molecules to associate into a double layer, a **bilayer,** in which only the polar ends of the surface molecules are exposed to the water **(Figure 2.14).** The nonpolar ends of the molecules associate in the interior of the bilayer, where they are not exposed to the water. Exclusion of their nonpolar regions by water is all that holds membranes together.

The membrane at the surface of cells prevents the watery solution inside the cell from mixing directly with the watery solution outside the cell. By doing so, the surface membrane, kept intact by nonpolar exclusion by water, maintains the internal environment and organization necessary for cellular life and its evolution.

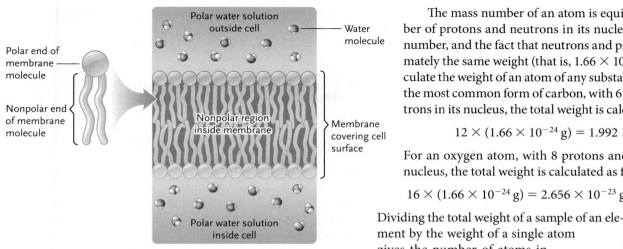

FIGURE 2.14 **Formation of the membrane covering the cell surface by lipid molecules.** Exclusion by polar water molecules forces the nonpolar ends of lipid molecules to associate into the bilayer that forms the membrane.
© Cengage Learning 2014

The Small Size and Polarity of Its Molecules Makes Water a Good Solvent

Because water molecules are small and strongly polar, they can penetrate or coat the surfaces of other polar and charged molecules and ions. The surface coat, called a **hydration layer,** reduces the attraction between the molecules or ions and promotes their separation and entry into a **solution,** where they are suspended individually, surrounded by water molecules. Once in solution, the hydration layer prevents the polar molecules or ions from reassociating. In such a solution, water is called the **solvent,** and the molecules of a substance dissolved in water are called the **solute.**

For example, when a teaspoon of table salt is added to water, water molecules quickly form hydration layers around the Na^+ and Cl^- ions in the salt crystals, reducing the attraction between the ions so much that they separate from the crystal and enter the surrounding water lattice as individual ions **(Figure 2.15).** If the water evaporates, the hydration layer is eliminated, exposing the strong positive and negative charges of the ions. The opposite charges attract and reestablish the ionic bonds that hold the ions in salt crystals. As the last of the water evaporates, all of the Na^+ and Cl^- ions reassociate, reestablishing the solid, crystalline form.

In the Cell, Chemical Reactions Involve Solutes Dissolved in Aqueous Solutions

In the cell, chemical reactions depend on solutes dissolved in aqueous solutions. To understand these reactions, you need to know the number of atoms and molecules involved. **Concentration** is the number of molecules or ions of a substance in a unit volume of space, such as a milliliter (mL) or liter (L). The number of molecules or ions in a unit volume cannot be counted directly but can be calculated indirectly by using the mass number of atoms as the starting point. The same method is used to prepare a solution with a known number of molecules per unit volume.

The mass number of an atom is equivalent to the number of protons and neutrons in its nucleus. From the mass number, and the fact that neutrons and protons are approximately the same weight (that is, 1.66×10^{-24} g), you can calculate the weight of an atom of any substance. For an atom of the most common form of carbon, with 6 protons and 6 neutrons in its nucleus, the total weight is calculated as follows:

$$12 \times (1.66 \times 10^{-24} \text{ g}) = 1.992 \times 10^{-23} \text{ g}$$

For an oxygen atom, with 8 protons and 8 neutrons in its nucleus, the total weight is calculated as follows:

$$16 \times (1.66 \times 10^{-24} \text{ g}) = 2.656 \times 10^{-23} \text{ g}$$

Dividing the total weight of a sample of an element by the weight of a single atom gives the number of atoms in the sample. Suppose you have a carbon sample that weighs 12 g—a weight in grams equal to the atom's mass number.

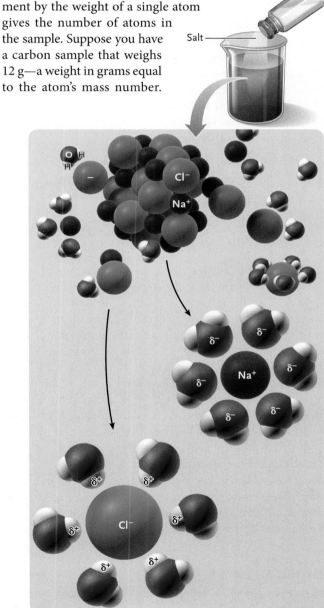

FIGURE 2.15 Water molecules forming a hydration layer around Na^+ and Cl^- ions, which promotes their separation and entry into solution.
© Cengage Learning 2014

(A weight in grams equal to the mass number is known as an **atomic weight** of an element.) Dividing 12 g by the weight of one carbon atom gives the following result:

$$\frac{12}{(1.992 \times 10^{-23} \text{ g})} = 6.022 \times 10^{23} \text{ atoms}$$

If you divide the atomic weight of oxygen (16 g) by the weight of one oxygen atom, you get the same result:

$$\frac{16}{(2.656 \times 10^{-23} \text{ g})} = 6.022 \times 10^{23} \text{ atoms}$$

In fact, dividing the atomic weight of any element by the weight of an atom of that element always produces the same number: 6.022×10^{23}. This number is called **Avogadro's number** after Amedeo Avogadro, the nineteenth-century Italian chemist who first discovered the relationship.

The same relationship holds for molecules. The **molecular weight** of any molecule is the sum of the atomic weights of all of the atoms in the molecule. For NaCl, the total mass number is $23 + 35 = 58$ (a sodium atom has 11 protons and 12 neutrons, and a chlorine atom has 17 protons and 18 neutrons). The weight of an NaCl molecule is therefore:

$$58 \times (1.66 \times 10^{-24} \text{ g}) = 9.628 \times 10^{-23} \text{ g}$$

Dividing a molecular weight of NaCl (58 g) by the weight of a single NaCl molecule gives:

$$\frac{58}{(9.628 \times 10^{-23} \text{ g})} = 6.022 \times 10^{23} \text{ molecules}$$

When concentrations are described, the atomic weight of an element or the molecular weight of a compound—the amount that contains 6.022×10^{23} atoms or molecules—is known as a **mole** (abbreviated **mol**). More strictly, chemists define the mole as the amount of a substance that contains as many atoms or molecules as there are atoms in exactly 12 g of carbon-12. As we saw above, the number of atoms in 12 g of carbon-12 is 6.022×10^{23}. The number of moles of a substance dissolved in 1 L of solution is known as the **molarity** (abbreviated M) of the solution. This relationship is highly useful in chemistry and biology because we know that two solutions having the same volume and molarity but composed of different substances will contain the same number of molecules of the substances.

Water has still other properties that contribute to its ability to sustain life, the most important being that its molecules separate into ions. These ions help maintain an environment inside living organisms that promotes the chemical reactions of life.

STUDY BREAK 2.4

1. How do hydrogen bonds between water molecules contribute to the properties of water?
2. How do a solute, a solvent, and a solution differ?

2.5 Water Ionization and Acids, Bases, and Buffers

The most critical property of water that is unrelated to its hydrogen-bond lattice is its ability to separate, or **dissociate,** to produce positively charged *hydrogen ions* (H^+, or protons) and *hydroxide ions* (OH^-):

$$H_2O \rightleftharpoons H^+ + OH^-$$

(The double arrow means that the reaction is **reversible**—depending on conditions, it may go from left to right or from right to left.) The proportion of water molecules that dissociates to release protons and hydroxide ions is small. However, because of the dissociation, water always contains some H^+ and OH^- ions.

Substances Act as Acids or Bases by Altering the Concentrations of H^+ and OH^- Ions in Water

In pure water, the concentrations of H^+ and OH^- ions are equal. However, adding other substances may alter the relative concentrations of H^+ and OH^-, making them unequal. Some substances, called **acids,** are proton donors that release H^+ (and anions) when they are dissolved in water, effectively increasing the H^+ concentration. For example, hydrochloric acid (HCl) dissociates into H^+ and Cl^- when dissolved in water:

$$HCl \rightleftharpoons H^+ + Cl^-$$

Other substances, called **bases,** are proton acceptors that reduce the H^+ concentration of a solution. Most bases dissociate in water into a hydroxide ion (OH^-) and a cation. The hydroxide ion can act as a base by accepting a proton (H^+) to produce water. For example, sodium hydroxide (NaOH) separates into Na^+ and OH^- ions when dissolved in water:

$$NaOH \rightarrow Na^+ + OH^-$$

The excess OH^- combines with H^+ to produce water:

$$OH^- + H^+ \rightarrow H_2O$$

thereby reducing the H^+ concentration.

Other bases do not dissociate to produce hydroxide ions directly. For example, ammonia (NH_3), a poisonous gas, acts as a base when dissolved in water by directly accepting a proton from water to produce an ammonium ion and releasing a hydroxide ion:

$$NH_3 + H_2O \rightarrow NH_4 + OH^-$$

The concentration of H^+ ions in a water solution, as compared with the concentration of OH^- ions, determines the **acidity** of the solution. Scientists measure acidity using a numerical scale from 0 to 14, called the **pH scale.** Because the number of H^+ ions in solution increases exponentially as the acidity increases, the scale is based on logarithms of this number to make the values manageable:

$$pH = -\log_{10}[H^+]$$

In this formula, the brackets indicate concentration in moles per liter of the substance within them. The negative of the logarithm is used to give a positive number for the pH value. For example, a pure water solution is *neutral*—neither acidic nor basic. That is, in a neutral water solution, the concentration of *both* H^+ and OH^- ions is 1×10^{-7} M (0.0000001 M), with the product of the two ion concentrations being constant in an aqueous solution at 25°C and given by $[H^+][OH^-] = (1 \times 10^{-7}) \times (1 \times 10^{-7}) = 1 \times 10^{-14}$. The $\log_{10}$ of 1×10^{-7} is -7. The negative of the logarithm -7 is 7. Thus, a neutral water solution with an H^+ concentration of 1×10^{-7} M has a pH of 7. *Acidic* solutions have pH values less than 7, with pH 0 being the value for the highly acidic 1 M hydrochloric acid (HCl); *basic* solutions have pH values greater than 7, with pH 14 being the value for the highly basic 1 M sodium hydroxide (NaOH) (basic solutions are also called *alkaline* solutions). Each whole number on the pH scale represents a value 10 times greater or less than the next number. Thus, a solution with a pH of 4 is 10 times more acidic than one with a pH of 5, and a solution with a pH of 6 is 100 times more acidic than a solution with a pH of 8. (The pH of many familiar solutions is shown in **Figure 2.16.**)

Acidity is important to cells because even small changes, on the order of 0.1 or even 0.01 pH unit, can drastically affect biological reactions. In large part, this effect reflects changes in the structure of proteins that occur when the water solution surrounding the proteins has too few or too many hydrogen ions. Consequently, all living organisms have elaborate systems that control their internal acidity by regulating H^+ concentration near the neutral value of pH 7.

Acidity is also important to the environment in which we live. Where the air is unpolluted, rainwater is only slightly acidic. However, in regions where certain pollutants are released into the air in large quantities by industry and automobile exhaust, the polluting chemicals combine with atmospheric water to produce **acid precipitation,** rain, sleet, snow, or fog with a pH below 6. Acid precipitation may have a pH as low as 3, about the same pH as that of vinegar. It can sicken and kill wildlife such as fishes and birds, as well as plants and trees (see also discussion in Chapter 55). Humans are also affected; acid precipitation can contribute to human respiratory diseases such as bronchitis and asthma.

Buffers Help Keep pH under Control

Living organisms control the internal pH of their cells with **buffers,** substances that compensate for pH changes by absorbing or releasing H^+. When H^+ ions are released in excess by biological reactions, buffers combine with them and remove them from the solution; if the concentration of H^+ decreases greatly, buffers release additional H^+ to restore the balance. Most buffers are weak acids or bases, or combinations of these substances that dissociate reversibly in water solutions to release or absorb H^+ or OH^-. (Weak acids, such as acetic acid, or weak bases, such as ammonia, are substances that release relatively few H^+ or OH^- ions in a water solution. Strong acids or bases are substances that dissociate extensively in a water solution. HCl is a strong acid; NaOH is a strong base.)

The buffering mechanism that maintains blood pH near neutral values is a primary example. In humans and many other animals, blood pH is buffered by a *carbonic acid–bicarbonate buffer system*. In water solutions, carbonic acid (H_2CO_3), which is a weak acid, dissociates readily into bicarbonate ions (HCO_3^-) and H^+:

$$H_2CO_3 \rightleftharpoons HCO_3^- + H^+$$

The reaction is reversible. If H^+ is present in excess, the reaction is pushed to the left—the excess H^+ ions combine with bicarbonate ions to form H_2CO_3. If the H^+ concentration declines below normal levels, the reaction is pushed to the right—H_2CO_3 disso-

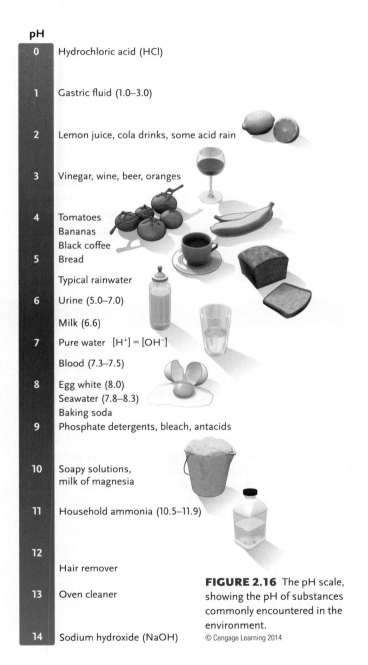

pH

0	Hydrochloric acid (HCl)
1	Gastric fluid (1.0–3.0)
2	Lemon juice, cola drinks, some acid rain
3	Vinegar, wine, beer, oranges
4	Tomatoes Bananas Black coffee
5	Bread Typical rainwater
6	Urine (5.0–7.0) Milk (6.6)
7	Pure water $[H^+] = [OH^-]$ Blood (7.3–7.5)
8	Egg white (8.0) Seawater (7.8–8.3) Baking soda
9	Phosphate detergents, bleach, antacids
10	Soapy solutions, milk of magnesia
11	Household ammonia (10.5–11.9)
12	Hair remover
13	Oven cleaner
14	Sodium hydroxide (NaOH)

FIGURE 2.16 The pH scale, showing the pH of substances commonly encountered in the environment.
© Cengage Learning 2014

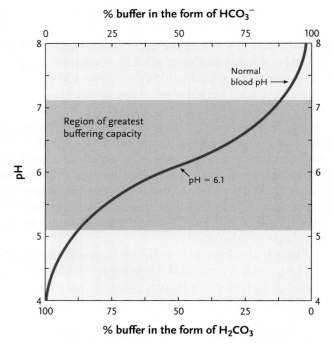

% buffer in the form of HCO₃⁻

Normal blood pH →

Region of greatest buffering capacity

pH = 6.1

% buffer in the form of H₂CO₃

FIGURE 2.17 Properties of the carbonic acid–bicarbonate buffer system. The colored zone is the pH range of greatest buffering capacity for this buffer system.

© Cengage Learning 2014

ciates into HCO_3^- and H^+, restoring the H^+ concentration. **Figure 2.17** graphs the pH of a solution (such as blood) as its relative proportions of H_2CO_3 and HCO_3^- change. The colored zone of the graph indicates the range of greatest *buffering capacity*, that is,

the range in which changes in the relative proportions of H_2CO_3 and HCO_3^- produce *little* change in the pH of the solution. Note, however, that at pH values lower than 5.1 and higher than 7.1, the slope of the curve is much greater. Here, changes in the relative proportions of H_2CO_3 and HCO_3^- produce a *large* change in the pH of the solution. Interestingly, the normal pH of blood is 7.4, which, as Figure 2.17 shows, is outside the region of greatest buffering capacity for this buffer system. In the body, other mechanisms help keep the blood pH relatively constant. (More on blood pH appears in Section 46.4.)

All buffers have curves similar to that of Figure 2.17. Each buffer has a specific range of greatest buffering capacity.

This chapter examined the basic structure of atoms and molecules and discussed the unusual properties of water that make it ideal for supporting life. The next chapter looks more closely at the structure and properties of carbon and at the great multitude of molecules based on this element.

STUDY BREAK 2.5

1. Distinguish between acids and bases. What are their properties?
2. Why are buffers important for living organisms?

THINK OUTSIDE THE BOOK

The H_2CO_3–HCO_3^- buffering system is only one of the mechanisms by which the human body maintains blood pH at a relatively constant level. Collaboratively, or on your own, research what happens to the body when blood pH becomes abnormal.

 UNANSWERED QUESTIONS

Bioremediation, the application of chemical and biological knowledge to decontaminate polluted environments, was introduced in *Why It Matters*. Among the contaminants of the environment is uranium, a radioactive heavy metal.

How can the effectiveness of uranium bioremediation by bacteria in soils and sediments be optimized?

Uranium contamination of the environment is a big concern in areas with uranium ore mining and processing activities. Mining tails in soils and sediments can lead to high concentrations of uranium in the groundwater, which can then be transported into nearby drinking water supplies. One bioremediation approach to remove uranium from groundwater is to use bacteria that exist naturally in the soils and sediments at the site. The metabolism of the bacteria transforms uranium from its soluble form to an insoluble form. In this way it is possible to immobilize uranium from spreading in the groundwater. The effectiveness of bioremediation in natural environments depends on many factors and is the subject of ongoing research. My laboratory at Pennsylvania State University is a participant in a U.S. Department of Energy (DOE) project studying bio-

remediation of uranium from a previous uranium ore mining and processing site near Rifle, Colorado. Uranium in the Rifle groundwater directly flows into the Colorado River and causes a significant environmental problem. Scientists have known for some time it is also possible to inject organic carbon such as acetate into the subsurface to stimulate the indigenous bacteria to transform the contaminants. My research group is researching how injected organic carbon is delivered to other locations together with ground water flow, and how the spatial distribution of different types of minerals affect the rate at which uranium is being remediated. We use flow-through column experiments, field experiments, as well as computational tools to quantify the uranium bioremediation rates and to predict the system behavior under different conditions. For example, we have found that the spatial distribution of important minerals, especially those that contain iron, play a critical role in determining the effectiveness of bioremediation. If the iron-containing minerals are all located at the vicinity of the injection wells, the amount of immobilized uranium can be more than two times larger than that in the case where the iron-containing minerals are distributed spatially in an even fashion.

The complexity of the subsurface system, where natural variability in hydrological, microbiologic, and geochemical properties exist, necessitate an interdisciplinary research approach to this bioremediation problem. The scientists in our laboratory are part of a team of microbiologists, geochemists, hydrogeologists, and geophysicists from various universities and national laboratories. As a team we study different aspects of uranium bioremediation including determining the type of bacteria that are important for uranium removal, the distribution of important minerals at the site, the movement of groundwater, and how all these factors determine the overall efficacy of bioremediation at the site.

THINK LIKE A SCIENTIST What other environmental factors might change the functioning of bacteria in the bioremediation of uranium?

Li Li is a professor in the Department of Energy and Mineral Engineering at the Pennsylvania State University, University Park. Her main research interests include understanding reactive transport processes in natural subsurface systems that are important to many applications. Learn more about her work at http://www.eme.psu.edu/faculty/lili.html.

Courtesy of Dr. Li Li

REVIEW KEY CONCEPTS

To access the course materials and companion resources for this text, please visit www.cengagebrain.com.

2.1 The Organization of Matter: Elements and Atoms

- Matter is anything that occupies space and has mass. Matter is composed of elements, each consisting of atoms of the same kind.

- Atoms combine chemically in fixed numbers and ratios to form the molecules of living and nonliving matter. Compounds are molecules in which the component atoms are different.

  Animation: Atomic number, mass number

2.2 Atomic Structure

- Atoms consist of an atomic nucleus that contains protons and neutrons surrounded by one or more electrons traveling in orbitals. Each orbital can hold a maximum of two electrons (Figure 2.3).

- All atoms of an element have the same number of protons, but the number of neutrons is variable. The number of protons in an atom is designated by its atomic number; the number of protons plus neutrons is designated by the mass number (Figure 2.4 and Table 2.1).

- Isotopes are atoms of an element with differing numbers of neutrons. The isotopes of an atom differ in physical but not chemical properties (Figure 2.4).

- Electrons surround an atomic nucleus in orbitals occupying energy levels that increase in discrete steps (Figures 2.5 and 2.6).

- The chemical activities of atoms are determined largely by the number of electrons in the outermost energy level. Atoms that have the outermost level filled with electrons are nonreactive, whereas atoms in which that level is not completely filled with electrons are reactive. Atoms tend to lose, gain, or share electrons to fill the outermost energy level.

  Animation: Subatomic particles

  Animation: Isotopes of hydrogen

2.3 Chemical Bonds and Chemical Reactions

- An ionic bond forms between atoms that gain or lose electrons in the outermost energy level completely, that is, between a positively charged cation and a negatively charged anion (Figure 2.7).

- A covalent bond is established by a pair of electrons shared between two atoms. If the electrons are shared equally, the covalent bond is nonpolar (Figure 2.8).

- If electrons are shared unequally in a covalent bond, the atoms carry partial positive and negative charges and the bond is polar (Figure 2.9).

- Polar molecules tend to associate with other polar molecules and to exclude nonpolar molecules. Polar molecules that associate readily with water are hydrophilic; nonpolar molecules excluded by water are hydrophobic.

- A hydrogen bond is a weak attraction between a hydrogen atom made partially positive by unequal electron sharing and another atom—usually oxygen, nitrogen, or sulfur—made partially negative by unequal electron sharing (Figure 2.10).

- Van der Waals forces, bonds even weaker than hydrogen bonds, can form when natural changes in the electron density of molecules produce regions of positive and negative charge, which cause the molecules to stick together briefly.

- Chemical reactions occur when molecules form or break chemical bonds. The atoms or molecules entering into a chemical reaction are the reactants, and those leaving a reaction are the products.

  Animation: Examples of hydrogen bonds

  Animation: Ionic bonding

  Animation: Models for methane

2.4 Hydrogen Bonds and the Properties of Water

- The hydrogen-bond lattice gives water unusual properties that are vital to living organisms, including high specific heat, boiling point, cohesion, and surface tension (Figures 2.12 and 2.13).

- The polarity of the water molecules in the hydrogen-bond lattice makes it difficult for nonpolar substances to penetrate the lattice. The distinct polar and nonpolar environments created by water are critical to the organization of cells (Figure 2.14).

- The polar properties of water allow it to form a hydration layer over the surfaces of polar and charged biological molecules, particularly proteins. Many chemical reactions depend on the special molecular conditions created by the hydration layer (Figure 2.15).

- The polarity of water allows ions and polar molecules to dissolve readily in water, making it a good solvent.

  Animation: Dissolution

  Animation: Structure of water

2.5 Water Ionization and Acids, Bases, and Buffers

- Acids are substances that increase the H^+ concentration by releasing additional H^+ as they dissolve in water; bases are substances that decrease the H^+ concentration by gathering H^+ or releasing OH^- as they dissolve.

- The relative concentrations of H^+ and OH^- in a water solution determine the acidity of the solution, which is expressed quantitatively as pH on a number scale ranging from 0 to 14. Neutral solutions, in which the concentrations of H^+ and OH^- are equal, have a pH of 7. Solutions with pH less than 7 have H^+ in excess and are acidic; solutions with pH greater than 7 have OH^- in excess and are basic or alkaline (Figure 2.16).

- The pH of living cells is regulated by buffers, which absorb or release H^+ to compensate for changes in H^+ concentration (Figure 2.17).

UNDERSTAND & APPLY

Test Your Knowledge

1. Which of the following statements about the mass number of an atom is *incorrect*?
   a. It has a unit defined as a dalton.
   b. On Earth, it equals the atomic weight.
   c. Unlike the atomic weight of an atom, it does not change when gravitational forces change.
   d. It equals the number of electrons in an atom.
   e. It is the sum of the protons and neutrons in the atomic nucleus.

2. Oxygen (O) is a(n) ___, while the oxygen we breathe (O_2) is a(n) ___, and the carbon dioxide we exhale is a(n) ___.
   a. compound; molecule; element
   b. atom; compound; element
   c. element; atom; molecule
   d. atom; element; molecule
   e. element; molecule; compound

3. The chemical activity of an atom:
   a. depends on the electrons in the outermost energy level.
   b. is increased when the outermost energy level is filled with electrons.
   c. depends on its $1s$ but not its $2s$ or $2p$ orbitals.
   d. is increased when valence electrons completely fill the outer orbitals.
   e. of oxygen prevents it from sharing its electrons with other atoms.

4. When electrons are shared equally between atoms, they form:
   a. a polar covalent bond.
   b. a nonpolar covalent bond.
   c. an ionic bond.
   d. a hydrogen bond.
   e. a van der Waals force.

5. Which of the following is *not* a property of water?
   a. It has a low boiling point compared with other molecules.
   b. It has a high heat of vaporization.
   c. Its molecules resist separation, a property called cohesion.
   d. It has the property of adhesion, the ability to stick to charged and polar groups in molecules.
   e. It can form hydrogen bonds to molecules below but not above its surface.

6. Which of the following is *not* a hydrophilic body fluid?
   a. blood
   b. sweat
   c. tears
   d. oil
   e. saliva

7. The water lattice:
   a. is formed from hydrophobic bonds.
   b. causes ice to be denser than water.
   c. causes water to have a relatively low specific heat.
   d. excludes nonpolar substances.
   e. is held together by hydrogen bonds that are permanent; that is, they never break and reform.

8. A hydrogen bond is:
   a. a strong attraction between hydrogen and another atom.
   b. a bond between a hydrogen atom already covalently bound to one atom and made partially negative by unequal electron sharing with another atom.
   c. a bond between a hydrogen atom already covalently bound to one atom and made partially positive by unequal electron sharing with another atom.
   d. weaker than van der Waals forces.
   e. exemplified by the two hydrogens covalently bound to oxygen in the water molecule.

9. If the water in a pond has a pH of 5, the hydroxide concentration would be
   a. 10^{-5} M.
   b. 10^{-10} M.
   c. 10^5 M.
   d. 10^9 M.
   e. 10^{-9} M.

10. Because of a sudden hormonal imbalance, a patient's blood was tested and shown to have a pH of 7.5. What does this pH value mean?
    a. This is more acidic than normal blood.
    b. It represents a weak alkaline fluid.
    c. This is caused by a release of large amounts of hydrogen ions into the system.
    d. The reaction $H_2CO_3 \rightarrow HCO_3^- + H^+$ is pushed to the left.
    e. This is probably caused by excess CO_2 in the blood.

Discuss the Concepts

1. Detergents allow particles of oil to mix with water. From the information presented in this chapter, how do you think detergents work?

2. What would living conditions be like on Earth if ice were denser than liquid water?

3. You place a metal pan full of water on the stove and turn on the heat. After a few minutes, the handle is too hot to touch but the water is only warm. How do you explain this observation?

4. You are studying a chemical reaction accelerated by an enzyme. H^+ forms during the reaction, but the enzyme's activity is lost at low pH. What could you include in the reaction mix to keep the enzyme's activity at high levels? Explain how your suggestion might solve the problem.

Design an Experiment

You know that adding NaOH to HCl results in the formation of common table salt, NaCl. You have a 0.5 *M* HCl solution. What weight of NaOH would you need to add to convert all of the HCl to NaCl? (**Note:** Chemical reactions have the potential to be dangerous. Please do not attempt to perform this reaction.)

Interpret the Data

The pH of human stomach acid ranges from 1.0 to 3.0, whereas a healthy esophagus has a pH of approximately 7.0. In gastroesophageal reflux disease (GERD), often called acid reflux, stomach acid flows backward from the stomach into the esophagus. Repeated episodes in which esophageal pH goes below 4.0, considered clinical acid reflux, can result in bleeding ulcers and damage to the esophageal lining. The data in the **Figure**, from a patient with GERD, show esophageal pH during a sleeping reflux event.

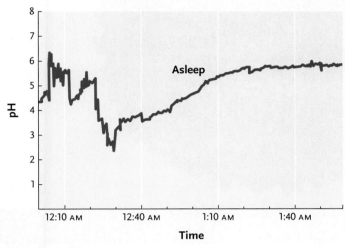

© Cengage Learning 2014

1. How many minutes does it take to go from the peak of the reflux event (when pH is most acidic) to when the reflux event is over?

2. What is the molar concentration of H^+ and OH^- ions: (a) during sleep after the reflux event; and (b) during the peak of the reflux event? Be sure to include the correct concentration units in your answer.

3. What is the change in concentration of H^+ and OH^- during the peak of the reflux attack compared to the clinical value of acid reflux? Be sure to include the correct concentration units in your answer.

Source: Based on T. Demeester et al. 1976. Patterns of gastroesophageal reflux in health and disease. *Annals of Surgery* 184:459–469.

Apply Evolutionary Thinking

What properties of water made the evolution of life possible?

3

STUDY OUTLINE

© SCIENCE PHOTO LIBRARY/AgeFotostock

Ferritin, a protein consisting of 24 subunits. It is the main iron-storage protein within cells of prokaryotes and eukaryotes.

Biological Molecules: The Carbon Compounds of Life

Why it matters . . . In the Pacific Northwest, vast forests of coniferous trees have survived another winter **(Figure 3.1).** With the arrival of spring, rising temperatures and water from melting snow stimulate renewed growth. Carbon dioxide (CO_2) from the air enters the needlelike leaves of the trees through microscopic pores. Using energy from sunlight, the trees combine the water and carbon dioxide into sugars and other carbon-based compounds through the process of photosynthesis. The lives of plants, and almost all other organisms, depend directly or indirectly on the products of photosynthesis.

The amount of CO_2 in the atmosphere is critical to photosynthesis. Researchers have found that CO_2 concentration changes with the seasons. It declines during spring and summer, when plants and other photosynthetic organisms withdraw large amounts of the gas from the air and convert it into sugars and other complex carbon compounds. It increases during fall and winter, when photosynthesis decreases and decomposers that release the gas as a metabolic by-product increase. Great quantities of CO_2 are also added to the atmosphere by forest fires and by the burning of coal, oil, gasoline, and other fossil fuels in automobiles, aircraft, trains, power plants, and other industries. The resulting increase in atmospheric CO_2 contributes to global warming.

The importance of atmospheric CO_2 to food production and world climate are just two examples of how carbon and its compounds are fundamental to the entire living world, from the structures and activities of single cells to physical effects that take place on a global scale. Carbon compounds form the structures of living organisms and take part in all biological reactions. They also serve as sources of energy for living organisms and as an energy resource for much of the world's industry—for example, coal and oil are the fossil remains of long-dead organisms. This chapter outlines the structures and functions of biological carbon compounds.

FIGURE 3.1 **Conifers around Mount Rainier in Washington State.** As is true of all other organisms, the structure, activities, and survival of these trees start with the carbon atom and its diverse molecular partners in organic compounds.

3.1 Formation and Modification of Biological Molecules

The bonding properties of carbon enable it to form an astounding variety of chain and ring structures that are the backbones of all biological molecules. The wide variety of carbon-based molecules has been responsible for the wide diversity of organisms that have evolved.

Carbon Chains and Rings Form the Backbones of All Biological Molecules

Collectively, molecules based on carbon are known as **organic molecules.** All other substances—those without carbon atoms in their structures—are **inorganic molecules.** A few of the smallest carbon-containing molecules that occur in the environment as minerals or atmospheric gases, such as CO_2, are also considered inorganic molecules. Outside of water, the four major classes of organic molecules—*carbohydrates, lipids, proteins,* and *nucleic acids*—form almost the entire substance of living organisms. They are discussed in turn in subsequent sections in this chapter.

In organic molecules, carbon atoms bond covalently to each other and to other atoms (chiefly hydrogen, oxygen, nitrogen, and sulfur) in molecules that range in size from a few atoms to thousands, or even millions, of atoms. Molecules consisting of carbon linked only to hydrogen atoms are called **hydrocarbons** (*hydro-* refers to hydrogen, not to water).

As discussed in Section 2.3, carbon has four unpaired outer electrons that it readily shares to complete its outermost energy level, forming four covalent bonds. The simplest hydrocarbon, CH_4 (methane), consists of a single carbon atom bonded to four hydrogen atoms (see Figure 2.8A and B). More complex hydrocarbons involve two or more carbon atoms arranged in a linear unbranched chain, a linear branched chain, or a structure with one or more rings. The number of bonds between neighboring carbon atoms diversifies the structures. A triple bond can only occur in a two-carbon hydrocarbon, but single and double bonds

A. Two-carbon hydrocarbons with single, double, and triple bonding

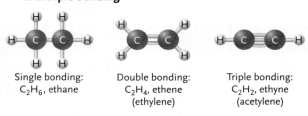

Single bonding: C_2H_6, ethane

Double bonding: C_2H_4, ethene (ethylene)

Triple bonding: C_2H_2, ethyne (acetylene)

B. Linear and branched hydrocarbon chains

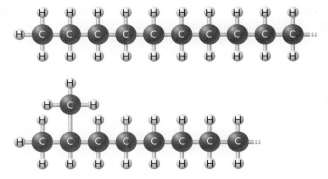

C. Hydrocarbon ring, in this case with double bonds

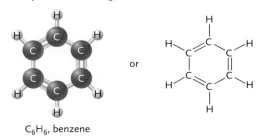

C_6H_6, benzene

or

FIGURE 3.2 **Examples of hydrocarbon structures.**
© Cengage Learning 2014

are found in both linear and ring hydrocarbons **(Figure 3.2).** All in all, there is almost no limit to the number of different hydrocarbon structures that carbon and hydrogen can form.

Functional Groups Confer Specific Properties to Biological Molecules

Carbohydrates, lipids, proteins, and nucleic acids contain particular small, reactive groups of atoms called **functional groups.** Each of those groups has specific chemical properties, which are then also found in the larger molecules containing them. Thus, the number and arrangement of functional groups in a larger molecule give that molecule its particular properties.

Functional groups can participate in biological reactions. The functional groups that enter most frequently into biological reactions are the **hydroxyl group** (—OH), **carbonyl group** ($>C$=O), **carboxyl group** (—COOH), **amino group** (—NH$_2$), **phosphate group** (—OPO$_3^{2-}$), and **sulfhydryl group** (—SH) **(Table 3.1).** The functional groups (boxed in blue in the table) are

TABLE 3.1	Common Functional Groups of Organic Molecules		
Functional Group (boxed in blue)	Major Classes of Molecules and Examples of Them		Properties
Hydroxyl R—OH	An oxygen atom linked to a hydrogen atom. In a molecule, it is linked to an R group on the other side.	**Alcohols** Ethyl alcohol (in alcoholic beverages)	Polar; confers polarity on the parts of the molecules that contain them. Hydrogen bonds with water facilitating dissolving of organic molecules. Enables an alcohol to form linkages with other organic molecules through dehydration synthesis reactions (see Figure 3.5A).
Carbonyl R—C=O | H R—C—R || O	Oxygen atom linked to an atom by a double bond. In aldehydes, the carbonyl group is linked to a carbon atom at the end of a carbon chain. In ketones, the carbonyl group is linked to a carbon atom in the interior of a carbon chain.	**Aldehydes** Acetaldehyde **Ketones** Acetone (a solvent)	Reactive parts of aldehydes and ketones, molecules that act as major building blocks of carbohydrates, and that also take part in the reactions supplying energy for cellular activities.
Carboxyl R—COOH or R—C || O | OH	A carbonyl group and a hydroxyl group combined.	**Carboxylic acids** (a type of organic acid) Acetic acid (in vinegar)	Gives organic molecules acidic properties because its —OH group readily releases the hydrogen as a proton (H^+) in aqueous solutions (such as in cells) (see Section 2.5) converting it from a non-ionized to an ionized form: $R-C(O)(OH) \rightleftharpoons R-C(O)(O^-) + H^+$
Amino R—NH₂ or R—N(H)(H)	A nitrogen atom bonded on one side to two hydrogen atoms. In a molecule it is linked to an R group on the other side.	**Amines** Alanine (an amino acid)	Readily acts as an organic base by accepting a proton (H^+) in aqueous solutions, converting it from a nonionized to an ionized form: $R-N(H)(H) + H^+ \rightleftharpoons R-\overset{+}{N}(H)(H)(H)$

(Continued)

linked by covalent bonds to other atoms in biological molecules, usually carbon atoms. The symbol *R* is often used to represent the chain of carbon atoms. A double bond, such as that in the carbonyl group, indicates that two pairs of electrons are shared between the carbon and oxygen atoms.

Isomers Have the Same Chemical Formula but Different Molecular Structures

Isomers are two or more molecules with the same chemical formula but different molecular structures. Molecules that are mirror images of one another are an example of **stereoisomers**. That

is, often one or more of the carbon atoms in an organic molecule link to four different atoms or functional groups. A carbon linked in this way is called an *asymmetric carbon*. Asymmetric carbons have important effects on the molecule because they can take either of two fixed positions in space with respect to other carbons in a carbon chain.

Figure 3.3 shows the stereoisomers of an amino acid. The chemical formula of each, as well as the connections between atoms and groups, is the same. The difference between the two forms is similar to the difference between your two hands. Although both hands have four fingers and a thumb, they are not identical; rather, they are mirror images of each other. That is,

TABLE 3.1 | **Common Functional Groups of Organic Molecules (Continued)**

Functional Group (boxed in blue)		Major Classes of Molecules and Examples of Them		Properties
Phosphate R—O—P with O⁻ groups	A central phosphorus bound to four oxygen atoms. In molecules, one of the oxygen atoms links to an R group.	**Organic phosphates** Glyceraldehyde-3-phosphate structure	Glyceraldehyde-3-phosphate (product of photosynthesis). Nucleotides and nucleic acids are also examples.	Molecules that contain phosphate groups react as weak acids because one or both —OH groups readily release their hydrogens as H^+ in aqueous solutions converting them from a nonionized to an ionized form: $$R—O—P(OH)(O)—OH \rightleftharpoons R—O—P(O^-)(O)—O^- + 2\,H^+$$ A phosphate group can bridge two organic building blocks to form a larger structure, for example DNA: Organic subunit —O—P(O⁻)(O)—O— Organic subunit Phosphate groups are added to or removed from biological molecules as part of reactions that conserve or release energy, or, for many proteins, to alter activity.
Sulfhydryl R—SH	A sulfur atom linked to a hydrogen atom. In a molecule, the other side is linked to an R group.	**Thiols** Mercaptoethanol	Mercaptoethanol	Easily converted into a covalent linkage, in which it loses its hydrogen atom as it binds. In many linking reactions, two sulfhydryl groups form a **disulfide linkage** (—S—S—): $R—SH + HS—R \rightarrow R—S—S—R + 2\,H^+ + 2\text{ electrons}$ disulfide linkage

when you hold your right hand in front of a mirror, the reflection looks like your left hand and vice versa. One of the stereoisomers is designated the L isomer (L for *laevus* = left) (see left side of Figure 3.3 with the left hand). The other stereoisomer is called the D isomer (D for *dexter* = right) (see right side of Figure 3.3 with the right hand).

The difference between L and D stereoisomers is critical to biological function. Typically, one of the two forms enters much more readily into cellular reactions; just as your left hand does not fit readily into a right-hand glove, enzymes (proteins that accelerate chemical reactions in living organisms) fit best to one of the two forms of a stereoisomer. For example, most of the enzymes that catalyze the biochemical reactions involving amino acids recognize the L stereoisomer. Many other kinds of biological molecules besides amino acids form stereoisomers. Most of the enzymes that catalyze the biochemical reactions involving sugars recognize the D stereoisomer, making this form much more common among cellular carbohydrates than L stereoisomers.

Structural isomers are two molecules with the same chemical formula but atoms that are connected in different ways. The sugars glucose and fructose are examples of structural isomers **(Figure 3.4).**

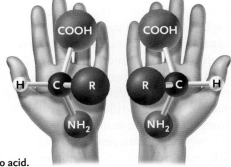

FIGURE 3.3
Stereoisomers of an amino acid.
Adapted from http://creationwiki.org/File:Chirality.jpg
"L isomer"   "D isomer"

A. Glucose
(an aldehyde)

B. Fructose
(a ketone)

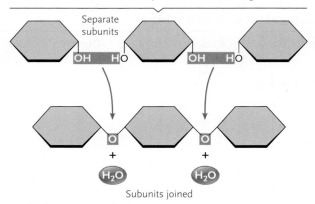

FIGURE 3.4 Glucose and fructose, structural isomers of a six-carbon sugar with the chemical formula $C_6H_{12}O_6$. **(A)** In glucose, the aldehyde isomer, the carbonyl group (shaded region) is located at the end of the carbon chain. **(B)** In fructose, the ketone isomer, the carbonyl group is located inside the carbon chain. For convenience, the carbons of the sugars are numbered, with 1 being the carbon at the end nearest the carbonyl group.
© Cengage Learning 2014

A Water Molecule Is Added or Removed in Many Reactions Involving Functional Groups

In many of the reactions that involve functional groups, the components of a water molecule, —H and —OH, are removed from or added to the groups as they interact. When the components of a water molecule are *removed* during a reaction (usually as part of the assembly of a larger molecule from smaller subunits), the reaction is called a **dehydration synthesis reaction** or **condensation reaction (Figure 3.5A).** For example, this type of reaction occurs when individual sugar molecules combine to form a starch molecule. In **hydrolysis,** the reverse reaction, the components of a water molecule are *added* to functional groups as molecules are broken into smaller subunits **(Figure 3.5B).** For example, the breakdown of a protein molecule into individual amino acids occurs by hydrolysis in the digestive processes of animals.

Of the functional groups, hydroxyl groups readily enter dehydration synthesis reactions and are formed as part of hydrolysis reactions. Carboxyl groups readily enter into dehydration synthesis reactions, giving up hydroxyl groups as organic molecules combine into larger assemblies. Amino groups also readily enter dehydration synthesis reactions, releasing H^+ as it links subunits into larger molecules. For example, the joining of amino acids in the synthesis of proteins involves a dehydration reaction involving the carboxyl group of one amino acid and the amino group of another amino acid (see Figure 3.19).

Many Carbohydrates, Proteins, and Nucleic Acids Are Macromolecules

Many carbohydrates, proteins, and nucleic acids are large *polymers* (*poly* = many; *mer* = unit). A **polymer** is a molecule assembled from subunit molecules called *monomers* into a

A. Dehydration synthesis reaction
Dehydration synthesis reactions remove the components of a water molecule as new covalent bonds join subunits into a larger molecule.

B. Hydrolysis reaction
Hydrolysis reactions add the components of a water molecule as covalent bonds are broken, splitting a molecule into smaller subunits.

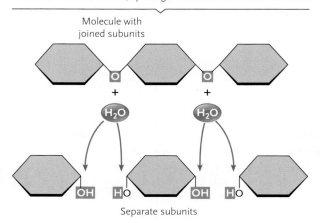

FIGURE 3.5 Dehydration synthesis and hydrolysis reactions.
© Cengage Learning 2014

chain by covalent bonds. The process of assembly of a polymer from monomers is called **polymerization.** The polymerization reactions are dehydration synthesis reactions. The opposite reactions—breakdown of polymers into monomers—occur by hydrolysis.

Each type of polymeric biological molecule contains one type of monomer. The monomers may be identical, or they may have chemical variations, depending on the molecule. The variations among monomer structures are responsible for the highly diverse and varied biological molecules found in living organisms. For instance, proteins are polymers consisting of amino acid monomers. There are 20 different amino acids, each with identical amino and carboxyl functional groups that enable them to undergo polymerization (see Figure 3.17), but each also with a different R group. The proteins assembled from these amino acids vary in number and organization in the polymer chains, resulting in a huge variety of proteins with different structures and, therefore, functions.

Somewhat arbitrarily, a single polymer molecule with a mass of 1,000 Da (daltons) or more is called a **macromolecule** (*macro* = large). By that criterion, many representatives of carbohydrates, proteins, and nucleic acids are macromolecules. In a number of instances, macromolecules interact to form even larger functional molecular structures in cells. For example, the ribosome, the cellular structure that plays the central role in the polymerization of amino acids into a protein chain, consists of several RNA macromolecules and many protein macromolecules.

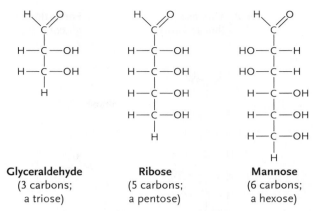

Glyceraldehyde
(3 carbons;
a triose)

Ribose
(5 carbons;
a pentose)

Mannose
(6 carbons;
a hexose)

FIGURE 3.6 **Some representative monosaccharides.** The triose, glyceraldehyde, takes part in energy-yielding reactions and photosynthesis. The pentose, ribose, is a component of RNA and of molecules that carry energy. The hexose, mannose, is a fuel substance and a component of glycolipids and glycoproteins.
© Cengage Learning 2014

3.2 Carbohydrates

Carbohydrates, the most abundant organic molecules in the world, serve many functions. Together with fats, they act as the major fuel substances providing chemical energy for cellular activities. The carbohydrate sucrose, common table sugar, is consumed in large quantities as an energy source in the human diet. Energy-providing carbohydrates are stored in plant cells as **starch** and in animal cells as **glycogen,** both consisting of long chains of repeating carbohydrate subunits linked end to end. Chains of carbohydrate subunits also form many structural molecules, such as **cellulose,** one of the primary constituents of plant cell walls.

Carbohydrates contain only carbon, hydrogen, and oxygen atoms, in an approximate ratio of 1 carbon:2 hydrogens:1 oxygen (CH_2O). The names of many carbohydrates end in *-ose*. The smallest carbohydrates, the **monosaccharides** (*mono* = one; *saccharum* = sugar), contain three to seven carbon atoms. For example, the monosaccharide glucose consists of a chain of six carbons and has the molecular formula $C_6H_{12}O_6$. Two monosaccharides polymerize to form a **disaccharide** such as sucrose, which is common table sugar. Carbohydrate polymers with more than 10 linked monosaccharide monomers are called **polysaccharides.** Starch, glycogen, and cellulose are common polysaccharides.

Monosaccharides Are the Structural Units of Carbohydrates

Carbohydrates occur either as monosaccharides or as polymers of monosaccharide units linked together. Monosaccharides are soluble in water, and most have a sweet taste. Of the monosaccharides, those that contain three carbons (*trioses*), five carbons (*pentoses*), and six carbons (*hexoses*) are most common in living organisms (**Figure 3.6**).

All monosaccharides can occur in the linear form shown in Figure 3.6. In this form, each carbon atom in the chain except one has both an —H and an —OH group attached to it. The remaining carbon is part of a carbonyl group, which may be located at the end of the carbon chain in the aldehyde position, or inside the chain in the ketone position, resulting in structural isomers (see Figure 3.4).

Monosaccharides with four or more carbons can fold back on themselves to assume a ring form. Folding into a ring occurs through a reaction between two functional groups in the same monosaccharide, as occurs in glucose (**Figure 3.7**). The ring form of most five- and six-carbon sugars is much more common in cells than the linear form.

In the ring form of many five- or six-carbon monosaccharides, including glucose, the carbon at the 1 position of the ring is asymmetric because its four bonds link to different groups of atoms. This asymmetry allows monosaccharides such as glucose to exist as two different stereoisomers. The glucose stereoisomer with an —OH group pointing below the plane of the ring is known as *alpha-glucose,* or *α-glucose;* the stereoisomer with an —OH group pointing above the plane of the ring is known as *beta-glucose,* or *β-glucose* (see Figure 3.7B). Other five- and six-carbon monosaccharide rings have similar α- and β-configurations.

The α- and β-rings of monosaccharides can give the polysaccharides assembled from them vastly different chemical properties. For example, starches, which are assembled from α-glucose units, are biologically reactive polysaccharides easily digested by animals; cellulose, which is assembled from β-glucose units, is relatively unreactive and, for most animals, completely indigestible.

A. Glucose (linear form)

B. Formation of glucose rings

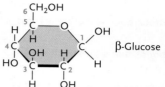

C. Simplified ring structure

α-Glucose

or

β-Glucose

D. Space-filling model

FIGURE 3.7 **Ring formation by glucose. (A)** Glucose in linear form. **(B)** The ring form of glucose is produced by a reaction between the aldehyde group at the 1 carbon and the hydroxyl group at the 5 carbon. The reaction produces two glucose stereoisomers, α- and β-glucose. If the ring is considered to lie in the plane of the page, the —OH group points below the page in α-glucose and upward from the page in β-glucose. In the ring form, the thicker lines along one side indicate that you are viewing the ring edge on. For simplicity, the group at the 6 carbon is shown as CH₂OH in this and later diagrams. **(C)** A commonly used, simplified representation of the glucose ring (in this case α-glucose), in which the Cs designating carbons of the ring are omitted. Other sugar rings similarly are drawn this way. **(D)** A space-filling model of glucose, showing the volumes occupied by the atoms. Carbon atoms are black, oxygen atoms are red, and hydrogen atoms are white.

© Cengage Learning 2014

Two Monosaccharides Link to Form a Disaccharide

Disaccharides typically are assembled from two monosaccharides covalently joined by a dehydration synthesis reaction. For example, the disaccharide maltose is formed by the linkage of two α-glucose molecules **(Figure 3.8A)** with oxygen as a bridge between the number 1 carbon of the first glucose unit and the 4 carbon of the second glucose unit. Bonds of this type, which commonly link monosaccharides into chains, are known as **glycosidic bonds**. A glycosidic bond between a 1 carbon and a 4 carbon is written in chemical shorthand as a 1→4 linkage; 1→2, 1→3, and 1→6 linkages are also common in carbohydrate chains. The linkages are designated as α or β depending on the orientation of the —OH group at the 1 carbon that forms the bond. In maltose, the —OH group is in the α position. Therefore, the link between the two glucose subunits of maltose is written as an α(1→4) linkage.

FIGURE 3.8 **Disaccharides.** A dehydration synthesis reaction involving two monosaccharides produces a disaccharide. The components of a water molecule (in blue) are removed from the monosaccharides as they join.

© Cengage Learning 2014

Maltose, sucrose, and lactose are common disaccharides. Maltose (see Figure 3.8A) is present in germinating seeds and is a major sugar used in the brewing industry. Sucrose, which contains a glucose and a fructose unit **(Figure 3.8B)**, is transported to and from different parts of leafy plants. It is probably the most plentiful sugar in nature. Table sugar is made by extracting and crystallizing sucrose from plants, such as sugar cane and sugar beets. Lactose, assembled from a glucose and a galactose unit **(Figure 3.8C)**, is the primary sugar of milk.

Monosaccharides Link in Longer Chains to Form Polysaccharides

Polysaccharides are the macromolecules formed by polymerization of monosaccharide monomers through dehydration synthesis reactions. The most common polysaccharides—the plant starches,

A. Formation of maltose
Maltose is assembled from two glucose molecules.

Glucose + Glucose → Maltose + H₂O

α(1→4) linkage

B. Sucrose
Sucrose is assembled from glucose and fructose.

α(1→2) linkage

Glucose unit   Fructose unit

C. Lactose
Lactose is assembled from galactose and glucose.

β(1→4) linkage

Galactose unit   Glucose unit

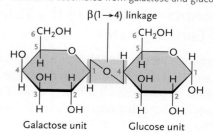

glycogen, and cellulose—are polymers of hundreds or thousands of glucose units. Other polysaccharides are built up from a variety of different sugar units. Polysaccharides may be linear, unbranched molecules, or they may contain one or more branches in which side chains of sugar units attach to a main chain.

Figure 3.9 shows four common polysaccharides. Plant starches include both linear, unbranched forms such as amylose **(Figure 3.9A)** and branched forms such as amylopectin. Glycogen **(Figure 3.9B)**, a more highly branched polysaccharide than amylopectin, can be assembled or disassembled readily to take up or

A. Amylose, a plant starch

Amylose, formed from α-glucose units joined end to end in α(1→4) linkages. The coiled structures are induced by the bond angles in the α-linkages.

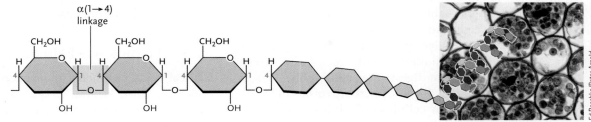

Amylose grains (purple) in plant root tissue

B. Glycogen, found in animal tissues

Glycogen, formed from glucose units joined in chains by α(1→4) linkages; side branches are linked to the chains by α(1→6) linkages (boxed in blue).

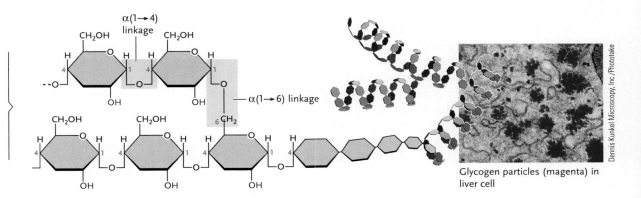

Glycogen particles (magenta) in liver cell

C. Cellulose, the primary fiber in plant cell walls

Cellulose, formed from glucose units joined end to end by β(1→4) linkages. Hundreds to thousands of cellulose chains line up side by side, in an arrangement reinforced by hydrogen bonds between the chains, to form cellulose microfibrils in plant cells.

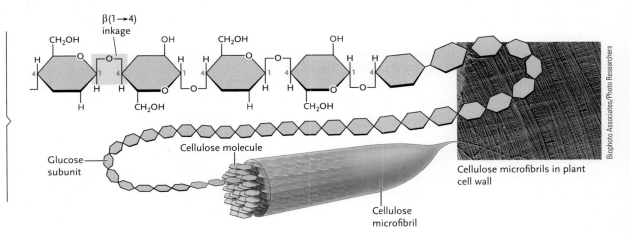

Glucose subunit

Cellulose molecule

Cellulose microfibril

Cellulose microfibrils in plant cell wall

D. Chitin, a reinforcing fiber in the external skeleton of arthropods and the cell walls of some fungi

Chitin, formed from β(1→4) linkages joining glucose units modified by the addition of nitrogen-containing groups. The external body armor of the tick is reinforced by chitin fibers.

FIGURE 3.9 Four common polysaccharides.
© Cengage Learning 2014

release glucose; it is stored in large quantities in the liver and muscle tissues of many animals.

Cellulose **(Figure 3.9C),** probably the most abundant carbohydrate on Earth, is an unbranched polysaccharide assembled from glucose monomers bound together by β-linkages. It is the primary structural fiber of plant cell walls; in this role, cellulose has been likened to the steel rods in reinforced concrete. Its tough fibers enable the cell walls of plants to withstand enormous weight and stress. Fabrics such as cotton and linen are made from cellulose fibers extracted from plant cell walls. Animals such as mollusks, crustaceans, and insects synthesize an enzyme that digests the cellulose they eat. In ruminant mammals, such as cows, microorganisms in the digestive tract break down cellulose. Cellulose passes unchanged through the human digestive tract as indigestible fibrous matter. Many nutritionists maintain that the bulk provided by cellulose fibers helps maintain healthy digestive function.

Chitin **(Figure 3.9D),** another tough and resilient polysaccharide, is assembled from glucose units modified by the addition of nitrogen-containing groups. Similar to the subunits of cellulose, the modified glucose monomers of chitin are held together by β-linkages. Chitin is the main structural fiber in the external skeletons and other hard body parts of arthropods such as insects, crabs, and spiders. It is also a structural material in the cell walls of fungi such as mushrooms, bread molds, and yeasts. Unlike cellulose, chitin is digested by enzymes that are widespread among microorganisms, plants, and many animals. In plants and animals, including humans and other mammals, chitin-digesting enzymes occur primarily as part of defenses against fungal infections. However, humans cannot digest chitin as a food source.

Polysaccharides also occur on the surfaces of cells, particularly in animals. These surface polysaccharides are attached to both the protein and lipid molecules in membranes. They help hold the cells of animals together and serve as recognition sites between cells.

STUDY BREAK 3.2

What is the difference between a monosaccharide, a disaccharide, and a polysaccharide? Give examples of each.

3.3 Lipids

Lipids are a diverse group of water-insoluble, primarily nonpolar biological molecules composed mostly of hydrocarbons. Some are large molecules, but they are not large enough to be considered macromolecules. Lipids also are not considered to be polymers.

As a result of their nonpolar character, lipids typically dissolve much more readily in nonpolar solvents, such as acetone and chloroform, than in water, the polar solvent of living organisms. Their insolubility in water underlies their ability to form cell membranes, the thin molecular films that create boundaries between and within cells.

In addition to forming membranes, some lipids are stored and used in cells as an energy source. Other lipids serve as hormones that regulate cellular activities. Three types of lipid molecules—*neutral lipids, phospholipids,* and *steroids*—occur most commonly in living organisms.

Neutral Lipids Are Familiar as Fats and Oils

Neutral lipids, commonly found in cells as energy-storage molecules, are called "neutral" because at cellular pH they have no charged groups; they are, therefore, nonpolar. There are two types of neutral lipids: **oils** and **fats.** Oils are liquid at biological temperatures, and fats are semisolid. Generally, neutral lipids are insoluble in water.

Almost all neutral lipids are formed by dehydration synthesis reactions involving glycerol and three *fatty acids* **(Figure 3.10).** The product is also called a **triglyceride.** Glycerol is a three-carbon alcohol with an —OH attached to each carbon. In its free state, glycerol is a polar, water soluble, sweet-tasting alcohol. The glycerol forms the backbone of the triglyceride. A **fatty acid** contains a single hydrocarbon chain with a carboxyl group (—COOH) at one end. The carboxyl group gives the fatty acid its acidic properties. In the synthesis of a triglyceride, three dehydration synthesis reactions occur, each involving the carboxyl group of one fatty acid and one of the hydroxyl groups of glycerol. A covalent bond formed between a carboxyl group and a hydroxyl group, as here, is called an **ester linkage.** In the formation of the three ester linkages, the polar groups of glycerol are eliminated, resulting in the nonpolar triglyceride.

The fatty acids in living organisms contain four or more carbons in their hydrocarbon chain, with the most common forms having even-numbered chains of 14 to 22 carbons. Only the shortest fatty acid chains are water-soluble. As chain length increases, fatty acids become progressively less water-soluble and more oily.

If the hydrocarbon chain of a fatty acid binds the maximum possible number of hydrogen atoms, so that only single bonds link the carbon atoms, the fatty acid is said to be **saturated** with hydrogen atoms (as in stearic acid in **Figure 3.11A**). If one or more double bonds link the carbons (**Figure 3.11B,** arrow), reducing the number of hydrogen atoms bound, the fatty acid is **unsaturated.** Fatty acids with one double bond are **monounsaturated;** those with more than one double bond are **polyunsaturated.**

Unsaturated fatty acid chains tend to bend or "kink" at a double bond (see Figures 3.11B and 3.14C). The kink makes the chains more disordered and thus more fluid at biological temperatures. Consequently, unsaturated fatty acids melt at lower temperatures than saturated fatty acids of the same length, and they generally have oily rather than fatty characteristics.

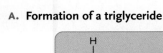

A. Formation of a triglyceride

Triglyceride

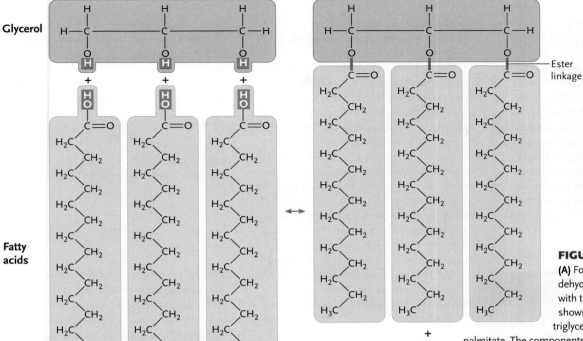

FIGURE 3.10 Triglycerides.
(A) Formation of a triglyceride by dehydration synthesis of glycerol with three fatty acids. The fatty acid shown is palmitic acid, and the triglyceride product is glyceryl palmitate. The components of a water molecule (in blue) are removed from the glycerol and fatty acids in each of the three bonds formed. **(B)** Space-filling model of the triglyceride, glyceryl palmitate.
© Cengage Learning 2014

In foods, saturated fatty acids are usually found in solid animal fats, such as butter, whereas unsaturated fatty acids are usually found in vegetable oils, such as liquid canola oil. Nonetheless, both solid animal fats and liquid vegetable oils contain some saturated and some unsaturated fatty acids.

The fatty acids linked to a glycerol may be different or the same. Different organisms usually have distinctive combinations of fatty acids in their triglycerides. As with individual fatty acids, triglycerides generally become less fluid as the length of their fatty acid chains increases; those with shorter chains remain liquid as oils at biological temperatures, and those with longer chains solidify as fats. The degree of saturation of the fatty acid chains also affects the fluidity of triglycerides—the more saturated, the less fluid the triglyceride. Plant oils are converted commercially to fats by *hydrogenation*—that is, adding hydrogen atoms to increase the degree of saturation, as in the conversion of vegetable oils to margarines and shortening.

Triglycerides are used widely as stored energy in animals. Gram for gram, they yield more than twice as much energy as carbohydrates do (see Chapter 47). Therefore, fats are an excellent source of energy in the diet. Storing the equivalent amount of energy as carbohydrates rather than fats would add more than 100 pounds to the weight of an average man or woman. A layer of fatty tissue just under the skin also serves as an insulating blanket in humans, other mammals, and birds. Triglycerides secreted from special glands in waterfowl and other birds

A. Stearic acid,
$CH_3(CH_2)_{16}COOH$

B. Oleic acid,
$CH_3(CH_2)_7CH = CH(CH_2)_7COOH$

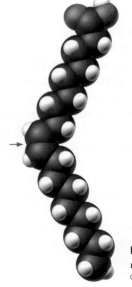

FIGURE 3.11 Fatty acids. **(A)** Stearic acid, a saturated fatty acid. **(B)** Oleic acid, an unsaturated fatty acid. An arrow marks the "kink" introduced by the double bond.
© Cengage Learning 2014

Fats, Cholesterol, and Coronary Artery Disease

Hardening of the arteries, or *atherosclerosis,* is a condition in which deposits of lipid and fibrous material called *plaque* build up in the walls of arteries, the vessels that supply oxygenated blood to body tissues. Plaque reduces the internal diameter of the arteries, restricting or even completely blocking the flow of blood. Blockage of the coronary arteries that supply oxygenated blood to the heart muscle **(Figure)** can severely impair heart function, a condition called *coronary heart disease.* In ex-treme cases, it can lead to destruction of heart muscle tissue, as occurs in a heart attack (myocardial infarction).

Your body requires a certain amount of cholesterol, but the liver normally makes enough to meet this demand. Additional cholesterol is made from fats taken in as food. Cholesterol is found in the blood bound to low-density lipoprotein (LDL) and high-density lipoprotein (HDL). LDL cholesterol is considered "bad" because of a positive correlation between its level in the blood and the risk for coronary heart disease. LDL cholesterol contributes to plaque formation as atherosclerosis proceeds. In contrast, HDL cholesterol is "good" because high levels appear to provide some protection against coronary heart disease. Simplifying, HDL cholesterol removes excess cholesterol from plaques in arteries, thereby reducing plaque buildup. The cholesterol that has been removed is transported by the HDL cholesterol to the liver where it is broken down.

Fats in food affect cholesterol levels in the blood. Diets high in saturated fats raise LDL cholesterol levels, but levels of HDL cholesterol are not affected by such a diet. Foods of animal origin typically contain saturated fats, and foods of plant origin typically contain unsaturated fats.

In the food industry, unsaturated vegetable oils are often processed to solidify the fats. The process, partial hydrogenation, adds hydrogen atoms to unsatu-rated sites, eliminating many double bonds and generating substances known as *trans* fatty acids (or *trans* fats). Usually the hydrogen atoms at a double bond are positioned on the same side of the carbon chain, producing a *cis* (Latin, "on the same side") fatty acid:

$$\begin{array}{ccc} H & H \\ | & | \\ -C & = & C- \end{array}$$

but in a *trans* (Latin, "across") fatty acid, the hydrogen atoms are on different sides of the chain at some double bonds:

$$\begin{array}{ccc} H \\ | \\ -C & = & C- \\ & & | \\ & & H \end{array}$$

Trans fatty acids are found in many vegetable shortenings, some margarines, cookies, cakes, doughnuts, and other foods made with or fried in partially hydrogenated fats.

Research from human feeding studies has shown that *trans* fatty acids raise LDL cholesterol levels nearly as much as saturated fatty acids do. More seriously, intake of *trans* fatty acids at levels found in a typical U.S. diet also appears to reduce HDL cholesterol levels. In addition, clinical studies have demonstrated a positive correlation between the intake of *trans* fatty acids and the occurrence of coronary heart disease. A regulation to add the *trans* fatty acid content to nutritional labels went into effect in the United States in January 2006.

Coronary artery

Atherosclerotic plaques

Cardiac muscle (heart muscle tissue)

Micrograph Louis L. Lainey

Atherosclerotic plaques (bright areas) in the coronary arteries of a patient with heart disease.

© Cengage Learning 2014

© iStockphoto/Keith Szafranski

FIGURE 3.12 Penguins of the Antarctic, one of several animals that have a thick, insulating layer of fatty tissue that contains triglycerides under the skin. An oily coating secreted by a gland near their tail and spread by their face and bill keeps their feathers watertight and dry.

help make their feathers water repellent (as in the penguins shown in **Figure 3.12**).

Unsaturated fats are considered healthier than saturated fats in the human diet. Saturated fats have been implicated in the development of atherosclerosis (see *Focus on Applied Research*), a disease in which arteries, particularly those serving the heart, become clogged with fatty deposits.

Fatty acids may also combine with long-chain alcohols or hydrocarbon structures to form **waxes,** which are harder and less greasy than fats. Insoluble in water, waxy coatings help keep skin, hair, or feathers of animals protected, lubricated, and pliable. In humans, earwax lubricates the outer ear canal and protects the eardrum. Honeybees use a wax secreted by glands in their abdomen to construct the comb in which larvae are raised and honey is stored.

Many plants secrete waxes that form a protective exterior layer, which greatly reduces water loss from the plants and resists invasion by infective agents such as bacteria and viruses. This waxy covering gives cherries, apples, and many other fruits their shiny appearance.

Phospholipids Provide the Framework of Biological Membranes

Phosphate-containing lipids called **phospholipids** are the primary lipids of cell membranes. In the most common phospholipids, glycerol forms the backbone of the molecule as in triglycerides, but only two of its binding sites are linked to fatty acids **(Figure 3.13)**. The third site is linked to a polar phosphate group, which binds to yet another polar unit. The end of the molecule containing the fatty acids is nonpolar and hydrophobic, and the end with the phosphate group is polar and hydrophilic.

In polar environments, such as a water solution, phospholipids assume arrangements in which only their polar ends are exposed to the water; their nonpolar ends collect together in a region that excludes water. One of these arrangements, the *bilayer,* is the structural basis of membranes, the organizing boundaries of all living cells (see Figure 2.14 and Chapter 6). In a bilayer, formed by a film of phospholipids just two molecules thick, the phospholipid molecules are aligned so that the polar groups face the surrounding water molecules at the surfaces of the bilayer. The hydrocarbon chains of the phospholipids are packed together in the interior of the bilayer, where they form a nonpolar, hydrophobic region that excludes water. The bilayer remains stable because, if disturbed, the hydrophobic, nonpolar hydrocarbon chains of the phospholipids become exposed to the surrounding watery solution, and the molecule returns to its normal bilayer arrangement.

In the origin of cells billions of years ago, researchers hypothesize that organic molecules assembled into aggregates that became bounded by lipid bilayer membranes to form primitive protocells. Protocells are considered key to the origin of life because life depends on reactions occurring in a controlled and sequestered environment; namely, the cell. Protocells are the presumed precursors of cells. (The origin of cells is discussed in more detail in Chapter 25.)

FIGURE 3.13 Phospholipid structure. (A) The arrangement of components in phospholipids. **(B)** Phosphatidyl ethanolamine, a common membrane phospholipid. **(C)** Space-filling model of phosphatidyl ethanolamine. The kink in the fatty acid chain on the right reflects a double bond at this position. **(D)** Diagram widely used to depict a phospholipid molecule in cell membrane diagrams. The sphere represents the polar end of the molecule, and the zigzag lines represent the nonpolar fatty acid chains. (The kink in the fatty acid chain is not depicted.)
© Cengage Learning 2014

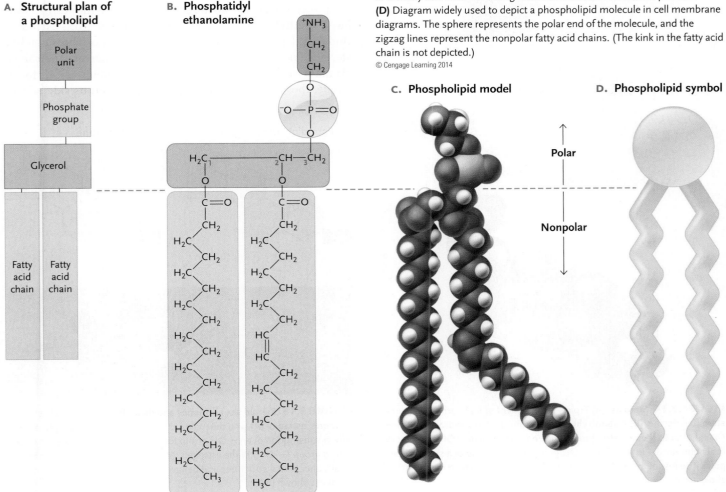

Steroids Contribute to Membrane Structure and Work as Hormones

Steroids are a group of lipids with structures based on a framework of four carbon rings **(Figure 3.14A).** Small differences in the side groups attached to the rings distinguish one steroid from another. The most abundant steroids, the **sterols,** have a single polar —OH group linked to one end of the ring framework and a complex, nonpolar hydrocarbon chain at the other end **(Figure 3.14B).** Although sterols are almost completely hydrophobic, the single hydroxyl group gives one end of the molecule a slightly polar, hydrophilic character. As a result, sterols also have dual solubility properties and, like phospholipids, tend to assume positions that satisfy these properties. In biological membranes, they line up beside the phospholipid molecules with their polar —OH group facing the membrane surface and their nonpolar ends buried in the nonpolar membrane interior.

Cholesterol (see **Figure 3.14B, C**) is an important component of the boundary membrane surrounding animal cells; similar sterols, called **phytosterols,** occur in plant cell membranes. Deposits derived from cholesterol also collect inside arteries in atherosclerosis (see *Focus on Applied Research*).

Other steroids, the *steroid hormones,* are important regulatory molecules in animals; they control development, behavior, and many internal biochemical processes. The sex hormones that control differentiation of the sexes and sexual behavior are primary examples of steroid hormones **(Figure 3.15).** Small differences in the functional groups of steroid hormones have vastly different effects in animals. For instance, the two key differences between the estrogen estradiol, the primary female sex hormone, and the androgen testosterone, the male sex hormone, are that estradiol has an —OH in the position where testosterone has an =O, and testosterone has a methyl group (—CH_3) that is absent from estradiol. Bodybuilders and other athletes sometimes use hormonelike steroids (anabolic-androgenic steroids) to increase their muscle mass (see *Focus on Basic Research* in Chapter 42). Unfortunately, these substances also produce numerous side effects,

A. Arrangement of carbon rings in a steroid

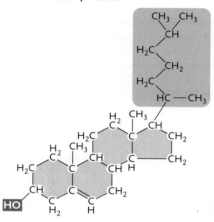

B. Cholesterol, a sterol

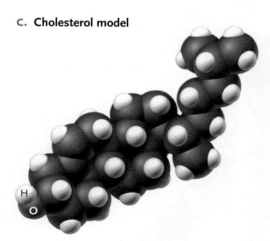

C. Cholesterol model

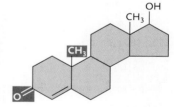

FIGURE 3.14 Steroids. (A) Typical arrangement of four carbon rings in a steroid molecule. **(B)** A sterol, cholesterol. Sterols have a hydrocarbon side chain linked to the ring structure at one end and a single —OH group at the other end (boxed in red). The —OH group makes its end of a sterol slightly polar. The rest of the molecule is nonpolar. **(C)** A space-filling model of cholesterol.
© Cengage Learning 2014

A. Estradiol, an estrogen   **B. Testosterone**

C.

FIGURE 3.15 Steroid sex hormones and their effects. The female sex hormone, estradiol **(A),** and the male sex hormone, testosterone **(B),** differ only in substitution of an —OH group for an oxygen and the absence of one methyl group (—CH_3) in the estrogen. Although small, these differences greatly alter sexual structures and behavior in animals, such as humans, and the ducks shown in **(C).**
© Cengage Learning 2014

including elevated cholesterol, elevated blood pressure, and acne. Other steroids occur as poisons in the venoms of toads and other animals.

Several other lipid types have structures unrelated to triglycerides, phospholipids, or steroids. Among these are *chlorophylls* and *carotenoids,* pigments that absorb light and participate in its conversion to chemical energy in plants (see Chapter 9). Lipid groups also combine with carbohydrates to form *glycolipids* and with proteins to form *lipoproteins.* Both glycolipids and lipoproteins form parts of cell membranes, where they perform vital structural and functional roles.

STUDY BREAK 3.3

What are the three most common lipids in living organisms? Distinguish between their structures.

3.4 Proteins

Proteins are macromolecules that perform many vital functions in living organisms **(Table 3.2).** Some provide structural support for cells; others, called **enzymes,** increase the rate of cellular reactions; still others impart movement to cells and cellular structures. Proteins also transport substances across biological membranes, serve as recognition and receptor molecules at cell surfaces, or regulate the activity of other proteins and DNA. Some proteins work as hormones or defend against foreign substances, such as infectious microorganisms. Many toxins and venoms are proteins.

All of the protein molecules that carry out these and other functions are fundamentally similar in structure. All are polymers consisting of one or more unbranched chains of monomers called amino acids. An **amino acid** is a molecule that contains both an amino and a carboxyl group. Although the most com-

TABLE 3.2	Major Protein Functions	
Protein Type	Function	Examples
Structural proteins	Support	Microtubule and microfilament proteins form supporting fibers inside cells; collagen and other proteins surround and support animal cells; cell wall proteins support plant cells.
Enzymatic proteins	Increase the rate of biological reactions	Among thousands of examples, DNA polymerase increases the rate of duplication of DNA molecules; RuBP (ribulose 1,5-bisphosphate) carboxylase/oxygenase increases the rates of the first synthetic reactions of photosynthesis; lipases and proteases in the digestive enzymes increase the rate of breakdown of fats and proteins, respectively.
Membrane transport proteins	Speed up movement of substances across biological membranes	Ion transporters move ions such as Na^+, K^+, and Ca^{2+} across membranes; glucose transporters move glucose into cells; aquaporins allow water molecules to move across membranes.
Motile proteins	Produce cellular movements	Myosin acts on microfilaments to produce muscle movements; dynein acts on microtubules to produce the whipping movements of sperm tails, flagella, and cilia (the last two are whiplike appendages on the surfaces of many eukaryotic cells); kinesin acts on microtubules of the cytoskeleton (the scaffolding of eukaryotic cells responsible for cellular movement, cell division, and the organization of organelles).
Regulatory proteins	Promote or inhibit the activity of other cellular molecules	Nuclear regulatory proteins turn genes on or off to control the activity of DNA; protein kinases add phosphate groups to other proteins to modify their activity.
Receptor proteins	Bind molecules at cell surface or within cell; some trigger internal cellular responses	Hormone receptors bind hormones at the cell surface or within cells and trigger cellular responses; cellular adhesion molecules help hold cells together by binding molecules on other cells; LDL receptors bind cholesterol-containing particles at cell surfaces.
Hormones	Carry regulatory signals between cells	Insulin regulates sugar levels in the bloodstream; growth hormone regulates cellular growth and division.
Antibodies	Defend against invading molecules and organisms	Antibodies recognize, bind, and help eliminate essentially any protein of infecting bacteria and viruses, and many other types of molecules, both natural and artificial.
Storage proteins	Hold amino acids and other substances in stored form	Ovalbumin is a storage protein of eggs; apolipoproteins hold cholesterol in stored form for transport through the bloodstream.
Venoms and toxins	Interfere with competing organisms	Ricin is a castor-bean protein that stops protein synthesis; bungarotoxin is a snake venom that causes muscle paralysis.

© Cengage Learning 2014

mon proteins contain 50 to 1,000 amino acids, some proteins found in nature have as few as 3 or as many as 50,000 amino acid units. Proteins range in shape from globular or spherical forms to elongated fibers, and they vary from soluble to completely insoluble in water solutions. Some proteins have single functions, whereas others have multiple functions.

Cells Assemble 20 Kinds of Amino Acids into Proteins by Forming Peptide Bonds

The cells of all organisms use 20 different amino acids as the building blocks of proteins **(Figure 3.16)**. Of these 20 amino acids, 19 have the same structural plan—a central carbon atom is attached to an amino group ($-NH_2$), a carboxyl group ($-COOH$), and a hydrogen atom:

$$H_2N-\underset{\underset{H}{|}}{\overset{\overset{R}{|}}{C}}-COOH$$

The remaining bond of the central carbon is linked to 1 of 19 different side groups represented by the R (see shaded regions in Figure 3.16); its usage for amino acids refers to a range from a single hydrogen atom to complex carbon-containing chains or rings. The remaining amino acid, proline, has a ring structure that includes the central carbon atom; that carbon bonds to a $-COOH$ group on one side and to an $=NH$ (imino) group that forms part of the ring at the other side (see Figure 3.16). (Strictly speaking, proline is an imino acid.) At the pH of the cell, the amino acids assume ionic forms: the amino group becomes $-NH_3^+$ ($-NH_2^+$ in the case of proline), and the carboxyl group becomes $-COO^-$.

FIGURE 3.16 The 20 amino acids used by cells to make proteins. The side group of each amino acid is boxed in brown. The amino acids are shown in the ionic forms in which they are found at the pH within the cell. Three-letter and one-letter abbreviations commonly used for the amino acids appear below each diagram. All amino acids assembled into proteins are in the L-form, one of two possible stereoisomers.
© Cengage Learning 2014

A. Nonpolar amino acids

Alanine Ala A · Valine Val V · Leucine Leu L · Isoleucine Ile I · Glycine Gly G · Phenylalanine Phe F · Tryptophan Trp W · Methionine Met M · Proline Pro P

B. Uncharged polar amino acids

Cysteine Cys C · Serine Ser S · Threonine Thr T · Tyrosine Tyr Y · Asparagine Asn N · Glutamine Gln Q

C. Negatively charged (acidic) polar amino acids

Aspartic acid Asp D · Glutamic acid Glu E

D. Positively charged (basic) polar amino acids

Lysine Lys K · Arginine Arg R · Histidine His H

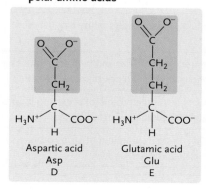

Differences in the side groups give the amino acids their individual properties. Some amino acids are nonpolar (see Figure 3.16A), and some are polar (see Figure 3.16B, C, D). Among the polar amino acids, some are uncharged (see Figure 3.16B), some are negatively charged (acidic) (see Figure 3.16C), and some are positively charged (basic) (see Figure 3.16D). Many of the side groups contain reactive functional groups, such as —NH₂, —OH, —COOH, or —SH, which may interact with atoms located elsewhere in the same protein or with molecules and ions outside the protein.

The sulfhydryl group (—SH) in the amino acid cysteine is particularly important in protein structure. The sulfhydryl groups in the side groups of two cysteines located in different regions of the same protein, or in different proteins, can react to produce disulfide linkages (—S—S—) (see Table 3.1). The linkages fasten amino acid chains together **(Figure 3.17)** and help hold proteins in their three-dimensional shape.

Overall, the varied properties and functions of proteins depend on the types and locations of the different amino acid side groups in their structures. The variations in the number and types of amino acids mean that the total number of possible proteins is extremely large.

Covalent bonds link amino acids into the chains of subunits that make proteins. The link, a **peptide bond,** is formed by a dehydration synthesis reaction between the amino group of one amino acid and the carboxyl group of a second as shown in **Figure 3.18.** In the cell, an amino acid chain always has an —NH₃⁺ group at one end, called the **N-terminal end,** and a —COO⁻ group at the other end, called the **C-terminal end.**

The chain of amino acids formed by sequential peptide bonds is a **polypeptide,** and it is only part of the complex structure of proteins. That is, once assembled, an amino acid chain may fold in various patterns, and more than one chain may combine to form a finished protein, adding to the structural and functional variability of proteins.

Proteins Have as Many as Four Levels of Structure

Proteins have up to four levels of structure, with each level imparting different characteristics and degrees of structural complexity to the molecule **(Figure 3.19)**. **Primary structure** is the particular and unique sequence of amino acids forming a polypeptide; **secondary structure** is produced by the twists and turns of the amino acid chain; and **tertiary structure** is the folding of

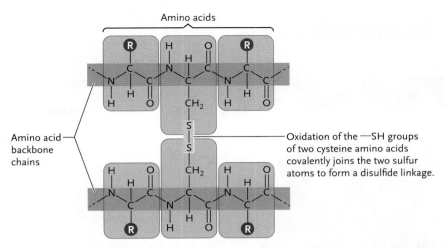

FIGURE 3.17 A disulfide linkage between two amino acid chains or two regions of the same chain. The linkage is formed by a reaction between the sulfhydryl groups (—SH) of cysteines. The circled Rs indicate the side groups of other amino acids in the chains. Figure 3.20 shows disulfide linkages in a real protein, and Figure 3.23 shows how disulfide linkages help maintain a protein's conformation.
© Cengage Learning 2014

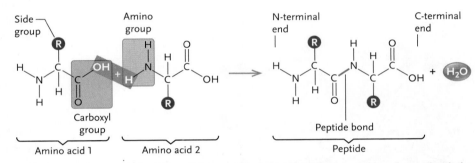

FIGURE 3.18 A peptide bond formed by reaction of the carboxyl group of one amino acid with the amino group of a second amino acid. The reaction is a typical dehydration synthesis reaction.
© Cengage Learning 2014

the amino acid chain, with its secondary structures, into the overall three-dimensional shape of a protein. All proteins have primary, secondary, and tertiary structures. **Quaternary structure,** when present, refers to the arrangement of polypeptide chains in a protein that is formed from more than one chain.

Primary Structure Is the Fundamental Determinant of Protein Form and Function

The primary structure of a protein—the sequence in which amino acids are linked—underlies the other, higher levels of structure. Changing even a single amino acid of the primary structure alters the secondary, tertiary, and quaternary structures to at least some degree and, by so doing, can alter or even destroy the biological functions of a protein. For example, substitution of a single amino acid in the blood protein hemoglobin produces an altered form responsible for sickle-cell anemia (see Chapter 15); a number of other blood disorders are caused by single amino acid substitutions in other parts of the protein.

Because primary structure is so fundamentally important, many years of research have been devoted to determining the

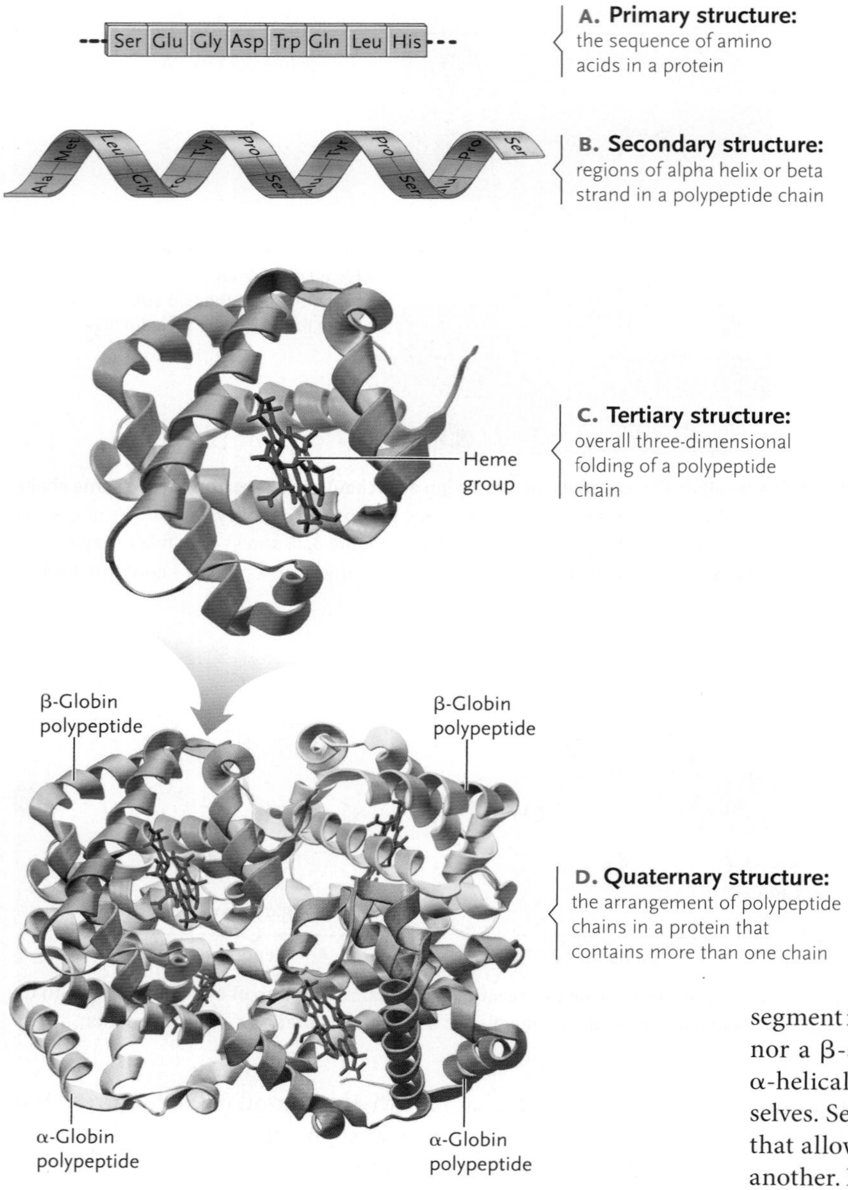

A. Primary structure: the sequence of amino acids in a protein

--- | Ser | Glu | Gly | Asp | Trp | Gln | Leu | His | ---

B. Secondary structure: regions of alpha helix or beta strand in a polypeptide chain

C. Tertiary structure: overall three-dimensional folding of a polypeptide chain

Heme group

β-Globin polypeptide

β-Globin polypeptide

D. Quaternary structure: the arrangement of polypeptide chains in a protein that contains more than one chain

α-Globin polypeptide

α-Globin polypeptide

FIGURE 3.19 The four levels of protein structure. The protein shown in **(C)** is one of the subunits of a hemoglobin molecule; the heme group (in red) is an iron-containing group that binds oxygen. **(D)** A complete hemoglobin molecule.
© Cengage Learning 2014

Twists and Other Arrangements of the Amino Acid Chain Form the Secondary Structure of a Protein

The amino acid chain of a protein, rather than being stretched out in linear form, is folded into arrangements that form the protein's secondary structure. Two highly regular secondary structures, the *alpha helix* and the *beta strand,* are particularly stable and make an amino acid chain resistant to bending. Most proteins have segments of both arrangements.

THE ALPHA HELIX In the **alpha (α) helix,** first identified by Linus Pauling and Robert Corey at the California Institute of Technology in 1951, the backbone of the amino acid chain is twisted into a regular, right-hand spiral **(Figure 3.21)**. The amino acid side groups extend outward from the twisted backbone. The structure is stabilized by regularly spaced hydrogen bonds (dotted lines in Figure 3.21) between atoms in the backbone.

Most proteins contain segments of α helix, which are rigid and rodlike, in at least some regions. Globular proteins usually contain several short α-helical segments that run in different directions, connected by segments of *random coil*. A random coil segment is a sequence of amino acids that has neither an α-helical nor a β-strand structure. It provides flexible sites that allow α-helical or β-strand segments to bend or fold back on themselves. Segments of random coil also commonly act as "hinges" that allow major parts of proteins to move with respect to one another. Fibrous proteins, such as the collagens, a major component of tendons, bone, and other extracellular structures in animals, typically contain one or more α-helical segments that run the length of the molecule, with few or no bendable regions of random coil.

THE BETA STRAND Pauling and Corey were also the first to identify the beta (β) strand as a major secondary protein structure. In a β strand, the amino acid chain zigzags in a flat plane rather than twisting into a coil. In many proteins, β strands are aligned side

amino acid sequence of proteins. Initial success came in 1953, when the English biochemist Frederick Sanger deduced the amino acid sequence of insulin, a protein-based hormone, using samples obtained from cows **(Figure 3.20)**. Now, the amino acid sequences of literally thousands of proteins have been determined. Knowledge of the primary structure of proteins can allow their three-dimensional structure and functions to be predicted and reveal relationships among proteins.

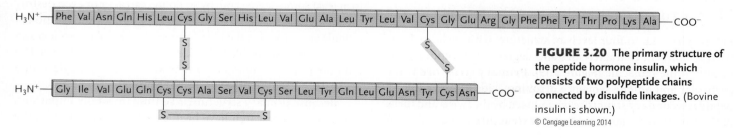

H_3N^+ — | Phe | Val | Asn | Gln | His | Leu | Cys | Gly | Ser | His | Leu | Val | Glu | Ala | Leu | Tyr | Leu | Val | Cys | Gly | Glu | Arg | Gly | Phe | Phe | Tyr | Thr | Pro | Lys | Ala | — COO^-

H_3N^+ — | Gly | Ile | Val | Glu | Gln | Cys | Cys | Ala | Ser | Val | Cys | Ser | Leu | Tyr | Gln | Leu | Glu | Asn | Tyr | Cys | Asn | — COO^-

FIGURE 3.20 The primary structure of the peptide hormone insulin, which consists of two polypeptide chains connected by disulfide linkages. (Bovine insulin is shown.)
© Cengage Learning 2014

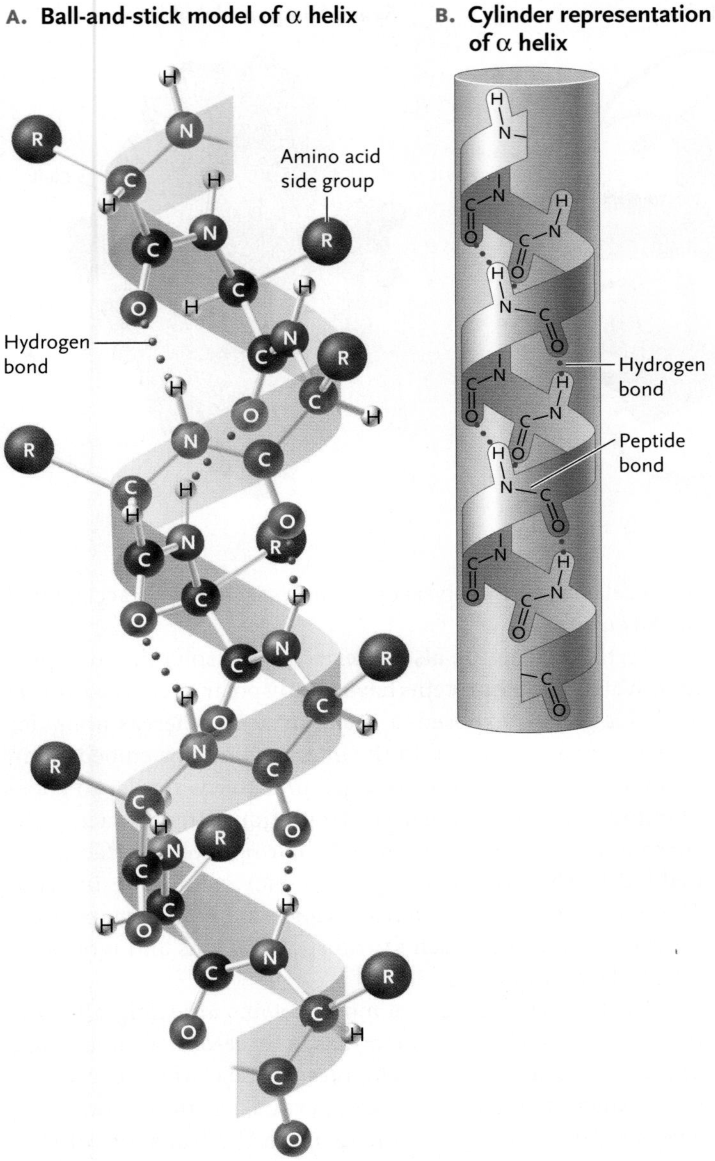

A. Ball-and-stick model of α helix

R

N

C

H

C

Amino acid side group

N

O

C

R

H

Hydrogen bond

N

C

R

H

R

C

N

O

H

C

N

R

O

C

H

N

C

R

O

H

R

C

N

C

O

N

R

C

H

R

C

N

O

H

C

O

B. Cylinder representation of α helix

H

N

N

H

C

Hydrogen bond

N

C

O

Peptide bond

N

C

O

C

N

O

C

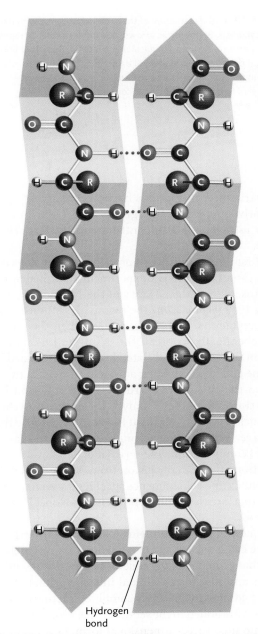

Hydrogen bond

FIGURE 3.21 The α helix, a type of secondary structure in proteins. **(A)** A model of the α helix showing atoms as spheres and covalent bonds as rods. The backbone of the amino acid chain is held in a spiral by hydrogen bonds formed at regular intervals. **(B)** The cylinder often is used to depict an α helix in protein diagrams, with peptide and hydrogen bonds also shown.
© Cengage Learning 2014

FIGURE 3.22 A β sheet formed by side-by-side alignment of two β strands, a type of secondary structure in proteins. The β strands are held together stably by hydrogen bonds. In this sheet, the β strands run in opposite directions, as shown by the arrows, which point in the direction of the C-terminal end of each polypeptide chain. Strands may also run in the same direction in a β sheet. Arrows alone often are used to represent β strands in protein diagrams.
© Cengage Learning 2014

by side in the same or opposite directions to form a structure known as a **beta (β) sheet (Figure 3.22).** Hydrogen bonds between adjacent β strands stabilize the sheet, making it a highly rigid structure. Beta sheets may lie in a flat plane or may twist into propeller- or barrel-like structures.

Beta strands and sheets occur in many proteins, usually in combination with α-helical segments. One notable exception is in the silk protein secreted by silk worms, which contains only β sheets. This exceptionally stable structure, reinforced by an extensive network of hydrogen bonds, underlies the unusually high tensile strength of silk fibers.

The Tertiary Structure of a Protein Is Its Overall Three-Dimensional Conformation

The tertiary structure of a protein is its overall three-dimensional shape, or **conformation (Figure 3.23).** The contents of α-helical and β-strand secondary structure segments, together with the number and position of disulfide linkages and hydrogen bonds, play the major roles in folding each protein into its

FIGURE 3.23 Tertiary structure of the protein lysozyme, with helices shown as cylinders, β strands as arrows, and random coils as ropes. Lysozyme is an enzyme found in nasal mucus, tears, and other body secretions; it destroys the cell walls of bacteria by breaking down molecules in the wall. Disulfide bonds are shown in red. A space-filling model of lysozyme is shown for comparison.
© Cengage Learning 2014

Lysozyme

Space-filling model of lysozyme

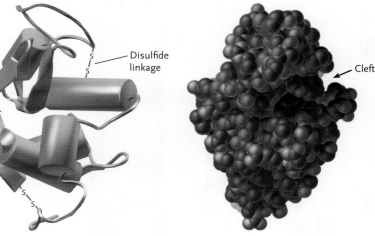

Disulfide linkage

Cleft

tertiary structure. Attractions between positively and negatively charged side groups and polar or nonpolar associations also contribute to the tertiary structure.

The first insight into how a protein assumes its tertiary structure came from a classic experiment published by Christian Anfinsen and Edgar Haber of the National Institutes of Health in 1962 **(Figure 3.24)**. The researchers studied ribonuclease, an enzyme that hydrolyzes RNA. When they treated the enzyme chemically to break the disulfide linkages holding the protein in its functional state, the protein unfolded and had no enzyme activity. Unfolding a protein from its active conformation so that it loses its structure and function is called **denaturation.** When they removed the denaturing chemicals, the ribonuclease slowly regained full activity because the disulfide linkages reformed, enabling the protein to reassume its functional conformation. The reversal of denaturation is called **renaturation.** The key conclusion from Anfinsen's experiment was that the amino acid sequence specifies the tertiary structure of a protein. Christian Anfinsen received a Nobel Prize in 1972 for this work.

The question of how proteins fold into their tertiary structure in the cell is the subject of contemporary research. Results indicate that proteins fold gradually as they are assembled—as successive amino acids are linked into the primary structure, the chain folds into increasingly complex structures. As the final amino acids are added to the sequence, the protein completes its folding into final three-dimensional form. One nagging question about this process is how proteins assume their correct tertiary structure among the different possibilities that may exist for a given amino acid sequence. For many proteins, "guide" proteins called **chaperone proteins** or **chaperonins** answer this question; they bind temporarily with newly synthesized proteins, directing their conformation toward the correct tertiary structure and inhibiting incorrect arrangements as the new proteins fold **(Figure 3.25)**.

Tertiary structure determines a protein's function. That is, a protein's tertiary structure buries some amino acid side groups in its interior and exposes others at the surface. The distribution and three-dimensional arrangement of the side groups, in combination with their chemical properties, determine the overall chemical activity of the protein. For example, the tertiary structure of the antibacterial enzyme lysozyme (see Figure 3.23) has a cleft that binds a polysaccharide found in bac-

terial cell walls; hydrolysis of the polysaccharide is accelerated by the enzyme.

Tertiary structure also determines the solubility of a protein. Water-soluble proteins have mostly polar or charged amino acid side groups exposed at their surfaces, whereas nonpolar side groups are clustered in the interior. Proteins embedded in nonpolar membranes are arranged in patterns similar to phospholipids, with their polar (hydrophilic) segments facing the surrounding watery solution and their nonpolar surfaces embedded in the nonpolar (hydrophobic) membrane interior. These dual-solubility proteins perform many important functions in membranes, such as transporting ions and molecules into and out of cells.

The tertiary structure of most proteins is flexible, allowing them to undergo limited alterations in three-dimensional shape known as **conformational changes.** These changes contribute to the function of many proteins, particularly those working as enzymes, in cellular movements or in the transport of substances across cell membranes.

As you learned earlier, chemical treatment can denature a protein in the test tube (see Figure 3.24). Excessive heat can also break the hydrogen bonds holding a protein in its natural conformation, causing it to denature and lose its biological activity. Denaturation is one of the major reasons few living organisms can tolerate temperatures greater than 45°C. Extreme changes in pH, which alter the charge of amino acid side groups and weaken or destroy ionic bonds, can also cause protein denaturation.

For some proteins, denaturation is permanent. A familiar example of a permanently denatured protein is a cooked egg white. In its natural form, the egg white protein albumin dissolves in water to form a clear solution. The heat of cooking denatures it permanently into an insoluble, whitish mass. For other proteins, such as ribonuclease (mentioned previously), denaturation is reversible; the proteins can renature and return to their functional form if the temperature or pH returns to normal values.

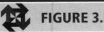 **FIGURE 3.24** | **Experimental Research**

Anfinsen's Experiment Demonstrating That the Amino Acid Sequence of a Protein Specifies Its Tertiary Structure

Question: What is the relationship between the amino acid sequence of a protein and its conformation?

Experiment: Anfinsen and Haber studied the 124-amino acid enzyme ribonuclease in the test tube. They knew that the native (functional) enzyme has four disulfide linkages between amino acids 26 and 84, 40 and 95, 58 and 110, and 65 and 72 (see figure). They treated the active enzyme with a mixture of urea and β-mercaptoethanol, which breaks disulfide linkages. They then removed the two chemicals and left the enzyme solution in air.

Results: The chemical treatment broke the four disulfide linkages, which caused the protein to denature and lose its enzyme activity. After the chemicals had been removed and the enzyme solution exposed to air, Anfinsen made the crucial observation that the protein renatured, slowly regaining enzyme activity. Ultimately the solution showed 90% of the activity of the native enzyme.

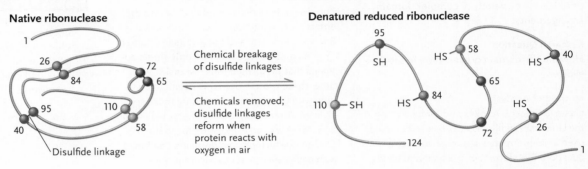

Native ribonuclease

Disulfide linkage

Chemical breakage of disulfide linkages

Chemicals removed; disulfide linkages reform when protein reacts with oxygen in air

Denatured reduced ribonuclease

He realized that oxygen from the air had reacted with the —SH groups of the denatured enzyme causing disulfide linkages to reform, and that the enzyme had spontaneously refolded into its native, active conformation. All physical and chemical properties of the refolded enzyme the researchers measured were the same as those of the native enzyme, confirming that the same disulfide bridges had formed as in the native enzyme.

Conclusion: Anfinsen concluded that the information for determining the three-dimensional shape of ribonuclease is in its amino acid sequence.

THINK LIKE A SCIENTIST If denatured ribonuclease renatures in the presence of a high concentration of urea, the renatured enzyme has physical and chemical properties similar to those of the native enzyme indicating that refolding had occurred, but enzyme activity is less than 1% of that of the native enzyme. Interpret this result.

Source: E. Haber and C. Anfinsen. 1962. Side-chain interactions governing the pairing of half-cystine residues in ribonuclease. *Journal of Biological Chemistry* 237:1839–1844.

Multiple Polypeptide Chains Form Quaternary Structure

Some complex proteins, such as hemoglobin and antibody molecules, have *quaternary structure*—that is, the presence and arrangement of two or more polypeptide chains (see Figure 3.19D). The same bonds and forces that fold single polypeptide chains into tertiary structures, including hydrogen bonds, polar and nonpolar attractions, and disulfide linkages, also hold the multiple polypeptide chains together. During the assembly of multichain proteins, chaperonins also promote correct association of the individual amino acid chains and inhibit incorrect formations.

Combinations of Secondary, Tertiary, and Quaternary Structure Form Functional Domains in Many Proteins

In many proteins, folding of the amino acid chain (or chains) produces distinct, large structural subdivisions called **domains (Figure 3.26A).** Often, one domain of a protein is connected to

A Big Bang in Protein Structure Evolution: How did the domain organization in proteins evolve?

Many proteins have distinct, large structural regions called domains (see Figure 3.26). In multifunctional proteins, different domains are responsible for the various functions. Further, proteins that have similar functions will have similar domains responsible for those functions. In other words, protein domains may be considered as units of structure and function that have combined to produce a wide variety of complex domain arrangements.

Research Question

How did the domain organization in proteins evolve?

Method of Analysis

Gustavo Caetano-Anollés and Minglei Wang at the University of Illinois at Urbana-Champaign used a bioinformatics approach to answer the question. Bioinformatics is a fusion of biology

with mathematics and computer science used to manage and analyze biological data (see Chapter 19). Typically, bioinformatics involves the analysis of DNA sequences and/or protein sequences in comprehensive databases. The researchers analyzed domain structure and organization in proteins encoded in hundreds of fully sequenced genomes. From the data they reconstructed phylogenetic trees of protein structures. These phylogenetic trees laid out a rough timeline for, and details of, the evolution of domain organization in proteins.

Conclusion

The researchers discovered that, before the emergence of the three taxonomic domains—Bacteria, Archaea, and Eukarya (see Chapter 1)—most proteins contained only single domains that performed multiple tasks. With time, these protein domains began to combine with one another, becoming more specialized. After a long period of gradual evolution, there was an explosion ("big bang") of change in which protein domains combined with each other or split apart to produce a

wide range of novel domains, each typically more specialized than its ancestors. The explosion in the evolution of domain organization is of particular interest because it coincided with the rapidly increasing diversity of the three taxonomic domains. Thereafter, the protein domains diverged markedly from each other as the diversification of life continued. Diversification of protein domains was most extensive in the Eukarya, allowing eukaryotic organisms to exhibit functions of their proteins not possible in other organisms.

THINK LIKE A SCIENTIST

1. Give an example of a multiple-domain protein that is mentioned in the text. What are its domains?
2. What sort of bioinformatics evidence would indicate that a domain in a protein had diverged?

Source: M. Wang and G. Caetano-Anollés. 2009. The evolutionary mechanics of domain organization in proteomes and the rise of modularity in the protein world. *Structure* 17:66–78.

---

another by a segment of random coil. The hinge formed by the flexible random coil allows domains to move with respect to one another. Hinged domains of this type are typical of proteins that produce motion and also occur in many enzymes. *Insights from the Molecular Revolution* discusses how the domain organization in proteins might have evolved.

Many proteins have multiple functions. For instance, the sperm surface protein SPAM1 (sperm adhesion molecule 1) plays multiple roles in mammalian fertilization. In proteins with multiple functions, individual functions are often located in different domains **(Figure 3.26B)**, meaning that domains are functional as well as structural subdivisions. Different proteins often share one or more domains with particular functions. For example, a type of domain that releases energy to power biological reactions appears in similar form in many enzymes and motile proteins. The appearance of similar domains in different proteins suggests that the proteins may have evolved through a mechanism that mixes existing domains into new combinations.

The three-dimensional arrangement of amino acid chains within and between domains also produces highly specialized regions called **motifs.** Several types of motifs, each with a specialized function, occur in proteins. Examples are presented in Section 16.2.

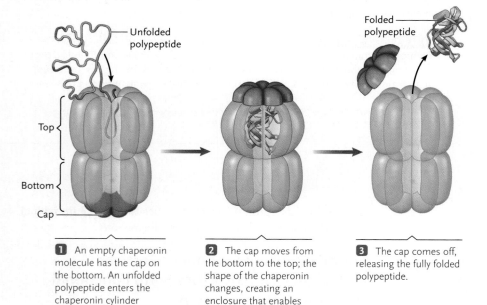

FIGURE 3.25 **Role of a chaperonin in folding a polypeptide.** The three parts of the chaperonin are the top and bottom, which form a cylinder, and the cap.
© Cengage Learning 2014

1 An empty chaperonin molecule has the cap on the bottom. An unfolded polypeptide enters the chaperonin cylinder at the top.

2 The cap moves from the bottom to the top; the shape of the chaperonin changes, creating an enclosure that enables the polypeptide to fold.

3 The cap comes off, releasing the fully folded polypeptide.

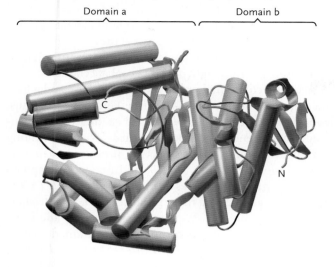

A. **Two domains in an enzyme that assembles DNA molecules**

Domain a   Domain b

C

N

B. **The same protein, showing the domain surfaces**

Domain a   Domain b

DNA molecule

Site that assembles DNA molecules (DNA polymerase site)

Site that corrects mistakes during DNA assembly (exonuclease site)

FIGURE 3.26 Domains in proteins. (A) Two domains in part of an enzyme that assembles DNA molecules in the bacterium *Escherichia coli.* The α helices are shown as cylinders, the β strands as arrows, and the random coils as ropes. "N" indicates the N-terminal end of the protein, and "C" indicates the C-terminal end. **(B)** The same view of the protein as in **(A),** showing the domain surfaces and functional sites.
© Cengage Learning 2014

Proteins Combine with Units Derived from Other Classes of Biological Molecules

We have already mentioned the linkage of proteins to lipids to form lipoproteins. Proteins also link with carbohydrates to form *glycoproteins,* which function as enzymes, antibodies, recognition and receptor molecules at the cell surface, and parts of extracellular supports such as collagen. In fact, most of the known proteins located at the cell surface or in the spaces between cells

are glycoproteins. Linkage of proteins to nucleic acids produces *nucleoproteins,* which form such vital structures as *chromosomes,* the structures that organize DNA inside cells, and *ribosomes,* which carry out the process of protein synthesis in the cell.

This section has demonstrated the importance of amino acid sequence to the structure and function of proteins and highlighted the great variability in proteins produced by differences in their amino acid sequence. The next section considers the nucleic acids, which store and transmit the information required to arrange amino acids into particular sequences in proteins.

STUDY BREAK 3.4

1. What gives amino acids their individual properties?
2. What is a peptide bond, and what type of reaction forms it?
3. What are functional domains of proteins, and how are they formed?

THINK OUTSIDE THE BOOK

Collaboratively or on your own, investigate whether chaperonins are the same in bacteria, archaeans, and eukaryotes.

3.5 Nucleotides and Nucleic Acids

Nucleic acids are another class of macromolecules, in this case, long polymers assembled from repeating monomers called *nucleotides.* The two types of nucleic acids are DNA and RNA. **DNA (deoxyribonucleic acid)** stores the hereditary information responsible for inherited traits in all eukaryotes and prokaryotes and in a large group of viruses. **RNA (ribonucleic acid)** is the hereditary molecule of another large group of viruses; in all organisms, one major type of RNA carries the instructions for assembling proteins from DNA to the sites where the proteins are made inside cells (see Chapter 15). Another major type of RNA forms part of ribosomes, the structural units that assemble proteins, and a third major type of RNA brings amino acids to the ribosomes for their assembly into proteins (see Chapter 15). A fourth major type of RNA is involved in regulating gene expression (see Chapter 16).

Nucleotides Consist of a Nitrogenous Base, a Five-Carbon Sugar, and One or More Phosphate Groups

A **nucleotide,** the monomer of nucleic acids, consists of three parts linked together by covalent bonds: (1) a **nitrogenous base** (a nitrogen-containing molecule that accepts protons), formed from rings of carbon and nitrogen atoms; (2) a five-carbon, ring-shaped sugar; and (3) one to three phosphate groups **(Figure 3.27).** The two types of nitrogenous bases are **pyrimidines,** with one carbon–nitrogen ring, and **purines,** with two carbon–nitrogen rings **(Figure 3.28).** Three pyrimidine bases—uracil (U), thymine (T), and cytosine (C)—and two purine bases—adenine (A) and guanine (G)—form parts of nucleic acids in cells.

A. Overall structural plan of a nucleotide

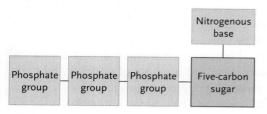

B. Chemical structures of nucleotides

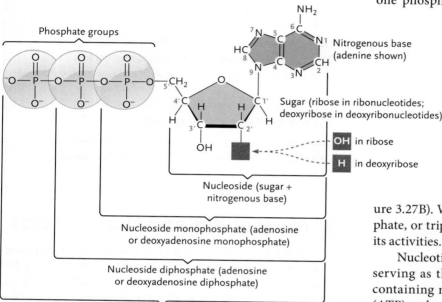

Phosphate groups

Nitrogenous base (adenine shown)

Sugar (ribose in ribonucleotides; deoxyribose in deoxyribonucleotides)

OH in ribose

H in deoxyribose

Nucleoside (sugar + nitrogenous base)

Nucleoside monophosphate (adenosine or deoxyadenosine monophosphate)

Nucleoside diphosphate (adenosine or deoxyadenosine diphosphate)

Nucleoside triphosphate (adenosine or deoxyadenosine triphosphate)

Other nucleotides:

Containing guanine: Guanosine or deoxyguanosine monophosphate, diphosphate, or triphosphate

Containing cytosine: Cytidine or deoxycytidine monophosphate, diphosphate, or triphosphate

Containing thymine: Thymidine monophosphate, diphosphate, or triphosphate

Containing uracil: Uridine monophosphate, diphosphate, or triphosphate

FIGURE 3.27 Nucleotide structure.
© Cengage Learning 2014

In nucleotides, the nitrogenous bases link covalently to either **deoxyribose** in DNA or **ribose** in RNA. Nucleotides containing deoxyribose are called **deoxyribonucleotides** and nucleotides containing ribose are called **ribonucleotides.** Deoxyribose and ribose are five-carbon sugars. The carbons of these two sugars are numbered with a prime symbol—1′, 2′, 3′, 4′, and 5′—to distinguish them from the carbons and nitrogens in the nitrogenous bases, which are numbered without primes (see Figure 3.27). The two sugars differ only in the chemical group bound to the 2′ carbon (boxed in red in Figure 3.27B): deoxyribose has an —H at this position and ribose has an —OH

group. The prefix *deoxy-* in deoxyribose indicates that oxygen is absent at this position in the DNA sugar. In individual, unlinked nucleotides, a chain of one, two, or three phosphate groups bonds to the ribose or deoxyribose sugar at the 5′ carbon; nucleotides are called monophosphates, diphosphates, or triphosphates according to the length of this phosphate chain.

A structure containing only a nitrogenous base and a five-carbon sugar is called a **nucleoside** (see Figure 3.27B). Thus, nucleotides are *nucleoside phosphates.* For example, the nucleoside containing adenine and ribose is called *adenosine.* Adding one phosphate group to this structure produces *adenosine monophosphate (AMP),* adding two phosphate groups produces *adenosine diphosphate (ADP),* and adding three produces *adenosine triphosphate (ATP).* The corresponding adenine–deoxyribose complexes are named *deoxyadenosine monophosphate (dAMP), deoxyadenosine diphosphate (dADP),* and *deoxyadenosine triphosphate (dATP).* The lowercase *d* in the abbreviations indicates that the nucleoside contains the deoxyribose form of the sugar. Equivalent names and abbreviations are used for the other nucleotides (see Figure 3.27B). Whether a nucleotide is a monophosphate, diphosphate, or triphosphate has fundamentally important effects on its activities.

Nucleotides perform many functions in cells in addition to serving as the building blocks of nucleic acids. Two ribose-containing nucleotides in particular, adenosine triphosphate (ATP) and guanosine triphosphate (GTP), are the primary mol-

Pyrimidines

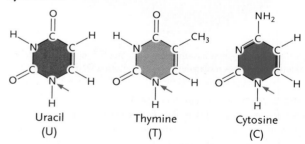

Uracil (U)

Thymine (T)

Cytosine (C)

Purines

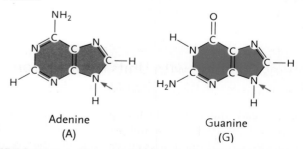

Adenine (A)

Guanine (G)

FIGURE 3.28 Pyrimidine and purine bases of nucleotides and nucleic acids. Red arrows indicate where the bases link to ribose or deoxyribose sugars in the formation of nucleotides.
© Cengage Learning 2014

FIGURE 3.29 Linkage of nucleotides to form the nucleic acids DNA and RNA. P is a phosphate group (see Figure 3.27). **(A)** In DNA, the bases adenine (A), thymine (T), cytosine (C), or guanine (G) may be bound at the base positions. The lilac zones are the sugar (deoxyribose)–phosphate backbones of the two polynucleotide strands of the DNA molecule. Dotted lines designate hydrogen bonds. **(B)** In RNA, A, G, C, or uracil (U) may be bound at the base positions. The light green zone is the sugar (ribose)–phosphate backbone of the polynucleotide strand of the RNA molecule.

© Cengage Learning 2014

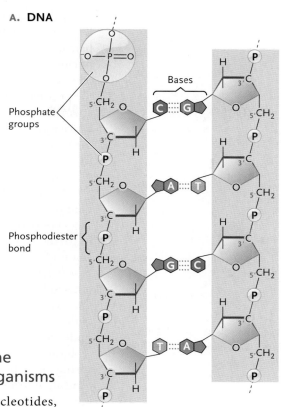

A. DNA

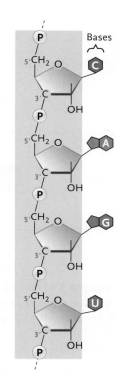

B. RNA

ecules that transport chemical energy from one reaction system to another; the same nucleotides regulate and adjust cellular activity. Molecules derived from nucleotides play important roles in biochemical reactions by delivering reactants or electrons from one system to another.

Nucleic Acids DNA and RNA Are the Informational Molecules of All Organisms

DNA and RNA consist of chains of nucleotides, *polynucleotide chains,* with one nucleotide linked to the next by a bridging phosphate group between the 5′ carbon of one sugar and the 3′ carbon of the next sugar in line; this linkage is called a **phosphodiester bond (Figure 3.29).** This arrangement of alternating sugar and phosphate groups forms the backbone of a nucleic acid chain. The nitrogenous bases of the nucleotides project from this backbone.

Each nucleotide of a DNA chain contains deoxyribose and one of the four bases A, T, G, or C. Each nucleotide of an RNA chain contains ribose and one of the four bases A, U, G, or C. Thymine and uracil differ only in a single functional group: in T a methyl (—CH$_3$) group is linked to the ring, but in U it is replaced by a hydrogen (see Figure 3.28). The differences in sugar and pyrimidine bases between DNA and RNA account for important differences in the structure and functions of these nucleic acids inside cells.

DNA Molecules in Cells Consist of Two Nucleotide Chains Wound Together

In cells, DNA takes the form of a **double helix,** first discovered by James D. Watson and Francis H. C. Crick in 1953, in collaboration with Maurice Wilkins and

FIGURE 3.30 The DNA double helix. **(A)** Arrangement of sugars, phosphate groups, and bases in the DNA double helix. The dotted lines between the bases designate hydrogen bonds. **(B)** Space-filling model of the DNA double helix. The paired bases, which lie in flat planes, are seen edge-on in this view.

© Cengage Learning 2014

Rosalind Franklin (see Chapter 14 for details of their discovery). The double helix they described consists of two nucleotide chains wrapped around each other in a spiral that resembles a twisted ladder **(Figure 3.30).** The sides of the ladder are the

A. DNA double helix, showing arrangement of sugars, phosphate groups, and bases

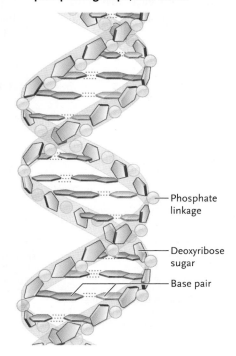

B. Space-filling model of DNA double helix

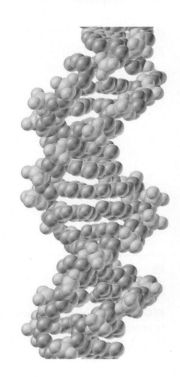

Phosphate linkage

Deoxyribose sugar

Base pair

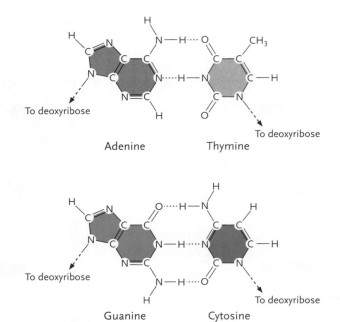

Adenine   Thymine

Guanine   Cytosine

FIGURE 3.31 The DNA base pairs A–T (adenine–thymine) and G–C (guanine–cytosine), as seen from one end of a DNA molecule. Dotted lines between the bases designate hydrogen bonds.
© Cengage Learning 2014

sugar–phosphate backbones of the two chains, which twist around each other in a right-handed direction to form the double spiral. The rungs of the ladder are the nitrogenous bases, which extend inward from the sugars toward the center of the helix. Each rung consists of a pair of nitrogenous bases held in a flat plane roughly perpendicular to the long axis of the helix. The two nucleotide chains of a DNA double helix are held together primarily by hydrogen bonds between the base pairs. Slightly more than 10 base pairs are packed into each turn of the double helix. A DNA double-helix molecule is also referred to as double-stranded DNA.

The space separating the sugar–phosphate backbones of a DNA double helix is just wide enough to accommodate a base pair that consists of one purine and one pyrimidine. Purine–purine base pairs are too wide and pyrimidine–pyrimidine pairs are too narrow to fit this space exactly. More specifically, of the possible purine–pyrimidine pairs, only two combinations, adenine with thymine, and guanine with cytosine, can form stable hydrogen bonds so that the base pair fits precisely within the double helix **(Figure 3.31)**. An adenine–thymine (A–T) pair forms two stabilizing hydrogen bonds; a guanine–cytosine (G–C) pair forms three.

As Watson and Crick pointed out in the initial report of their discovery, the formation of A–T and G–C pairs allows the sequence of one nucleotide chain to determine the sequence of its partner in the double helix. Thus, wherever a T occurs on one chain of a DNA double helix, an A occurs opposite it on the other chain; wherever a C occurs on one chain, a G occurs on the other side (see Figure 3.29). That is, the nucleotide sequence of one chain is said to be *complementary* to the nucleotide sequence of the other chain. The complementary nature of the two chains underlies the processes when DNA molecules are copied—replicated—to pass hereditary information from parents to offspring and when RNA copies are made of DNA molecules to transmit information within cells (see Chapter 14). In DNA replication, one nucleotide chain is used as a **template** for the assembly of a complementary chain according to the A–T and G–C base-pairing rules **(Figure 3.32)**.

RNA Molecules Are Usually Single Nucleotide Chains

In contrast to DNA, RNA molecules exist largely as single, rather than double, polynucleotide chains in living cells. That is, RNA is typically single-stranded. However, RNA molecules can fold and twist back on themselves to form double-helical regions. The patterns of these fold-back double helices are as vital to RNA func-

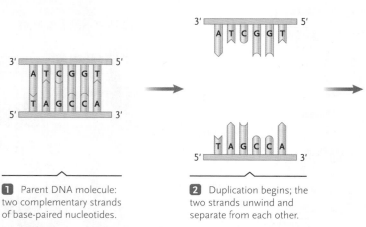

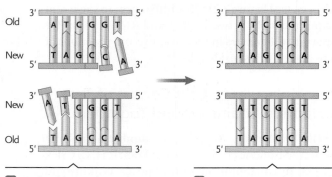

1 Parent DNA molecule: two complementary strands of base-paired nucleotides.

2 Duplication begins; the two strands unwind and separate from each other.

3 Each "old" strand serves as a template for addition of bases according to the A–T and G–C base-pairing rules.

4 Bases positioned on each old strand are joined together into a "new" strand. Each half-old, half-new DNA molecule is an exact duplicate of the parent molecule.

FIGURE 3.32 How complementary base pairing allows DNA molecules to be replicated precisely.
© Cengage Learning 2014

tion as the folding of amino acid chains is to protein function. "Hybrid" double helices, which consist of an RNA chain paired with a DNA chain, are formed temporarily when RNA copies are made of DNA chains (see Chapter 15). In the RNA–RNA or hybrid RNA–DNA helices, U in RNA takes over the pairing functions of T, forming A–U rather than A–T base pairs.

The description of nucleic acid molecules in this section, with the discussions of carbohydrates, lipids, and proteins in earlier sections, completes our survey of the major classes of or-

ganic molecules found in living organisms. The next chapter discusses the functions of molecules in one of these classes, the enzymatic proteins, and the relationships of energy changes to the biological reactions speeded by enzymes.

STUDY BREAK 3.5

1. What is the monomer of a nucleic acid macromolecule?
2. What are the chemical differences between DNA and RNA?

UNANSWERED QUESTIONS

How are the enzymes that synthesize fatty acids and cholesterol regulated in the cells?

The cells of humans and animals contain thousands of different lipids that form the lipid bilayer membranes that bound cells and subcellular organelles. Several of the signaling molecules or hormones that regulate metabolic processes also derive from lipids. Despite their ubiquitous functions, we know very little about how the body controls the level of each lipid substance to ensure a concentration that is adequate for function but guards against overaccumulation. Overaccumulation can produce disorders ranging from brain degeneration to heart attacks.

One area of recent progress concerns the mechanism that controls the body's production of two of the most abundant lipids, cholesterol and fatty acids. Much of this work has been carried out in a laboratory that is led jointly by Joseph L. Goldstein and myself at the University of Texas Southwestern Medical School in Dallas, Texas. Several years ago our group discovered sterol regulatory element binding proteins (SREBPs), which are transcription factors. (Transcription factors are regulatory proteins that bind to specific sequences in DNA and activate transcription of nearby genes. Transcription is the first step in the process of gene expression, in which the DNA sequence of a gene is copied into an RNA molecule. This RNA sequence later directs assembly of the protein's amino acid sequence in translation. See Chapters 15 and 16.) The SREBPs selectively activate several dozen genes, including all of the ones necessary for the production of cholesterol or its uptake from plasma.

SREBPs are synthesized initially on membranes of the endoplasmic reticulum (ER). To reach the nucleus, the SREBPs must be transported from the ER to the cell's Golgi complex where they are cleaved by proteases to release a fragment that enters the nucleus and activates transcription. Transport of SREBPs from ER to Golgi requires the protein Scap, which binds SREBPs immediately after their synthesis.

Why are SREBPs attached to ER membranes, and why do cells go through this elaborate transport process to release the active fragments? The reason is simple: it allows the synthesis of membrane cholesterol to be regulated by the concentration of cholesterol in the ER membrane. When cholesterol builds up in ER membranes, the cholesterol binds to

Scap, changing its conformation and blocking the transport of SREBPs to the Golgi. The SREBPs are no longer processed by the Golgi proteases, and the amount of the fragment that goes to the nucleus decreases. Synthesis of cholesterol is diminished, and it remains low until the cholesterol content of ER membranes falls, upon which the SREBPs are again transported to the Golgi where they activate the genes required for cholesterol synthesis.

The elucidation of the SREBP processing pathway is a step forward, but many questions remain unanswered. Many of these center on the precise way in which Scap senses the cholesterol level in membranes. Also, we must determine the mechanisms that control the other constituents of cell membranes, primarily phospholipids. Only then will we understand how our cells manage to create membranes with the precise chemical and physical properties that allow them to perform their vital functions.

THINK LIKE A SCIENTIST Defects in SREBP regulation contribute to common diseases, ranging from heart attacks to obesity and diabetes. How might better understanding of the SREBP regulatory mechanism contribute to treatment of diabetes?

Courtesy of Dr. Michael S. Brown

Courtesy of Dr. Joseph L. Goldstein

Dr. Michael S. Brown and **Dr. Joseph L. Goldstein,** both of the University of Texas Southwestern Medical School, discovered the low-density lipoprotein (LDL) receptor. They shared many awards for this work, including the Nobel Prize for Medicine or Physiology. Dr. Brown is Paul J. Thomas Professor of Molecular Genetics and Director of the Jonsson Center for Molecular Genetics; Dr. Goldstein is chairman of the Department of Molecular Genetics. To learn more about their research, go to http://www4.utsouthwestern.edu/moleculargenetics/pages/brown/lab.html.

REVIEW KEY CONCEPTS

To access the course materials and companion resources for this text, please visit www.cengagebrain.com.

3.1 Formation and Modification of Biological Molecules

- Carbon atoms readily share electrons, allowing each carbon atom to form four covalent bonds with other carbon atoms or atoms of other elements. The resulting extensive chain and ring structures form the backbones of diverse organic compounds (Figure 3.2).

- The structure and behavior of organic molecules, as well as their linkage into larger units, depend on the chemical properties of functional groups. Particular combinations of functional groups determine whether an organic molecule is an alcohol, aldehyde, ketone, or acid (Table 3.1).

- Isomers have the same chemical formula but different molecular structures. Molecules that are mirror images of one another are an example of stereoisomers. Structural isomers are two molecules with atoms arranged in different ways (Figure 3.4).

- In a dehydration synthesis reaction, the components of a water molecule are removed as subunits assemble. In hydrolysis, the components of a water molecule are added as subunits are broken apart (Figure 3.5).

- Many carbohydrates, proteins, and nucleic acids are large polymers of monomer subunits. Polymerization reactions assemble monomers into polymers. Polymers larger than 1,000 daltons in mass are considered to be macromolecules.

Animation: Dehydration synthesis and hydrolysis

3.2 Carbohydrates

- Carbohydrates are molecules in which carbon, hydrogen, and oxygen occur in the approximate ratio 1:2:1.

- Monosaccharides are carbohydrate subunits that contain three to seven carbons (Figures 3.5–3.7).

- Monosaccharides have D and L stereoisomers. Typically, one of the two forms is used in cellular reactions because it has a molecular shape that can be recognized by the enzyme accelerating the reaction, whereas the other form does not.

- Two monosaccharides join to form a disaccharide; greater numbers form polysaccharides (Figures 3.8 and 3.9).

Animation: Structure of starch and cellulose

3.3 Lipids

- Lipids are hydrocarbon-based, water-insoluble, nonpolar molecules. Biological lipids include neutral lipids, phospholipids, and steroids.

- Neutral lipids, which are primarily energy-storing molecules, have a glycerol backbone and three fatty acid chains (Figures 3.10 and 3.11).

- Phospholipids are similar to neutral lipids except that a phosphate group and a polar organic unit substitute for one of the fatty acids (Figure 3.13). In polar environments (such as a water solution), phospholipids orient with their polar end facing the water and their nonpolar ends clustered in a region that excludes water. This orientation underlies the formation of bilayers, the structural framework of biological membranes.

- Steroids, which consist of four carbon rings carrying primarily nonpolar groups, function chiefly as components of membranes and as hormones in animals (Figures 3.14 and 3.15).

- Lipids link with carbohydrates to form glycolipids and with proteins to form lipoproteins, both of which play important roles in cell membranes.

Animation: Fatty acids

Animation: Structure of a Phospholipid

Animation: Triglyceride formation

3.4 Proteins

- Proteins are assembled from 20 different amino acids. Amino acids have a central carbon to which is attached an amino group, a carboxyl group, a hydrogen atom, and a side group that differs in each amino acid (Figure 3.16).

- Peptide bonds between the amino group of one amino acid and the carboxyl group of another amino acid link amino acids into chains (Figure 3.18).

- A protein may have four levels of structure. Its primary structure is the linear sequence of amino acids in a polypeptide chain; secondary structure is the arrangement of the amino acid chain into α helices or β strands and sheets; tertiary structure is the protein's overall conformation. Quaternary structure is the number and arrangement of polypeptide chains in a protein (Figures 3.19–3.23).

- In many proteins, combinations of secondary, tertiary, and quaternary structure form functional domains (Figure 3.26).

- Proteins combine with lipids to produce lipoproteins, with carbohydrates to produce glycoproteins, and with nucleic acids to form nucleoproteins.

Animation: Peptide bond formation

Animation: Secondary and tertiary structure

3.5 Nucleotides and Nucleic Acids

- A nucleotide consists of a nitrogenous base, a five-carbon sugar, and one to three phosphate groups (Figures 3.27 and 3.28).

- Nucleotides are linked into nucleic acid chains by covalent bonds between their sugar and phosphate groups. Alternating sugar and phosphate groups form the backbone of a nucleic acid chain (Figure 3.29).

- There are two nucleic acids: DNA and RNA. DNA contains nucleotides with the nitrogenous bases adenine (A), thymine (T), guanine (G), or cytosine (C) linked to the sugar deoxyribose; RNA contains nucleotides with the nitrogenous bases adenine, uracil (U), guanine, or cytosine linked to the sugar ribose (Figures 3.27–3.29).

- In a DNA double helix, two nucleotide chains wind around each other like a twisted ladder, with the sugar–phosphate backbones of the two chains forming the sides of the ladder and the nitrogenous bases forming the rungs of the ladder (Figure 3.30).

- A–T and G–C base pairs mean that the sequences of the two nucleotide chains of a DNA double helix are complements of each other. Complementary pairing underlies the processes that replicate DNA and copy RNA from DNA (Figures 3.31 and 3.32).

Animation: DNA structure

Animation: Subunits of DNA

Animation: DNA Replication in Detail

UNDERSTAND & APPLY

Test Your Knowledge

1. Which functional group has a double bond and forms organic acids?
   a. carboxyl
   b. amino
   c. hydroxyl
   d. carbonyl
   e. sulfhydryl

2. Which of the following characteristics is *not* common to carbohydrates, lipids, and proteins?
   a. They are composed of a carbon backbone with functional groups attached.
   b. Monomers of these molecules undergo dehydration synthesis to form polymers.
   c. Their polymers are broken apart by hydrolysis.
   d. The backbones of the polymers are primarily polar molecules.
   e. The molecules are held together by covalent bonding.

3. Cellulose is to carbohydrate as:
   a. amino acid is to protein.
   b. lipid is to fat.
   c. collagen is to protein.
   d. nucleic acid is to DNA.
   e. nucleic acid is to RNA.

4. Maltose, sucrose, and lactose differ from one another:
   a. because not all contain glucose.
   b. because not all of them exist in ring form.
   c. in the number of carbons in the sugar.
   d. in the number of hexose monomers involved.
   e. by the linkage of the monomers.

5. Lipids that are liquid at room temperature:
   a. are fats.
   b. contain more hydrogen atoms than lipids that are solids at room temperature.
   c. if polyunsaturated, contain several double bonds in their fatty acid chains.
   d. lack glycerol.
   e. are not stored in cells as triglycerides.

6. Which of the following statements about steroids is *false*?
   a. They are classified as lipids because, like lipids, they are nonpolar.
   b. They can act as regulatory molecules in animals.
   c. They are composed of four carbon rings.
   d. They are highly soluble in water.
   e. Their most abundant form is as sterols.

7. The term *secondary structure* refers to a protein's:
   a. sequence of amino acids.
   b. structure that results from local interactions between different amino acids in the chain.
   c. interactions with a second protein chain.
   d. interaction with a chaperonin.
   e. interactions with carbohydrates.

8. The first and major effect in denaturation of proteins is that:
   a. peptide bonds break.
   b. α helices unwind.
   c. β sheet structures unfold.
   d. tertiary structure is changed.
   e. quaternary structures disassemble.

9. In living systems:
   a. proteins rarely combine with other macromolecules.
   b. enzymes are always proteins.
   c. proteins are composed of 24 amino acids.
   d. chaperonins inhibit protein movement.
   e. a protein domain refers to the place in the cell where proteins are synthesized and function.

10. RNA differs from DNA because:
    a. RNA may contain the pyrimidine uracil, and DNA does not.
    b. RNA is always single-stranded when functioning, and DNA is always double-stranded.
    c. the pentose sugar in RNA has one less O atom than the pentose sugar in DNA.
    d. RNA is more stable and is broken down by enzymes less easily than DNA.
    e. RNA is a much larger molecule than DNA.

Discuss the Concepts

1. Identify the following structures as a carbohydrate, fatty acid, amino acid, or polypeptide:

   a. $H_3N^+ - \overset{\overset{\displaystyle R}{\displaystyle |}}{\underset{\underset{\displaystyle H}{\displaystyle |}}{C}} - COO^-$ (The R indicates an organic group.)

   b. $C_6H_{12}O_6$
   c. $(glycine)_{20}$
   d. $CH_3(CH_2)_{16}COOH$

2. Lipoproteins are relatively large, spherical clumps of protein and lipid molecules that circulate in the blood of mammals. They are like suitcases that move cholesterol, fatty acid remnants, triglycerides, and phospholipids from one place to another in the body. Given what you know about the insolubility of lipids in water, which of the three kinds of lipids would you predict to be on the outside of a lipoprotein clump, bathed in the fluid portion of blood?

3. The shapes of a protein's domains often give clues to its functions. For example, protein HLA (human leukocyte antigen) is a type of recognition protein on the outer surface of all vertebrate body cells. Certain cells of the immune system use HLAs to distinguish self (the body's own cells) from nonself (invading cells). Each HLA protein has a jawlike region that can bind to molecular parts of an invader. It thus alerts the immune system that the body has been invaded. Speculate on what might happen if a mutation makes the jawlike region misfold.

4. Explain how polar and nonpolar groups are important in the structure and functions of lipids, proteins, and nucleic acids.

Design an Experiment

A clerk in a health food store tells you that natural vitamin C extracted from rose hips is better for you than synthetic vitamin C. Given your understanding of the structure of organic molecules, how would you respond? Design an experiment to test whether the rose hips and synthetic vitamin C preparations differ in their effects.

Interpret the Data

Cholesterol does not dissolve in blood. It is carried through the bloodstream by lipoproteins, as described in *Focus on Applied Research*. Low-density lipoprotein (LDL) carries cholesterol to body tissues, such as artery walls, where it can form health-endangering deposits. LDL is often called "bad" cholesterol. High-density lipoprotein (HDL) carries cholesterol away from tissues to the liver for disposal; it is often called "good" cholesterol.

In 1990, Ronald Mensink and Martijn Katan tested the effects of different dietary fats on blood lipoprotein levels. They placed 59 men and women on a diet in which 10% of their daily energy intake consisted of *cis* fatty acids, *trans* fatty acids, or saturated fats. (All subjects were tested on each of the diets.) Blood LDL and HDL levels of the subjects were measured after three weeks on the diet. The **Table** displays the averaged results, shown in mg/dL (milligrams per deciliter) of blood.

Effect of Diet on Lipoprotein Levels (mg/dL)

	cis fatty acids	*trans* fatty acids	Saturated fats	Optimal level
LDL	103	117	121	<100
HDL	55	48	55	>40
LDL-to-HDL ratio	1.87	2.43	2.2	<2

© Cengage Learning 2014

1. In which group was the level of LDL ("bad" cholesterol) highest and in which group was the level of HDL ("good" cholesterol) lowest?

2. An elevated risk of heart disease has been correlated with increasing LDL-to-HDL ratios. Which group had the highest LDL-to-HDL ratio?

3. Rank the three diets from best to worst according to their potential effect on cardiovascular health.

Source: R. P. Mensink and M. B. Katan. 1990. Effect of dietary trans fatty acids on high-density and low-density lipoprotein cholesterol levels in healthy subjects. *New England Journal of Medicine* 323:439–445.

Apply Evolutionary Thinking

How do you think the primary structure (amino acid) sequence of proteins could inform us about the evolutionary relationships of proteins?

Fly caught by a leaf of the English sundew *(Drosera anglica)*. Enzymes secreted by the hairs on the leaf digest trapped insects, providing nutrients to the plant.

Matthijs Wetterauw/shutterstock.com

Energy, Enzymes, and Biological Reactions

Why it matters . . . Earth is a cold place, at least when it comes to chemical reactions. Life cannot survive at the high temperatures routinely used in most laboratories and industrial plants for chemical synthesis. Instead, life relies on substances called *catalysts,* which speed up the rates of reaction without the need for an increase in temperature. The acceleration of a reaction by a catalyst is called *catalysis.* Most of the catalysts in biological systems are proteins called *enzymes.*

How good are enzymes at increasing the rate of a reaction? Richard Wolfenden and his colleagues at the University of North Carolina experimentally measured the rates for a range of uncatalyzed and enzyme-catalyzed biological reactions. They found the greatest difference between the uncatalyzed rate and the enzyme-catalyzed rate for a reaction that removes a phosphate group from a molecule.

In the cell, a group of enzymes called phosphatases catalyze the removal of phosphate groups from a number of molecules, including proteins. The reversible addition and removal of a phosphate group from particular proteins is a central mechanism of intracellular communication in almost all cells. In a cell using a phosphatase enzyme, the phosphate removal reaction takes approximately 10 milliseconds (msec). Wolfenden's research group calculated that in an aqueous environment such as a cell, without an enzyme, the phosphate removal reaction would take over 1 trillion (10^{12}) years to occur. This exceeds the current estimate for the age of the universe! By contrast, the enzyme-catalyzed reaction is 21 orders of magnitude (10^{21}) faster.

For most reactions, the difference between the uncatalyzed rate and the enzyme-catalyzed rate is many millions of times. Because life requires temperatures that are relatively low (below 100°C), life as we know would not have evolved without enzymes to speed up the rates of reactions.

Enzymes are key players in **metabolism**—the biochemical modification and use of organic molecules and energy to support the activities of life. Metabolism, which occurs only in living organisms, comprises thousands of biochemical reactions that accomplish the special activities we associate with life, such as growth, reproduction, movement, and the ability to respond to stimuli.

Understanding how biological reactions occur and how enzymes work requires knowledge of the basic laws of chemistry and physics. All reactions, whether they occur inside living organisms or in the outside, inanimate world, obey the same chemical and physical laws that operate everywhere in the universe. These fundamental laws are our starting point for an exploration of the nature of energy and how cells use it to conduct their activities.

4.1 Energy, Life, and the Laws of Thermodynamics

Life, like all chemical and physical activities, is an energy-driven process. Energy cannot be measured or weighed directly. We can detect it only through its effects on matter, including its ability to move objects against opposing forces, such as friction, gravity, or pressure, or to push chemical reactions toward completion. Therefore, energy is most conveniently defined as *the capacity to do work*. Even when you are asleep, cells of your muscles, brain, and other parts of your body are at work and using energy.

Energy Exists in Different Forms and States

Energy can exist in many different forms, including heat, chemical, electrical, mechanical, and radiant energy. Visible light, infrared and ultraviolet light, gamma rays, and X-rays are all types of radiant energy. Although the forms of energy are different, energy can be converted readily from one form to another. For example, chemical energy is transformed into electrical energy in a flashlight battery, and electrical energy is transformed into light and heat energy in the flashlight bulb. In green plants, the radiant energy of sunlight is transformed into chemical energy in the form of complex sugars and other organic molecules (see Chapter 9).

All forms of energy can exist in one of two states: kinetic and potential. **Kinetic energy** (*kinetikos* = putting in motion) is the energy of an object because it is in motion. Examples of everyday objects that possess kinetic energy are waves in the ocean, a hit baseball, and a falling rock. Examples from the natural world are electricity (which is a flow of electrons), photons of light, and heat. The movement present in kinetic energy is useful because it can perform work by making other objects move. **Potential energy** is stored energy, that is, the energy an object has because of its location or chemical structure. A boulder on the top of a hill has potential energy because of its position in Earth's gravitational field. Chemical energy, nuclear energy, gravitational energy, and stored mechanical energy are forms of potential energy.

Potential energy can be converted to kinetic energy and vice versa. Consider a roller coaster. When the train descends from its maximum height, its potential energy is converted into kinetic energy, and the coaster accelerates. On the next hill, kinetic energy is converted back to potential energy and the coaster slows. In the ups and downs of the ride, the coaster's potential energy and kinetic energy interchange continuously but, importantly, the sum of potential energy and kinetic energy remains constant.

The Laws of Thermodynamics Describe the Energy Flow in Natural Systems

The study of energy and its transformations is called **thermodynamics.** Quantitative research by chemists and physicists in the nineteenth century regarding energy flow between systems and the surroundings led to the formulation of two fundamental laws of thermodynamics that apply equally to living cells and to stars and galaxies. These laws allow us to predict whether reactions of any kind, including biological reactions, can occur. That is, if particular groups of molecules are placed together, are they likely to react chemically and change into different groups of molecules? The laws also give us the information needed to trace energy flows in biological reactions: They allow us to estimate the amount of energy released or required as a reaction proceeds.

When discussing thermodynamics, scientists refer to a *system,* which is the object under study. A system is whatever we define it to be—a single molecule, a cell, a planet. Everything outside a system is its *surroundings.* The universe, in this context, is the total of the system and the surroundings. There are three types of systems: isolated, closed, and open. An *isolated system* does not exchange matter or energy with its surroundings **(Figure 4.1A)**. A perfectly insulated Thermos flask is an example of an isolated system. A *closed system* can exchange energy but not matter with its surroundings **(Figure 4.1B).** Earth is a closed system. It takes in a great amount of energy from the sun and releases heat, but essentially no matter is exchanged between Earth and the rest of the universe (barring the odd space probe, of course). An *open system* can exchange both energy and matter with its surroundings **(Figure 4.1C)**. All living organisms are open systems.

The First Law of Thermodynamics Addresses the Energy Content of Systems and Their Surroundings

The **first law of thermodynamics** states that *energy can be transformed from one form to another or transferred from one place to another, but it cannot be created or destroyed.* That is, in any process that involves an energy change, the *total amount of energy in a system and its surroundings remains constant.* This law is also called the *principle of conservation of energy.*

A. Isolated system: does not exchange matter or energy with its surroundings

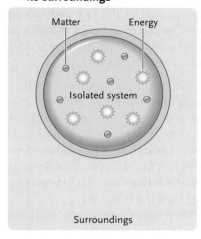

B. Closed system: exchanges energy with its surroundings

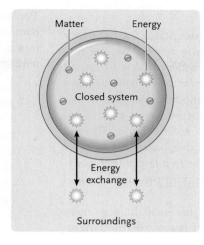

C. Open system: exchanges both energy and matter with its surroundings

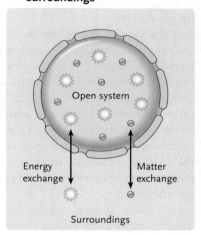

If energy can be neither created nor destroyed, what is the ultimate source of the energy we and other living organisms use? For almost all organisms, the ultimate source is the sun (**Figure 4.2**). Plants capture the kinetic energy of light radiating from the sun by absorbing it and converting it to the chemical potential energy of complex organic molecules—primarily sugars, starches, and lipids (see Chapter 9). These substances are used as fuels by the plants themselves, by animals that feed on plants, and by organisms (such as fungi and bacteria) that break down the bodies of dead organisms. The chemical potential energy stored in sugars and other organic molecules is used for growth, reproduction, and other work of living organisms.

Eventually, most of the solar energy absorbed by green plants is converted into heat energy as the activities of life take place. Heat (a form of kinetic energy) is largely unusable by living organisms; as a result, most of the heat released by the reactions of living organisms radiates to their surroundings, and then from Earth into space.

How does the principle of conservation of energy apply to biochemical reactions? Molecules have both kinetic and potential energy. Kinetic energy for molecules above absolute zero (−273°C) is reflected in the constant motion of the molecules, whereas potential energy for molecules is the energy contained in the arrangement of atoms and chemical bonds. The energy content of reacting systems provides part of the information required to predict the likelihood and direction of chemical reactions. Usually, the energy content of the reactants in a chemical reaction (see Section 2.3) is larger than the energy content of the products. Thus, reactions usually progress to a state in which the products have *minimum energy content*. When this is the case, the difference in energy content between reactants and products in the reacting system is released to the surroundings.

FIGURE 4.2
Energy flow from the sun to photosynthetic organisms (here, colonies of the green alga *Volvox*), which capture the kinetic radiant energy of sunlight and convert it to potential chemical energy in the form of complex organic molecules.

Radiant energy lost from the sun

Light energy absorbed by photosynthetic organisms and converted into potential chemical energy

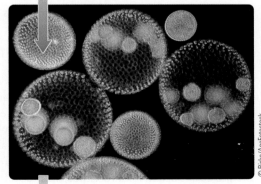

Heat energy lost from photosynthetic organisms

The Second Law of Thermodynamics Considers Changes in the Degree of Order in Reacting Systems

The second law of thermodynamics explains why, as any energy change occurs, the objects (matter) involved in the change typically become more disordered. (Your room and the kitchen at home are probably the best examples of this phenomenon.) You know from experience that it takes energy to straighten out (decrease) the disorder (as when you clean up your room).

The **second law of thermodynamics** states this tendency toward disorder formally, in terms of a system and its surroundings: in any process in which a system changes from an initial to a final state, *the total disorder of a system and its surroundings always increases.* In thermodynamics, disorder is called **entropy.** If the system and its surroundings are defined as the entire universe, the second law means that as changes occur anywhere in the universe, the total disorder or entropy of the universe constantly increases. As the first law of thermodynamics asserts, however, the total energy in the universe does not change.

At first glance, living organisms—which are open systems—appear to violate the second law of thermodynamics. As a fertilized egg develops into an adult animal, it becomes more highly ordered (decreases its entropy) as it synthesizes organic molecules from less complex substances. However, the entropy of the whole system—the surroundings as well as the organism—must be considered as growth proceeds. For the fertilized egg—the initial state—its surroundings include all the carbohydrates, fats, and other complex organic molecules the animal will use as it develops into an adult. For the adult—the final state—the surroundings include the animal's waste products (water, carbon dioxide, and many relatively simple organic molecules), which are collectively much less complex than the organic molecules used as fuels. When the total reactants, including all the nutrients, and the total products, including all the waste materials, are tallied, the total change satisfies both laws of thermodynamics—the total energy content remains constant, and the entropy of the system and its surroundings increases.

STUDY BREAK 4.1

1. What are kinetic energy and potential energy?
2. In thermodynamics, what is meant by an isolated system, a closed system, and an open system?

4.2 Free Energy and Spontaneous Reactions

Applying the first and second laws of thermodynamics together allows us to predict whether any particular chemical or physical reaction will occur without an input of energy. Such reactions are called **spontaneous reactions** in thermodynamics. In this usage, the word *spontaneous* means only that a reaction will occur—it does not describe the rate of a reaction. Spontaneous reactions may proceed very slowly, such as the formation of rust on a nail, or very quickly, such as a match bursting into flame.

Energy Content and Entropy Contribute to Making a Reaction Spontaneous

Two factors related to the first and second laws of thermodynamics must be taken into account to determine whether a reaction is spontaneous: (1) the change in energy content of a system; and (2) its change in entropy.

1. *Reactions tend to be spontaneous if the products have less potential energy than the reactants.* The potential energy in a system is its **enthalpy** (*en* = to put into; *-thalpein* = to heat), symbolized by *H*. Reactions that release energy are termed **exothermic**—the products have less potential energy than the reactants. Reactions that absorb energy are termed **endothermic**—the products have more potential energy than the reactants.

    When natural gas burns, methane reacts spontaneously with oxygen producing carbon dioxide and water. This reaction is exothermic, producing a large quantity of heat, as the products have less potential energy than the reactants. In a system composed of a glass of ice cubes in water melting at room temperature, the system absorbs energy from its surroundings and the water has greater potential energy than the ice. The process of ice melting is endothermic, yet it is spontaneous. Clearly, some other factor besides potential energy is involved.

2. *Reactions tend to be spontaneous when the products are less ordered (more random) than the reactants.* Reactions tend to occur spontaneously if the entropy of the products is greater than the entropy of the reactants.

    Consider again the glass of ice in water. It is an increase in entropy that makes the melting of the ice a spontaneous process at room temperature. Molecules of ice are far more ordered (possess lower entropy) than molecules of water moving around randomly (see Section 2.4).

We have now learned that a combination of factors, *energy content* and *entropy,* contributes to determining whether a chemical reaction proceeds spontaneously. These two concepts are used in the next section to explore quantitatively how to predict whether or not a reaction proceeds spontaneously.

The Change in Free Energy Indicates Whether a Reaction Is Spontaneous

Recall from the second law of thermodynamics that energy transformations are not 100% efficient; some of the energy is lost as an increase in entropy. How much energy is available? The portion of a system's energy that is available to do work is called **free energy,** which is symbolized as *G* in recognition of the physicist Josiah Willard Gibbs, who developed the concept. In living organisms, free energy accomplishes the chemical and physical work involved in activities such as the synthesis of molecules, movement, and reproduction.

The change in free energy, ΔG ($\Delta G = G_{\text{final state}} - G_{\text{initial state}}$; Δ, pronounced delta, means change in), can be calculated for any chemical reaction from the formula

$$\Delta G = \Delta H - T\Delta S$$

where ΔH is the change in enthalpy, T is the absolute temperature in degrees Kelvin (K, where K = °C + 273.16), and ΔS is the change in entropy.

For a reaction to be spontaneous, ΔG must be negative. As the above formula tells us, both the entropy and the enthalpy of a reaction can influence the overall ΔG. In some processes, such as the combustion of methane, the large loss of potential energy, negative enthalpy (ΔH), dominates in making a reaction spontaneous. In other reactions, such as the melting of ice at room temperature, a decrease in order (ΔS increases) dominates. Once we know what the ΔG is for a reaction, we can determine if the reaction will proceed spontaneously.

Spontaneous Reactions Typically Reach an Equilibrium Point Rather than Going to Completion

In many spontaneous biological reactions, the reactants may not convert completely to products even though the reactions have a negative ΔG. Instead, the reactions run in the direction of completion (toward reactants or toward products) until they reach the equilibrium point, a state of balance between the opposing factors pushing the reaction in either direction. At the equilibrium point, both reactants and products are present and the reactions typically are reversible.

Consider as an example the chemical reaction in which glucose-1-phosphate is converted into glucose-6-phosphate, starting with 0.02 M glucose-1-phosphate as the reactant (**Figure 4.3**). The reaction proceeds spontaneously until there is 0.019 M glucose-6-phosphate (product) and 0.001 M of glucose-1-phosphate (reactant) in the solution. In fact, regardless of the amounts of each you start with, the reaction reaches a point at which there is 95% glucose-6-phosphate and 5% glucose-1-phosphate. This is the point of *chemical equilibrium*: the reaction does not stop, but the rate of the forward reaction equals the rate of the reverse reaction. As a system moves toward equilibrium, its free energy becomes progressively lower and reaches its lowest point when the system achieves equilibrium ($\Delta G = 0$). You can think of a reaction as an energy valley with the equilibrium point being at the bottom. To move away from equilibrium requires free energy and thus will not be spontaneous.

The point of equilibrium of a reaction is related to its ΔG. The more negative the ΔG, the further toward completion the reaction will move before equilibrium is established. If the reaction shown in Figure 4.3 had a positive ΔG, the reaction would run in reverse toward glucose-1-phosphate. Many reactions have a ΔG that is near zero and are thus readily **reversible** by adjusting the concentration of products and re-

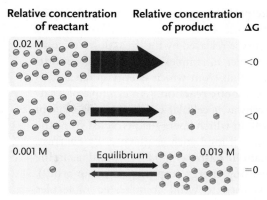

FIGURE 4.3 **The equilibrium point of a reaction.** No matter what quantities of glucose-1-phosphate and glucose-6-phosphate are dissolved in water, when equilibrium is reached, there is 95% glucose-6-phosphate (product) and 5% glucose-1-phosphate (reactant). At equilibrium, the number of reactant molecules being converted to product molecules equals the number of product molecules being converted back to reactant molecules. The reaction at the equilibrium point is reversible; it may be made to run to the right (forward) by adding more reactants, or to the left (reverse) by adding more products.
© Cengage Learning 2014

actants slightly. Reversible reactions are written with a double arrow:

$$A + B \rightleftharpoons C + D$$
$$\text{reactants} \qquad \text{products}$$

The reaction in Figure 4.3 is an isolated system and, over time, equilibrium is reached, with ΔG becoming zero. However, many individual reactions in living organisms never reach an equilibrium point because living systems are open; thus, the supply of reactants is constant and, as products are formed, they do not accumulate but become the reactants of another reaction. In fact, the ΔG of life is always negative as organisms constantly take in energy-rich molecules (or light, if they are photosynthetic) and use them to do work. Organisms reach equilibrium, with $\Delta G = 0$, only when they die.

Metabolic Pathways Consist of Exergonic and Endergonic Reactions

Based on the free energy of reactants and products, every reaction can be placed into one of two groups. An **exergonic reaction** (*ergon* = work) **(Figure 4.4A)** is one in that releases free energy—the ΔG is negative because the products contain less free energy than the reactants. In an **endergonic reaction (Figure 4.4B)**, the products contain more free energy than the reactants; therefore, ΔG is positive. The reactants involved in endergonic reactions need to gain free energy from the surroundings to form the products of the reaction.

In metabolism, individual reactions tend to be part of a *metabolic pathway,* a series of reactions in which the products of one reaction are used immediately as the reactants for the next reaction in the series. In one type of metabolic pathway, called a catabolic pathway (*cata* = downward, as in the sense of a rock releasing energy as it rolls down a hill), energy is released by the breakdown of complex molecules to simpler compounds. (An individual reaction from which energy is released is called a **catabolic reaction.**) An example of a catabolic pathway is cellular respiration, the topic of Chapter 8, in which energy is extracted from the breakdown of food such as glucose. By contrast, in an **anabolic pathway** (*ana* = upward, as in the sense of using energy to push a rock up a hill), energy is used to build complicated molecules from simpler ones; these pathways are often called *biosynthetic pathways.* (An individual reaction that requires energy input is called an **anabolic reaction,** or a **biosynthetic reaction.**) Examples of anabolic pathways include photosynthesis, the topic of Chapter 9, as well as the synthesis of macromolecules such as proteins and nucleic acids.

The overall ΔG of a catabolic pathway is negative, whereas the overall ΔG of an anabolic pathway is positive. Any one pathway consists of a number of individual reactions, each of which may have a positive or negative ΔG. However, when you sum the individual reaction ΔG values for a catabolic pathway, the overall free energy is negative, and for an anabolic reaction, the overall free energy is positive.

A. Exergonic reaction: free energy is released, products have less free energy than reactants, and the reaction proceeds spontaneously

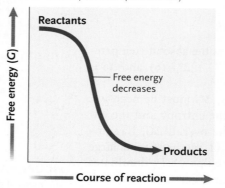

B. Endergonic reaction: free energy is gained, products have more free energy than reactants, and the reaction is not spontaneous

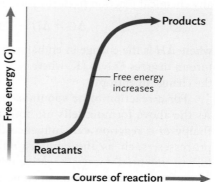

FIGURE 4.4 Exergonic (A) and endergonic (B) reactions. An endergonic reaction proceeds only if energy is supplied by an exergonic reaction.
© Cengage Learning 2014

STUDY BREAK 4.2

1. What two factors must be taken into account to determine if a reaction will proceed spontaneously?
2. What is the relation between ΔG and the concentrations of reactants and products at the equilibrium point of a reaction?
3. Distinguish between exergonic and endergonic reactions, and between catabolic and anabolic reactions. How are the two categories of reactions related?

4.3 Adenosine Triphosphate (ATP): The Energy Currency of the Cell

Many reactions within cells involve the assembly of complex molecules from more simple components. As we learned in the previous section, these reactions have a positive ΔG and are called endergonic; they may be part of both catabolic and anabolic pathways. How the cell supplies energy to drive these endergonic reactions is highly conserved among all forms of life and involves the nucleotide adenosine triphosphate (ATP).

ATP Hydrolysis Releases Free Energy

ATP is the best example of a molecule that contains large amounts of free energy because of its high-energy phosphate bonds. ATP consists of the five-carbon sugar ribose linked to the nitrogenous base adenine and a chain of three phosphate groups **(Figure 4.5A).**

Much of the potential energy of ATP is associated with the three phosphate groups. Removal of one or two of the three phosphate groups is a spontaneous reaction that releases large amounts of free energy. (As you will learn in Chapter 15, ATP is also a building block in the synthesis of RNA.)

The breakdown of ATP is a hydrolysis reaction **(Figure 4.5B)** and results in the formation of adenosine diphosphate (ADP) and a molecule of inorganic phosphate (Ⓟ$_i$).

$$ATP + H_2O \rightarrow ADP + P_i$$

$$\Delta G = -7.3 \text{ kcal/mol}$$

ADP can be further hydrolyzed to adenosine monophosphate (AMP), but this releases somewhat less free energy than the hydrolysis of ATP.

Phosphate Groups from ATP Hydrolysis Couple Reactions

The hydrolysis of ATP in water in a test tube is an exergonic reaction that releases free energy that simply warms up the water. How do living cells couple the hydrolysis of ATP to an endergonic reaction so that energy is not simply wasted as heat?

The answer is a process called **energy coupling,** the use of an exergonic reaction—ATP hydrolysis—to drive an endergonic reaction. In other words, the two reactions become coupled. In a coupled reaction, ATP is hydrolyzed and its terminal phosphate group is transferred to the reactant molecule of the endergonic reaction. The addition of a phosphate group to a molecule is called **phosphorylation,** and the modified molecule is said to have been *phosphorylated.* Phosphorylation makes a molecule less stable (more reactive) than when it is unphosphorylated. Energy coupling requires the action of an enzyme to bring the ATP and reactant molecule into close association. The enzyme has a specific site on it that binds both the ATP and the reactant molecule, allowing for transfer of the phosphate group.

A. Chemical structures of AMP, ADP, and ATP

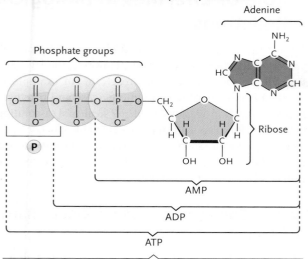

With one phosphate group, the molecule is known as AMP; with two phosphates, the molecule is called ADP. Each added phosphate packs additional potential chemical energy into the molecular structure.

B. Hydrolysis reaction removing a phosphate group from ATP

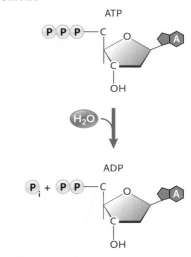

FIGURE 4.5 ATP, the primary molecule that couples energy-requiring reactions to energy-releasing reactions in living organisms. (P_i is the symbol used in this book for inorganic phosphate.)
© Cengage Learning 2014

Energy coupling is used, for example, in the reaction in which ammonia (NH_3) is added to glutamic acid, an amino acid with one amino group, to produce glutamine, an amino acid with two amino groups. The overall reaction is **(Figure 4.6A)**:

$$\text{glutamic acid} + NH_3 \rightarrow \text{glutamine} + H_2O$$

$$\Delta G = 13.4 \text{ kcal/mol}$$

The positive value for ΔG indicates that the reaction cannot proceed spontaneously. Both glutamic acid and glutamine are used in the assembly of proteins. This reaction is common in most cells. The product, glutamine, is a donor of nitrogen for other reactions in the cell.

A. Without ATP, reaction is not spontaneous because ΔG is positive

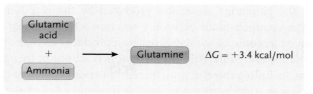

B. ATP hydrolysis is an exergonic reaction

C. Coupled with ATP hydrolysis, the glutamine synthesis reaction is spontaneous because net ΔG is negative

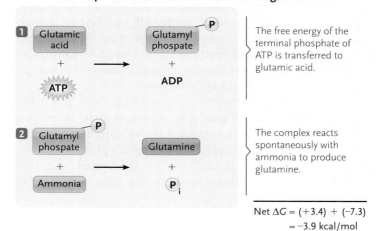

Net ΔG = (+3.4) + (−7.3)
= −3.9 kcal/mol

FIGURE 4.6 Energy coupling using ATP in the synthesis of glutamine from glutamic acid and ammonia.
© Cengage Learning 2014

Cells carry out this endergonic reaction by using energy released from ATP hydrolysis **(Figure 4.6B)**. The coupled reaction is shown in **Figure 4.6C**. As a first step, glutamic acid is phosphorylated using the phosphate group removed from ATP, forming glutamyl phosphate:

$$\text{glutamic acid} + \text{ATP} \rightarrow \text{glutamyl phosphate} + \text{ADP}$$

The ΔG for this step is negative, making the reaction spontaneous, but much less free energy is released than in the hydrolysis of ATP to ADP + P_i. In the second step, glutamyl phosphate reacts with NH_3:

$$\text{glutamyl phosphate} + NH_3 \rightarrow \text{glutamine} + P_i$$

This second step also has a negative ΔG value and is spontaneous.

Even though the reaction proceeds in two steps, it is usually written as one reaction, with a combined negative ΔG value:

$$\text{glutamic acid} + NH_3 + \text{ATP} \rightarrow \text{glutamine} + \text{ADP} + P_i$$

$$\Delta G = -3.9 \text{ kcal/mol}$$

Because ΔG is negative, the coupled reaction is spontaneous and releases energy. The difference between −3.9 kcal/mol and the

−7.3 kcal/mol released by hydrolyzing ATP to ADP + P_i (that is, +3.4 kcal/mol) represents potential chemical energy transferred to the glutamine molecules produced by the reaction. That is, the coupling of an exergonic reaction—ATP hydrolysis—to an endergonic biosynthesis reaction produces an overall reaction that is exergonic. All the endergonic reactions of living organisms, including those of growth, reproduction, movement, and response to stimuli, are made possible by coupling reactions in this way.

Cells Also Couple Reactions to Regenerate ATP

Coupling reactions occur continuously in living cells, consuming a tremendous amount of ATP. How do cells generate that ATP? ATP is a renewable resource that is synthesized by recombining ADP and P_i. Although ATP hydrolysis is an exergonic reaction, ATP synthesis from ADP and P_i is an energy-requiring, endergonic reaction. The energy for ATP synthesis comes from exergonic reactions that involve breakdown of complex molecules that contain an abundance of free energy: carbohydrates, proteins, and fats in food.

The continual hydrolysis and resynthesis of ATP is called the **ATP/ADP cycle (Figure 4.7)**. Approximately 10 million ATP molecules are hydrolyzed and resynthesized each second in a typical cell, illustrating that this cycle operates at an astonishing rate. In fact, if ATP were not regenerated from ADP and P_i, the average human would use an estimated 75 kg of ATP per day.

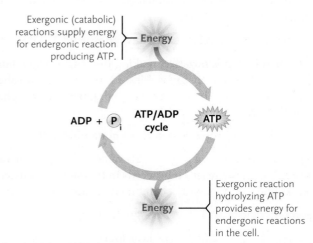

FIGURE 4.7 The ATP/ADP cycle that couples reactions releasing free energy and reactions requiring free energy.
© Cengage Learning 2014

4.4 Role of Enzymes in Biological Reactions

The laws of thermodynamics are useful because they can tell us if a process will occur spontaneously. However, the laws do not tell us anything about the rate of a reaction. For example, even though the breakdown of sucrose into glucose and fructose is a spontaneous reaction with a ΔG of −7 kcal/mol, a solution of sucrose will sit for years without any detectable glucose or fructose forming. That is, *spontaneous reactions do not necessarily proceed rapidly*. In this section, we discuss how the speed of a reaction can be altered by enzymes.

Activation Energy Represents a Kinetic Barrier for a Reaction

In our example above, what prevents sucrose from being converted rapidly into glucose and fructose? Chemical reactions require bonds to break and new bonds to form. For bonds to be broken, they must be strained or otherwise made less stable so that breakage can occur. To get reacting molecules into a less stable (more unstable) state requires a small input of energy. Thus, even though a reaction is spontaneous (negative ΔG), the reaction will not start unless a relatively small boost of energy is added **(Figure 4.8A)**. This initial energy investment required to

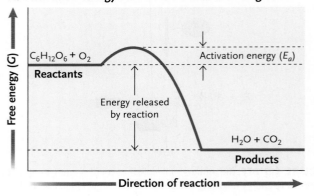

A. Activation energy barrier in the oxidation of glucose

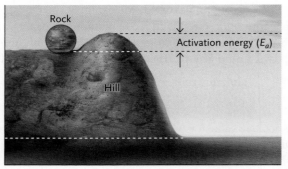

B. "Activation energy" barrier in the movement of a rock downhill

FIGURE 4.8 Activation energy (E_a).
© Cengage Learning 2014

start a reaction is called the **activation energy,** symbolized E_a. Molecules that gain the necessary activation energy occupy what is called the *transition state,* where bonds are unstable and are ready to be broken.

A rock resting in a depression at the top of a hill provides a physical example of activation energy **(Figure 4.8B).** The rock will not roll downhill spontaneously, even though its position represents considerable potential energy and the total "reaction"—the downward movement of the rock—is spontaneous and releases free energy. In this example, the activation energy is the effort required to raise the rock over the rim of the depression and start its downhill roll.

What provides the activation energy in chemical reactions? The molecules taking part in chemical reactions are in constant motion at temperatures above absolute zero. Periodically, reacting molecules gain enough energy to reach the transition state. For a solution of sucrose, the number of molecules that reach the transition state at any one time is very small. However, if a significant number of reactant molecules reach the transition state, the free energy that is released may be enough to get the remaining reactants to the transition state. For example, in the chemistry lab, heat commonly provides the energy needed for reactant molecules to get to the transition state and, therefore, to speed up the rate of a reaction. In biology, however, using heat to speed up a reaction is problematic for two reasons: (1) high temperatures destroy the structural components of cells, particularly proteins; and (2) an increase in temperature would speed up all possible chemical reactions in a cell, not just the specific reactions that are part of metabolism.

Enzymes Accelerate Reactions by Reducing the Activation Energy

How can you increase the rate of a reaction without raising the temperature? The answer is that you can use a **catalyst,** a chemical agent that accelerates the rate of a reaction without itself being changed by the reaction. The process of accelerating a reaction with a catalyst is called **catalysis,** and we say that the chemical agent responsible *catalyzes* the reaction. The most common biological catalysts are proteins called **enzymes** (*enzym* = in yeast).

The activation energy for a reaction represents a kinetic barrier that prevents spontaneous reactions from proceeding quickly. The greater the activation energy barrier, the slower the reaction will proceed. Enzymes increase the rate of reaction by lowering this barrier—by lowering the activation energy of the reaction **(Figure 4.9).** Because the rate of a reaction is proportional to the number of reactant molecules that can acquire enough energy to get to the transition state, enzymes make it possible for a greater proportion of reactant molecules to attain the activation energy.

Although enzymes lower the activation energy of a reaction, as shown in Figure 4.9, they do not alter the change in free energy (ΔG) of the reaction. The free energy values of the reactants and products are the same; the only difference is the path the reaction takes.

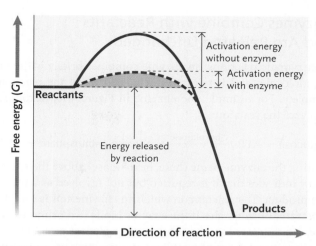

FIGURE 4.9 **Effect of enzymes in reducing the activation energy (E_a).** The reduction allows biological reactions to proceed rapidly at the relatively low temperatures that can be tolerated by living organisms.

© Cengage Learning 2014

Let us be clear about what enzymes do and do not do with regard to biological reactions. By lowering the activation energy, enzymes DO speed up the rate of spontaneous (exergonic) reactions. However, enzymes DO NOT supply free energy to a reaction. Therefore, enzymes CANNOT make an endergonic reaction proceed spontaneously. ATP hydrolysis can be coupled to an endergonic reaction to make it proceed spontaneously but, alone, an enzyme cannot. Finally, enzymes DO NOT change the ΔG of a reaction.

Cells have thousands of different enzymes. They vary from relatively small molecules, with single polypeptide chains containing as few as 100 amino acids, to large complexes that include many polypeptide chains totaling thousands of amino acids. It is the three-dimensional structure of a protein—its *conformation*—that determines the protein's function. Conformation is determined by the number of polypeptides in the protein and the amino acid sequence of each polypeptide. (The three-dimensional structure of proteins is described in Section 3.4.) Each enzyme has a specific protein structure that has evolved to catalyze a specific reaction.

Different enzymes are found in all areas of the cell, from the aqueous cell solution to the cell membranes. Other enzymes are released to catalyze reactions outside the cell. For example, enzymes that catalyze reactions breaking down food molecules are released from cells into the digestive cavity in all animals.

The majority of enzymes have names ending in *-ase.* The rest of the name typically relates to the substrate of the enzyme or to the type of reaction with which the enzyme is associated. For example, enzymes that break down proteins are called *proteinases* or *proteases.*

Enzymes are not the only biological molecules capable of accelerating reaction rates. Some RNA molecules (see Section 4.6) also have this capacity. As we do in this book, most biologists reserve the term *enzyme* for protein molecules that can accelerate reaction rates and call the RNA molecules with this capacity *ribozymes.*

Enzymes Combine with Reactants and Are Released Unchanged

In enzymatic reactions, an enzyme combines briefly with reacting molecules and is released unchanged when the reaction is complete. For example, the enzyme in **Figure 4.10**, hexokinase, catalyzes the reaction:

$$\text{glucose} + \text{ATP} \xrightarrow{\text{hexokinase}} \text{glucose-6-phosphate} + \text{ADP}$$

Writing the enzyme name (here, hexokinase) above the reaction arrow indicates that it is required but not involved as a reactant or a product. The reactant on which an enzyme acts is called the **substrate**, or substrates if the enzyme binds two or more molecules. In this reaction, the substrates are glucose and ATP.

Each type of enzyme catalyzes the reaction of a single type of substrate molecule or a group of closely related molecules. This **enzyme specificity** explains why a typical cell needs about 4,000 different enzymes to function properly. Notice in Figure 4.10 that the enzyme is much larger than the substrate. Moreover, the substrate interacts with only a very small region of the enzyme called the **active site**, the place on the enzyme where catalysis occurs. The active site is usually a pocket or groove that is formed when the enzyme protein folds into its functional three-dimensional shape.

When the substrate binds initially at the active site of the enzyme, both the enzyme and substrate molecules are distorted, which stabilizes the substrate molecule in the transition state and makes its chemical bonds ready for reaction. Figure 4.10 illustrates the change in enzyme conformation on substrate binding, a phenomenon called *induced fit.*

Once an enzyme–substrate complex is formed, catalysis occurs, with the enzyme converting the substrate into one or more products. Because enzymes are released unchanged after a reaction, enzyme molecules cycle repeatedly through reactions, combining with reactants and releasing products **(Figure 4.11)**. Depending on the enzyme, the rate at which reactants are bound and catalyzed and at which products are released varies from 100 times to 10 million times per second. These high rates of catalysis mean that a small number of enzyme molecules can catalyze large numbers of reactions.

Some enzymes consist of polypeptide chains only. However, many enzymes require a **cofactor**, a nonprotein group that binds precisely to the enzyme, for catalytic activity. The role of a cofactor in an enzyme's catalytic activity varies with the cofactor and the enzyme. Some cofactors are metallic ions, including iron, copper, magnesium, zinc, and manganese. Other cofactors are small organic molecules called **coenzymes**, which are often derived from vitamins. Some coenzymes bind loosely to enzymes; others—called *prosthetic groups*—bind tightly.

Enzymes Reduce the Activation Energy by Stabilizing the Transition State

How do enzymes reduce the activation energy of a reaction? Recall that substrate molecules need to be in the transition state for catalysis to occur. Enzymes stabilize the transition state, doing so through three major mechanisms:

1. Enzymes *bring the reacting molecules together.* Reacting molecules can assume the transition state only when they collide. Binding of reactant molecules to an enzyme's active site brings them close together in the correct orientation for catalysis to take place.

2. Enzymes *expose the reactant molecules to altered charge environments that promote catalysis.* In some systems, the active site of the enzyme contains ionic groups whose positive or negative charges alter the substrate in a way that favors catalysis.

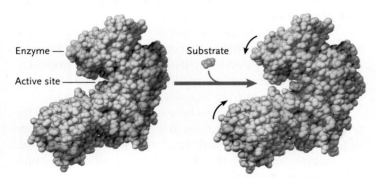

FIGURE 4.10 Combination of an enzyme, hexokinase (in blue), with its substrate, glucose (in yellow). Hexokinase catalyzes the phosphorylation of glucose to form glucose-6-phosphate. The phosphate group that is transferred to glucose is not shown. Note how the enzyme undergoes a conformational change (an induced fit), closing the active site more tightly as it binds the substrate.
© Cengage Learning 2014

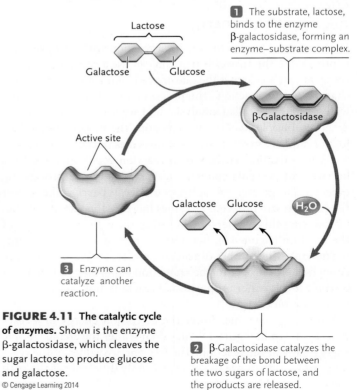

FIGURE 4.11 The catalytic cycle of enzymes. Shown is the enzyme β-galactosidase, which cleaves the sugar lactose to produce glucose and galactose.
© Cengage Learning 2014

1 The substrate, lactose, binds to the enzyme β-galactosidase, forming an enzyme–substrate complex.

2 β-Galactosidase catalyzes the breakage of the bond between the two sugars of lactose, and the products are released.

3 Enzyme can catalyze another reaction.

3. Enzymes *change the shape of the substrate molecules.* As mentioned previously, the active site can distort substrate molecules into a conformation that mimics the transition state.

Regardless of the mechanism, the binding of the substrate molecule(s) to the active site results in stabilization of the substrate in the transition state conformation. Substrate molecules do attain the transition state in the absence of an enzyme, but that is a rare event. The inclusion of an enzyme enables many more molecules to reach the transition state more rapidly. Fundamentally, this is why an enzyme speeds up the rate of a reaction.

STUDY BREAK 4.4

1. How do enzymes increase the rates of the reaction they catalyze?
2. Can enzymes alter the ∆G of a reaction?

THINK OUTSIDE THE BOOK

Use the Internet to determine which enzymes are involved in meat tenderizers, the production of soft cheeses such as brie and camembert, and the production of yogurt. For each enzyme you identify, give the organism that makes it, as well as the reaction catalyzed and how that reaction relates to its use in food.

4.5 Conditions and Factors That Affect Enzyme Activity

Several conditions can alter enzyme activity, including changes in the concentration of substrate and other molecules that can bind to enzymes. In addition, a number of control mechanisms modify enzyme activity, thereby adjusting reaction rates to meet a cell's requirements for chemical products. Changes in pH and temperature can also have a significant effect on enzyme activity.

Enzyme and Substrate Concentrations Influence the Rate of Catalysis

Biochemists use a wide range of approaches to study enzymes. These include molecular tools to study the structure and regulation of the gene encoding the enzyme and sophisticated computer algorithms to model the three-dimensional structure of the enzyme itself.

FIGURE 4.12 Effect of increasing enzyme concentration **(A)** or substrate concentration **(B)** on the rate of an enzyme-catalyzed reaction.
© Cengage Learning 2014

The most fundamental and central approach has been to determine the rate of an enzyme-catalyzed reaction and how it changes in response to altering certain experimental parameters. Typically this requires isolating the enzyme from a cell, incubating it in an appropriate buffered solution, and supplying the reaction mixture with substrate. With these constituents, a researcher can then determine the rate of catalysis, usually by measuring the rate at which the product of the reaction is formed.

As shown in **Figure 4.12A**, in the presence of excess substrate (that is, at high concentrations), the rate of catalysis is proportional to the amount of enzyme. As enzyme concentration increases, the rate of product formation increases. In this system, the rate of the reaction is limited by the amount of enzyme in the reaction mixture. What happens to the reaction rate if we keep the concentration of enzyme constant at some intermediate level and change the concentration of substrate from low to high? As shown in **Figure 4.12B**, at very low concentrations, substrate molecules collide so infrequently with enzyme molecules that the reaction proceeds slowly. As the substrate concentration increases, the reaction rate initially increases as enzyme and substrate molecules collide more frequently. But, as the enzyme molecules approach the maximum rate at which they can combine with reactants and release products, increasing substrate concentration has a smaller and smaller effect, and the rate of reaction eventually levels off. When the enzymes are cycling as rapidly as possible, further increases in substrate concentration have no effect on the reaction rate. At this point, the enzymes are said to be **saturated** with the substrate (the saturation level is shown by a horizontal dashed line in Figure 4.12B).

Cells Adjust Enzyme Activity to Meet Their Needs for Reaction Products

Cells adjust the activity of many enzymes upward or downward to meet their needs for reaction products. Several mechanisms are used in this regulation, including: (1) competitive and noncompetitive inhibition; (2) a form of noncompetitive control called *allosteric regulation;* and (3) covalent modification of enzyme structure by the addition or removal of chemical groups.

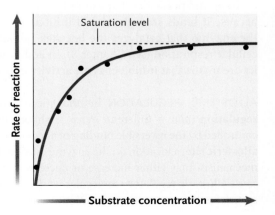

A. Rate of reaction as function of enzyme concentration (substrate at high concentration)

Rate of reaction / Enzyme concentration

B. Rate of reaction as function of substrate concentration (enzyme amount constant)

Saturation level

Rate of reaction / Substrate concentration

COMPETITIVE AND NONCOMPETITIVE INHIBITION Reversible inhibition of enzyme activity serves as an important mechanism of metabolism regulation. A typical cell contains thousands of enzymes, and for many enzymes that synthesize a specific molecule, usually another enzyme exists that catalyzes its breakdown. If both of these enzymes were active in the same cell compartment at the same time, the two metabolic pathways they catalyze would run simultaneously in opposite directions and have no overall effect other than using up energy. To prevent this, the cell is able to regulate enzyme activity in such a way that not all enzymes are active at the same time.

Many enzymes are regulated by natural inhibitors. These *enzyme inhibitors* are nonsubstrate molecules that bind to an enzyme and decrease its activity, thereby lowering the rate at which an enzyme can catalyze a reaction. Control by the inhibitors changes enzyme activity precisely to meet the needs of the cell for the products of the reaction catalyzed by the enzyme.

Regulation of enzyme activity by inhibitors may be competitive or noncompetitive depending on the inhibitor. In **competitive inhibition,** an inhibitor competes with the normal substrate for binding to an enzyme's active site **(Figure 4.13A).** That is, competitive inhibitors have shapes resembling the normal substrate closely enough to fit into and occupy the active site, thereby blocking access for the normal substrate and slowing the reaction rate. If the concentration of the inhibitor is high enough, the reaction may stop completely. Competitive inhibitors are useful in enzyme research because their structure helps identify the region of a normal substrate that binds to an enzyme.

In **noncompetitive inhibition,** an inhibitor binds to an enzyme at a site other than its active site and changes the conformation of the enzyme so that the ability of the active site to bind substrate efficiently is reduced **(Figure 4.13B).**

Inhibitors (competitive or noncompetitive) differ with respect to how strongly they bind to enzymes. In *reversible inhibition,* the binding of an inhibitor to an enzyme is weak, and when the inhibitor releases, enzyme activity returns to normal. In *irreversible inhibition,* an inhibitor binds so strongly to an enzyme through the formation of covalent bonds that the enzyme is completely disabled. Irreversible inhibition can only be overcome if the cell synthesizes more of the enzyme.

Not surprisingly, many irreversible inhibitors that act on critical enzymes are toxic to the cell. They include a wide variety of drugs and pesticides. For example, cyanide is a potent poison because it binds strongly to and inhibits cytochrome oxidase, the enzyme that catalyzes the last step of electron transfer in cellular respiration (see Chapter 8). In addition, many antibiotics are toxins that inhibit enzyme activity in bacteria.

ALLOSTERIC REGULATION In the mechanism of **allosteric regulation** (*allo* = different; *stereo* = shape), enzyme activity is controlled by the reversible binding of a regulatory molecule to the **allosteric site,** a location on the enzyme outside the active site. The mechanism may either increase or decrease enzyme activity. Because allosteric regulatory molecules work by binding to sites separate from the active site, their action is noncompetitive.

A. Competitive inhibition

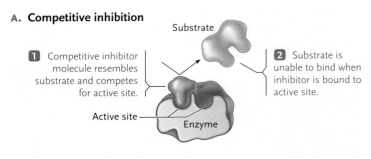

1 Competitive inhibitor molecule resembles substrate and competes for active site.

Active site

Enzyme

Substrate

2 Substrate is unable to bind when inhibitor is bound to active site.

B. Noncompetitive inhibition

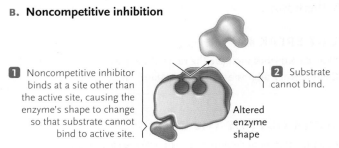

1 Noncompetitive inhibitor binds at a site other than the active site, causing the enzyme's shape to change so that substrate cannot bind to active site.

Altered enzyme shape

2 Substrate cannot bind.

FIGURE 4.13 How competitive (A) and noncompetitive (B) inhibitors reduce enzyme activity.
© Cengage Learning 2014

Enzymes controlled by allosteric regulation typically have two alternate conformations controlled from the allosteric site. In one conformation, called the *high-affinity state* (the active form), the enzyme binds strongly to its substrate; in the other conformation, *the low-affinity state* (the inactive form), the enzyme binds the substrate weakly or not at all. Binding with regulatory substances may induce either state: binding an **allosteric activator** converts it from the low- to high-affinity state and therefore increases enzyme activity, and binding an **allosteric inhibitor** converts an allosteric enzyme from the high- to low-affinity state and therefore decreases enzyme activity **(Figure 4.14).**

Frequently, allosteric inhibitors are a product of the metabolic pathway they regulate. If the product accumulates in excess, its effect as an inhibitor automatically slows or stops the enzymatic reaction producing it, typically by inhibiting the enzyme that catalyzes the first reaction of the pathway. If the product becomes too scarce, the inhibition is reduced and its production increases. Regulation of this type, in which the product of a reaction acts as a regulator of the reaction, is termed **feedback inhibition** (also called **end-product inhibition**). Feedback inhibition prevents cellular resources from being wasted in the synthesis of molecules made at intermediate steps of the pathway.

The biochemical pathway that makes the amino acid isoleucine from threonine is an example of feedback inhibition. The pathway proceeds in five steps, each catalyzed by an enzyme **(Figure 4.15).** The end product of the pathway, isoleucine, is an allosteric inhibitor of the first enzyme of the pathway, threonine deaminase. If the cell makes more isoleucine than it needs, isoleucine binds reversibly with threonine deaminase at the allosteric site, converting the enzyme to the low-affinity state and inhibiting its ability to combine with threonine, the substrate for the first reac-

Allosteric activation

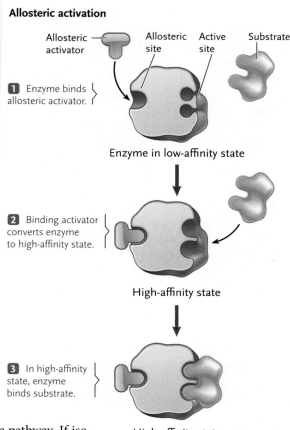

Enzyme in low-affinity state

1 Enzyme binds allosteric activator.

2 Binding activator converts enzyme to high-affinity state.

High-affinity state

3 In high-affinity state, enzyme binds substrate.

High-affinity state

Allosteric inhibition

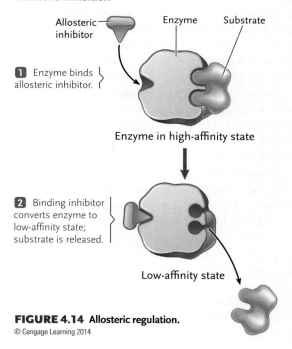

Enzyme in high-affinity state

1 Enzyme binds allosteric inhibitor.

2 Binding inhibitor converts enzyme to low-affinity state; substrate is released.

Low-affinity state

FIGURE 4.14 Allosteric regulation.
© Cengage Learning 2014

---

tion in the pathway. If isoleucine levels drop too low, the allosteric site of threonine deaminase is vacated, the enzyme is converted to the high-affinity state, and isoleucine production increases.

REGULATION BY CHEMICAL MODIFICATION Many key enzymes are regulated by chemical linkage to other substances, typically ions, functional groups such as phosphate or methyl groups, or units derived from nucleotides. The regulatory substances induce folding changes in the enzyme that increase or decrease its activity.

For example, chemical modification by the addition or removal of phosphate groups is a highly significant mechanism of cellular regulation that is used by all organisms from bacteria to humans. Typically, regulatory phosphate groups derived from ATP or other nucleotides are added to the regulated enzymes by other enzymes known as protein kinases. The addition of a phosphate group (phosphorylation, as you learned earlier) either increases or decreases enzyme activity or activates or deactivates the enzyme, depending on the particular enzyme and where the phosphate group is added to the enzyme.

Removal of phosphate groups—*dephosphorylation*—reverses the effects of phosphorylation. Dephosphorylation is carried out by a different group of enzymes called *protein phosphatases*. The balance between phosphorylation and dephosphorylation of the enzymes modified by the protein kinases and protein phosphatases closely regulates cellular activity, often as a part of the response to external signal molecules (see Chapter 7).

pH and Temperature Are Key Factors Affecting Enzyme Activity

The activity of most enzymes is altered strongly by changes in pH and temperature. Characteristically, enzymes reach maximal activity within a narrow range of pH or temperature; at levels outside this range, enzyme activity decreases. These effects produce a typically peaked curve when enzyme activity is plotted, with the peak where pH or temperature produces maximal activity.

FIGURE 4.15 **Feedback inhibition in the pathway that produces isoleucine from threonine.** If the product of the pathway, isoleucine, accumulates in excess, it slows or stops the pathway by acting as an allosteric inhibitor of the enzyme that catalyzes the first step in the pathway.
© Cengage Learning 2014

Threonine

⊖ → Enzyme 1 (threonine deaminase)

Intermediate A

Enzyme 2

Intermediate B

Enzyme 3

Intermediate C

Enzyme 4

Intermediate D

Enzyme 5

Feedback inhibition

Isoleucine

EFFECTS OF pH CHANGES Typically, each enzyme has an optimal pH where it operates at peak efficiency in speeding the rate of its biochemical reaction **(Figure 4.16)**. On either side of this pH optimum, the rate of the catalyzed reaction decreases. The effects become more extreme at pH values farther from the optimum, until the rate drops to zero. An enzyme's dependence on pH typically is caused by ionizable amino acids. A change in pH from the optimal value alters the charges of those amino acids, which modifies the conformation of the protein and eventually causes denaturation of the enzyme.

Most enzymes have an optimum of about pH 7, near the pH of the cellular contents. Enzymes that are secreted from cells may have pH optima farther from neutrality. For example, pepsin, a protein-digesting enzyme secreted into the stomach, has an optimum of pH 1.5, close to the acidity of stomach contents. Similarly, trypsin, also a protein-digesting enzyme, has an optimum of about pH 8, allowing it to function well in the somewhat alkaline contents of the intestine where it is secreted.

EFFECTS OF TEMPERATURE CHANGES The effects of temperature changes on enzyme activity reflect two distinct processes:

1. Temperature has a general effect on chemical reactions of all kinds. As the temperature rises, the rate of chemical reactions typically increases. This effect reflects increases in the kinetic motion of all molecules, with more frequent and stronger collisions as the temperature rises.

2. Temperature has an effect on all proteins, including enzymes. As the temperature rises, the kinetic motions of the amino acid chains of an enzyme increase, along with the strength and frequency of collisions between enzymes and surrounding molecules. At some point, these disturbances become strong enough to denature the enzyme: the hydrogen bonds and other forces that maintain its three-dimensional structure break, making the enzyme unfold and lose its function.

The two effects of temperature act in opposition to each other to produce characteristic changes in the rate of enzymatic catalysis **(Figure 4.17)**. In the range of 0° to about 40°C, the reaction rate doubles for every 10°C increase in temperature. Above 40°C, the increasing kinetic motion begins to unfold the enzyme, reducing the rate of increase in enzyme activity. At some point, as temperature continues to rise, the unfolding causes the reaction rate to level off at a peak. Further increases cause such extensive unfolding that the reaction rate decreases rapidly to zero. For most enzymes, the peak in activity lies between 40° and 50°C; the drop-off becomes steep at 55°C and falls to zero at about 60°C. Thus, the rate of an enzyme-catalyzed reaction peaks at a temperature at which kinetic motion is greatest but no significant unfolding of the enzyme has occurred.

Although most enzymes have a temperature optimum between 40° and 50°C, some have activity peaks below or above this range. For example, the enzymes of maize (corn) pollen function best near 30°C and undergo steep reductions in activity above 32°C. As a result, environmental temperatures above 32°C can seriously inhibit the growth of corn crops. Many animals living in frigid regions have enzymes with much lower temperature optima than average. For example, the enzymes of arctic snow fleas are most active at −10°C. At the other extreme are the enzymes of archaeans that live in hot springs, which are so resistant to denaturation that they remain active at temperatures of 85°C or more.

STUDY BREAK 4.5

1. Why do enzyme-catalyzed reactions reach a saturation level when substrate concentration is increased?
2. What is the difference between competitive and noncompetitive inhibition?
3. Why will the activity of an enzyme eventually decrease to zero as the temperature rises?

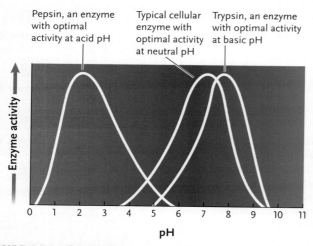

FIGURE 4.16 Effects of pH on enzyme activity. An enzyme typically has an optimal pH at which it is most active; at pH values above or below the optimum, the rate of enzyme activity drops off. At extreme pH values, the rate drops to zero.
© Cengage Learning 2014

Pepsin, an enzyme with optimal activity at acid pH

Typical cellular enzyme with optimal activity at neutral pH

Trypsin, an enzyme with optimal activity at basic pH

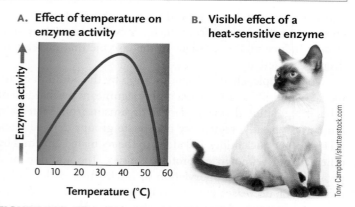

A. Effect of temperature on enzyme activity

B. Visible effect of a heat-sensitive enzyme

FIGURE 4.17 Effect of temperature on enzyme activity. (A) As the temperature rises, the rate of the catalyzed reaction increases proportionally until the temperature reaches the point at which the enzyme begins to denature. The rate drops off steeply as denaturation progresses and becomes complete. **(B)** Visible effects of environmental temperature on enzyme activity in Siamese cats. The fur on the extremities—ears, nose, paws, and tail—contains more dark brown pigment (melanin) than the rest of the body. A heat-sensitive enzyme controlling melanin production is denatured in warmer body regions, so dark pigment is not produced and fur color is lighter.
© Cengage Learning 2014

Ribozymes: Can RNA catalyze peptide bond formation in protein synthesis?

Harry Noller's experiment showed that if proteins were removed from ribosomes, the remaining ribosomal RNA (rRNA) molecules could still catalyze the central reaction of protein synthesis, linkage of amino acids into chains via peptide bonds. However, his work did not eliminate the possibility that undetectable small amounts of ribosomal proteins in the preparations might be catalyzing peptide bond formation.

Research Question

Can RNA catalyze peptide bond formation in protein synthesis?

Experiment

Research by Biliang Zhang and Thomas R. Cech of the University of Colorado Boulder showed that rRNA synthesized artificially and, therefore, never exposed to ribosomal proteins could catalyze the formation of peptide bonds between amino acids (see Figure 3.18).

1. The researchers first synthesized an extremely large pool of RNA molecules in the test tube. Parts of the RNA sequence were the same in every molecule (blue in the **Figure**) and parts differed randomly from molecule to molecule (green in the **Figure**), but did not vary in length. The investigators linked the amino acid phenylalanine (Phe) to the 5′ end of each RNA molecule by a disulfide (—S—S—) bond (**Figure**, left side).

2. To the pool of RNAs, they added the amino acid methionine (Met) linked to the

nucleotide AMP (see Figure 3.27). (In the cell, single amino acids linked to AMP are used in the pathway that makes proteins.) The methionine–AMP was "tagged" by linking the small molecule biotin to it (**Figure**, left side).

3. They allowed the molecules to react, hypothesizing that some of the RNA molecules would have the right sequence to act as a ribozyme and form a peptide bond between the methionine and the phenylalanine (**Figure**, right side).

To determine if peptide bonds had formed, Zhang and Cech poured the reaction mixture through a column packed with chemical beads that bind to biotin. Such binding trapped any RNA molecules that were able to catalyze the joining of the two amino acids, whereas unreactive RNA molecules flowed out the bottom of the column. The RNA molecules with the biotin tag were then washed from the column and separated from the linked amino acids by adding a reagent that breaks disulfide bonds.

4. The ribozyme RNA molecules washed from the column were analyzed and refined.

Eventually, the researchers obtained ribozymes that catalyzed peptide bond formation at rates 100,000 times faster than the same reaction occurring spontaneously without a catalyst.

Conclusion

Zhang and Cech's experiments confirmed a feature of ribozyme activity that is critical to the role proposed for these RNA-based catalysts in the primitive RNA world—their ability to catalyze formation of the fundamental linkage tying amino acids together in proteins. Thus, during the evolution of life, proteins could have been made first in quantity by RNA, with no requirement for either DNA or enzymatic proteins.

THINK LIKE A SCIENTIST Deletion of 20 nucleotides of the constant region from either the 5′ or the 3′ end of the active ribozymes identified resulted in a dramatic decrease in catalytic activity. Interpret this result.

Source: B. Zhang and T. R. Cech. 1997. Peptide bond formation by in vitro selected ribozymes. *Nature* 390:96–100.

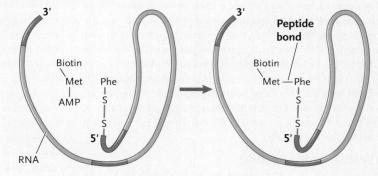

4.6 RNA-Based Biological Catalysts: Ribozymes

In 1981, biochemist Thomas R. Cech of the University of Colorado at Boulder discovered a group of RNA molecules that appeared to be capable of accelerating the rate of certain biological reactions without being changed by the reactions. This discovery was a great surprise to the scientific community. Further work demonstrated that these RNA-based catalysts, now called **ribozymes,** are part of the biochemical machinery of all cells. Cech and another scientist, Yale University biochemist Sidney Altman, received the Nobel Prize in 1989 for their research establishing that ribozymes are essential cellular catalysts.

Most of the known ribozymes speed the cutting and splicing reactions that remove surplus segments from RNA molecules as part of their conversion into finished form. Some have other functions, however. For example, Harry F. Noller and his coworkers at the University of California, Santa Cruz found that ribosomes, the cell structures that assemble amino acids into proteins, can still link amino acids together even if their proteins are removed. After the proteins are extracted, only RNA molecules are left in the ribosomes, indicating that a ribozyme catalyzes this central reaction of protein synthesis. (See Harry Noller's accounting of some of this work in the Unanswered Question of Chapter 15.) After Noller's discovery, Cech and his colleague, Biliang Zhang, confirmed that ribozymes can actually catalyze this reaction (see *Insights from the Molecular Revolution* for an outline of Cech and Zhang's experiment).

Ribozymes provide a possible solution to a long-standing "chicken-or-egg" paradox about the evolution of life: did proteins or nucleic acids come first in evolution? It is difficult to understand how DNA could exist without the enzymatic proteins required for its duplication. At the same time, it is difficult to understand how enzymes could exist without nucleic acids, which contain the information required to make them. Ribozymes offer a way around this dilemma because they could have acted as *both* enzymes and informational molecules when cellular life first appeared. The earliest forms of life therefore might have inhabited an "RNA world" in which neither DNA nor proteins played critical roles (see Chapter 25). If so, ribozymes—the most recently discovered biological catalysts—may have existed for the longest time.

This chapter concludes our survey of the chemical underpinnings of biology. In the next chapter, we survey the structure of cells, the fundamental units into which biological molecules are organized and where molecules interact to produce the characteristics of life.

STUDY BREAK 4.6

What is a ribozyme, and how does it fit the definition of an enzyme?

UNANSWERED QUESTIONS

You learned in this chapter that RNA molecules can, like proteins, fold into complex three-dimensional structures, bind substrate molecules, and catalyze chemical reactions. You also learned the widely accepted *RNA world hypothesis,* which states that early forms of life had a stage in which RNA served both as genome and as the only genome-encoded catalyst. The RNA world hypothesis is based on several lines of evidence, one of which is that the concept of the RNA world solves the "chicken-or-egg" paradox described at the end of Section 4.6. Other pieces of evidence for the RNA world hypothesis are *molecular fossils*—molecules that exist in today's cells and whose existence is most easily explained by an earlier stage of life, in which RNA was the dominant type of molecule. For example, the ribosome, the machinery that synthesizes proteins by translation, is a catalytic RNA (ribozyme) (see *Insights from the Molecular Evolution,* Chapter 15). A different confirmation of the RNA world hypothesis comes from *in vitro* selection experiments. For example, catalytic RNAs have been isolated from large pools of random RNA sequences. Such man-made ribozymes are able to catalyze RNA polymerization, which means that it should be possible to form a self-replicating system of catalytic RNAs—an RNA world.

How did the RNA world originate?

There are at least three problems concerning how the first RNA world organism originated:

1. It is hard to explain how the components necessary for the synthesis of RNA could have been generated in sufficient purity and concentration in a prebiotic soup. However, our current understanding of chemistry is focused on pure systems; therefore, it is possible that we lack a good understanding of how reactions can proceed in mixtures with many different molecules. Recently, the lab of John Sutherland at the University of Manchester, United Kingdom, found a synthetic pathway for the synthesis of nucleotides that appears more plausible for a prebiotic soup, but we are a long way from developing a consistent pathway for the prebiotic synthesis of RNA.

2. RNA is not very stable chemically. The phosphodiester bonds in RNA can hydrolyze easily, depending on chemical and physical conditions such as the concentration of magnesium ions, the pH, and the temperature.

3. No catalytic RNA (ribozyme) has been made or found that could serve as the self-replicating core of an RNA world organism. The best human-made ribozymes that can catalyze RNA polymerization to facilitate self-replication have a size in the range of 200 nucleotides. However, there are 4^{200} possibilities to arrange 200 nucleotides in their sequence, which means that the probability that this ribozyme (or other, related sequences) could emerge from a prebiotic soup by chance is astronomically unlikely.

A possible solution to these three problems is that a different form of life existed before the RNA world. This pre-RNA world would not have used RNA but a related molecule with an easier prebiotic synthesis, a higher chemical stability, and a higher chance of finding catalytic molecules in a pool of random sequences. Possible candidates for such molecules are TNA (threose nucleic acid), GNA (glycerol nucleic acid), and PNA (peptide nucleic acid). These variants differ from RNA in their backbone, which is not composed of ribose and phosphate (as in the case of RNA) but of threose and phosphate (for TNA), glycerol and phosphate (for GNA), or amino acids (for PNA). Although none of these molecules has been found in cellular nuclei, they are called nucleic acids because they contain the same nucleobases as RNA. The molecular geometry of these molecules allows them to base pair with RNA, which would allow the transfer of genetic information from a pre-RNA world to the RNA world. One problem in this transition is that the genomic information of a pre-RNA world may not have been useful for an RNA world because the different chemistries lead to structural differences in three-dimensionally folded polymers. The next years of research will show whether these molecules have the potential to generate a pre-RNA world organism.

THINK LIKE A SCIENTIST Could a single catalytic RNA molecule replicate itself? Keep in mind that it needs to template the synthesis of every nucleotide in its sequence by base-pairing, which includes the nucleotides of its catalytic center.

Ulrich Müller is Assistant Professor at the University of California, San Diego. He tries to generate an RNA world organism in the lab to find out how an RNA world could have looked like. To learn more about Dr. Müller's research go to http://www-chem.ucsd.edu/faculty/

REVIEW KEY CONCEPTS

To access the course materials and companion resources for this text, please visit www.cengagebrain.com.

4.1 Energy, Life, and the Laws of Thermodynamics

- Energy is the capacity to do work. Kinetic energy is the energy of motion; potential energy is energy stored in an object because of its location or chemical structure. Energy may be readily converted between potential and kinetic states.

- Thermodynamics is the study of energy flow between a system and its surroundings during chemical and physical reactions. A system that does not exchange energy or matter with its surroundings is an isolated system. A system that exchanges energy but not matter with its surroundings is a closed system. A system that exchanges both energy and matter with its surroundings is an open system (Figure 4.1).

- The first law of thermodynamics states that the total amount of energy in a system and its surroundings remains constant. The second law states that in any process involving a spontaneous (possible) change from an initial to a final state, the total entropy (disorder) of the system and its surroundings always increases.

4.2 Free Energy and Spontaneous Reactions

- A spontaneous reaction is one that will occur without the input of energy from the surroundings. A spontaneous reaction releases free energy—energy that is available to do work.

- The free energy equation, $\Delta G = \Delta H - T\Delta S$, states that the free energy change, ΔG, is influenced by two factors: the change in enthalpy (potential energy in a system) and the change in entropy of the system as a reaction goes to completion.

- Factors that oppose the completion of spontaneous reactions, such as the relative concentrations of reactants and products, produce an equilibrium point at which reactants are converted to products and products are converted back to reactants, at equal rates (Figure 4.3).

- Organisms reach equilibrium ($\Delta G = 0$) only when they die.

- Reactions with a negative ΔG are spontaneous; they release free energy and are known as exergonic reactions. Reactions with a positive ΔG require free energy and are known as endergonic reactions (Figure 4.4).

- Metabolism is the biochemical modification and use of energy in the synthesis and breakdown of organic molecules. A catabolic reaction releases the potential energy of a molecule in breaking it down to a simpler molecule (ΔG is negative). An anabolic (biosynthetic) reaction uses energy to convert a simple molecule to a more complex molecule (ΔG is positive). Typically, individual reactions operate in metabolic pathways. Individual reactions in a particular pathway can be catabolic or anabolic; it is the sum of the reactions that makes the pathway catabolic or anabolic.

Animation: Energy changes in chemical work

Animation: Chemical equilibrium

4.3 Adenosine Triphosphate (ATP): The Energy Currency of the Cell

- The hydrolysis of ATP releases free energy that can be used as a source of energy for the cell (Figure 4.5).

- A cell can couple the exergonic reaction of ATP hydrolysis to make an otherwise endergonic (anabolic) reaction proceed spontaneously. These coupling reactions require enzymes (Figure 4.6).

- The ATP used in coupling reactions is replenished by reactions that link ATP synthesis to catabolic reactions. ATP thus cycles between reactions that release free energy and reactions that require free energy (Figure 4.7).

4.4 Role of Enzymes in Biological Reactions

- What prevents many exergonic reactions from proceeding rapidly is that they need to overcome an energy barrier (the activation energy, E_a) to get to the transition state (Figure 4.8).

- Enzymes are catalysts that greatly speed the rate at which spontaneous reactions occur because they lower the activation energy (Figure 4.9).

- Enzymes usually are specific: they catalyze reactions of only a single type of molecule or a group of closely related molecules (Figure 4.10).

- Catalysis occurs at the active site, which is the site where the enzyme binds to the substrate (reactant molecule). After combining briefly with the substrate, the enzyme is released unchanged when the reaction is complete (Figure 4.11).

- Many enzymes require a cofactor, a nonprotein group that binds to the enzyme, for catalytic activity. Some cofactors are ions; others are small organic molecules called coenzymes. Some coenzymes bind loosely to enzymes whereas others, called prosthetic groups, bind tightly.

- Enzymes reduce the activation energy by inducing the transition state of the reaction, from which the reaction can move easily in the direction of either products or reactants.

- Three major mechanisms contribute to enzymatic catalysis by reducing the activation energy: (1) enzymes bring reacting molecules together; (2) enzymes expose reactant molecules to altered charge environments that promote catalysis; and (3) enzymes change the shape of a substrate molecule.

Animation: How catalase works

Animation: Enzymes and their role in lowering activation energy

Animation: Induced fit model

4.5 Conditions and Factors That Affect Enzyme Activity

- When substrate is abundant, the rate of a reaction is proportional to the amount of enzyme. At a fixed enzyme concentration, the rate of a reaction increases with substrate concentration until the enzyme becomes saturated with reactants. At that point, further increases in substrate concentration do not increase the rate of the reaction (Figure 4.12).

- Many cellular enzymes are regulated by nonsubstrate molecules called inhibitors. Competitive inhibitors interfere with reaction rates by combining with the active site of an enzyme; noncompetitive inhibitors combine with sites elsewhere on the enzyme (Figure 4.13).

- Allosteric regulation resembles noncompetitive inhibition except that regulatory molecules may either increase or decrease enzyme activity. Allosteric regulation often carries out feedback inhibition, in which a product of an enzyme-catalyzed pathway acts as an allosteric inhibitor of the first enzyme in the pathway (Figures 4.14 and 4.15).

- Many key enzymes are regulated by chemical modification, by substances such as ions and certain functional groups. The modifications change enzyme conformation resulting in increased or decreased activity.

- Typically, an enzyme has optimal activity at a certain pH and a certain temperature; at pH and temperature values above and below the optimum, the reaction rate falls off (Figures 4.16 and 4.17).

Animation: Allosteric activation

Animation: Allosteric inhibition

4.6 RNA-Based Biological Catalysts: Ribozymes

- RNA-based catalysts called ribozymes speed some types of biological reactions; these include cutting and splicing reactions in which surplus segments are removed from RNA molecules and linking reactions that combine amino acids into polypeptide chains.

UNDERSTAND & APPLY

Test Your Knowledge

1. The capacity to do work best defines:
   a. a metabolic pathway.
   b. entropy.
   c. kinetic or potential energy.
   d. a chemical equilibrium.
   e. thermodynamics.

2. The assembly of proteins from amino acids is best described as:
   a. a conversion of kinetic energy to potential energy reaction.
   b. an entropy reaction.
   c. a catabolic reaction.
   d. an anabolic reaction.
   e. an energy free reaction.

3. When two glucose molecules react to form maltose:
   a. the reaction represents a negative ΔG.
   b. free energy had to be available to allow the reaction to proceed.
   c. the reaction is exothermic.
   d. it supports the second law of thermodynamics, which states there is tendency of the universe toward disorder.
   e. the resulting product has less potential energy than the reactants.

4. When glucose reacts with ATP to form glucose-6-phosphate:
   a. the synthesis of glucose-6-phosphate is exergonic.
   b. ADP is at a higher energy level than ATP.
   c. glucose-6-phosphate is at a higher energy level than glucose.
   d. because ATP donates a phosphate to glucose, this is not a coupled reaction.
   e. the reaction is spontaneous.

5. In the following graph:

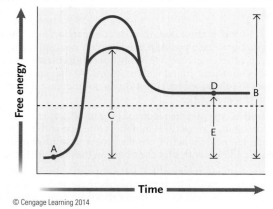

© Cengage Learning 2014

   a. A represents the product.
   b. B represents the energy of activation when enzymes are present.
   c. C is the free energy difference between A and D.
   d. C is the energy of activation without enzymes.
   e. E is the difference in free energy between the reactant and the products.

6. Which of the following methods is *not* used by enzymes to increase the rate of reactions?
   a. covalent bonding with the substrate at their active site
   b. bringing reacting molecules into close proximity
   c. orienting reactants into positions to favor transition states
   d. changing charges on reactants to hasten their reactivity
   e. increasing fit of enzyme and substrate that reduces the energy of activation

7. In an enzymatic reaction:
   a. the enzyme leaves the reaction chemically unchanged.
   b. if the enzyme molecules approach maximal rate, and the substrate is continually increased, the rate of the reaction does not reach saturation.
   c. in the stomach, enzymes would have an optimal activity at a neutral pH.
   d. increasing temperature above the optimal value slows the reaction rate.
   e. the least important level of organization for an enzyme is its tertiary structure.

8. Which of the following statements about the allosteric site is true?
   a. The allosteric site is a second active site on a substrate in a metabolic pathway.
   b. The allosteric site on an enzyme can allow the product of a metabolic pathway to inhibit that enzyme and stop the pathway.
   c. When the allosteric site of an enzyme is occupied, the reaction is irreversible and the enzyme cannot react again.
   d. An allosteric activator prevents binding at the active site.
   e. An enzyme that possesses allosteric sites does not possess an active site.

9. Which of the following statements about inhibition is true?
   a. Allosteric inhibitors and allosteric activators are competitive for a given enzyme.
   b. If an inhibitor binds the active site, it is considered noncompetitive.
   c. If an inhibitor binds to a site other than the active site, this is competitive inhibition.
   d. A noncompetitive inhibitor is believed to change the shape of the enzyme, making its active site inoperable.
   e. Competitive inhibition is usually not reversible.

10. Which of the following statements is *incorrect*?
    a. Ribozymes can link amino acids to form protein.
    b. Ribozymes can act as enzymes.
    c. Ribozymes can act as informational molecules.
    d. Ribozymes are suggested as the first molecules of life.
    e. Ribozymes are proteins.

Discuss the Concepts

1. Trees become more complex as they develop spontaneously from seeds to adults. Does this process violate the second law of thermodynamics? Why or why not?

2. Trace the flow of energy through your body. What products increase the entropy of you and your surroundings?

3. You have found a molecular substance that accelerates the rate of a particular reaction. What kind of information would you need to demonstrate that this molecular substance is an enzyme?

4. The addition or removal of phosphate groups from ATP is a fully reversible reaction. In what way does this reversibility facilitate the use of ATP as a coupling agent for cellular reactions?

5. Researchers once hypothesized that an enzyme and its substrate fit together like a lock and key but that the products do not fit the enzyme. Examine this idea with respect to reversible reactions.

Design an Experiment

Succinate dehydrogenase is part of the cellular biochemical machinery for breaking down sugars, fatty acids, and amino acids into carbon dioxide and water, with the capture of their chemical energy as ATP. Suppose you are measuring the activity of this enzyme extracted from cells in test-tube reactions. You find that the rate of the reaction converting succinate to fumarate catalyzed by succinate dehydrogenase is inhibited by the addition of malonate to the reaction mixture. Design an experiment that will tell you whether malonate is acting as a competitive or a noncompetitive inhibitor.

Interpret the Data

The postsynaptic density 95 (PSD-95) protein plays a key role in mammalian nervous system responses by concentrating and organizing receptors on the nerve cells. Future drugs that change these receptors could change learning and memory at the cellular level. A key part of the interaction between PSD-95 and the proposed drugs is the thermodynamics of binding between them.

The thermodynamics of binding between PSD-95 and two different small polypeptides (similar to those in the proposed drugs) are shown in the graphs. Graph A shows data for a parent polypeptide, and graph B shows the data for a mutated polypeptide (with some altered amino acids) of the same length. On the x axis is the ratio (in moles) between the amount of bound polypeptide and PSD-95. When the molar ratio is very near zero on the x axis in each graph, all of the PSD-95 molecules available are bound by the polypeptides. The molar ratio can be manipulated by injecting small amounts of polypeptide into the solution, and the change in energy content of the injectant (ΔH measured in kcal/mole) is measured with each accumulating injection.

A. Parent polypeptide

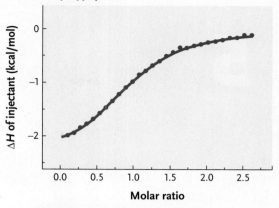

B. Mutated polypeptide

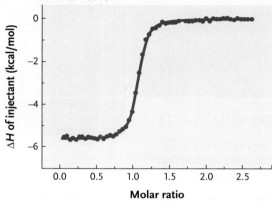

Reprinted (adapted) from D. Saro et al. 2007. A thermodynamic ligand binding study of the third PDZ domain (PDZ3) from the mammalian neuronal protein PSD-95. *Biochemistry* 46:6340–6352. Copyright 2007 American Chemical Society.

1. If $T\Delta S$ for the parent polypeptide (graph A) is 4.0 kcal/mol and for the mutated polypeptide (graph B) is 3.5 kcal/mol, what is the ΔG for each polypeptide binding to the protein?

2. Using the ΔG values calculated in question 1, state whether each polypeptide binding is endergonic or exergonic, spontaneous or not spontaneous, and which would yield more free energy on binding to PSD-95.

3. If the mutated polypeptide represented a drug that needed to compete with the parent polypeptide in a nerve cell, would this mutated polypeptide be very effective?

Source: From D. Saro et al. 2007. A thermodynamic ligand binding study of the third PDZ domain (PDZ3) from the mammalian neuronal protein PSD-95. *Biochemistry* 46:6340–6352.

Apply Evolutionary Thinking

If RNA appeared first in evolution, establishing an RNA world, which do you think would evolve next: DNA or proteins? Why?

5

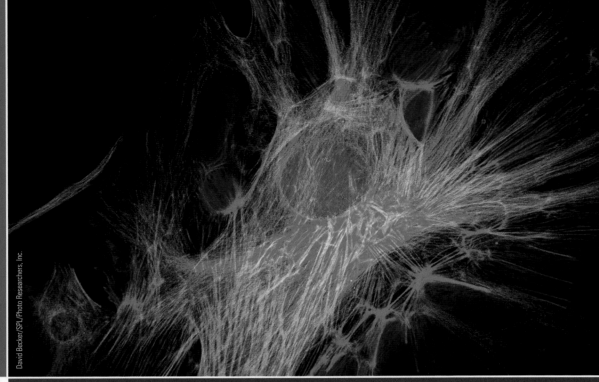

David Becker/SPL/Photo Researchers, Inc.

Cells fluorescently labeled to visualize their internal structure (confocal light micrograph). Cell nuclei are shown in blue and parts of the cytoskeleton in red and green.

The Cell: An Overview

Why it matters . . . In the mid-1600s, Robert Hooke, Curator of Instruments for the Royal Society of England, was at the forefront of studies applying the newly invented light microscopes to biological materials. When Hooke looked at thinly sliced cork from a mature tree through a microscope, he observed tiny compartments **(Figure 5.1A)**. He gave them the Latin name *cellulae,* meaning "small rooms"—hence, the origin of the biological term *cell.* Hooke was actually looking at the walls of dead cells, which is what cork consists of.

Reports of cells also came from other sources. By the late 1600s, Anton van Leeuwenhoek **(Figure 5.1B),** a Dutch shopkeeper, observed "many very little animalcules, very prettily a-moving," using a single-lens microscope of his own construction. Leeuwenhoek discovered and described diverse protists (see Chapter 27), sperm cells, and even bacteria, organisms so small that they would not be seen by others for another two centuries.

In the 1820s, improvements in microscopes brought cells into sharper focus. Robert Brown, an English botanist, noticed a discrete, spherical body inside some cells; he called it a *nucleus.* In 1838, a German botanist, Matthias Schleiden, speculated that the nucleus had something to do with the development of a cell. The following year, zoologist Theodor Schwann of Germany expanded Schleiden's idea to propose that all animals and plants consist of cells that contain a nucleus. He also proposed that even when a cell forms part of a larger organism, it has an individual life of its own. However, an important question remained: Where do cells come from? A decade later, the German physiologist Rudolf Virchow answered this question. From his studies of cell growth and reproduction, Virchow proposed that cells arise only from preexisting cells by a process of division.

FIGURE 5.1 **Investigations leading to the first descriptions of cells. (A)** The cork cells drawn by Robert Hooke and the compound microscope he used to examine them. **(B)** Anton van Leeuwenhoek holding his microscope, which consisted of a single, small sphere of glass fixed in a holder. He viewed objects by holding them close to one side of the glass sphere and looking at them through the other side.

Thus, by the middle of the nineteenth century, microscopic observations had yielded three profound generalizations, which together constitute what is now known as the **cell theory:**

1. All organisms are composed of one or more cells.
2. The cell is the basic structural and functional unit of all living organisms.
3. Cells arise only from the division of preexisting cells.

These tenets were fundamental to the development of biological science.

This chapter provides an overview of our current understanding of the structure and functions of cells, emphasizing both the similarities among all cells and some of the most basic differences among cells of various organisms. The variations in cells that help make particular groups of organisms distinctive are discussed in later chapters. This chapter also introduces some of the modern microscopes that enable us to learn more about cell structure.

5.1 Basic Features of Cell Structure and Function

As the basic structural and functional units of all living organisms, cells carry out the essential processes of life. They contain highly organized systems of molecules, including the nucleic acids DNA and RNA, which carry hereditary information and direct the manufacture of cellular molecules. Cells use chemical molecules or light as energy sources for their activities.

Cells also respond to changes in their external environment by altering their internal reactions. Further, cells duplicate and pass on their hereditary information as part of cellular reproduction. All these activities occur in cells that, in most cases, are invisible to the naked eye.

Some types of organisms, including almost all bacteria and archaeans; some protists, such as amoebas; and some fungi, such as yeasts, are unicellular. Each of these cells is a functionally independent organism capable of carrying out all activities necessary for its life. In more complex multicellular organisms, including plants and animals, the activities of life are divided among varying numbers of specialized cells. However, individual cells of multicellular organisms are potentially capable of surviving by themselves if placed in a chemical medium that can sustain them.

If cells are broken open, the property of life is lost: they are unable to grow, reproduce, or respond to outside stimuli in a coordinated, potentially independent fashion. This fact confirms the second tenet of the cell theory: life as we know it does not exist in units more simple than individual cells. *Viruses,* which consist only of a nucleic acid molecule surrounded by a protein coat, cannot carry out most of the activities of life. Their only capacity is to infect living cells and direct them to make more virus particles of the same kind. (Viruses are discussed in Chapter 17.)

Cells Are Small and Are Visualized Using a Microscope

Cells assume a wide variety of forms in different prokaryotes and eukaryotes **(Figure 5.2)**. Individual cells range in size from tiny bacteria to an egg yolk, a single cell that can be several centimeters in diameter. Yet, all cells are organized according to the same basic plan, and all have structures that perform similar activities.

Most cells are too small to be seen by the unaided eye: humans cannot see objects smaller than about 0.1 mm in diameter. The smallest bacteria have diameters of about 0.5 μm (a micrometer is 1,000 times smaller than a millimeter). The cells of multicellular animals range from about 5 to 30 μm in diameter. Your red blood cells are 7 to 8 μm across—a string of 2,500 of these cells is needed to span the width of your thumbnail. Plant cells range from about 10 μm to a few hundred micrometers in diameter. (**Figure 5.3** explains the units of measurement used in biology to study molecules and cells.)

To see cells and the structures within them biologists use **microscopy,** a technique for producing visible images of objects, biological or otherwise, that are too small to be seen by the human eye **(Figure 5.4)**. The instrument of microscopy is the **microscope.** The two common types of microscopes are **light microscopes,** which use light to illuminate the specimen (the

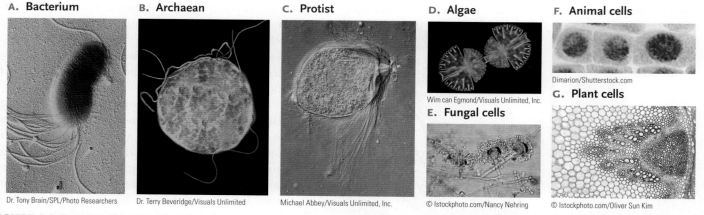

A. Bacterium
Dr. Tony Brain/SPL/Photo Researchers

B. Archaean
Dr. Terry Beveridge/Visuals Unlimited

C. Protist
Michael Abbey/Visuals Unlimited, Inc.

D. Algae
Wim can Egmond/Visuals Unlimited, Inc.

E. Fungal cells
© Istockphoto.com/Nancy Nehring

F. Animal cells
Dimarion/Shutterstock.com

G. Plant cells
© Istockphoto.com/Oliver Sun Kim

FIGURE 5.2 Examples of the varied kinds of cells: **(A)** and **(B)** are prokaryotes, the others are eukaryotes. **(A)** A bacterial cell with flagella, *Pseudomonas fluorescens*. **(B)** An archaean, the extremophile *Sulfolobus acidocaldarius*. **(C)** *Trichonympha*, a protist that lives in a termite's gut. **(D)** Two cells of *Micrasterias*, an algal protist. **(E)** Fungal cells of the bread mold *Aspergillus*. **(F)** Animal cells. **(G)** Cells in the stem of a sunflower, *Helianthus annuus*.

object being viewed), and **electron microscopes,** which use electrons to illuminate the specimen. Different types of microscopes give different magnification and resolution of the specimen. Just as for a camera or a pair of binoculars, **magnification** is the ratio of the object as viewed to its real size, usually given as something like 1,200×. **Resolution** is the minimum distance two points in the specimen can be separated and still be seen as two points. Resolution depends primarily on the wavelength of light or electrons used to illuminate the specimen; the shorter the wavelength, the better the resolution. Hence, electron microscopes have higher resolution than light microscopes. Biologists choose the type of microscopy technique based on what they need to see in the specimen; selected examples are shown in Figure 5.4.

While cells vary in size, there is an upper limit to cell size due to the change in the surface area-to-volume ratio of an object as its size increases **(Figure 5.5)**. For example, doubling the diameter of a cell increases its volume by eight times but increases its surface area by only four times. The significance of this relationship is that the volume of a cell determines the amount of chemical activity that can take place within it, whereas the surface area determines the amount of substances that can be exchanged between the inside of the cell and the outside environment. Nutrients must constantly enter cells, and wastes must constantly leave; however, past a certain point, increasing the diameter of a cell gives a surface area that is insufficient to maintain an adequate nutrient–waste exchange for its entire volume.

Some cells increase their ability to exchange materials with their surroundings by flattening or by developing surface folds or extensions that increase their surface area. For example, human intestinal cells have closely packed, fingerlike extensions that increase their surface area, which greatly enhances their ability to absorb digested food molecules.

Cells Have a DNA-Containing Central Region That Is Surrounded by Cytoplasm

All cells are bounded by the **plasma membrane,** a bilayer made of lipids with embedded protein molecules **(Figure 5.6)**. (The chemical structure of the lipid bilayer was described in Section 3.3.) The lipid bilayer is a hydrophobic barrier to the passage of water-soluble substances, but selected water-soluble substances can penetrate cell membranes through transport

1 centimeter (cm) = 1/100 meter or 0.4 inch	**Unaided human eye**
	3 cm — Chicken egg (the "yolk")
1 millimeter (mm) = 1/1,000 meter	
	1 mm — Frog egg, fish egg
1 micrometer (µm) = 1/1,000,000 meter	**Light microscopes**
	100 µm — Human egg
	10–100 — Typical plant cell
	5–30 — Typical animal cell
	2–10 — Chloroplast
	1–5 — Mitochondrion
	5 — *Anabaena* (cyanobacterium)
	1 — *Escherichia coli*
1 nanometer (nm) = 1/1,000,000,000 meter	**Electron microscopes**
	100 nm — Large virus (HIV, influenza virus)
	25 — Ribosome
	7–10 — Cell membrane (thickness)
	2 — DNA double helix (diameter)
	0.1 — Hydrogen atom

1 meter = 10^2 cm = 10^3 mm = 10^6 µm = 10^9 nm

FIGURE 5.3 **Units of measure and the ranges in which they are used in the study of molecules and cells.** The vertical scale in each box is logarithmic.
© Cengage Learning 2014

FIGURE 5.4   **Research Method**

Light and Electron Microscopy

Purpose: In biology, microscopy is used to view organisms, cells, and structures within cells in their natural state or after being treated (stained) so that specific structures can be seen more clearly. All of the photographs of cells and cell structures in this book were made using microscopy.

Protocol: A light microscope uses a beam of light to illuminate the specimen and forms a magnified image of the specimen with glass lenses. An electron microscope uses a beam of electrons to illuminate the specimen and forms an image with magnetic fields. Electron microscopy provides higher resolution and higher magnification than light microscopy.

Light microscopy
Micrographs are of the protist *Paramecium*.

Electron microscopy
Micrographs are of the green alga *Scenedesmus*.

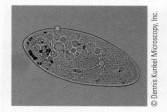

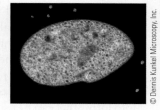

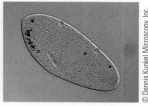

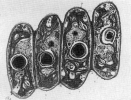

Bright field microscopy: Light passes directly through the specimen. Many cell structures have insufficient contrast to be discerned. Staining with a dye is used to enhance contrast in a specimen, as shown here, but this treatment usually fixes and kills the cells.

Dark field microscopy: Light illuminates the specimen at an angle, and only light scattered by the specimen reaches the viewing lens of the microscope. This gives a bright image of the cell against a black background.

Phase-contrast microscopy: Differences in refraction (the way light is bent) caused by variations in the density of the specimen are visualized as differences in contrast. Otherwise invisible structures are revealed with this technique, and living cells in action can be photographed or filmed.

Transmission electron microscopy (TEM): A beam of electrons is focused on a thin section of a specimen in a vacuum. Electrons that pass through form the image; structures that scatter electrons appear dark. TEM is used primarily to examine structures within cells. Various staining and fixing methods are used to highlight structures of interest.

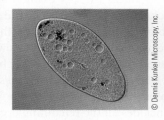

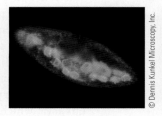

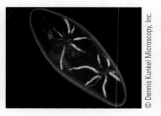

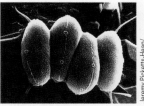

Nomarski (differential interference contrast): Similar to phase-contrast microscopy, special lenses enhance differences in density, giving a cell a 3D appearance.

Fluorescence microscopy: Different structures or molecules in cells are stained with specific fluorescent dyes. The stained structures or molecules fluoresce when the microscope illuminates them with ultraviolet light, and their locations are seen by viewing the emitted visible light.

Confocal laser scanning microscopy: Lasers scan across a fluorescently stained specimen, and a computer focuses the light to show a single plane through the cell. This provides a sharper 3D image than other light microscopy techniques.

Scanning electron microscopy (SEM): A beam of electrons is scanned across a whole cell or organism, and the electrons excited on the specimen surface are converted to a 3D-appearing image.

Interpreting the Results: Different techniques of light and electron microscopy produce images that reveal different structures or functions of the specimen. A micrograph is a photograph of an image formed by a microscope.

FIGURE 5.5 Relationship between surface area and volume.

The surface area of an object increases as a square of the linear dimension, whereas the volume increases as a cube of that dimension.
© Cengage Learning 2014

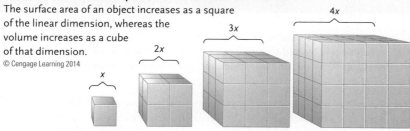

Total surface area	$6x^2$	$6(2x)^2 = 24x^2$	$6(3x)^2 = 54x^2$	$6(4x)^2 = 96x^2$
Total volume	x^3	$(2x)^3 = 8x^3$	$(3x)^3 = 27x^3$	$(4x)^3 = 64x^3$
Surface area/ volume ratio	6:1	3:1	2:1	1.5:1

protein channels. The selective movement of ions and water-soluble molecules through the transport proteins maintains the specialized internal ionic and molecular environments required for cellular life. (Membrane structure and functions are discussed further in Chapter 6.)

A central region of all cells contains DNA molecules, which store hereditary information. The hereditary information is organized in the form of *genes*—segments of DNA that code for individual proteins. The central region also contains proteins that help maintain the DNA structure and enzymes that duplicate DNA and copy its information into RNA.

All the parts of the cell between the plasma membrane and the central region comprise the **cytoplasm.** The cytoplasm contains the *organelles,* the *cytosol,* and the *cytoskeleton.* The **organelles** ("little organs") are small, organized structures important for cell function. The **cytosol** is an aqueous (water) solution containing ions and various organic molecules. The **cytoskeleton** is a protein-based framework of filamentous structures that, among other things, helps maintain proper cell shape and plays key roles in cell division and chromosome segregation from cell generation to cell generation. The cytoskeleton was once thought to be specific to eukaryotes, but recent research has shown that all major eukaryotic cytoskeletal proteins have functional equivalents in prokaryotes.

Many of the cell's vital activities occur in the cytoplasm, including the synthesis and assembly of most of the molecules required for growth and reproduction (except those made in the central region) and the conversion of chemical and light energy into forms that can be used by cells. The cytoplasm also conducts stimulatory signals from the outside into the cell interior and carries out chemical reactions that respond to these signals.

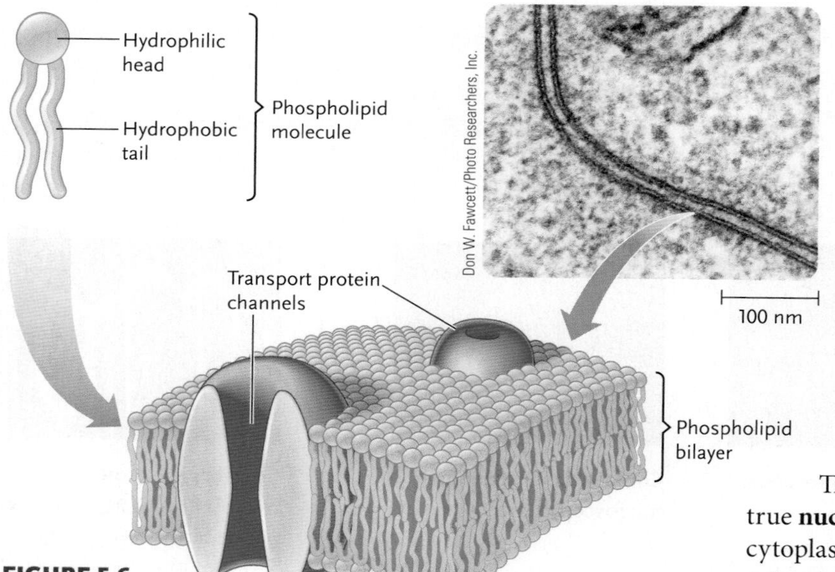

Cells Occur in Prokaryotic and Eukaryotic Forms, Each with Distinctive Structures and Organization

Organisms fall into two fundamental groups, prokaryotes and eukaryotes, based on the organization of their cells. **Prokaryotes** (*pro* = before; *karyon* = nucleus) make up two domains of organisms, the Bacteria and the Archaea. The DNA-containing central region of prokaryotic cells, the **nucleoid,** has no boundary membrane separating it from the cytoplasm. Many species of bacteria contain few if any internal membranes, but a number of other bacterial species contain extensive internal membranes.

The **eukaryotes** (*eu* = true) make up the Domain Eukarya, which includes all the remaining organisms. The DNA-containing central region of eukaryotic cells, a true **nucleus,** is separated by membranes from the surrounding cytoplasm. The cytoplasm of eukaryotic cells typically contains extensive membrane systems that form organelles with their own distinct environments and specialized functions. As in prokaryotes, a plasma membrane surrounds eukaryotic cells as the outer limit of the cytoplasm.

The remainder of this chapter surveys the components of prokaryotic and eukaryotic cells in more detail.

Hydrophilic head
Hydrophobic tail
Phospholipid molecule

Transport protein channels

Don W. Fawcett/Photo Researchers, Inc.

100 nm

Phospholipid bilayer

FIGURE 5.6

The plasma membrane, which forms the outer limit of a cell's cytoplasm.

The plasma membrane consists of a phospholipid bilayer, an arrangement of phospholipids two molecules thick, which provides the framework of all biological membranes. Water-soluble substances cannot pass through the phospholipid part of the membrane. Instead, they pass through protein channels in the membrane; two proteins that transport substances across the membrane are shown. Other types of proteins are also associated with the plasma membrane. (Inset) Electron micrograph showing the plasma membranes of two adjacent animal cells.
© Cengage Learning 2014

STUDY BREAK 5.1

What is the plasma membrane, and what are its main functions?

5.2 Prokaryotic Cells

Most prokaryotic cells are relatively small, usually not much more than a few micrometers in length and a micrometer or less in diameter. A typical human cell is about ten times larger in diameter and over 8,000 times larger in volume than an average prokaryotic cell.

The three shapes most common among prokaryotes are spherical, rodlike, and spiral. *Escherichia coli (E. coli),* a normal inhabitant of the mammalian intestine that has been studied extensively as a model organism in genetics, molecular biology, and genomics research, is rodlike in shape. **Figure 5.7** shows an electron micrograph (EM) and diagram of *E. coli* to illustrate the basic features of prokaryotic cell structure. More detail about prokaryotic cell structure and function, as well as about the diversity of prokaryotic organisms, is presented in Chapter 26.

The genetic material of prokaryotes is located in the nucleoid; in an electron microscope, that region of the cell is seen to contain a highly folded mass of DNA (see Figure 5.7). For most species, the DNA is a single, circular molecule that unfolds when released from the cell. This DNA molecule is called the **prokaryotic chromosome.** (Chapter 17 discusses the genetics of prokaryotes.)

Individual genes in the DNA molecule encode the information required to make proteins. This information is copied into a type of RNA molecule called *messenger RNA (mRNA)* (described in more detail in Chapter 15). Small, roughly spherical particles in the cytoplasm, the **ribosomes,** use the information in the mRNA to assemble amino acids into proteins (described in more detail in Chapter 15). A prokaryotic ribosome consists of a large and a small subunit, each formed from a combination of *ribosomal RNA (rRNA)* and protein molecules.

Each prokaryotic ribosome contains three types of rRNA molecules, which are also copied from the DNA, and more than 50 proteins.

In almost all prokaryotes, the plasma membrane is surrounded by a rigid external layer of material, the **cell wall,** which ranges in thickness from 15 to 100 nm or more (a nanometer is one-billionth of a meter). The cell wall provides rigidity to prokaryotic cells and helps protect the cell from physical damage. In many prokaryotic cells, the wall is coated with an external layer of polysaccharides called the **glycocalyx** (a "sugar coating" from *glykys* = sweet; *calyx* = cup or vessel). When the glycocalyx is diffuse and loosely associated with the cells, it is a **slime layer;** when it is gelatinous and attached more firmly to cells, it is a **capsule.** The glycocalyx helps protect prokaryotic cells from physical damage and desiccation, and may enable a cell to attach to a surface, such as other prokaryotic cells (as in forming a colony), eukaryotic cells (as in *Streptococcus pneumoniae* attaching to lung cells), or nonliving substrates (such as rocks).

The plasma membrane itself performs several vital functions in prokaryotes. Besides transporting materials into and out of the cells, it contains most of the molecular systems that metabolize food molecules into the chemical energy of ATP. In photosynthetic prokaryotes, the molecules that absorb light energy and convert it to the chemical energy of ATP are also associated with the plasma membrane or with internal, saclike membranes derived from the plasma membrane.

Many prokaryotic species contain few if any internal membranes; in such cells, most cellular functions occur either on the plasma membrane or in the cytoplasm. But some prokaryotes have more extensive internal membrane structures. For example, photosynthetic bacteria and archaeans have complex layers of intracellular membranes formed by invagi-

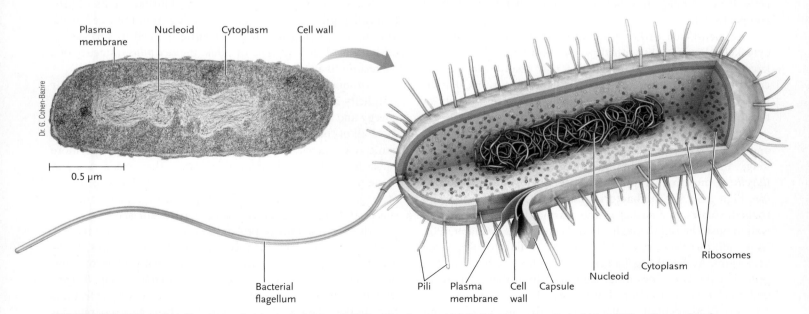

Plasma membrane · Nucleoid · Cytoplasm · Cell wall

Dr. G. Cohen-Bazire

0.5 μm

Bacterial flagellum · Pili · Plasma membrane · Cell wall · Capsule · Nucleoid · Cytoplasm · Ribosomes

FIGURE 5.7 Prokaryotic cell structure. An electron micrograph (left) and a diagram (right) of the bacterium *Escherichia coli*. The pili extending from the cell wall attach bacterial cells to other cells of the same species or to eukaryotic cells as a part of infection. A typical *E. coli* has four flagella.
© Cengage Learning 2014

An Old Kingdom in a New Domain: Do archaeans define a distinct domain of life?

Many archaeans live in extreme environments that can be tolerated by no other organisms, suggesting that they might belong in a distinct domain of life. For example, *Methanococcus jannaschii* was first found in an oceanic hot water vent at a depth of more than 2,600 m (8,500 feet). It can live at temperatures as high as 94°C, which is almost the temperature of boiling water, and can tolerate pressures as high as 200 times the pressure of air at sea level.

Research Question

Are archaeans a distinct domain of life?

Experiment

In 1996, Carol J. Bult, Carl R. Woese, J. Craig Venter, and 37 other scientists at the Institute for Genomic Research published the complete genomic DNA sequence of *Methanococcus jannaschii.* (The sequence was obtained using techniques outlined in Chapter 19.) Using computer algorithms, the scientists compared the sequence with the already known genome sequences of several bacteria and of brewer's yeast *(Saccharomyces cerevisiae),* the first eukaryote whose genome was sequenced completely.

Results

The researchers found genes coding for 1,738 proteins in the *Methanococcus* genome. Of these, only 38% were related to genes coding for known proteins in either bacteria or eukaryotes. The remaining 62% have no known relatives in organisms of those two groups.

Some features of *Methanococcus* DNA are typically prokaryotic. Its single, circular chromosome is in a nucleoid, which is not bounded by a membrane. Its protein-coding genes are organized into functional groups called *operons,* each having several genes copied as a unit into a single mRNA molecule (see discussion in Section 16.1). By contrast, each protein-coding gene in eukaryotes is copied into a separate mRNA molecule. Some of the proteins encoded in *Methanococcus* DNA, including enzymes active in energy metabolism, membrane transport, and cell division, are similar to those of bacteria. Other pro-teins encoded in the *Methanococcus* DNA are similar to those of eukaryotes, including enzymes and other proteins that carry out DNA replication and the copying of genes into mRNA.

Conclusion

Methanococcus has a majority of genes that are unique, some that are typically bacterial, and some that are typically eukaryotic. This finding supports the proposal, first advanced by Woese, that *Methanococcus* and its archaean relatives are a separate domain of life, the Archaea, with the Bacteria and the Eukarya as the other domains. Together, Bacteria and Archaea are the prokaryotes. Woese's three-domain system is used in this book.

THINK LIKE A SCIENTIST What result would have indicated that *Methanococcus jannaschii* is a member of the Domain Bacteria rather than a member of the Domain Archaea?

Source: C. J. Bult et al. 1996. Complete genome sequence of the methanogenic archaeon, *Methanococcus jannaschii. Science* 273:1058–1073.

nations of the plasma membrane on which photosynthesis takes place. And members of the bacterial phylum Planctomycetes have complex internal membranes that form distinct compartments.

As mentioned earlier, prokaryotic cells have filamentous cytoskeletal structures with functions similar to those in eukaryotes. Prokaryotic cytoskeletons play important roles in creating and maintaining the proper shape of cells, in cell division and, for certain prokaryotes, in determining polarity of the cells.

Many bacteria and archaeans can move through liquids and across wet surfaces. Most commonly they do so using long, threadlike protein fibers called **flagella** (singular, *flagellum,* meaning whip), which extend from the cell surface (see Figure 5.2A). The **bacterial flagellum,** which is helically shaped, rotates in a socket in the plasma membrane and cell wall to push the cell through a liquid medium (see Chapter 26). In *E. coli,* for instance, rotating bundles of flagella propel the bacterium. Archaeal flagella function similarly to bacterial flagella, but the two types differ significantly in their structures and mechanisms of action. Both types of prokaryotic flagella are also fundamentally different from the much larger and more complex flagella of eukaryotic cells, which are described in Section 5.3.

Some bacteria and archaeans have hairlike shafts of protein called **pili** (singular, *pilus*) extending from their cell walls. The main function of pili is attaching the cell to surfaces or other cells. A special type of pilus, the *sex pilus,* attaches one bacterium to another during conjugation (see Chapter 17).

Although prokaryotic cells appear relatively simple, their simplicity is deceptive. Most can use a variety of substances as energy and carbon sources, and they are able to synthesize almost all of their required organic molecules from simple inorganic raw materials. In many respects, prokaryotes are more versatile biochemically than eukaryotes. Their small size and metabolic versatility are reflected in their abundance; prokaryotes vastly outnumber all other types of organisms and live successfully in almost all regions of Earth's surface.

The two domains of prokaryotes, the Bacteria and the Archaea, share many biochemical and molecular features. However, the archaeans also share some features with eukaryotes and have other characteristics that are unique to their group. *Insights from the Molecular Revolution* describes the discovery of features that support the classification of the Archaea as a separate do-

FIGURE 5.8 **Research Method**

Cell Fractionation

Purpose: Cell fractionation partitions cells into fractions containing a single cell component, such as mitochondria or ribosomes. Once isolated, the cell component can be disassembled by the same general techniques to analyze its structure and function.

Protocol:

1. Break open intact cells by sonication (high-frequency sound waves), grinding in fine glass beads, or exposure to detergents that disrupt plasma membranes.

2. Use sequential centrifugations at increasing speeds to separate and purify cell structures. The spinning centrifuge drives cellular structures to the bottom of tube at a rate that depends on their shape and density. With each centrifugation, the largest and densest components are isolated and concentrated into a pellet; the remaining solution, the supernatant, is drawn off and can be centrifuged again at higher speed.

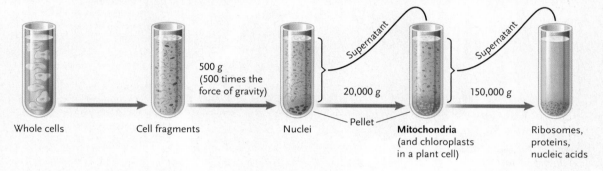

Whole cells   Cell fragments   500 g (500 times the force of gravity)   Nuclei   Supernatant   20,000 g   Pellet   **Mitochondria** (and chloroplasts in a plant cell)   Supernatant   150,000 g   Ribosomes, proteins, nucleic acids

3. Resuspend the pellet containing the isolated cell components and subfractionate using the same general techniques to examine the components of organelles.

Interpreting the Results: Many of the cell or organelle subfractions generated by cell fractionation retain their biological activity, making them useful in studies of various cellular processes. For example, cell fractionation is used to determine the cellular location of a protein or biological reaction, such as whether it is free in the cytosol or associated with a membrane.

---

main. As shown in Figure 1.11B, evolutionary divergence from the common ancestor of all living organisms resulted in the ancestral lineage of present-day members of the Domain Bacteria and the common ancestral lineage of present-day members of the Domain Archaea and members of the Domain Eukarya.

STUDY BREAK 5.2

Where in a prokaryotic cell is DNA found? How is that DNA organized?

5.3 Eukaryotic Cells

The domain of the eukaryotes, Eukarya, is divided into four major groups: the protists (see Chapter 27), fungi (see Chapter 30), animals (see Chapters 31 and 32), and plants (see Chapters 28 and 29). The rest of the chapter focuses on the cell components that are common to all or large groups of eukaryotic organisms.

Eukaryotic Cells Have a True Nucleus and Cytoplasmic Organelles Enclosed within a Plasma Membrane

The cells of all eukaryotes have a true nucleus enclosed by membranes. The cytoplasm surrounding the nucleus contains a system of membranous organelles, each specialized to carry out one or more major functions of energy metabolism and molecular synthesis, storage, and transport. The cytosol, the cytoplasmic solution surrounding the organelles, participates in energy metabolism and molecular synthesis and performs specialized functions in support and motility. Researchers have discovered much about the structures and functions of various cellular organelles by using the research method of *cell fractionation* to isolate them and purify them **(Figure 5.8).**

The eukaryotic plasma membrane carries out various functions through several types of embedded proteins. Some of these proteins form channels through the plasma membrane that transport substances into and out of the cell (see Chapter 6).

A. Diagram of an animal cell, highlighting the major organelles and their primary locations (Not shown are flagellae and cilia, which are found in some cells.)

Present in animal cells but not plant cells
Centrosomes with centrioles
Lysosomes
Cilia

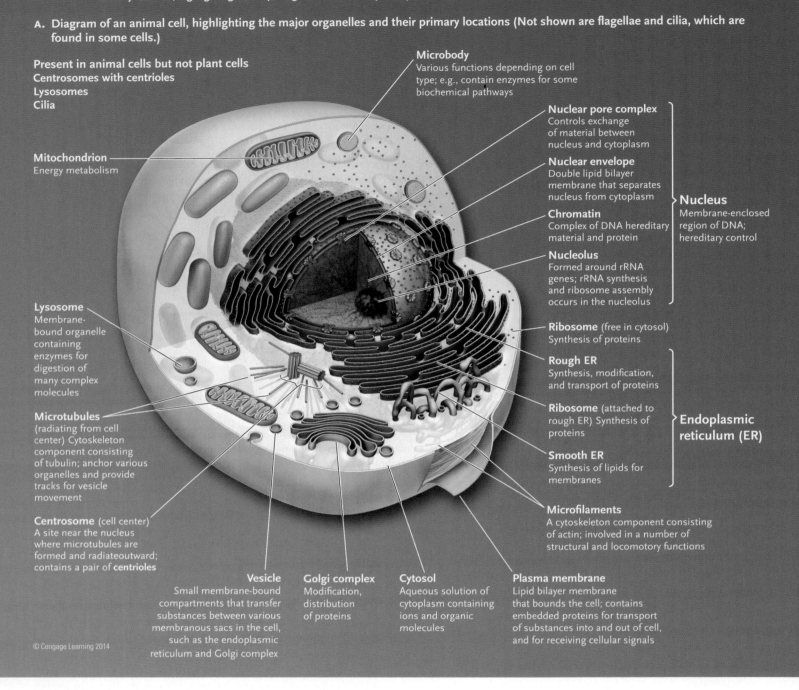

Microbody
Various functions depending on cell type; e.g., contain enzymes for some biochemical pathways

Nuclear pore complex
Controls exchange of material between nucleus and cytoplasm

Nuclear envelope
Double lipid bilayer membrane that separates nucleus from cytoplasm

Chromatin
Complex of DNA hereditary material and protein

Nucleolus
Formed around rRNA genes; rRNA synthesis and ribosome assembly occurs in the nucleolus

Nucleus
Membrane-enclosed region of DNA; hereditary control

Ribosome (free in cytosol)
Synthesis of proteins

Rough ER
Synthesis, modification, and transport of proteins

Ribosome (attached to rough ER) Synthesis of proteins

Smooth ER
Synthesis of lipids for membranes

Endoplasmic reticulum (ER)

Mitochondrion
Energy metabolism

Lysosome
Membrane-bound organelle containing enzymes for digestion of many complex molecules

Microtubules
(radiating from cell center) Cytoskeleton component consisting of tubulin; anchor various organelles and provide tracks for vesicle movement

Centrosome (cell center)
A site near the nucleus where microtubules are formed and radiateoutward; contains a pair of **centrioles**

Microfilaments
A cytoskeleton component consisting of actin; involved in a number of structural and locomotory functions

Vesicle
Small membrane-bound compartments that transfer substances between various membranous sacs in the cell, such as the endoplasmic reticulum and Golgi complex

Golgi complex
Modification, distribution of proteins

Cytosol
Aqueous solution of cytoplasm containing ions and organic molecules

Plasma membrane
Lipid bilayer membrane that bounds the cell; contains embedded proteins for transport of substances into and out of cell, and for receiving cellular signals

© Cengage Learning 2014

Other proteins in the plasma membrane act as receptors; they recognize and bind specific signal molecules in the cellular environment and trigger internal responses (see Chapter 7). In some eukaryotes, particularly animals, plasma membrane proteins recognize and adhere to molecules on the surfaces of other cells. Yet other plasma membrane proteins are important markers in the immune system, labeling cells as "self," that is, belonging to the organism. Therefore, the immune system can identify cells without those markers as being foreign, most likely *pathogens* (disease-causing organisms or viruses) (see Chapter 45).

A supportive cell wall surrounds the plasma membrane of fungal, plant, and many protist cells. Because the cell wall lies outside the plasma membrane, it is an *extracellular* structure (*extra* = outside). Although animal cells do not have cell walls, they also form extracellular material with supportive and other functions (see Section 5.5).

Figure 5.9A presents a diagram of a representative animal cell and **Figure 5.9B** presents a diagram of a representative plant cell to show where the nucleus, cytoplasmic organelles, and other structures are located. The following sections discuss the

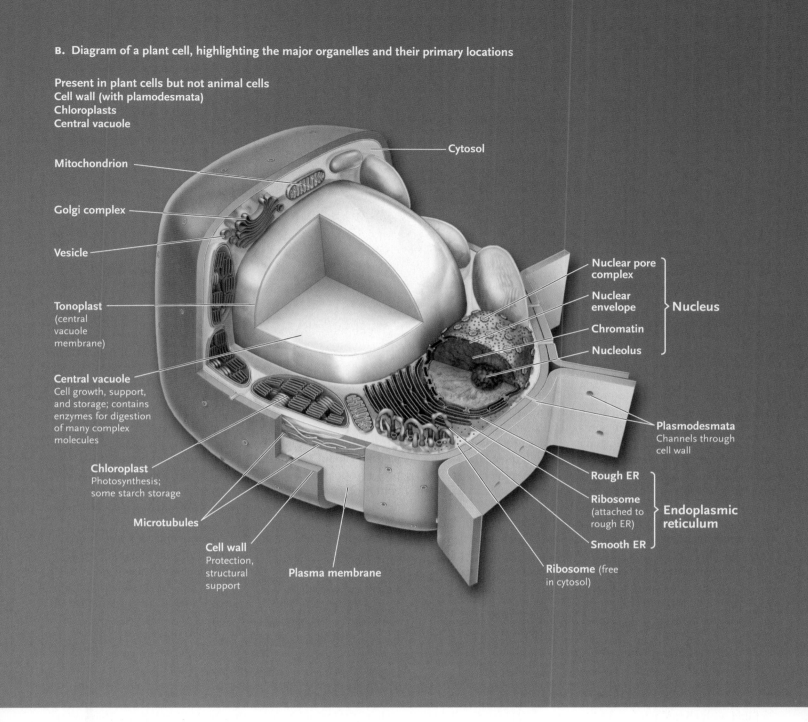

Present in plant cells but not animal cells
Cell wall (with plamodesmata)
Chloroplasts
Central vacuole

Cytosol

Mitochondrion

Golgi complex

Vesicle

Nuclear pore complex

Nuclear envelope

} Nucleus

Chromatin

Nucleolus

Tonoplast
(central
vacuole
membrane)

Central vacuole
Cell growth, support,
and storage; contains
enzymes for digestion
of many complex
molecules

Plasmodesmata
Channels through
cell wall

Chloroplast
Photosynthesis;
some starch storage

Rough ER

Ribosome
(attached to
rough ER)

} Endoplasmic
reticulum

Microtubules

Smooth ER

Cell wall
Protection,
structural
support

Plasma membrane

Ribosome (free
in cytosol)

structure and function of eukaryotic cell parts in more detail, beginning with the nucleus.

The Eukaryotic Nucleus Contains Much More DNA Than the Prokaryotic Nucleoid

The nucleus (see Figures 5.9A and 5.9B) is separated from the cytoplasm by the **nuclear envelope,** which consists of two lipid bilayer membranes, one just inside the other and separated by a narrow space **(Figure 5.10).** A network of protein filaments called

lamins lines and reinforces the inner surface of the nuclear envelope in animal cells. Lamins are a type of intermediate filament (see later in this section). Evolutionarily unrelated proteins line the inner surface of the nuclear envelope in protists, fungi, and plants and carry out the same function.

Embedded in the nuclear envelope are many hundreds of nuclear pore complexes. A **nuclear pore complex** is a large, octagonally symmetric, cylindrical structure formed of many types of proteins, called *nucleoporins*. Probably the largest protein complex in the cell, it exchanges components between the

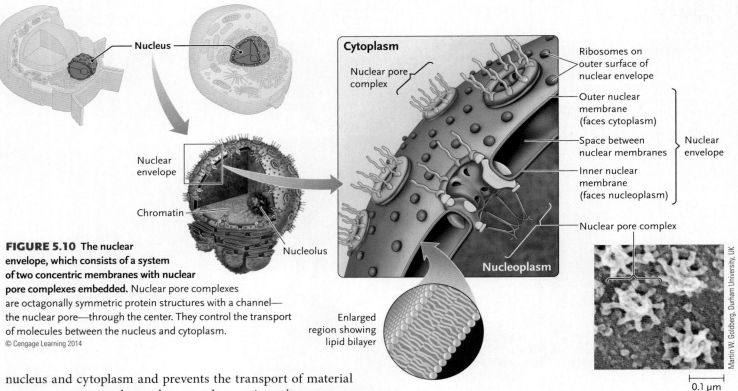

FIGURE 5.10 The nuclear envelope, which consists of a system of two concentric membranes with nuclear pore complexes embedded. Nuclear pore complexes are octagonally symmetric protein structures with a channel—the nuclear pore—through the center. They control the transport of molecules between the nucleus and cytoplasm.

© Cengage Learning 2014

Labels in figure: Nucleus · Nuclear envelope · Chromatin · Nucleolus · Cytoplasm · Nuclear pore complex · Ribosomes on outer surface of nuclear envelope · Outer nuclear membrane (faces cytoplasm) · Space between nuclear membranes · Inner nuclear membrane (faces nucleoplasm) · Nuclear envelope · Nuclear pore complex · Nucleoplasm · Enlarged region showing lipid bilayer · 0.1 μm

Martin W. Goldberg, Durham University, UK

nucleus and cytoplasm and prevents the transport of material not meant to cross the nuclear membrane. A *nuclear pore*—a channel through the nuclear pore complex—is the path for the assisted exchange of large molecules such as proteins and RNA molecules with the cytoplasm, whereas small molecules simply pass through unassisted. A protein or RNA molecule (called the *cargo*) associates with a transport protein acting as a chaperone to shuttle the cargo through the pore.

Some proteins—for instance, the enzymes for replicating and repairing DNA—must be imported into the nucleus to carry out their functions. Proteins to be imported into the nucleus are distinguished from those that function in the cytosol by the presence of a special, short amino acid sequence called a **nuclear localization signal.** A specific protein in the cytosol recognizes and binds to the signal and moves the protein containing it to the nuclear pore complex where it is then transported through the pore into the nucleus. **Figure 5.11** shows how researchers discovered the nuclear localization signal.

The liquid or semiliquid substance within the nucleus is called the **nucleoplasm.** Most of the space inside the nucleus is filled with **chromatin,** a combination of DNA and proteins. By contrast with most prokaryotes, most of the hereditary information of a eukaryote is distributed among several to many linear DNA molecules in the nucleus. Each individual DNA molecule with its associated proteins is a **eukaryotic chromosome.** The terms *chromatin* and *chromosome* are similar but have distinct meanings. *Chromatin* refers to any collection of eukaryotic DNA molecules with their associated proteins. *Chromosome* refers to one complete DNA molecule with its associated proteins.

Eukaryotic nuclei contain much more DNA than do prokaryotic nucleoids. For example, the entire complement of 46 chromosomes in the nucleus of a human cell has a total DNA length of about 2 meters (m), compared with about 1,500 μm in prokaryotic cells with the most DNA. Some eukaryotic cells contain even more DNA; for example, a single frog or salamander nucleus, although of microscopic diameter, is packed with about 10 m of DNA.

A eukaryotic nucleus also contains one or more **nucleoli** (singular, *nucleolus*), which look like irregular masses of small fibers and granules (see Figure 5.9). These structures form around the genes coding for the rRNA molecules of ribosomes. Within the nucleolus, the information in rRNA genes is copied into rRNA molecules, which combine with proteins to form ribosomal subunits. The ribosomal subunits then leave the nucleoli and exit the nucleus through the nuclear pore complexes to enter the cytoplasm, where they join to form complete ribosomes on mRNAs.

The genes for most of the proteins that the organism can make are found within the chromatin, as are the genes for specialized RNA molecules such as rRNA molecules. Expression of these genes is carefully controlled as required for the function of each cell. The other proteins in the cell are specified by genes in the DNA of mitochondria and chloroplasts. Mitochondria and chloroplasts are discussed later in the chapter.

Eukaryotic Ribosomes Are Either Free in the Cytosol or Attached to Membranes

Like prokaryotic ribosomes, a eukaryotic ribosome consists of a large and a small subunit **(Figure 5.12)**. However, the structures of bacterial, archaeal, and eukaryotic ribosomes, al-

FIGURE 5.11 | **Experimental Research**

Discovery of the Nuclear Localization Signal

Question: How are proteins that are imported into the nucleus identified by the import machinery?

Experiment: Alan Smith and his colleagues at the National Institute for Medical Research, Mill Hill, London, studied a viral protein that normally is found in the nucleus after the virus infects a cell. They mutated the 708-amino-acid protein, changing one or more specific amino acids in the protein or deleting segments of the protein, and determined whether the alterations affected the location of the viral protein in rodent or monkey cells in culture.

Results: The researchers obtained the following results:

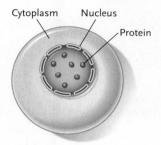

Normal protein: Localized to nucleus.

The amino acid at position 128 (thought to be important for the protein to bind to DNA) was mutated from lysine to threonine: Mutated protein localized in cytoplasm. The researchers interpreted this result to mean that the mutated amino acid was important for localizing the protein to the nucleus. Mutating other amino acids in the same region of the protein impaired import of the protein into the nucleus, but did not abolish it.

Deleting amino acids 1 to 126 (or any part of that region) or 136 to 708 (or any part of that region): Protein localized to nucleus, meaning that amino acids in those regions are not important for nuclear localization.

Deleting amino acids 127 to 133: Protein localized to cytoplasm, meaning that this amino acid sequence is necessary for nuclear localization of the viral protein. Other deletions involving parts of this region gave the same result.

Conclusion: By mutating the viral protein sequence, the researchers identified a seven-amino-acid segment of the protein, amino acids 127 to 133, that is necessary for localization of the protein to the nucleus. In follow-up experiments, they added this amino acid sequence to a cellular enzyme protein normally found only in the cytoplasm and determined that the modified protein localized to the nucleus. Therefore, the seven-amino-acid sequence is a nuclear localization signal. Continuing research has shown that this sequence is only the first example of similar sequences in other nuclear proteins. Thus, the identification of nuclear localization signals was a key step toward understanding the import of proteins into the nucleus.

THINK LIKE A SCIENTIST At the time of the discovery of the nuclear localization signal, the nature of the molecular system for the transport of nuclear proteins was not known. Imagine you are at that time and speculate on the molecular nature of the transport system.

Sources: D. Kalderon, W. D. Richardson, A. F. Markham, and A. E. Smith. 1984. Sequence requirements for nuclear location of simian virus 40 large-T antigen. *Nature* 311:33–38; D. Kalderon, B. L. Roberts, W. D. Richardson, and A. E. Smith. 1984. A short amino acid sequence able to specify nuclear location. *Cell* 39:499–509.

though similar, are not identical. In general, eukaryotic ribosomes are larger than either bacterial or archaeal ribosomes; they contain four types of rRNA molecules and more than 80 proteins. Their function is identical to that of prokaryotic ribosomes: They use the information in mRNA to assemble amino acids into proteins.

Some eukaryotic ribosomes are suspended freely in the cytosol; others are attached to membranes. Proteins made on free ribosomes in the cytosol may remain in the cytosol, pass through the nuclear pores into the nucleus, or become parts of mitochondria, chloroplasts, the cytoskeleton, or other cytoplasmic structures. Proteins that enter the nucleus become part of chromatin, line the nuclear envelope (the lamins), or remain in solution in the nucleoplasm.

Many ribosomes are attached to membranes. Some ribosomes are attached to the nuclear envelope, but most are attached to a network of membranes in the cytosol called the *endoplasmic reticulum* (ER) (described in more detail next). The proteins made on ribosomes attached to the ER follow a special path to other organelles within the cell.

An Endomembrane System Divides the Cytoplasm into Functional and Structural Compartments

Eukaryotic cells are characterized by an **endomembrane system** (*endo* = within), a collection of interrelated internal membranous sacs that divide the cell into functional and structural compartments. The membranes of the endomembrane system are lipid bilayers.

The endomembrane system has a number of functions, including the synthesis and modification of proteins and their transport into membranes and organelles or to the outside of the cell, the synthesis of lipids, and the detoxification of some toxins. The membranes of the system are connected either directly in the physical sense or indirectly by **vesicles,** which are small membrane-bound compartments that transfer sub-

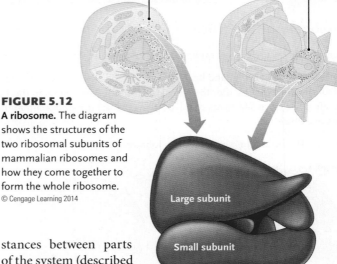

FIGURE 5.12
A ribosome. The diagram shows the structures of the two ribosomal subunits of mammalian ribosomes and how they come together to form the whole ribosome.
© Cengage Learning 2014

Ribosome

Ribosome

Large subunit

Small subunit

stances between parts of the system (described later in Figure 5.17).

The components of the endomembrane system include the nuclear envelope, endoplasmic reticulum, Golgi complex, lysosomes, vesicles, and plasma membrane. The plasma membrane and the nuclear envelope are discussed earlier in this chapter. The functions of the other organelles are described in the following sections.

ENDOPLASMIC RETICULUM The **endoplasmic reticulum** (ER) is an extensive interconnected network (*reticulum* = little net) of membranous channels and vesicles called **cisternae** (singular, *cisterna*). Each cisterna is formed by a single membrane that surrounds an enclosed space called the **ER lumen (Figure 5.13)**. The ER occurs in two forms:

1. The **rough ER** (see Figure 5.13A) gets its name from the ribosomes that stud its outer surface. The proteins made on ribosomes attached to the ER enter the ER lumen,

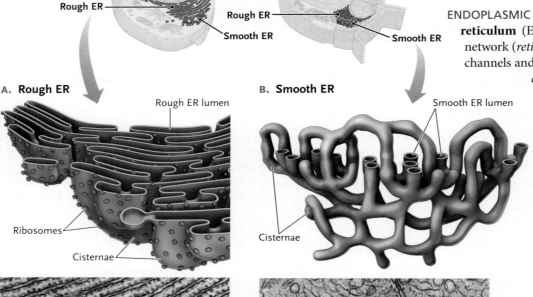

Rough ER
Rough ER
Smooth ER
Smooth ER

A. Rough ER

Rough ER lumen

Ribosomes

Cisternae

Vesicle budding from rough ER   Ribosome

B. Smooth ER

Smooth ER lumen

Cisternae

Smooth ER lumen

0.5 μm

Don W. Fawcett/Visuals Unlimited, Inc.

Don W. Fawcett/Visuals Unlimited, Inc.

FIGURE 5.13 **The endoplasmic reticulum.** **(A)** Rough ER, showing the ribosomes that stud the membrane surfaces facing the cytoplasm. Proteins synthesized on these ribosomes enter the lumen of the rough ER where they are modified chemically and then begin their path to their final destinations in the cell. **(B)** Smooth ER membranes. Among their functions are the synthesis of lipids for cell membranes, and enzymatic conversion of certain toxic molecules to safer molecules.
© Cengage Learning 2014

where they fold into their final form. Chemical modifications of these proteins, such as addition of carbohydrate groups to produce glycoproteins, occur in the lumen. The proteins are then delivered to other regions of the cell within small vesicles that pinch off from the ER, travel through the cytosol, and join with the organelle that performs the next steps in their modification and distribution. For most of the proteins made on the rough ER, the next destination is the Golgi complex, which packages and sorts them for delivery to their final destinations.

The outer membrane of the nuclear envelope is closely related in structure and function to the rough ER, to which it is connected. This membrane is also a "rough" membrane, studded with ribosomes attached to the surface facing the cytoplasm. The proteins made on these ribosomes enter the space between the two nuclear envelope membranes. From there, the proteins can move into the ER and on to other cellular locations.

2. The **smooth ER** (see Figure 5.13B) is so called because its membranes have no ribosomes attached to their surfaces. The smooth ER has various functions in the cytoplasm, including synthesis of lipids that become part of cell membranes. In some cells, such as those of the liver, smooth ER membranes contain enzymes that convert drugs, poisons, and toxic by-products of cellular metabolism into substances that can be tolerated or more easily removed from the body.

The rough and smooth ER membranes are often connected, making the entire ER system a continuous network of interconnected channels in the cytoplasm. The relative proportions of rough and smooth ER reflect cellular activities in protein and lipid synthesis. Cells that are highly active in making proteins to be released outside the cell, such as pancreatic cells that make digestive enzymes, are packed with rough ER but have relatively little smooth ER. By contrast, cells that primarily synthesize lipids or break down toxic substances are packed with smooth ER but contain little rough ER.

GOLGI COMPLEX Camillo Golgi, a late-nineteenth-century Italian neuroscientist and Nobel laureate, discovered the **Golgi complex.** The Golgi complex consists of a stack of flattened, membranous sacs (without attached ribosomes) known as cisternae **(Figure 5.14)**. In most cells, the complex looks like a stack of cupped pancakes, and like pancakes, they are separate sacs, not interconnected as the ER cisternae are. Typically there are between three and eight cisternae. The number and size of Golgi complexes can vary with cell type and the metabolic activity of the cell. Some cells have a single complex, whereas cells highly active in secreting proteins from the cell can have hundreds of complexes. Golgi complexes are usually located near concentrations of rough ER membranes, between the ER and the plasma membrane.

The Golgi complex receives proteins that were made in the ER and transported to the complex in vesicles. When the vesicles contact the *cis* face of the complex (which faces the nucleus), they fuse with the Golgi membrane and release their

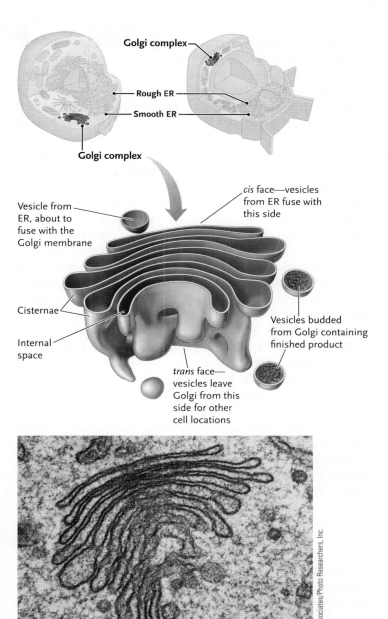

FIGURE 5.14 The Golgi complex.
© Cengage Learning 2014

contents directly into the cisternal (see Figure 5.14). Within the Golgi complex, the proteins are chemically modified, for example, by removing segments of the amino acid chain, adding small functional groups, or adding lipid or carbohydrate units. The modified proteins are transported within the Golgi to the *trans* face of the complex (which faces the plasma membrane), where they are sorted into vesicles that bud off from the margins of the Golgi (see Figure 5.14). The content of a vesicle is kept separate from the cytosol by the vesicle membrane. Researchers have proposed three quite different models for how proteins move through the Golgi complex. The mechanism is a subject of active current research.

The Golgi complex regulates the movement of several types of proteins. Some are secreted from the cell, others become embedded in the plasma membrane as integral membrane proteins, and yet others are placed in lysosomes. The modifications of the proteins within the Golgi complex include adding "zip codes" to the proteins, which tags them for sorting to their final destinations. For instance, proteins secreted from the cell are transported to the plasma membrane in **secretory vesicles,** which release their contents to the exterior by **exocytosis (Figure 5.15A).** In this process, a secretory vesicle fuses with the plasma membrane and spills the vesicle contents to the outside. The contents of secretory vesicles vary, including signaling molecules such as peptide hormones and neurotransmitters (see Chapter 7), waste products or toxic substances, and enzymes (such as from cells lining the intestine). The membrane of a vesicle that fuses with the plasma membrane becomes part of the plasma membrane. In fact, this process is used to expand the surface of the cell during cell growth.

Vesicles also may form by the reverse process, called **endocytosis,** which brings molecules into the cell from the exterior **(Figure 5.15B).** In this process, the plasma membrane forms a pocket, which bulges inward and pinches off into the cytoplasm as an **endocytic vesicle.** Once in the cytoplasm, endocytic vesicles, which contain segments of the plasma membrane as well as proteins and other molecules, are carried to the Golgi complex or to other destinations such as lysosomes in animal cells. The substances carried to the Golgi complex are sorted and placed into vesicles for routing to other locations, which may include lysosomes. Those routed to lysosomes are digested into molecular subunits that may be recycled as building blocks for the biological molecules of the cell. Exocytosis and endocytosis are discussed in more detail in Chapter 6.

LYSOSOMES **Lysosomes** (*lys* = breakdown; *some* = body) are small, membrane-bound vesicles that contain more than 30 hydrolytic enzymes for the digestion of many complex molecules, including proteins, lipids, nucleic acids, and polysaccharides **(Figure 5.16).** The cell recycles the subunits of these molecules. Lysosomes are found in animals, but not in plants. The functions of lysosomes in plants are carried out by the central vacuole (see Section 5.4). Depending on the contents they are digesting, lysosomes assume a variety of sizes and shapes instead of a uniform structure as is characteristic of other organelles. Most commonly, lysosomes are small (0.1–0.5 μm in diameter) oval or spherical bodies. A human cell contains about 300 lysosomes.

Lysosomes are formed by budding from the Golgi complex. Their hydrolytic enzymes are synthesized in the rough ER, modified in the lumen of the ER to identify them as being bound for a lysosome, transported to the Golgi complex in a vesicle, and then packaged in the budding lysosome.

The pH within lysosomes is acidic (pH ~5) and is significantly lower than the pH of the cytosol (pH ~7.2). The hydrolytic enzymes in the lysosomes function optimally at the acidic pH within the organelle, but they do not function well at the

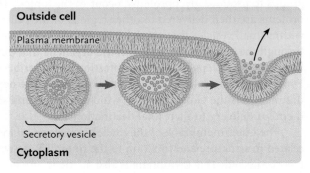

A. Exocytosis: A secretory vesicle fuses with the plasma membrane, releasing the vesicle contents to the cell exterior. The vesicle membrane becomes part of the plasma membrane.

Outside cell

Plasma membrane

Secretory vesicle

Cytoplasm

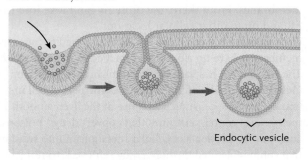

B. Endocytosis: Materials from the cell exterior are enclosed in a segment of the plasma membrane that pockets inward and pinches off as an endocytic vesicle.

Endocytic vesicle

FIGURE 5.15 Exocytosis and endocytosis.
© Cengage Learning 2014

pH of the cytosol; this difference reduces the risk to the viability of the cell should the enzymes be released from the vesicle.

Lysosomal enzymes can digest several types of materials. They digest food molecules entering the cell by endocytosis when an endocytic vesicle fuses with a lysosome. In a process called *autophagy,* they digest organelles that are not functioning correctly. A membrane surrounds the defective organelle, forming a large vesicle that fuses with one or more lysosomes; the organelle then is degraded by the hydrolytic enzymes. They also play a role in **phagocytosis,** a process in which some types of cells engulf bacteria or other cellular debris to break them down.

Lysosome

Lysosome containing ingested material

FIGURE 5.16
A lysosome.
© Cengage Learning 2014

Don W. Fawcett/Photo Researchers, Inc.

The ER and Golgi complex are part of the endomembrane system, which releases proteins and other substances to the cell exterior and gathers materials from outside the cell.

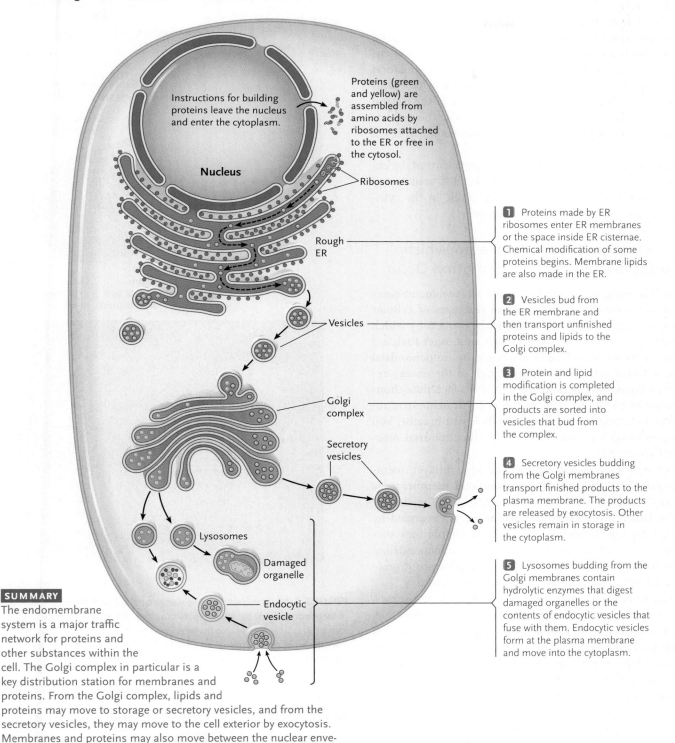

Instructions for building proteins leave the nucleus and enter the cytoplasm.

Proteins (green and yellow) are assembled from amino acids by ribosomes attached to the ER or free in the cytosol.

Nucleus

Ribosomes

Rough ER

Vesicles

Golgi complex

Secretory vesicles

Lysosomes

Damaged organelle

Endocytic vesicle

1 Proteins made by ER ribosomes enter ER membranes or the space inside ER cisternae. Chemical modification of some proteins begins. Membrane lipids are also made in the ER.

2 Vesicles bud from the ER membrane and then transport unfinished proteins and lipids to the Golgi complex.

3 Protein and lipid modification is completed in the Golgi complex, and products are sorted into vesicles that bud from the complex.

4 Secretory vesicles budding from the Golgi membranes transport finished products to the plasma membrane. The products are released by exocytosis. Other vesicles remain in storage in the cytoplasm.

5 Lysosomes budding from the Golgi membranes contain hydrolytic enzymes that digest damaged organelles or the contents of endocytic vesicles that fuse with them. Endocytic vesicles form at the plasma membrane and move into the cytoplasm.

SUMMARY

The endomembrane system is a major traffic network for proteins and other substances within the cell. The Golgi complex in particular is a key distribution station for membranes and proteins. From the Golgi complex, lipids and proteins may move to storage or secretory vesicles, and from the secretory vesicles, they may move to the cell exterior by exocytosis. Membranes and proteins may also move between the nuclear envelope and the endomembrane system. Proteins and other materials that enter cells by endocytosis also enter the endomembrane system to travel to the Golgi complex for sorting and distribution to other locations. Lysosomes, originating as vesicles from the Golgi complex, digest damaged organelles or the contents of endocytic vesicles.

© Cengage Learning 2014

THINK LIKE A SCIENTIST Consider an animal cell that is secreting a particular protein. You treat cells with a chemical and find that the protein no longer is secreted. What possible steps could be affected by the chemical?

These cells include the white blood cells known as *phagocytes,* which play an important role in the immune system (see Chapter 45). Phagocytosis produces a large vesicle that contains the engulfed materials until lysosomes fuse with the vesicle and release the hydrolytic enzymes necessary for degrading them.

In certain human genetic diseases known as *lysosomal storage diseases,* one of the hydrolytic enzymes normally found in the lysosome is absent. As a result, the substrate of that enzyme accumulates in the lysosomes, and this accumulation eventually interferes with normal cellular activities. An example is Tay-Sachs disease, which is a fatal disease of the central nervous system caused by the failure to synthesize the enzyme needed for hydrolysis of fatty acid derivatives found in brain and nerve cells.

We have completed our survey of the components of the endomembrane system, including their interrelated structures and functions. **Figure 5.17** illustrates the vesicle traffic in the cytoplasm that involves the endomembrane system.

Mitochondria Are the Organelles in Which Some Reactions of Cellular Respiration Occur

Mitochondria (singular, *mitochondrion*) are the membrane-bound organelles in which some of the reactions of cellular respiration occurs. *Cellular respiration* is the process by which energy-rich molecules such as sugars, fats, and other fuels are broken down to water and carbon dioxide by mitochondrial reactions, with the release of energy. Much of the energy released by the breakdown is captured in ATP. In fact, mitochondria generate most of the ATP of the cell. Mitochondria require oxygen for cellular respiration—when you breathe, you are taking in oxygen primarily for your mitochondrial reactions (see Chapter 8).

Mitochondria are enclosed by two lipid bilayer membranes **(Figure 5.18).** The **outer mitochondrial membrane** is smooth and covers the outside of the organelle. The surface area of the **inner mitochondrial membrane** is expanded by folds called **cristae** (singular, *crista*). Both membranes surround the innermost compartment of the mitochondrion, called the **mitochondrial matrix.** The ATP-generating reactions of mitochondria occur in the cristae and matrix.

The mitochondrial matrix also contains DNA, ribosomes, and other molecular machinery. Proteins encoded by the DNA are components of the enzyme machinery for the reactions of cellular respiration carried out by mitochondria. Other protein components of the enzyme machinery are encoded by nuclear genes and are imported into the organelle.

Mitochondrial DNA, ribosomes, and other molecular machinery resemble the equivalent structures in bacteria. These and other similarities led to the *endosymbiotic theory,* which proposes that mitochondria may have originated from a mutually advantageous relationship between an ingested prokaryotic cell and the prokaryotic cell that ingested it. The ingested prokaryotic cell evolved over time to become mitochondria. Chapter 25 discusses the endosymbiotic theory in more detail.

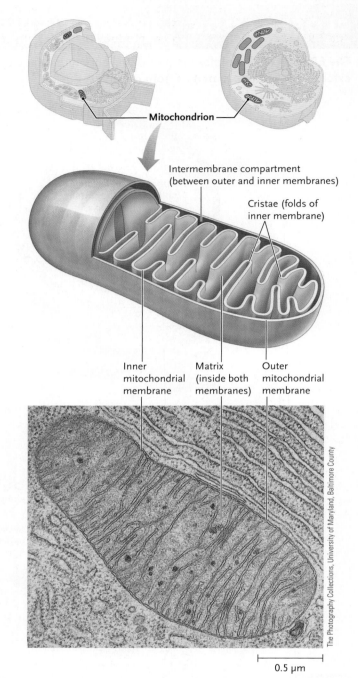

FIGURE 5.18 Mitochondria. The electron micrograph shows a mitochondrion from bat pancreas, surrounded by cytoplasm containing rough ER. Cristae extend into the interior of the mitochondrion as folds from the inner mitochondrial membrane. The darkly stained granules inside the mitochondrion are probably lipid deposits.
© Cengage Learning 2014

Microbodies Carry Out Vital Reactions That Link Metabolic Pathways

Microbodies are small, relatively simple membrane-bound organelles found in various forms in essentially all eukaryotic cells. They consist of a single boundary lipid bilayer membrane that encloses a collection of enzymes and other proteins **(Figure 5.19)**. Recent research has shown that the ER is involved in microbody production. Proteins and phospholipids are continuously imported into microbodies. The phospholipids are used for new membrane synthesis, leading to growth of the microbody. Division of a microbody then produces new microbodies.

Microbodies have various functions that are often specific to an organism or cell type. Commonly, they contain enzymes that conduct preparatory or intermediate reactions linking major biochemical pathways. For example, the series of reactions that allows cells to use fats as an energy source begins in microbodies and continues in mitochondria. Beginning or intermediate steps in the breakdown of some amino acids and alcohols also take place in microbodies, including about half of the ethyl alcohol that humans consume. Many types of microbodies produce as a by-product the toxic substance hydrogen peroxide (H_2O_2), which is broken down into water and oxygen by the enzyme *catalase*. Microbodies with this reaction are often termed **peroxisomes.**

Microbodies in plants convert oils or fats to sugars that can be used directly for energy-releasing reactions in mitochondria or for reactions that require sugars as chemical building blocks. These microbody reactions are particularly important in plant embryos that develop from oily seeds, such as those of the peanut or soybean. Depending on the particular reaction pathways they carry out, plant microbodies are called peroxisomes, *glyoxysomes,* or *glycosomes.*

The Cytoskeleton Supports and Moves Cell Structures

The characteristic shape and internal organization of each type of cell is maintained in part by its cytoskeleton, the interconnected system of protein tubes and fibers that extends throughout the cytoplasm. The cytoskeleton also reinforces the plasma membrane and functions in movement, both of structures within the cell and of the cell as a whole. It is most highly developed in animal cells, in which it fills and supports the cytoplasm from the plasma membrane to the nuclear envelope. Although cytoskeletal structures are also present in plant cells, the fibers and tubes of the system are less prominent; much of cellular support in plants is provided by the cell wall and a large central vacuole (described in Section 5.4).

The cytoskeleton of animal and plant cells contains structural elements of three major types: *microtubules, intermediate filaments,* and *microfilaments* **(Figure 5.20)**. Microtubules are the largest cytoskeletal elements, and microfilaments are the smallest. Each cytoskeletal element is assembled from proteins—microtubules from *tubulins,* intermediate filaments from a large and varied group of *intermediate filament proteins,* and microfilaments from *actins.*

Microtubules (Figure 5.20A) are tubes with an outer diameter of about 25 nm and an inner diameter of about 15 nm.

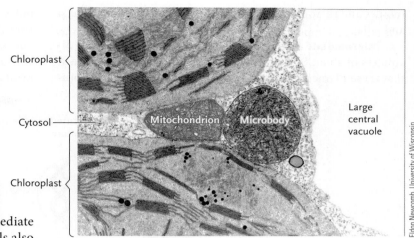

FIGURE 5.19 A microbody in the cytoplasm of a tobacco leaf cell. The EM has been colorized to make the structures easier to identify.

They vary widely in length from less than 200 nm to several micrometers. The wall of the microtubule consists of 13 protein filaments arranged side by side. Each filament is a linear polymer of tubulin dimers organized head-to-tail in each filament, each dimer consisting of one α-tubulin and one β-tubulin subunit. The end of a microtubule filament with α-tubulin subunits is the + (plus) end; the other end with the β-tubulin subunits is the − (minus) end, at the ends of the filaments. Microtubules change their lengths as required by their functions. This is seen readily in animal cells that are changing shape. Microtubules change length by the addition or removal of tubulin dimers; this occurs asymmetrically, with dimers adding or detaching more rapidly at the + end than at the − end. The lengths of microtubules are tightly regulated in the cell.

Many of the cytoskeletal microtubules in animal cells are formed and radiate outward from a site near the nucleus termed the **cell center** or **centrosome** (see Figure 5.9). At its midpoint are two short, barrel-shaped structures also formed from microtubules called the **centrioles** (see Figure 5.24). Often, intermediate filaments also extend from the cell center, apparently held in the same radiating pattern by linkage to microtubules.

Microtubules have various functions, as listed in Figure 5.20A. One of those functions is moving animal cells. Animal cell movements are generated by "motor" proteins that push or pull against microtubules or microfilaments, much as our muscles produce body movements by acting on bones of the skeleton. One end of a motor protein is firmly fixed to a cell structure such as a vesicle or to a microtubule or microfilament. Using energy from ATP hydrolysis, the other end "walks" along another microtubule or microfilament by making an attachment, forcefully swiveling a short distance, and then releasing **(Figure 5.21)**. The motor proteins that walk along microtubules are called *dyneins* and *kinesins,* and ones that walk along microfilaments are called *myosins.* Some cell movements, such as the whipping motions of sperm tails, depend entirely on microtubules and their motor proteins (see later discussion of flagella and cilia).

Intermediate filaments (Figure 5.20B) are fibers with diameters of about 8 to 12 nm. ("Intermediate" signifies, in fact, that these filaments are intermediate in size between microtubules and microfilaments.) Found only in multicellular organisms, intermediate filaments occur singly, in parallel bundles, and in interlinked networks, either alone or in combination with microtubules, microfilaments, or both. Intermediate filaments consist of proteins that are tissue-specific in their com-

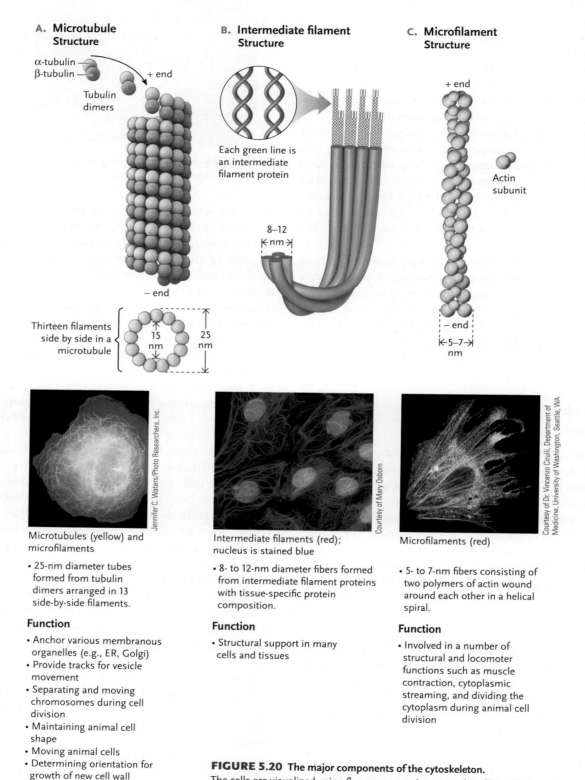

A. Microtubule Structure

α-tubulin
β-tubulin
Tubulin dimers
+ end
− end

Thirteen filaments side by side in a microtubule
15 nm
25 nm

Microtubules (yellow) and microfilaments

• 25-nm diameter tubes formed from tubulin dimers arranged in 13 side-by-side filaments.

Function

• Anchor various membranous organelles (e.g., ER, Golgi)
• Provide tracks for vesicle movement
• Separating and moving chromosomes during cell division
• Maintaining animal cell shape
• Moving animal cells
• Determining orientation for growth of new cell wall during plant cell division

B. Intermediate filament Structure

Each green line is an intermediate filament protein
8–12 nm

Intermediate filaments (red); nucleus is stained blue

• 8- to 12-nm diameter fibers formed from intermediate filament proteins with tissue-specific protein composition.

Function

• Structural support in many cells and tissues

C. Microfilament Structure

+ end
Actin subunit
− end
5–7 nm

Microfilaments (red)

• 5- to 7-nm fibers consisting of two polymers of actin wound around each other in a helical spiral.

Function

• Involved in a number of structural and locomoter functions such as muscle contraction, cytoplasmic streaming, and dividing the cytoplasm during animal cell division

Jennifer C. Waters/Photo Researchers, Inc.

Courtesy of Mary Osborn

Courtesy of Dr. Vincenzo Cirulli, Department of Medicine, University of Washington, Seattle, WA

FIGURE 5.20 The major components of the cytoskeleton.
The cells are visualized using fluorescence microscopy (see Figure 5.4).
© Cengage Learning 2014

FIGURE 5.21

The microtubule motor protein kinesin.
(A) Structure of the end of a kinesin molecule that "walks" along a microtubule, with α-helical segments shown as spirals and β strands as flat ribbons. **(B)** How a kinesin molecule walks along the surface of a molecule by alternately attaching and releasing its "feet."
© Cengage Learning 2014

A. "Walking" end of a kinesin molecule

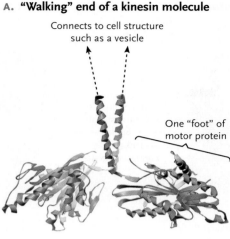

Connects to cell structure such as a vesicle

One "foot" of motor protein

B. How a kinesin molecule "walks"

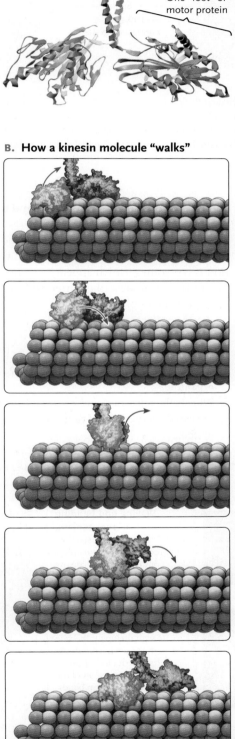

position. Despite the molecular diversity of intermediate filaments, however, they all play similar roles in the cell, providing structural support in many cells and tissues. For example, the nucleus in epithelial cells is held within the cell by a basketlike network of intermediate filaments made of keratins. Keratins are also found in animal hair, nails, and claws.

Microfilaments (Figure 5.20C) are thin protein fibers 5 to 7 nm in diameter that consist of two polymers of actin subunits wound around each other in a long helical spiral. The actin subunits are asymmetric in shape, and they are all oriented in the same way in the polymer chains of a microfilament. Thus, as for microtubules, the two ends are designated + (plus) and − (minus). And as for microtubules, growth and disassembly occur more rapidly at the + end than at the − end.

Microfilaments occur in almost all eukaryotic cells and are involved in a number of structural and locomotor processes (see Figure 5.20C). They are best known as one of the two components of the contractile elements in muscle fibers of vertebrates (the roles of myosin and microfilaments in muscle contraction are discussed in Chapter 43). Microfilaments are involved in the actively flowing motion of cytoplasm called *cytoplasmic streaming,* which can transport nutrients, proteins, and organelles in both animal and plant cells, and which is responsible for amoeboid movement. When animal cells divide, microfilaments are responsible for dividing the cytoplasm (see Chapter 10 for further discussion).

Flagella Propel Cells, and Cilia Move Materials over the Cell Surface

Flagella and *cilia* (singular, *cilium*) are elongated, slender, motile structures that extend from the cell surface. They are identical in structure except that cilia are usually shorter than flagella and occur on cells in greater numbers. Whiplike or oarlike movements of a flagellum propel a cell through a watery medium, and cilia move fluids over the cell surface.

A bundle of microtubules extends from the base to the tip of a flagellum or cilium **(Figure 5.22).** In the bundle, a circle of nine double microtubules surrounds a central pair of single microtubules, forming what is known as the 9 + 2 complex. Dynein motor proteins slide the microtubules of the 9 + 2 complex over each other to produce the movements of a flagellum or cilium **(Figure 5.23).**

Flagella and cilia arise from the centrioles. These barrel-shaped structures contain a bundle of microtubules similar to the 9 + 2 complex, except that the central pair of microtubules is missing and the outer circle is formed from a ring of nine triple rather than double microtubules (compare Figure 5.23 and **Figure 5.24**). During the formation of a flagellum or cilium, a centriole moves to a position just under the plasma membrane. Then two of the three microtubules of each triplet grow outward from one end of the centriole to form the ring of nine double microtubules. The two central microtubules of the 9 + 2 complex also grow from the end of the centriole, but without direct connection to any centriole microtubules. The centriole remains at the innermost end of a flagellum or cilium when its develop-

A. Eukaryotic flagellum　　　**B. Cross section of flagellum**　　　**C. Micrograph of flagellum**

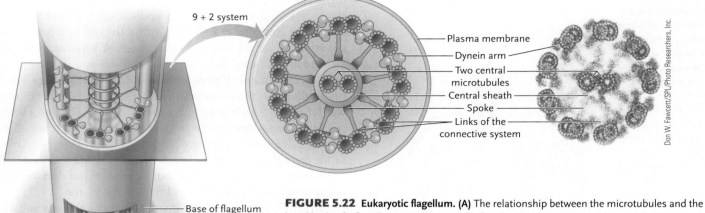

9 + 2 system

Plasma membrane
Dynein arm
Two central microtubules
Central sheath
Spoke
Links of the connective system

Don W. Fawcett/SPL/Photo Researchers, Inc.

Base of flagellum or cilium

Plasma membrane (cell surface)

Basal body or centriole

FIGURE 5.22 Eukaryotic flagellum. **(A)** The relationship between the microtubules and the basal body of a flagellum. **(B)** Diagram of a flagellum in cross section, showing the 9 + 2 system of microtubules. The spokes and connecting links hold the system together. **(C)** Electron micrograph of a flagellum in cross section; individual tubulin molecules are visible in the microtubule walls.
© Cengage Learning 2014

ment is complete as the **basal body** of the structure (see Figure 5.22).

Cilia and flagella are found in protozoa and algae, and many types of animal cells have flagella—the tail of a sperm cell is a flagellum—as do the reproductive cells of some plants. In humans, cilia cover the surfaces of cells lining cavities or tubes in some parts of the body. For example, cilia on cells lining the ventricles (cavities) of the brain circulate fluid through the brain, and cilia in the oviducts conduct eggs from the ovaries to the uterus. Cilia covering cells that line the air passages of the lungs sweep out mucus containing bacteria, dust particles, and other contaminants.

Although the purpose of the eukaryotic flagellum is the same as that of prokaryotic flagella, the genes that encode the components of the flagellar apparatus of cells of Bacteria, Archaea, and Eukarya are different in each case. Thus, as mentioned earlier in the chapter, the three types of flagella are anal-

ogous, not homologous, structures, and they must have evolved independently.

With a few exceptions, the cell structures described so far in this chapter occur in all eukaryotic cells. The major exception is lysosomes, which appear to be restricted to animal cells. The next section describes three additional structures that are characteristic of plant cells.

STUDY BREAK 5.3

1. Where in a eukaryotic cell is DNA found? How is that DNA organized?
2. What is the nucleolus, and what is its function?
3. Explain the structure and function of the endomembrane system.
4. What is the structure and function of a mitochondrion?
5. What is the structure and function of the cytoskeleton?

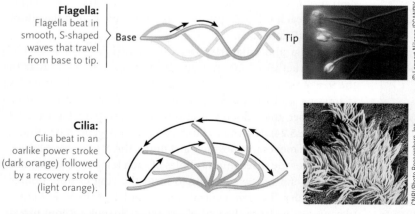

Flagella:
Flagella beat in smooth, S-shaped waves that travel from base to tip.

Base　　Tip

© Lennart Nilsson/SCANPIX

Cilia:
Cilia beat in an oarlike power stroke (dark orange) followed by a recovery stroke (light orange).

CNRI/Photo Researchers, Inc.

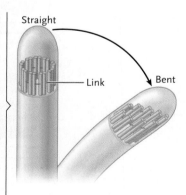

Waving and bending mechanism:
The waves and bends are produced by dynein motor proteins, which slide the microtubule doublets over each other. An examination of the tip of a bent cilium or flagellum shows that the doublets extend farther toward the tip on the side toward the bend, confirming that the doublets actually slide as the shaft of the cilium or flagellum bends.

Straight
Link
Bent

FIGURE 5.23 Flagellar and ciliary beating patterns. The micrographs show a few human sperm, each with a flagellum (top), and cilia from the lining of an airway in the lungs (bottom).
© Cengage Learning 2014

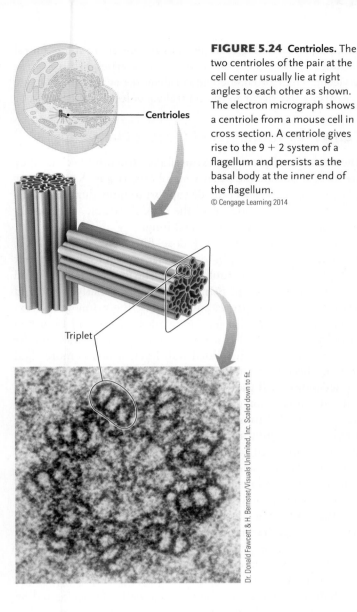

FIGURE 5.24 Centrioles. The two centrioles of the pair at the cell center usually lie at right angles to each other as shown. The electron micrograph shows a centriole from a mouse cell in cross section. A centriole gives rise to the 9 + 2 system of a flagellum and persists as the basal body at the inner end of the flagellum.
© Cengage Learning 2014

Centrioles

Triplet

Dr. Donald Fawcett & H. Bernstet/Visuals Unlimited, Inc. Scaled down to fit.

THINK OUTSIDE THE BOOK >

On your own or collaboratively, explore the Internet and the research literature to develop an outline of the molecular steps that a protein with a nuclear localization signal follows for nuclear import (that is, being transported through a nuclear pore complex).

5.4 Specialized Structures of Plant Cells

Chloroplasts, a large and highly specialized central vacuole, and cell walls give plant cells their distinctive characteristics, but these structures also occur in some other eukaryotes—chloroplasts in algal protists and cell walls in algal protists and fungi.

Chloroplasts Are Biochemical Factories Powered by Sunlight

Chloroplasts (*chloro* = yellow–green), the sites of photosynthesis in plant cells, are members of a family of plant organelles known collectively as **plastids.** Other members of the family include amyloplasts and chromoplasts. **Amyloplasts** (*amylo* = starch) are colorless plastids that store starch, a product of photosynthesis. They occur in great numbers in the roots or tubers of some plants, such as the potato. **Chromoplasts** (*chromo* = color) contain red and yellow pigments and are responsible for the colors of ripening fruits or autumn leaves.

All plastids contain DNA genomes and molecular machinery for gene expression and the synthesis of proteins on ribosomes. For example, chloroplast-encoded proteins are components of the enzyme machinery for photosynthesis and for some of the proteins of chloroplast ribosomes. The other protein components of the enzyme machinery and of the ribosomes are encoded by nuclear genes and are imported into the organelle.

Chloroplasts, like mitochondria, are usually lens- or disc-shaped and are surrounded by a smooth **outer boundary membrane** and an **inner boundary membrane,** which lies just inside the outer membrane **(Figure 5.25).** These two boundary membranes completely enclose an inner compartment, the stroma. Within the stroma is a third membrane system that consists of flattened, closed sacs called thylakoids. In higher plants, the thylakoids are stacked, one on top of another, forming structures called **grana** (singular, *granum*).

The thylakoid membranes contain molecules that absorb light energy and convert it to chemical energy in photosynthesis. The primary molecule absorbing light is *chlorophyll,* a green pigment that is present in all chloroplasts. The chemical energy is used by enzymes in the stroma to make carbohydrates and other complex organic molecules from water, carbon dioxide, and other simple inorganic precursors. The organic molecules produced in chloroplasts, or from biochemical building blocks made in chloroplasts, are the ultimate food source for most organisms. (The physical and biochemical reactions of chloroplasts are described in Chapter 9.)

The chloroplast stroma contains DNA and ribosomes that resemble those of certain photosynthetic bacteria. Because of these similarities, chloroplasts, like mitochondria, are believed to have originated from ancient prokaryotes that became permanent residents of the eukaryotic cells ancestral to the plant lineage (see Chapter 25 for further discussion).

Central Vacuoles Have Diverse Roles in Storage, Structural Support, and Cell Growth

Central vacuoles (see Figure 5.9B; also see Chapter 34) are large vesicles identified as distinct organelles of plant cells because they perform specialized functions unique to plants. In a mature plant cell, 90% or more of the cell's volume may be occupied by

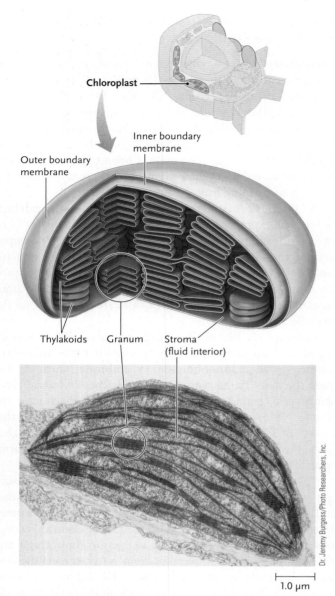

FIGURE 5.25 Chloroplast structure. The electron micrograph shows a maize (corn) chloroplast.
© Cengage Learning 2014

Labels on figure:
Chloroplast
Inner boundary membrane
Outer boundary membrane
Thylakoids   Granum   Stroma (fluid interior)
Dr. Jeremy Burgess/Photo Researchers, Inc.
1.0 μm

one or more large central vacuoles. The remainder of the cytoplasm and the nucleus of these cells are restricted to a narrow zone between the central vacuole and the plasma membrane. The pressure within the central vacuole supports the cells (see Chapter 6).

The membrane that surrounds the central vacuole, the **tonoplast,** contains transport proteins that move substances into and out of the central vacuole. As plant cells mature, they grow primarily by increases in the pressure and volume of the central vacuole.

Central vacuoles conduct other vital functions. They store salts, organic acids, sugars, storage proteins, pigments, and, in some cells, waste products. Pigments concentrated in the vacuoles produce the colors of many flowers. Enzymes capable of breaking down biological molecules are present in some central vacuoles, giving them some of the properties of lysosomes. Molecules that provide chemical defenses against pathogenic organisms also occur in the central vacuoles of some plants.

Cell Walls Support and Protect Plant Cells

The cell walls of plants are extracellular structures because they are located outside the plasma membrane (**Figure 5.26;** also see Chapter 33). Cell walls provide support to individual cells, contain the pressure produced in the central vacuole, and protect cells against invading bacteria and fungi.

Cell walls consist of cellulose fibers (see Figure 3.9C), which give tensile strength to the walls, embedded in a network of highly branched carbohydrates. The initial cell wall laid down by a plant cell, the **primary cell wall,** is relatively soft and flexible. As the cell grows and matures, the primary wall expands and additional layers of cellulose fibers and branched carbohydrates are laid down between the primary wall and the plasma membrane. The added wall layer, which is more rigid and may become many times thicker than the primary wall, is the **secondary cell wall.** In woody plants and trees, secondary cell walls are reinforced by *lignin,* a hard, highly resistant substance assembled from complex alcohols, surrounding the cellulose fibers. Lignin-impregnated cell walls are actually stronger than reinforced concrete by weight; hence, trees can grow to substantial size, and the wood of trees is used extensively in human cultures to make many structures and objects, including houses, tables, and chairs.

The walls of adjacent cells are held together by a layer of gel-like polysaccharides called the **middle lamella,** which acts as an intercellular glue (see Figure 5.26). The polysaccharide material of the middle lamella, called *pectin,* is extracted from some plants and used to thicken jams and jellies.

Both primary and secondary cell walls are perforated by minute channels, the **plasmodesmata** (singular, *plasmodesma;* see Figure 5.26 and Figure 5.9B). A typical plant cell has between 1,000 and 100,000 plasmodesmata connecting it to abutting cells. These cytosol-filled channels are lined by plasma membranes, so that connected cells essentially all have one continuous surface membrane. Most plasmodesmata also contain a narrow tubelike structure derived from the smooth endoplasmic reticulum of the connected cells. Plasmodesmata allow ions and small molecules to move directly from one cell to another through the connecting cytosol, without having to penetrate the plasma membranes or cell walls. Proteins and nucleic acids move through some plasmodesmata using energy-dependent processes.

Cell walls also surround the cells of fungi and algal protists. Carbohydrate molecules form the major framework of cell walls in most of these organisms, as they do in plants. In some, the wall fibers contain chitin (see Figure 3.9D) instead of cellulose. Details of cell wall structure in the algal protists and fungi, as well as in different subgroups of the plants, are presented in later chapters devoted to these organisms.

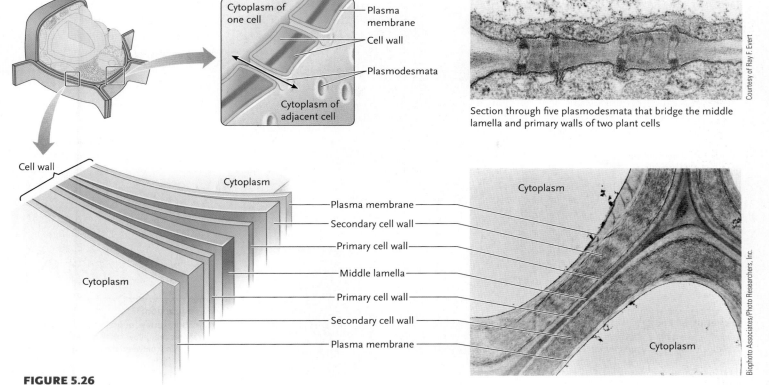

Section through five plasmodesmata that bridge the middle lamella and primary walls of two plant cells

Courtesy of Ray F. Evert

Biophoto Associates/Photo Researchers, Inc.

Labels (upper diagram):
Cytoplasm of one cell
Plasma membrane
Cell wall
Plasmodesmata
Cytoplasm of adjacent cell

Labels (lower diagram):
Cell wall
Cytoplasm
Plasma membrane
Secondary cell wall
Primary cell wall
Middle lamella
Primary cell wall
Secondary cell wall
Plasma membrane
Cytoplasm
Cytoplasm

FIGURE 5.26

Cell wall structure in plants.

The upper right diagram and electron micrograph show plasmodesmata, which form openings in the cell wall that directly connect the cytoplasm of adjacent cells. The lower diagram and electron micrograph show the successive layers in the cell wall between two plant cells that have laid down secondary wall material.

© Cengage Learning 2014

As noted earlier, animal cells do not form rigid, external, layered structures equivalent to the walls of plant cells. However, most animal cells secrete extracellular material and have other structures at the cell surface that play vital roles in the support and organization of animal body structures. The next section describes these and other surface structures of animal cells.

STUDY BREAK 5.4

1. What is the structure and function of a chloroplast?
2. What is the function of the central vacuole in plants?

5.5 The Animal Cell Surface

Animal cells have specialized structures that help hold cells together, produce avenues of communication between cells, and organize body structures. Molecular systems that perform these functions are organized at three levels: individual **cell adhesion molecules** bind cells together, more complex **cell junctions** seal the spaces between cells and provide direct communication between cells, and the **extracellular matrix (ECM)** supports and protects cells and provides mechanical linkages, such as those between muscles and bone.

Cell Adhesion Molecules Organize Animal Cells into Tissues and Organs

Cell adhesion molecules are glycoproteins embedded in the plasma membrane. They help maintain body form and structure in animals ranging from sponges to the most complex invertebrates and vertebrates. Cell adhesion molecules bind to specific molecules on other cells. Most cells in solid body tissues are held together by many different cell adhesion molecules.

Connections between cells formed by cell adhesion molecules only become permanent as an embryo develops into an adult. Cancer cells typically lose these adhesions, allowing them to break loose from their original locations, migrate to new locations, and form additional tumors.

Some bacteria and viruses—such as the virus that causes the common cold—target cell adhesion molecules as attachment sites during infection. Cell adhesion molecules are also partly responsible for the ability of cells to recognize one another as being part of the same individual or foreign. For example, rejection of organ transplants in mammals results from an immune response triggered by the foreign cell-surface molecules.

Cell Junctions Reinforce Cell Adhesions and Provide Avenues of Communication

Cell junctions reinforce cell adhesions in cells of adult animals. Three types of cell junctions are common in animal tissues **(Figure 5.27)**. **Anchoring junctions** form buttonlike spots, or belts,

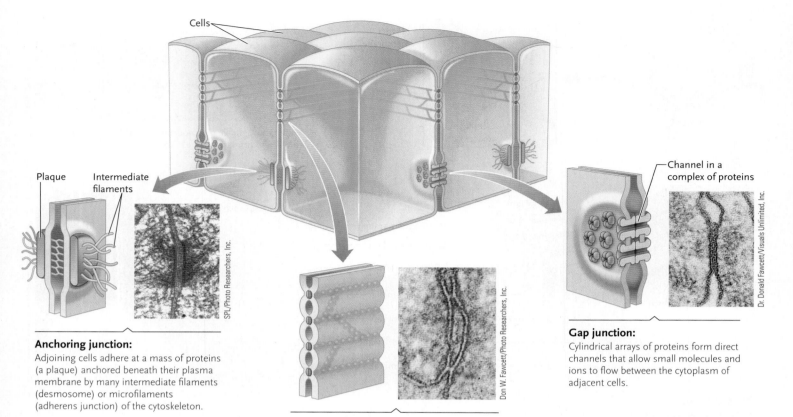

Anchoring junction:
Adjoining cells adhere at a mass of proteins (a plaque) anchored beneath their plasma membrane by many intermediate filaments (desmosome) or microfilaments (adherens junction) of the cytoskeleton.

Plaque   Intermediate filaments

SPL/Photo Researchers, Inc.

Cells

Channel in a complex of proteins

Dr. Donald Fawcett/Visuals Unlimited, Inc.

Gap junction:
Cylindrical arrays of proteins form direct channels that allow small molecules and ions to flow between the cytoplasm of adjacent cells.

Tight junction:
Tight connections form between adjacent cells by fusion of plasma membrane proteins on their outer surfaces. A complex network of junction proteins makes a seal tight enough to prevent leaks of ions or molecules between cells.

Don W. Fawcett/Photo Researchers, Inc.

FIGURE 5.27 Anchoring junctions, tight junctions, and gap junctions, which connect cells in animal tissues. Anchoring junctions reinforce the cell-to-cell connections made by cell adhesion molecules, tight junctions seal the spaces between cells, and gap junctions create direct channels of communication between animal cells.
© Cengage Learning 2014

that run entirely around cells, "welding" adjacent cells together. For some anchoring junctions known as **desmosomes,** intermediate filaments anchor the junction in the underlying cytoplasm; in other anchoring junctions known as **adherens junctions,** microfilaments are the anchoring cytoskeletal component. Anchoring junctions are most common in tissues that are subject to stretching, shear, or other mechanical forces—for example, heart muscle, skin, and the cell layers that cover organs or line body cavities and ducts.

Tight junctions, as the name indicates, are regions of tight connections between membranes of adjacent cells (see Figure 5.27). The connection is so tight that it can keep particles as small as ions from moving between the cells in the layers.

Tight junctions seal the spaces between cells in the cell layers that cover internal organs and the outer surface of the body, or the layers that line internal cavities and ducts. For example, tight junctions between cells that line the stomach, intestine, and bladder keep the contents of these body cavities from leaking into surrounding tissues.

A tight junction is formed by direct fusion of proteins on the outer surfaces of the two plasma membranes of adjacent cells. Strands of the tight junction proteins form a complex network that gives the appearance of stitch work holding the cells together. Within a tight junction, the plasma membrane is not joined continuously; instead, there are regions of intercellular space. Nonetheless, the network of junction proteins is sufficient to make the tight cell connections characteristic of these junctions.

Gap junctions open direct channels that allow ions and small molecules to pass directly from one cell to another (see Figure 5.27). Hollow protein cylinders embedded in the plasma membranes of adjacent cells line up and form a sort of pipeline that connects the cytoplasm of one cell with the cytoplasm of the next. The flow of ions and small molecules through the channels provides almost instantaneous communication between animal cells, similar to the communication that plasmodesmata provide between plant cells.

In vertebrates, gap junctions occur between cells within almost all body tissues, but not between cells of different tissues. These junctions are particularly important in heart muscle tissues and in the smooth muscle tissues that form the uterus, where their pathways of communication allow cells of the organ to operate as a coordinated unit. Although most nerve tissues do not have gap junctions, nerve cells in dental pulp are connected by gap junctions; they are responsible for the discomfort you feel if your teeth are disturbed or damaged, or when a dentist pokes a probe into a cavity.

The Extracellular Matrix Organizes the Cell Exterior

Many types of animal cells are embedded in an ECM that consists of proteins and polysaccharides secreted by the cells themselves **(Figure 5.28)**. The primary function of the ECM is protection and support. The ECM forms the mass of skin, bones, and tendons; it also forms many highly specialized extracellular structures such as the cornea of the eye and filtering networks in the kidney. The ECM also affects cell division, adhesion, motility, and embryonic development, and it takes part in reactions to wounds and disease.

Glycoproteins are the main component of the ECM. In most animals, the most abundant ECM glycoprotein is *collagen,* which forms fibers with great tensile strength and elasticity. In vertebrates, the collagens of tendons, cartilage, and bone are the most abundant proteins of the body, making up about half of the total body protein by weight. (Collagens and their roles in body structures are described in further detail in Chapter 38.)

The consistency of the matrix, which may range from soft and jellylike to hard and elastic, depends on a network of proteoglycans that surrounds the collagen fibers. *Proteoglycans* are glycoproteins that consist of small proteins noncovalently attached to long polysaccharide molecules. Matrix consistency depends on the number of interlinks in this network, which determines how much water can be trapped in it. For example, cartilage, which contains a high proportion of interlinked glycoproteins, is relatively soft. Tendons, which are almost pure collagen, are tough and elastic. In bone, the glycoprotein network that surrounds collagen fibers is impregnated with mineral crystals, producing a dense and hard—but still elastic—structure that is about as strong as fiberglass or reinforced concrete.

Yet another class of glycoproteins is *fibronectins,* which aid in organizing the ECM and help cells attach to it. Fibronectins

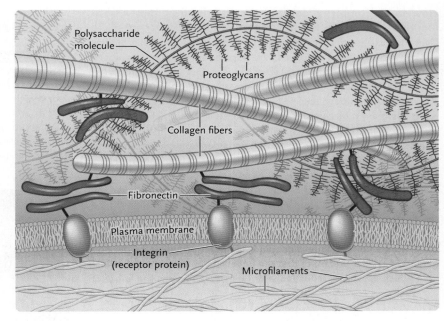

FIGURE 5.28 Components of the extracellular matrix in an animal cell.
© Cengage Learning 2014

bind to receptor proteins called *integrins* that span the plasma membrane. On the cytoplasmic side of the plasma membrane, the integrins bind to microfilaments of the cytoskeleton. Integrins integrate changes outside and inside the cell by communicating changes in the ECM to the cytoskeleton.

Having laid the groundwork for cell structure and function in this chapter, we next take up further details of individual cell structures, beginning with the roles of cell membranes in transport in the next chapter.

STUDY BREAK 5.5

1. Distinguish between anchoring junctions, tight junctions, and gap junctions.
2. What is the structure and function of the extracellular matrix?

 UNANSWERED QUESTIONS

The field of cell biology seeks to understand the properties and behaviors of cells, including their growth and division, shape and movement, subcellular organization and transport systems, and interactions and communication with each other and the environment. Although cell biology research has dramatically enhanced our basic understanding of cells, many fundamental questions remain to be answered.

How do eukaryotic cells balance assembly and disassembly pathways to maintain a complex cellular and subcellular structure?

Eukaryotic cells have a high degree of organizational complexity in the form of numerous subcellular organelles and structures. Maintenance of this complexity depends on a delicate balance between the pathways that promote the formation and disassembly of each structure. How is

this balance achieved? In my lab, we are working to answer this question in part by studying the interaction between bacterial and viral pathogens and their eukaryotic host cells. Many pathogens invade, move within, and spread between host cells by manipulating the assembly and disassembly of cytoskeletal polymers. Because pathogens exploit normal cellular pathways that regulate assembly and disassembly, they are very useful tools for identifying key molecular players and determining their mechanistic roles. For example, our studies of the intracellular actin-based movement of the bacterial pathogens *Listeria monocytogenes* and *Rickettsia parkeri* have helped define cellular molecules and pathways that control the assembly and organization of the actin cytoskeleton. Continued progress in addressing this question will come from studying pathogens and their host cells, and from reconstructing subcellular processes outside of the cell.

How do the complex properties of cells arise from the functions and interactions of their molecular components? Over the past few decades, inspired by advances in molecular biology and genetics, most cell biologists have pursued a reductionist approach by determining the detailed role of individual genes and proteins in cell structure and behavior. This has been very successful, and together with advances in genome sequencing technology, it has enabled cell biologists to catalog many molecules that perform key roles in the cell. In the past few years, many cell biologists have embraced a more holistic approach called *systems biology,* which is focused on understanding how the complex properties of cells and subcellular systems arise from the properties and interactions of their individual parts. For instance, Jonathan Weismann's lab at the University of California, San Francisco has studied the translation, localization, and abundance of thousands of proteins in the yeast *Saccharomyces cerevisiae,* revealing insights into global mechanisms that regulate protein abundance and function. Systems cell biologists often work with mathematicians and computer scientists to develop models of cellular processes that can be refined by experimental validation. Which approach, reductionist or holistic, represents the future of cell biology? Both do, as each relies on the other to generate insights and hypotheses that move the field forward.

THINK LIKE A SCIENTIST There are several major approaches cell biologists use to investigate complex cell structures and processes. One is to perturb the activity of genes or proteins and assess the effect on cellular function. Another is to build or reconstitute the structure or process in a test tube. What are strengths and weaknesses of each approach? What types of questions can be answered using one or the other? Can you think of technical challenges that might arise for each one?

Courtesy of Matthew Welch

Matthew Welch is a professor of Molecular and Cell Biology at the University of California, Berkeley. His research interests include cytoskeleton dynamics and microbial pathogenesis. Learn more about his work at http://mcb.berkeley.edu/labs/welch.

REVIEW KEY CONCEPTS

To access the course materials and companion resources for this text, please visit www.cengagebrain.com.

5.1 Basic Features of Cell Structure and Function

- According to the cell theory: (1) all living organisms are composed of cells; (2) cells are the structural and functional units of life; and (3) cells arise only from the division of preexisting cells.

- Cells of all kinds are divided internally into a central region containing the genetic material and the cytoplasm, which consists of the cytosol, the cytoskeleton, and organelles and is bounded by the plasma membrane.

- The plasma membrane is a lipid bilayer in which transport proteins are embedded (Figure 5.6).

- In the cytoplasm, proteins are made, most of the other molecules required for growth and reproduction are assembled, and energy absorbed from the surroundings is converted into energy usable by the cell.

Animation: How an electron microscope works

Animation: How a light microscope works

Animation: Cell membranes

5.2 Prokaryotic Cells

- Prokaryotic cells are surrounded by a plasma membrane and, in most groups, are enclosed by a cell wall. The genetic material, typically a single, circular DNA molecule, is located in the nucleoid. The cytoplasm contains masses of ribosomes (Figure 5.7).

Animation: Typical prokaryotic cell

5.3 Eukaryotic Cells

- Eukaryotic cells have a true nucleus, which is separated from the cytoplasm by the nuclear envelope perforated by nuclear pores. A plasma membrane forms the outer boundary of the cell. Other membrane systems enclose specialized compartments as organelles in the cytoplasm (Figure 5.9).

- The eukaryotic nucleus contains chromatin, a combination of DNA and proteins. A specialized segment of the chromatin forms the nucleolus, where ribosomal RNA molecules are made and combined with ribosomal proteins to make ribosomes. The nuclear envelope contains nuclear pore complexes with pores that allow passive or assisted transport of molecules between the nucleus and the cytoplasm. Proteins destined for the nucleus contain a short amino acid sequence called a nuclear localization signal (Figures 5.10 and 5.11).

- Eukaryotic cytoplasm contains ribosomes (Figure 5.12), an endomembrane system, mitochondria, microbodies, the cytoskeleton, and some organelles specific to certain organisms. The endomembrane system includes the nuclear envelope, ER, Golgi complex, lysosomes, vesicles, and plasma membrane.

- The endoplasmic reticulum (ER) occurs in two forms, as rough and smooth ER. The ribosome-studded rough ER makes proteins that become part of cell membranes or are released from the cell. Smooth ER synthesizes lipids and breaks down toxic substances (Figure 5.13).

- The Golgi complex chemically modifies proteins made in the rough ER and sorts finished proteins to be secreted from the cell, embedded in the plasma membrane, or included in lysosomes (Figures 5.14, 5.15, and 5.17).

- Lysosomes, specialized vesicles that contain hydrolytic enzymes, digest complex molecules such as food molecules that enter the cell by endocytosis, cellular organelles that are no longer functioning correctly, and engulfed bacteria and cell debris (Figure 5.16).

- Mitochondria carry out cellular respiration, the conversion of fuel molecules into the energy of ATP (Figure 5.18).

- Microbodies conduct the initial steps in fat breakdown and other reactions that link major biochemical pathways in the cytoplasm (Figure 5.19).

- The cytoskeleton is a supportive structure built from microtubules, intermediate filaments, and microfilaments. Motor proteins walking along microtubules and microfilaments produce most movements of animal cells (Figures 5.20 and 5.21).

- Motor protein-controlled sliding of microtubules generates the movements of flagella and cilia. Flagella and cilia arise from centrioles (Figures 5.22–5.24).

Animation: Process of secretion

Animation: Cytoskeletal components

Animation: Common eukaryotic organelles

Animation: Flagella structure

Animation: Structure of a mitochondrion

Animation: Motor proteins

Animation: Nuclear envelope

Animation: The endomembrane system

Animation: Generalized Animal Cell

Animation: Structure of a mitochondrion

Animation: Generalized Plant Cell

5.4 Specialized Structures of Plant Cells

- Plant cells contain all the eukaryotic structures found in animal cells except for lysosomes. They also contain three structures not found in animal cells: chloroplasts, a central vacuole, and a cell wall (Figure 5.9B).
- Chloroplasts contain pigments and molecular systems that absorb light energy and convert it to chemical energy. The chemical energy is used inside the chloroplasts to assemble carbohydrates and other organic molecules from simple inorganic raw materials (Figure 5.25).

- The large central vacuole, which consists of a tonoplast enclosing an inner space, develops pressure that supports plant cells, accounts for much of cellular growth by enlarging as cells mature, and serves as a storage site for substances including waste materials (Figure 5.9B).
- A cellulose cell wall surrounds plant cells, providing support and protection. Plant cell walls are perforated by plasmodesmata, channels that provide direct pathways of communication between the cytoplasm of adjacent cells (Figure 5.26).

Animation: Plant cell walls

Animation: Sites of photosynthesis

Animation: Structure of a chloroplast

5.5 The Animal Cell Surface

- Animal cells have specialized surface molecules and structures that function in cell adhesion, communication, and support.
- Cell adhesion molecules bind to specific molecules on other cells. The adhesions organize and hold together cells of the same type in body tissues.
- Cell adhesions are reinforced by various junctions. Anchoring junctions hold cells together. Tight junctions seal together the plasma membranes of adjacent cells, preventing ions and molecules from moving between the cells. Gap junctions open direct channels between the cytoplasm of adjacent cells (Figure 5.27).
- The extracellular matrix, formed from collagen proteins embedded in a matrix of branched glycoproteins, functions primarily in cell and body protection and support but also affects cell division, motility, embryonic development, and wound healing (Figure 5.28).

Animation: Animal cell junctions

UNDERSTAND & APPLY

Test Your Knowledge

1. You are examining a cell from a crime scene using an electron microscope. It contains ribosomes, DNA, a plasma membrane, a cell wall, and mitochondria. What type of cell is it?
   a. lung cell
   b. plant cell
   c. prokaryotic cell
   d. cell from the surface of a human fingernail
   e. sperm cell

2. A prokaryote converts food energy into the chemical energy of ATP on/in its:
   a. chromosome.
   b. flagella.
   c. ribosomes.
   d. cell wall.
   e. plasma membrane.

3. Eukaryotic and prokaryotic ribosomes are similar in that:
   a. both contain a small subunit, but only eukaryotes contain a large subunit.
   b. both contain the same number of proteins.
   c. both use mRNA to assemble amino acids into proteins.
   d. both contain the same number of types of rRNA.
   e. both produce proteins that can pass through pores into the nucleus.

4. Which of the following structures does *not* require an immediate source of energy to function?
   a. central vacuoles
   b. cilia
   c. microtubules
   d. microfilaments
   e. microbodies

5. Which of the following structures is *not* used in eukaryotic protein manufacture and secretion?
   a. ribosome
   b. lysosome
   c. rough ER
   d. secretory vesicle
   e. Golgi complex

6. Which of the following are glycoproteins whose function is affected by the common cold virus?
   a. plasmodesmata
   b. desmosomes
   c. cell adhesion molecules
   d. flagella
   e. cilia

7. An electron micrograph shows that a cell has extensive amounts of rough ER throughout. One can deduce from this that the cell is:
   a. synthesizing and metabolizing carbohydrates.
   b. synthesizing and secreting proteins.
   c. synthesizing ATP.
   d. contracting.
   e. resting metabolically.

8. Which of the following contributes to the sealed lining of the digestive tract to keep food inside it?
   a. a central vacuole that stores proteins
   b. tight junctions formed by direct fusion of proteins
   c. gap junctions that communicate between cells of the stomach lining and its muscular wall
   d. desmosomes forming buttonlike spots or a belt to keep cells joined together
   e. plasmodesmata that help cells communicate their activities

9. Which of the following statements about proteins is correct?
   a. Proteins are transported to the rough ER for use within the cell.
   b. Lipids and carbohydrates are added to proteins by the Golgi complex.
   c. Proteins are transported directly into the cytosol for secretion from the cell.
   d. Proteins that are to be stored by the cell are moved to the rough ER.
   e. Proteins are synthesized in vesicles.

10. Which of the following is *not* a component of the cytoskeleton?
    a. microtubules
    b. actins
    c. microfilaments
    d. cilia
    e. cytokeratins

Discuss the Concepts

1. Many compound microscopes have a filter that eliminates all wavelengths except that of blue light, thereby allowing only blue light to pass through the microscope. Use the spectrum of visible light (see Figure 9.4) to explain why the filter improves the resolution of light microscopes.

2. Explain why aliens invading Earth are not likely to be giant cells the size of humans.

3. An electron micrograph of a cell shows the cytoplasm packed with rough ER membranes, a Golgi complex, and mitochondria. What activities might this cell concentrate on? Why would large numbers of mitochondria be required for these activities?

4. Assuming that mitochondria evolved from bacteria that entered cells by endocytosis, what are the likely origins of the outer and inner mitochondrial membranes?

5. Researchers have noticed that some men who were sterile because their sperm cells were unable to move also had chronic infections of the respiratory tract. What might be the connection between these two symptoms?

Design an Experiment

The unicellular alga *Chlamydomonas reinhardtii* has two flagella assembled from tubulin proteins. If a researcher changes the pH from approximately neutral (their normal growing condition) to pH 4.5, *Chlamydomonas* cells spontaneously lose their flagella. After the cells are returned to neutral pH, they regrow the flagella—a process called reflagellation. Assuming that you have deflagellated *Chlamydomonas* cells, devise experiments to answer the following questions:

1. Do new tubulin proteins need to be made for reflagellation to occur, or is there a reservoir of proteins in the cell?

2. Is the production of new mRNA for the tubulin proteins necessary for reflagellation?

3. What is the optimal pH for reflagellation?

Interpret the Data

Investigators studying protein changes during aging examined enzyme activity in cells extracted from the nematode worm *Caenorhabditis elegans*. The cell extracts were treated to conserve enzyme activity, although the investigators noted that some proteins were broken down by the extraction procedure. The extracts were centrifuged, and seven fractions were collected in sequence to isolate the location of activity by protease enzymes called cathepsins. Examine the activity profiles in the **Figure**. In which fraction and, hence, in which eukaryotic cellular structure are these enzymes most active?

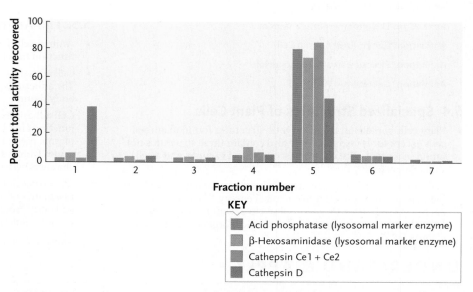

KEY

	Acid phosphatase (lysosomal marker enzyme)
	β-Hexosaminidase (lysosomal marker enzyme)
	Cathepsin Ce1 + Ce2
	Cathepsin D

FIGURE Distribution of enzyme activity in fractions from centrifugation of an organelle pellet. The fractions are numbered 1 to 7 from the top to the bottom of the centrifuge tube. Fraction 1 contains cytosolic contents and is the supernatant, and fraction 7 contains cellular debris and membrane fragments.

Reprinted from G. J. Sarkis et al. 1988. Decline in protease activities with age in the nematode *Caenorhabiditis elegans*. *Mechanisms of Ageing and Development* 45:191–201. Copyright 1988, with permission from Elsevier.

Source: G. J. Sarkis et al. 1988. Decline in protease activities with age in the nematode *Caenorhabiditis elegans*. *Mechanisms of Ageing and Development* 45:191–201.

Apply Evolutionary Thinking

What aspects of cell structure suggest that prokaryotes and eukaryotes share a common ancestor in their evolutionary history?

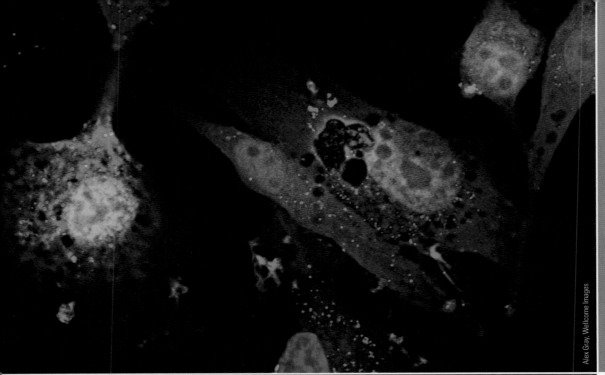

Endocytosis in cancer cells (confocal micrograph). The red spots are fluorescent spheres used to follow the process of endocytosis; some of the spheres have been taken up by cells.

Membranes and Transport

Why it matters . . . Cystic fibrosis (CF) is one of the most common genetic diseases. It affects approximately 1 in 4,000 children born in the United States. People with CF experience a progressive impairment of lung and gastrointestinal function. Although the treatment of CF patients is slowly improving, their average lifespan remains under 40 years. CF is caused by mutations in a single gene that codes for a protein so vital that a change in a particular amino acid has life-altering consequences. What is this protein, and what role does it play in the body?

To maintain their internal environment, the cells of all organisms must exchange molecules and ions constantly with the fluid environment that surrounds them. What makes this possible is the **plasma membrane,** the thin layer of lipids and proteins that separates a cell from its surroundings. Some of the proteins—called transport proteins—move particular ions and molecules, including water, in a directed way across the membrane. Different transport proteins move different molecules or ions. One such transport protein is cystic fibrosis transmembrane conductance regulator, or CFTR, which is found in the plasma membrane of epithelial cells **(Figure 6.1).** Epithelial cells are organized into sheetlike layers that form coverings and linings, which are typically exposed to water, air, or fluids within the body. Epithelial cells line the passageways and ducts of the lungs, liver, pancreas, intestines, reproductive system, and skin. CFTR pumps chloride ions out of those cells, and water follows the ions producing a thin, watery film over the epithelial tissue surface. Mucus can slide easily over the tissue because of the moist surface. (Epithelial cells are described in more detail in Chapter 38.)

Particular mutations in the *CFTR* gene result in a CFTR molecule that transports chloride ions poorly or not at all. The most common mutation is a small deletion in the gene that removes one amino acid from the CFTR protein. If an individual has one normal and one mutant copy of the gene, there are enough working CFTR molecules for normal chloride transport. However, if

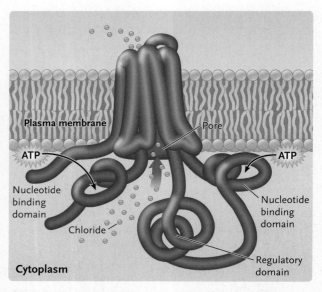

FIGURE 6.1 Molecular model of cystic fibrosis transmembrane conductance regulator (CFTR), a chloride ion transport protein embedded in the plasma membrane. The nucleotide binding domains bind ATP which provides energy for ion transport by the protein.
© Cengage Learning 2014

an individual has two mutant copies of the gene, all CFTR molecules are mutant, and chloride transport is defective—the individual has CF. Not enough chloride ions are transported out of the cell, so not enough water leaves either. As a result, mucus sticks to the drier epithelial tissue, building up into a thick mass. In the respiratory tract, the abnormally thick mucus means much diminished protection from bacteria, which leads to infections. Chronic lung infections produce progressive damage to the respiratory tract; at some point, lung function becomes insufficient to support life and the CF patient dies.

For a person to have CF, one mutant copy of the gene must be inherited from each parent. The frequency of CF individuals in the United States varies among groups—1/2,500 to 1/3,500 for Caucasian Americans, 1/17,000 for African Americans, and 1/31,000 for Asian Americans. Recent evidence suggests that the higher frequency among Caucasian Americans may be a result of natural selection in their European ancestors, where having a single mutant copy conferred resistance to diseases that were common at the time. Most treatments center on managing the symptoms, and include pounding the back or chest to dislodge mucus, and antibiotic or other pharmaceutical treatments to control infections. Research is ongoing to develop more effective treatments. One interesting research approach is to develop pharmaceuticals that target the defective CFTR itself. Basic research has located the disease-causing mutations in the CFTR molecule, so researchers may be able to design treatments that change the transport protein to a more functional form.

The plasma membrane, in which the CFTR molecule is located, forms the outer boundary of every living cell, and encloses the intracellular contents. The plasma membrane also regulates the passage of molecules and ions between the inside and outside of the cell, helping to determine the cell's composi-

tion. Within eukaryotic cells, membranes surrounding internal organelles play similar roles, creating environments that differ from the surrounding cytosol.

The structure and function of biological membranes are the focus of this chapter. We first consider the structure of membranes and then examine how membranes transport substances selectively in and out of cells and organelles. Other roles of membranes, including recognition of molecules on other cells, adherence to other cells or extracellular materials, and reception of molecular signals such as hormones, are the subjects of Chapters 7, 42, and 45 in this book.

6.1 Membrane Structure and Function

A watery fluid medium—or aqueous solution—bathes both surfaces of all biological membranes. The membranes are also fluid, but they are kept separate from their surroundings by the properties of the lipid and protein molecules from which they are formed.

Biological Membranes Contain Both Lipid and Protein Molecules

Biological membranes consist of lipids and proteins assembled into a thin film. The proportions of lipid and protein molecules in membranes vary, depending on the functions of the membranes in the cells.

MEMBRANE LIPIDS *Phospholipids* and *sterols* are the two major types of lipids in membranes (see Section 3.3). Phospholipids have a polar (electrically charged) end containing a phosphate group linked to one of several alcohols or amino acids, and a nonpolar (uncharged) end containing two nonpolar fatty-acid tails. **Figure 6.2A** shows an example of a phospholipid, phosphatidylcholine, in which the polar head contains choline. The polar end is hydrophilic—it "prefers" being in an aqueous environment—and the nonpolar end is hydrophobic—it "prefers" being in an environment from which water is excluded. In other words, phospholipids have dual solubility properties.

In an aqueous medium, phospholipid molecules satisfy their dual solubility properties by assembling into a **bilayer**—a layer two molecules thick **(Figure 6.2B)**. In a bilayer, the polar ends of the phospholipid molecules are located at the surfaces, where they face the aqueous media. The nonpolar fatty-acid chains arrange themselves end to end in the membrane interior, in a nonpolar region that excludes water.

At low temperatures, the phospholipid bilayer freezes into a semisolid, gel-like state **(Figure 6.2C)**. When a phospholipid bilayer sheet is shaken in water, it breaks and spontaneously forms small *vesicles* **(Figure 6.2D)**. Vesicles consist of a spherical shell of phospholipid bilayer enclosing a small droplet of water.

Membrane sterols also have dual solubility properties. As explained in Section 3.4, these molecules have nonpolar car-

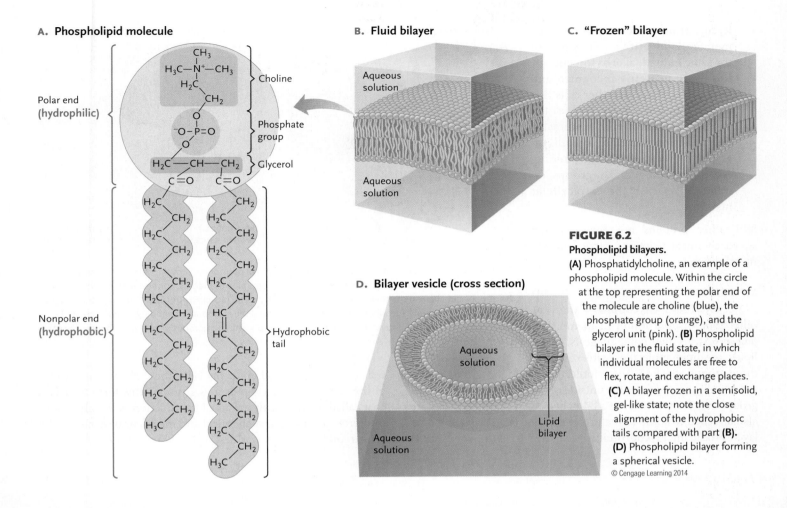

A. Phospholipid molecule

Polar end (hydrophilic)

Choline
Phosphate group
Glycerol

Nonpolar end (hydrophobic)

Hydrophobic tail

B. Fluid bilayer

Aqueous solution

Aqueous solution

C. "Frozen" bilayer

D. Bilayer vesicle (cross section)

Aqueous solution

Aqueous solution

Lipid bilayer

FIGURE 6.2
Phospholipid bilayers.
(A) Phosphatidylcholine, an example of a phospholipid molecule. Within the circle at the top representing the polar end of the molecule are choline (blue), the phosphate group (orange), and the glycerol unit (pink). **(B)** Phospholipid bilayer in the fluid state, in which individual molecules are free to flex, rotate, and exchange places. **(C)** A bilayer frozen in a semisolid, gel-like state; note the close alignment of the hydrophobic tails compared with part **(B)**. **(D)** Phospholipid bilayer forming a spherical vesicle.
© Cengage Learning 2014

bon rings with a nonpolar side chain at one end and a single polar group (an —OH group) at the other end. In biological membranes, sterols pack into membranes alongside the phospholipid hydrocarbon chains, with only the polar end extending into the polar membrane surface **(Figure 6.3).** The predominant sterol of animal cell membranes is **cholesterol,** which is

important for keeping the membranes fluid. A variety of sterols, called *phytosterols,* are found in plants. The high similarity of bilayer membranes in all cells—prokaryotic and eukaryotic—is evidence that the basic structure of membranes evolved during the earliest stages of life on Earth, and has been preserved ever since.

MEMBRANE PROTEINS Membrane proteins also have hydrophilic and hydrophobic regions that give them dual solubility properties. The hydrophobic regions of membrane proteins are formed by segments of the amino acid chain with hydrophobic side groups. These hydrophobic segments are often wound into α helices, which span the membrane bilayer **(Figure 6.4).** (Protein structural elements are described in Section 3.4.) The hydrophobic segments are connected by loops of hydrophilic amino acids that extend into the polar regions at the membrane surfaces (for example, see Figure 6.4).

Each type of membrane has a characteristic group of proteins that is responsible for its specialized functions:

• **Transport proteins** form channels that allow selected polar molecules and ions to pass across a membrane.

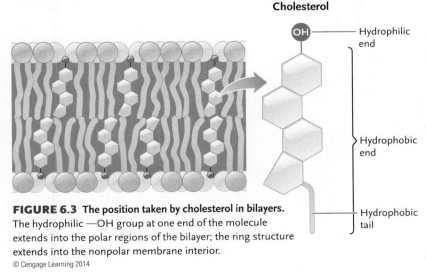

Cholesterol

Hydrophilic end

Hydrophobic end

Hydrophobic tail

FIGURE 6.3 **The position taken by cholesterol in bilayers.**
The hydrophilic —OH group at one end of the molecule extends into the polar regions of the bilayer; the ring structure extends into the nonpolar membrane interior.
© Cengage Learning 2014

- **Recognition proteins** in the plasma membrane identify a cell as part of the same individual or as foreign.
- **Receptor proteins** recognize and bind molecules from other cells that act as chemical signals, such as the peptide hormone insulin in animals.
- **Cell adhesion proteins** bind cells together by recognizing and binding receptors or chemical groups on other cells or on the extracellular matrix.

Still other proteins are enzymes that speed chemical reactions carried out by membranes.

MEMBRANE GLYCOLIPIDS AND GLYCOPROTEINS **Glycolipids** are another type of lipid found in membranes **(Figure 6.5)**. As their name suggests, glycolipids are lipid molecules with carbohydrate groups attached. These molecules are found in the part of the membrane that faces the outside of the cell. The carbohydrate groups may form a linear or a branched chain. Carbohydrates are also attached to some of the proteins in the exterior-facing lipid layer, producing **glycoproteins** (see Figure 6.5). In many animal cells, such as intestinal epithelial cells, the carbohydrate groups of the cell surface glycolipids and glycoproteins form a surface coat called the **glycocalyx.** Part of the function of the glycocalyx is to protect cells against chemical and mechanical damage. (The prokaryotic glycocalyx was described in Section 5.2.)

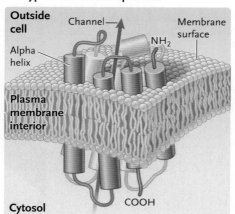
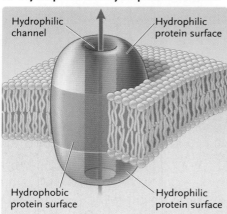

A. Typical membrane protein

B. Hydrophilic and hydrophobic surfaces

FIGURE 6.4 **Structure of membrane proteins. (A)** Typical membrane protein, bacteriorhodopsin, showing the membrane-spanning alpha-helical segments (blue cylinders), connected by flexible loops of the amino acid chain at the membrane surfaces. **(B)** The same protein as in **(A)** in a diagram that shows hydrophilic (blue) and hydrophobic (orange) surfaces and the membrane-spanning channel created by this protein. Bacteriorhodopsin absorbs light energy in plasma membranes of photosynthetic archaeans.
© Cengage Learning 2014

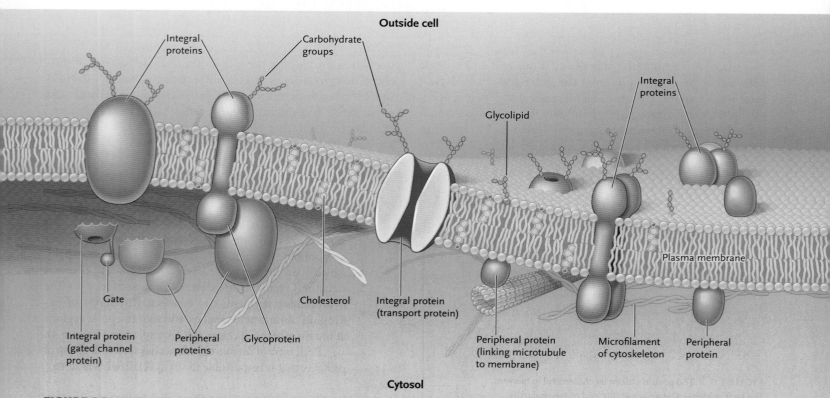

FIGURE 6.5 **Membrane structure according to the fluid mosaic model, in which integral membrane proteins are suspended individually in a fluid bilayer.** Peripheral proteins are attached to integral proteins or membrane lipids mostly on the cytoplasmic side of the membrane (shown only on the inner surface in the figure). In the plasma membrane, carbohydrate groups of membrane glycoproteins and glycolipids face the cell exterior.
© Cengage Learning 2014

Keeping Membranes Fluid at Cold Temperatures

Biological membranes of all organisms are evolutionarily conserved. The fluid state of biological membranes is critical to membrane function and, therefore, vital to cellular life. When membranes freeze, the phospholipids form a semisolid gel in which they are unable to move (see Figure 6.2C), and proteins become locked in place. Freezing can kill cells by impeding vital membrane functions such as transport.

Many eukaryotic organisms, including higher plants, animals, and some protists, adapt to colder temperatures by changing membrane lipids. Experiments have shown that, in animals with body temperatures that fluctuate with environmental temperature, such as fish, amphibians, and reptiles, both the proportion of double bonds in membrane phospholipids and the cholesterol content are increased at lower temperatures. How do these changes affect membrane fluidity? Double bonds in unsaturated fatty acids introduce "kinks" in their hydrocarbon chain (see Figure 3.14); the kinks help bilayers stay fluid at lower temperatures by interfering with packing of the hydrocarbons.

Cholesterol depresses the freezing point by interfering with close packing of membrane phospholipids.

All of these membrane changes also occur in mammals that enter hibernation in cold climates, thereby preventing their membranes from freezing. The resistance to freezing allows the nerve cells of a hibernating mammal to remain active so that the animal can maintain basic body functions and respond, although sluggishly, to external stimuli. In active, nonhibernating mammals, membranes freeze into the gel state at about 15°C.

The Fluid Mosaic Model Explains Membrane Structure

Membrane structure has been and continues to be a subject for research. The current view of membrane structure is based on the fluid mosaic model advanced by S. Jonathan Singer and Garth L. Nicolson at the University of California, San Diego, in 1972 (see Figure 6.5). The **fluid mosaic model** proposes that the membrane consists of a fluid phospholipid bilayer in which proteins are embedded and float freely. This model revolutionized how scientists think about membrane structure and function.

The "fluid" part of the fluid mosaic model refers to the phospholipid molecules, which vibrate, flex back and forth, spin around their long axis, move sideways, and exchange places within the same bilayer half. Only rarely does a phospholipid flip-flop between the two layers. Phospholipids exchange places within a layer millions of times a second, making the phospholipid molecules in the membrane highly dynamic. Membrane fluidity is critical to the functions of membrane proteins and allows membranes to accommodate, for example, cell growth, motility, and surface stresses.

Lipid composition and temperature affect the fluidity of membranes. For example, shorter fatty acid chains or a greater proportion of unsaturated fatty acids reduces the ability of the hydrophobic tails to interact with each other. As a result, the membrane remains fluid at lower temperatures. This ability of membranes to remain fluid at lower temperatures is important for organisms whose temperature fluctuates with that of the environment. Such organisms adapt to the lower temperature by synthesizing shorter tails and with more double bonds.

Cholesterol also plays an important role in enabling membrane fluidity over a range of temperature. As you learned, cholesterol packs into the membrane alongside the phospholipid molecules (see Figure 6.3). At higher temperatures, such as 37°C (the body temperature of mammals), cholesterol makes the membrane less fluid. Without cholesterol, the membrane would become too fluid and would become leaky, allowing ions to cross in an uncontrolled manner. This leaking disrupts the function of the cell and it is likely to die. As described previously, cholesterol reduces membrane fluidity at high temperatures, thereby providing some protection. At low temperatures, cholesterol has the opposite effect: it prevents the hydrophobic tails from coming together and crystallizing, thereby maintaining membrane fluidity. (See *Focus on Basic Research* for a description of other strategies that organisms use to keep their membranes from freezing at low temperatures.)

The "mosaic" part of the fluid mosaic model refers to the membrane proteins, most of which float individually in the fluid lipid bilayer, like icebergs in the sea. Membrane proteins are larger than membrane lipids, and those that move do so much more slowly than do lipids. A number of membrane proteins are attached to the cytoskeleton. These proteins either are immobile or move in a directed fashion, such as along cytoskeletal filaments.

Membrane proteins are oriented across the membrane so that particular functional groups and active sites face either the inside or the outside membrane surface. The inside and outside halves of the bilayer also contain different mixtures of phospholipids. These differences make biological membranes *asymmetric* and give their inside and outside surfaces different functions.

Proteins that are embedded in the phospholipid bilayer are termed **integral proteins** (see Figure 6.5). Essentially all transport, receptor, recognition, and cell adhesion proteins that give membranes their specific functions are integral membrane proteins.

Other proteins, called **peripheral proteins** (see Figure 6.5), are held to membrane surfaces by noncovalent bonds—hydrogen bonds and ionic bonds—formed with the polar parts of integral membrane proteins or membrane lipids. Most peripheral proteins are on the cytoplasmic side of the membrane. Some peripheral proteins are parts of the cytoskeleton, such as mi-

FIGURE 6.6 **Experimental Research**

The Frye–Edidin Experiment Demonstrating That the Phospholipid Bilayer Is Fluid

Question: Do membrane proteins move in the phospholipid bilayer?

Experiment: Frye and Edidin labeled membrane proteins on cultured human and mouse cells with fluorescent dyes, red for human proteins and green for mouse proteins. Human and mouse cells were then fused and the pattern of fluorescence was followed under a microscope.

Results: After 40 minutes, the fluorescence pattern showed that the human and mouse membrane proteins had mixed completely.

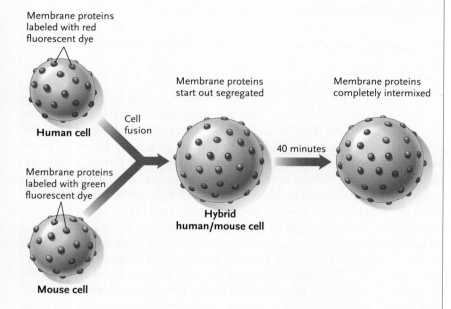

Membrane proteins labeled with red fluorescent dye

Human cell

Membrane proteins labeled with green fluorescent dye

Mouse cell

Cell fusion

Membrane proteins start out segregated

Hybrid human/mouse cell

40 minutes

Membrane proteins completely intermixed

Conclusion: The rapid mixing of membrane proteins in the hybrid human/mouse cells showed that membrane proteins move in the phospholipid bilayer, indicating that the membrane is fluid.

THINK LIKE A SCIENTIST An alternative explanation for the observed intermixing of membrane proteins is new synthesis of the proteins and the insertion of those proteins all over the plasma membrane to replace proteins that are broken down. Frye and Edidin did some experiments to determine whether the intermixing was due to the fluid nature of the phospholipid bilayer or to new protein synthesis. In one experiment, they incubated the hybrid cells with three different protein synthesis inhibitors. They observed that the membrane proteins were completely mixed after the incubation period. In another experiment, the researchers performed the experiment at a variety of temperatures below that of the initial experiment, which was performed at 37°C. They observed that the lower the temperature, the less intermixing of membrane proteins occurred. Essentially no intermixing occurred at 15°C and below. How do these results support the fluid phospholipid bilayer model?

Source: L. D. Frye and M. Edidin. 1970. The rapid intermixing of cell surface antigens after formation of mouse–human heterokaryons. *Journal of Cell Science* 7:319–335.

crotubules, microfilaments, or intermediate filaments, or proteins that link the cytoskeleton together. These structures hold some integral membrane proteins in place. For example, this anchoring constrains many types of receptors to the sides of cells that face body surfaces, cavities, or tubes.

The Fluid Mosaic Model Is Supported Fully by Experimental Evidence

The novel ideas of a fluid membrane and a flexible mosaic arrangement of proteins and lipids challenged an accepted model in which a relatively rigid, stable membrane was coated on both sides with proteins arranged like jam on bread. Researchers tested the new model with intensive research. The experimental evidence from that research supports every major hypothesis of the model completely: that membrane lipids are arranged in a bilayer; that the bilayer is fluid; that proteins are suspended individually in the bilayer; and that the arrangement of both membrane lipids and proteins is asymmetric.

EVIDENCE THAT MEMBRANES ARE FLUID In 1970, L. David Frye and Michael A. Edidin grew human cells and mouse cells separately in tissue culture. Then they added antibodies that bound to either human or mouse membrane proteins **(Figure 6.6)**. The anti-human antibodies were attached to dye molecules that fluoresce red under ultraviolet light, and the anti-mouse antibodies to molecules that fluoresce green. The researchers then fused the two cells. The red and green tagged proteins started out in separate halves of the fused cell, but over a period of 40 minutes they gradually intermixed, demonstrating that the proteins were floating freely in a fluid membrane.

Based on the measured rates at which molecules mix in biological membranes, the membrane bilayer appears to be about as fluid as light machine oil, such as the lubricants you might use around the house to oil a door hinge or a bicycle.

EVIDENCE FOR MEMBRANE ASYMMETRY AND INDIVIDUAL SUSPENSION OF PROTEINS An experiment that used membranes prepared for electron microscopy by the **freeze-fracture technique** confirmed that the membrane is a bilayer with proteins suspended in it individually and that the arrangement of membrane lipids and proteins is asymmetric **(Figure 6.7)**. In this technique, experimenters freeze a block of cells rapidly by dipping it

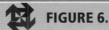 **FIGURE 6.7** | **Research Method**

Freeze Fracture

Purpose: Quick-frozen cells are fractured to split apart lipid bilayers for analysis of the membrane interior.

Protocol:

1. The specimen is frozen quickly in liquid nitrogen and then fractured by a sharp blow by a knife edge.

2. The fracture may travel over membrane surfaces as it passes through the specimen, or it may split membrane bilayers into inner and outer halves as shown here.

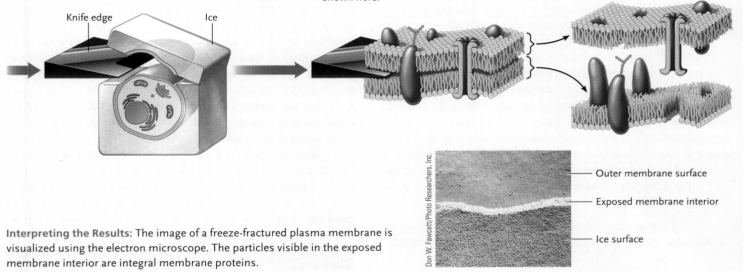

Knife edge · Ice

Outer membrane surface

Exposed membrane interior

Ice surface

Don W. Fawcett/Photo Researchers, Inc.

Interpreting the Results: The image of a freeze-fractured plasma membrane is visualized using the electron microscope. The particles visible in the exposed membrane interior are integral membrane proteins.

© Cengage Learning 2014

---

in, for example, liquid nitrogen. Then they fracture the block by giving it a blow from the sharp edge of a microscopic knife. Often, the fracture splits bilayers into inner and outer halves, exposing the hydrophobic membrane interior. In the electron microscope, the split membranes appear as smooth layers in which individual particles the size of proteins are embedded (see Figure 6.7C).

Recent research has shown that the asymmetry of lipids and proteins in membranes is not the result of free mixing of the components. Rather, both lipids and proteins tend to form ordered arrangements in membranes, and their movements may be restricted. In fact, many cellular processes require a specific structural organization of membrane lipids and proteins. Cell communication, the topic of the next chapter, is an example of such a process.

Membranes Have Several Functions

Membranes perform a diverse array of functions. As you will see, often the membrane protein defines the function of a membrane or membrane segment.

- Membranes define the boundaries of cells and, in eukaryotes, the boundaries of compartments (for example, the nucleus, mitochondria, and chloroplasts).

- Membranes are permeability barriers, permitting regulated control of the contents of cells compared with the extracellular environments.

- Some membranes have enzyme activities that are the properties of the membrane proteins. For instance, the particular enzymes found in or on their membranes define the specific properties of many eukaryotic organelles, such as the endoplasmic reticulum, Golgi complex, lysosome, mitochondrion, and chloroplast (see Sections 5.3 and 5.4). Enzymes on mitochondrial and chloroplast internal membranes, for example, play essential roles in converting the chemical energy in energy-rich nutrients to ATP in cellular respiration, and the conversion of light energy to the chemical energy of ATP in photosynthesis (see Sections 5.3 and 5.4, and Chapters 8 and 9).

- Membrane-spanning channel proteins form channels that selectively transport specific ions or water through the membrane. Proteins that form channels for ions may either allow the ions to pass freely, or may regulate their passage. For example, specific channel proteins control the passage of individual ions such as Na^+, K^+, Ca^{2+}, and Cl^- (see next section).

- Membrane-spanning carrier proteins bind to specific substances and transport them across the membrane. An example is the carrier protein for transporting glucose into cells (see next section).

- Some membrane proteins serve as receptors that recognize and bind specific molecules in the extracellular environment. Depending on the system, binding serves as the first step in bringing the substance into the cell, or it activates the receptor and triggers a series of molecular events within the cell that leads to a cellular response. For instance, some hormones bind to receptors in the plasma membrane and cause the cell to change its gene activity (discussed in Chapter 7).

- Membranes have electrical properties as a result of an uneven distribution of particular ions inside and outside of the cell. These electrical properties can serve as a mechanism of signal conduction when a cell receives an electrical, chemical, or mechanical stimulus. For instance, neurons (nerve cells) and muscle cells conduct electrical signals by using the electrical properties of membranes (discussed in Chapter 39).

- Some membrane proteins facilitate cell adhesion and cell-to-cell communication. For example, in gap junctions particular plasma membrane proteins of adjacent cells line up and form pipelines between the two cells (see Section 5.5 and Figure 5.27).

In the remainder of the chapter we focus on the functions of membranes in transporting substances into and out of the cell.

STUDY BREAK 6.1

1. Describe the fluid mosaic model for membrane structure.
2. Give two examples each of integral proteins and peripheral proteins.

6.2 Functions of Membranes in Transport: Passive Transport

Transport is the controlled movement of ions and molecules from one side of a membrane to the other. Membrane proteins are the molecules responsible for transport. Typically the movement is *directional;* that is, some ions and molecules consistently move into cells, whereas others move out of cells. Transport is also *specific;* that is, only certain ions and molecules move directionally across membranes. Transport is critical to the ionic and molecular organization of cells, and with it, the maintenance of cellular life.

Transport occurs by two mechanisms:

1. **Passive transport** depends on concentration differences on the two sides of a membrane (concentration = number of molecules or ions per unit volume). Passive transport moves ions and molecules across the membrane *with* the concentration gradient; that is, from the side with the higher concentration to the side with the lower concentration. The difference in concentration provides the energy for this form of transport.

2. **Active transport** moves ions or molecules *against* the concentration gradient; that is, from the side with the lower concentration to the side with the higher concentration. Active transport uses energy obtained directly or indirectly by breaking down ATP. **Table 6.1** compares the properties of passive and active transport.

Passive Transport Is Based on Diffusion

Passive transport is a form of **diffusion,** the net movement of ions or molecules from a region of higher concentration to a region of lower concentration. Diffusion depends on the constant motion of ions or molecules at temperatures above absolute zero ($-273°C$). For instance, if you add a drop of food dye to a container of clear water, the dye molecules, and therefore the color, will spread or *diffuse* from their initial center of high concentration until they are distributed evenly. At this point, the water has an even color.

Diffusion involves a *net* movement of molecules or ions. Molecules and ions actually move in all directions at all times in a solution as a result of thermal (heat) energy. But when there is a concentration difference, termed a **concentration gradient,** more molecules or ions move from the area of higher

TABLE 6.1	Characteristics of Transport Mechanisms		
	Passive Transport		
Characteristic	Simple Diffusion	Facilitated Diffusion	Active Transport
Membrane component responsible for transport	Lipids, water	Proteins, water	Proteins
Binding of transported substance	No	Yes	Yes
Energy source	Concentration gradients	Concentration gradients	ATP hydrolysis or concentration gradients
Direction of transport	With gradient of transported substance	With gradient of transported substance	Against gradient of transported substance
Specificity for molecules or molecular classes	Nonspecific	Specific	Specific
Saturation at high concentrations of transported molecules	No	Yes	Yes

© Cengage Learning 2014

concentration to areas of lower concentration than in the opposite direction. Even after their concentration is the same in all regions, molecules or ions still move constantly from one space to another, but there is no net change in concentration on either side. This condition is an example of a *dynamic equilibrium* (*dynamic* with respect to the continuous movement, and *equilibrium* with respect to the exact balance between opposing forces).

Substances Move Passively through Membranes by Simple or Facilitated Diffusion

Hydrophobic (nonpolar) molecules are able to dissolve in the lipid bilayer of a membrane and move through it freely. By contrast, the hydrophobic core of the membrane impedes the movement of hydrophilic molecules such as ions and polar molecules; thus, their passage is blocked or slow. Membranes that affect diffusion in this way are said to be **selectively permeable.**

TRANSPORT BY SIMPLE DIFFUSION A few small substances diffuse through the lipid part of a biological membrane. With one major exception—water—these substances are nonpolar inorganic gases such as O_2, N_2, and CO_2 and nonpolar organic molecules such as steroid hormones. This type of transport, which depends solely on molecular size and lipid solubility, is **simple diffusion** (see Table 6.1).

Water is a strongly polar molecule. Nevertheless, water molecules are small enough to slip through momentary spaces created between the hydrocarbon tails of phospholipid molecules as they flex and move in a fluid bilayer. This type of water movement across the membrane is relatively slow.

TRANSPORT BY FACILITATED DIFFUSION Many polar and charged molecules such as water, amino acids, sugars, and ions diffuse across membranes with the help of transport proteins, a mechanism termed **facilitated diffusion.** The transport proteins enable polar and charged molecules to avoid interaction with the hydrophobic lipid bilayer (see Table 6.1).

Facilitated diffusion is specific in that the membrane proteins involved transport certain polar and charged molecules, but not others. Facilitated diffusion is also dependent on concentration gradients. That is, proteins aid the transport of polar and charged molecules through membranes, but a favorable concentration gradient provides the energy for transport. Transport stops if the gradient falls to zero.

Two Groups of Transport Proteins Carry Out Facilitated Diffusion

The proteins that carry out facilitated diffusion are integral membrane proteins that extend entirely through the membrane. Two types of transport proteins are involved in facilitated diffusion:

1. **Channel proteins** form hydrophilic channels in the membrane through which water and ions can pass **(Figure 6.8A).**

The channel "facilitates" the diffusion of molecules through the membrane by providing an avenue. For example, facilitated diffusion of water through membranes occurs through specialized water channels called **aquaporins** (see Figure 6.8A(1)). A billion molecules of water per second can move through an aquaporin channel. How the molecules move is fascinating. Each water molecule is severed from its hydrogen-bonded neighbors as it is handed off to a succession of hydrogen-bonding sites on the aquaporin protein in the channel. Peter Agre at Johns Hopkins University in Baltimore, Maryland, received a Nobel Prize in 2003 for his discovery of aquaporins. (Aquaporins in plants are discussed in Chapter 34.)

Other channel proteins, **ion channels,** facilitate the transport of ions such as sodium (Na^+), potassium (K^+), calcium (Ca^{2+}), and chlorine (Cl^-). Ion channels occur in all eukaryotes. Most ion channels are **gated channels;** that is, they switch between open, closed, or intermediate states. For instance, the gates may open or close in response to changes in voltage across the membrane, or by binding signal molecules. The opening or closing involves changes in the protein's three-dimensional shape. In animals, voltage-gated ion channels are used in nerve conduction and the control of muscle contraction (see Chapters 39 and 43); Figure 6.8A(2) shows a voltage-gated K^+ channel.

Gated ion channels perform functions that are vital to survival, as illustrated by the effects of hereditary defects in the channels. For example, as you learned in *Why It Matters,* the lethal genetic disease *cystic fibrosis* (CF) results from a fault in a Cl^- channel.

2. **Carrier proteins** also form passageways through the lipid bilayer **(Figure 6.8B).** Carrier proteins each bind a specific single solute, such as glucose or an amino acid, and transport it across the lipid bilayer. (Glucose is also transported by active transport, as described in the next section.) Because a single solute is transferred in this carrier-mediated fashion, the transfer is called *uniport transport.* In performing the transport step, the carrier protein undergoes conformational changes that progressively move the solute-binding site from one side of the membrane to the other, thereby transporting the solute. This property distinguishes carrier protein function from channel protein function.

Facilitated diffusion by carrier proteins can become *saturated* when there are too few transport proteins to handle all the solute molecules. For example, if glucose is added at higher and higher concentrations to the solution that surrounds an animal cell, the rate at which glucose passes through the membrane at first increases proportionately with the increase in concentration. However, at some point, as the glucose concentration is increased still further, the increase in the rate of transport slows. Eventually, further increases in concentration cause no additional rise in the rate of transport—the transport mechanism is saturated. By contrast, saturation does not occur for simple diffusion.

A. Channel protein

Channel proteins form hydrophilic channels in the membrane through which water and ions can move.

1 Aquaporin

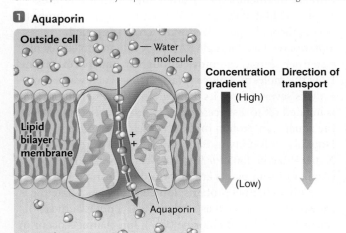

An aquaporin is a water channel. Water molecules move through the channel by being handed off to a succession of hydrogen-bonding sites on the channel in this protein.

2 K⁺ voltage-gated channel

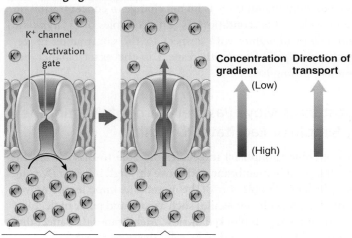

With normal voltage across the membrane, the activation gate of the K⁺ channel is closed and K⁺ cannot move across the membrane.

In response to a voltage change across the membrane, the activation gate of the K⁺ channel opens, and K⁺ moves with its concentration gradient from the cytoplasm to outside the cell.

B. Carrier protein

Carrier proteins each bind a single solute and transport it across the lipid bilayer. During the transport step, the carrier protein undergoes conformational changes that progressively move the solute-binding site from one side of the membrane to the other, thereby transporting the solute. Shown is the transport of glucose.

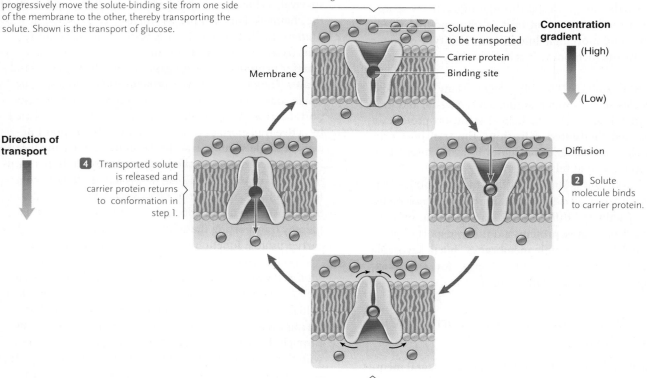

1 Carrier protein is in conformation so that binding site is exposed toward region of higher concentration.

2 Solute molecule binds to carrier protein.

3 In response to binding, carrier protein changes conformation so that binding site is exposed to region of lower concentration.

4 Transported solute is released and carrier protein returns to conformation in step 1.

FIGURE 6.8 **Transport proteins for facilitated diffusion.**
(A) Channel protein: (1) aquaporin; (2) K⁺ voltage-gated channel. **(B)** Carrier protein.
© Cengage Learning 2014

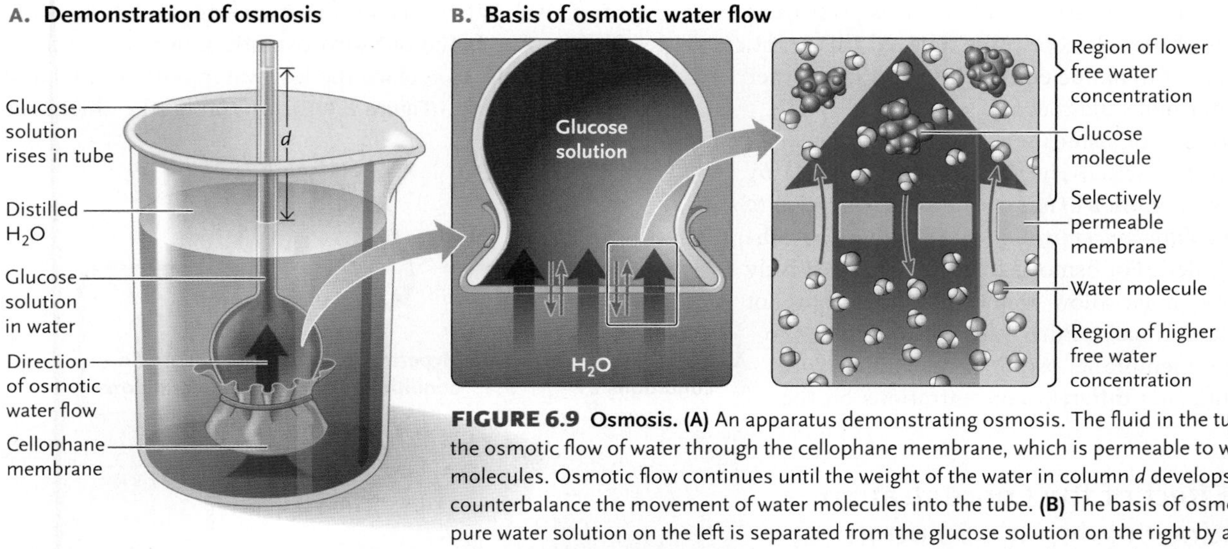

A. Demonstration of osmosis

- Glucose solution rises in tube
- Distilled H₂O
- Glucose solution in water
- Direction of osmotic water flow
- Cellophane membrane

B. Basis of osmotic water flow

Glucose solution

H₂O

- Region of lower free water concentration
- Glucose molecule
- Selectively permeable membrane
- Water molecule
- Region of higher free water concentration

FIGURE 6.9 Osmosis. **(A)** An apparatus demonstrating osmosis. The fluid in the tube rises because of the osmotic flow of water through the cellophane membrane, which is permeable to water but not to glucose molecules. Osmotic flow continues until the weight of the water in column d develops enough pressure to counterbalance the movement of water molecules into the tube. **(B)** The basis of osmotic water flow. The pure water solution on the left is separated from the glucose solution on the right by a membrane permeable to water but not to glucose. The free water concentration on the glucose side is lower than on the water-only side because water molecules are associated with the glucose molecules. That is, water molecules are in greater concentration on the bottom than on the top. Although water molecules move in both directions across the membrane (small red arrows), there is a net upward movement of water (blue arrows), with the water's concentration gradient.

© Cengage Learning 2014

Because the proteins that perform facilitated diffusion are specific, cells can control the kinds of molecules and ions that pass through their membranes by regulating the types of transport proteins in their membranes. As a result, each type of cellular membrane, and each type of cell, has its own group of transport proteins and passes a characteristic group of substances by facilitated diffusion. The kinds of transport proteins present in a cell ultimately depend on the activity of genes in the cell nucleus.

STUDY BREAK 6.2

1. What is the difference between passive and active transport?
2. What is the difference between simple and facilitated diffusion?

6.3 Passive Water Transport and Osmosis

As discussed earlier, water can follow concentration gradients and diffuse passively across membranes in response. It diffuses directly both through the membrane and aquaporins. The passive transport of water, called **osmosis,** occurs constantly in living cells. Inward or outward movement of water by osmosis develops forces that can cause cells to swell and burst or shrink and shrivel up. Much of the energy budget of many cell types, particularly in animals, is spent counteracting the inward or outward movement of water by osmosis.

Osmosis Can Be Demonstrated in a Purely Physical System

The apparatus shown in **Figure 6.9A** is a favorite laboratory demonstration of osmosis. It consists of an inverted thistle tube (so named because its shape resembles a thistle flower) tightly sealed at its lower end by a sheet of cellophane. The tube is filled

with a solution of glucose molecules in water and is suspended in a beaker of distilled water. The cellophane acts as a selectively permeable membrane because its pores are large enough to admit water molecules but not glucose. At the start of the experiment, the position of the tube is set so the level of the liquid in the tube is at the same level as the distilled water in the beaker. Almost immediately, the level of the solution in the tube begins to rise, eventually reaching a maximum height above the liquid in the beaker.

The liquid rises in the tube because water moves by osmosis from the beaker into the thistle tube. The movement occurs passively, in response to a concentration gradient in which the water molecules are more concentrated in the beaker than inside the thistle tube. The basis for the gradient is shown in **Figure 6.9B.** The glucose molecules are more concentrated on one side of the selectively permeable membrane. On this side, association of water molecules with those solute molecules reduces the amount of water available to cross the membrane. Thus, although initially there is an equal apparent water concentration on each side of the membrane, there is a difference in the *free water* concentration—that is, the water available to move across the membrane. Specifically, the concentration of free water molecules is lower on the glucose side than on the pure water side. In response, more water molecules from the pure water side will hit the pores in the membrane than from the solute side, producing a net movement of water from the pure water side to the glucose solution side. Osmosis is the net diffusion of water molecules through a selectively permeable membrane in response to a gradient of this type.

The solution stops rising in the tube when the pressure created by the weight of the raised solution exactly balances the tendency of water molecules to move from the beaker into the tube in response to the concentration gradient. This pressure is the **osmotic pressure** of the solution in the tube. At this point, the system is in a state of dynamic equilibrium and no further net movement of water molecules occurs.

A formal definition for osmosis is *the net movement of water molecules across a selectively permeable membrane by passive diffusion, from a solution of lesser solute concentration to a solution of greater solute concentration* (the *solute* is the substance dissolved in water). For osmosis to occur, the selectively permeable membrane must allow water molecules, but not molecules of the solute, to pass. Pure water does not need to be on one side of the membrane; osmotic water movement also occurs if a solute is at different concentrations on the two sides.

Tonicity Is the Effect of the Concentration of Nonpenetrating Solutes in a Solution on Cell Volume

The solution surrounding a cell can affect cell volume—whether the cell remains the same size, shrinks, or swells—depending on the relative concentrations of nonpenetrating solutes surrounding the cell and within the cell. *Nonpenetrating solutes* are proteins and other molecules that cannot pass through a membrane impermeable to them but that membrane is freely permeable to water. The effect a solution has on cell volume when the solution surrounds the cell is the **tonicity** of the solution (*tonos* = tension or tone). Tonicity is a property of a solution with respect to a particular membrane. That is, the same surrounding solution can have different effects on different cell types.

If the solution surrounding a cell contains nonpenetrating solutes at lower concentrations than in the cell, the solution is said to be **hypotonic** to the cell (*hypo* = under or below). When a cell is in a hypotonic solution, water enters by osmosis and the cell tends to swell **(Figure 6.10A)**. Animal cells (for instance, red blood cells) in a hypotonic solution may actually swell to the point of bursting. However, in most plant cells, strong walls prevent the cells from bursting in a hypotonic solution. In most land plants, the cells at the surfaces of roots are surrounded by almost pure water, which is hypotonic to the cells and tissues of the root. As a result, water flows from the surrounding soil into the root cells by osmosis. The osmotic pressure developed by the inward flow contributes part of the force required to raise water from the roots to the leaves of the plant. Osmosis also drives water into cells of the stems and leaves of plants. The resulting osmotic pressure, called **turgor pressure,** pushes the cells tightly against their walls and supports the softer tissues against the force of gravity. (Turgor pressure in plants is described more in Chapter 34.)

If the solution that surrounds a cell contains nonpenetrating solutes at higher concentrations than in the cell, the outside solution is said to be **hypertonic** to the cells (*hyper*

= over or above). When a cell is in a hypertonic solution, water leaves by osmosis. If the outward osmotic movement exceeds the capacity of cells to replace the lost water, both animal and plant cells will shrink **(Figure 6.10B)**. In plants, the shrinkage

FIGURE 6.10 Tonicity and osmotic water movement. The diagrams show what happens when a cellophane bag filled with a 2 *M* sucrose solution is placed in a **(A)** hypotonic, **(B)** hypertonic, or **(C)** isotonic solution. They also show the corresponding effects of these three types of solutions on animal and plant cells.
© Cengage Learning 2014

and loss of internal osmotic pressure under these conditions causes stems and leaves to wilt. In extreme cases, plant cells shrink so much that they retract from their walls, a condition known as **plasmolysis.**

In animals, ions, proteins, and other molecules are concentrated in extracellular fluids, as well as inside cells, so that the concentration of water inside and outside cells is usually equal or **isotonic** (*iso* = the same; see **Figure 6.10C**). To keep fluids on either side of the plasma membrane isotonic, animal cells must constantly use energy to pump Na^+ from inside to outside by active transport (see Section 6.4); otherwise, water would move inward by osmosis and cause the cells to burst. For animal cells, an isotonic solution is usually optimal, whereas for plant cells, an isotonic solution results in some loss of turgor. The mechanisms by which plants and animals balance their water content by regulating osmosis are discussed in Chapters 34 and 48, respectively.

Passive transport, driven by concentration gradients, accounts for much of the movement of water, ions, and many types of molecules into or out of cells. In addition, all cells transport some ions and molecules against their concentration gradients by active transport (see the next section).

STUDY BREAK 6.3

1. What conditions are required for osmosis to occur?
2. Explain the effect of a hypertonic solution that surrounds animal cells.

6.4 Active Transport

Facilitated diffusion accelerates the movement of substances across cellular membranes, but it is limited to transport with a concentration gradient. To transport substances across a membrane against a concentration gradient requires **active transport,** a process that requires energy input.

The three main functions of active transport in cells and organelles are: (1) uptake of essential nutrients from the fluid surrounding cells even when their concentrations are lower than in cells; (2) removal of secretory or waste materials from cells or organelles even when the concentration of those materials is higher outside the cells or organelles; and (3) maintenance of essentially constant intracellular concentrations of H^+, Ca^{2+}, Na^+, and K^+.

Because ions are charged molecules, active transport of ions may contribute to voltage—an electrical potential difference—across the plasma membrane, called a **membrane potential.** The unequal distribution of ions across the membrane created by passive transport also contributes to the voltage. Neurons and muscle cells use the membrane potential in a specialized way. That is, in response to electrical, chemical, mechanical, and certain other types of stimuli, their membrane potential changes rapidly and transiently. In nerve cells, for example, this type of transport is the basis for transmission of a nerve impulse (discussed in Chapter 39).

In short, active transport usually establishes differences in solute concentrations or of voltage across membranes that are important for cell or organelle function. By contrast, passive diffusion and facilitated diffusion act mostly to move substances across membranes in the direction toward equalizing their concentrations on each side.

Active Transport Requires a Direct or Indirect Input of Energy Derived from ATP Hydrolysis

There are two kinds of active transport:

1. In **primary active transport,** the same protein that transports a substance also hydrolyzes ATP to power the transport directly.
2. In **secondary active transport,** the transport is driven indirectly by ATP hydrolysis. That is, the transporter proteins do not break down ATP but use instead a favorable concentration gradient of ions, built up by primary active transport, as their energy source for active transport of a different ion or molecule.

Other features of active transport resemble facilitated diffusion (listed in Table 6.1). Both processes depend on membrane transport proteins, both are specific, and both can be saturated. In both, the transport proteins are carrier proteins that change their conformation as they function.

Primary Active Transport Moves Positively Charged Ions across Membranes

The primary active transport pumps all move positively charged ions—H^+, Ca^{2+}, Na^+, and K^+—across membranes **(Figure 6.11).** The gradients of positive ions established by primary active transport pumps underlie functions that are essential for cellular life. For example, H^+ **pumps** (also called **proton pumps**) move H^+ hydrogen ions (protons) across membranes, temporarily binding a phosphate group removed from ATP during the pumping cycle. Proton pumps have various functions. For example, proton pumps in the plasma membrane in prokaryotes, plants, and fungi generate membrane potential. Proton pumps in lysosomes of animals and vacuoles of plants and fungi keep the pH within the organelle low, activating the enzymes contained within them.

The Ca^{2+} **pump** (or **calcium pump**) is distributed widely among eukaryotes. It pushes Ca^{2+} from the cytoplasm to the cell exterior, and also from the cytosol into the vesicles of the endoplasmic reticulum (ER). As a result, Ca^{2+} concentration is typically high outside cells and inside ER vesicles and low in the cytosol. This Ca^{2+} gradient is used universally among eukaryotes as a regulatory control of cellular activities as diverse as secretion, microtubule assembly, and muscle contraction. For the process of muscle contraction in animals, calcium pumps release stored Ca^{2+} from vesicles inside a muscle fiber (a single cell in a muscle) which initiates a series of steps leading to contraction of the fiber (see Chapter 43). In plants, a calcium pump is involved in pollen growth and fertilization.

In primary active transport, the carrier protein that transports a substance also hydrolyzes ATP to power the transport directly. Primary active transport pumps all positively charged ions across membranes against their concentration gradients.

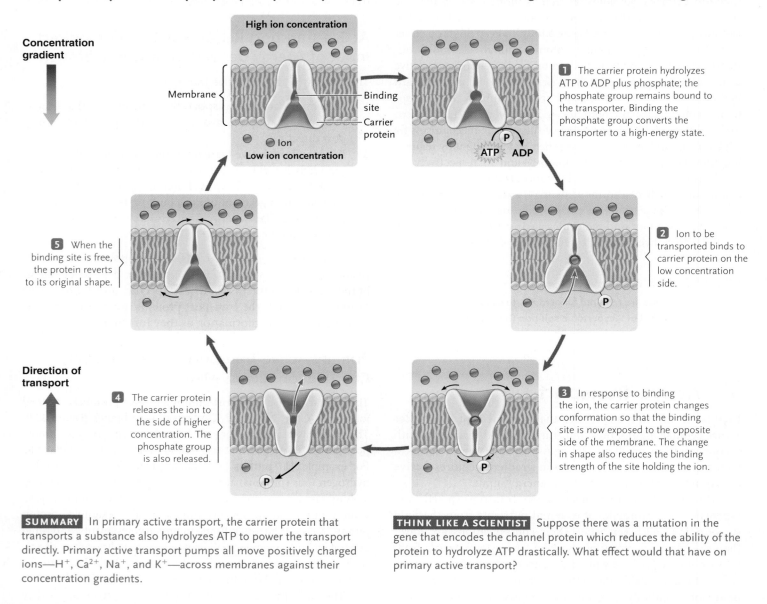

SUMMARY In primary active transport, the carrier protein that transports a substance also hydrolyzes ATP to power the transport directly. Primary active transport pumps all move positively charged ions—H⁺, Ca²⁺, Na⁺, and K⁺—across membranes against their concentration gradients.

THINK LIKE A SCIENTIST Suppose there was a mutation in the gene that encodes the channel protein which reduces the ability of the protein to hydrolyze ATP drastically. What effect would that have on primary active transport?

© Cengage Learning 2014

The **Na⁺/K⁺ pump** (also known as the **sodium–potassium pump or Na⁺/K⁺-ATPase**), located in the plasma membrane of all animal cells, pushes three Na⁺ out of the cell and two K⁺ into the cell in the same pumping cycle **(Figure 6.12)**. As a result, positive charges accumulate in excess outside the membrane, and the inside of the cell becomes negatively charged with respect to the outside. This creates a membrane potential measuring from about −50 to −200 millivolts (mV; 1 millivolt = 1/1,000th of a volt), with the minus sign indicat-ing that the charge inside the cell is negative versus the outside. That is, there is both a concentration difference (of the ions) and an electrical charge difference on the two sides of the membrane, constituting what is called an **electrochemical gradient.** Electrochemical gradients store energy that is used for other transport mechanisms. For instance, an electro-chemical gradient across a nerve cell membrane drives the movement of ions involved in nerve impulse transmission (see Chapter 39).

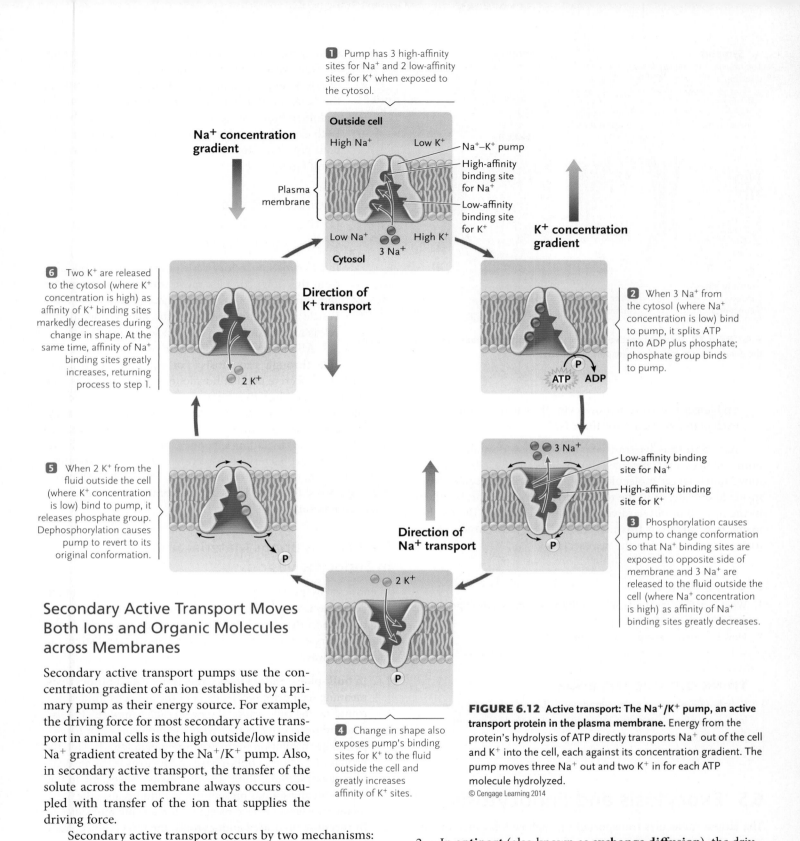

1 Pump has 3 high-affinity sites for Na$^+$ and 2 low-affinity sites for K$^+$ when exposed to the cytosol.

Na$^+$ concentration gradient

Outside cell

High Na$^+$ Low K$^+$

Na$^+$–K$^+$ pump

High-affinity binding site for Na$^+$

Plasma membrane

Low-affinity binding site for K$^+$

Low Na$^+$ High K$^+$

3 Na$^+$

Cytosol

K$^+$ concentration gradient

Direction of K$^+$ transport

6 Two K$^+$ are released to the cytosol (where K$^+$ concentration is high) as affinity of K$^+$ binding sites markedly decreases during change in shape. At the same time, affinity of Na$^+$ binding sites greatly increases, returning process to step 1.

2 K$^+$

2 When 3 Na$^+$ from the cytosol (where Na$^+$ concentration is low) bind to pump, it splits ATP into ADP plus phosphate; phosphate group binds to pump.

P

ATP   ADP

5 When 2 K$^+$ from the fluid outside the cell (where K$^+$ concentration is low) bind to pump, it releases phosphate group. Dephosphorylation causes pump to revert to its original conformation.

P

3 Na$^+$

Low-affinity binding site for Na$^+$

High-affinity binding site for K$^+$

3 Phosphorylation causes pump to change conformation so that Na$^+$ binding sites are exposed to opposite side of membrane and 3 Na$^+$ are released to the fluid outside the cell (where Na$^+$ concentration is high) as affinity of Na$^+$ binding sites greatly decreases.

P

Direction of Na$^+$ transport

2 K$^+$

P

4 Change in shape also exposes pump's binding sites for K$^+$ to the fluid outside the cell and greatly increases affinity of K$^+$ sites.

Secondary Active Transport Moves Both Ions and Organic Molecules across Membranes

Secondary active transport pumps use the concentration gradient of an ion established by a primary pump as their energy source. For example, the driving force for most secondary active transport in animal cells is the high outside/low inside Na$^+$ gradient created by the Na$^+$/K$^+$ pump. Also, in secondary active transport, the transfer of the solute across the membrane always occurs coupled with transfer of the ion that supplies the driving force.

Secondary active transport occurs by two mechanisms:

1. In **symport** (also called **cotransport**), the solute moves through the membrane channel in the same direction as the driving ion **(Figure 6.13A)**. Sugars, such as glucose, and amino acids are examples of molecules actively transported into cells by symport.

FIGURE 6.12 Active transport: The Na$^+$/K$^+$ pump, an active transport protein in the plasma membrane. Energy from the protein's hydrolysis of ATP directly transports Na$^+$ out of the cell and K$^+$ into the cell, each against its concentration gradient. The pump moves three Na$^+$ out and two K$^+$ in for each ATP molecule hydrolyzed.
© Cengage Learning 2014

2. In **antiport** (also known as **exchange diffusion**), the driving ion moves through the membrane channel in one direction, providing the energy for the active transport of another molecule through the membrane in the opposite direction **(Figure 6.13B)** In many cases, ions are exchanged by antiport. For example, in many tissues, a Na$^+$/Ca^{2+}

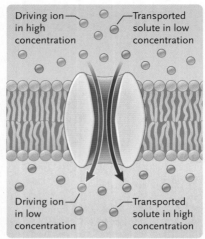

A. Symport
The transported solute moves in the same direction as the gradient of the driving ion.

Driving ion in high concentration — Transported solute in low concentration

Driving ion in low concentration — Transported solute in high concentration

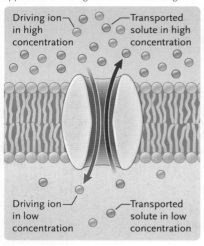

B. Antiport
The transported solute moves in the direction opposite from the gradient of the driving ion.

Driving ion in high concentration — Transported solute in high concentration

Driving ion in low concentration — Transported solute in low concentration

FIGURE 6.13 Secondary active transport: a concentration gradient of an ion is used as the energy source for active transport of a solute.
© Cengage Learning 2014

exchanger is used to remove cytosolic calcium from cells, exchanging one Ca^{2+} for three Na^+.

Active and passive transport move ions and smaller hydrophilic molecules across cellular membranes. Cells can also move much larger molecules or aggregates of molecules from inside to outside, or in the reverse direction, by including them in the inward or outward vesicle traffic of the cell. The mechanisms that carry out this movement—exocytosis and endocytosis—are discussed in the next section.

STUDY BREAK 6.4

1. What is active transport? What is the difference between primary and secondary active transport?
2. How is a membrane potential generated?

THINK OUTSIDE THE BOOK >

Collaboratively or on your own, draw a diagram showing how ions and solutes exchange in a symport system and an antiport system in the intestinal tract.

6.5 Exocytosis and Endocytosis

The largest molecules transported through cellular membranes by passive and active transport are in the size range of amino acids or monosaccharides such as glucose. Eukaryotic cells import and export larger molecules by exocytosis and endocytosis (introduced in Section 5.3). The export of materials by exocytosis primarily carries secretory proteins and some waste materials from the cytoplasm to the cell exterior. Import by endocyto-

sis may carry proteins, larger aggregates of molecules, or even whole cells from the outside into the cytoplasm. Exocytosis and endocytosis also contribute to the back-and-forth flow of membranes between the endomembrane system and the plasma membrane. Both exocytosis and endocytosis require energy; thus, both processes stop if the ability of a cell to make ATP is inhibited.

Exocytosis Releases Molecules to the Outside of the Cell by Means of Secretory Vesicles

In exocytosis, secretory vesicles originated by budding from the Golgi complex move through the cytoplasm and contact the plasma membrane **(Figure 6.14A)**. The vesicle membrane fuses with the plasma membrane, releasing the contents of the vesicle to the cell exterior.

All eukaryotic cells secrete materials to the outside through exocytosis. For example, in animals, glandular cells secrete peptide hormones or milk proteins, and cells that line the digestive tract secrete mucus and digestive enzymes. Plant cells, fungal cells, and bacterial cells use exocytosis to secrete proteins and other macromolecules associated with the cell wall, including enzymes, proteins, and carbohydrates. Fungi and bacteria also secrete enzymes by exocytosis to digest nutrients in their environments. Finally, all organisms use exocytosis to place integral membrane proteins in the plasma membrane.

Endocytosis Brings Materials into Cells in Endocytic Vesicles

In endocytosis, proteins and other substances are trapped in pit-like depressions that bulge inward from the plasma membrane. The depression then pinches off as an endocytic vesicle. Endocytosis occurs in most eukaryotic cells by one of two distinct but related pathways:

1. In **bulk-phase endocytosis** (sometimes called **pinocytosis,** meaning "cell drinking") a drop of the aqueous fluid surrounding the cell—called the *extracellular fluid* (ECF)—is taken into the cell together with any molecules that happen to be in solution in the water **(Figure 6.14B)**. In fact, the primary function of bulk-phase endocytosis is the absorption of extracellular fluid. The process is nonspecific in that it takes in any solutes present in the fluid because the membrane lacks surface receptors for specific molecules.

2. In **receptor-mediated endocytosis,** the target molecules to be taken in are bound to receptor proteins on the outer cell surface **(Figure 6.14C, D)**. The receptors, which are integral proteins of the plasma membrane, recognize and bind only certain molecules in the solution that surrounds the cell, which makes this type of endocytosis highly specific. After binding their target molecules, the receptors

A. Exocytosis: vesicle joins plasma membrane, releases contents

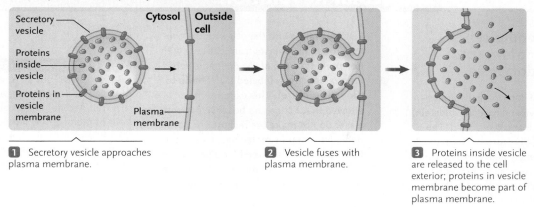

1 Secretory vesicle approaches plasma membrane.

2 Vesicle fuses with plasma membrane.

3 Proteins inside vesicle are released to the cell exterior; proteins in vesicle membrane become part of plasma membrane.

B. Bulk-phase endocytosis (pinocytosis): vesicle imports water and other substances from outside cell

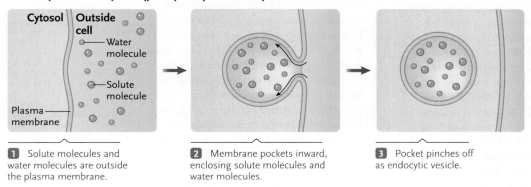

1 Solute molecules and water molecules are outside the plasma membrane.

2 Membrane pockets inward, enclosing solute molecules and water molecules.

3 Pocket pinches off as endocytic vesicle.

C. Receptor-mediated endocytosis: vesicle imports specific molecules

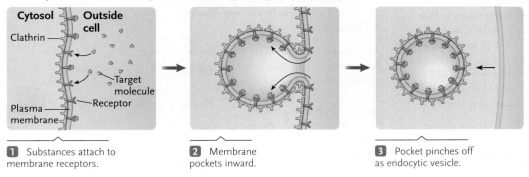

1 Substances attach to membrane receptors.

2 Membrane pockets inward.

3 Pocket pinches off as endocytic vesicle.

D. Micrographs of stages of receptor-mediated endocytosis shown in C

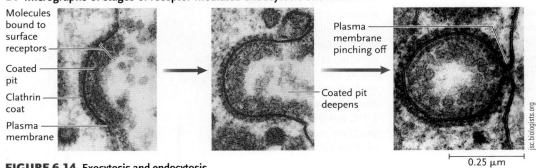

0.25 μm

jsc.biologists.org

FIGURE 6.14 Exocytosis and endocytosis.

© Cengage Learning 2014

Research Serendipity:
The discovery of receptor-mediated endocytosis

When they were still in training, two physicians, Michael Brown and Joseph Goldstein, treated two sisters, 6 and 8 years old. The children were dying of recurrent heart attacks brought on by extremely high cholesterol levels in their blood. Brown and Goldstein performed groundbreaking research to determine how such young children develop a condition that usually appears only in late middle or old age.

As you have learned, cholesterol is essential for keeping cell membranes fluid. The blood transports cholesterol to cells that need it. But blood cholesterol can cause *atherosclerosis* (thickening of the arteries as a result of the buildup of fatty materials such as cholesterol), which can be lethal. To lessen this danger, the body packages cholesterol with proteins to form lipoproteins such as low-density lipoprotein (LDL).

Patients with familial hypercholesteremia (FH) have a higher than normal concentration of LDL in their blood and often experience atherosclerosis and heart attacks early in life. Individuals with one copy of the mutated gene responsible for the disease have about twice the normal level of LDL, and begin to have heart attacks at 30 to 40 years old. Individuals like the two sisters, with two copies of the mutated gene, have 6- to 10-fold higher than normal levels of LDL in their blood, and they often have heart attacks in childhood.

Research Question

What is the molecular basis of familial hypercholesteremia?

Experiments

Brown and Goldstein of the University of Texas Southwestern Medical School, Dallas, Texas, carried out two key experiments:

1. The researchers cultured human cells (skin fibroblasts) from normal individuals and found that radioactively labeled LDL bound strongly to the cells. They interpreted this result to mean that the cells had specific surface receptors for LDL. Once bound to the receptor, the LDL was taken into the cell with the cholesterol. In fibroblasts from patients with two mutant forms of the *FH* gene, very little radioactively labeled LDL bound to the cells **(Figure)**. This experiment showed that FH occurs because patients either do not make the receptor or have defective receptor molecules. As a result, LDL accumulates in the blood because it cannot be taken into the cells. Further research has confirmed that a mutation in the gene for this receptor is responsible for FH.

2. To determine how the receptor–LDL complex enter cells, the two researchers, with Richard G. W. Anderson, used the same fibroblast system. In this case, they bound ferritin to the LDL. Ferritin is a protein that binds iron, which is electron dense, making it possible to detect the LDL visually as black dots under the electron microscope. When they incubated fibroblast cells from normal individuals with ferritin-labeled LDL, the investigators saw that the ferritin collected at short segments of the plasma membrane, which appeared to be indented and coated on both sides by "fuzzy material." These regions we now know to be clathrin-coated pits, and they are the cellular entry site for the receptor–LDL complex.

Conclusion

In a major step toward answering the question of what causes FH, Goldstein and Brown had discovered that LDL enters cells using a specific receptor on the cell surface, which is absent or reduced in FH patients. Serendipitously, they had also discovered the answer to broader question: How do cells take in specific molecules that cannot pass through the membrane? They had discovered receptor-mediated endocytosis (see Figures 6.14C and D). The researchers received the Nobel Prize for their discovery in 1986.

THINK LIKE A SCIENTIST The researchers also carried out an important variation of their first key experiment. Specifically, they cultured fibroblast cells in the presence both of radioactively labeled LDL and of a 50-fold excess of unlabeled LDL. The result of this experiment was that only a limited amount of radioactivity became associated with the cells, far less than in the experiment described in the box. Interpret this result, and explain whether or not it supports their overall conclusions.

Sources: R. G. W. Anderson, J. L. Goldstein, and M. S. Brown. 1976. Localization of low density lipoprotein receptors on plasma membrane of normal human fibroblasts and their absence in cells from a familial hypercholesteremia homozygote. *Proceedings of the National Academy of Sciences USA* 73:2434–2438; R. G. W. Anderson, J. L. Goldstein, and M. S. Brown. 1977. A mutation that impairs the ability of lipoprotein receptors to localize in coated pits on the cell surface of human fibroblasts. *Nature* 270:695–699; M. S. Brown and J. L. Goldstein. 1974. Familial hypercholesteremia: Defective binding of lipoproteins to cultured fibroblasts associated with impaired regulation of 3-hydroxy-3-methylglutaryl coenzyme A reductase activity. *Proceedings of the National Academy of Sciences USA* 71:788–792.

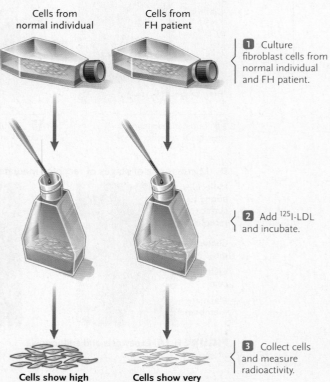

Cells from normal individual | Cells from FH patient

1 Culture fibroblast cells from normal individual and FH patient.

2 Add ^{125}I-LDL and incubate.

3 Collect cells and measure radioactivity.

Cells show high level of radioactivity | **Cells show very little radioactivity**

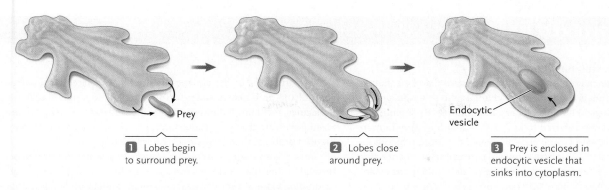

1 Lobes begin to surround prey.

2 Lobes close around prey.

3 Prey is enclosed in endocytic vesicle that sinks into cytoplasm.

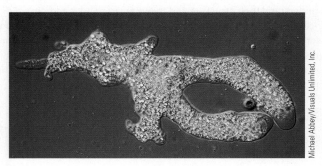

Michael Abbey/Visuals Unlimited, Inc.

FIGURE 6.15 Phagocytosis, in which lobes of the cytoplasm extend outward and surround a cell targeted as prey. The micrograph shows the protist *Chaos carolinense* engulfing a single-celled alga by phagocytosis (corresponding to step 2 in the diagram); white blood cells called phagocytes carry out a similar process in mammals.
© Cengage Learning 2014

collect into a depression in the plasma membrane called a **coated pit** because a network of proteins (called **clathrin**) coat and reinforce the cytoplasmic side. With the target molecules attached, the pits deepen and pinch free of the plasma membrane to form endocytic vesicles. Once in the cytoplasm, an endocytic vesicle rapidly loses its clathrin coat and may fuse with a lysosome. The enzymes within the lysosome then digest the contents of the vesicle, breaking them down into smaller molecules useful to the cell. These molecular products—for example, amino acids and monosaccharides—enter the cytoplasm by crossing the vesicle membrane via transport proteins. The membrane proteins are recycled to the plasma membrane.

Mammalian cells take in many substances by receptor-mediated endocytosis, including peptide hormones such as insulin, growth factors, enzymes, antibodies, blood proteins, iron, and vitamin B_{12}. Some viruses exploit the receptor-mediated exocytosis mechanism to enter cells. For instance, HIV, the virus that causes AIDS, binds to membrane receptors that function normally to internalize a needed molecule. The receptors that bind these substances to the plasma membrane are present in thousands to hundreds of thousands of copies. For example, a mammalian cell plasma membrane has about 20,000 receptors for *low-density lipoprotein (LDL)*. LDL, a complex of lipids and proteins, is the way cholesterol moves through the bloodstream. When LDL binds to its receptor on the membrane, it is taken into the cell by receptor-mediated endocytosis. Then, by the steps described earlier, the LDL is broken down within the cell and cholesterol is released into the cytoplasm. *Insights from the Molecular Revolution* describes the discovery of receptor-mediated endocytosis.

Some specialized cells, such as certain white blood cells (*phagocytes*) in the bloodstream, or protists such as *Amoeba proteus,* can take in large aggregates of molecules, cell parts, or even whole cells by a process related to receptor-mediated endocytosis. The process, called **phagocytosis** (meaning "cell eating"), begins when surface receptors bind molecules on the substances to be taken in **(Figure 6.15).** Cytoplasmic lobes then extend, surround, and engulf the materials, forming a pit that pinches off and sinks into the cytoplasm as a large endocytic vesicle. Enzymes then digest the materials as in receptor-mediated endocytosis, and the cell permanently sequesters any remaining residues into storage vesicles or expels them by exocytosis as wastes.

Working together, exocytosis and endocytosis constantly cycle membrane segments between the internal cytoplasm and the cell surface. The balance of the two mechanisms maintains the surface area of the plasma membrane at controlled levels.

Thus, through the combined mechanisms of passive transport, active transport, exocytosis, and endocytosis, cells maintain their internal concentrations of ions and molecules and exchange larger molecules such as proteins with their surroundings. The next chapter explores cell communication through intercellular chemical messengers. Many of these messengers act through binding to specific proteins embedded in the plasma membrane.

STUDY BREAK 6.5

1. What is the mechanism of exocytosis?
2. What is the difference between bulk-phase endocytosis and receptor-mediated endocytosis?

How do aquaporin channels function?

The discovery of water channels along with the structural and mechanistic studies of ion channels fundamentally changed the scientific understanding of how biological fluids cross cell membranes. By serendipity, we discovered the protein referred to as AQP1 (aquaporin-1) in human red blood cells, and isolated, purified, and cloned the protein before its function was identified. AQP1 is now known to permit movement of water across cell membranes by osmosis. Cell biological determinations established the sites in humans where AQP1 is expressed, thereby predicting its physiological significance and also predicting several diseases states. For example, in the kidney, AQP1 allows the reabsorption of water from primary urine; in capillaries, AQP1 facilitates the reabsorption of tissue edema (swelling because of fluid accumulation); and in the area of the brain where cerebrospinal fluid is synthesized, AQP1 permits secretion of that fluid. Humans with mutations in the gene encoding AQP1 that cause the absence of AQP1 are characterized by inability to concentrate urine despite prolonged thirsting.

The discovery of AQP1 quickly led to the discovery of several other mammalian aquaporins. In humans, AQP2 resides in the final segment of the kidney where the neurohormone antidiuretic hormone (also known as vasopressin; see Chapter 44) regulates water reabsorption by controlling exocytosis and endocytosis of vesicles containing AQP2 channels. Genetic defects in AQP2 result in severe nephrogenic diabetes insipidus—a disease where children must drink gallons of fluid every day, because they make large volumes of dilute urine. AQP4 exists in multiple sites, including the perivascular membranes of astroglial cells in brain. This protein has been linked to epileptic seizures and brain edema after injury. AQP5 is present in secretory glands and is responsible for sweat, tears, and saliva. Other aquaporins are necessary for eye lens homeostasis; defects in AQP0 are linked to congenital cataracts.

The structures of the aquaporins are highly related. The narrowest span of the pore allows water to move rapidly in single file. All larger molecules, including hydrated ions, are blocked by their greater diameter. Fixed positively charged residues in the pore serve as barriers to movement of protons, thereby restricting passage of protons.

Thus, the mysterious process of transcellular water movement occurs through molecular water channels. Together these proteins form a "plumbing system" for cells. Research on the aquaporin family is ongoing to understand their structure and their known cellular and physiological functions in detail. Where defects in aquaporins are implicated in clinical disorders, that understanding will fuel new research into treatments or cures. Research is also being done to identify other cellular and physiological roles of individual aquaporins, as well as to search for new members of the family in humans and in other organisms.

THINK LIKE A SCIENTIST Why do you think it is functional that the water channels in the cell membrane formed by aquaporins restrict the passage of protons?

Courtesy of Peter Agre

Dr. Peter Agre is director of the Johns Hopkins Malaria Research Institute. In 2003 he and Roderick McKinnon of Rockefeller University were awarded the Nobel Prize in Chemistry "for discoveries concerning channels in cell membranes." In 2009 Dr. Agre became president of the American Association for the Advancement of Science. To learn more about his research, visit http://www.jhsph.edu/faculty/directory/profile/4671/Agre/Peter.

REVIEW KEY CONCEPTS

To access the course materials and companion resources for this text, please visit www.cengagebrain.com.

6.1 Membrane Structure and Function

- Both membrane phospholipids and membrane proteins have hydrophobic and hydrophilic regions, giving them dual solubility properties.
- Membranes are based on a fluid phospholipid bilayer, with the polar regions of the phospholipids at the surfaces of the bilayer and their nonpolar tails in the interior (Figures 6.2–6.5).
- Membrane proteins are suspended individually in the bilayer, with their hydrophilic regions at the membrane surfaces and their hydrophobic regions in the interior (Figures 6.4 and 6.5).
- The lipid bilayer forms the structural framework of membranes and is a barrier to the passage of most water-soluble molecules.
- Proteins embedded in the phospholipid bilayer perform most membrane functions, including transport of selected hydrophilic substances, recognition, signal reception, cell adhesion, and metabolism.

- Integral membrane proteins are embedded deeply in the bilayer, whereas peripheral membrane proteins associate with membrane surfaces (Figure 6.5).
- Membranes are asymmetric—different arrangements of membrane lipids and proteins occur in the two bilayer halves.
- Membranes have diverse functions, including defining the boundaries of cells and of internal compartments, acting as permeability barriers, and facilitating electric signal conduction. Membrane proteins also show diverse activities, acting as enzymes, channel proteins, carrier proteins, receptors, and cell adhesion molecules.

Animation: Lipid bilayer organization

Animation: Protein drift in Plasma Membranes

Animation: Cell membranes

6.2 Functions of Membranes in Transport: Passive Transport

- Passive transport depends on diffusion, the net movement of molecules from a region of higher concentration to a region of lower concentration. It does not require cells to expend energy (Table 6.1).

- Simple diffusion is the passive transport of substances across the lipid portion of cellular membranes. It proceeds most rapidly for small molecules that are soluble in lipids (Table 6.1).
- Facilitated diffusion is the diffusion of polar and charged molecules through a membrane aided by transport proteins in the membrane. It follows concentration gradients, is specific for certain substances, and becomes saturated at high concentrations of the transported substance (Figure 6.8 and Table 6.1).
- Most proteins that carry out facilitated diffusion of ions are controlled by "gates" that open or close their transport channels (Figure 6.8).

Animation: Passive transport

6.3 Passive Water Transport and Osmosis

- Osmosis is the net diffusion of water molecules across a selectively permeable membrane in response to differences in the concentration of solute molecules (Figure 6.9). Water moves from hypotonic (lower solute concentrations) to hypertonic solutions (higher solute concentrations). When the solutions on each side are isotonic, net osmotic movement of water ceases (Figure 6.10).

Animation: Osmosis

Animation: Tonicity

6.4 Active Transport

- Active transport moves substances against their concentration gradients and requires cells to expend energy. It depends on membrane proteins, is specific for certain substances, and becomes saturated at high concentrations of the transported substance (Table 6.1).

- Active transport proteins are either primary transport pumps, which directly use ATP for energy, or secondary transport pumps, which use favorable concentration gradients of positively charged ions, created by primary transport pumps, as their energy source (Figures 6.11–6.12).
- Secondary active transport may occur by symport, in which the transported substance moves in the same direction as the concentration gradient that provides energy, or by antiport, in which the transported substance moves in the direction opposite to the concentration gradient that provides energy (Figure 6.13).

Animation: Active Transport

Animation: Plasma Membranes: Active Transport

6.5 Exocytosis and Endocytosis

- Large molecules and particles are moved out of and into cells by exocytosis and endocytosis. The mechanisms allow substances to leave and enter cells without directly passing through the plasma membrane (Figure 6.14).
- In exocytosis, a vesicle carrying secreted materials contacts and fuses with the plasma membrane on its cytoplasmic side. The fused vesicle membrane releases the vesicle contents to the cell exterior (Figure 6.14A).
- In endocytosis, materials on the cell exterior are enclosed in a segment of the plasma membrane that pockets inward and pinches off on the cytoplasmic side as an endocytic vesicle. The two forms of endocytosis are bulk-phase (pinocytosis) and receptor-mediated endocytosis. Most of the materials that enter cells are digested into molecular subunits small enough to be transported across the vesicle membranes (Figures 6.14B–D).

Animation: Membrane cycling

Animation: Phagocytosis

UNDERSTAND & APPLY

Test Your Knowledge

1. In the fluid mosaic model:
   a. plasma membrane proteins orient their hydrophilic sides toward the internal bilayer.
   b. phospholipids often flip-flop between the inner and outer layers.
   c. the mosaic refers to proteins attached to the underlying cytoskeleton.
   d. the fluid refers to the phospholipid bilayer.
   e. the mosaic refers to the symmetry of the internal membrane proteins and sterols.

2. Which of the following statements is false? Proteins in the plasma membrane can:
   a. transport ions.
   b. transport chloride ions when there are two mutant copies of the cystic fibrosis transmembrane conductance regulator gene.
   c. recognize self versus foreign molecules.
   d. allow adhesion between the same tissue cells or cells of different tissues.
   e. combine with lipids or sugars to form complex macromolecules.

3. The freeze-fracture technique demonstrated:
   a. that the plasma membrane is a bilayer with individual proteins suspended in it.
   b. that the plasma membrane is fluid.
   c. that the arrangement of membrane lipids and proteins is symmetric.

   d. that proteins are bound to the cytoplasmic side but not embedded in the lipid bilayer.
   e. the direction of movement of solutes through the membrane.

4. In the following figure, assume that the setup was left unattended. Which of the following statements is correct?

Selectively permeable membrane	
Inside a cell	**Outside fluids**
Solvent 95%	Solvent 98%
Solute 5%	Solute 2%

© Cengage Learning 2014

   a. The relation of the cell to its environment is isotonic.
   b. The cell is in a hypertonic environment.
   c. The net flow of solvent is into the cell.
   d. The cell will soon shrink.
   e. Diffusion can occur here but not osmosis.

5. Which of the following statements is true for the figure in question 4?
   a. The net movement of solutes is into the cell.
   b. There is no concentration gradient.
   c. There is a potential for plasmolysis.
   d. The solvent will move against its concentration gradient.
   e. If this were a plant cell, turgor pressure would be maintained.

6. Using the principle of diffusion, a dialysis machine removes waste solutes from a patient's blood. Imagine blood runs through a cylinder wherein diffusion can occur across an artificial selectively permeable membrane to a saline solution on the other side. Which of the following statements is correct?
   a. Solutes move from lower to higher concentration.
   b. The concentration gradient is lower in the patient's blood than in the saline solution wash.
   c. The solutes are transported through a symport in the blood cell membrane.
   d. The saline solution has a lower concentration gradient of solute than the blood.
   e. The waste solutes are actively transported from the blood.

7. A characteristic of carrier molecules in a primary active transport pump is that:
   a. they cannot transport a substance and also hydrolyze ATP.
   b. they retain their same shape as they perform different roles.
   c. their primary role is to move negatively charged ions across membranes.
   d. they move Na^+ into a neural cell and K^+ out of the same cell.
   e. they act to establish an electrochemical gradient.

8. A driving ion moving through a membrane channel in one direction gives energy to actively transport another molecule in the opposite direction. What is this process called?
   a. facilitated diffusion
   b. exchange diffusion
   c. symport transport
   d. primary active transport pump
   e. cotransport

9. Phagocytosis illustrates which phenomenon?
   a. receptor-mediated endocytosis
   b. bulk-phase endocytosis
   c. exocytosis
   d. pinocytosis
   e. cotransport

10. Place in order the following events of receptor-mediated endocytosis.
    (1) Clathrin coat disappears.
    (2) Receptors collect in a coated pit covered with clathrin on the cytoplasmic side.
    (3) Receptors recognize and bind specific molecules.
    (4) Endocytic vesicle may fuse with lysosome whereas receptors are recycled to the cell surface.
    (5) Pits deepen and pinch free of plasma membrane to form endocytic vesicles.
    a. 4, 1, 2, 5, 3
    b. 2, 1, 3, 5, 4
    c. 3, 2, 5, 1, 4
    d. 4, 1, 5, 2, 3
    e. 3, 1, 2, 4, 5

Discuss the Concepts

1. The bacterium *Vibrio cholerae* causes cholera, a disease characterized by severe diarrhea that may cause infected people to lose up to 20 L of fluid in a day. The bacterium enters the body when someone drinks contaminated water. It adheres to the intestinal lining, where it causes cells of the lining to release sodium and chloride ions. Explain how this release is related to the massive fluid loss.

2. Irrigation is widely used in dryer areas of the United States to support agriculture. In those regions, the water evaporates and leaves behind deposits of salt. What problems might these salt deposits cause for plants?

3. In hospitals, solutions of glucose with a concentration of 0.3 M can be introduced directly into the bloodstream of patients without tissue damage by osmotic water movement. The same is true of NaCl solutions, but these must be adjusted to 0.15 M to be introduced without damage. Explain why one solution is introduced at 0.3 M and the other at 0.15 M.

Design an Experiment

Design an experiment to determine the concentration of NaCl (table salt) in water that is isotonic to potato cells. Use only the following materials: a knife, small cookie cutters, and a balance.

Interpret the Data

Some cancer cells are insensitive to typical chemotherapy. Research into the mechanisms underlying this insensitivity uncovered an ability by these cells to "pump" the treatment drug out of the cell against its concentration gradient. Additional drugs have been developed that inhibit the pump, thus trapping the chemotherapeutic agent inside to promote cancer cell destruction.

The **Figure** shows what happens when two types of cells are treated with a ^{3}H-labeled anti-cancer drug, paclitaxel.

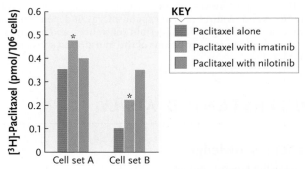

T. Shen et al. 2009. Imatinib and nilotinib reverse multidrug resistance in cancer cells by inhibiting the efflux activity of the MRP7 (ABCC10). *PLoS ONE* 4(10):e7520. doi:10.1371/journal.pone.0007520.

1. Which set of cells (A or B) would be described as resistant to the cancer treatment? Explain your answer. What type of transport are the resistant cells using?

2. Two additional drugs, imatinib and nilotinib, are evaluated for their ability to overcome the cancer cells' ability to "pump out" the chemotherapeutic agent. An asterisk (*) indicates a statistically significant difference from the cells receiving paclitaxel alone. Do the additional drugs seem to be effective in overcoming the pump? Which set of graphs (A or B) best supports your answer? Explain your answer.

Source: T. Shen et al. 2009. Imatinib and nilotinib reverse multidrug resistance in cancer cells by inhibiting the efflux activity of the MRP7 (ABCC10). *PLoS ONE* 4(10):e7520. doi:10.1371/journal.pone.0007520.

Apply Evolutionary Thinking

What evidence would convince you that membranes and active transport mechanisms evolved from an ancestor common to both prokaryotes and eukaryotes?

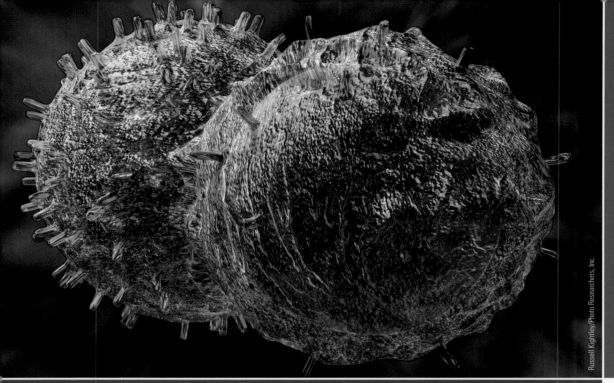

Russell Kightley/Photo Researchers, Inc.

A B cell and a T cell communicating by direct contact in the human immune system (computer image). Cell communication coordinates the cellular defense against disease.

Cell Communication

Why it matters . . . Hundreds of aircraft approach and leave airports traveling at various speeds, altitudes, and directions. How are all these aircraft kept separate, and routed to and from their airports safely and efficiently? The answer lies in a highly organized system of controllers, signals, and receivers. As the aircraft arrive and depart, they follow directions issued by air traffic controllers. Although thousands of different messages are traveling through the airspace, each pilot has a radio receiver tuned to a frequency specific for only that aircraft. The flow of directing signals, followed individually by each aircraft in the vicinity, keeps the traffic unscrambled and moving safely.

An equivalent system of signals and tuned receivers evolved hundreds of millions of years ago, as one of the developments that made multicellular life possible. Within a multicellular organism, the activities of individual cells are directed by *signal molecules* such as hormones that are released by certain controlling cells. Although the *controlling cells* release many signals, each receiving cell—the *target cell*—has *receptors* that are "tuned" to recognize only one or a few of the many signal molecules that circulate in its vicinity; other signal molecules pass by without effect because the cell has no receptors for them.

When a target cell binds a signal molecule, it modifies its internal activities in accordance with the signal, coordinating its functions with the activities of other cells of the organism. The responses of the target cell may include changes in gene activity, protein synthesis, transport of molecules across the plasma membrane, metabolic reactions, secretion, movement, division, or even "suicide"—that is, the programmed death of the receiving cell. As part of its response, a cell may itself become a controller by releasing signal molecules that modify the activity of other cell types. A signal molecule triggers a cellular pathway that results in a response in the target cell. The series of steps from signal molecule to response is a *signaling pathway.* The total network of signaling pathways allows multicellular organisms to grow, develop, reproduce, and compensate for environmental changes in an internally coordinated fashion. Maintaining the internal environment within a narrow tolerable range is *homeostasis.* The system of communication between cells through signaling pathways is called **cell signaling.** Research in cell signaling is a highly important field of biology, motivated by the desire to understand the growth, development, and function of organisms.

This chapter describes the major pathways that form parts of the cell communication system based on both surface and internal receptors, including the links that tie the different response pathways into fully integrated networks.

7.1 Cell Communication: An Overview

Cell communication is essential to orchestrate the activities of cells in multicellular organisms, and also takes place among single-celled organisms. In this chapter we focus on the principles of cell communication in animals, and in subsequent chapters you will see how similar principles apply to plants, fungi, and even bacteria and archaea.

Cell Communication in Animals May Involve Nearby or Distant Cells

Communication is critical for the function and survival of cells that compose a multicellular animal. For example, the ability of cells to communicate with one another in a regulated way is responsible for the controlled growth and development of an animal, as well as the integrated activities of its tissues and organs.

Cells communicate with one another in three ways:

1. **By direct contact.** In communication by direct contact, adjacent cells have direct channels linking their cytoplasms. In this rapid means of communication, small molecules and ions exchange directly between the two cytoplasms. In animal cells, the direct channels of communication are *gap junctions,* the specialized connections between the cytoplasms of adjacent cells (see Section 5.5). The main role of gap junctions is to synchronize metabolic activities or electrical signals between cells in a tissue. For example, gap junctions play a key role in spreading electrical signals from one cell to the next in cardiac muscle. In plant cells, the direct channels of communication are plasmodesmata (see Section 5.4 and Chapter 34). Small molecules moving between adjacent cells in plants include certain plant hormones that regulate growth (see Chapter 37). In this way, plant hormones are distributed to other cells.

    Cells can also communicate directly through *cell–cell recognition.* In this process, animal cells with particular membrane-bound cell-surface molecules dock with one another, initiating communication between the cells. For

example, cell–cell recognition of this kind activates particular cells in a mammal's immune system in order to mount an immune response (see the figure at the start of this chapter, and Figures 45.5 and 45.10).

2. **By local signaling.** In local signaling, a cell releases a signal molecule that diffuses through the *extracellular fluid* (the aqueous fluid surrounding and between the cells) and causes a response in nearby target cells. Here the effect of cell signaling is local, so the signal molecule is called a *local regulator* and the process is called *paracrine regulation* (**Figure 7.1A;** and see Chapter 42). In some cases the local regulator acts on the same cell that produces it, and this is called *autocrine regulation* (**Figure 7.1B;** and see Chapter 42). For example, many of the growth factors that regulate cell division are local regulators that act in both a paracrine and autocrine fashion.

3. **By long-distance signaling.** In this form of communication, a controlling cell secretes a long-distance signaling molecule called a **hormone** (*hormaein* = to excite), which produces a response in target cells that may be far from the controlling cell (**Figure 7.1C;** and see Chapter 42). This method is the most common means of cell communication. Hormones are found in both animals and plants. In animals, hormones secreted by controlling cells enter the circulatory system where they travel to target cells elsewhere in the body. For example, in response to stress, cells of a mammal's adrenal glands (located on top of the kidneys)—the controlling cells—secrete the hormone epinephrine (also known as *adrenaline*) into the bloodstream. Epinephrine acts on target cells to increase the amount of glucose in the blood. In plants, most hormones travel to target cells by moving through cells rather than by moving through vessels. Some plant hormones are gases that diffuse through the air to the target tissues. The actions of plant hormones are discussed in Chapter 37.

Cell communication by long-distance signaling is the focus of this chapter, and we will use the epinephrine example to illustrate the principles involved. In the 1950s, Earl Sutherland and his research team at Case Western Reserve University, Cleveland, Ohio, wanted to understand how the hormone epinephrine activates the enzyme *glycogen phosphorylase.* In the liver, this enzyme catalyzes the breakdown of glycogen—a polymer of glucose molecules—into glucose molecules, which are then released into the bloodstream. The overall effect of this response to epinephrine secretion is to supply energy to the major muscles responsible for locomotion—the body is now ready for physical activity or to handle stress.

Sutherland's key experiments are shown in **Figure 7.2.** He demonstrated that enzyme activation did not involve epinephrine directly but required an unknown (at the time) cellular factor. Sutherland called the hormone the *first messenger* in the system and the unknown cellular factor the *second messenger.* He proposed the following chain of reactions: epinephrine (the first messenger) stimulates the membrane fraction of the cell to

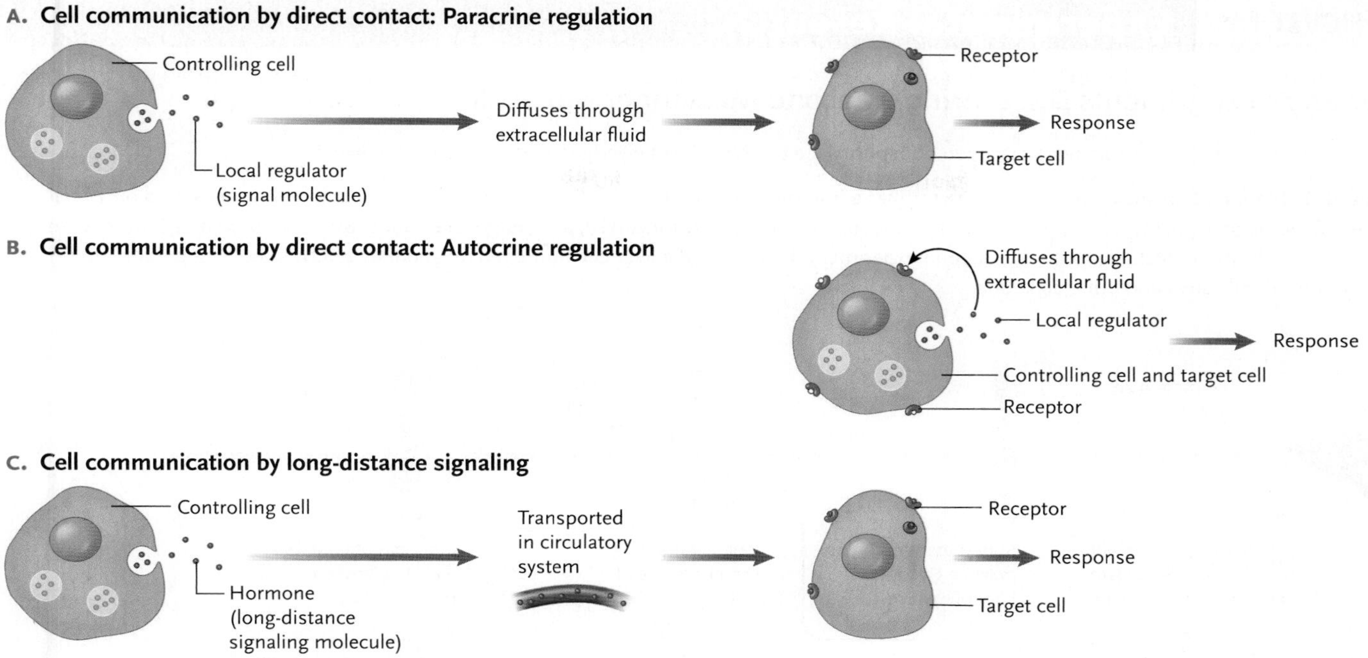

A. Cell communication by direct contact: Paracrine regulation

Controlling cell

Local regulator (signal molecule)

Diffuses through extracellular fluid

Receptor

Response

Target cell

B. Cell communication by direct contact: Autocrine regulation

Diffuses through extracellular fluid

Local regulator

Controlling cell and target cell

Receptor

Response

C. Cell communication by long-distance signaling

Controlling cell

Hormone (long-distance signaling molecule)

Transported in circulatory system

Receptor

Response

Target cell

FIGURE 7.1 Cell communication by local signaling (A, B) and by long-distance signaling (C).
© Cengage Learning 2014

produce a second messenger molecule, which activates the glycogen phosphorylase for conversion of glycogen to glucose. Later in the chapter, we return to this topic and describe the natures and functions of second messenger molecules. Sutherland was awarded a Nobel Prize in 1971 for his discoveries concerning the mechanisms of the action of hormones.

Sutherland's discovery was critical to understanding the mechanism of action of epinephrine and, in fact, of many other hormones. His work also illustrates how this type of long-distance cell signaling operates: a controlling cell releases a signal molecule that causes a response (affects the function) in target cells. Target cells process the signal in three sequential steps **(Figure 7.3)**:

1. **Reception.** Reception is the binding of a signal molecule with a specific receptor of target cells (see Figure 7.3, step 1). Target cells have receptors that are specific for the signal molecule, which distinguishes them from cells that do not respond to the signal molecule. The signals themselves may be polar (charged, hydrophilic) molecules or nonpolar (hydrophobic) molecules, and their receptors are shaped to recognize and bind them specifically. Receptors for polar signal molecules are embedded in the plasma membrane with a binding site for the signal molecule on the cell surface (see Figure 7.3). Epinephrine, the first messenger in Sutherland's research, is a polar hormone that is recognized by a surface receptor embedded in the plasma membrane of target cells. Receptors for nonpolar molecules are located within the cell (described later in the chapter; see Figure 7.14). In this case, the nonpolar signal molecule passes freely through the

plasma membrane and interacts with its receptor within the cell. Steroid hormones such as testosterone and estrogen are examples of nonpolar signal molecules.

2. **Transduction.** Transduction is the process of changing the signal into the form necessary to cause the cellular response (see Figure 7.3, step 2). The initial signal binds to and activates the receptor, changing it to a form that initiates transduction. Transduction typically involves a cascade of reactions that include several different molecules, referred to as a *signaling cascade*. For example, with respect to Sutherland's work, after epinephrine binds to its surface receptor, the signal is transmitted through the plasma membrane into the cell to another protein, which, in turn, causes the production of numerous small second messenger molecules. As we shall see later, both proteins and second messengers can be part of the signaling cascade that results in triggering a cellular response.

3. **Response.** Last, the transduced signal causes a specific cellular response. That response depends on the signal and the receptors of the target cell. In Sutherland's work, the response was the activation of the enzyme glycogen phosphorylase. The active enzyme catalyzes the conversion of stored glycogen to glucose, which is the response to the signal delivered by epinephrine.

The sequence of reception, transduction, and response is common to all the signaling pathways we will encounter in this chapter, although they vary greatly in detail. Importantly, all signaling pathways have "off switches" that serve to stop the cellular response when no such response is needed (see later in the chapter).

FIGURE 7.2 | **Experimental Research**

Sutherland's Experiments Discovering a Second Messenger Molecule

Question: How does epinephrine activate glycogen phosphorylase to break down glycogen into glucose in the liver?

Experiment: Sutherland had shown in one experiment that a homogenate (disrupted cells, consisting of cytoplasm, membranes, and other cell components) would activate glycogen phosphorylase if incubated with epinephrine, ATP, and magnesium ions. Sutherland then set out to learn more about the activation mechanism. In this second experiment, he prepared a liver cell homogenate and then centrifuged it, separating the cytoplasm from membranes and other cell debris.

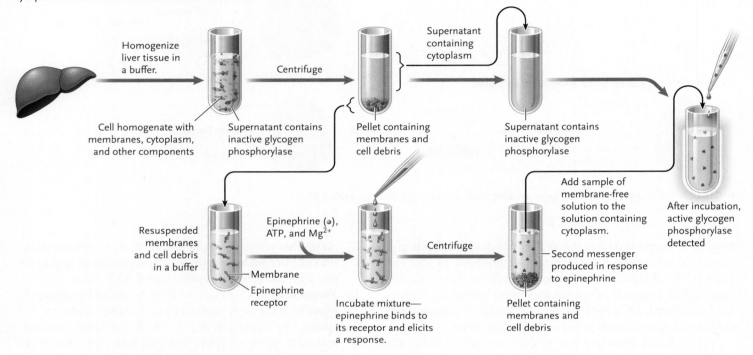

The cytoplasm was moved to a new tube, and the pellet with the membranes and cell debris was resuspended in a buffer. Neither the cytoplasm nor the membrane fractions had active glycogen phosphorylase. Next, he added epinephrine, ATP, and magnesium ions to the resuspended membranes and incubated the mixture. Centrifuging the mixture pelleted the membranes to the bottom of the tube. A sample of the supernatant (the membrane-free solution above the pellet) was added to the solution containing cytoplasm and the mixture was incubated.

Result: Active glycogen phosphorylase was detected in the mixture.

Conclusion: Sutherland had shown that the response to the hormone epinephrine—the activation of glycogen phosphorylase—does not involve epinephrine directly, but requires another cellular factor. He named the factor the *second messenger*, with the hormone itself being the *first messenger*.

THINK LIKE A SCIENTIST The researchers did some experiments to investigate the nature of the second messenger molecule. In one of their experiments, they showed that the second messenger remained active after heating in boiling water for 3 minutes. Interpret that result.

Source: T. W. Rall, E. W. Sutherland, and J. Berthet. 1957. The relationship of epinephrine and glucagon to liver phosphorylase. IV. Effect of epinephrine and glucagon on the reactivation of phosphorylase in liver homogenates. *Journal of Biological Chemistry* 224:463–475.

Cell Communication Is Evolutionarily Ancient

A number of cell communication properties are ancient, evolutionarily speaking. That is, mechanisms for one cell to signal to another cell and elicit a response most likely existed in unicellular organisms before the evolution of multicellularity. For instance, research with present-day bacteria has shown that a number of species alter their patterns of gene expression in response to changes in population density. In this process of *quorum sensing*, bacteria release signal molecules in increasing concentration as cell density increases. The molecules are sensed by the cells in the population, and each cell then responds to adapt to the changing environment.

In the unicellular eukaryote, yeast, sexual mating begins when one cell secretes a hormone that is recognized by a cell of

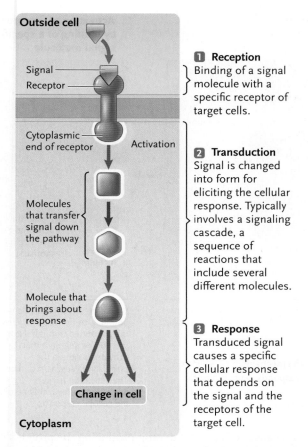

Outside cell

Signal

Receptor

1 **Reception**
Binding of a signal molecule with a specific receptor of target cells.

Cytoplasmic end of receptor

Activation

2 **Transduction**
Signal is changed into form for eliciting the cellular response. Typically involves a signaling cascade, a sequence of reactions that include several different molecules.

Molecules that transfer signal down the pathway

Molecule that brings about response

3 **Response**
Transduced signal causes a specific cellular response that depends on the signal and the receptors of the target cell.

Change in cell

Cytoplasm

FIGURE 7.3 The three stages of signal transduction: reception, transduction, and response (shown for a signal transduction system using a surface receptor).
© Cengage Learning 2014

a different "sex," signaling that the two cells are compatible for mating. In multicellular eukaryotes, complex cell communication pathways coordinate the activities of multiple cell types. Some protein components of these pathways are found in both prokaryotes and eukaryotes, indicating that they are evolutionarily ancient. Other proteins in the pathways appeared only after eukaryotes evolved. For instance, protein kinases—enzymes that add phosphate groups to other proteins to control their activity—are one of the largest families of proteins in eukaryotes, yet they are absent in prokaryotes. Scientists believe that the evolution of protein kinases was an important step in the development of multicellularity.

Beyond individual components, entire cell signaling pathways are conserved between organisms. For example, one pathway for cell growth control is conserved between *Drosophila* (the fruit fly) and humans, indicating that the pathway is at least 800 million years old. In short, the principles of cell communication are similar in unicellular and multicellular organisms, and some components are shared between them, but there is no single evolutionary root for the pathways involved. Many of the examples in this chapter focus on the systems working in animals, particularly in mammals, from which most of our knowledge of cell

communication has been developed. (The plant communication and control systems are described in more detail in Chapter 37.) This discussion begins with a few fundamental principles that underlie the often complex networks of cell communication.

STUDY BREAK 7.1

What accounts for the specificity of a cellular response to a signal molecule?

7.2 Cell Communication Systems with Surface Receptors

Cell communication systems using surface receptors have three components: (1) the extracellular signal molecules released by controlling cells; (2) the surface receptors on target cells that receive the signals; and (3) the internal response pathways triggered when receptors bind a signal. Surface receptors in mammals and other vertebrates recognize and bind polar, water-soluble signal molecules. These molecules are released by controlling cells and enter the extracellular fluid, and then pass into the blood circulation (in animals with a circulatory system). Two major types of polar signal molecules are polar hormones (hormones that are not steroids) and neurotransmitters. As you learned earlier, epinephrine is an example of a polar hormone, as are peptide hormones, which, as a group, affect all body systems (see Chapter 42). For example, insulin regulates sugar levels in blood. Neurotransmitters are molecules released by neurons that trigger activity in other neurons or other cells in the body; they include small peptides, individual amino acids or their derivatives, and other chemical substances (see Chapter 39).

Surface Receptors Are Integral Membrane Glycoproteins

The surface receptors that recognize and bind signal molecules are all glycoproteins—proteins with attached carbohydrate chains (see Section 3.4). They are integral membrane proteins that extend entirely through the plasma membrane **(Figure 7.4A)**. Typically the signal-binding site of the receptor is the part of the protein that extends from the outer membrane surface, and which is folded in a way that closely fits the signal molecule. The fit is specific, so a particular receptor binds only one type of signal molecule or a closely related group of signal molecules.

A signal molecule brings about specific changes in cells to which it binds. When a signal molecule binds to a surface receptor, the molecular structure of that receptor changes so that it transmits the signal through the plasma membrane, activating the cytoplasmic end of the receptor. The activated receptor then initiates the first step in a cascade of molecular events—the signaling cascade—that triggers the cellular response (**Figure 7.4B**).

Animal cells typically have hundreds to thousands of surface receptors that represent many receptor types. Receptors for a specific peptide hormone may number from 500 to as many as 100,000 or more per cell. Different cell types contain distinct combinations of receptors, allowing them to react individually to the polar signal molecules in the extracellular fluid. The combination of surface receptors on particular cell types is not fixed but changes as cells develop. Changes also occur as normal cells are transformed into cancer cells.

The Signaling Molecule Bound by a Surface Receptor Triggers Response Pathways within the Cell

Signal transduction pathways triggered by surface receptors are common to all animal cells. At least parts of the pathways are also found in protists, fungi, and plants. Signal transduction involving surface receptors has three characteristics:

1. Binding of a signal molecule to a surface receptor is sufficient to trigger the cellular response—the signal molecule does not enter the cell. For example, experiments have shown that: (1) a signal molecule produces no response if it is injected directly into the cytoplasm; and (2) unrelated molecules that mimic the structure of the normal extracellular signal molecule can trigger or block a full cellular response as long as they can bind to the recognition site of the receptor. In fact, many medical conditions are treated with drugs that are signal molecule mimics. For example, propranolol, a beta-blocker drug, inhibits receptors involved in contractions of certain muscles, and is used to reduce the strength of cardiac contractions and to reduce blood pressure.

2. The signal is relayed inside the cell by **protein kinases,** enzymes that transfer a phosphate group from ATP to one or more sites on particular proteins (see Section 4.5). As shown in **Figure 7.5,** protein kinases often act in a chain catalyzing a series of phosphorylation reactions called a *phosphorylation cascade,* to pass a signal along. The first kinase catalyzes phosphorylation of the second, which then becomes active and phosphorylates the third kinase, which then becomes active, and so on. The last protein in the cascade is the *target protein.* Phosphorylation of a target protein stimulates or inhibits its activity depending on the particular protein. This change in activity brings about the cellular response. For example, phosphorylating a target protein that regulates whether a set of genes are turned on or off

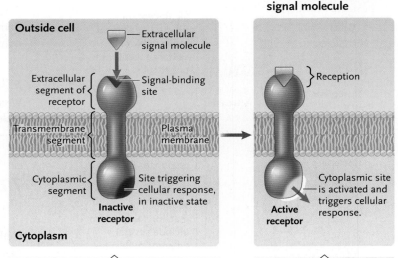

A. Surface receptor

Outside cell

Extracellular signal molecule

Extracellular segment of receptor — Signal-binding site

Transmembrane segment

Plasma membrane

Cytoplasmic segment — Site triggering cellular response, in inactive state

Inactive receptor

Cytoplasm

B. Activation of receptor by binding of a specific signal molecule

}Reception

Cytoplasmic site is activated and triggers cellular response.

Active receptor

A surface receptor has an extracellular segment with a site that recognizes and binds a particular signal molecule.

When the signal molecule is bound, a conformational change is transmitted through the transmembrane segment that activates a site on the cytoplasmic segment of receptor. The activation triggers a reaction pathway that results in the cellular response.

FIGURE 7.4 **The mechanism by which a surface receptor responds when it binds a signal molecule.**
© Cengage Learning 2014

could cause cells to start or stop producing the proteins the genes encode. The change in this set of proteins then causes a response related to the functions of those proteins.

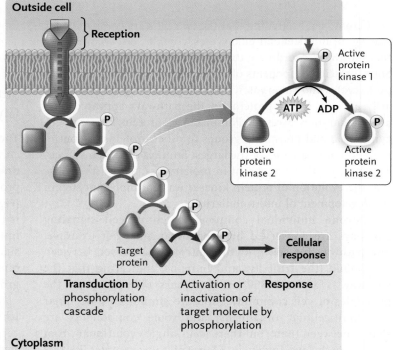

Outside cell

}**Reception**

P Active protein kinase 1

ATP ADP

Inactive protein kinase 2

Active protein kinase 2

Target protein

Cellular response

Transduction by phosphorylation cascade

Activation or inactivation of target molecule by phosphorylation

Response

Cytoplasm

FIGURE 7.5 **Phosphorylation, a key reaction in many signaling pathways.**
© Cengage Learning 2014

The effects of protein kinases in signal transduction pathways are balanced or reversed by another group of enzymes called **protein phosphatases,** which remove phosphate groups from target proteins. Unlike the protein kinases, which are active only when a surface receptor binds a signal molecule, most of the protein phosphatases are continuously active in cells. By continually removing phosphate groups from target proteins, the protein phosphatases quickly shut off a signal transduction pathway if its signal molecule is no longer bound at the cell surface.

Two scientists, Edwin Krebs and Edmond Fischer at the University of Washington, Seattle, first discovered that protein kinases add phosphate groups to control the activities of key proteins in cells and obtained evidence showing that protein phosphatases reverse these phosphorylations. Krebs and Fischer, who began their experiments in the 1950s, received a Nobel Prize in 1992 for their discoveries concerning reversible protein phosphorylation.

3. An increase in the magnitude of each step occurs as a signal transduction pathway proceeds, a phenomenon called **amplification (Figure 7.6).** Amplification occurs because many of the proteins that carry out individual steps in the pathways, including the protein kinases, are enzymes. Once activated, each enzyme can activate hundreds of proteins including other enzymes that enter the next step in the pathway. Generally, the more enzyme-catalyzed steps in a response pathway, the greater the amplification. As a result, just a few extracellular signal molecules binding to their receptors can produce a full internal response. For similar reasons, amplification also occurs for signal transduction pathways that involve internal receptors.

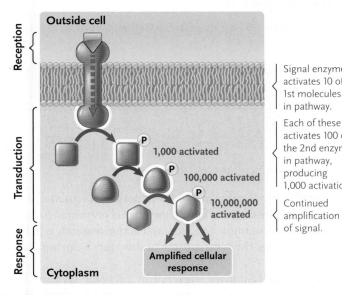

FIGURE 7.6 Amplification in signal transduction.
© Cengage Learning 2014

Signal Transduction Pathways Have "Off Switches"

Once signal molecules are released into the body's circulation, they remain for only a certain time. Either they are broken down at a steady rate by enzymes in organs such as the liver, or they are excreted by the kidneys. The removal process ensures that the signal molecules are active only as long as controlling cells are secreting them.

As signal transduction runs its course, the receptors and their bound signal molecules are removed from the target cell surface by endocytosis (see Section 6.5). Both the receptor and its bound signal molecule may be degraded in lysosomes after entering the cell. Alternatively, the receptors may be separated from the signal molecules and recycled to the cell surface, whereas the signal molecules are degraded as above. Thus, surface receptors participate in an extremely lively cellular "conversation" with moment-to-moment shifts in the information.

Next, you will see how the three hallmarks of surface receptor pathways (surface receptor, kinase cascade, amplification) play out in two large surface receptor families: the receptor tyrosine kinases and the G-protein–coupled receptors.

STUDY BREAK 7.2

1. What are protein kinases, and how are they involved in signal transduction pathways?
2. How is amplification accomplished in a signal transduction pathway?

7.3 Surface Receptors with Built-in Protein Kinase Activity: Receptor Tyrosine Kinases

One major type of surface receptors, the **receptor tyrosine kinases,** have their own protein kinase activity on the cytoplasmic end of the protein. Binding of a signal molecule to this type of receptor turns on the receptor's built-in protein kinase, which leads to activation of the receptor. The activated receptor then initiates a signaling cascade, which results in a cellular response.

For this type of receptor, initiation of transduction occurs when two receptor molecules each bind a signal molecule in the reception step, move together in the membrane, and assemble into a dimer (a pair of monomers bonded together) (**Figure 7.7,** step 1). The protein kinases of each receptor monomer are activated by dimer formation, and they phosphorylate the partner monomer in the dimer, a process called *autophosphorylation* (Figure 7.7, step 2). The phosphorylation is of tyrosine amino acids, which gives this type of receptors their name. The multiple phosphorylations activate many different sites on the dimer. When a signaling protein binds to an activated site, it initiates a transduction

FIGURE 7.7 The action of a receptor tyrosine kinase, a receptor type with built-in protein kinase activity.

1 Signal molecules bind and two receptors assemble into a dimer

When no signal molecules are bound, the receptor monomers are separate and their cytoplasmic protein kinase sites are inactive. When a signal molecule binds, the receptors assemble into dimers.

2 Activation of protein kinases and autophosphorylation of the receptor

Conformational changes then activate protein kinases on the receptor monomers, leading to phosphorylation of tyrosine amino acids.

3 Transduction and cellular responses

Signaling proteins bind to the activated receptor and become activated. Each then initiates a transduction pathway that produces a cellular response.

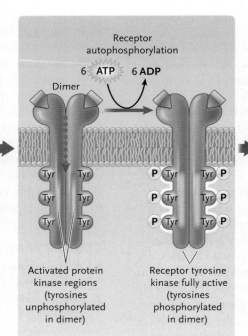

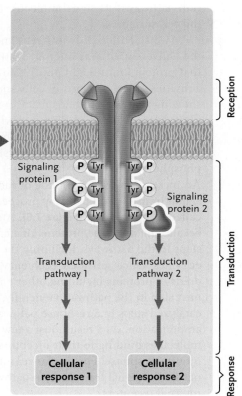

SUMMARY Binding of a signal molecule causes receptor monomers to form a dimer that becomes active by autophosphorylation. Signaling proteins bind to the activated receptor and become activated, each initiating a transduction pathway that produces a cellular response. That is, binding of a signal molecule to a receptor tyrosine kinase can initiate multiple signal transduction pathways and cause multiple cellular responses.

THINK LIKE A SCIENTIST Suppose a mutation in a gene for a receptor tyrosine kinase receptor causes the receptor to form dimers and activate without having a signal molecule bind. What would be the effect of that mutation?

© Cengage Learning 2014

pathway leading to a cellular response (Figure 7.7, step 3). Since different receptor tyrosine kinases bind different combinations of signaling proteins, the receptors initiate different responses.

Receptor tyrosine kinases are found in all multicellular animals, but not in plants or fungi. Fifty-eight genes in the human genome encode receptor tyrosine kinases. In mammals, more generally, receptor tyrosine kinases fall into about 20 different families, all related to one another in structure and amino acid sequence. Each family member is presumed to have diverged from a common ancestor, and active research is being done to determine the mechanisms of divergence during their evolution.

The cellular responses triggered by receptor tyrosine kinases are among the most important processes of animal cells. For example, the receptor tyrosine kinases binding the peptide hormone *insulin,* a regulator of carbohydrate metabolism, trigger diverse cellular responses, including effects on glucose uptake, the rates of many metabolic reactions, and cell growth and division. (The insulin receptor is exceptional because it is permanently in a tetrameric [four-monomer] form.) Other receptor tyrosine kinases bind growth factors, including *epidermal growth factor, platelet-derived growth factor,* and *nerve growth factor,* which are important peptides.

Hereditary defects in the insulin receptor are responsible for some forms of *diabetes,* a disease in which glucose accumulates in the blood because it cannot be absorbed in sufficient quantity by body cells. The cells with faulty receptors do not respond to insulin's signal to add glucose receptors to take up glucose. (The role of insulin in glucose metabolism and diabetes is discussed further in Chapter 42.)

STUDY BREAK 7.3

1. How does a receptor tyrosine kinase become activated?
2. Once activated fully, how does a receptor tyrosine kinase bring about a cellular response?

THINK OUTSIDE THE BOOK

Using a specific example, outline how an alteration in a receptor tyrosine kinase can contribute to the development of a cancer.

7.4 G-Protein–Coupled Receptors

A second large family of surface receptors, known as the **G-protein–coupled receptors,** respond to a signal by activating an inner membrane protein called a G protein, which is closely associated with the cytoplasmic end of the receptor. G proteins are so named because they bind the guanine nucleotides GDP

(guanosine diphosphate) and GTP (guanosine triphosphate). G-protein–coupled receptors are found in animals (both multicellular and unicellular forms), plants, fungi, and certain protists. Researchers have identified thousands of different G-protein–coupled receptors in mammals, including thousands involved in recognizing and binding odor molecules as part of the mammalian sense of smell, light-activated receptors in the eye, and many receptors for hormones and neurotransmitters. Almost all of the receptors of this group are large glycoproteins built up from a single polypeptide chain anchored in the plasma membrane by seven segments of the amino acid chain that zigzag back and forth across the membrane seven times (**Figure 7.8**).

Unlike receptor tyrosine kinases, G-protein–coupled receptors lack built-in protein kinase activity.

G Proteins Are Key Molecular Switches in Second-Messenger Pathways

In signal transduction pathways controlled by G-protein–coupled receptors, the extracellular signal molecule is called the **first messenger.** The binding of the first messenger to the receptor activates it (**Figure 7.9,** step 1). Coupled to the receptor is a G protein, which is called a molecular switch because it switches between an inactive form with GDP bound to it (step 2), and an active form in which GDP is replaced by GTP. When the first messenger activates the receptor, it activates the G protein by causing it to release GDP and bind GTP (step 3). The GTP-bound subunit of the G protein breaks off and binds to a plasma membrane–associated enzyme called the **effector** (steps 4 and 5), activating it. The activated effector now generates one or more internal, nonprotein signal molecules called **second messengers** (step 6). The second messengers directly or indirectly activate protein kinases, which elicit the cellular response by adding phosphate groups to specific target proteins (step 7).

The separate protein kinases of these pathways all add phosphate groups to serine or threonine amino acids in their target proteins, which typically are:

- enzymes that catalyze steps in metabolic pathways.
- ion channels in the plasma and other membranes.
- regulatory proteins that control gene activity and cell division.

The pathway from first messengers to target proteins is common to all G-protein–coupled receptors.

As long as a G-protein–coupled receptor is bound to a first messenger, the receptor keeps the G protein active. The activated G protein, in turn, keeps the effector active in generating second messengers. If the first messenger is released from the receptor, or if the receptor is taken into the cell by endocytosis, GTP is hydrolyzed to GDP, which inactivates the G protein. As a result, the effector becomes inactive, turning "off" the response pathway.

Cells can make a variety of G proteins, with each type activating a different cellular response. Alfred G. Gilman at the University of Virginia, Charlottesville, and Martin Rodbell at

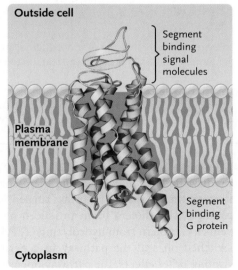

FIGURE 7.8 Structure of the G-protein–coupled receptors, which activate separate protein kinases. These receptors have seven transmembrane α-helical segments that zigzag across the plasma membrane. Binding of a signal molecule at the cell surface, by inducing changes in the positions of some of the helices, activates the cytoplasmic end of the receptor.
© Cengage Learning 2014

Outside cell

Segment binding signal molecules

Plasma membrane

Segment binding G protein

Cytoplasm

FIGURE 7.9 Response pathways activated by G-protein–coupled receptors, in which protein kinase activity is separate from the receptor.

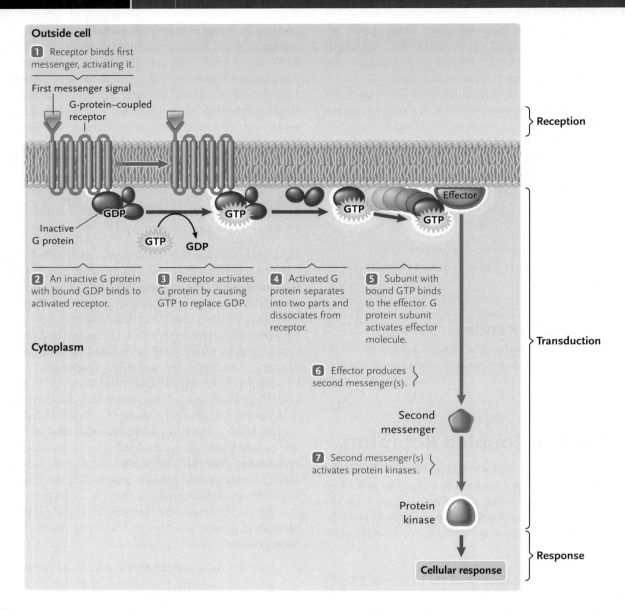

Outside cell

1 Receptor binds first messenger, activating it.

First messenger signal

G-protein–coupled receptor

Reception

Inactive G protein

GDP

GTP

GDP

GTP

GTP

GTP

Effector

2 An inactive G protein with bound GDP binds to activated receptor.

3 Receptor activates G protein by causing GTP to replace GDP.

4 Activated G protein separates into two parts and dissociates from receptor.

5 Subunit with bound GTP binds to the effector. G protein subunit activates effector molecule.

Cytoplasm

6 Effector produces second messenger(s).

Transduction

Second messenger

7 Second messenger(s) activates protein kinases.

Protein kinase

Response

Cellular response

SUMMARY The signal molecule is the first messenger. The effector is an enzyme that generates one or more internal signal molecules called second messengers. The second messengers directly or indirectly activate the protein kinases of the pathway, leading to the cellular response. In sum, the entire control pathway operates through the following sequence:

first messenger → receptor → G proteins → effector → protein kinases → target proteins

THINK LIKE A SCIENTIST What would be the possible effects of a malfunction of a G-protein–coupled receptor?

© Cengage Learning 2014

the National Institutes of Health, Bethesda, Maryland, received a Nobel Prize in 1994 for their discovery of G proteins and the role of these proteins in signal transduction in cells.

The importance of G proteins to cellular metabolism is underscored by the fact that they are targets of toxins released by some infecting bacteria. Typically, the toxins are enzymes that

modify the G proteins, making them continuously active and keeping their response pathways turned "on" at high levels. For example, the cholera toxin produced by *Vibrio cholerae* prevents a G protein from hydrolyzing GTP, keeping the G protein switched on and the pathway in a permanently active state. Among other effects, the pathway opens ion channels in intes-

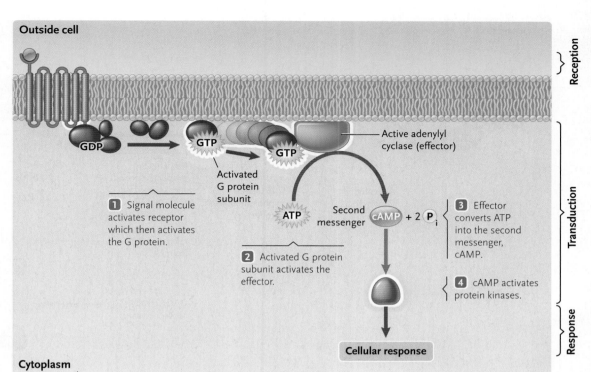

tinal cells, causing severe diarrhea through a massive release of salt and water from the body into the intestinal tract. Unless the resulting dehydration of the body is relieved, death can result quickly.

Two Major G-Protein–Coupled Receptor–Response Pathways Involve Different Second Messengers

Activated G proteins bring about a cellular response through two major receptor–response pathways in which different effectors generate different second messengers. One pathway involves the second messenger **cyclic AMP** (**cAMP**—cyclic 3′,5′-adenosine monophosphate), a relatively small, water-soluble molecule derived from ATP **(Figure 7.10).** The effector that produces cAMP is the enzyme *adenylyl cyclase,* which converts ATP to cAMP **(Figure 7.11).** cAMP diffuses through the cytoplasm and activates protein kinases that add phosphate groups to target proteins. The other pathway involves two second messengers: **inositol triphosphate (IP₃)** and **diacylglycerol (DAG).** The effector of this pathway, an enzyme called *phospholipase C,* produces both of these second messengers by breaking down a membrane phospholipid **(Figure 7.12).**

IP_3 is a small, water-soluble molecule that diffuses rapidly through the cytoplasm. DAG is hydrophobic; it remains and functions in the plasma membrane.

The primary effect of IP_3 in animal cells is to activate transport proteins in the endoplasmic reticulum (ER), which release Ca^{2+} stored in the ER into the cytoplasm. The released Ca^{2+}, either alone or in combination with DAG, activates a protein kinase cascade that brings about the cellular effect. As *Focus on Basic Research* describes, researchers developed techniques to detect Ca^{2+} release inside cells for studying cell signaling.

Both major G-protein–coupled receptor–response pathways are balanced by reactions that constantly eliminate their second messengers. For example, cAMP is quickly converted to AMP (5′-adenosine monophosphate) by *phosphodiesterase,* an enzyme that is continuously active in the cytoplasm (see Figure 7.11). The rapid elimination of the second messengers provides another highly effective off switch for the pathways,

FIGURE 7.11 cAMP. The second messenger, cAMP, is made from ATP by adenylyl cyclase and is broken down to AMP by phosphodiesterase.
© Cengage Learning 2014

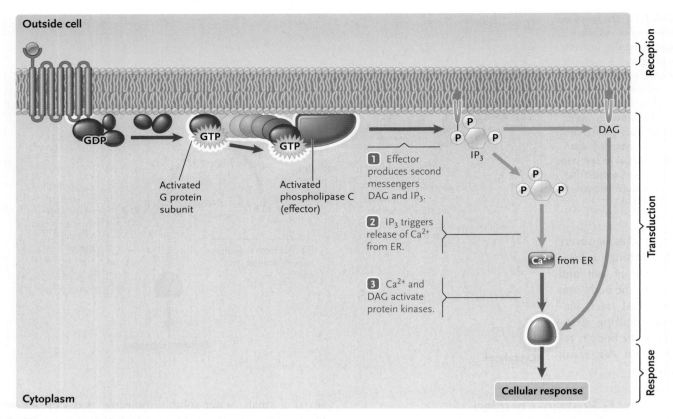

FIGURE 7.12 The operation of IP₃/DAG receptor–response pathways. Two second messengers, IP₃ and DAG, are produced by the pathway. IP₃ opens Ca²⁺ channels in ER membranes, releasing the ion into the cytoplasm. The Ca²⁺, with DAG in some cases, directly or indirectly activates the protein kinases of the pathway, which add phosphate groups to target proteins to initiate the cellular response.
© Cengage Learning 2014

ensuring that protein kinases are inactivated quickly if the receptor becomes inactive. Still another off switch is provided by protein phosphatases that remove the phosphate groups added to proteins by the protein kinases.

As in the receptor tyrosine kinase pathways, the activities of the pathways controlled by cAMP and IP₃/DAG second messengers are also stopped by endocytosis of receptors and their bound extracellular signals. As with all cell signaling pathways, cells vary in their response to cAMP or IP₃/DAG pathways depending on the type of G-protein–coupled receptors on the cell surface and the kinds of protein kinases present in the cytoplasm.

The cAMP second-messenger pathway is found in animals and fungi. In plants, cAMP may be involved in germination and in some plant defensive responses (see Chapter 37), although the pathways are not well understood. The IP₃/DAG second-messenger pathway is universally distributed among eukaryotic organisms, including both vertebrate and invertebrate animals, fungi, and plants. In plants, IP₃ releases Ca²⁺ primarily from the large central vacuole rather than from the ER.

SPECIFIC EXAMPLES OF CYCLIC AMP PATHWAYS Many polar hormones act as first messengers for cAMP pathways in mam-

mals and other vertebrates. The receptors that bind these hormones control such varied cellular responses as the uptake and oxidation of glucose, glycogen breakdown or synthesis, ion transport, the transport of amino acids into cells, and cell division.

For example, a cAMP pathway is involved in regulating the level of glucose, the fundamental fuel of cells. When the level of blood glucose falls too low in mammals, cells in the pancreas release the peptide hormone glucagon. Glucagon triggers a cAMP receptor–response pathway (see Figure 7.10) in liver cells, which stimulates them to break down glycogen into glucose units that pass from the liver cells into the bloodstream. When the level of blood glucose is excessive, the opposite occurs: the enzyme *glycogen synthase* joins glucose units to produce glycogen.

SPECIFIC EXAMPLES OF IP₃/DAG PATHWAYS The IP₃/DAG-response pathways are also activated by a large number of polar hormones (including growth factors) and neurotransmitters, leading to responses as varied as sugar and ion transport, glucose oxidation, cell growth and division, and movements such as smooth muscle contraction.

Among the mammalian hormones that activate the pathways are antidiuretic hormone, angiotensin, and norepi-

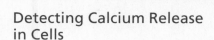

Detecting Calcium Release in Cells

Because calcium ions are used as a control element in all eukaryotic cells, it was important to develop techniques for detecting Ca^{2+} when it is released into the cytosol. One technique uses substances that release a burst of light when they bind the ion. *Aequorin,* a protein produced by jellyfish, ctenophores, and many other luminescent organisms, is such a substance. Aequorin is injected into the cytoplasm of cells using microscopic needles, and it releases light when IP_3 opens Ca^{2+} channels in the ER, causing an increase in cytosolic Ca^{2+} concentration.

Artificially made, water-soluble molecules called *fura-2* and *quin-2* are also used as indicators of Ca^{2+} release. These molecules fluoresce (emit light) when exposed to ultraviolet (UV) light, and the wavelength of the fluorescence differs depending on whether the molecules are bound to or free of Ca^{2+}. Therefore, the amount of Ca^{2+} released into the cytosol can be quantified by measuring the amount of fluorescence at each of the two wavelengths. Experimentally, investigators combine the molecules with a hydrophobic organic molecule that allows them to pass directly through the plasma membrane. After fura-2 or quin-2 is inside the cell, cellular enzymes remove the added organic group, releasing the Ca^{2+} indicators into the cytosol.

In a typical experiment designed to follow steps in the IP_3/DAG pathway, an investigator might want to know whether a given hormone triggers the pathway in a group of cells. The investigator first adds aequorin or quin-2 to the cells, and then the hormone. If the cells emit a bright flash of light, it is a good indication that the hormone triggers the IP_3/DAG pathway.

Experiments using these methods and others have revealed the many cellular processes controlled by Ca^{2+} concentration inside cells, including cellular response pathways, cell movements, assembly and disassembly of the cytoskeleton, secretion, and endocytosis.

---

nephrine. Antidiuretic hormone, also known as vasopressin, helps the body conserve water by reducing the output of urine. Angiotensin helps maintain blood volume and pressure (see Chapter 48). Norepinephrine (also known as noradrenaline), together with epinephrine, brings about the fight-or-flight response in threatening or stressful situations (see Chapter 42).

Many growth factors operate through IP_3/DAG pathways. Defects in the receptors or other parts of the pathways that lead to higher-than-normal levels of DAG in response to growth factors are often associated with the progression of some forms of cancer. This is because DAG, in turn, causes an overactivity of the protein kinases responsible for stimulating cell growth and division. Also, plant substances in a group called *phorbol esters* resemble DAG so closely that they can promote cancer in animals by activating the same protein kinases.

IP_3/DAG pathways have also been linked to mental disease, particularly *bipolar disorder* (previously called *manic depression*), in which patients experience periodic changes in mood. Lithium has been used for many years as a therapeutic agent for bipolar disorder. Recent research has shown that lithium reduces the activity of IP_3/DAG pathways that release neurotransmitters; among them are some neurotransmitters that take part in brain function. Lithium also relieves cluster headaches and premenstrual tension, suggesting that IP_3/DAG pathways may be linked to these conditions as well.

In plants, IP_3/DAG pathways control responses to conditions such as water loss and changes in light intensity or salinity. Plant hormones—relatively small, nonprotein molecules such as the *cytokinins* (derivatives of the nucleotide base adenine)—act as first messengers activating some of the IP_3/DAG pathways of these organisms. Cytokinins are discussed in Chapter 37.

Some Signaling Pathways Combine a Receptor Tyrosine Kinase with the G Protein Ras

Some pathways important in gene regulation link certain receptor tyrosine kinases to a specific type of G protein called Ras. When the receptor tyrosine kinase receives a signal (**Figure 7.13,** step 1), it activates by autophosphorylation (step 2). Adapter proteins then bind to the phosphorylated receptor and bridge to Ras, causing GTP to bind to, and activate, Ras (step 3). The activated Ras sets in motion a phosphorylation cascade that involves a series of three enzymes known as *mitogen-activated protein kinases* (MAP kinases; step 4). The last MAP kinase in the cascade, when activated, enters the nucleus (step 5) and phosphorylates other proteins, which then change the expression of certain genes, particularly activating those involved in cell division (step 6). (A *mitogen* is a substance that controls cell division, hence the name of the kinases.)

Changes in gene expression can have far-reaching effects on the cell, such as determining whether a cell divides or how frequently it divides. The Ras proteins are of major interest to investigators because of their role in linking receptor tyrosine kinases to gene regulation, as well as their major roles in the development of many types of cancer when their function is altered.

Both the Ras proteins and the MAP kinases are widely distributed among eukaryotes. Ras has been detected in eukaryotic organisms ranging from yeasts to humans and higher plants. Similarly, MAP kinases have been identified in eukaryotes as diverse as yeasts, roundworms, insects, humans, and plants. *Insights from the Molecular Revolution* presents evidence that influenza virus uses a MAP kinase cascade to aid its propagation in infected cells.

In this section, we have surveyed major response pathways linked to surface receptors that bind peptide hormones, growth

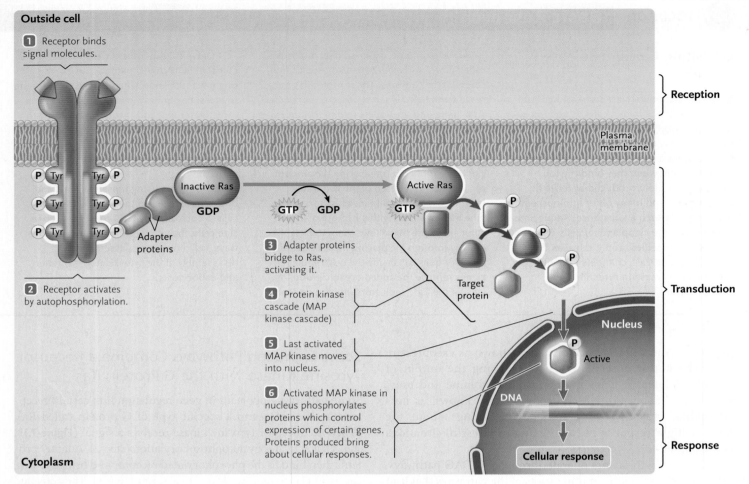

Outside cell

1 Receptor binds signal molecules.

} **Reception**

Plasma membrane

P Tyr   Tyr P

P Tyr   Tyr P

P Tyr   Tyr P

Adapter proteins

Inactive Ras

GDP

GTP   GDP

Active Ras

GTP

P

P

P

Target protein

2 Receptor activates by autophosphorylation.

3 Adapter proteins bridge to Ras, activating it.

4 Protein kinase cascade (MAP kinase cascade)

5 Last activated MAP kinase moves into nucleus.

6 Activated MAP kinase in nucleus phosphorylates proteins which control expression of certain genes. Proteins produced bring about cellular responses.

Cytoplasm

} **Transduction**

Nucleus

P   Active

DNA

} **Response**

Cellular response

FIGURE 7.13 The pathway from receptor tyrosine kinases to gene regulation, including the G protein, Ras, and MAP kinase.
© Cengage Learning 2014

factors, and neurotransmitters. We now turn to the other major type of signal receptor: the internal receptors binding signal molecules—primarily steroid hormones—that penetrate through the plasma membrane.

7.5 Pathways Triggered by Internal Receptors: Steroid Hormone Receptors

Cells of many types have internal receptors that respond to signals arriving from the cell exterior. Unlike the signal molecules that bind to surface receptors, these signals, primarily steroid hormones, penetrate through the plasma membrane to trigger response pathways inside the cells. The internal receptors, called **steroid hormone receptors,** are typically control proteins that turn on (sometimes off) specific genes when they are activated by binding a signal molecule.

Steroid Hormones Have Widely Different Effects That Depend on Relatively Small Chemical Differences

Steroid hormones are relatively small, nonpolar molecules derived from cholesterol, with a chemical structure based on four carbon rings (see Figure 3.14). Steroid hormones combine with hydrophilic carrier proteins that mask their hydrophobic groups and hold them in solution in the blood and extracellular fluids. When a steroid hormone–carrier protein complex contacts the surface of a cell, the hormone is released and penetrates directly through the plasma membrane. On the cytoplasmic side, the hormone binds to its internal receptor.

The various steroid hormones differ only in the side groups attached to their carbon rings. Although the differences are small, they are responsible for highly distinctive effects. For example, the male and female sex hormones of

Virus Infections and Cell Signaling Pathways: Does influenza virus propagation involve a cellular MAP kinase cascade?

Epidemic outbreaks of influenza occur almost every year. Some influenza epidemics become pandemics, epidemics of new strains that spread through many parts of the world. One particularly virulent pandemic, caused by a viral strain called influenzavirus A/H1N1, was the so-called Spanish Flu of 1918 to 1919. It infected an estimated 500 million people, of whom 50 to 100 million died.

An influenza virus consists of an RNA genome within a protein coat (the genomes of all living organisms are DNA). Like all viruses, influenza viruses have a limited number of genes and, therefore, must manipulate host cell functions in order to propagate. When an influenza virus infects a cell, it elicits responses in the cell to combat the infection. To be successful in its infection, the virus must overcome these antiviral activities. Various viruses support their propagation by activating MAP kinase cascades in cells they infect. A particular MAP kinase signaling pathway, called the Raf/MEK/ERK cascade, has a major role in regulating cell growth and proliferation.

Research Question: Does influenza A virus propagation involve the Raf/MEK/ERK MAP kinase cascade?

Experiments: Stephan Pleschka of the University of Geissen, Germany, Stephan Ludwig of the University of Würzburg, Germany, and other researchers performed two experiments with mammalian cells in culture to answer the question **(Figure)**: (1) They tested whether the viruses activated the Raf/MEK/ERK pathway in infected cells that either were untreated or treated with U0126, a substance that inhibits MEK, the middle kinase in the Raf/MEK/ERK MAP kinase cascade; (2) In a similar experiment, they then tested whether the virus would still proliferate if the experimenters shut down the Raf/MEK/ERK pathway with U0126.

Results: Experiment 1 showed that influenza A virus activates the Raf/MEK/ERK MAP kinase cascade and that U0126 could block that activation (see Figure). Experiment 2 showed that U0126 reduces the production of progeny viruses by about 80% compared with cultures not treated with the inhibitor.

Conclusion: The experiments showed that influenza A virus propagation requires virus-induced signaling through that kinase cascade pathway. More recent experiments have shown that specific inhibition of the Raf/MEK/ERK signaling cascade causes a marked impairment in the growth of all influenza A viruses tested. The targeting of signaling pathways that are essential for virus propagation has become the basis for clinical research with the goal of developing effective antiviral drugs.

THINK LIKE A SCIENTIST How might you show that an inhibitor of the Raf/MEK/ERK MAP kinase signaling cascade, such as U0126, is an effective antiviral drug?

Sources: S. Pleschka et al. 2001. Influenza virus propagation is impaired by inhibition of the Raf/MEK/ERK signaling cascade. *Nature Cell Biology* 3:301–305; S. Ludwig. 2009. Targeting cell signalling pathways to fight the flu: towards a paradigm change in anti-influenza therapy. *Journal of Antimicrobial Chemotherapy* 64:1–4.

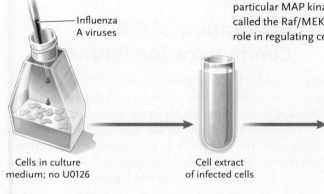

Influenza A viruses

Cells in culture medium; no U0126

Cell extract of infected cells

Experiment 1: Analyze sample of extract for ERK activity: ERK activity was detected, indicating that MAP kinase cascade had been activated.

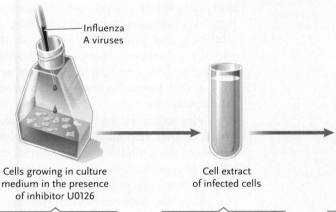

Influenza A viruses

Cells growing in culture medium in the presence of inhibitor U0126

Cell extract of infected cells

Experiment 2: Analyze sample of extract for ERK activity: Very little ERK activity was detected, indicating that U0126 could block activation of the MAP kinase cascade by the virus.

The researchers infected mammalian cells growing in culture with influenza A virus and incubated them for a few hours in either the absence (top) or presence (bottom) of chemical U0126, an inhibitor of the MAP kinase cascade.

The scientists broke open cells, centrifuged to remove cell debris, and collected the cell extract.

Samples of the extract were analyzed for the presence of ERK, the last kinase in the cascade.

FIGURE Experiments testing whether influenza A viruses activate the MAP kinase cascade.

mammals, testosterone and estrogen, respectively, which are responsible for many of the structural and behavioral differences between male and female mammals, differ only in minor substitutions in side groups at two positions (see Figure 3.15). The differences cause the hormones to be recognized by different receptors, which activate specific group of genes leading to development of individuals as males or females.

The Response of a Cell to Steroid Hormones Depends on Its Internal Receptors and the Genes They Activate

Steroid hormone receptors are proteins with two major domains (**Figure 7.14**). One domain recognizes and binds a specific steroid hormone. The other domain interacts with the regions of target genes that control the expression of those genes. When a steroid hormone combines with the hormone-binding domain, the gene activation domain changes shape, thus enabling the complex to bind to the control regions of the target genes that the hormone affects. For most steroid hormone receptors, binding of the activated receptor to a gene control region activates that gene.

Steroid hormones, like polar hormones, are released by cells in one part of an organism and are carried by the organism's circulation to other cells. Whether a cell responds to a steroid hormone depends on whether it has a receptor for the hormone within the cell. The type of response depends on the genes that are recognized and turned on (or off) by an activated receptor. Depending on the receptor type and the particular genes it recognizes, even the same steroid hormone can have highly varied effects on different cells. (The effects of steroid hormones are described in more detail in Chapter 42.)

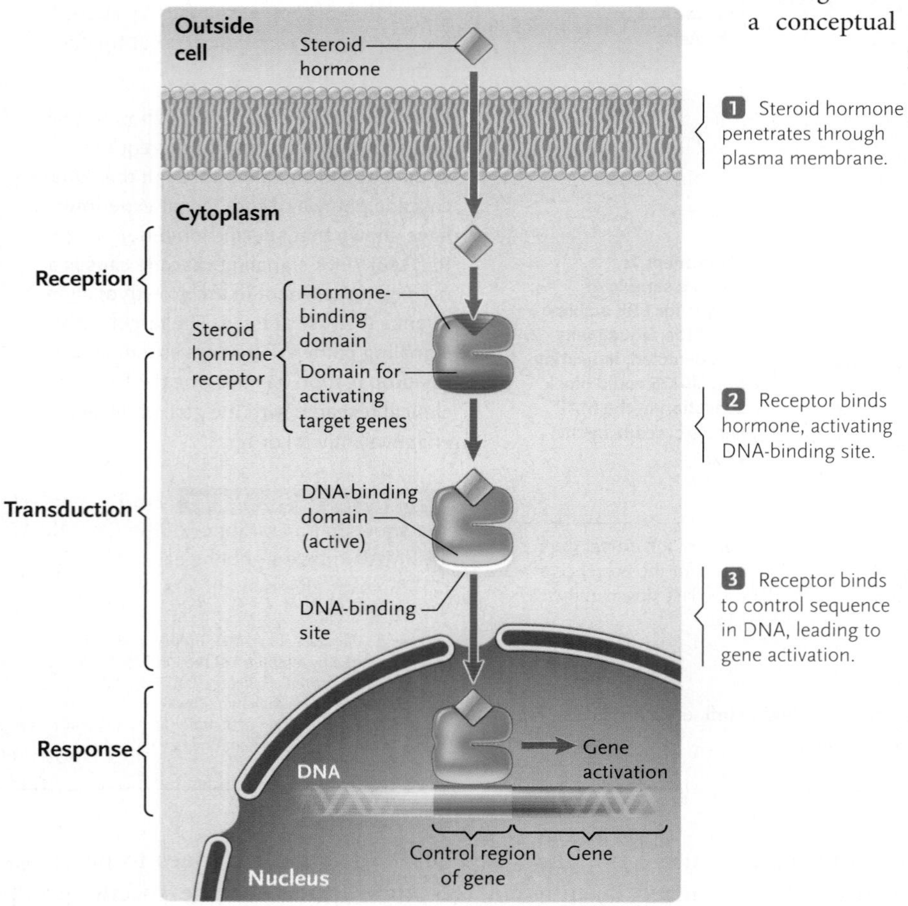

Taken together, the various types of receptor tyrosine kinases, G-protein–coupled receptors, and steroid hormone receptors, prime cells to respond to a stream of specific signals that continuously fine-tune their function. How are the signals integrated within the cell and organism to produce harmony rather than chaos? The next section shows how the various signal pathways are integrated into a coordinated response.

STUDY BREAK 7.5

1. What distinguishes a steroid receptor from a receptor tyrosine kinase receptor or a G-protein–coupled receptor?
2. By what means does a specific steroid hormone result in a specific cellular response?

7.6 Integration of Cell Communication Pathways

Cells are under the continual influence of many simultaneous signal molecules. The cell signaling pathways may operate independently, or communicate with one another to integrate their responses to cellular signals coming from different controlling cells. The interpathway interaction is called **cross-talk;** a conceptual example that involves two second-messenger pathways is shown in **Figure 7.15**. For example, a protein kinase in one pathway might phosphorylate a site on a target protein in another signal transduction pathway, activating or inhibiting that protein, depending on the site of the phosphorylation. The cross-talk can be extensive, resulting in a complex network of interactions between cell communication pathways.

Cross-talk often leads to modifications of the cellular responses controlled by the pathways. Such modifications fine-tune the effects of combinations of signal molecules binding to the receptors of a cell. For example, cross-talk between second-messenger pathways is involved in particular types of olfactory (smell) signal transduction in rats and probably in many other animals, including humans. The two pathways involved are activated on stimulation with distinct odors. One pathway involves cAMP as the second messenger, and the other involves IP_3. However, the two olfactory second-messenger pathways do not work independently; rather, they operate in an antagonistic way. That is, experimentally blocking key enzymes of one signal transduction cascade inhibits that pathway, while simultaneously augmenting the activity of the other path-

FIGURE 7.14 Pathway of gene activation by steroid hormone receptors.

© Cengage Learning 2014

Labels in figure:
- Outside cell — Steroid hormone
- **1** Steroid hormone penetrates through plasma membrane.
- Cytoplasm
- Reception
- Steroid hormone receptor — Hormone-binding domain; Domain for activating target genes
- **2** Receptor binds hormone, activating DNA-binding site.
- Transduction — DNA-binding domain (active); DNA-binding site
- **3** Receptor binds to control sequence in DNA, leading to gene activation.
- Response — DNA; Gene activation
- Control region of gene; Gene
- Nucleus

FIGURE 7.15 Cross-talk, the interaction between cell communication pathways to integrate the responses to signal molecules.
© Cengage Learning 2014

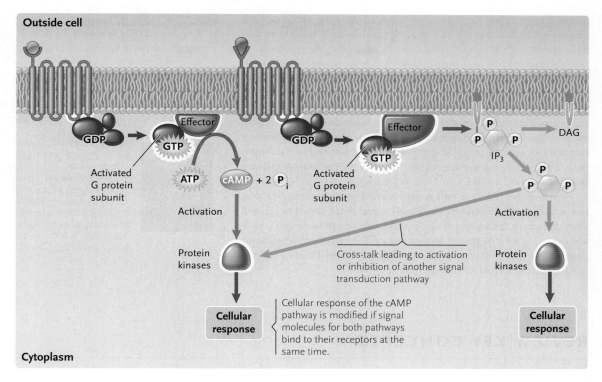

way. The cross-talk may be a way to refine an animal's olfactory sensory perception by helping discriminate different odor molecules more effectively.

Direct channels of communication may also be involved in a cross-talk network. For example, gap junctions between the cytoplasms of adjacent cells admit ions and small molecules, including the Ca^{2+}, cAMP, and IP_3 second messengers released by the receptor–response pathways. (Gap junctions are discussed in Section 5.5.) Thus, one cell that receives a signal through its surface receptors can transmit the signal to other cells in the same tissue via the connecting gap junctions, thereby coordinating the functions of those cells. For instance, cardiac muscle cells are connected by gap junc-

tions, and the Ca^{2+} flow regulates coordinated muscle fiber contractions.

The entire system integrating cellular response mechanisms, tied together by many avenues of cross-talk between individual pathways, creates a sensitively balanced control mechanism that regulates and coordinates the activities of individual cells into the working unit of the organism.

STUDY BREAK 7.6

What cell communication pathways might be integrated in a cross-talk network?

UNANSWERED QUESTIONS

How does cross-talk between signaling pathways influence a behavior?

As you learned in this chapter, cell signaling by the sex hormones of animals is responsible for many of the structural and behavioral characteristics of males and females. Another type of cell signaling, called neural signaling, is also important in reproductive behavior. Currently, many laboratories are actively investigating possible cross-talk between these two major types of cell signaling. Specifically, researchers are investigating the cellular processes by which steroid hormones that are involved in mammalian reproductive behavior act on neurons. During the estrous cycle of many animals, including female rats, guinea pigs, hamsters, and mice, the ovarian hormones estradiol and progesterone regulate the expression of reproductive behaviors via cellular processes, including binding to steroid hormone receptors in neurons that are involved in the behaviors. But do neural signals also activate steroid hormone receptors?

The steroid hormone receptor model presented in this chapter is that an intracellular steroid hormone receptor becomes activated when the steroid hormone binds to it. This mechanism, which involves both estrogen receptors (for estradiol) and progestin receptors (for progesterone) in the brain, is consistent with a great deal of the research on the cellular mechanisms of hormonal regulation of reproductive behaviors. However, research based on work of Shaila Mani and collaborators has now shown that the regulation of reproductive behaviors involves cross-talk between neurotransmitter signaling pathways and steroid hormone receptors. Although steroid hormones bind directly to the steroid hormone receptors, neurotransmitters, including dopamine, acting via second-messenger pathways, can also activate steroid hormone receptors in the absence of a hormone. In addition, my research group at the University of Massachusetts, Amherst, has shown that when a male rat attempts to mate with a female rat, the mating stimulation somehow activates the female's neural progestin receptors, presumably by a process that involves the release of particular neurotransmitters onto neurons containing the receptors. In fact, although it had always been thought that progesterone is required to facilitate the full expression of sexual behaviors in female rats, stimulation by the male can substitute for progesterone. Mating stimulation induces reproductive behavior similar to that induced by the secretion of steroid hormones. That is, how a male behaves toward a female alters neurotransmitter release in her brain, presumably then activating steroid hormone receptors in some neurons. This activation results in neuronal changes, many of which are the same as those caused by the hormone secretions from the female's ovaries.

How does this hormone-independent steroid hormone receptor activation occur? In which neurons would you expect these events occur, and what characteristics would you expect of the neurons (for example, inputs and outputs)? What might regulate the process? The results of experiments designed to answer these questions will give valuable insights into the mechanisms of steroid hormone action in the brain.

THINK LIKE A SCIENTIST

1. Why do you think the male's behavior can substitute for the hormone in facilitating sexual behavior?

2. Progestin receptor regulation of sexual behavior is a useful model for studying the interaction between neurotransmitters and steroid hormone receptors. Would you expect this cross-talk mechanism to be limited to sexual behavior, or do you think it might come into play with other hormone-regulated behaviors?

Courtesy of Jeffrey Blaustein

Jeffrey Blaustein is a professor in the Neuroscience and Behavior Program and is a member of the Center for Neuroendocrine Studies at the University of Massachusetts Amherst. His research interests are in the many ways in which the environment can influence hormonal processes in the brain resulting in changes in behavior. In recent years, the interest of his group has expanded to the influences of stress around the time of puberty on response to ovarian hormones in adulthood. To learn more about the work of his research group, go to http://www.umass.edu/cns/blaustein.

REVIEW KEY CONCEPTS

To access the course materials and companion resources for this text, please visit www.cengagebrain.com.

7.1 Cell Communication: An Overview

- Cells communicate with one another by direct contact, local signaling, and long-distance signaling.

- In long-distance signaling, a controlling cell releases a signal molecule that causes a response of target cells. Target cells process the signal in three steps: reception, transduction, and response. This process is called signal transduction (Figure 7.3).

- Some cell communication properties are evolutionarily ancient. In some cases, entire cell signaling pathways are conserved between distantly related organisms.

7.2 Cell Communication Systems with Surface Receptors

- Cell communication systems based on surface receptors have three components: (1) extracellular signal molecules; (2) surface receptors that receive the signals; and (3) internal response pathways triggered when receptors bind a signal.

- The systems based on surface receptors respond to polar hormones and neurotransmitters. Polar hormones include peptide hormones and growth factors, which affect cell growth, division, and differentiation. Neurotransmitters include small peptides, individual amino acids or their derivatives, and other chemical substances.

- Surface receptors are integral membrane proteins that extend through the plasma membrane. Binding a signal molecule induces a molecular change in the receptor that activates its cytoplasmic end (Figure 7.4).

- Many cellular response pathways operate by activating protein kinases, which add phosphate groups that stimulate or inhibit the activities of the target proteins, bringing about the cellular response (Figure 7.5). Protein phosphatases that remove phosphate groups from target proteins reverse the response. In addition, receptors are removed by endocytosis when signal transduction has run its course.

- Each step of a response pathway catalyzed by an enzyme is amplified, because each enzyme can activate hundreds or thousands of proteins that enter the next step in the pathway. Through amplification, a few signal molecules can bring about a full cellular response (Figure 7.6).

  Animation: Signal transduction

7.3 Surface Receptors with Built-in Protein Kinase Activity: Receptor Tyrosine Kinases

- When receptor tyrosine kinases bind a signal molecule, the protein kinase site is activated and adds phosphate groups to tyrosines in the receptor itself activating those sites. When a signaling protein binds to an activated site, it initiates a transduction pathway leading to a cellular response. The binding of different combinations of signaling proteins to different tyrosine kinases produces different responses (Figure 7.7).

7.4 G-Protein–Coupled Receptors

- In the pathways activated by G-protein–coupled receptors, binding of the extracellular signal molecule (the first messenger) activates a site on the cytoplasmic end of the receptor (Figure 7.8). The activated receptor turns on a G protein, which acts as a molecular switch. The G protein is active when it is bound to GTP and inactive when it is bound to GDP (Figure 7.9).

- The active G protein switches on the effector, an enzyme that generates small internal signal molecules called second messengers. The second messengers activate the protein kinases of the pathway (Figure 7.9).

- In one of the two major pathways triggered by G-protein–coupled receptors, the effector, adenylyl cyclase, generates cAMP as second messenger. cAMP activates specific protein kinases (Figures 7.10 and 7.11).

- In the other major pathway, the activated effector, phospholipase C, generates two second messengers, IP_3 and DAG. IP_3 activates transport proteins in the ER, which release stored Ca^{2+}. The Ca^{2+} alone, or with DAG, activates specific protein kinases that phosphorylate their target proteins (Figure 7.12).

- Both the cAMP and IP$_3$/DAG pathways are balanced by reactions that constantly eliminate their second messengers. Both pathways are also stopped by protein phosphatases that remove phosphate groups from target proteins and by endocytosis of receptors and their bound signals.

- Mutated systems can turn on the pathways permanently, contributing to the progression of some forms of cancer.

- Some pathways important in gene regulation link certain receptor tyrosine kinases to a specific G protein called Ras. When the receptor binds a signal molecule, it phosphorylates itself, and adapter proteins then bind, bridging to Ras, activating it. Activated Ras turns on the MAP kinase cascade. The last MAP kinase in the cascade phosphorylates target proteins in the nucleus, which turn on specific genes (Figure 7.13). Many of those genes control cell division.

Animation: Response pathways activated by G-protein–coupled receptors

7.5 Pathways Triggered by Internal Receptors: Steroid Hormone Receptors

- Steroid hormones penetrate through the plasma membrane to bind to receptors within the cell, activating the receptors. The internal receptors are regulatory proteins that turn on specific genes, producing the cellular response (Figure 7.14).

- Steroid hormone receptors have a domain that recognizes and binds a specific steroid hormone and a domain that interacts with the controlling regions of target genes. A cell responds to a steroid hormone only if it has an internal receptor for the hormone, and the type of response depends on the genes that are turned on by an activated receptor (Figure 7.14).

Animation: Hormones and target cell receptors

7.6 Integration of Cell Communication Pathways

- In cross-talk, cell signaling pathways communicate with one another to integrate responses to cellular signals. Cross-talk may result in a complex network of interactions between cell communication pathways. Cross-talk often modifies the cellular responses controlled by the pathways, fine-tuning the effects of combinations of signal molecules binding to a cell (Figure 7.15).

- In animals, inputs from other cellular response systems, including cell adhesion molecules and molecules arriving through gap junctions, also can be involved in the cross-talk network.

UNDERSTAND & APPLY

Test Your Knowledge

1. In signal transduction, which of the following is *not* a target protein?
   a. proteins that regulate gene activity
   b. hormones that activate the receptor
   c. enzymes of pathways
   d. transport proteins
   e. enzymes of cell reactions

2. Which of the following could *not* elicit a signal transduction response?
   a. a signal molecule injected directly into the cytoplasm
   b. a virus mimicking a normal signal molecule
   c. a peptide hormone
   d. a steroid hormone
   e. a neurotransmitter

3. A cell that responds to a signal molecule is distinguished from a cell that does not respond by the fact that it has:
   a. a cell adhesion molecule.
   b. cAMP.
   c. a first messenger molecule.
   d. a receptor.
   e. a protein kinase.

4. The mechanism to activate an immune cell to make an antibody involves signal transduction using tyrosine kinases. Place in order the following series of steps to activate this function.
   (1) The activated receptor phosphorylates cytoplasmic proteins.
   (2) Conformational change occurs in the receptor tyrosine kinase.
   (3) Cytoplasmic protein crosses the nuclear membrane to activate genes.
   (4) An immune hormone signals the immune cell.
   (5) Activation of protein kinase site(s) adds phosphates to the receptor to activate it.
   a. 2, 1, 4, 3, 5
   b. 5, 3, 4, 2, 1
   c. 4, 1, 5, 2, 3
   d. 4, 2, 5, 1, 3
   e. 2, 5, 3, 4, 1

5. Which of the following describes the ability of enzymes, involving few surface receptors, to activate thousands of molecules in a stepwise pathway?
   a. autophosphorylation
   b. second-messenger enhancement
   c. amplification
   d. ion channel regulation
   e. G protein turn-on

6. Which of the following is *incorrect* about pathways activated by G-protein–coupled receptors?
   a. The extracellular signal is the first messenger.
   b. When activated, plasma membrane-bound G protein can switch on an effector.
   c. Second messengers enter the nucleus.
   d. ATP converts to cAMP to activate protein kinases.
   e. Protein kinases phosphorylate molecules to change cellular activity.

7. Which of the following would *not* inhibit signal transduction?
   a. Phosphate groups are removed from proteins.
   b. Endocytosis acts on receptors and their bound signals.
   c. Receptors and signals separate.
   d. Receptors and bound signals enter lysosomes.
   e. Autophosphorylation targets the cytoplasmic portion of the receptor.

8. An internal receptor binds both a signal molecule and controlling region of a gene. What type of receptor is it?
   a. protein
   b. steroid
   c. IP$_3$/DAG
   d. receptor tyrosine kinase
   e. switch protein

9. Place in order the following steps for the normal activity of a Ras protein.
   - (1) Ras turns on the MAP kinase cascade.
   - (2) Adaptor proteins connect phosphorylated tyrosine on a receptor to Ras.
   - (3) GTP activates Ras by binding to it, displacing GDP.
   - (4) The last MAP kinase in the cascade phosphorylates proteins in the nucleus that activate genes.
   - (5) Receptor tyrosine kinase binds a signal molecule and is activated.
   - a. 1, 2, 3, 4, 5
   - b. 2, 3, 5, 1, 4
   - c. 5, 2, 3, 1, 4
   - d. 2, 3, 1, 5, 4
   - e. 4, 1, 5, 3, 2

10. Which of the following does *not* exemplify cross-talk?
   - a. a protein kinase in one pathway that phosphorylates a site on a target protein in another signal transduction pathway
   - b. modifications of cellular responses controlled by pathways
   - c. two second-messenger pathways interacting
   - d. olfactory sensory perception
   - e. signal transduction pathways controlled by G-protein–coupled receptors

Discuss the Concepts

1. Describe the possible ways in which a G-protein–coupled receptor pathway could become defective and not trigger any cellular responses.

2. Is providing extra insulin an effective cure for an individual who has diabetes that is caused by a hereditary defect in the insulin receptor? Why or why not?

3. There are molecules called GTP analogs that resemble GTP so closely that they can be bound by G proteins. However, they cannot be hydrolyzed by cellular GTPases. What differences in effect would you expect if you inject GTP or a nonhydrolyzable GTP analog into a liver cell that responds to glucagon?

4. Why do you suppose cells evolved internal response mechanisms using switching molecules that bind GTP instead of ATP?

Design an Experiment

How would you set up an experiment to determine whether a hormone receptor is located on the cell surface or inside the cell?

Interpret the Data

In most individuals with cystic fibrosis, the 508th amino acid of the CFTR protein (a phenylalanine) is missing. A CFTR protein with this change is synthesized correctly, and it can transport ions correctly, but it never reaches the plasma membrane to do its job.

Sergei Bannykh and his coworkers developed a procedure to measure the relative amounts of the CFTR protein localized in different regions of the cell. They compared the pattern of CFTR distribution in normal cells with the pattern in CFTR-mutated cells. A summary of their results is shown in the **Figure**.

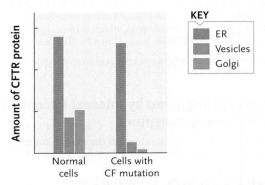

FIGURE Comparison of the amounts of CFTR protein associated with endoplasmic reticulum (blue), vesicles traveling from ER to Golgi (green), and Golgi bodies (orange). The patterns of CFTR distribution in normal cells, and the cells with the most common cystic fibrosis mutation, were compared.

© Cengage Learning 2014

1. Which organelle contains the least amount of CFTR protein in normal cells? In CF cells? Which contains the most?

2. In which organelle is the amount of CFTR protein in CF cells closest to the amount in normal cells?

3. Where is the mutated CFTR protein getting held up?

Apply Evolutionary Thinking

Based on their distributions among different groups of organisms, which signaling pathway is the oldest?

Mitochondrion (colorized TEM). Mitochondria are the sites of cellular respiration.

8

Harvesting Chemical Energy: Cellular Respiration

Why it matters . . . In the early 1960s, Swedish physician Rolf Luft mulled over some odd symptoms of a patient. The young woman felt weak and too hot all the time (with a body temperature of up to 38.4°C). Even on the coldest winter days she never stopped perspiring, and her skin was always flushed. She was also underweight (40 kg), despite consuming about 3,500 calories per day.

Luft inferred that his patient's symptoms pointed to a metabolic disorder. Her cells were very active, but much of their activity was being dissipated as metabolic heat. Tests that measured her metabolic rate—the amount of energy her body was expending—showed the patient's oxygen consumption was the highest ever recorded, about twice the normal rate.

Luft also examined a tissue sample from the patient's skeletal muscles under a microscope. He found that her muscle cells contained many more mitochondria—the ATP-producing organelles of the cell—than are normally present in muscle cells. In addition, her mitochondria were abnormally shaped and their interiors were packed to an abnormal degree with cristae, the infoldings of the inner mitochondrial membrane (see Section 5.3). Other studies showed that the mitochondria were engaged in cellular respiration—their prime function—but little ATP was being generated.

The disorder, now called *Luft syndrome,* was the first disorder to be linked directly to a defective mitochondrion. By analogy, someone with this disorder functions like a city with half of its power plants shut down. Skeletal and heart muscles, the brain, and other hardworking body parts with high energy demands are hurt the most by the inability of mitochondria to provide enough energy for metabolic demands. More than 100 mitochondrial disorders are now known. Interestingly, to this day, the genetic defect responsible for Luft syndrome is not known.

Defective mitochondria also contribute to many age-related problems, including type 1 diabetes, atherosclerosis, and amyotrophic lateral sclerosis (ALS, also called Lou Gehrig disease), as well as Parkinson, Alzheimer, and Huntington diseases.

Clearly, human health depends on mitochondria that are structurally sound and that function properly. More broadly, every animal, plant, fungus, and most protists depend on mitochondria that are functioning correctly to grow and survive.

In mitochondria, ATP forms as part of the reactions of cellular respiration. **Cellular respiration** is the collection of metabolic reactions within cells that breaks down food molecules to produce energy in the form of ATP. ATP fuels nearly all the reactions that keep cells, and organisms, metabolically active. In eukaryotes and many prokaryotes, oxygen is a reactant in the ATP-producing process. This form of cellular respiration is **aerobic respiration** (*aero* = air, *bios* = life). In some prokaryotes, a molecule other than oxygen, such as sulfate or nitrate, is used in the ATP-producing process. This form of cellular respiration is **anaerobic respiration** (*an* = without). The discussion of the reactions of cellular respiration in this chapter concerns aerobic respiration except where noted.

The primary source of the food molecules broken down in cellular respiration is *photosynthesis,* which is described in Chapter 9. Photosynthesis is the process in which light energy is captured and used to split water into hydrogen and oxygen. The hydrogen from the water is combined with carbon dioxide to synthesize carbohydrates and other organic molecules. The other product of the splitting of water is oxygen, a molecule needed for cellular respiration. Photosynthesis occurs in most plants, many protists, and some prokaryotes.

Together, cellular respiration and photosynthesis are the major biological steps of the carbon cycle, which is the global circulation of carbon atoms. The carbon cycle is described in Section 54.3. Atmospheric carbon in the carbon cycle is mostly in the form of CO_2, which is a product of cellular respiration.

8.1 Overview of Cellular Energy Metabolism

Electron-rich food molecules synthesized by photosynthetic organisms are used by the organisms themselves and by other organisms that ingest the photosynthetic organisms, or parts of them. The electrons are removed from fuel substances, such as sugars, and donated to other molecules, such as oxygen, that act as electron acceptors. In the process, some of the energy of the electrons is released and used to drive the synthesis of ATP. ATP provides energy for most of the energy-consuming activities in the cell. Thus, life and its systems are driven by a cycle of electron flow that is powered by light in photosynthesis and oxidation in cellular respiration **(Figure 8.1)**.

Coupled Oxidation and Reduction Reactions Produce the Flow of Electrons for Energy Metabolism

The partial or full loss of electrons (e^-) from a substance is termed an **oxidation,** and the substance from which the electrons are lost—called the *electron donor*—is said to be **oxidized.** The partial or full gain of electrons to a substance is termed a **reduction,** and the substance that gains the electrons—called the *electron acceptor*—is said to be **reduced.** A simple mnemonic to remember the direction of electron transfer is OIL RIG: Oxidation Is Loss (of electrons), Reduction Is Gain (of electrons). The term *oxidation* was originally used to describe the reaction that occurs when fuel substances are burned in air, in which oxygen directly accepts electrons removed from the fuels. However, although the term *oxidation* suggests that oxygen is involved in electron loss, most cellular oxidations occur without the direct participation of oxygen. The term *reduction* refers to the decrease in positive electrical charge that occurs when electrons, which are negatively charged, are added to a substance. Although the term *reduction* suggests that the energy level of molecules is decreased when they accept electrons, molecules typically gain energy from added electrons.

FIGURE 8.1 The flow of energy from sunlight to ATP. **(A)** Photosynthesis occurs in plants, many protists, and some prokaryotes; **(B)** cellular respiration (aerobic respiration) occurs in all eukaryotes, including plants, and in some prokaryotes.
© Cengage Learning 2014

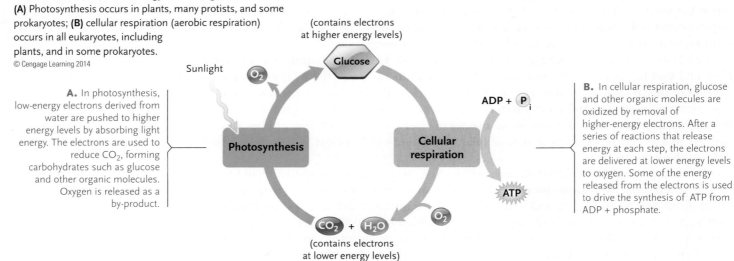

A. In photosynthesis, low-energy electrons derived from water are pushed to higher energy levels by absorbing light energy. The electrons are used to reduce CO_2, forming carbohydrates such as glucose and other organic molecules. Oxygen is released as a by-product.

(contains electrons at higher energy levels)

Sunlight

Glucose

ADP + P_i

Photosynthesis

Cellular respiration

ATP

CO_2 + H_2O

(contains electrons at lower energy levels)

B. In cellular respiration, glucose and other organic molecules are oxidized by removal of higher-energy electrons. After a series of reactions that release energy at each step, the electrons are delivered at lower energy levels to oxygen. Some of the energy released from the electrons is used to drive the synthesis of ATP from ADP + phosphate.

Oxidation and reduction are coupled reactions that remove electrons from a donor molecule and simultaneously add them to an acceptor molecule. Such coupled oxidation–reduction reactions are also called **redox reactions;** a generalized redox reaction can be written as:

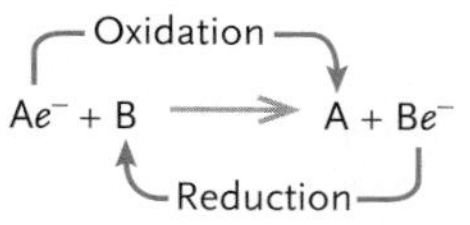

$$\overset{\text{Oxidation}}{\underset{\text{Reduction}}{Ae^- + B \longrightarrow A + Be^-}}$$

In a redox reaction, electrons release some of their energy as they pass from a donor molecule to an acceptor molecule. This free energy (see Section 4.1) is available for cellular work, such as ATP synthesis. ATP is the primary agent that couples exergonic and endergonic reactions in the cell to facilitate the synthesis of complex molecules (see Section 4.2). The harnessing of energy into a useful form such as ATP when electrons move from a higher-energy state to a lower-energy state is analogous to what happens when water moves through turbines at a dam and produces electricity.

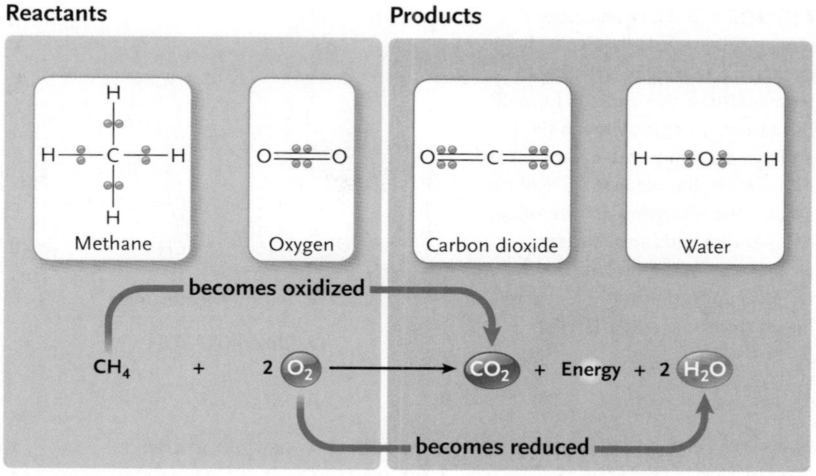

As you have just learned, oxidation and reduction are defined with respect to the gain or loss of electrons. You will see in the reactions described later in the chapter that electron movement is associated with H atoms. Recall from Section 2.3 that a hydrogen atom, H, consists of a proton and an electron: $H = H^+ + e^-$. Therefore, the transfer of a hydrogen atom involves the transfer of an electron. As a result, when a molecule loses a hydrogen atom, it becomes oxidized.

The gain or loss of an electron in a redox reaction is not always complete. That is, in some redox reactions, electrons are transferred completely from one atom to another, whereas in others the degree to which electrons are shared between atoms changes. (Sharing of electrons in covalent bonds is described in Section 2.3.) The condition of electron sharing is said to involve a relative loss or gain of electrons; most redox reactions in the electron transfer system discussed later in the chapter are of this type. The redox reaction between methane and oxygen (the burning of natural gas in air) that produces carbon dioxide and water illustrates a change in the degree of electron sharing **(Figure 8.2)**. The dots in the figure indicate the positions of the electrons involved in the covalent bonds of the reactants and products.

Compare the reactant methane with the product CO_2. In methane, the covalent electrons are shared essentially equally between bonded C and H atoms because C and H are almost equally electronegative. In CO_2, electrons are closer to the O atoms than to the C atom in the C=O bonds because O atoms are highly electronegative. Overall, this means that the C atom has partially "lost" its shared electrons in the reaction. In short, methane has been oxidized. Now compare the oxygen reactant with the product water. In the O_2 molecule, the two O atoms share their electrons equally. The oxygen reacts with the hydrogen from methane, producing water in which the electrons are

Reactants   **Products**

FIGURE 8.2 Relative loss and gain of electrons in a redox reaction, the burning of methane (natural gas) in oxygen. Compare the positions of the electrons in the covalent bonds of reactants and products. In this redox reaction, methane is oxidized because the carbon atom has partially lost its shared electrons, and oxygen is reduced because the oxygen atoms have partially gained electrons.
© Cengage Learning 2014

closer to the O atom than to the H atoms. This means that each O atom has partially "gained" electrons; in short, oxygen has been reduced. Because of this, the reaction between CH_4 and O_2 releases a lot of heat energy as the electrons in the C—H bonds of CH_4 move closer to the electronegative atoms that form CO_2. The more electronegative an atom is, the greater the force that holds the electrons to that atom and therefore the greater the energy required to lose an electron.

Electrons Flow from Fuel Substances to Final Electron Acceptors

The energy of the electrons removed during cellular oxidations originates in the reactions of photosynthesis (see Figure 8.1A). During photosynthesis, electrons derived from water are pushed to very high energy levels using energy from the absorption of light. These higher-energy electrons, together with H^+ from water, are combined with carbon dioxide to form sugar molecules and then are removed by the oxidative reactions that release energy for cellular activities (see Figure 8.1B). As electrons pass to acceptor molecules, they lose much of their energy. Some of this energy drives the synthesis of ATP from ADP and P_i (a phosphate group from an inorganic source) (see Section 4.2).

The total amount of energy obtained from electrons flowing through cellular oxidative pathways depends on the difference between their high energy level in fuel substances and the lower energy level in the molecule that acts as the *final acceptor* for electrons, that is, the last molecule reduced in cellular pathways. The lower the energy level in the final acceptor, the greater the yield of energy for cellular activities. Oxygen is the final acceptor in the most efficient and highly developed form of cellular oxidation: cellular respiration (see Figure 8.1B). The

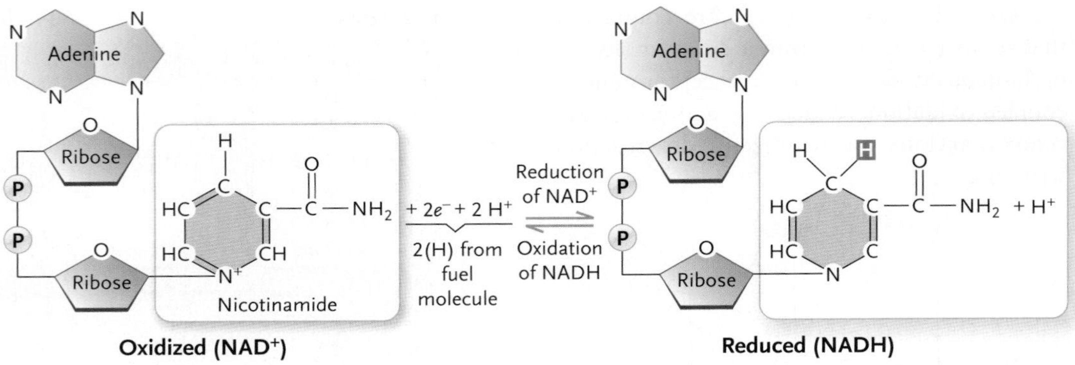

FIGURE 8.3 Electron carrier NAD⁺. When a fuel molecule is oxidized, releasing two hydrogen atoms, NAD⁺, the oxidized form of the carrier, accepts a proton (H⁺) and two electrons and is transformed into NADH, the reduced form of the carrier. The nitrogenous base (blue) of NAD that adds and releases electrons and protons is nicotinamide, which is derived from the vitamin niacin (nicotinic acid).

© Cengage Learning 2014

Oxidized (NAD⁺) ⟷ Reduced (NADH)

$+ 2e^- + 2 H^+$

Reduction of NAD⁺

2(H) from fuel molecule

Oxidation of NADH

very low energy level of the electrons added to oxygen allows a maximum output of energy for ATP synthesis. As part of the final reduction, oxygen combines with protons and electrons to form water.

In Cellular Respiration, Cells Make ATP by Oxidative Phosphorylation

Cellular respiration includes both the reactions that transfer electrons from organic molecules (fuel molecules, commonly known as food) to oxygen and the reactions that make ATP. These reactions are often written in a summary form that uses glucose ($C_6H_{12}O_6$) as the initial reactant:

$$C_6H_{12}O_6 + 6 O_2 + 32 ADP + 32 P_i \rightarrow 6 H_2O + 6 CO_2 + 32 ATP$$

In this overall reaction, electrons and protons are transferred from glucose to oxygen, forming water, and the carbons left after this transfer are released as carbon dioxide. ATP synthesis by the addition of P_i to ADP is the key, and final, step of this reaction. As discussed in Section 4.2, *phosphorylation* is the term for a reaction that adds a phosphate group to a substance such as ADP. How we derive the 32 ATP molecules is explained later in this chapter.

The oxidation of fuel molecules in cellular respiration is catalyzed by a number of *dehydrogenase* enzymes that facilitate the transfer of electrons from a fuel molecule to a molecule that acts as an *electron carrier*. The most common electron carrier is the coenzyme **nicotinamide adenine dinucleotide (NAD⁺: Figure 8.3).** (Coenzymes are discussed in Section 4.4.) As its name indicates, NAD⁺ is a type of nucleotide (see Section 3.5). In cellular respiration, dehydrogenases remove two H atoms from a substrate molecule and transfer the two electrons, but only one of the protons, to NAD⁺ (the oxidized form), resulting in its complete reduction to NADH (the reduced form). The other proton is released. Later you will learn that the potential energy carried in NADH is used in the synthesis of ATP.

The entire process of cellular respiration can be divided into three stages **(Figure 8.4):**

1. In **glycolysis,** enzymes break a molecule of glucose (contains six carbon atoms) into two molecules of pyruvate (an organic compound with a backbone of three carbon

atoms). Some NADH is produced from NAD⁺, and some ATP is synthesized by **substrate-level phosphorylation,** an enzyme-catalyzed reaction that transfers a phosphate group from a substrate to ADP **(Figure 8.5).**

2. In **pyruvate oxidation,** enzymes convert the three-carbon pyruvate into a two-carbon acetyl group that enters the **citric acid cycle** where it is completely oxidized to carbon dioxide. Some NADH is produced from NAD⁺, and some

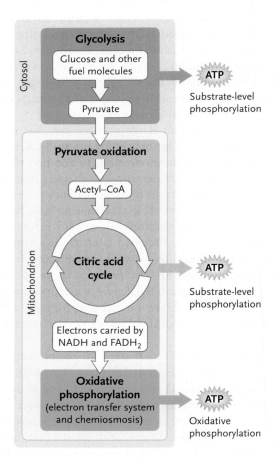

FIGURE 8.4 The three stages of cellular respiration: (1) glycolysis; (2) pyruvate oxidation and the citric acid cycle; and (3) oxidative phosphorylation, which includes the electron transfer system and chemiosmosis.

© Cengage Learning 2014

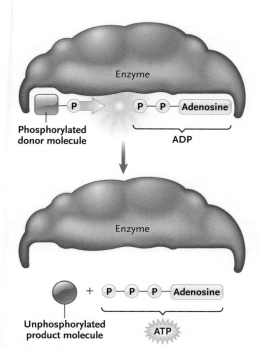

FIGURE 8.5 Substrate-level phosphorylation. A phosphate group is transferred from a high-energy donor directly to ADP, forming ATP.
© Cengage Learning 2014

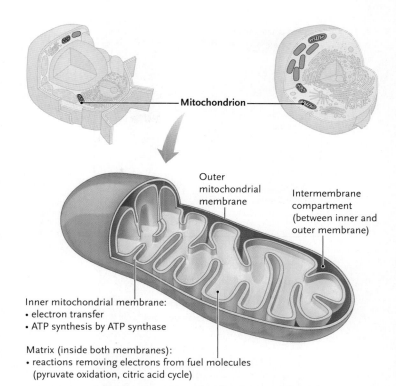

FIGURE 8.6 Membranes and compartments of a mitochondrion. Label lines that end in a dot indicate a compartment enclosed by the membranes.
© Cengage Learning 2014

ATP is synthesized during the citric acid cycle by substrate-level phosphorylation.

3. In **oxidative phosphorylation,** high-energy electrons produced from stages 1 and 2 are delivered to oxygen by a sequence of electron carriers in the **electron transfer system.** Free energy released by the electron flow then generates an H^+ gradient in a process called **chemiosmosis.** The enzyme **ATP synthase** uses the H^+ gradient as the energy source to make ATP from ADP and P_i.

In eukaryotes, most of the reactions of cellular respiration occur in various regions of the mitochondrion **(Figure 8.6)**; only glycolysis is located in the cytosol. Pyruvate oxidation and the citric acid cycle take place in the mitochondrial matrix. The inner mitochondrial membrane houses the electron transfer system and the ATP synthase enzymes. Transport proteins, located primarily in the inner membrane, control the substances that enter and leave mitochondria.

In prokaryotes, glycolysis, pyruvate oxidation, and the citric acid cycle all take place in the cytosol. The other reactions of cellular respiration occur on the plasma membrane.

The next three sections examine the three stages of cellular respiration in turn.

STUDY BREAK 8.1

1. Distinguish between oxidation and reduction.
2. Distinguish between cellular respiration and oxidative phosphorylation.

> **THINK OUTSIDE THE BOOK**
>
> Collaboratively or individually, find techniques for cell fractionation and prepare an outline of the steps needed to isolate and purify mitochondria from an animal cell.

8.2 Glycolysis: Splitting the Sugar in Half

Glycolysis, the first series of oxidative reactions that remove electrons from cellular fuel molecules, takes place in the cytosol of all organisms. In glycolysis (*glykys* = sweet, *lysis* = breakdown), sugars such as glucose are partially oxidized and broken down into smaller molecules, and a relatively small amount of ATP is produced. Glycolysis is also known as the Embden–Meyerhof pathway in honor of Gustav Embden and Otto Meyerhof, two German physiological chemists who (separately) made the most important contributions to determining the sequence of reactions in the pathway. Meyerhof received a Nobel Prize in 1922 for his work.

Glycolysis starts with the six-carbon sugar glucose and produces two molecules of the three-carbon organic substance *pyruvate* or *pyruvic acid* in 10 sequential enzyme-catalyzed reactions. (The *-ate* suffix indicates the ionized form of an organic acid such as pyruvate, in which the carboxyl group —COOH dissociates to —COO$^-$ + H$^+$ which is usual

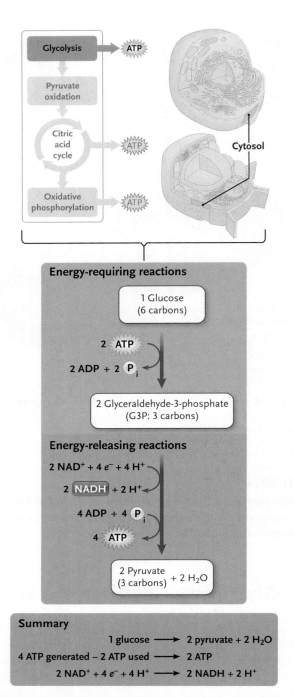

Energy-requiring reactions

1 Glucose
(6 carbons)

2 ATP

2 ADP + 2 P$_i$

2 Glyceraldehyde-3-phosphate
(G3P: 3 carbons)

Energy-releasing reactions

$2 NAD^+ + 4 e^- + 4 H^+$

$2 \boxed{NADH} + 2 H^+$

$4 ADP + 4 P_i$

4 ATP

2 Pyruvate
(3 carbons) + 2 H$_2$O

Summary

1 glucose $\longrightarrow$ 2 pyruvate + 2 H$_2$O

4 ATP generated − 2 ATP used $\longrightarrow$ 2 ATP

$2 NAD^+ + 4 e^- + 4 H^+ \longrightarrow 2 NADH + 2 H^+$

FIGURE 8.7 **Overall reactions of glycolysis.** Glycolysis splits glucose (six carbons) into pyruvate (three carbons) and yields ATP and NADH.
© Cengage Learning 2014

under cellular conditions.) Pyruvate still contains many electrons that can be removed by oxidation, and it is the primary fuel substance for the second stage of cellular respiration.

The Reactions of Glycolysis Include Energy-Requiring and Energy-Releasing Steps

The initial steps of glycolysis are *energy-requiring reactions* (**Figure 8.7,** red arrow)—2 ATP are hydrolyzed. The steps convert a molecule of the six-carbon glucose into two molecules of the

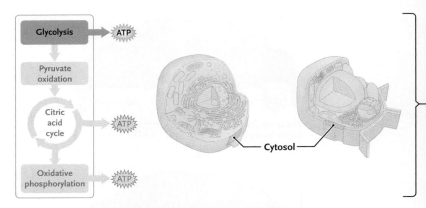

FIGURE 8.8 **Reactions of glycolysis, which occur in the cytosol.** Because two molecules of G3P are produced in reaction 5, all the reactions from 6 to 10 are doubled (not shown). The names of the enzymes that catalyze each reaction are in rust.
© Cengage Learning 2014

three-carbon phosphorylated compound, glyceraldehyde-3-phosphate (G3P). In the subsequent *energy-releasing reactions* of glycolysis (Figure 8.7, blue arrow), each G3P molecule is converted to a molecule of pyruvate. In these reactions, NAD$^+$ is reduced to NADH, and 4 ATP molecules are produced by substrate-level phosphorylation. Energywise, glycolysis results in a net gain of energy per glucose molecule of 2 ATP and 2 NADH. The end product of the glycolysis pathway is two molecules of pyruvate for each input molecule of glucose. The two molecules of pyruvate contain all of the six carbon atoms of the glucose molecule.

The reactions of glycolysis are shown in detail in **Figure 8.8.** Two redox reactions occur in glycolysis, the first in reaction 6, and the second in reaction 9.

For each molecule of glucose that enters the pathway, the energy-requiring reactions 1 to 5 generate 2 molecules of G3P using 2 ATP, and the energy-releasing reactions 6 to 10 convert the 2 molecules of G3P to 2 molecules of pyruvate, producing 4 ATP and 2 NADH. The net reactants and products of glycolysis in equation form are:

$$1 \text{ glucose} + 2 ADP + 2 P_i + 2 NAD^+ + 4 e^- + 4 H^+ \rightarrow$$
$$2 \text{ pyruvate} + 2 ATP + 2 NADH + 2 H^+ + 2 H_2O$$

Each ATP molecule produced in the energy-releasing steps of glycolysis—steps 7 and 10 (see Figure 8.7)—results from *substrate-level phosphorylation* (see Figure 8.5). In step 7 the molecule donating a phosphate group to ADP is 1,3-bisphosphoglycerate, and in step 10 the donor molecule is phosphoenolpyruvate (PEP).

Glycolysis Is Regulated at Key Points

The rate of sugar oxidation by glycolysis is closely regulated by several mechanisms to match the cell's need for ATP. For example, if excess ATP is present in the cytosol, it binds to *phosphofructokinase,* the enzyme that catalyzes reaction 3 in Figure 8.8, inhibiting its activity. This is an example of feedback inhibition (see Section 4.5 and Figure 4.15). The result-

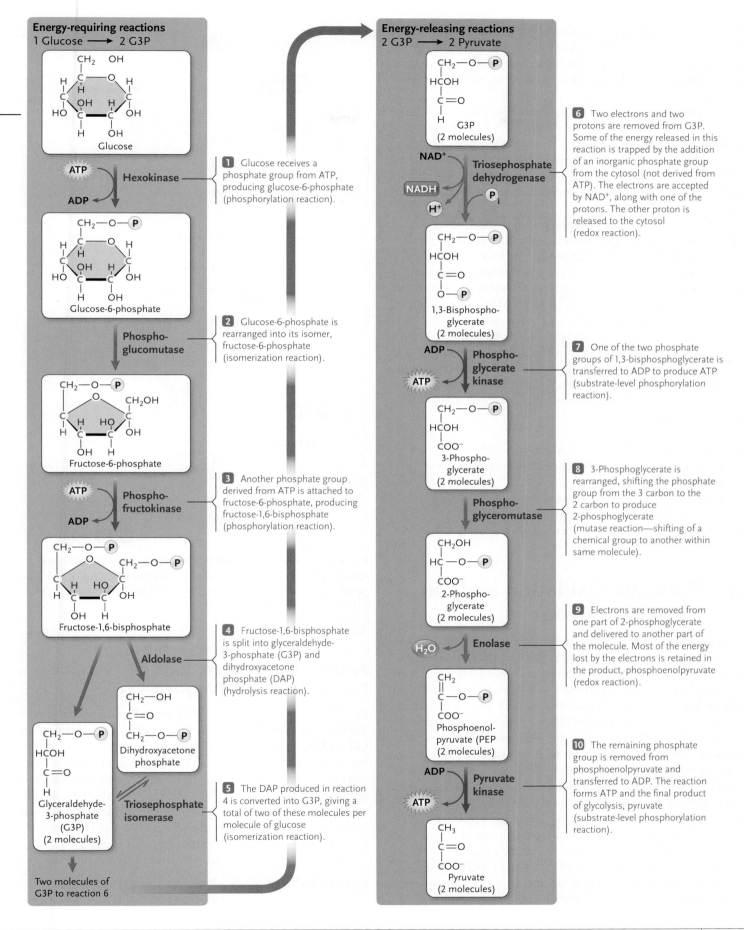

Energy-requiring reactions
1 Glucose ⟶ 2 G3P

Glucose

ATP → Hexokinase → ADP

1 Glucose receives a phosphate group from ATP, producing glucose-6-phosphate (phosphorylation reaction).

Glucose-6-phosphate

Phospho-glucomutase

2 Glucose-6-phosphate is rearranged into its isomer, fructose-6-phosphate (isomerization reaction).

Fructose-6-phosphate

ATP → Phospho-fructokinase → ADP

3 Another phosphate group, derived from ATP is attached to fructose-6-phosphate, producing fructose-1,6-bisphosphate (phosphorylation reaction).

Fructose-1,6-bisphosphate

Aldolase

4 Fructose-1,6-bisphosphate is split into glyceraldehyde-3-phosphate (G3P) and dihydroxyacetone phosphate (DAP) (hydrolysis reaction).

Dihydroxyacetone phosphate

Glyceraldehyde-3-phosphate (G3P) (2 molecules)

Triosephosphate isomerase

5 The DAP produced in reaction 4 is converted into G3P, giving a total of two of these molecules per molecule of glucose (isomerization reaction).

Two molecules of G3P to reaction 6

Energy-releasing reactions
2 G3P ⟶ 2 Pyruvate

G3P (2 molecules)

NAD⁺ → Triosephosphate dehydrogenase

NADH
H⁺
Pᵢ

6 Two electrons and two protons are removed from G3P. Some of the energy released in this reaction is trapped by the addition of an inorganic phosphate group from the cytosol (not derived from ATP). The electrons are accepted by NAD⁺, along with one of the protons. The other proton is released to the cytosol (redox reaction).

1,3-Bisphospho-glycerate (2 molecules)

ADP → Phospho-glycerate kinase → ATP

7 One of the two phosphate groups of 1,3-bisphosphoglycerate is transferred to ADP to produce ATP (substrate-level phosphorylation reaction).

3-Phospho-glycerate (2 molecules)

Phospho-glyceromutase

8 3-Phosphoglycerate is rearranged, shifting the phosphate group from the 3 carbon to the 2 carbon to produce 2-phosphoglycerate (mutase reaction—shifting of a chemical group to another within same molecule).

2-Phospho-glycerate (2 molecules)

H₂O → Enolase

9 Electrons are removed from one part of 2-phosphoglycerate and delivered to another part of the molecule. Most of the energy lost by the electrons is retained in the product, phosphoenolpyruvate (redox reaction).

Phosphoenol-pyruvate (PEP) (2 molecules)

ADP → Pyruvate kinase → ATP

10 The remaining phosphate group is removed from phosphoenolpyruvate and transferred to ADP. The reaction forms ATP and the final product of glycolysis, pyruvate (substrate-level phosphorylation reaction).

Pyruvate (2 molecules)

ing decrease in the concentration of the product of reaction 3, fructose-1,6-bisphosphate, slows or stops the subsequent reactions of glycolysis. Thus, glycolysis does not oxidize fuel substances needlessly when there is an adequate supply of ATP.

If energy-requiring activities then take place in the cell, ATP concentration in the cytosol decreases, and ADP concentration increases. As a result, ATP is released from phosphofructokinase, relieving inhibition of the enzyme. In addition, ADP activates the enzyme and, therefore, the rates of glycolysis and ATP production increase proportionately as cellular activities convert ATP to ADP.

NADH also inhibits phosphofructokinase. This inhibition slows glycolysis if excess NADH is present, such as when oxidative phosphorylation has been slowed by limited oxygen supplies. The systems that regulate phosphofructokinase and other enzymes of glycolysis closely balance the rate of the pathway to produce adequate supplies of ATP and NADH without oxidizing excess quantities of glucose and other sugars.

Our discussion of the oxidative reactions that supply electrons now moves from the cytosol to mitochondria, the site of pyruvate oxidation and the citric acid cycle. These reactions complete the breakdown of fuel substances into carbon dioxide and provide most of the electrons that drive electron transfer and ATP synthesis.

STUDY BREAK 8.2

1. What are the energy-requiring and energy-releasing steps of glycolysis?
2. What is the redox reaction in glycolysis?
3. How is ATP synthesized in glycolysis?
4. Why is phosphofructokinase a target for inhibition by ATP?

8.3 Pyruvate Oxidation and the Citric Acid Cycle

Glycolysis produces pyruvate molecules in the cytosol, and an active transport mechanism moves them into the mitochondrial matrix where pyruvate oxidation and the citric acid cycle proceed. An overview of these two processes is presented in **Figure 8.9.** Oxidation of pyruvate generates CO_2, acetyl–coenzyme A (acetyl–CoA), and NADH in a redox reaction involving NAD^+. The acetyl group of acetyl–CoA enters the **citric acid cycle.** As the citric acid cycle turns, every available electron carried into the cycle from pyruvate oxidation is transferred to NAD^+ or to another nucleotide-based molecule, *flavin adenine dinucleotide* (FAD; the reduced form is $FADH_2$). With each turn of the cycle, substrate-level phosphorylation produces 1 ATP. The combined action of pyruvate oxidation and the citric acid cycle oxidizes the three-carbon products of glycolysis completely to carbon dioxide. The 3 NADH and 1 $FADH_2$ produced for each acetyl–CoA during this stage carry high-energy electrons to the electron transfer system in the mitochondrion.

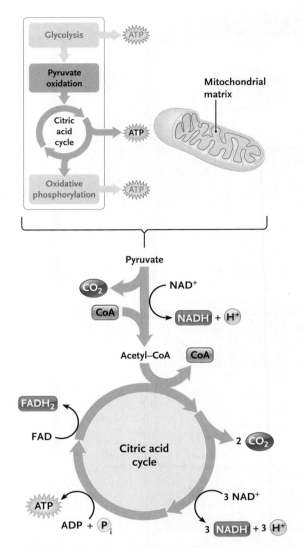

FIGURE 8.9 Overall reactions of pyruvate oxidation and the citric acid cycle. Each turn of the cycle oxidizes an acetyl group of acetyl–CoA to 2 CO_2. Acetyl–CoA, NAD^+, FAD, ADP, and P_i enter the cycle; CoA, NADH, $FADH_2$, ATP, and CO_2 are released as products.
© Cengage Learning 2014

Pyruvate Oxidation Produces the Two-Carbon Fuel of the Citric Acid Cycle

In pyruvate oxidation (also called **pyruvic acid oxidation**), a multienzyme complex removes the $—COO^-$ from pyruvate as CO_2 and then oxidizes the remaining two-carbon fragment of pyruvate to an acetyl group ($CH_3CO—$) **(Figure 8.10).** Two electrons and two protons are released by these reactions; the electrons and one proton are accepted by NAD^+, reducing it to NADH, and the other proton is released as free H^+. The acetyl group is transferred to the nucleotide-based carrier *coenzyme A* (CoA). As acetyl–CoA, it carries acetyl groups to the citric acid cycle.

In summary, the pyruvate oxidation reaction is:

pyruvate + CoA + NAD^+ →

acetyl–CoA + NADH + H^+ + CO_2

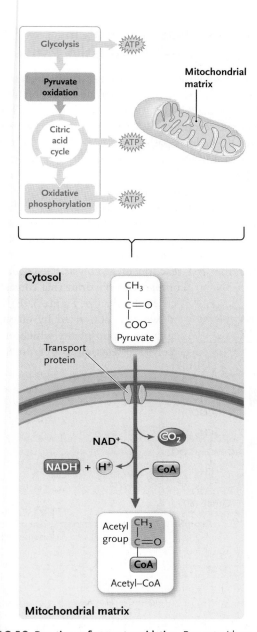

FIGURE 8.10 Reactions of pyruvate oxidation. Pyruvate (three carbons) is oxidized to an acetyl group (two carbons) and CO_2. NAD^+ accepts two electrons and one proton removed in the oxidation. The acetyl group, carried by CoA, is the fuel for the citric acid cycle.
© Cengage Learning 2014

Because each glucose molecule that enters glycolysis produces two molecules of pyruvate, all the reactants and products in this equation are *doubled* when pyruvate oxidation is considered a continuation of glycolysis.

The Citric Acid Cycle Oxidizes Acetyl Groups Completely to CO_2

The reactions of the citric acid cycle **(Figure 8.11)** oxidize acetyl groups completely to CO_2 and synthesize some ATP molecules by substrate-level phosphorylation (see Figure 8.5). The citric

acid cycle gets its name from citrate, the product of the first reaction of the cycle. It is also called the **tricarboxylic acid cycle** or the **Krebs cycle,** named after Hans Krebs, a German-born scientist who worked out the majority of the reactions in the cycle in research he conducted in England beginning in 1932. How did Krebs discover the citric acid cycle? Using minced pigeon breast muscle (a tissue used in flight that has a very high rate of cellular respiration), he demonstrated that the four-carbon organic acids succinate, fumarate, malate, and oxaloacetate stimulated the consumption of oxygen by the tissue. He also showed that the oxidation of pyruvate by the muscle tissue is stimulated by the six-carbon organic acids citrate and isocitrate, and the five-carbon acid α-ketoglutarate. While several other researchers pieced together segments of the reaction series, Krebs was the first to reason that the organic acids were linked into a cycle of reactions rather than a linear series. Krebs was awarded a Nobel Prize in 1953 for his elucidation of the citric acid cycle.

The citric acid cycle has eight reactions, each catalyzed by a specific enzyme. All of the enzymes are located in the mitochondrial matrix except the enzyme for reaction 6, succinate dehydrogenase, which is bound to the inner mitochondrial membrane on the matrix side. In a complete turn of the cycle, one two-carbon acetyl unit is consumed and two molecules of CO_2 are released (at reactions 3 and 4), thereby completing the conversion of all the C atoms originally in glucose to CO_2. The CoA molecule that carried the acetyl group to the cycle is released and participates again in pyruvate oxidation to pick up another acetyl group. Electron pairs are removed at each of four oxidations in the cycle (reactions 3, 4, 6, and 8). Three of the oxidations use NAD^+ as the electron acceptor, producing 3 NADH, and one uses FAD, producing 1 $FADH_2$. Substrate-level phosphorylation generates 1 ATP as part of reaction 5. Therefore, the net reactants and products of one turn of the citric acid cycle are:

$$1 \text{ acetyl–CoA} + 3 \text{ NAD}^+ + 1 \text{ FAD} + 1 \text{ ADP} + 1 \text{ P}_i + 2 \text{ H}_2\text{O} \rightarrow$$
$$2 \text{ CO}_2 + 3 \text{ NADH} + 1 \text{ FADH}_2 + 1 \text{ ATP} + 3 \text{ H}^+ + 1 \text{ CoA}$$

Because one molecule of glucose is converted to two molecules of pyruvate by glycolysis and each molecule of pyruvate is converted to one acetyl group, all the reactants and products in this equation are *doubled* when the citric acid cycle is considered a continuation of glycolysis and pyruvate oxidation.

Most of the energy released by the four oxidations of the cycle is associated with the high-energy electrons carried by the 3 NADH and 1 $FADH_2$. These high-energy electrons enter the electron transfer system, where their energy is used to make most of the ATP produced in cellular respiration.

Like glycolysis, the citric acid cycle is regulated at several steps to match its rate to the cell's requirements for ATP. For example, the enzyme that catalyzes the first reaction of the citric acid cycle, *citrate synthase,* is inhibited by elevated ATP concentrations. The inhibitions automatically slow or stop the cycle when ATP production exceeds the demands of the cell and, by doing so, conserve cellular fuels.

FIGURE 8.11 Reactions of the citric acid cycle.

FIGURE 8.11 Reactions of the citric acid cycle. Acetyl–CoA, NAD$^+$, FAD, ADP, and P$_i$ enter the cycle; CoA, NADH, FADH$_2$, ATP, and CO$_2$ are released as products. The CoA released in reaction 1 can cycle back for another turn of pyruvate oxidation. Enzyme names are in rust.

© Cengage Learning 2014

Carbohydrates, Fats, and Proteins Can Function as Electron Sources for Oxidative Pathways

In addition to glucose and other six-carbon sugars, reactions leading from glycolysis through pyruvate oxidation also oxidize a wide range of carbohydrates, lipids, and proteins. These molecules enter the reaction pathways at various points. **Figure 8.12** summarizes the cellular pathways involved. It shows the central role of CoA in funneling acetyl groups from different pathways into the citric acid cycle and of the mitochondrion as the site where most of these groups are oxidized.

Carbohydrates such as sucrose and other disaccharides are easily broken into monosaccharides such as glucose and fructose that enter glycolysis at early steps. Starch (see Figure 3.9A) is hydrolyzed by digestive enzymes into individual glucose molecules, which enter the first reaction of glycolysis. Glycogen, a more complex carbohydrate that consists of glucose subunits (see Figure 3.9B), is broken down and converted by enzymes into glucose-6-phosphate, which enters glycolysis at reaction 2 of Figure 8.8.

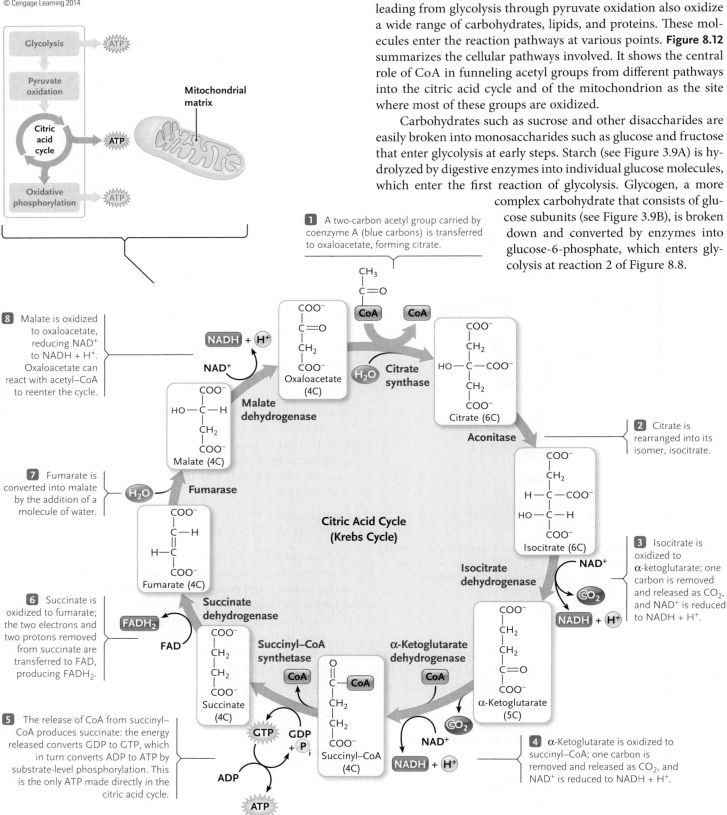

1 A two-carbon acetyl group carried by coenzyme A (blue carbons) is transferred to oxaloacetate, forming citrate.

8 Malate is oxidized to oxaloacetate, reducing NAD$^+$ to NADH + H$^+$. Oxaloacetate can react with acetyl–CoA to reenter the cycle.

7 Fumarate is converted into malate by the addition of a molecule of water.

6 Succinate is oxidized to fumarate; the two electrons and two protons removed from succinate are transferred to FAD, producing FADH$_2$.

5 The release of CoA from succinyl–CoA produces succinate: the energy released converts GDP to GTP, which in turn converts ADP to ATP by substrate-level phosphorylation. This is the only ATP made directly in the citric acid cycle.

2 Citrate is rearranged into its isomer, isocitrate.

3 Isocitrate is oxidized to α-ketoglutarate; one carbon is removed and released as CO$_2$, and NAD$^+$ is reduced to NADH + H$^+$.

4 α-Ketoglutarate is oxidized to succinyl–CoA; one carbon is removed and released as CO$_2$, and NAD$^+$ is reduced to NADH + H$^+$.

Citric Acid Cycle (Krebs Cycle)

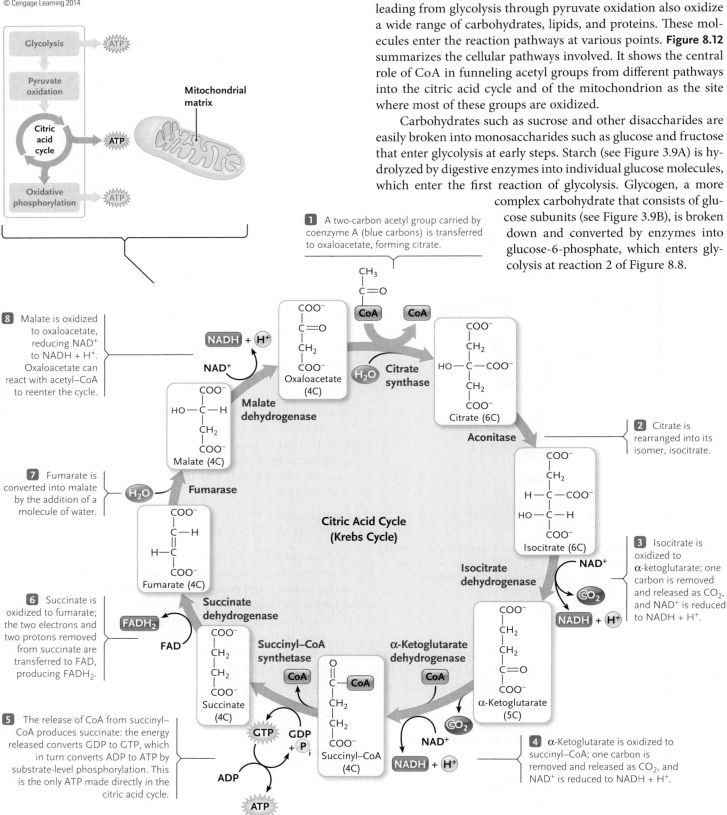

Among the fats, triglycerides (see Figure 3.10) are major sources of electrons for ATP synthesis. Before entering the oxidative reactions, the triglycerides are hydrolyzed into glycerol and individual fatty acids. The glycerol is converted to G3P and enters glycolysis at reaction 6 of Figure 8.8, in the ATP-producing portion of the pathway. The fatty acids—and many other types of lipids—are split into two-carbon fragments, which enter the citric acid cycle as acetyl–CoA. The energy released by the oxidation of fats, by weight, is comparatively high—a little more than twice the energy of oxidation of carbohydrates or proteins. This fact explains why fats are an excellent source of energy in the diet.

Proteins are hydrolyzed to amino acids before oxidation. During oxidation, the amino group is removed, and the remainder of the molecule enters the pathway of carbohydrate oxidation as either pyruvate, acetyl units carried by CoA, or intermediates of the citric acid cycle. For example, the amino acid alanine is converted into pyruvate, leucine is converted into acetyl units, and phenylalanine is converted into fumarate. Fumarate enters the citric acid cycle at reaction 7 of Figure 8.11.

STUDY BREAK 8.3

Summarize the fate of pyruvate molecules produced by glycolysis.

8.4 Oxidative Phosphorylation: The Electron Transfer System and Chemiosmosis

From the standpoint of ATP synthesis, the most significant products of glycolysis, pyruvate oxidation, and the citric acid cycle are the many high-energy electrons removed from fuel molecules and picked up by the carrier molecules NAD^+ or FAD as a result of redox reactions. These electrons are released by the reduced form of these carriers into the electron transfer system of mitochondria (in eukaryotes).

The **mitochondrial electron transfer system** consists of a series of electron carriers that alternately pick up and release electrons, and ultimately transfer them to their final acceptor, oxygen. As the electrons flow through the system, they release free energy, which is used to build a gradient of H^+ across the inner mitochondrial membrane. The gradient goes from a high concentration of H^+ in the intermembrane compartment to a low concentration of H^+ in the matrix. The H^+ gradient supplies the energy that drives ATP synthesis by mitochondrial ATP synthase.

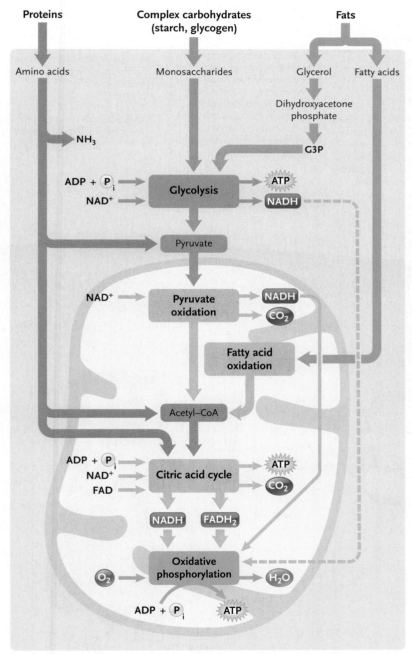

FIGURE 8.12 **Major pathways that oxidize carbohydrates, fats, and proteins.** Reactions that occur in the cytosol are shown against a tan background; reactions that occur in mitochondria are shown inside the organelle. CoA funnels the products of many oxidative pathways into the citric acid cycle.

© Cengage Learning 2014

FIGURE 8.13 Oxidative phosphorylation: the mitochondrial electron transfer system and chemiosmosis.

Oxidative phosphorylation involves the electron transfer system (steps 1–6), and chemiosmosis by ATP synthase (steps 7–9). Blue arrows indicate electron flow; red arrows indicate H$^+$ movement.

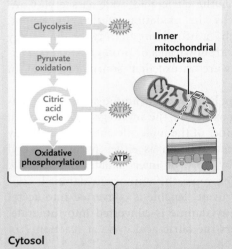

1 Complex I picks up high-energy electrons from NADH and conducts them via two electron carriers, FMN (flavin mononucleotide) and an Fe/S (iron–sulfur) protein, to ubiquinone.

2 Complex II oxidizes FADH$_2$ to FAD; the two electrons released are transferred to ubiquinone, and the two protons released go into the matrix. Electrons that pass to ubiquinone by the complex II reaction bypass complex I of the electron transfer system.

3 Complex III accepts electrons from ubiquinone and transfers them through the electron carriers in the complex— cytochrome b, an Fe/S protein, and cytochrome c_1—to cytochrome c, which is free in the intermembrane space.

4 Complex IV accepts electrons from cytochrome c and delivers them via electron carriers cytochromes a and a_3 to oxygen. Four protons are added to a molecule of O$_2$ as it accepts four electrons, forming 2 H$_2$O.

5 As electrons move through the electron transfer system, they release free energy. Part of the released energy is lost as heat, but some is used by the mitochondrion to transport H$^+$ across the inner mitochondrial membrane from the matrix to the inter membrane compartment at complexes I, III, and IV.

6 The resulting H$^+$ gradient supplies the energy that drives ATP synthesis by ATP synthase.

7 Because of the gradient, H$^+$ flows across the inner membrane and into the matrix through a channel in the ATP synthase.

8 The flow of H$^+$ activates ATP synthase, making the headpiece and stalk rotate.

9 As a result of changes in shape and position as it turns, the headpiece catalyzes the synthesis of ATP from ADP and P$_i$.

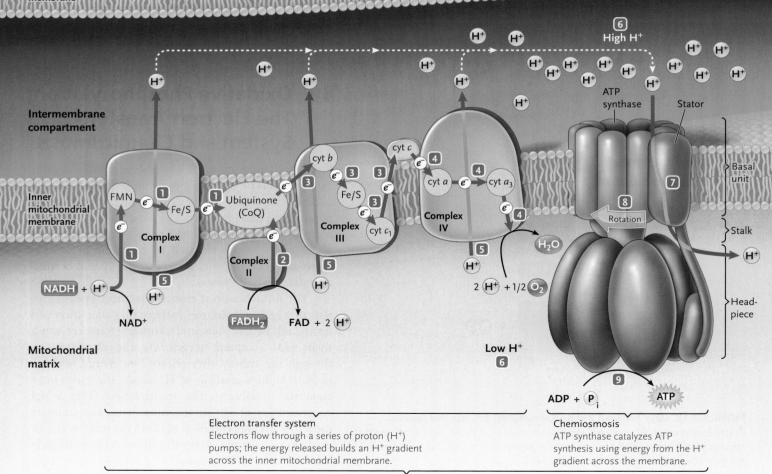

Electron transfer system
Electrons flow through a series of proton (H$^+$) pumps; the energy released builds an H$^+$ gradient across the inner mitochondrial membrane.

Chemiosmosis
ATP synthase catalyzes ATP synthesis using energy from the H$^+$ gradient across the membrane.

Oxidative phosphorylation

SUMMARY Three major protein complexes serve as electron carriers in the mitochondrial electron transfer system. Energy from electron flow through the complexes is used by the complexes to pump H^+ from the mitochondrial matrix into the intermembrane compartment. The resulting H^+ gradient supplies the energy that drives ATP synthesis by the membrane-embedded ATP synthase.

THINK LIKE A SCIENTIST Using cell fractionation techniques (see Figure 5.8), you can isolate intact mitochondria from cells. Assume that the outer mitochondrial membrane, but not the inner mitochondrial membrane is permeable to H^+. You place isolated mitochondria in a low-pH environment. The H^+ concentration will increase in which part of the mitochondrion? What effect will that increase have on ATP production?

© Cengage Learning 2014

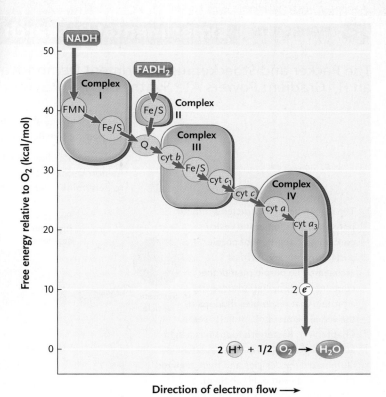

FIGURE 8.14 Organization of the mitochondrial electron transfer system from high to low free energy. Electrons flow spontaneously from one molecule to the next in the series.
© Cengage Learning 2014

In the Electron Transfer System, Electrons Flow through Protein Complexes in the Inner Mitochondrial Membrane

The mitochondrial electron transfer system includes three major protein complexes, which serve as electron carriers (**Figure 8.13**). These protein complexes, numbered I, III, and IV, are integral membrane proteins located in the inner mitochondrial membrane. In addition, a smaller complex, complex II, is bound to the inner mitochondrial membrane on the matrix side. Associated with the system are two small, highly mobile electron carriers, *cytochrome c* and *ubiquinone* (also known as coenzyme Q, or CoQ), which shuttle electrons between the major complexes.

Cytochromes are proteins with a heme prosthetic group that contains an iron atom. (Heme is an iron-containing group that binds oxygen; see Figure 3.19. A prosthetic group is a cofactor that binds tightly to a protein or enzyme; see Section 4.4.) The iron atom accepts and donates electrons. One cytochrome in particular, cytochrome *c,* was important historically in determining evolutionary relationships. Because the amino acid sequence of cytochrome *c* has been conserved by evolution, comparisons of cytochrome *c* amino acid sequences from a wide variety of eukaryotic species contributed to the construction of phylogenetic trees (see Figure 24.11). Present-day phylogeny studies typically use DNA or RNA sequences (see Chapter 24).

Electrons flow through the major protein complexes as shown in Figure 8.13, steps 1 to 4. Note that complex II, a succinate dehydrogenase complex, catalyzes two reactions. One is reaction 6 of the citric acid cycle, the conversion of succinate to fumarate (see Figure 8.11). In that reaction, FAD accepts two protons and two electrons and is reduced to $FADH_2$. The other reaction is shown in Figure 8.13, step 2.

The poison cyanide does its deadly work by blocking the transfer of electrons from complex IV to oxygen. The gas car-

bon monoxide inhibits complex IV activity, leading to abnormalities in mitochondrial function. In this way, the carbon monoxide in tobacco smoke contributes to the development of diseases associated with smoking.

What is the driving force for electron flow through the protein complexes of the electron transfer system? **Figure 8.14** shows that the individual electron carriers of the system are organized specifically from high to low free energy. Any single component has a higher affinity for electrons than the preceding carrier in the series has. Overall, molecules such as NADH contain an abundance of free energy and can be oxidized readily, whereas O_2, the terminal electron acceptor of the series, can be reduced easily. As a consequence of this organization, electron movement through the system is spontaneous, releasing free energy.

Three Major Electron Transfer Complexes Pump H^+ across the Inner Mitochondrial Membrane

Using energy from electron flow, the proteins of complexes I, III, and IV pump (actively transport) H^+ (protons) from the matrix to the intermembrane compartment of the mitochondrion (see Figure 8.13, step 5). The result is an H^+ gradient with a high concentration of H^+ in the intermembrane compartment and a low concentration of H^+ in the matrix (step 6). Because protons carry a positive charge, the asymmetric distribution of protons generates an electrical and chemical gradient across the inner mitochondrial

The Racker and Stoeckenius Experiment Demonstrating That an H$^+$ Gradient Powers ATP Synthesis by ATP Synthase

Question: Does an H$^+$ gradient power ATP synthase-catalyzed ATP synthesis, thereby supporting Mitchell's chemiosmotic hypothesis?

Experiment: Efraim Racker of Cornell University and Walther Stoeckenius of University of California, San Francisco, made membrane vesicles that had a proton pump and ATP synthase to determine whether proton-motive force drives ATP synthesis.

1. The researchers constructed synthetic phospholipid membrane vesicles containing a segment of purple surface membrane from an archaean. The purple membrane contained only bacteriorhodopsin, a protein that resembles rhodopsin, the visual pigment of animals (see Chapter 41). Bacteriorhodopsin is a light-activated proton pump. The researchers illuminated the vesicles and then analyzed the concentration of H$^+$ in them.

Cutaway view of a synthetic vesicle

Result: H$^+$ is pumped into the vesicles, creating an H$^+$ gradient.

2. The researchers next made synthetic vesicles containing ATP synthase from both bacteriorhodopsin and bovine heart mitochondria. The ATP synthase molecule was oriented so that the ATP-synthesizing headpiece was on the outside of the vesicles. They added ADP and P$_i$ to the medium containing the vesicles and tested whether ATP was produced in the dark and after a period of illumination.

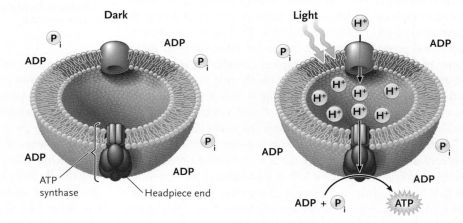

Result: In the dark, no ATP was synthesized.

Result: In the light, ATP was synthesized.

Together, these results showed that light activated the bacteriorhodopsin to produce an H$^+$ gradient, with H$^+$ moving from the outside to the inside of the vesicle (like the movement from the mitochondrial matrix to the intermembrane compartment in Figure 8.13), and that the energy from the H$^+$ gradient drove ATP synthesis by ATP synthase.

Conclusion: An H$^+$ gradient—and, therefore, proton-motive force—powers ATP synthesis by ATP synthase. The results support Mitchell's chemiosmotic hypothesis for ATP synthesis in mitochondria.

THINK LIKE A SCIENTIST In the experiment described, ATP was synthesized in the light. What do you predict would happen if the vesicles were initially exposed to light and then they were placed in the dark?

Source: E. Racker and W. Stoeckenius. 1974. Reconstitution of purple membrane vesicles catalyzing light-driven proton uptake and adenosine triphosphate formation. *Journal of Biological Chemistry* 249:662–663.

membrane, with the intermembrane compartment more positively charged than the matrix. The combination of a proton gradient and voltage gradient across the membrane produces stored energy known as the **proton-motive force.** This force contributes energy for ATP synthesis, as well as for the cotransport of substances to and from mitochondria (see Section 6.4).

Chemiosmosis Powers ATP Synthesis by an H⁺ Gradient

Within the mitochondrion, ATP is synthesized by ATP synthase, an enzyme embedded in the inner mitochondrial membrane. In 1961, British scientist Peter Mitchell of Glynn Research Laboratories proposed that mitochondrial electron transfer produces an H⁺ gradient and that the gradient powers ATP synthesis by ATP synthase. He called this pioneering model the **chemiosmotic hypothesis;** the process is commonly called *chemiosmosis* (see Figure 8.13). At the time, this hypothesis was a radical proposal because most researchers thought that the energy of electron transfer was stored as a high-energy chemical intermediate. No such intermediate was ever found, and eventually, Mitchell's hypothesis was supported by the results of many experiments, one of which is described in **Figure 8.15.** Mitchell received a Nobel Prize in 1978 for his model and supporting research.

How does ATP synthase use the H⁺ gradient to power ATP synthesis in chemiosmosis? ATP synthase consists of a *basal unit* embedded in the inner mitochondrial membrane connected by a *stalk* to a *headpiece* located in the matrix. A peripheral stalk called a *stator* bridges the basal unit and headpiece (see Figure 8.13). ATP synthase functions like an active transport ion pump. In Chapter 6, we described active transport pumps that use the energy created by hydrolysis of ATP to ADP and P$_i$ to transport ions across membranes against their concentration gradients (see Figure 6.11). However, if the concentration of an ion is very high on the side toward which it is normally transported, the pump runs in reverse—that is, the ion is transported backward through the pump, and the pump adds phosphate to ADP to generate ATP. That is how ATP synthase operates in mitochondrial membranes. Proton-motive force moves protons in the intermembrane space through the channel in the enzyme's basal unit, down their concentration gradient, and into the matrix (see Figure 8.13, step 7). The flow of H⁺ powers ATP synthesis by the headpiece; this phosphorylation reaction is chemiosmosis (steps 8 and 9). ATP synthase occurs in similar form and works in the same way in mitochondria, chloroplasts, and prokaryotes capable of oxidative phosphorylation.

Thirty-Two ATP Molecules Are Produced for Each Molecule of Glucose Oxidized Completely to CO₂ and H₂O

How many ATP molecules are produced as electrons flow through the mitochondrial electron transfer system? The most recent research indicates that approximately 2.5 ATP are synthe-

sized as a pair of electrons released by NADH travels through the entire electron transfer pathway to oxygen. The shorter pathway, followed by an electron pair released from FADH$_2$ by complex II to oxygen, synthesizes about 1.5 ATP. (Some accounts of ATP production round these numbers to 3 and 2 molecules of ATP, respectively.)

These numbers allow us to estimate the total amount of ATP that would be produced by the complete oxidation of glucose to CO₂ and H₂O if the entire H⁺ gradient produced by electron transfer is used for ATP synthesis **(Figure 8.16).** During glycolysis, substrate-level phosphorylation produces 2 ATP and 2 NADH. In pyruvate oxidation, 2 NADH are produced from the two molecules of pyruvate.

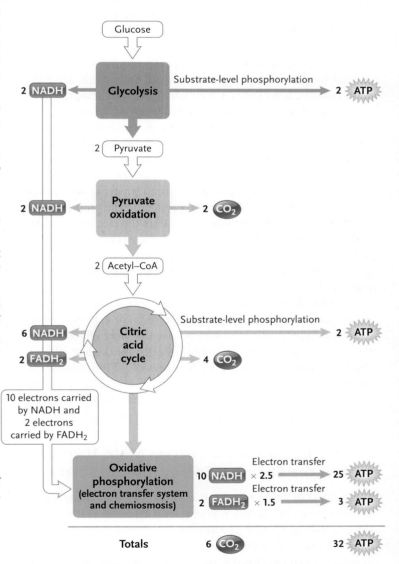

FIGURE 8.16 Summary of ATP production from the complete oxidation of a molecule of glucose. The total of 32 ATP assumes that electrons carried from glycolysis by NADH are transferred to NAD⁺ inside mitochondria. If the electrons from glycolysis are instead transferred to FAD inside mitochondria, total production will be 30 ATP.

© Cengage Learning 2014

The subsequent citric acid cycle turns twice for each molecule of glucose that enters glycolysis, yielding a total of 2 ATP produced by substrate-level phosphorylation, as well as 6 NADH, 2 FADH$_2$, and 4 CO$_2$.

The combination of glycolysis, pyruvate oxidation, and the citric acid cycle has the following summary reaction:

$$\text{glucose} + 4\,\text{ADP} + 4\,\text{P}_i + 10\,\text{NAD}^+ + 2\,\text{FAD} \rightarrow$$
$$4\,\text{ATP} + 10\,\text{NADH} + 10\,\text{H}^+ + 2\,\text{FADH}_2 + 6\,\text{CO}_2$$

Therefore, high-energy electrons carried by 10 NADH molecules (2 from glycolysis, 2 from pyruvate oxidation, and 6 from the citric acid cycle) and 2 FADH$_2$ molecules (from the citric acid cycle) enter the electron transfer system of the mitochondrion (see Figure 8.16). Assuming 2.5 molecules of ATP generated per NADH molecule and 1.5 molecules of ATP generated per FADH$_2$ molecule as discussed earlier, this gives a total of 32 ATP molecules for each molecule of glucose oxidized completely to CO$_2$ and H$_2$O.

The total of 32 ATP assumes that the two pairs of electrons carried by the 2 NADH reduced in glycolysis each drive the synthesis of 2.5 ATP when traversing the mitochondrial electron transfer system. However, because NADH cannot penetrate the mitochondrial membranes, its electrons are transferred into the mitochondrion by one of two shuttle systems. The more efficient shuttle mechanism transfers the electrons to NAD$^+$ as the acceptor inside mitochondria. These electron pairs, when passed through the electron transfer system, result in the synthesis of 2.5 ATP each, producing the grand total of 32 ATP. The less efficient shuttle transfers the electrons to FAD as the acceptor inside mitochondria. These electron pairs, when passed through the electron transfer system, result in the synthesis of only 1.5 ATP each and produce a grand total of 30 ATP instead of 32.

Which shuttle system predominates depends on the particular species and the cell types involved. For example, heart, liver, and kidney cells in mammals use the more efficient shuttle system; skeletal muscle and brain cells use the less efficient shuttle system. Regardless, the numbers of ATP produced are idealized, because mitochondria also use the H$^+$ gradient to drive cotransport; any of the energy in the gradient used for this activity would reduce ATP production proportionately.

Cellular Respiration Conserves More Than 30% of the Chemical Energy of Glucose in ATP

The process of cellular respiration is not 100% efficient; it does not convert all the chemical energy of glucose to ATP. By using the estimate of 32 ATP produced for each molecule of glucose oxidized under ideal conditions, we can estimate the overall efficiency of cellular glucose oxidation. Efficiency refers to the percentage of the chemical energy of glucose conserved as ATP energy.

Under standard conditions, including neutral pH (pH = 7) and a temperature of 25°C, the hydrolysis of ATP to ADP yields about 7.0 kilocalories per mole (kcal/mol). Assuming that complete glucose oxidation produces 32 ATP, the total energy conserved in ATP production would be about 224 kcal/mol. By contrast, if glucose is simply burned in air, it releases 686 kcal/mol. On this basis, the efficiency of cellular glucose oxidation would be about 33% (224/686 × 100 = about 33%). This value is considerably better than that of most devices designed by human engineers—for example, the engine of an automobile extracts only about 25% of the energy in the fuel it burns.

The chemical energy released by cellular oxidations that is not captured in ATP synthesis is released as heat. In mammals and birds, this source of heat maintains body temperature at a constant level. In certain mammalian tissues, including *brown fat* (see Chapter 48), the inner mitochondrial membranes contain *uncoupling proteins* (UCPs) that make the inner mitochondrial membrane "leaky" to H$^+$. As a result, electron transfer runs without building an H$^+$ gradient or synthesizing ATP and releases all the energy extracted from the electrons as heat. Brown fat with UCPs occurs in significant quantities in hibernating mammals and in very young mammals, including human infants. (*Insights from the Molecular Revolution* describes research showing that some plants also use UCPs in mitochondrial membranes to heat tissues.)

THINK OUTSIDE THE BOOK

A few human genetic diseases result from mutations that affect mitochondrial function. Collaboratively or individually, find an example of such a disease and research how the genetic mutation disrupts mitochondrial function and leads to the disease symptoms.

8.5 Fermentation

When oxygen is plentiful, electrons carried by the 2 NADH produced by glycolysis are passed to the electron transfer system inside mitochondria, and the released energy drives ATP synthesis. When oxygen is absent or limited, the electrons may be used in fermentation. In **fermentation,** electrons carried by NADH are transferred to an organic acceptor molecule rather than to the electron transfer system. This transfer converts the NADH to NAD$^+$, which is required to accept electrons in reaction 6 of glycolysis (see Figure 8.8). As a result, glycolysis continues to supply ATP by substrate-level phosphorylation. The amount of ATP that is generated by fermentation is far less than is produced by cellular respiration.

Two types of fermentation reactions exist: lactate fermentation and alcoholic fermentation **(Figure 8.17)**. **Lactate**

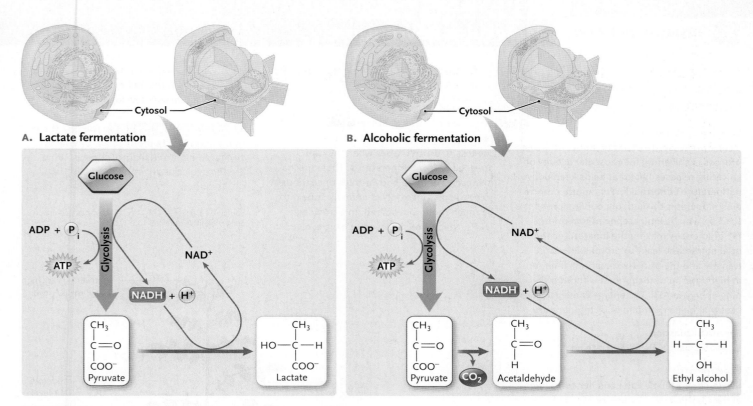

A. Lactate fermentation

Cytosol

Glucose

ADP + P_i

Glycolysis

ATP

NADH + H^+

NAD$^+$

$\begin{array}{c} CH_3 \\ | \\ C=O \\ | \\ COO^- \end{array}$
Pyruvate

$\begin{array}{c} CH_3 \\ | \\ HO-C-H \\ | \\ COO^- \end{array}$
Lactate

B. Alcoholic fermentation

Cytosol

Glucose

ADP + P_i

Glycolysis

ATP

NADH + H^+

NAD$^+$

$\begin{array}{c} CH_3 \\ | \\ C=O \\ | \\ COO^- \end{array}$
Pyruvate

CO_2

$\begin{array}{c} CH_3 \\ | \\ C=O \\ | \\ H \end{array}$
Acetaldehyde

$\begin{array}{c} CH_3 \\ | \\ H-C-H \\ | \\ OH \end{array}$
Ethyl alcohol

FIGURE 8.17 Fermentation reactions that produce (A) lactate and (B) ethyl alcohol. The fermentations, which occur in the cytosol, convert NADH to NAD$^+$, allowing the electron carrier to cycle back to glycolysis. This process keeps glycolysis running, with continued production of ATP.
© Cengage Learning 2014

fermentation converts pyruvate into lactate (Figure 8.17A). This reaction occurs in many bacteria, in some plant tissues, and in certain animal tissues such as skeletal muscle cells. When vigorous contraction of muscle cells calls for more oxygen than the circulation can supply, lactate fermentation takes place. For example, lactate accumulates in the leg muscles of a sprinter during a 100-meter race. The lactate temporarily stores electrons, and when the oxygen content of the muscle cells returns to normal levels, the reverse of the reaction in Figure 8.17A regenerates pyruvate and NADH. The pyruvate can then be used in the second stage of cellular respiration, and the NADH contributes its electron pair to the electron transfer system. Lactate is also the fermentation product of some bacteria; the sour taste of buttermilk, yogurt, and dill pickles is a sign of their activity.

Alcoholic fermentation (Figure 8.17B) occurs in some plant tissues, in certain invertebrates and protists, in certain bacteria, and in some single-celled fungi such as yeasts. In this reaction, pyruvate is converted into ethyl alcohol (which has two carbons) and carbon dioxide in a two-step series that also converts NADH into NAD$^+$. Alcoholic fermentation by yeasts has widespread commercial applications. Bakers use the yeast *Saccharomyces cerevisiae* to make bread dough rise. They mix the yeast with a small amount of sugar and blend the mixture into the dough, where oxygen levels are low. As the yeast cells convert the sugar into ethyl alcohol and carbon dioxide, the gaseous CO_2 expands and creates bubbles that cause the dough

to rise. Oven heat evaporates the alcohol and causes further expansion of the bubbles, producing a light-textured product. Alcoholic fermentation is also the mainstay of beer and wine production. Fruits are a natural home to wild yeasts—for instance, the dusty appearance of the surface of grapes is yeast. For example, winemakers rely on a mixture of wild and cultivated yeasts to produce wine. Alcoholic fermentation also occurs naturally in the environment; for example, overripe or rotting fruit frequently will start to ferment, and birds that eat the fruit may become too drunk to fly.

Fermentation is a lifestyle for some organisms. In bacteria and fungi that lack the enzymes and factors to carry out oxidative phosphorylation, fermentation is the only source of ATP. These organisms are called **strict anaerobes.** In general, strict anaerobes require an oxygen-free environment; they cannot utilize oxygen as a final electron acceptor. Among these organisms are the bacteria that cause botulism, tetanus, and some other serious diseases. For example, the bacterium that causes botulism (*Clostridium botulinum*) thrives in the oxygen-free environment of improperly sterilized canned foods. It is the absence of oxygen in canned foods that prevents the growth of most other microorganisms.

Other organisms, called **facultative anaerobes,** can switch between fermentation and full oxidative pathways, depending on the oxygen supply. Facultative anaerobes include *Escherichia coli,* a bacterium that inhabits the digestive tract of humans; the *Lactobacillus* bacteria used to produce buttermilk and yogurt; and *S. cerevisiae,* the yeast used in brewing, wine making, and baking. Many cell types in higher organisms, including vertebrate muscle cells, are also facultative anaerobes.

Hot Potatoes: Do plants use uncoupling proteins to generate heat?

Mammals use several biochemical and molecular processes to maintain body heat. One process is shivering; the muscular activity of shivering releases heat that helps keep body temperature at normal levels. Another mechanism operates through uncoupling proteins (UCPs), which eliminate the mitochondrial H^+ gradient by making the inner mitochondrial membrane leaky to protons. Electron transfer and the oxidative reactions then run at high rates in mitochondria without trapping energy in ATP. The energy is released as heat that helps maintain body temperature.

Research Question

Do plants use UCPs to generate heat?

Experiments

Research by Maryse Laloi and her colleagues at the Max Planck Institute for Molecular Plant Physiology in Germany shows that some tissues in plants may use the same process involving UCPs to generate heat. Their evidence is as follows:

1. Potato plants (*Solanum tuberosum*) have a gene with a DNA sequence similar to that of a mammalian UCP gene. The potato gene protein is clearly related to the mammalian protein, and also has the same overall three-dimensional structure.

2. The researchers used the DNA of the potato UCP gene to probe for the presence of messenger RNA (mRNA), the gene product that specifies the amino acid sequence of proteins in the cytoplasm, with the results shown in **Figure 1**. These results indicate that the potato UCP gene is active at different levels in various plant tissues, suggesting that certain tissues naturally need warming for optimal function.

3. Laloi and her coworkers then tested whether exposing potato plants to cold temperatures could induce greater synthesis of the UCP mRNA (**Figure 2**). The cold treatment resulted in an increase in UCP mRNA level in the leaves.

Conclusion

The research indicates that potato plants probably use the mitochondrial uncoupling process to warm tissues when they are stressed by low temperatures. Thus, mechanisms for warming body tissues once thought to be the province only of animals appear to be much more widespread. In particular, UCPs, which were believed to have evolved in relatively recent evolutionary times with the appearance of birds and mammals, may be a much more ancient development.

THINK LIKE A SCIENTIST The box shows that potato plants produce a UCP that is similar to a mammalian UCP gene. What are the evolutionary implications of that observation?

Source: M. Laloi et al. 1997. A plant cold-induced uncoupling protein. *Nature* 389:135–136.

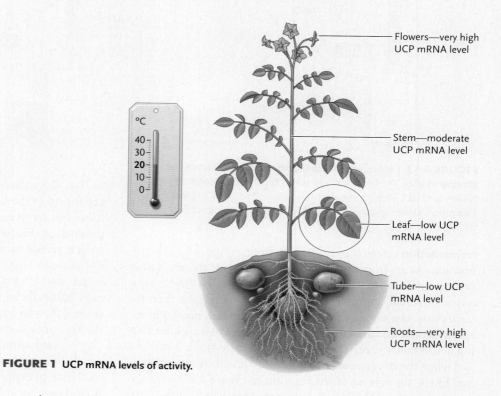

FIGURE 1 UCP mRNA levels of activity.

- Flowers—very high UCP mRNA level
- Stem—moderate UCP mRNA level
- Leaf—low UCP mRNA level
- Tuber—low UCP mRNA level
- Roots—very high UCP mRNA level

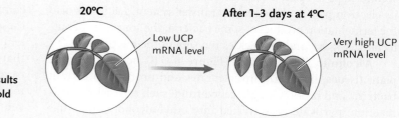

20°C — Low UCP mRNA level

After 1–3 days at 4°C — Very high UCP mRNA level

FIGURE 2 Results of exposure to cold temperature.

© Cengage Learning 2014

Some prokaryotic and eukaryotic cells are **strict aerobes**—that is, they have an absolute requirement for oxygen to survive and are unable to live solely by fermentations. Vertebrate brain cells are key examples of strict aerobes.

Fermentation differs from anaerobic respiration, the form of cellular respiration used by some prokaryotes for ATP production (see *Why It Matters* for this chapter). In fermentation the electrons carried by NADH are transferred to an organic acceptor molecule, while in anaerobic respiration the electrons are transferred to an electron transfer system. However, in contrast to aerobic respiration, in anaerobic respiration the electrons are transferred through the electron transfer system to a molecule other than oxygen, such as sulfate.

This chapter traced the flow of high-energy electrons from fuel molecules to ATP. As part of the process, the fuels are broken into molecules of carbon dioxide and water. The next chapter shows how photosynthetic organisms use these inorganic raw materials to produce organic molecules in a process that pushes the electrons back to high energy levels by absorbing the energy of sunlight.

STUDY BREAK 8.5

What is fermentation, and when does it occur? What are the two types of fermentation?

UNANSWERED QUESTIONS

Glycolysis and energy metabolism are crucial for the normal functioning of an animal. Research of many kinds is being conducted in this area, such as characterizing the molecular components of the reactions in detail and determining how they are regulated. The goal is to generate comprehensive models of cellular respiration and its regulation. Following is a specific example of ongoing research related to human disease caused by defects in cellular respiration.

How do mitochondrial proteins change in patients with Alzheimer disease?

Alzheimer disease (AD) is an age-dependent, irreversible, neurodegenerative disorder in humans. Symptoms include a progressive deterioration of cognitive functions and, in particular, a significant loss of memory. Neuropathologically, AD is characterized by the presence of extracellular amyloid plaques, intracellular neurofibrillary tangles, and synaptic and neuronal loss. Reduced brain metabolism occurs early in the onset of AD. One of the mechanisms for this physiological change appears to be damage to or reduction of key mitochondrial components, including enzymes of the citric acid cycle and the oxidative phosphorylation system. However, the complete scope of mitochondrial protein changes has not been established, nor have detailed comparisons been made of mitochondrial protein changes among AD patients.

To begin to address these questions, research is being carried out in my laboratory at The University of Texas at Dallas to analyze quantitatively the complete set of mitochondrial proteins in healthy versus AD brains. The results will show the changes that occur in the mitochondrial proteome in the two tissues. This approach is called quantitative comparative proteomic profiling. (Proteomics is the complete characterization of the proteins present in a cell, cell compartment, tissue, organ, or organism; see Chapter 19. The set of proteins identified in such a study is called the proteome.) We are using a transgenic mouse model of AD in the first stage of this research; that is, the mice have been genetically engineered with altered genes so that they develop AD. (The generation of transgenic organisms is described in Chapter 18.)

The results of our experiments have demonstrated the levels that many mitochondrial proteins are altered in the brains of transgenic AD mice. Interestingly, both down-regulated and up-regulated mitochondrial proteins were identified in AD brains. These dysregulated mitochondrial proteins participate in many different metabolic functions, including the citric acid cycle, oxidative phosphorylation, pyruvate metabolism, fatty acid oxidation, ketone body metabolism, metabolite transport, oxidative stress, mitochondrial protein synthesis, mitochondrial protein import, and cell growth and apoptosis (a type of programmed cell death; see Chapter 45). We have also determined that these changes in the mitochondrial proteome occurs early in AD before the development of significant plaque and tangle pathologies. Future experiments will be directed toward examining changes in the mitochondrial proteome in the brains of human AD patients. Ultimately, the results of our experiments may lead to the development of treatments that can slow or halt the progression of AD in humans.

THINK LIKE A SCIENTIST What significance do you attribute to the statement in this essay that changes in levels of many mitochondrial proteins occurs early in AD before the development of significant plaque and tangle pathologies? Develop a scientific hypothesis based on this statement.

Gail A. Breen is an associate professor in the Department of Molecular and Cell Biology at The University of Texas at Dallas. Her current research focuses on mitochondrial biogenesis and the role of mitochondria in neurodegenerative diseases, such as Alzheimer disease. To learn more about Dr. Breen's research go to http://www.utdallas.edu/biology/faculty/breen.html.

REVIEW KEY CONCEPTS

To access the course materials and companion resources for this text, please visit www.cengagebrain.com.

8.1 Overview of Cellular Energy Metabolism

- Plants and almost all other organisms obtain energy for cellular activities through cellular respiration, the process of transferring electrons from donor organic molecules to a final acceptor molecule such as oxygen; the energy that is released drives ATP synthesis (Figure 8.1).

- Oxidation–reduction reactions, called redox reactions, partially or completely transfer electrons from donor to acceptor atoms; the donor is oxidized as it releases electrons, and the acceptor is reduced (Figure 8.2).

- Cellular respiration occurs in three stages: (1) in glycolysis, glucose is converted to two molecules of pyruvate through a series of enzyme-catalyzed reactions. ATP is produced by substrate-level phosphorylation, an enzyme-catalyzed reaction that transfers a phosphate group from a substrate to ADP; (2) in pyruvate oxidation and the citric acid cycle, pyruvate is converted to an acetyl compound that is oxidized completely to carbon dioxide; and (3) in oxidative phosphorylation, which is comprised of the electron transfer system and chemiosmosis, high-energy electrons produced from the first two stages pass through the transfer system, with much of their energy being used to establish an H^+ gradient across the membrane that drives the synthesis of ATP from ADP and P_i (Figures 8.4 and 8.5).

- In eukaryotes, most of the reactions of cellular respiration occur in mitochondria. In prokaryotes, glycolysis, pyruvate oxidation, and the citric acid cycle occur in the cytosol, while the rest of cellular respiration occurs on the plasma membrane (Figure 8.6).

Animation: Cellular Respiration

Animation: Structure of a mitochondrion

8.2 Glycolysis: Splitting the Sugar in Half

- In glycolysis, which occurs in the cytosol, glucose (six carbons) is oxidized into two molecules of pyruvate (three carbons each). Electrons removed in the oxidations are delivered to NAD^+, producing NADH. The reaction sequence produces a net gain of 2 ATP, 2 NADH, and 2 pyruvate molecules for each molecule of glucose oxidized (Figures 8.7 and 8.8).

Animation: Glycolysis

8.3 Pyruvate Oxidation and the Citric Acid Cycle

- In pyruvate oxidation, which occurs inside mitochondria, 1 pyruvate (three carbons) is oxidized to 1 acetyl group (two carbons) and 1 carbon dioxide (CO_2). Electrons removed in the oxidation are accepted by 1 NAD^+ to produce 1 NADH. The acetyl group is transferred to coenzyme A, which carries it to the citric acid cycle (Figure 8.10).

- In the citric acid cycle, acetyl groups are oxidized completely to CO_2. Electrons removed in the oxidations are accepted by NAD^+ or FAD, and substrate-level phosphorylation produces ATP. For each acetyl group oxidized by the cycle, 2 CO_2, 1 ATP, 3 NADH, and 1 $FADH_2$ are produced (Figure 8.11).

Animation: Alternative energy sources

8.4 Oxidative Phosphorylation: The Electron Transfer System and Chemiosmosis

- Electrons are passed from NADH and $FADH_2$ to the electron transfer system, which consists of four protein complexes and two smaller shuttle carriers. As the electrons flow from one carrier to the next through the system, some of their energy is used by the complexes to pump protons across the inner mitochondrial membrane (Figures 8.13 and 8.14).

- The three major protein complexes (I, III, and IV) pump H^+ from the matrix to the intermembrane compartment, generating an H^+ gradient with a high concentration in the intermembrane compartment and a low concentration in the matrix (Figure 8.13).

- The H^+ gradient produced by the electron transfer system is used by ATP synthase as an energy source for synthesis of ATP from ADP and P_i. The ATP synthase is embedded in the inner mitochondrial membrane together with the electron transfer system (Figure 8.13).

- An estimated 2.5 ATP are synthesized as each electron pair travels from NADH to oxygen through the mitochondrial electron transfer system; about 1.5 ATP are synthesized as each electron pair travels through the system from $FADH_2$ to oxygen. Using these totals gives an efficiency of more than 30% for the utilization of energy released by glucose oxidation if the H^+ gradient is used only for ATP production (Figure 8.16).

Animation: The mitochondrial electron transfer system and oxidative phosphorylation

8.5 Fermentation

- Lactate fermentation and alcoholic fermentation are reaction pathways that deliver electrons carried from glycolysis by NADH to organic acceptor molecules, thereby converting NADH back to NAD^+. The NAD^+ is required to accept electrons generated by glycolysis, allowing glycolysis to supply ATP by substrate-level phosphorylation (Figure 8.17).

Animation: The fermentation reactions

UNDERSTAND & APPLY

Test Your Knowledge

1. What is the final acceptor for electrons in cellular respiration?
   a. oxygen
   b. ATP
   c. carbon dioxide
   d. hydrogen
   e. water

2. In glycolysis:
   a. free oxygen is required for the reactions to occur.
   b. ATP is used when glucose and fructose-6-phosphate are phosphorylated, and ATP is synthesized when 3-phosphoglycerate and pyruvate are formed.
   c. the enzymes that move phosphate groups on and off the molecules are uncoupling proteins.
   d. the product with the highest potential energy in the pathway is pyruvate.
   e. the end product of glycolysis moves to the electron transfer system.

3. Which of the following statements about phosphofructokinase is *false*?
   a. It is located and has its main activity in the inner mitochondrial membrane.
   b. It can be inhibited by NADH to slow glycolysis.
   c. It can be inactivated by ATP at an inhibitory site on its surface.
   d. It can be activated by ADP at an excitatory site on its surface.
   e. It can cause ADP to form.

4. Which of the following statements is *false*? Imagine that you ingested three chocolate bars just before sitting down to study this chapter. Most likely:
   a. your brain cells are using ATP.
   b. there is no deficit of the initial substrate to begin glycolysis.
   c. the respiratory processes in your brain cells are moving atoms from glycolysis through the citric acid cycle to the electron transfer system.
   d. after a couple of hours, you change position and stretch to rest certain muscle cells, which removes lactate from these muscles.
   e. after 2 hours, your brain cells are oxygen-deficient.

5. If ADP is produced in excess in cellular respiration, this excess ADP will:
   a. bind glucose to turn off glycolysis.
   b. bind glucose-6-phosphate to turn off glycolysis.
   c. bind phosphofructokinase to turn on or keep glycolysis turned on.
   d. cause lactate to form.
   e. increase oxaloacetate binding to increase NAD^+ production.

6. Which of the following statements is *false*? In cellular respiration:
   a. one molecule of glucose can produce about 32 ATP.
   b. oxygen combines directly with glucose to form carbon dioxide.
   c. a series of energy-requiring reactions is coupled to a series of energy-releasing reactions.
   d. NADH and $FADH_2$ allow H^+ to be pumped across the inner mitochondrial membrane.
   e. the electron transfer system occurs in the inner mitochondrial membrane.

7. You are reading this text while breathing in oxygen and breathing out carbon dioxide. The carbon dioxide arises from:
   a. glucose in glycolysis.
   b. NAD^+ redox reactions in the mitochondrial matrix.
   c. NADH redox reactions on the inner mitochondrial membrane.
   d. $FADH_2$ in the electron transfer system.
   e. the oxidation of pyruvate, isocitrate, and α-ketoglutarate in the citric acid cycle.

8. In the citric acid cycle:
   a. NADH and H^+ are produced when α-ketoglutarate is both produced and metabolized.
   b. ATP is produced by oxidative phosphorylation.
   c. to progress from a four-carbon molecule to a six-carbon molecule, CO_2 enters the cycle.
   d. $FADH_2$ is formed when succinate is converted to oxaloacetate.
   e. the cycle "turns" once for each molecule of glucose metabolized.

9. For each NADH produced from the citric acid cycle, about how many ATP are formed?
   a. 38
   b. 36
   c. 32
   d. 2.5
   e. 2.0

10. In the 1950s, a diet pill that had the effect of "poisoning" ATP synthase was tried. The person taking it could not use glucose and "lost weight"—and ultimately his or her life. Today, we know that the immediate effect of poisoning ATP synthase is:
    a. ATP would not be made in the electron transfer system.
    b. H^+ movement across the inner mitochondrial membrane would increase.
    c. more than 32 ATP could be produced from a molecule of glucose.
    d. ADP would be united with phosphate more readily in the mitochondria.
    e. ATP would react with oxygen.

Discuss the Concepts

1. Why do you think nucleic acids are not oxidized extensively as a cellular energy source?

2. A hospital patient was regularly found to be intoxicated. He denied that he was drinking alcoholic beverages. The doctors and nurses made a special point to eliminate the possibility that the patient or his friends were smuggling alcohol into his room, but he was still regularly intoxicated. Then, one of the doctors had an idea that turned out to be correct and cured the patient of his intoxication. The idea involved the patient's digestive system and one of the oxidative reactions covered in this chapter. What was the doctor's idea?

Design an Experiment

There are several ways to measure cellular respiration experimentally. For example, CO_2 and O_2 gas sensors measure changes over time in the concentration of carbon dioxide or oxygen, respectively. Design two experiments to test the effects of changing two different variables or conditions (one per experiment) on the respiration of a research organism of your choice.

Interpret the Data

As CO_2 concentrations increase in the atmosphere, biologists continue to explore the role of respiration from plants as a small, but potentially important contribution beyond fossil fuel combustion. The data in the **Table** were collected from the leaf of a sagebrush plant from a semiarid ecosystem in Wyoming, enclosed in a chamber that measures the rate of CO_2 exchange. The respiration rate is the amount of CO_2 in micromoles lost by the leaf per square meter per second, which results in the negative numbers. The temperature values are from the leaves as they are heated or cooled during the measurements.

Observation	Temperature (°C)	Respiration Rate ($\mu mol/m^2/s$)
1	25	−2.0
2	30	−2.7
3	35	−4.1
4	40	−5.8
5	20	−1.3
6	15	−1.0
7	10	−0.7

© Cengage Learning 2014

1. Make a graph of the data, with temperature on the x axis and respiration rate on the y axis.

2. The Q10 value of respiration is the increase in respiration, expressed as the ratio of the higher rate to the lower rate, with a 10°C change in temperature. What is the approximate (whole number) Q10 of respiration for this sagebrush leaf?

3. What describes the relationship between temperature and respiration, a line or a curve? Does the Q10 that you calculated in 2 suggest a line or a curve?

4. How might the predicted increase in temperature due to elevated CO_2 concentrations impact respiration of sagebrush leaves? Do you think the data presented here are all that is needed to predict this impact?

Source: Data based on unpublished research by Brent Ewers, University of Wyoming.

Apply Evolutionary Thinking

Which of the two phosphorylation mechanisms, oxidative phosphorylation or substrate-level phosphorylation, is likely to have appeared first in evolution? Why?

9

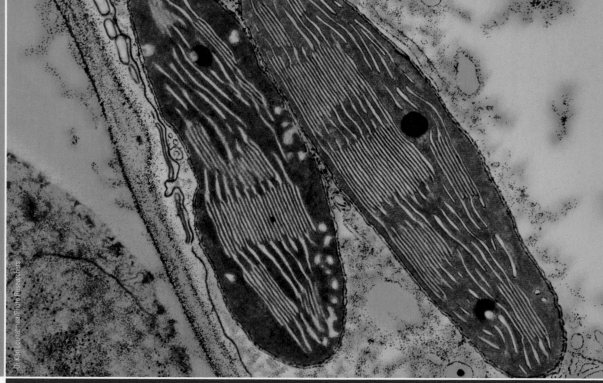

Chloroplasts in the leaf of the pea plant *Pisum sativum* (colorized TEM). The light-dependent reactions of photosynthesis take place within the thylakoids of the chloroplasts (thylakoid membranes are shown in yellow).

Photosynthesis

Why it matters . . . By the late 1880s, scientists realized that green algae and plants use light as a source of energy to make organic molecules. This conversion of light energy to chemical energy in the form of sugar and other organic molecules is called **photosynthesis.** The scientists also knew that these organisms release oxygen as part of their photosynthetic reactions. Among these scientists was a German botanist, Theodor Engelmann, who was curious about the particular colors of light used in photosynthesis. Was green light the most effective in promoting photosynthesis or were other colors more effective?

Engelmann used a light microscope and a glass prism to find the answer to this question. Yet his experiment stands today as a classic one, both for the fundamental importance of his conclusion and for the simple but elegant methods he used to obtain it. Engelmann placed a strand of a filamentous green alga on a glass microscope slide along with water containing aerobic bacteria (bacteria that require oxygen to survive). He adjusted the prism so that it split a beam of visible light into its separate colors, which spread like a rainbow across the strand **(Figure 9.1).** After a short time, he noticed that the bacteria had

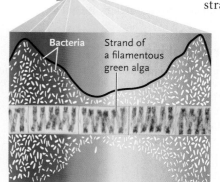

FIGURE 9.1 Engelmann's 1882 experiment revealing the action spectrum of light used in photosynthesis by a filamentous green alga. Using a glass prism, Engelmann broke up a beam of light into a spectrum of colors, which were cast across a microscope slide with a strand of the alga in water containing aerobic bacteria. The bacteria clustered along the algal strand in the regions where oxygen was released in greatest quantity—the regions in which photosynthesis proceeded at the greatest rate. Those regions corresponded to the colors (wavelengths) of light being absorbed most effectively by the alga—violet, blue, and red.

© Cengage Learning 2014

formed large clusters in the blue and violet light at one end of the strand, and in the red light at the other end. Very few bacteria were found in the green light. Evidently, violet, blue, and red light caused the most oxygen to be released, and Engelmann concluded that these colors of light—rather than green—were used most effectively in photosynthesis.

Engelmann used the distribution of bacteria in the light to construct a curve called an *action spectrum* for the wavelengths of light that fell on the alga; it showed the relative effect of each color of light on photosynthesis (black curve in Figure 9.1). Engelmann's results were so accurate that an action spectrum obtained with modern equipment fits closely with his bacterial distribution. However, his results were controversial for some 60 years, until instruments that enabled direct measurements of the effects of specific wavelengths of light on photosynthesis became available.

Scientists now know that photosynthetic organisms, which include plants, some protists (the algae), and some archaeans and bacteria, absorb the radiant energy of sunlight and convert it into chemical energy. The organisms use the chemical energy to convert simple inorganic raw materials—water, carbon dioxide (CO_2) from the air, and inorganic minerals from the soil—into complex organic molecules. Photosynthesis is still not completely understood, so it remains a subject of active research today.

This chapter begins with an overview of the photosynthetic reactions. We then examine light and light absorption and the reactions that use absorbed energy to make organic molecules from inorganic substances. This chapter focuses on oxygenic (oxygen-generating) photosynthesis in plants and green algae; other eukaryotic photosynthesizers have individual variations on the process (see Chapter 27). Prokaryotic photosynthesis is described in Chapter 26.

9.1 Photosynthesis: An Overview

Plants and other photosynthetic organisms are the *primary producers* of Earth; they convert the energy of sunlight into chemical energy and use this chemical energy to assemble simple inorganic raw materials into complex organic molecules. Primary producers use some of the organic molecules they make as an energy source for their own activities. But they also serve—directly or indirectly—as a food source for *consumers,* the animals that live by eating plants or other animals. Eventually, the bodies of both primary producers and consumers provide chemical energy for bacteria, fungi, and other *decomposers.*

Photosynthesizers and other organisms that make all of their required organic molecules from CO_2 and other inorganic sources such as water are called **autotrophs** (*autos* = self, *trophos* = feeding). Autotrophs that use light as the energy source to make organic molecules by photosynthesis are called **photoautotrophs.** Consumers and decomposers, which need a source of organic molecules to survive, are called **heterotrophs** (*hetero* = different).

As the pathway of energy flows from the sun through plants (primary producers) and animals to decomposers, the organic molecules made by photosynthesis are broken down into inorganic molecules again, and the chemical energy captured in photosynthesis is released as heat energy. Because the reactions capturing light energy are the first step in this pathway, photosynthesis is the vital link between the energy of sunlight and the vast majority of living organisms.

Electrons Play a Primary Role in Photosynthesis

Photosynthesis proceeds in two stages, each involving multiple reactions **(Figure 9.2).** In the first stage, the **light-dependent reactions,** the energy of sunlight is absorbed and converted into chemical energy in the form of two substances: ATP and NADPH. ATP is the main energy source for plant cells (as it is for all types of living cells), and NADPH (nicotinamide adenine dinucleotide phosphate) carries electrons that are pushed to high energy levels by absorbed light. In the second stage of photosynthesis, the **light-independent reactions** (also called the *Calvin cycle*), these electrons are used as a source of energy to convert inorganic CO_2 to an organic form. The conversion process, called **CO_2 fixation,** is a reduction reaction, in which electrons are added to CO_2. As part of the reduction, protons are also added to CO_2 (reduction and oxidation are discussed in Section 8.1). With the added electrons and protons (H^+), CO_2 is converted to a carbohydrate that contains carbon, hydrogen, and oxygen atoms in the ratio 1 C : 2 H : 1 O.

$$CO_2 + H^+ + e^- \rightarrow (CH_2O)_n$$

Carbohydrate units are often symbolized as $(CH_2O)_n$, with the "*n*" indicating that different carbohydrates are formed from different multiples of the carbohydrate unit.

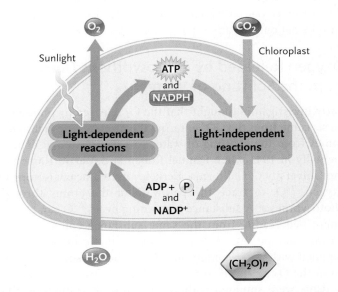

FIGURE 9.2 The light-dependent and light-independent reactions of photosynthesis, and their interlinking reactants and products. Both series of reactions occur in the chloroplasts of plants and algae.
© Cengage Learning 2014

In plants, algae, and one group of photosynthetic bacteria (the cyanobacteria), the source of electrons and protons for CO_2 fixation is water (H_2O), the most abundant substance on Earth. Oxygen (O_2) generated from the splitting of the water molecule is released into the environment as a by-product of photosynthesis:

$$2 H_2O \rightarrow 4 H^+ + 4 e^- + O_2$$

Thus, plants, algae, and cyanobacteria use three resources that are readily available—sunlight, water, and CO_2—to produce almost all the organic matter on Earth, and to supply the oxygen of our atmosphere.

In organisms that are able to split water, the two reactions shown above are combined and multiplied by six to produce a six-carbon carbohydrate such as glucose:

$$6 CO_2 + 12 H_2O \rightarrow C_6H_{12}O_6 + 6 O_2 + 6 H_2O$$

Note that water appears on both sides of the equation; it is both consumed as a reactant and generated as a product in photosynthesis.

Although glucose is the major product of photosynthesis, other monosaccharides, disaccharides, polysaccharides, lipids, and amino acids are also produced indirectly. In fact, all the organic molecules of plants are assembled as direct or indirect products of photosynthesis.

The relationships between the light-dependent and light-independent reactions are summarized in Figure 9.2. Notice that the ATP and NADPH produced by the light-dependent reactions, along with CO_2, are the reactants of the light-independent reactions. The ADP, inorganic phosphate (P_i), and $NADP^+$ produced by the light-independent reactions, along with H_2O, are the reactants for the light-dependent reactions. The light-dependent and light-independent reactions thus form a cycle in which the net inputs are H_2O and CO_2, and the net outputs are organic molecules and O_2.

Oxygen Released by Photosynthesis Derives from the Splitting of Water

Early investigators thought that the O_2 released by photosynthesis came from the CO_2 entering the process. The fact that it comes from the splitting of water was demonstrated experimentally in 1941 when Samuel Ruben and Martin Kamen of the University of California, Berkeley used a heavy isotope of oxygen, ^{18}O, to trace the pathways of the atoms through photosynthesis. A substance containing heavy ^{18}O can be distinguished readily from the same substance containing the normal isotope, ^{16}O. When a photosynthetic organism was supplied with water containing ^{18}O, the heavy isotope showed up in the O_2 given off in photosynthesis. However, if the organisms were supplied with carbon dioxide containing ^{18}O, the heavy isotope showed up in the carbohydrate and water molecules assembled during the reactions—but not in the oxygen gas. This experiment, and similar experiments using different isotopes, revealed where each atom of the reactants end up in products:

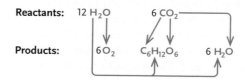

The water-splitting reaction probably developed even before oxygen-consuming organisms appeared, evolving first in photosynthetic bacteria that resembled present-day cyanobacteria. The oxygen released by the reaction profoundly changed the atmosphere. It allowed for aerobic respiration in which oxygen serves as the final acceptor for electrons removed in cellular oxidations. The existence of all animals depends on the oxygen provided by the water-splitting reaction of photosynthesis.

In Eukaryotes, Photosynthesis Takes Place in Chloroplasts

In eukaryotes, the photosynthetic reactions take place in the chloroplasts of plants and algae; in cyanobacteria, the reactions are distributed between the plasma membrane and the cytosol.

Chloroplasts from individual algal and plant groups differ in structural details. The chloroplasts of plants and green algae are formed from three membranes that enclose three compartments inside the organelles (**Figure 9.3**). (Chloroplast structure is also described in Section 5.4.) An *outer membrane* covers the entire surface of the organelle. An *inner membrane* lies just inside the outer membrane. Between the outer and inner membranes is an *intermembrane compartment*. The fluid within the compartment formed by the inner membrane is the *stroma*. Within the stroma is the third membrane system, the *thylakoid membranes*, which form flattened, closed sacs called *thylakoids*. The space enclosed by a thylakoid is called the *thylakoid lumen*.

In green algae and higher plants, thylakoids are arranged into stacks called *grana* (singular, *granum*; shown in Figure 9.3). The grana are interconnected by flattened, tubular membranes called *stromal lamellae*. The stromal lamellae probably link the thylakoid lumens into a single continuous space within the stroma.

The thylakoid membranes and stromal lamellae house the molecules that carry out the light-dependent reactions of photosynthesis, which include the pigments, electron transfer carriers, and ATP synthase enzymes for ATP production. The light-independent reactions are concentrated in the stroma.

In higher plants, the CO_2 required for photosynthesis diffuses to cells containing chloroplasts after entering the plant through *stomata* (singular, *stoma*), small pores in the surface of the leaves (particularly the undersurface) and stems. (Stomata are described in Section 28.1, and are shown in Figures 9.14 and 28.3.) The O_2 produced in photosynthesis diffuses from the cells and exits through the stomata, as does the H_2O. The water and minerals required for photosynthesis are absorbed by the roots and transported to cells containing chloroplasts through

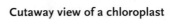

FIGURE 9.3 The membranes and compartments of chloroplasts.
© Cengage Learning 2014

tubular conducting cells. The organic products of photosynthesis are distributed to all parts of the plant by other conducting cells (see Chapter 36).

STUDY BREAK 9.1

1. What are the two stages of photosynthesis?
2. In which organelle does photosynthesis take place in plants? Where in that organelle are the two stages of photosynthesis carried out?

THINK OUTSIDE THE BOOK

Scientists have been working to develop an artificial version of photosynthesis that can be used to produce liquid fuels from CO_2 and H_2O. Collaboratively or individually, find an example of research on artificial photosynthesis and prepare an outline of how the system works or is anticipated to work.

9.2 The Light-Dependent Reactions of Photosynthesis

In this section we discuss the light-dependent reactions (also referred to more simply as the light reactions), in which light energy is converted to chemical energy. The light-dependent reactions involve two main processes: (1) light absorption; and (2) synthesis of NADPH and ATP. We will describe each of these processes in turn. To keep the bigger picture in perspective, you may find it useful to refer periodically to the summary of photosynthesis shown in Figure 9.2.

Electrons in Pigment Molecules Absorb Light Energy in Photosynthesis

The first process in photosynthesis is light absorption. What is light? Visible light is a form of radiant energy. It makes up a small part of the **electromagnetic spectrum (Figure 9.4)**, which ranges from radio waves to gamma rays. The various forms of electromagnetic radiation differ in *wavelength*—the horizontal distance between the crests of successive waves. Radio waves have wavelengths in the range of 10 meters to hundreds of kilometers, and gamma rays have wavelengths in the range of one hundredth to one millionth of a nanometer. The average wavelength of radiowaves for an FM radio station, for example, is 3 m.

The radiation humans detect as visible light has wavelengths between about 700 nm, seen as red light, and 400 nm, seen as blue light. We see the entire spectrum of wavelengths from 700 to 400 nm com-

Cutaway of a small section from the leaf

Leaf's upper surface

Photosynthetic cells

CO_2

O_2

Stomata (through which O_2 and CO_2 are exchanged with the atmosphere)

One of the photosynthetic cells, with green chloroplasts

Large central vacuole

Nucleus

Cutaway view of a chloroplast

Outer membrane

Inner membrane

Thylakoids
• light absorption by chlorophylls and carotenoids
• electron transfer
• ATP synthesis by ATP synthase

Stroma (space around thylakoids)
• light-independent reactions

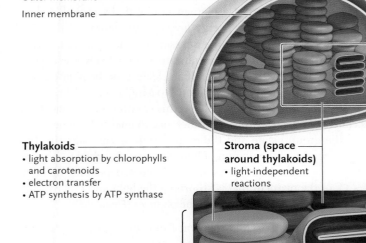

Granum

Stromal lamella

Thylakoid lumen

Thylakoid membrane

A. Range of the electromagnetic spectrum

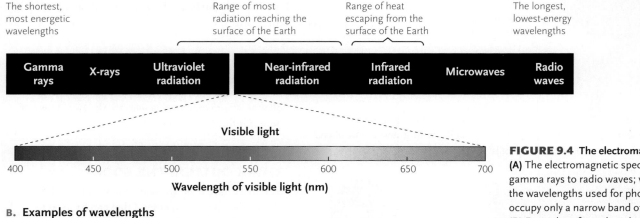

The shortest, most energetic wavelengths

Range of most radiation reaching the surface of the Earth

Range of heat escaping from the surface of the Earth

The longest, lowest-energy wavelengths

| Gamma rays | X-rays | Ultraviolet radiation | Near-infrared radiation | Infrared radiation | Microwaves | Radio waves |

Visible light

400   450   500   550   600   650   700

Wavelength of visible light (nm)

FIGURE 9.4 **The electromagnetic spectrum.**
(A) The electromagnetic spectrum ranges from gamma rays to radio waves; visible light and the wavelengths used for photosynthesis occupy only a narrow band of the spectrum. **(B)** Examples of wavelengths, showing the difference between the longest and shortest wavelengths of visible light.
© Cengage Learning 2014

B. Examples of wavelengths

400-nm wavelength

700-nm wavelength

bined together as white light. The energy of light interacts with matter in elementary particles called *photons*. Each photon is a discrete unit that contains a fixed amount of energy. That energy is inversely proportional to its wavelength: the shorter the wavelength, the greater the energy of a photon.

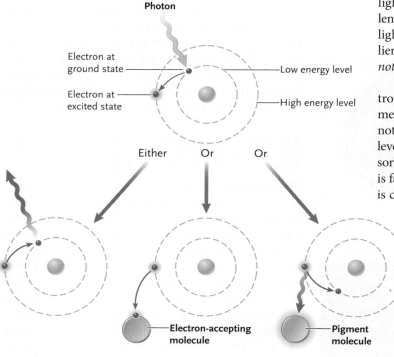

Photon is absorbed by an excitable electron that moves from a relatively low energy level to a higher energy level.

Photon

Electron at ground state — Low energy level

Electron at excited state — High energy level

Either   Or   Or

Electron-accepting molecule

Pigment molecule

The electron returns to its ground state by emitting a less energetic photon (fluorescence) or releasing energy as heat.

The higher-energy electron is accepted by an electron-accepting molecule, the primary acceptor.

The electron returns to its ground state, and the energy released transfers to a neighboring pigment molecule.

In photosynthesis, light is absorbed by molecules of green pigments called **chlorophylls** (*chloros* = yellow-green, *phyllon* = leaf) and yellow-orange pigments called **carotenoids** (*carota* = carrot). Pigment molecules such as chlorophyll appear colored to an observer because they absorb the energy of visible light at certain wavelengths and transmit or reflect other wavelengths. The color of a pigment is produced by the transmitted or reflected light. Plants look green because chlorophyll absorbs blue and red light most strongly, and transmits or reflects most of the wavelengths in between; we see the reflected light as green. This green light, as demonstrated by Engelmann's experiment described earlier in *Why It Matters,* is the combination of wavelengths that are *not used* by the plants in photosynthesis.

Light is absorbed in a pigment molecule by excitable electrons occupying certain energy levels in the atoms of the pigments (see Section 2.2 for the discussion on energy levels). When not absorbing light, these electrons are at a relatively low energy level known as the *ground state*. If an electron in the pigment absorbs the energy of a photon, it jumps to a higher energy level that is farther from the atomic nucleus. This condition of the electron is called the *excited state*. The difference in energy level between the ground state and the excited state is equivalent to the energy of the photon of light that was absorbed.

One of three events then occurs, depending on both the pigment that is absorbing the light and other molecules in the vicinity of the pigment **(Figure 9.5):**

- The excited electron from the pigment molecule returns to its ground state, releasing its energy either as heat, or as an emission of light of a longer wavelength than the absorbed light, a process called *fluorescence.*

FIGURE 9.5 **Alternative effects of light absorbed by a pigment molecule.**
© Cengage Learning 2014

A. Chlorophyll structure

B. Carotenoid structure

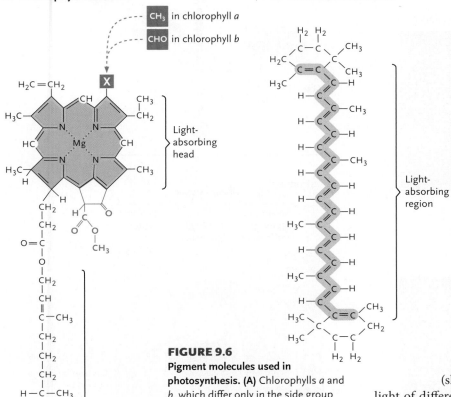

FIGURE 9.6

Pigment molecules used in photosynthesis. (A) Chlorophylls *a* and *b*, which differ only in the side group attached at the X. Light-absorbing electrons are distributed among the bonds shaded in orange. The chlorophylls are similar in structure to the cytochromes, which occur in both the chloroplast and mitochondrial electron transfer systems. **(B)** In carotenoids, the light-absorbing electrons are distributed in a series of alternating double and single bonds in the backbone of these pigments.

© Cengage Learning 2014

- The excited electron is transferred from the pigment molecule to a nearby electron-accepting molecule called a *primary acceptor.*

- The energy of the excited electron, but not the electron itself, is transferred to a neighboring pigment molecule. This transfer excites the second molecule, while the first molecule returns to its ground state. Very little energy is lost in this energy transfer.

Chlorophylls Are the Main Light Receptor Pigments for Photosynthesis

Chlorophylls are the major photosynthetic pigments in plants, green algae, and cyanobacteria. They absorb photons and transfer excited electrons to primary acceptor molecules. In the transfer, the chlorophyll is oxidized because it loses an electron,

and the primary acceptor is reduced because it gains an electron. Closely related molecules, the *bacteriochlorophylls,* carry out the same functions in other photosynthetic bacteria. Carotenoids are accessory pigments that absorb light energy at a different wavelength than those absorbed by chlorophylls.

Chlorophylls and carotenoids are bound to proteins that are embedded in photosynthetic membranes. In plants and green algae, they are located in the thylakoid membranes of chloroplasts; in photosynthetic bacteria, they are located in the plasma membrane.

Molecules of the chlorophyll family **(Figure 9.6A)** have a carbon ring structure with a magnesium atom bound at the center. The ring is attached to a long, hydrophobic side chain. The main types of chlorophyll are chlorophyll *a* and chlorophyll *b*. Chlorophyll *a* and chlorophyll *b* differ only in one side group that is attached to a carbon of the ring structure (shown in Figure 9.6A).

A chlorophyll molecule contains a network of electrons capable of absorbing light (shaded in orange in Figure 9.6A). The amount of light of different wavelengths that is absorbed by a pigment is called an **absorption spectrum;** it is usually shown as a graph in which the height of the curve at any wavelength indicates the amount of light absorbed. **Figure 9.7A** shows the absorption spectra for chlorophylls *a* and *b*.

The carotenoids are built on a long backbone that typically contains 40 carbon atoms **(Figure 9.6B)**. Carotenoids may expand the range of wavelengths used for photosynthesis because they absorb different wavelengths that chlorophyll does not absorb. A more important role for carotenoids than light capture for photosynthesis is in *photoprotection,* the protection of photosynthetic organisms against potentially harmful photooxidative processes. Carotenoids transmit or reflect other wavelengths in combinations that appear yellow, orange, red, or brown, depending on the type of carotenoid. The carotenoids contribute to the red, orange, and yellow colors of vegetables and fruits and to the brilliant colors of autumn leaves, in which the green color is lost when the chlorophylls break down.

The light absorbed by the chlorophylls is the main driver of the reactions of photosynthesis. Plotting the effectiveness of light of each wavelength in driving photosynthesis produces a graph called the **action spectrum** of photosynthesis **(Figure 9.7B** shows the action spectrum of higher plants). The action spectrum is usually determined by measuring the amount of O_2 released by photosynthesis carried out at different wavelengths of visible light, as Engelmann did indirectly in his experiment described in *Why It Matters* (compare Figures 9.1 and 9.7B).

In all eukaryotic photosynthesizers, a specialized chlorophyll *a* molecule passes excited electrons to the primary acceptor. Other chlorophyll molecules, along with carotenoids, act as

A. The absorption spectra of chlorophylls *a* and *b*

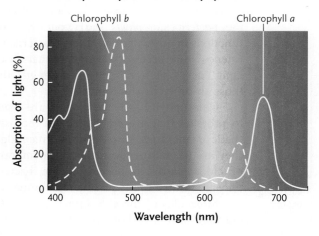

B. The action spectrum in higher plants

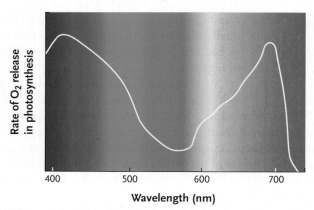

FIGURE 9.7 The absorption spectra of chlorophylls *a* and *b* (A) and the action spectrum of photosynthesis (B) in higher plants.
© Cengage Learning 2014

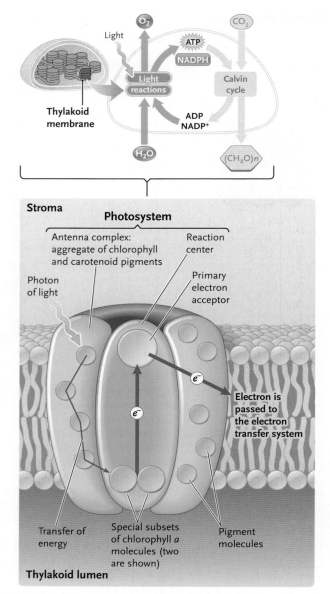

FIGURE 9.8 Major components of a photosystem: a group of pigments forming an antenna complex (light-harvesting complex) and a reaction center. Light energy absorbed anywhere in the antenna complex is conducted to special subsets of chlorophyll *a* molecules in the reaction center (two molecules are shown for illustration). The absorbed light is converted to chemical energy when an excited electron from the chlorophyll *a* is transferred to a stable orbital in a primary acceptor, also in the reaction center. High-energy electrons are passed out of the photosystem to the electron transfer system. The blue arrows show the path of energy flow.
© Cengage Learning 2014

accessory pigments that pass their energy to chlorophyll *a*. Light energy that is absorbed by the entire collection of chlorophyll and carotenoid molecules in chloroplasts is passed to the specialized chlorophyll *a* molecules that are directly involved in transforming light into chemical energy.

The Photosynthetic Pigments Are Organized into Photosystems in Chloroplasts

The light-absorbing pigments are organized with proteins and other molecules into large complexes called **photosystems (Figure 9.8),** which are embedded in thylakoid membranes and stromal lamellae. The photosystems are the sites at which light is absorbed and converted into chemical energy.

Plants, green algae, and cyanobacteria have two types of these complexes, called **photosystems II** and **I,** which carry out different parts of the light-dependent reactions. Photosystem II, in addition, is closely linked to a group of enzymes that car-

ries out the initial reaction of splitting water into electrons, protons, and oxygen. (Note: As you will learn soon, photosystem II functions before photosystem I in the light-dependent reactions. However, photosystem I was named because it was discovered first; the systems were given their numbers before their order of use in the reactions was worked out.)

Each photosystem consists of two closely associated components: an **antenna complex** (also called a *light-harvesting complex*), and a **reaction center.** The antenna complex contains an aggregate of many chlorophyll pigments and a number of carotenoid pigments. The chlorophyll molecules are anchored in the complex by being bound to specific membrane proteins. In this form, they are arranged efficiently to optimize the capture of light energy.

The reaction center contains special subsets of chlorophyll *a* molecules complexed with proteins. The chlorophyll *a* molecules in the reaction center of photosystem II are called *P680* (*P* = pigment) because they absorb light optimally at a wavelength of 680 nm. Those in the reaction center of photosystem I are called *P700* because they absorb light optimally at a wavelength of 700 nm.

Light energy in the form of photons is absorbed by the pigment molecules of the antenna complex. This absorbed light energy reaches P680 and P700 in the reaction center where it is captured in the form of an excited electron that is passed to a primary acceptor molecule. That electron is passed to the electron transfer system, which carries electrons away from the primary acceptor.

Some components of the electron transfer system are located within the photosystems and other components are separate.

Electron Flow from Water to NADP⁺ Leads to the Synthesis of NADPH and ATP

In the second main process of the light-dependent reactions, the electrons obtained from the splitting of water (two electrons per molecule of water; see Section 9.1) are used for the synthesis of NADPH. These electrons, which were pushed to higher levels by light energy, pass through an electron transfer system consisting of a series of electron carriers that are arranged in a chain **(Figure 9.9).** The electron carriers of the photosynthetic system consist of nonprotein organic groups that pick up and release the electrons traveling through the system. The carriers include the same types that act in mitochondrial electron transfer—cytochromes, quinones, and iron–sulfur centers (discussed in Section 8.4).

The electron carriers are alternately reduced and oxidized as they pick up and release electrons in sequence. Electrons

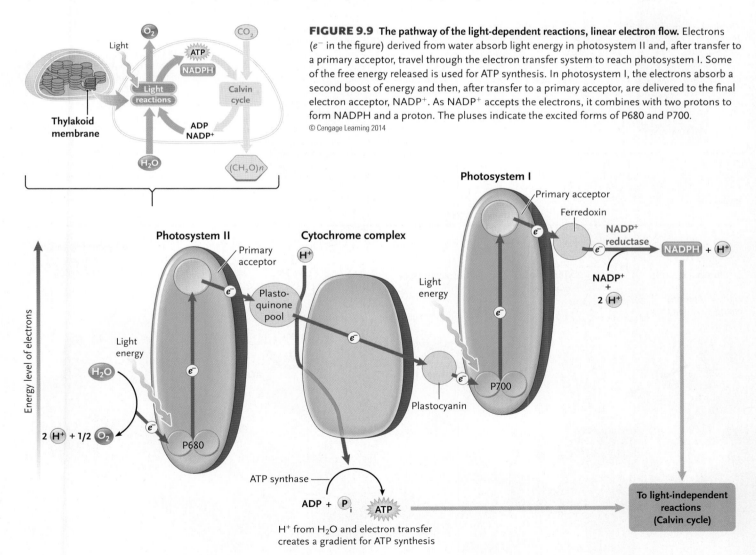

FIGURE 9.9 **The pathway of the light-dependent reactions, linear electron flow.** Electrons (e^- in the figure) derived from water absorb light energy in photosystem II and, after transfer to a primary acceptor, travel through the electron transfer system to reach photosystem I. Some of the free energy released is used for ATP synthesis. In photosystem I, the electrons absorb a second boost of energy and then, after transfer to a primary acceptor, are delivered to the final electron acceptor, NADP⁺. As NADP⁺ accepts the electrons, it combines with two protons to form NADPH and a proton. The pluses indicate the excited forms of P680 and P700.
© Cengage Learning 2014

FIGURE 9.10 The chloroplast electron transfer system (steps 1–9) and chemiosmosis (steps 10–12), illustrating the synthesis of NADPH and ATP by the noncyclic electron flow pathway.

The components of the electron transfer system and chemiosmosis are located in the chloroplast thylakoid membrane.

1 Electrons from the water-splitting reaction system are accepted one at a time by a P680 chlorophyll *a* in the reaction center of photosystem II. As P680 accepts the electrons, they are raised to the excited state, using energy passed to the reaction center from the light-absorbing pigment molecules in the antenna complex. The excited electrons are immediately transferred to the primary acceptor of photosystem II, a modified form of chlorophyll *a*.

2 The electrons flow through a short chain of carriers within photosystem II and then transfer to the mobile carrier plastoquinone. The plastoquinones form a "pool" of molecules within the thylakoid membranes.

3 Plastoquinones pass the electrons to the cytochrome complex. As it accepts and releases electrons, the cytochrome complex pumps H^+ from the stroma into the thylakoid lumen. Those protons drive ATP synthesis (see step 8).

4 Electrons now pass to the mobile carrier *plastocyanin*, which shuttles electrons between the cytochrome complex and photosystem I.

5 Electrons flow to a P700 chlorophyll *a* in the reaction center of photosystem I, where they are excited to high energy levels again by absorbing more light energy. The excited electrons are transferred from P700 to the primary acceptor of photosystem I, a modified chlorophyll *a*.

6 After passage through carriers within photosystem I, the electrons are transferred to the iron-sulfur protein ferredoxin, which acts as a mobile carrier.

7 The ferredoxin transfers the electrons, still at very high energy levels, to $NADP^+$, the final acceptor of the noncyclic pathway. $NADP^+$ is reduced to NADPH by $NADP^+$ reductase. Electron transfer is now complete.

8 Proton pumping by the plastoquinones and the cytochrome complex, as described in step 3, creates a concentration gradient of H^+ with the high concentration within the thylakoid lumen and the low concentration in the stroma.

9 The H^+ gradient supplies the energy that drives ATP synthesis by ATP synthase.

10 Due to the gradient, H^+ ions flow across the inner membrane into the matrix through a channel in the ATP synthase.

11 The flow of H^+ activates ATP synthase, making the headpiece and stalk rotate.

12 As a result of changes in shape and position as it turns, the headpiece catalyzes the synthesis of ATP from ADP and P_i.

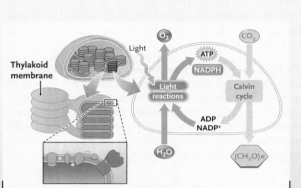

Thylakoid membrane — Light — O_2 — CO_2 — ATP — NADPH — Light reactions — Calvin cycle — ADP — $NADP^+$ — H_2O — $(CH_2O)n$

Stroma

Electron transfer system — Chemiosmosis

To light-independent reactions (Calvin cycle)

Photosystem II — H^+ — Cytochrome complex — Photosystem I — H^+ — Low H^+ **9**

Light energy — Antenna complex — Primary acceptor — Pigment molecules — H^+ — Light energy — Ferredoxin — $2\ H^+ + NADP^+$ — $H^+ +$ NADPH — ATP

6 — e^- — e^- — **7** — H^+

NADP$^+$ reductase — ADP + P_i — **12**

2 — e^-

1 — P680 — **8** — **3** — e^- — **5** — P700 — **5** — Stator — **10** — Rotation — **11**

1 — Plastoquinone — **4** — Plastocyanin — e^- — e^- — e^- — ATP synthase

e^- — Water-splitting complex

H_2O — $2\ H^+ + 1/2\ O_2$ — H^+ — H^+ — H^+ — H^+ — H^+ — H^+ — H^+

H^+ — H^+ — H^+ — High H^+ **9** — H^+ — H^+ — H^+ — H^+ — H^+ — Thylakoid membrane

H^+ — H^+

Thylakoid lumen

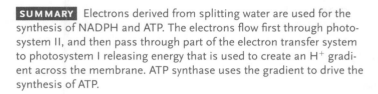

SUMMARY Electrons derived from splitting water are used for the synthesis of NADPH and ATP. The electrons flow first through photosystem II, and then pass through part of the electron transfer system to photosystem I releasing energy that is used to create an H^+ gradient across the membrane. ATP synthase uses the gradient to drive the synthesis of ATP.

THINK LIKE A SCIENTIST How many phospholipid bilayers separate the cytosol from the stroma?

© Cengage Learning 2014

from water first flow through photosystem II, becoming excited to a higher energy level in P680 through absorbed light energy (see Figure 9.9). The electrons then flow "downhill" in energy level through the part of the electron transfer system connecting photosystems II and I.

This consists of a pool of molecules of the electron carrier *plastoquinone,* a *cytochrome complex,* and the mobile carrier protein *plastocyanin.* As electrons pass through the system they release free energy at each transfer from a donor to an acceptor molecule. Some of this free energy is used to create a gradient of H^+ across the membrane. The gradient provides the energy source for ATP synthesis, just as it does in mitochondria.

The electrons then pass to photosystem I, where they are excited a second time in P700 through absorbed light energy. The high-energy electrons enter a short part of the electron transfer system that leads to the final acceptor of the chloroplast system, $NADP^+$. The enzyme $NADP^+$ reductase reduces $NADP^+$ to NADPH by using two electrons and two protons from the surrounding water solution, and by releasing one proton. This pathway is frequently called **linear electron flow** because electrons travel in a one-way direction from H_2O to $NADP^+$; it is sometimes called the *Z scheme* because of the zigzag-like changes in electron energy level (shown by the blue arrows in Figure 9.9).

Figure 9.10 shows how the electron transfer and ATP synthesis systems for the light-dependent reactions are organized in the thylakoid membrane and lays out the linear electron pathway for NADPH and ATP synthesis. The electron transfer system involves steps 1 to 9, and the chemiosmotic synthesis of ATP involves steps 10 to 12.

The flow of electrons through the electron transfer system leads to the generation of an H^+ gradient across the inner membrane (steps 1–8). That gradient is enhanced by the addition of two protons to the lumen for each water molecule split, and by the removal of one proton from the stroma for each NADPH molecule synthesized. Because protons carry a positive charge, an electrical gradient forms across the thylakoid membrane, with the lumen more positively charged than the stroma. The

combination of a proton gradient and a voltage gradient across the membrane produces stored energy known as the *proton-motive force* (also discussed for cellular respiration in Section 8.4), which contributes energy for ATP synthesis by ATP synthase. Just as for the mitochondrial ATP synthase, the chloroplast ATP synthase is embedded in the same membranes as the electron transfer system. Protons flow through a membrane channel from the thylakoid lumen to the stroma along their concentration gradient (step 10). Free energy is released as H^+ moves through the channel; it powers synthesis of ATP from ADP and P_i by the ATP synthase (steps 11–12). This process of using an H^+ gradient to power ATP synthesis—*chemiosmosis*—is the same as that used for ATP synthesis in mitochondria (see Section 8.4).

The overall yield of the linear electron flow pathway is one molecule of NADPH and one molecule of ATP for each pair of electrons produced from the splitting of water. The synthesis of ATP coupled to the transfer of electrons energized by photons of light is called **photophosphorylation.** This process is analogous to oxidative phosphorylation in mitochondria (see Section 8.4), except that in chloroplasts light provides the energy for establishing the proton gradient.

Comparing the linear pathway with the mitochondrial electron transfer system (shown in Figure 8.13) reveals that the pathway from the plastoquinones through plastocyanin in chloroplasts is essentially the same as the pathway from the ubiquinones through cytochrome c in mitochondria. The similarities between the two pathways indicate that the electron transfer system is an ancient evolutionary development that became adapted to both photosynthesis and oxidative phosphorylation.

Electron Flow Can Also Drive ATP Synthesis by Flowing Cyclically around Photosystem I

Linear electron flow produces ATP and NADPH + H^+. In some cases, however, photosystem I works independently of photosystem II in a circular process called **cyclic electron flow (Figure 9.11).** In this process, electrons pass through the cytochrome complex and plastocyanin to the P700 chlorophyll *a* in the reaction center of photosystem I where they are excited by light energy. The electrons then flow from photosystem I to the mobile carrier ferredoxin, but rather than being used for $NADP^+$ reduction by $NADP^+$ reductase, they flow back to P700 in the cytochrome complex. The electrons again pass to plastocyanin and on to photosystem I where they receive another energy boost from light energy, and so the cycle continues. Each time electrons flow around the cycle, more H^+ is pumped across the thylakoid membranes, driving ATP synthesis in the way already described. The net result of cyclic electron flow is that light energy is converted into the chemical energy of ATP *without* the production of NADPH.

Cyclic electron flow was observed in higher plant chloroplasts over 50 years ago. However, researchers still debate whether it is a real physiological process. Recent experimental results, including studies of mutants that appear to lack the cy-

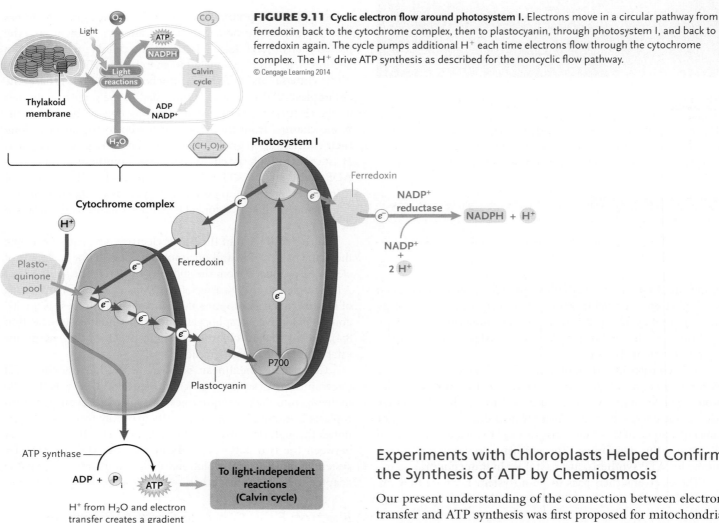

FIGURE 9.11 Cyclic electron flow around photosystem I. Electrons move in a circular pathway from ferredoxin back to the cytochrome complex, then to plastocyanin, through photosystem I, and back to ferredoxin again. The cycle pumps additional H⁺ each time electrons flow through the cytochrome complex. The H⁺ drive ATP synthesis as described for the noncyclic flow pathway.
© Cengage Learning 2014

clic electron flow pathway, support the hypothesis that the pathway plays an important role in the responses of plants to stress. That is, because cyclic electron flow produces ATP but no NADPH, the plants avoid risky overproduction of reducing power when stressed. Excess reducing power can be damaging because of the potential for some of the extra electrons "leaking out" of the system to form reactive oxygen species such as superoxide. Reactive oxygen species are toxic and can lead to the oxidative destruction of cells.

Photosynthesis occurs also in some groups of bacteria. In those organisms the photosynthetic electron transfer system components are embedded in membranes, but there are no chloroplasts like those of plants. Cyanobacteria, the only prokaryotes that produce oxygen by photosynthesis, contain both photosystems II and I and normally carry out the light-dependent reactions using the linear electron flow pathway. They are also capable of using the cyclic electron flow pathway involving photosystem I. Photosynthesis is also carried out by some archaea.

Experiments with Chloroplasts Helped Confirm the Synthesis of ATP by Chemiosmosis

Our present understanding of the connection between electron transfer and ATP synthesis was first proposed for mitochondria in Mitchell's chemiosmotic hypothesis (discussed in Section 8.4). Several experiments have shown that the same mechanism operates in chloroplasts. In one experiment, André Jagendorf and Ernest Uribe isolated chloroplasts from cells, broke them open, and treated the broken chloroplasts to create an H⁺ gradient across the thylakoid membrane. In the dark, ATP was made, indicating that the gradient, and not light-generated electron transfer, powers ATP synthesis.

Our description of photosynthesis to this point shows how the light-dependent reactions generate NADPH and ATP, which provide the reducing power and chemical energy required to produce organic molecules from CO_2. The next section follows NADPH and ATP through the light-independent reactions and shows how the organic molecules are produced.

STUDY BREAK 9.2

1. What is the difference in function between the chlorophyll *a* molecules in the antenna complexes and the chlorophyll *a* molecules in the reaction centers of the photosystems?
2. How is NADPH made in the linear electron flow pathway?
3. What is the difference between the linear electron flow pathway and the cyclic electron flow pathway?

Two-Dimensional Paper Chromatography and the Calvin Cycle

Beginning in 1945, Melvin Calvin, Andrew A. Benson, and their colleagues at the University of California, Berkeley combined CO_2 labeled with the radioactive carbon isotope ^{14}C (discussed in *Focus on Applied Research* in Chapter 2) with a technique called *two-dimensional paper chromatography* to trace the pathways of the light-independent reactions in a green alga of the genus *Chlorella*. The researchers exposed actively photosynthesizing *Chlorella* cells to the labeled carbon dioxide. Then, at various times, they removed cells and placed them in hot alcohol, which instantly stopped all the photosynthetic reactions of the algae.

They extracted radioactive carbohydrates from the cells and used two-dimensional paper chromatography to separate and to identify them chemically **(Figure)**.

By analyzing the labeled molecules revealed by the two-dimensional chromatography technique in the extracts prepared from *Chlorella* cells under different conditions, Calvin and his colleagues were able to reconstruct the reactions of the Calvin cycle. For example, in carbohydrate extracts made within a few seconds after the cells were exposed to the labeled CO_2, most of the radioactivity was found in 3PGA, indicating that it is one of the earliest products of photosynthesis. In extracts made after longer periods of exposure to the label, radioactivity showed up in G3P and in more complex substances, including a

variety of six-carbon sugars, sucrose, and starch. The researchers also examined the effect of reducing the amount of CO_2 available to the *Chlorella* cells so that photosynthesis worked slowly even in bright light. Under these conditions, RuBP accumulated in the cells, suggesting that it is the first substance to react with CO_2 in the light-independent reactions and that it accumulates if CO_2 is in short supply. Most of the intermediate compounds between CO_2 and six-carbon sugars were identified in similar experiments.

Using this information, Calvin and his colleagues pieced together the light-independent reactions of photosynthesis and showed that they formed a continuous cycle. Melvin Calvin was awarded a Nobel Prize in 1961 for his work on the assimilation of carbon dioxide in plants.

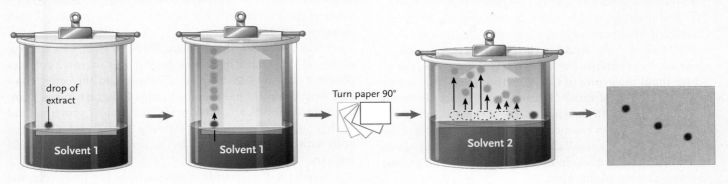

1 A drop of extract containing radioactive carbohydrates is placed at one corner of a piece of chromatography paper. The paper is placed in a jar with its edge touching a solvent. (Calvin used a water solution of butyl alcohol and propionic acid.)

2 The extracted molecules in the spot dissolve and are carried upward through the paper as the solvent rises. The rates of movement of the molecules vary according to their molecule size and solubility. The resulting vertical line of spots is the first dimension of the two-dimensional technique.

3 The paper is dried, turned 90°, and touched to a second solvent (Calvin used a water solution of phenol for this part of the experiment). As this solvent moves through the paper, the molecules again migrate upward from the spots produced by the first dimension, but at rates different from their mobility in the first solvent. This step, the second dimension of the two-dimensional technique, separates molecules that, although different, had produced a single spot in the first solvent because they had migrated at the same rate. The individual spots are identified by comparing their locations with the positions of spots made by known molecules when the "knowns" are run through the same procedure.

4 The paper is dried and covered with a sheet of photographic film. Radioactive molecules expose the film in spots over their locations in the paper. Developing the film reveals the locations of the radioactive spots. The spots on the film are compared with the spots on the paper to identify the molecules that were radioactive.

9.3 The Light-Independent Reactions of Photosynthesis

The electrons carried from the light-dependent reactions by NADPH retain much of the energy absorbed from sunlight. These electrons provide the reducing power required to fix CO_2 into carbohydrates and other organic molecules in the light-independent reactions. The ATP generated in the light-dependent reactions supplies additional energy for the light-independent reactions. The reactions using NADPH and ATP to fix CO_2 occur in a circuit

known as the **Calvin cycle,** named for its discoverer, Melvin Calvin. *Focus on Basic Research* describes the experiments Calvin and his colleagues used to elucidate the light-independent reactions.

The Calvin Cycle Uses NADPH, ATP, and CO_2 to Generate Carbohydrates

The light-independent reactions of the Calvin cycle use CO_2, ATP, and NADPH as inputs. For three input molecules of CO_2, one of which is used in each of three turns of the cycle, the key product is one molecule of the three-carbon carbohydrate molecule glyceraldehyde-3-phosphate (G3P). The G3P is used in reactions to synthesize glucose and a number of other organic molecules. In plants, the Calvin cycle takes place entirely in the chloroplast stroma (see Figure 9.3). The Calvin cycle also occurs in most photosynthetic prokaryotes, where it takes place in the cytoplasm. We focus here on plants.

Figure 9.12A tracks the carbon atoms through one turn of the Calvin cycle, and **Figure 9.12B** summarizes the reactions of three turns of the cycle, starting with three molecules of CO_2 and resulting in the release of one molecule of the product, G3P.

1. **Carbon fixation.** The first phase of the Calvin cycle, *carbon fixation*, involves the cycle's key reaction in which each input molecule of CO_2 is added to one molecule of ribulose 1,5-bisphosphate (RuBP), a five-carbon sugar, forming a transient six-carbon molecule that is cleaved to produce two three-carbon molecules of 3-phosphoglycerate (3PGA) (Figure 9.12B, reaction 1). This reaction, which fixes CO_2 into organic form, is catalyzed by the *carboxylase* activity of the key enzyme of the Calvin cycle, **RuBP carboxylase/oxygenase** (abbreviated as **rubisco**). For three turns of the cycle, the three input molecules of CO_2 (3 carbons) reacting with three molecules of RuBP (15 carbons) produce six molecules of 3PGA (18 carbons). Because the product of the carbon fixation reaction is a three-carbon molecule, the Calvin cycle is also called the C_3 **pathway,** and plants that initially fix carbon in this way are termed C_3 **plants.** Most plants are of this kind.

2. **Reduction.** In phase 2, *reduction,* reactions raise the energy level of 3PGA by the addition of a phosphate group transferred from ATP (Figure 9.12B, reaction 2) and electrons from NADPH (Figure 9.12B, reaction 3) to produce G3P, another three-carbon molecule. (The ATP and NADPH used are products of the light-dependent reactions.) For three turns of the cycle, six molecules of 3PGA (18 carbons) produce six molecules of G3P (18 carbons). One of these G3P molecule exits the cycle as a net product of the three turns and is used as the primary building block for reactions synthesizing the six-carbon glucose and many other organic molecules in chloroplasts. The other five molecules of G3P are used to regenerate RuBP in the next phase of the cycle.

3. **Regeneration.** In phase 3, *regeneration,* the G3P molecules generated by three turns of the cycle that do not exit the cycle are used to produce RuBP. First, G3P enters a complex series of reactions (Figure 9.12B, reaction 4) that yields the five-carbon sugar ribulose 5-phosphate. Then, in the final reaction of the cycle (Figure 9.12B, reaction 5), a phosphate group is transferred from ATP to regenerate the RuBP used in the first reaction. For three turns of the cycle, five molecules of G3P (15 carbons) produce three molecules of ribulose 5-phosphate (15 carbons) which then produce three molecules of RuBP (15 carbons).

Providing enough (CH_2O) units to make a six-carbon carbohydrate such as glucose requires six turns of the cycle.

In sum, for three turns of the Calvin cycle resulting in the net synthesis of one molecule of the product, G3P, 9 ATP and 6 NADPH are used. As mentioned earlier, those ATP and NADPH molecules derive from the light-dependent reactions of photosynthesis.

G3P Is the Starting Point for Synthesis of Many Other Organic Molecules

The net G3P formed by three turns of the Calvin cycle is the starting point for the production of a wide variety of organic molecules. More complex carbohydrates such as glucose and other monosaccharides are made from G3P by reactions that, in effect, reverse the first half of glycolysis. Glucose is a six-carbon sugar and G3P is a three-carbon sugar, which means that six turns of the Calvin cycle are needed for the synthesis of each molecule of glucose.

Once produced, the monosaccharides enter biochemical pathways that make disaccharides such as sucrose, polysaccharides such as starches and cellulose, and the other complex carbohydrates found in cell walls. Other pathways manufacture amino acids, fatty acids and lipids, proteins, and nucleic acids. The reactions forming these products occur both within chloroplasts and in the surrounding cytosol and nucleus.

Sucrose, a disaccharide of glucose linked to fructose, is the main form in which the products of photosynthesis circulate from cell to cell in vascular plants. Organic nutrients are stored in most vascular plants as sucrose or starch, or as a combination of the two in proportions that depend on the plant species. Sugar cane and sugar beets, which contain stored sucrose in high concentrations, are the main sources of the sucrose we use as table sugar.

Rubisco Is the Key Enzyme of the World's Food Economy

Rubisco, the enzyme that catalyzes the first reaction of the Calvin cycle, is unique to photosynthetic organisms. By catalyzing CO_2 fixation, it provides the source of organic molecules for most of the world's organisms—the enzyme converts about 100 billion tons of CO_2 into carbohydrates annually. There are so many rubisco molecules in chloroplasts that the enzyme may make up 50% or more of the total protein of plant leaves. As such, it is also the world's most abundant protein, estimated to total some 40 million tons worldwide, equivalent to about 6 kg for every human.

Rubisco has essentially the same overall structure in almost all photosynthetic organisms: eight copies each of a large and a small polypeptide, joined together in a 16-subunit struc-

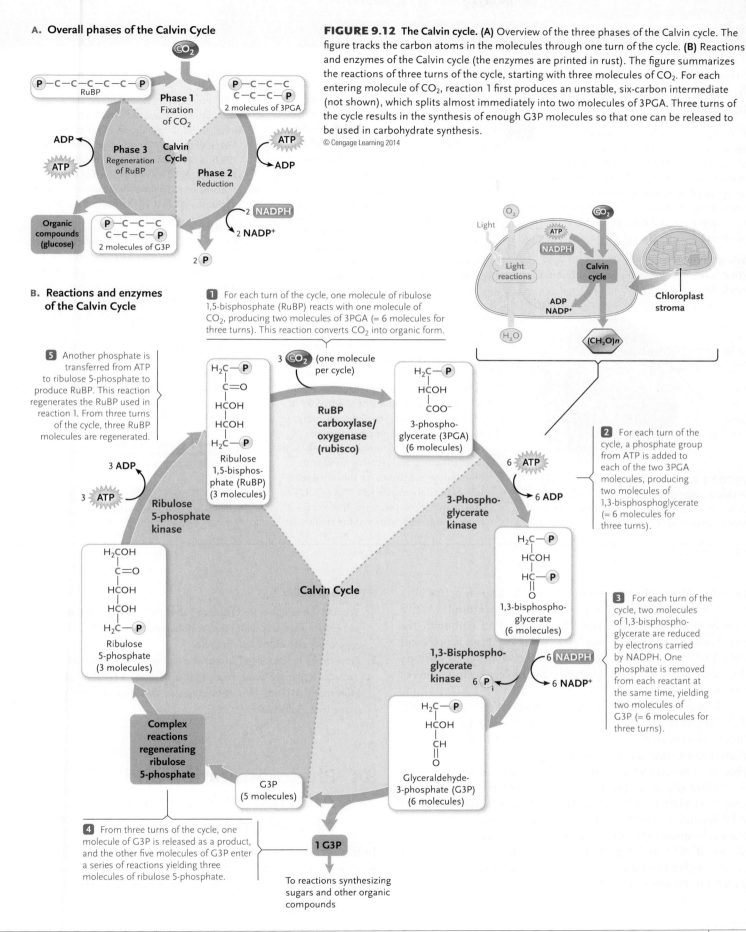

A. Overall phases of the Calvin Cycle

FIGURE 9.12 The Calvin cycle. (A) Overview of the three phases of the Calvin cycle. The figure tracks the carbon atoms in the molecules through one turn of the cycle. **(B)** Reactions and enzymes of the Calvin cycle (the enzymes are printed in rust). The figure summarizes the reactions of three turns of the cycle, starting with three molecules of CO_2. For each entering molecule of CO_2, reaction 1 first produces an unstable, six-carbon intermediate (not shown), which splits almost immediately into two molecules of 3PGA. Three turns of the cycle results in the synthesis of enough G3P molecules so that one can be released to be used in carbohydrate synthesis.
© Cengage Learning 2014

Phase 1
Fixation of CO_2

Phase 3
Regeneration of RuBP

Calvin Cycle

Phase 2
Reduction

Organic compounds (glucose)

B. Reactions and enzymes of the Calvin Cycle

1 For each turn of the cycle, one molecule of ribulose 1,5-bisphosphate (RuBP) reacts with one molecule of CO_2, producing two molecules of 3PGA (= 6 molecules for three turns). This reaction converts CO_2 into organic form.

5 Another phosphate is transferred from ATP to ribulose 5-phosphate to produce RuBP. This reaction regenerates the RuBP used in reaction 1. From three turns of the cycle, three RuBP molecules are regenerated.

Chloroplast stroma

$(CH_2O)n$

Light reactions

Calvin cycle

3 CO_2 (one molecule per cycle)

RuBP carboxylase/oxygenase (rubisco)

3-phospho-glycerate (3PGA) (6 molecules)

2 For each turn of the cycle, a phosphate group from ATP is added to each of the two 3PGA molecules, producing two molecules of 1,3-bisphosphoglycerate (= 6 molecules for three turns).

Ribulose 1,5-bisphos-phate (RuBP) (3 molecules)

Ribulose 5-phosphate kinase

3 ADP

3 ATP

6 ATP

6 ADP

3-Phospho-glycerate kinase

Ribulose 5-phosphate (3 molecules)

Calvin Cycle

1,3-bisphospho-glycerate (6 molecules)

3 For each turn of the cycle, two molecules of 1,3-bisphospho-glycerate are reduced by electrons carried by NADPH. One phosphate is removed from each reactant at the same time, yielding two molecules of G3P (= 6 molecules for three turns).

1,3-Bisphospho-glycerate kinase

6 NADPH

6 P_i

6 NADP⁺

Complex reactions regenerating ribulose 5-phosphate

G3P (5 molecules)

Glyceraldehyde-3-phosphate (G3P) (6 molecules)

4 From three turns of the cycle, one molecule of G3P is released as a product, and the other five molecules of G3P enter a series of reactions yielding three molecules of ribulose 5-phosphate.

1 G3P

To reactions synthesizing sugars and other organic compounds

INSIGHTS FROM THE | Molecular Revolution

Small but Pushy: What is the function of the small subunit of rubisco?

Rubisco is a 16-subunit protein with eight copies of a large polypeptide and eight copies of a small polypeptide **(Figure).** We noted that all the active sites of rubisco appear to be on the large polypeptide subunit of the enzyme. Even so, 99% of the catalytic activity of the enzyme is lost if the small subunit is removed.

Research Question

What is the function of the small subunit of rubisco?

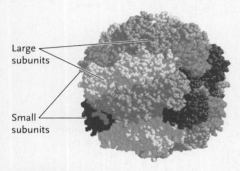

FIGURE A model for the structure of rubisco in chloroplasts of higher plants. The large subunits are shown in gray and white, and the small subunits in blue and orange.

Large subunits

Small subunits

Experiment

Betsy A. Read and F. Robert Tabita of The Ohio State University set out to answer this question using molecular techniques. They hypothesized that the structure of a specific region of the small subunit was critical to its function. To test this hypothesis, the investigators used DNA cloning techniques (see Chapter 18) to produce five versions of the small subunit, each with a different amino acid substituted for the normal one at five different positions in the protein. The locations chosen for the substitutions were believed to be important for the function of the enzyme. Then, they examined the effects of the substitutions on enzyme activity.

Results

1. One of the modified small subunits, which had glutamine substituted for arginine at position 88 in the amino acid sequence, was unable to assemble with the large subunit to form a complete enzyme complex. Thus, the arginine at position 88 is essential for normal enzyme assembly.

2. The four remaining modified versions of the small subunit assembled normally with large subunits, and all recognized and bound their substrates for the initial reaction of the Calvin cycle (see Figure 9.12) as ably as the normal enzyme. Therefore, these four modifications of the small sub-

unit had no effect on the specificity of the enzyme.

3. The investigators next checked the rates at which the enzymes with the four modified small subunits catalyzed CO_2 fixation. Three of the four modified enzymes ran the first reaction of the Calvin cycle at a rate of 35% or less of the rate of the normal enzyme. The most active worked only about half as fast as the normal enzyme.

Conclusion

The small subunit has a very significant effect on rubisco's rate of catalysis. The effect is critically important when considered in the context of the comparatively slow reaction rate of the normal enzyme. The enzyme's multiple form—eight copies of each subunit, massed together, all doing the same thing—and the very large amount of the enzyme packed into leaves compensate for the slow rate. Evidently, the small subunit evolved as yet another way to compensate for the enzyme's slow action, by pushing the large subunit to do its job faster.

THINK LIKE A SCIENTIST How many genes encode the rubisco enzyme?

Source: B. A. Read and F. R. Tabita. 1992. Amino acid substitutions in the small subunit of ribulose-1,5-bisphosphate carboxylase/oxygenase that influence catalytic activity of the holoenzyme. *Biochemistry* 31:519–525.

© Cengage Learning 2014

---

ture. The large subunit contains all of the known active sites where substrates, including CO_2 and RuBP, can bind. (Active sites of enzymes are described in Section 4.4.) Although the small subunit has no active sites, it is still essential for efficient operation of the enzyme. *Insights from the Molecular Revolution* describes experiments to determine the molecular functions of the small subunit.

Rubisco is also the key regulatory site of the Calvin cycle. The enzyme is stimulated by both NADPH and ATP; as long as these substances are available from the light-dependent reactions, the enzyme is active and the light-independent reactions proceed. During the daytime, when sunlight powers the light-dependent reactions, the abundant NADPH and ATP supplies keep the Calvin cycle running. In darkness, when NADPH and ATP become unavailable, the enzyme is inhibited and the Calvin cycle slows or stops. Similar controls based on the availability of ATP and NADPH also regulate the enzymes that catalyze other reactions of the Calvin cycle, including reactions 2 and 3 in Figure 9.12B.

STUDY BREAK 9.3

1. What is the reaction that rubisco catalyzes? Why is rubisco the key enzyme for producing the world's food, and how is it the key regulatory site of the Calvin cycle?

2. How many molecules of carbon dioxide must enter the Calvin cycle for a plant to produce a sugar containing 12 carbon atoms? How many ATP and NADPH molecules would be required to make that molecule?

9.4 Photorespiration and Alternative Processes of Carbon Fixation

In this section we consider *photorespiration* in plants, a mechanism with some features similar to aerobic respiration that uses O_2 as the first step in a pathway to generate CO_2. We also describe two alternative processes of carbon fixation, the C_4 pathway and the CAM pathway.

Photorespiration Produces Carbon Dioxide That Is Used by the Calvin Cycle

Rubisco, RuBP carboxylase/oxygenase, is so named because it is both a carboxylase and an oxygenase. That is, the enzyme has an active site to which either CO_2 or O_2 can bind. When CO_2 binds, rubisco acts as a carboxylase fixing the carbon of CO_2 by combining it with RuBP to produce 3PGA in phase 1 of the Calvin cycle (**Figure 9.13;** and see Figure 9.12). Overall, the carboxylase reaction of rubisco results in carbon gain.

When O_2 binds, rubisco acts as an oxygenase converting RuBP to one molecule of 3PGA and one molecule of a two-carbon compound, phosphoglycolate (see Figure 9.13). The 3PGA is used in the Calvin cycle within the chloroplast. The phosphoglycolate is hydrolyzed to its non-phosphorylated derivative, glycolate, which exits the chloroplasts and diffuses into peroxisomes (see Section 5.3). Glycolate is a toxic molecule, and it is broken down in the peroxisomes by reactions that release a molecule of CO_2, which can then be used for carbon fixation. In contrast to the carboxylase reaction, no carbon is fixed during the oxygenase reaction and there is no carbon gain as is seen for the carboxylase reaction. That is, in the oxygenase reaction, CO_2 is released, a net loss of carbon. But, in order to grow, all organisms must gain carbon.

The entire process from the oxygenase reaction of rubisco to the release of CO_2 is called **photorespiration.** The term was coined because the process occurs in the presence of light and, like aerobic respiration, it requires oxygen, and produces carbon dioxide and water. However, unlike aerobic respiration, photorespiration does not generate ATP.

The extent to which photorespiration occurs in a plant depends on the relative concentrations of CO_2 and O_2 in the leaves. Gas exchange—CO_2 in and O_2 out—between the air and the cells within the leaf occurs through the stomata of the leaves and stem. If gas exchange occurs normally, the relative concentrations of CO_2 and O_2 within the leaf favor CO_2 fixation by the Calvin cycle. That is, even though the CO_2 concentration is lower than that of O_2, rubisco has a ten-times greater affinity for CO_2 than it does for O_2. However, at higher concentrations of O_2 within the leaf, oxygen acts as a competitive inhibitor of the enzyme, and this favors the reaction of RuBP with O_2 rather than with CO_2: photorespiration occurs.

Photorespiration activity occurs more as temperatures rise due to the mechanism used to limit water loss from leaves. That is, the surface of the leaf is covered by a waxy *cuticle* that prevents water loss. The cuticle also prevents the rapid diffusion of gases, such as CO_2, into the leaf. Gas exchange occurs predominantly through the stomata. The plant regulates the size of stomata from fully closed to fully open so as to balance the demands for gas exchange with the need to minimize water loss. In particular, plants that are adapted to hot, dry climates are

A. Carboxylation reaction of rubisco—carbon gained

$$RuBP + CO_2 \xrightarrow{\text{Rubisco}} 2 \text{ 3-phosphoglycerate (3PGA)}$$

B. Oxygenation reaction of rubisco—carbon lost

$$RuBP + O_2 \xrightarrow{\text{Rubisco}} 1 \text{ 3PGA} + 1 \text{ Phosphoglycolate}$$

To reactions releasing CO_2

FIGURE 9.13 Reactions of rubisco. **(A)** Carboxylation reaction, the first reaction of the Calvin cycle, in which carbon from CO_2 is fixed to produce 3-phosphoglycerate. This activity of rubisco results in carbon gain. **(B)** Oxygenation reaction, the first step of photorespiration, in which O_2 instead of CO_2 is a substrate for the enzyme, resulting in the synthesis of 3PGA and phosphoglycolate. Phosphoglycolate then is used in reactions that ultimately release CO_2. This activity of rubisco results in carbon loss.
© Cengage Learning 2014

faced with a constant dilemma: their stomata must be open to let in CO_2 for the Calvin cycle, but their stomata should be closed to conserve water. A similar situation occurs for non-adapted plants when environmental conditions become hot and dry. With the stomata closed, photosynthesis rapidly consumes the CO_2 in the leaf and produces O_2, which accumulates in the chloroplasts and photorespiration occurs.

For a plant with high photorespiration rates at elevated temperatures, as much as 50% of the carbon fixed by the Calvin cycle may be lost because only one three-carbon 3PGA molecule is produced instead of two. Unfortunately, many economically important crop plants become seriously impaired by high photorespiration rates when temperatures rise; among them are the C_3 plants rice, barley, wheat, soybeans, tomatoes, and potatoes.

Alternative Processes of Carbon Fixation Minimize Photorespiration

Many plant species that live in hot, dry environments have evolved alternative processes of carbon fixation that minimize photorespiration and, therefore, its negative effect on photosynthesis. C_4 plants use the *C_4 pathway* to fix CO_2 into a four-carbon molecule, *oxaloacetate,* while CAM plants use the *CAM pathway* at night to fix carbon in the form of oxaloacetate. These special pathways precede the C_3 pathway (Calvin cycle) rather than replacing it.

C_4 **PLANTS AND THE** C_4 **PATHWAY** At least 19 families of flowering plants are C_4 plants. They include crabgrass and several crops important agriculturally, including corn and sugarcane. C_4 plants have a characteristic leaf anatomy **(Figure 9.14A):** photosynthetic *mesophyll cells* are tightly associated with specialized, chloroplast-rich *bundle sheath cells,* which encircle the veins of

A. C₄ plant leaf cross section

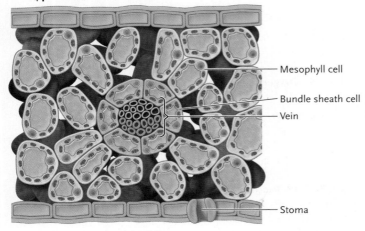

Mesophyll cell

Bundle sheath cell

Vein

Stoma

B. C₃ plant leaf cross section

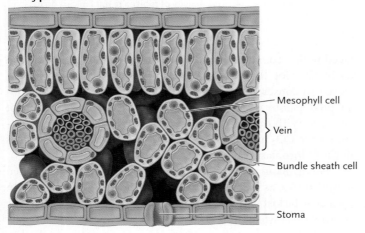

Mesophyll cell

Vein

Bundle sheath cell

Stoma

FIGURE 9.14 Leaf anatomy in (A) C₄ plants and (B) C₃ plants.
© Cengage Learning 2014

the leaf. In a C₃ plant leaf **(Figure 9.14B),** the bundle sheath cells have few chloroplasts and low concentrations of rubisco. These cells generate ATP by cyclic electron flow (see Figure 9.11).

C₄ plants use the *C₄ pathway* for carbon fixation. In the **C₄ pathway,** CO_2 that has diffused into the leaf through the stomata initially is fixed in mesophyll cells by combining it with the three-carbon *phosphoenolpyruvate (PEP)* to produce the four-carbon oxaloacetate **(Figure 9.15A).** The C₄ pathway gets its name because its first product is a four-carbon molecule rather than a three-carbon molecule, as in the C₃ pathway. The carbon fixation reaction producing oxaloacetate is catalyzed by *PEP carboxylase.* This enzyme has a much greater affinity for CO_2 than rubisco does and, unlike rubisco, it has no oxygenase activity.

As the C₄ pathway continues, the oxaloacetate is next reduced to the four-carbon *malate* by electrons transferred from NADPH. Malate diffuses into bundle sheath cells, the sites of the reactions of the Calvin cycle. (In C₃ plants, the Calvin cycle takes place in the mesophyll cells.) The malate is then oxidized to the three-carbon pyruvate, releasing CO_2, which is used in

the rubisco-catalyzed first step of the Calvin cycle. The ability to use malate to donate CO_2 in this way, leads to a higher $CO_2:O_2$ ratio in the bundle sheath cells at the time when rubisco is functional in the Calvin cycle, namely in the daylight. The higher $CO_2:O_2$ ratio means that photorespiration is extremely low or negligible in C₄ plants. The pyruvate produced diffuses into mesophyll cells where it is converted back into PEP in a reaction that consumes ATP.

In sum, in C₄ plants the C₄ pathway takes place in the mesophyll cells, while the Calvin cycle occurs in the bundle sheath cells (see Figure 9.15A). That is, carbon fixation and the Calvin cycle are separated spatially in different cell types.

If the C₄ pathway is so effective at reducing photorespiration, why is the pathway not used by all plants? The answer is that the C₄ pathway has an additional energy requirement: for each turn of the C₄ pathway cycle, one ATP molecule is hydrolyzed to regenerate PEP from pyruvate. This adds an energy requirement of six ATP molecules for each G3P produced by the Calvin cycle. However, as mentioned earlier, photorespiration potentially can decrease carbon fixation efficiency by as much as 50% in hot environments, so the additional ATP requirement is worthwhile. Moreover, hot environments typically receive a lot of sunshine; thus, the additional ATP requirement can be met easily by increasing the output of the light-dependent reactions.

C₄ plants also perform better where it is dry. Because PEP carboxylase has a very high affinity only for CO_2, C₄ plants are more efficient at fixing CO_2 than are C₃ plants. Therefore, they do not have to keep their stomata open for as long as a C₃ plant does under the same conditions. Because this reduces water loss, C₄ plants are much better suited to arid conditions.

CAM PLANTS AND THE CAM PATHWAY The acronym CAM stands for **crassulacean acid metabolism;** the name comes from the Crassulaceae family (stonecrops), where the pathway was first described. CAM plants live in very dry environments and have several evolutionary adaptations that enable them to survive the arid conditions. Besides succulents (water-storing plants) such as the stonecrops, CAM plants include some members of at least 25 plant families, including the cactus family, the lily family, and the orchid family.

Exactly as with the C₄ pathway in C₄ plants, in the **CAM pathway** of CAM plants CO_2 is initially fixed to oxaloacetate in a reaction catalyzed by PEP carboxylase **(Figure 9.15B).** The CO_2 produced by the oxidation of malate is used in the rubisco-catalyzed first step of the Calvin cycle. But, whereas carbon fixation and the Calvin cycle run in different cell types in C₄ plants, in CAM plants carbon fixation and the Calvin cycle *both* occur in mesophyll cells but are separated temporally. That is, they run at different times, initial carbon fixation at night and the Calvin cycle during the day.

CAM plants live in regions that are hot and dry during the day and cool at night. The stomata on their fleshy leaves or stems open only at night to minimize water loss. When they open, O_2 generated by photosynthesis is released, and CO_2 en-

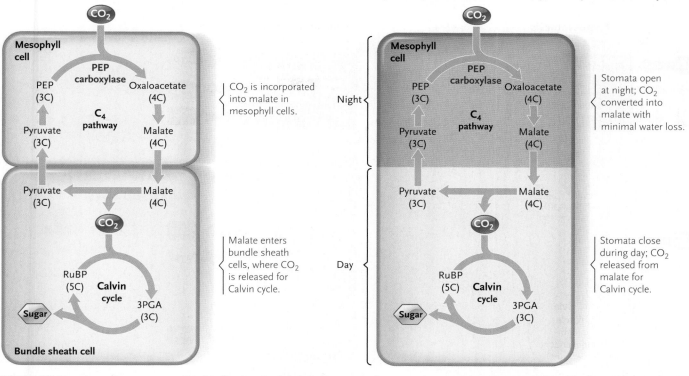

A. C₄ pathway in C₄ plants—spatial separation of steps

Mesophyll cell

PEP carboxylase

PEP (3C) → Oxaloacetate (4C)

C₄ pathway

Pyruvate (3C) ← Malate (4C)

CO₂ is incorporated into malate in mesophyll cells.

Pyruvate (3C) ← Malate (4C)

CO₂

RuBP (5C) **Calvin cycle** 3PGA (3C)

Sugar

Malate enters bundle sheath cells, where CO₂ is released for Calvin cycle.

Bundle sheath cell

B. CAM pathway in CAM plants—temporal separation of steps

Mesophyll cell

PEP carboxylase

Night

PEP (3C) → Oxaloacetate (4C)

C₄ pathway

Pyruvate (3C) ← Malate (4C)

Stomata open at night; CO₂ converted into malate with minimal water loss.

Day

Pyruvate (3C) ← Malate (4C)

CO₂

RuBP (5C) **Calvin cycle** 3PGA (3C)

Sugar

Stomata close during day; CO₂ released from malate for Calvin cycle.

FIGURE 9.15 Two alternative processes of carbon fixation to minimize photorespiration. In each case, carbon fixation produces the four-carbon oxaloacetate, which is processed to generate CO_2 that feeds into the Calvin (C_3) cycle. **(A)** In C_4 plants, carbon fixation and the Calvin cycle occur in different cell types: carbon fixation by the C_4 pathway takes place in mesophyll cells, while the Calvin cycle takes place in bundle sheath cells. **(B)** In CAM plants, carbon fixation and the Calvin cycle occur at different times in mesophyll cells: carbon fixation by the C_4 pathway takes place at night, while the Calvin cycle takes place during the day.
© Cengage Learning 2014

ters. The CO_2 is fixed into malate in the mesophyll cells by the CAM pathway; malate accumulates throughout the night and is stored in large cell vacuoles. During the day, the stomata close to reduce water loss in the hot conditions; this also cuts off the exchange of gases with the atmosphere. Malate now moves from the vacuole to the chloroplasts, where its oxidation to pyruvate generates CO_2 that is used by the Calvin cycle.

STUDY BREAK 9.4

1. When does photorespiration occur? What are the reactions of photorespiration, and what are the energetic consequences of the process?
2. What is the C_4 pathway, and how does it enable C_4 plants to circumvent photorespiration?
3. How are carbon fixation and the Calvin cycle different in C_4 plants and CAM plants?

9.5 Photosynthesis and Cellular Respiration Compared

This section addresses the similarities between photosynthesis and cellular respiration and the evolution of the two processes.

Photosynthesis and Cellular Respiration Are Similar Processes with Reactions That Essentially Are the Reverse of Each Other

A popular misconception is that photosynthesis occurs in plants, and cellular respiration occurs in animals only. In fact, both processes occur in plants, with photosynthesis confined to tissues containing chloroplasts and cellular respiration taking place in all cells. **Figure 9.16** presents side-by-side schematics of photosynthesis and cellular respiration to highlight their similarities and points of connection. Note that their overall reactions are basically the reverse of each other. That is, the reactants of photosynthesis—CO_2 and H_2O—are the products of cellular respiration, and the reactants of cellular respiration—glucose and O_2—are the products of photosynthesis. Both processes have key phosphorylation reactions involving an electron transfer system—photophosphorylation in photosynthesis and oxidative phosphorylation in cellular respiration—followed by the chemiosmotic synthesis of ATP. Further, G3P is found in the pathways of both processes. In photosynthesis, it is a product of the Calvin cycle and is used for the synthesis of sugars and other organic fuel molecules. In cellular respiration, it is an intermediate generated in glycolysis in the conversion of glucose to pyruvate. Thus, G3P is used by anabolic pathways when it is generated by photosynthesis, and it is a product of a catabolic pathway in cellular respiration.

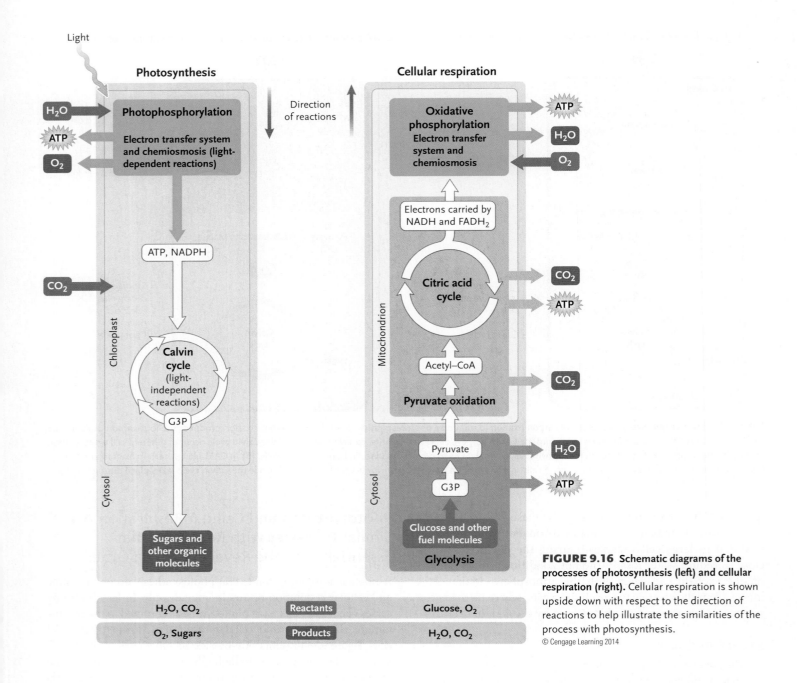

FIGURE 9.16 Schematic diagrams of the processes of photosynthesis (left) and cellular respiration (right). Cellular respiration is shown upside down with respect to the direction of reactions to help illustrate the similarities of the process with photosynthesis.
© Cengage Learning 2014

Evolution of Photosynthesis and Cellular Respiration

Oxygen (O_2) is a product of photosynthesis and is essential for cellular respiration. The early atmosphere of the earth, more than 3.5 billion years ago, had very little O_2. Prokaryotes that derived their energy from chemical compounds present on the early earth were the only organisms that lived at the time. By 2.7 billion years ago, some prokaryotic cells had acquired the ability to use energy from sunlight, a process that eventually evolved into early forms of photosynthesis, and they released oxygen into the atmosphere as a by-product.

We do not know the details of how photosynthesis first evolved. Studies of modern prokaryotes indicate that the evolution of photosynthesis was not a single event but rather a network of multiple events over long periods of time, giving rise to many different types of photosynthesis. Photosynthetic activity by prokaryotes between 2.4 and 2.0 billion years ago increased atmospheric oxygen to about 2%. Photosynthetic eukaryotes emerged about 850 million years ago, and by 300 to 350 million years ago they had contributed to increasing atmospheric oxygen to about 30% to 35%. The current level of oxygen in our atmosphere is about 20%.

The ancient accumulation of oxygen made cellular respiration possible. We do not know exactly how cellular respiration evolved but we do know that it evolved first in ancient prokaryotes through complex processes, long before any eukaryotes or multicellular organisms existed. The evolution of photosynthesis and respiration in prokaryotes was the foundation of eukaryotic evolution and the emergence of multicellular life. As you have learned, all eukaryotes carry out cellular respiration, and some also carry out photosynthesis. According to the endosymbiotic theory, ancient cells engulfed bacteria in which cellular respiration was already well developed, resulting in cells living within cells (see Chapter 25). These engulfed bacteria eventually evolved into the mitochondria present in eukaryotic cells. In a second event, cells already containing early mitochondria engulfed photosynthetic bacteria, which eventually evolved into chloroplasts. These cells were the ancestors of plants and other photosynthetic eukaryotes.

In this chapter, you have seen how photosynthesis supplies the organic molecules used as fuels by almost all the organisms of the world. It is a story of electron flow: electrons, pushed to high energy levels by the absorption of light energy, are added to CO_2, which is fixed into carbohydrates and other fuel molecules. The high-energy electrons are then removed from the fuel molecules by the oxidative reactions of cellular respiration, which use the released energy to power the activities of life. Among the most significant of these activities are cell growth and division, the subjects of the next chapter.

STUDY BREAK 9.5

How are the reactants and products of photosynthesis and cellular respiration related?

UNANSWERED QUESTIONS

Why is photosynthesis so inefficient, and what (if anything) can we do about it?

Photosynthesis is considered by many to be the most important biological process on Earth. Photosynthesis involves the highest energy processes of life; it is the process where (by far) most of the energy in our ecosystem is captured. All other biological processes are exergonic (they lose the energy captured by photosynthesis) and thus all other processes involve less energy than photosynthesis. It is also the process where, by far, the most energy in our ecosystem is *lost*. In fact, photosynthetic organisms dissipate (dump purposely) a large part of absorbed light energy to prevent the buildup of reactive oxygen species (intermediates of photosynthesis) that can damage the plant. At full sunlight, regulatory dissipation can involve more than 75% of absorbed light energy. Consequently, typical agricultural crops store only about 0.1% to 0.5% of their absorbed solar energy in the form of biomass. Interestingly, some plants and many green algae can store up to 10 times as much of this energy. Research is being done to understand why they are so efficient, and whether crop or biofuel plants can be engineered to produce more biomass.

How is the efficiency of photosynthesis regulated?

My laboratory at Washington State University is interested in how plants set their photosynthetic "strategies" and thus how they control light capture and electron and proton transfer reactions to balance their competing needs for efficient photosynthesis while avoiding toxic side reactions. As you have learned, energy conversion by the chloroplast involves the capture of light energy and the channeling of that energy through an electron transfer system with the eventual synthesis of NADPH and ATP. At high concentrations, many of the intermediates produced in this energy conversion can potentially destroy the photosynthetic apparatus, a phenomenon called photoinhibition. To prevent such damage, the effi-ciency of some of the photosystem components is decreased in a process called photoprotection. High temperatures lower the efficiency of photosynthesis, however. Evidence from a range of studies indicates that the balance between protection against photoinhibition and photosynthetic efficiency is important in enabling plants to acclimate to environmental changes, but it also limits their productivity. We have found that ATP synthase acts as a major regulator of photosynthesis, sensing the metabolic status of the chloroplast, and in response, regulating the proton-motive force by restricting proton flow out of the lumen. In turn, the proton-motive force controls the photoprotective dissipation of light energy and electron transfer. Among the major unanswered questions are: what controls the ATP synthase, what specifically does it sense, and how does this mechanism differ in high- and low-productivity plants? Answering these questions will illuminate how photosynthesis determines plant growth and survival. In addition, the technology developed as part of the research may lead to applications in plant breeding, particularly for improving energy storage by biofuels and food crops.

THINK LIKE A SCIENTIST Find some specific examples of plants with higher- and lower-efficiency photosynthesis. Why do you think some plants have higher photosynthetic rates than others? What are some potential selective advantages of lower rates of photosynthesis?

David Kramer

David Kramer is a professor and fellow at the Institute of Biological Chemistry (IBC) at Washington State University. Kramer's research group aims to understand how photosynthesis works, what specific reactions limit its productivity, and how these limitations can be overcome to redirect more output toward readily useable biofuels. Learn more about his research at http://ibc.wsu.edu/research/kramer/index.htm.

REVIEW KEY CONCEPTS

To access the course materials and companion resources for this text, please visit www.cengagebrain.com.

9.1 Photosynthesis: An Overview

- In photosynthesis, plants, algae, and photosynthetic prokaryotes use the energy of sunlight to drive synthesis of organic molecules from simple inorganic raw materials. The organic molecules are used by the photosynthesizers themselves as fuels; they also form the primary energy source for heterotrophs.

- The two overall stages of photosynthesis are the light-dependent and light-independent reactions. In eukaryotes, both stages take place inside chloroplasts (Figures 9.2 and 9.3).

- Photosynthesizers use light energy to push electrons to elevated energy levels. In eukaryotes and many prokaryotes, water is split as the source of the electrons for this process, and oxygen is released to the environment as a by-product.

- The high-energy electrons provide an indirect energy source for ATP synthesis and also for CO_2 fixation, in which CO_2 is fixed into organic substances by the addition of both electrons and protons.

Animation: T. Englemann's experiment

Animation: Photosynthesis Bio Experience 3D

Animation: Sites of photosynthesis

9.2 The Light-Dependent Reactions of Photosynthesis

- In the light-dependent reactions of photosynthesis, light energy is converted to chemical energy when electrons, excited by absorption of light in a pigment molecule, are passed from the pigment to a stable orbital in a primary acceptor molecule (Figure 9.5).

- Chlorophylls are the major photosynthetic pigments in plants, green algae, and cyanobacteria. Carotenoids are accessory pigments that absorb light energy at a different wavelength than those absorbed by chlorophylls. Carotenoids may expand the range of wavelengths used for photosynthesis, but their more important role is in photoprotection (Figures 9.6 and 9.7).

- In organisms that split water as their electron source, the pigments are organized with proteins into two photosystems. Special subsets of chlorophyll *a* pass excited electrons to primary acceptor molecules in the photosystems (Figure 9.8).

- Electrons obtained from splitting water are used for the synthesis of NADPH and ATP. In the linear electron flow pathway, electrons first flow through photosystem II, becoming excited there to a higher energy level, and then pass through an electron transfer system to photosystem I releasing energy that is used to create an H^+ gradient across the membrane. The gradient is used by ATP synthase to drive synthesis of ATP. The net products of the light-dependent reactions are ATP, NADPH, and oxygen (Figures 9.9 and 9.10).

- Electrons can also flow cyclically around photosystem I and the electron transfer system, building the H^+ concentration and allowing extra ATP to be produced, but no NADPH (Figure 9.11).

Animation: Noncyclic pathway of electron flow

Animation: Harvesting photo energy

9.3 The Light-Independent Reactions of Photosynthesis

- In the light-independent reactions of photosynthesis, CO_2 is reduced and converted into organic substances by the addition of electrons and hydrogen carried by the NADPH produced in the light-dependent reactions. ATP, also derived from the light-dependent reactions, provides additional energy. The key enzyme of the light-independent reactions is rubisco (RuBP carboxylase/oxygenase), which catalyzes the reaction that combines CO_2 into organic compounds (Figure 9.12).

- In the process, NADPH is oxidized to $NADP^+$, and ATP is hydrolyzed to ADP and phosphate. These products of the light-independent reactions cycle back as inputs to the light-dependent reactions.

- The Calvin cycle produces surplus molecules of G3P, which are the starting point for synthesis of glucose, sucrose, starches, and other organic molecules. The light-independent reactions take place in the chloroplast stroma in eukaryotes and in the cytoplasm of photosynthetic prokaryotes.

Animation: Calvin cycle

9.4 Photorespiration and Alternative Processes of Carbon Fixation

- When oxygen concentrations are high relative to CO_2 concentrations, rubisco acts as an oxygenase, catalyzing the combination of RuBP with O_2 rather than CO_2 and forming toxic products that cannot be used in photosynthesis. The toxic products are eliminated by reactions that release carbon as CO_2, greatly reducing the efficiency of photosynthesis. The entire process is called photorespiration because it uses oxygen and releases CO_2 (Figure 9.13).

- Some plants use alternative processes of carbon fixation that minimize photorespiration. C_4 plants use the C_4 pathway to first fix CO_2 into the four-carbon oxaloacetate in mesophyll cells (the site of the Calvin cycle in C_3 plants), and then produce CO_2 for the Calvin cycle in bundle sheath cells. CAM plants use the CAM pathway also to first fix CO_2 into oxaloacetate and then to generate CO_2 for the Calvin cycle. Here both the carbon fixation and the Calvin cycle occur in mesophyll cells, but they are separated by time; initial carbon fixation occurs at night and the Calvin cycle occurs during the day (Figures 9.14 and 9.15).

Animation: C_3-C_4 comparison

Animation: Carbon-fixing adaptations

9.5 Photosynthesis and Cellular Respiration Compared

- Photosynthesis occurs in the cells of plants that contain chloroplasts, whereas cellular respiration occurs in all cells. The overall reactions of the two processes are essentially the reverse of each other, with the reactants of one being the products of the other (Figure 9.16).

Test Your Knowledge

1. An organism exists for long periods by using only CO_2 and H_2O. It could be classified as a(n):
   a. herbivore.
   b. carnivore.
   c. decomposer.
   d. autotroph.
   e. heterotroph.

2. During the light-dependent reactions:
   a. CO_2 is fixed.
   b. NADPH and ATP are synthesized using electrons derived from splitting water.
   c. glucose is synthesized.
   d. water is split and the electrons generated are used for glucose synthesis.
   e. photosystem I is unlinked from photosystem II.

3. Which of the following is a correct step in the light-dependent reactions of the Z system?
   a. Light is absorbed at P700, and electrons flow through a pathway to $NADP^+$, the final acceptor of the linear pathway.
   b. Electrons flow from photosystem II to water.
   c. $NADP^+$ is oxidized to NADPH as it accepts electrons.
   d. Water is degraded to activate P680.
   e. Electrons pass through a thylakoid membrane to create energy to pump H^+ through the cytochrome complex.

4. The light-dependent reactions of photosynthesis resemble aerobic respiration as both:
   a. synthesize NADPH.
   b. synthesize NADH.
   c. require electron transfer systems to synthesize ATP.
   d. require oxygen as the final electron acceptor.
   e. have the same initial energy source.

5. The molecules that link the light-dependent and light-independent reactions are:
   a. ADP and H_2O.
   b. RuBP and CO_2.
   c. cytochromes and water.
   d. G3P and RuBP.
   e. ATP and NADPH.

6. You bite into a spinach leaf. Which one of the following is true?
   a. You are getting 50% of the protein in the leaf in the form of ribulose 1,5-bisphosphate carboxylase.
   b. The major pigment you are ingesting is a carotenoid.
   c. The water in the leaf is a product of the light-independent reactions.
   d. Any energy from the leaf you can use directly is in the form of ATP.
   e. The spinach most likely was grown in an area with a low CO_2 concentration.

7. Animal metabolism and plant metabolism are related in that:
   a. plants carry out photosynthesis and animals carry out respiration.
   b. G3P is found in the metabolic pathways of both animals and plants.
   c. G3P is used by catabolic pathways when it is generated by photosynthesis, and it is a product of an anabolic pathway in cellular respiration.
   d. light drives electron excitation.
   e. the reactants of photosynthesis drive cellular respiration in animals.

8. Which of the following statements about the C_4 cycle is *incorrect*?
   a. CO_2 initially combines with PEP.
   b. PEP carboxylase catalyzes a reaction to produce oxaloacetate.
   c. Oxaloacetate transfers electrons from NADPH and is reduced to malate.
   d. Less ATP is used to run the C_4 cycle than the C_3 cycle.
   e. The cycle runs when O_2 concentration is high.

9. In one turn of the Calvin cycle, one molecule of CO_2 generates:
   a. 6 ATP.
   b. 6 NADH.
   c. 6 ATP and 6 NADPH.
   d. one (CH_2O) unit of carbohydrate.
   e. one molecule of glucose.

10. The oxygen released by photosynthesis comes from:
    a. CO_2.
    b. H_2O.
    c. light.
    d. NADPH.
    e. electrons.

Discuss the Concepts

1. Suppose a garden in your neighborhood is filled with red, white, and blue petunias. Explain the floral colors in terms of which wavelengths of light are absorbed and reflected by the petals.

2. About 200 years ago, Jan Baptista van Helmont tried to determine the source of raw materials for plant growth. To do so, he planted a young tree weighing 5 pounds in a barrel filled with 200 pounds of soil. He watered the tree regularly. After 5 years, he again weighed the tree and the soil. At that time the tree weighed 169 pounds, 3 ounces, and the soil weighed 199 pounds, 14 ounces. Because the tree's weight had increased so much, and the soil's weight had remained about the same, he concluded that the tree gained weight as a result of the water he had added to the barrel. Analyze his conclusion in terms of the information you have learned from this chapter.

3. Like other accessory pigments, the carotenoids extend the range of wavelengths absorbed in photosynthesis. They also protect plants from a potentially lethal process known as *photooxidation*. This process begins when excitation energy in chlorophylls drives the conversion of oxygen into free radicals, substances that can damage organic compounds and kill cells. When plants that cannot produce carotenoids are grown in light, they bleach white and die. Given this observation, what molecules in the plants are likely to be destroyed by photooxidation?

4. What molecules would you have to provide a plant, theoretically speaking, for it to make glucose in the dark?

Design an Experiment

Space travelers of the future land on a planet in a distant galaxy, where they find populations of a carbon-based life form. The beings on this planet are of a vibrantly purple color. The travelers suspect that the beings secure the energy necessary for survival by a process similar to photosynthesis on Earth. How might they go about testing this conclusion?

Interpret the Data

Photosynthesis directly opposes respiration in determining how plants influence atmospheric CO_2 concentrations. When a leaf is in the light, both photosynthesis and respiration are occurring simultaneously. The data in the **Table** were collected from the leaf of a sagebrush plant that was enclosed in a chamber that measures the rate of CO_2 exchange. The same leaf was used to collect the data in *Interpret the Data* in Chapter 8. Respiration is shown as a negative and photosynthesis as a positive rate of CO_2 exchange. The net photosynthesis rate is the amount of CO_2 (in micromoles per square meter per second) assimilated by the leaf while respiration is occurring; a positive value indicates more photosynthesis is occurring than respiration. The light exposed to the leaf is quantified as the number of photons in the 400 to 700 nm wavelength, the Photosynthetic Photon Flux Density (PPFD); 2,000 μmol/m²/s is equivalent to the amount of light occurring at midday in full sun.

Observation	Photosynthetic Photon Flux Density (PPFD) (μmol/m²/s)	Net Photosynthesis (μmol/m²/s)
1	2,000	9.1
2	1,500	8.4
3	1,250	8.2
4	1,000	7.4
5	750	6.3
6	500	4.8
7	250	2.2
8	0	−2.0

1. Why is net photosynthesis negative when PPFD is zero? Looking at the respiration data from Chapter 8 *Interpret the Data*, at what temperature do you think these data were collected?

2. Make a graph of the data, with PPFD on the *x* axis and net photosynthesis rate on the *y* axis.

3. Is the relationship between PPFD and net photosynthesis best described as linear or curved? Why do you think the data have this type of relationship?

4. How might the predicted increase in temperature due to elevated CO_2 concentrations in the atmosphere impact net photosynthesis? Hint: combine the data from this question with the data from the table in *Interpret the Data* in Chapter 8. Do you think the data presented in both of these questions are sufficient to explain this impact?

Source: Data based on unpublished research by Brent Ewers, University of Wyoming.

Apply Evolutionary Thinking

If global warming raises the temperature of our climate significantly, will C_3 plants or C_4 plants be favored by natural selection? How will global warming change the geographical distributions of plants?

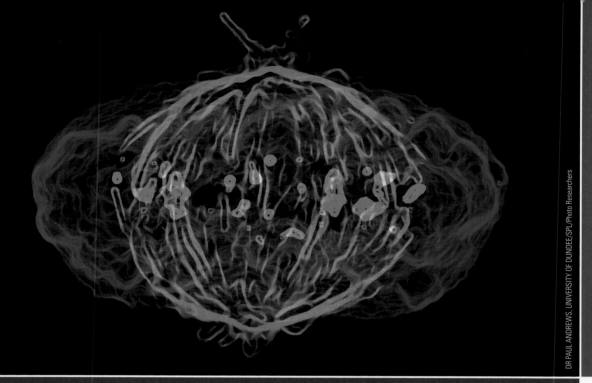

A cell in mitosis (fluorescence micrograph). The spindle (red) is separating copies of the cell's chromosomes (green) prior to cell division.

10

Cell Division and Mitosis

Why it matters . . . As the rainy season recedes in Northern India, rice paddies and other flooded areas begin to dry out. The resulting shallow, seasonal pools have provided an environment of slow-moving warm water for zebrafish *(Danio rerio)* to spawn **(Figure 10.1).** Over the past few months, many millions of cell divisions have changed the single-celled fertilized eggs into the complex multicellular tissues and organs of these small, boldly striped fish. Most cells in the adult fish have stopped dividing and are dedicated to particular functions.

Moving into the fast-running streams that feed the Ganges River, the young zebrafish often encounter larger predators such as knifefish *(Notopterus notopterus)*. Imagine that a knifefish attacks a zebrafish, tearing off part of a fin before its prey escapes. Within a week, the entire zebrafish fin will regenerate—skin, nerves, muscles, bones, and related tissues. The regeneration occurs because cells that were not growing and dividing are suddenly stimulated to grow and divide in a highly regulated way.

As a model organism for vertebrate development, the zebrafish is a popular tool for researchers to identify the stages of regeneration at the molecular level. In the first step of regeneration in injured zebrafish, existing skin cells migrate to close the wound and prevent bleeding. Then, cells just under the new skin transform into "regeneration cells" that form a temporary tissue called a blastema. The blastema cells begin to grow and divide. The large numbers of daughter cells produced are capable of maturing into new nerve, blood vessel, muscle, and bone cells in response to signal proteins that are produced by the skin. Once the regenerated fin has reached its normal size and shape, the new cells stop growing and dividing.

Since multicellular organisms are made almost entirely of cells and their products, understanding organismal development and structure at its most fundamental level by studying the regulation of cell division is necessary. Which conditions stimulate cells to divide? Which make them stop? This chapter focuses on how cell division occurs, and how that process is regulated.

FIGURE 10.1 Zebrafish *(Danio rerio).*

10.1 The Cycle of Cell Growth and Division: An Overview

A prokaryotic cell such as *Escherichia coli* or a single-celled eukaryotic microorganism such as baker's yeast *(Saccharomyces cerevisiae)* grows and divides as long as environmental conditions allow. But a cell inside the brain of a horse or in a yellow rose petal may neither grow nor divide. That is, in multicellular eukaryotes, cell division is under strict control to develop and maintain a mature body consisting of different subpopulations of cells. Most mature cells in multicellular organisms divide infrequently, if at all. However, the tissues of animals, plants, and other multicellular organisms also contain small populations of cells that are always actively dividing. The new cells are required, for instance, for growth (hair, new leaves), and replacement of cells lost to wear (blood cells, cells lining the intestine) and tear (wounds, virus infections).

Before dividing, most cells enter a period of growth in which they synthesize proteins, lipids, and carbohydrates and, during one particular stage, replicate their nuclear DNA. After this growth period, the nuclei divide and, usually, **cytokinesis** *(cyto* = cell; *kinesis* = movement)—the division of the cytoplasm—follows, partitioning nuclei into daughter cells. Each daughter nucleus contains one copy of the replicated parent DNA. This sequence of events—the period of growth followed by nuclear division and cytokinesis—is known as the **cell cycle.** The nuclear division part of the cell cycle is **mitosis.**

The Products of Mitosis Are Genetic Duplicates of the Dividing Cell

Mitosis serves the purpose of partitioning the replicated DNA equally and precisely, generating daughter cells, which are genetic copies of the parent cell. This is accomplished by three interrelated systems:

1. A master program of molecular checks and balances that ensures an orderly and timely progression through the cell cycle

2. Within the overall regulation of the cell cycle, a process of DNA synthesis that replicates each DNA chromosome into two copies with almost perfect fidelity (see Chapter 14)

3. A structural and mechanical web of interwoven "cables" and "motors" of the mitotic cytoskeleton that separates the replicated DNA molecules precisely into the daughter cells

However, at a particular stage of the life cycle of sexually reproducing organisms, a cell division process called *meiosis* produces some cells that are genetically different from the parent cells. **Meiosis** produces daughter nuclei that are different in that they have one half the number of chromosomes the parental nucleus had. Also, the mechanisms involved in producing the daughter nuclei produce arrangements of genes on chromosomes that are different from those in the parent cell (see Chapter 11). The cells that are the products of meiosis may function as gametes in animals (fusing with other gametes to make a zygote) and as spores in plants and many fungi (dividing by mitosis).

This chapter concentrates on the mechanical and regulatory aspects of mitosis in eukaryotes and cell division in prokaryotes; meiosis and its role in eukaryotic sexual reproduction are addressed in Chapter 11. We begin our discussion with **chromosomes,** the nuclear units of genetic information divided and distributed by mitotic cell division.

Chromosomes Are the Genetic Units That Are Partitioned by Mitosis

In all eukaryotes, the hereditary information within the nucleus is distributed among individual, linear DNA molecules. These DNA molecules are bound to proteins that stabilize the DNA molecules, assist in packaging DNA during cell division, and influence the expression of individual genes. In a cell, each *chromosome* *(chroma* = color; *soma* = body; **Figure 10.2**) is composed of one of these linear DNA molecules along with its associated proteins.

Most eukaryotic cells have two copies of each type of chromosome in their nuclei, so their chromosome complement is said to be **diploid,** or $2n$. For example, human body cells have 23 pairs of different chromosomes, for a diploid number of 46 chromosomes ($2n = 46$). The two chromosomes of each pair in a diploid cell are called **homologous chromosomes**—they have the same genes, in the same order in the DNA of the chromosomes. One member of the pair derives from the maternal parent and the other from the paternal parent. Some eukaryotes, mostly fungi, have only one copy of each type of chromosome in their nuclei throughout much of their life cycle, so their chromosome complement is said to be **haploid,** or n. For example, cells of the orange bread mold *Neurospora crassa* are haploid ($n = 7$) throughout much of the organism's life cycle. Baker's yeast *(S. cerevisiae),* a model organism that helped illuminate regulatory aspects of the cell cycle, is an example of an organism that

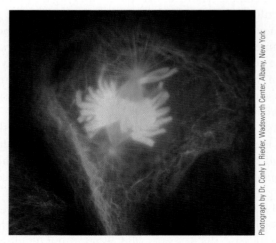

FIGURE 10.2 Eukaryotic chromosomes (blue) in a dividing animal cell. At this stage of division, the chromosomes have duplicated and are highly compacted compared to their extended state between cell divisions. The chromosomes are attached to the spindle (green), a structure formed from microtubules, which serves to partition the two sets of chromosomes to the two daughter cells as cell division proceeds.

Photograph by Dr. Conly L. Rieder, Wadsworth Center, Albany, New York can grow as haploid cells ($n = 16$) or as diploid cells ($2n = 32$). Still others, such as many plant species, have three, four, or even more complete sets of chromosomes in each cell. The number of chromosome sets is called the **ploidy** of a cell or species.

Before a cell divides in mitosis, duplication of each chromosome produces two identical copies of each chromosome called **sister chromatids.** Duplication of a chromosome involves replicating the DNA molecule it contains, plus doubling the proteins that are bound to the DNA to stabilize it. Newly formed sister chromatids are held together tightly by *sister chromatid cohesion,* in which proteins called *cohesins* encircle the sister chromatids along their length. During mitosis, the cohesins are removed and the sister chromatids are separated, with one of each pair going to each of the two daughter nuclei. *As a result of this precise division, each daughter nucleus receives exactly the same number and types of chromosomes, and contains the same genetic information, as the parent cell that entered the division.* The equal distribution of daughter chromosomes into each of the two daughter cells that result from cell division is called **chromosome segregation.**

The accuracy of chromosome replication and segregation in the mitotic cell cycle creates a group of genetically identical cells—**clones** of the original cell. Since all the diverse cell types of a complex multicellular organism arose by mitosis from a single zygote, they all contain the same genetic information. Forensic scientists rely on this feature of organisms when, for instance, they match the genetic profile of a small amount of tissue (for example, DNA from blood left at the scene of a crime) with a DNA sample from a suspect. In the laboratory, cells may be grown in **cell cultures,** living cells grown in laboratory vessels. Many types of prokaryotic and eukaryotic cells can be cultured in a growth medium optimized for the organism.

10.2 The Mitotic Cell Cycle

Growth and division of both diploid and haploid cells occurs in the mitotic cell cycle.

Interphase Extends from the End of One Mitosis to the Beginning of the Next Mitosis

If we set the formation of a new daughter cell as the beginning of the mitotic cell cycle, then the first and longest stage is **interphase (Figure 10.3).** Three phases of the cell cycle comprise interphase:

1. **G_1 phase,** in which the cell carries out its function, and in some cases grows
2. **S phase,** in which DNA replication and chromosome duplication occur
3. **G_2 phase,** a brief gap in the cell cycle during which cell growth continues and the cell prepares for mitosis (the fourth phase of the cell cycle; also called *M phase*) and cytokinesis

Usually, G_1 is the only phase of the cell cycle that varies in length. The other phases are typically uniform in length within a species. Thus, whether cells divide rapidly or slowly depends primarily on the length of G_1. Once DNA replication begins, most mammalian cells take about 10 to 12 hours to proceed through the S phase, about 4 to 6 hours to go through G_2, and about 1 hour or less to complete mitosis.

G_1 is also the stage in which many cell types stop dividing. Cells that are not destined to divide immediately enter a shunt from G_1 called the **G_0 phase.** In some cases, a cell in G_0 may start dividing again by reentering G_1. Some cells never resume the cell cycle; for example, most cells of the human nervous system stop dividing once they are fully mature.

Internal regulatory controls trigger each phase of the cell cycle, ensuring that the processes of one phase are completed successfully before the next phase can begin. Various internal mech-

G_2 refers to the second gap in which there is no DNA synthesis. During G_2, the cell continues to synthesize RNAs and proteins, including those for mitosis, and it continues to grow. The end of G_2 marks the end of interphase; mitosis then begins.

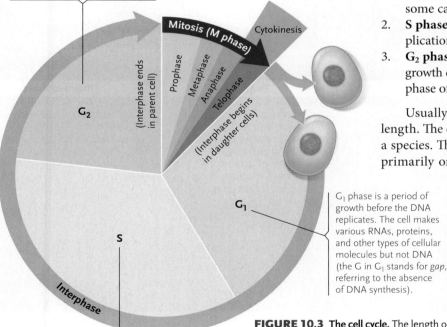

G_1 phase is a period of growth before the DNA replicates. The cell makes various RNAs, proteins, and other types of cellular molecules but not DNA (the G in G_1 stands for *gap,* referring to the absence of DNA synthesis).

If the cell is going to divide, DNA replication begins. During S phase, the cell duplicates each chromosome, including both the DNA and the chromosomal proteins, and it also continues synthesis of other cellular molecules.

FIGURE 10.3 The cell cycle. The length of G_1 varies, but for a given cell type, the timing of S, G_2, and mitosis is usually relatively uniform. Cytokinesis (segment at 2 o'clock) usually begins while mitosis is in progress and reaches completion as mitosis ends. Cells in a state of division arrest enter a shunt from G_1 called G_0 (not shown).
© Cengage Learning 2014

FIGURE 10.4 The stages of mitosis.

Light micrographs show mitosis in an animal cell (whitefish embryo). Diagrams show mitosis in an animal cell with two pairs of chromosomes. As you study these diagrams, consider that each diploid human cell, although only 40 to 50 μm in diameter, contains 2 meters of DNA distributed among 23 pairs of chromosomes.

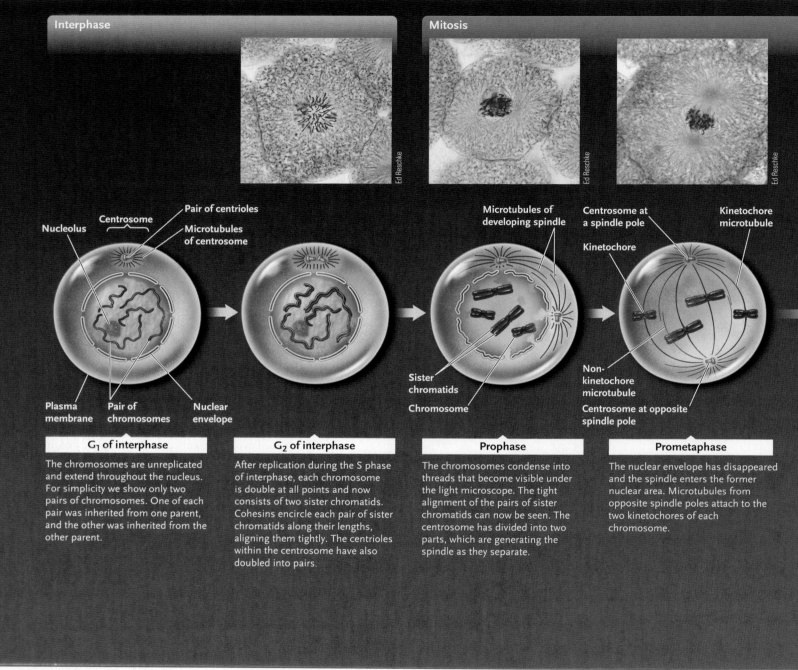

Interphase

Mitosis

Ed Reschke

Ed Reschke

Ed Reschke

Nucleolus
Centrosome
Pair of centrioles
Microtubules of centrosome

Microtubules of developing spindle

Centrosome at a spindle pole
Kinetochore microtubule

Kinetochore

Sister chromatids

Chromosome

Plasma membrane
Pair of chromosomes
Nuclear envelope

Non-kinetochore microtubule

Centrosome at opposite spindle pole

G₁ of interphase

The chromosomes are unreplicated and extend throughout the nucleus. For simplicity we show only two pairs of chromosomes. One of each pair was inherited from one parent, and the other was inherited from the other parent.

G₂ of interphase

After replication during the S phase of interphase, each chromosome is double at all points and now consists of two sister chromatids. Cohesins encircle each pair of sister chromatids along their lengths, aligning them tightly. The centrioles within the centrosome have also doubled into pairs.

Prophase

The chromosomes condense into threads that become visible under the light microscope. The tight alignment of the pairs of sister chromatids can now be seen. The centrosome has divided into two parts, which are generating the spindle as they separate.

Prometaphase

The nuclear envelope has disappeared and the spindle enters the former nuclear area. Microtubules from opposite spindle poles attach to the two kinetochores of each chromosome.

© Cengage Learning 2014

anisms also regulate the overall number of cycles that a cell goes through. These internal controls may be subject to various external influences such as other cells or viruses, as well as signal molecules, including hormones, growth factors, and death signals.

After Interphase, Mitosis Proceeds in Five Stages

Once it begins, mitosis proceeds continuously, without significant pauses or breaks. However, for convenience in study, biologists separate mitosis into five sequential stages: **prophase**

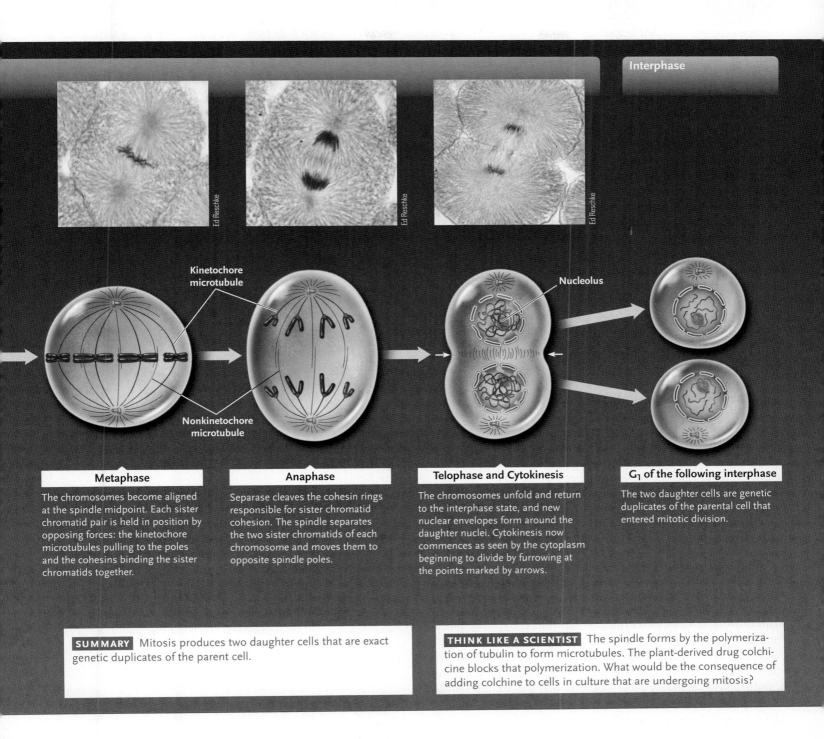

Kinetochore
microtubule

Nucleolus

Nonkinetochore
microtubule

Metaphase

The chromosomes become aligned at the spindle midpoint. Each sister chromatid pair is held in position by opposing forces: the kinetochore microtubules pulling to the poles and the cohesins binding the sister chromatids together.

Anaphase

Separase cleaves the cohesin rings responsible for sister chromatid cohesion. The spindle separates the two sister chromatids of each chromosome and moves them to opposite spindle poles.

Telophase and Cytokinesis

The chromosomes unfold and return to the interphase state, and new nuclear envelopes form around the daughter nuclei. Cytokinesis now commences as seen by the cytoplasm beginning to divide by furrowing at the points marked by arrows.

G_1 of the following interphase

The two daughter cells are genetic duplicates of the parental cell that entered mitotic division.

SUMMARY Mitosis produces two daughter cells that are exact genetic duplicates of the parent cell.

THINK LIKE A SCIENTIST The spindle forms by the polymerization of tubulin to form microtubules. The plant-derived drug colchicine blocks that polymerization. What would be the consequence of adding colchine to cells in culture that are undergoing mitosis?

(*pro* = before), **prometaphase** (*meta* = between), **metaphase, anaphase** (*ana* = back), and **telophase** (*telo* = end). **Figures 10.4** and **10.5** show the process of mitosis in an animal cell and in a plant cell.

PROPHASE During **prophase,** the greatly extended chromosomes that were replicated during interphase begin to *condense* into compact, rodlike structures (see chromatin packaging in Chapter 14). Each diploid human cell, although only about 40 to

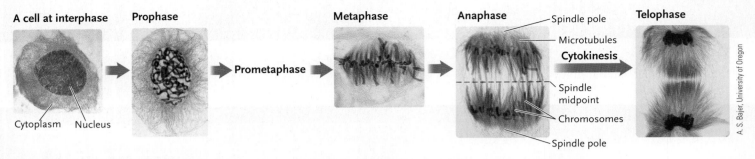

A cell at interphase   **Prophase**   **Metaphase**   **Anaphase**   Spindle pole   **Telophase**

Prometaphase   Microtubules   **Cytokinesis**

Spindle midpoint

Chromosomes

Cytoplasm   Nucleus   Spindle pole

A. S. Bajer, University of Oregon

FIGURE 10.5 Mitosis in a plant cell. The chromosomes are stained blue; the spindle microtubules are stained red.

50 μm in diameter, contains *2 meters* of DNA distributed among 23 pairs of chromosomes. Condensation during prophase packs these long DNA molecules into units small enough to be divided successfully during mitosis. As they condense, the chromosomes appear as thin threads under the light microscope. The word *mitosis* (*mitos* = thread) is derived from this thread-like appearance.

While condensation is in progress, the nucleolus becomes smaller and eventually disappears in most species. The disappearance reflects a substantial reduction in RNA synthesis, including the ribosomal RNA made in the nucleolus.

In the cytoplasm, the mitotic **spindle** (**Figure 10.6;** see also Figure 10.11), the structure that will later separate the chromatids, begins to form between the two centrosomes as they start migrating toward the opposite ends of the cell, where they will form the **spindle poles.** The spindle develops as two bundles of microtubules that radiate from the two spindle poles.

PROMETAPHASE At the end of prophase, the nuclear envelope breaks down, marking the beginning of **prometaphase.** Bundles of spindle microtubules grow from centrosomes at the *opposite spindle poles* toward the center of the cell. Some of the developing spindle enters the former nuclear area.

Each chromosome is still in a duplicated state made of two identical sister chromatids held together throughout their lengths, a phenomenon known as **sister chromatid adhesion.** By this time, a complex of several proteins, a **kinetochore,** has formed on each chromatid at the **centromere,** a region located at a particular position in a given chromosome. The centromere is often narrower than the rest of the chromosome. **Kinetochore microtubules** originating from the spindle poles bind to the kinetochores. These connections determine the outcome of mitosis, because they attach the sister chromatids of each chromosome to microtubules that lead to the opposite spindle poles (see Figure 10.6). Microtubules that do not attach to kinetochores—the **nonkinetochore microtubules**—overlap those from the opposite spindle pole.

METAPHASE During **metaphase,** the spindle reaches its final form and the spindle microtubules move the chromosomes into alignment at the spindle midpoint, also called the *metaphase plate.* The chromosomes complete their condensation in this stage. The pattern of condensation gives each chromosome a

characteristic shape, determined by the location of the centromere and the length of the chromatid arms.

Only when the chromosomes are all assembled at the spindle midpoint, with the two sister chromatids of each one attached to microtubules leading to opposite spindle poles, can metaphase give way to actual separation of chromatids in anaphase.

The length of each metaphase chromosome and the position of its centromere gives it a characteristic appearance. The complete set of metaphase chromosomes, arranged according to size and centromere position, forms the **karyotype** of a given species. In many cases, the karyotype is so distinctive that a species can

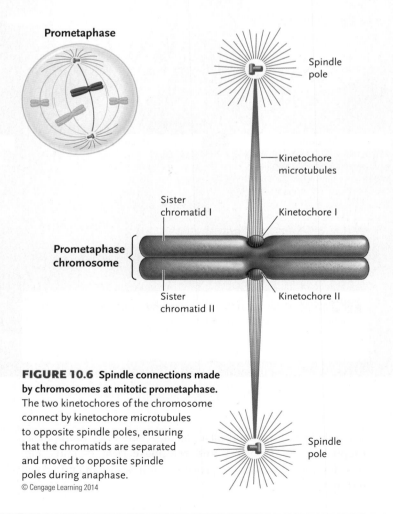

Prometaphase

Spindle pole

Kinetochore microtubules

Sister chromatid I

Kinetochore I

Prometaphase chromosome

Sister chromatid II

Kinetochore II

Spindle pole

FIGURE 10.6 Spindle connections made by chromosomes at mitotic prometaphase. The two kinetochores of the chromosome connect by kinetochore microtubules to opposite spindle poles, ensuring that the chromatids are separated and moved to opposite spindle poles during anaphase.
© Cengage Learning 2014

FIGURE 10.7 | Research Method

Preparing a Human Karyotype

Purpose: A karyotype is a display of chromosomes of an organism arranged in pairs. A normal karyotype has a characteristic appearance for each species. Examination of the karyotype of the chromosomes from a particular individual indicates whether the individual has a normal set of chromosomes or whether there are abnormalities in number or appearance of individual chromosomes. A normal karyotype can be used to indicate the species.

Protocol:

1. Add sample to culture medium that has stimulator for growth and division of cells (white blood cells in the case of blood). Incubate at 37°C. Add colchicine, which blocks the formation of microtubules. As a result, the spindle does not form and this causes mitosis to arrest at metaphase.

2. Stain the cells so that the chromosomes are distinguished. Some stains produce chromosome-specific banding patterns, as shown in the photograph below.

3. View the stained cells under a microscope equipped with a digital imaging system and take a digital photograph. A computer processes the photograph to arrange the chromosomes in pairs and numbers them according to size and shape.

Interpreting the Results: The karyotype is evaluated with respect to the scientific question being asked. For example, it may identify a particular species, or it may indicate whether or not the chromosome set of a human (fetus, child, or adult) is normal or aberrant.

Pair of homologous chromosomes

Pair of sister chromatids closely aligned side-by-side by sister chromatid cohesion

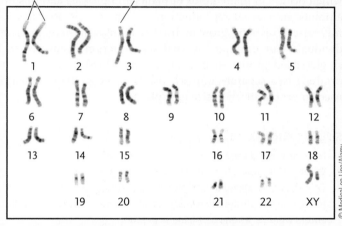

© Medical-on-Line/Alamy

be identified from this characteristic alone. **Figure 10.7** shows how human chromosomes are prepared for analysis as a karyotype.

ANAPHASE The proper alignment of chromosomes in metaphase triggers anaphase. An enzyme, *separase,* is activated and cleaves the cohesin rings around the pairs of sister chromatids. This cancels the force opposing the pull of sister chromatids to opposite poles. Thus, during **anaphase,** the spindle separates sister chromatids and pulls them to opposite spindle poles. The first signs of chromosome movement can be seen at the centromeres, where tension developed by the spindle pulls the kinetochores toward the poles. The movement continues until the separated chromatids, now called *daughter chromosomes,* have reached the two poles. At this point, chromosome segregation has been completed.

TELOPHASE During **telophase,** the spindle disassembles and the chromosomes at each spindle pole decondense, returning to the extended state typical of interphase. As decondensation pro-

ceeds, the nucleolus reappears, RNA transcription resumes, and a new nuclear envelope forms around the chromosomes at each pole producing the two daughter nuclei. At this point, nuclear division is complete.

Mitosis produces two daughter nuclei, each with identical sets of chromosomes compared to the parental cell. Cytokinesis, the division of the cytoplasm, typically follows the nuclear division stage of mitosis, and produces two daughter cells each with one of the two daughter nuclei. Cytokinesis proceeds by different pathways in the various kingdoms of eukaryotic organisms. In animals, protists, and many fungi, a groove, the **furrow,** girdles the cell and gradually deepens until it cuts the cytoplasm into two parts **(Figure 10.8).** In plants, a new cell wall, called the **cell plate,** forms between the daughter nuclei and grows laterally until it divides the cytoplasm in two **(Figure 10.9).** In both cases, the plane of cytoplasmic division is determined by the layer of microtubules that persist at the former spindle midpoint.

FIGURE 10.8 Cytokinesis by furrowing. The micrograph shows a furrow developing in the first division of a fertilized egg cell.
© Cengage Learning 2014

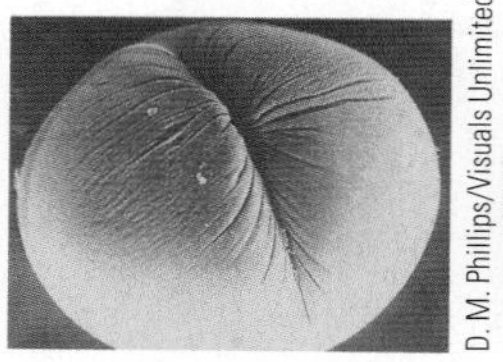

Contractile ring of microfilaments (see Section 5.5) just inside the plasma membrane

1 The microfilaments slide together, tightening the ring and constricting the cell. The constriction forms a groove—the furrow—in the plasma membrane.

2 Continued constriction causes the furrow to deepen gradually, much like the tightening of a drawstring.

3 Furrowing continues until the daughter nuclei are enclosed in separate cells. At the same time, the cytoplasmic division distributes organelles and other structures (which also have increased in number) approximately equally between the cells.

D. M. Phillips/Visuals Unlimited

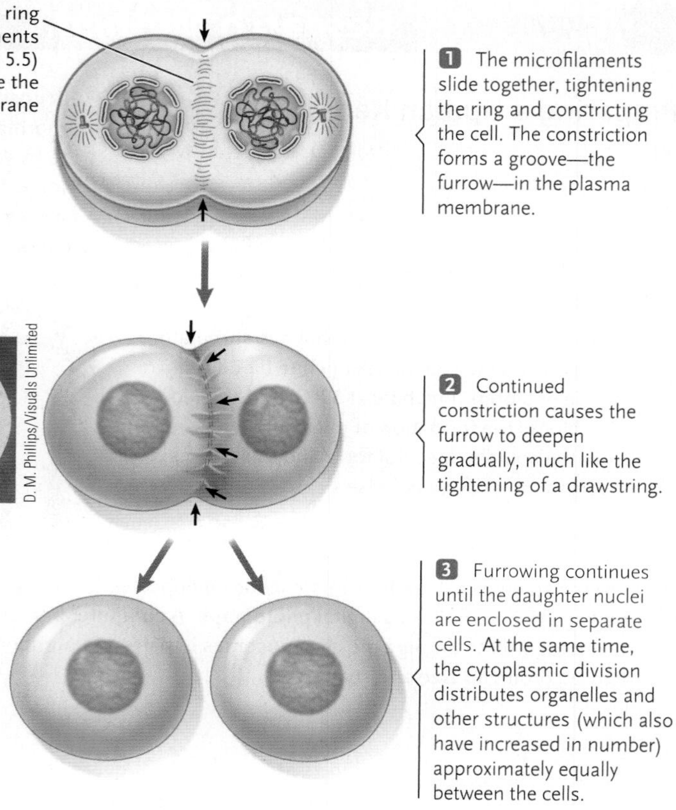

The Mitotic Cell Cycle Is Significant for Both Development and Reproduction

The mitotic cycle of interphase, nuclear division, and cytokinesis accounts for the growth of multicellular eukaryotes from single initial cells, such as a fertilized egg, to fully developed adults. Mitosis also serves as a mechanism of organismal reproduction called **vegetative** or **asexual reproduction,** which occurs in many kinds of plants and protists and in some animals. In asexual reproduction, daughter cells produced by mitotic cell division grow by further mitosis into complete individuals. For example, asexual reproduction occurs when a single-celled protist such as an amoeba divides by mitosis to produce two separate individuals, or when a stem cutting is used to propagate an entire new plant.

STUDY BREAK 10.2

1. In what order do the stages of mitosis occur?
2. What is the importance of centromeres to mitosis?
3. Colchicine, an alkaloid extracted from plants, prevents the formation of spindle microtubules. What would happen if a cell enters mitosis when colchicine is present?

10.3 Formation and Action of the Mitotic Spindle

The mitotic spindle is central to both mitosis and cytokinesis. The spindle is made up of microtubules and their motor proteins, and its activities depend on their changing patterns of organization during the cell cycle.

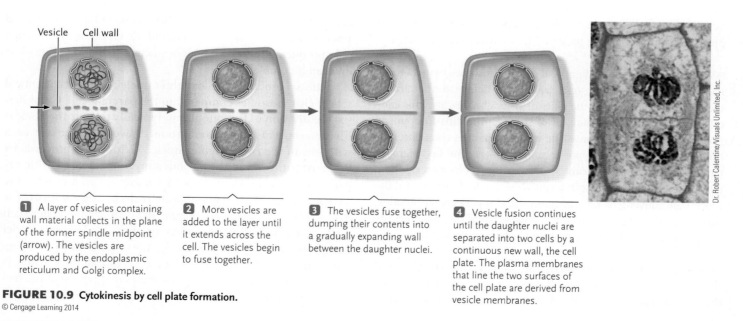

Vesicle   Cell wall

1 A layer of vesicles containing wall material collects in the plane of the former spindle midpoint (arrow). The vesicles are produced by the endoplasmic reticulum and Golgi complex.

2 More vesicles are added to the layer until it extends across the cell. The vesicles begin to fuse together.

3 The vesicles fuse together, dumping their contents into a gradually expanding wall between the daughter nuclei.

4 Vesicle fusion continues until the daughter nuclei are separated into two cells by a continuous new wall, the cell plate. The plasma membranes that line the two surfaces of the cell plate are derived from vesicle membranes.

Dr. Robert Calentine/Visuals Unlimited, Inc.

FIGURE 10.9 Cytokinesis by cell plate formation.
© Cengage Learning 2014

Microtubules form a major part of the interphase cytoskeleton of eukaryotic cells. (Section 5.3 outlines the patterns of microtubule organization in the cytoskeleton.) As mitosis approaches, the microtubules disassemble from their interphase arrangement and reorganize into the spindle, which grows until it fills almost the entire cell. This reorganization follows one of two pathways in different organisms, depending on the presence or absence of a *centrosome* during interphase. However, once organized, the basic function of the spindle is the same, regardless of whether a centrosome is present or not.

Animals and Plants Form Spindles in Different Ways

Figure 10.10 shows the **centrosome,** a site near the nucleus from which microtubules radiate outward in all directions, and its role in spindle formation. The centrosome is the main **microtubule organizing center (MTOC)** of animal cells and many protists. The centrosome contains a pair of **centrioles,** arranged at right angles to each other.

When the nuclear envelope breaks down at the end of prophase, the spindle (see Figure 10.10, step 4) moves into the region

formerly occupied by the nucleus and continues growing until it fills the cytoplasm. The microtubules that extend from the centrosomes also grow in length and extent, producing radiating arrays that appear starlike under the light microscope. Initially named by early microscopists, **asters** (*aster* = star) are the centrosomes at the spindle tips, which form the poles of the spindle. By separating the duplicated centrioles, the spindle ensures that, when the cytoplasm divides during cytokinesis, the daughter cells each receive a pair of centrioles.

Angiosperms (flowering plants) and most gymnosperms, such as conifers, lack centrosomes and centrioles. In these organisms, the spindle forms from microtubules that assemble in all directions from multiple MTOCs surrounding the entire nucleus (see prophase in Figure 10.5). Then, when the nuclear envelope breaks down at the end of prophase, the spindle moves into the former nuclear region.

Mitotic Spindles Move Chromosomes by a Combination of Two Mechanisms

When fully formed at metaphase, the spindle may contain from hundreds to many thousands of microtubules, depending on the species **(Figure 10.11).** As you learned earlier in the chapter, these microtubules are divided into two groups (see Figure 10.4):

1. *Kinetochore microtubules* connect the spindle poles to kinetochores, the complexes of proteins that form at centromeres during prometaphase.
2. *Nonkinetochore microtubules* extend between the spindle poles without connecting to chromosomes; at the spindle midpoint, the microtubules from one pole overlap with microtubules from the opposite pole.

Prophase

FIGURE 10.10 **The centrosome and its role in spindle formation.**
© Cengage Learning 2014

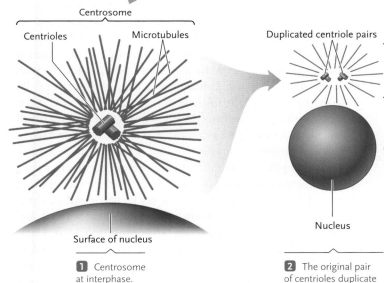

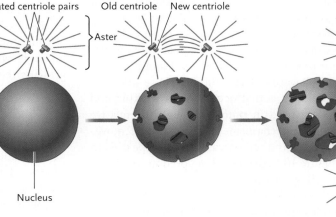

1 Centrosome at interphase.

2 The original pair of centrioles duplicate during the S phase of the cell cycle, producing two pairs of centrioles.

3 As prophase begins, the centrosome separates into two parts, each containing one "old" and one "new" centriole—one centriole of the original pair and its copy.

4 The duplicated centrosomes, containing the centrioles, continue to separate until they reach opposite sides of the nucleus. The microtubules between them lengthen and increase in number. By late prophase, the early spindle is complete, consisting of the separated centrosomes and a large mass of microtubules between them.

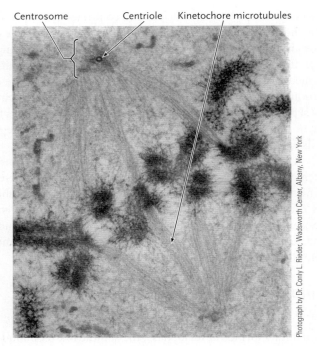

Centrosome   Centriole   Kinetochore microtubules

Photograph by Dr. Conly L. Rieder, Wadsworth Center, Albany, New York

FIGURE 10.11 **A fully developed spindle in a mammalian cell.** Only microtubules connected to chromosomes have been caught in the plane of this section. One of the centrioles is visible in cross section in the centrosome at the top of the micrograph.

Figure 10.12 describes an experiment which demonstrated that chromosomes can move by sliding along kinetochore microtubules toward the poles. Mechanistically this movement involves motor proteins in the kinetochores "walking" the chromosomes along microtubules **(Figure 10.13).** The tubulin subunits of the kinetochore microtubules disassemble as the kinetochores pass along them; as a result, the microtubules become shorter as the movement progresses. The movement is similar to pulling yourself, hand over hand, up a rope as it falls apart behind you.

Chromosomes can also move toward the poles by a mechanism in which motor proteins at the spindle poles pull kinetochore microtubules polewards, disassembling those microtubules into tubulin subunits as that occurs. Both walking and pulling mechanisms are used in mitosis, though the relative contributions of the two mechanisms to chromosome movement varies among species and cell types. (The cell type used in the experiment of Figure 10.12 used walking predominantly.)

As a dividing cell goes through anaphase, it elongates (compare Metaphase and Anaphase in Figure 10.4). Nonkinetochore microtubules are responsible for this cell elongation in two ways:

1. Motor proteins on overlapping nonkinetochore microtubules walk in opposite directions, thereby reducing the extent of overlap at the spindle midpoint. This activity pushes the spindle poles further apart.
2. The nonkinetochore microtubules also push the poles apart by growing in length as they slide along. This activity

also maintains the overlap of pairs of nonkinetochore microtubules at the spindle midpoint even as the motor proteins work to reduce the overlap.

STUDY BREAK 10.3

1. How does spindle formation differ in animals and plants?
2. How do mitotic spindles move chromosomes?

10.4 Cell Cycle Regulation

In this section we discuss experimental evidence for (and the operation of) regulatory mechanisms that control the mitotic cell cycle.

Cell Fusion Experiments and Studies of Yeast Mutants Identified Molecules that Control the Cell Cycle

The first insights into how the cell cycle is regulated came from experiments by Robert T. Johnson and Potu N. Rao at the University of Colorado Medical Center, Denver, published in 1970. They fused human HeLa cells (a type of cancer cell that can be grown in cell culture) that were in different stages of the cell cycle and determined whether one nucleus could influence the other **(Figure 10.14).** Their results suggested that specific molecules in the cytoplasm cause the progression of cells from G_1 to S, and from G_2 into M.

Some key research using baker's yeast, *S. cerevisiae,* helped to identify these cell cycle control molecules and contributed to our general understanding of how the cell cycle is regulated. (*Focus on Model Research Organisms* describes yeast and its role in research in more detail.) In particular, Leland Hartwell of the Fred Hutchinson Cancer Center, Seattle, investigated yeast mutants that become stuck at some point in the cell cycle, but only when they are cultured at a high temperature. By growing the mutant cells initially at the standard temperature, and then shifting the cells to the higher temperature, Hartwell was able to use time-lapse photomicroscopy to see if and when growth and division were affected. In this way he isolated many *cell division cycle,* or *cdc* mutants. By examining the mutants, he could identify the stage in the cell cycle where each mutant type was blocked by noting whether nuclei had divided, chromosomes had condensed, the mitotic spindle had formed, cytokinesis had occurred, and so on. Using this approach, Hartwell identified many genes that code for proteins involved in yeast's cell cycle and hypothesized where in the cycle these proteins operated. As might be expected, some of the proteins were involved in DNA replication, but a number of others were shown to function in cell cycle regulation. Hartwell received a Nobel Prize in 2001 for his discovery.

Paul Nurse of the Imperial Cancer Research Fund, London, carried out similar research with the fission yeast,

FIGURE 10.12 | **Experimental Research**

Movement of Chromosomes during Anaphase of Mitosis

Question: How do chromosomes move during anaphase of mitosis?

Experiment: One hypothesis for how chromosomes move during anaphase of mitosis was that the kinetochore microtubules moved, pulling chromosomes to the poles. An alternative hypothesis was that chromosomes move by sliding over or along kinetochore microtubules. To test the hypotheses, G. J. Gorbsky and his colleagues made regions of the kinetochore microtubules visibly distinct.

1. Kinetochore microtubules were combined with a dye molecule that bleaches when it is exposed to light.

2. The region of the spindle between the kinetochores and the poles was exposed to a microscopic beam of light that bleached a narrow stripe across the microtubules. The bleached region could be seen with a light microscope and analyzed as anaphase proceeded.

Results: The bleached region remained at the same distance from the pole as the chromosomes moved toward the pole.

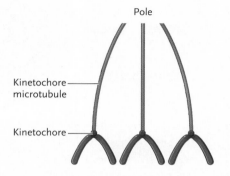

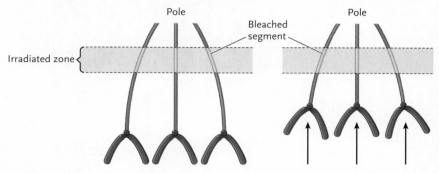

Conclusion: The results support the hypothesis that chromosomes move by sliding over or along kinetochore microtubules.

THINK LIKE A SCIENTIST What result would have supported the hypothesis that, during anaphase of mitosis, the kinetochore microtubules moved, pulling chromosomes to the poles?

Source: G. J. Gorbsky, P. J. Sammak, and G. G. Borisy. 1987. Chromosomes moved poleward in anaphase along stationary microtubules that coordinately disassemble from their kinetochore microtubules. *Journal of Cell Biology* 104:9–18.

Schizosaccharomyces pombe, a species that divides by fission rather than budding. He identified a gene called *cdc2* that encodes a protein needed for the cell to progress from G_2 to M. Nurse also made the breakthrough discovery that all eukaryotic cells studied have counterparts of the yeast *cdc2* gene, implying that this gene originated early during eukaryotic evolution and has played an essential role in cell cycle regulation in all eukaryotes since that time. The protein product of *cdc2* is a protein kinase, an enzyme that catalyzes the phosphorylation of a target protein. (Recall from Section 7.2 that phosphorylation of proteins by protein kinases can activate or inactivate proteins.) That discovery was pivotal in determining how cell cycle regulation occurs. Paul Nurse received a Nobel Prize in 2001 for his discovery.

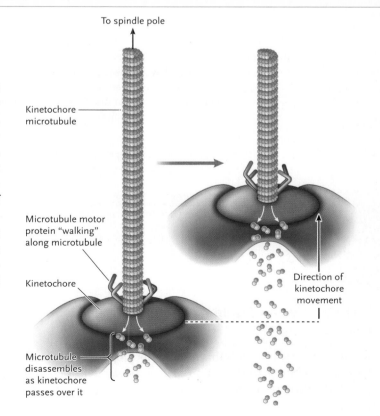

FIGURE 10.13 Microtubule motor proteins "walking" the kinetochore of a chromosome along a microtubule.

FIGURE 10.14 **Experimental Research**

Demonstrating the Existence of Molecules Controlling the Cell Cycle by Cell Fusion

Question: Do molecules in the cytoplasm direct the progression through the cell cycle?

Experiment: Johnson and Rao fused human HeLa cells at different stages of the cell cycle. Cell fusion produces a single cell with two separate nuclei. The researchers allowed the fused cells to grow and determined whether one nucleus influenced the other in terms of progression through the cell cycle.

1. Fusion of cell in S phase with cell in G_1 phase.   **2.** Fusion of cell in M (mitosis) with cell in any other stage.

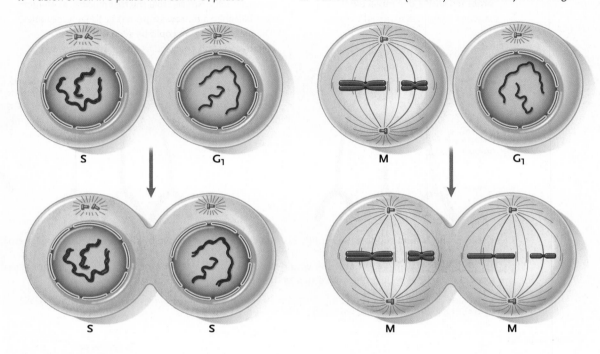

Result: DNA synthesis quickly began in the original G_1 nucleus. Normally, the G_1 nucleus would not have initiated DNA synthesis until it reached S phase itself, which could have been several hours later. The result suggested that one or more molecules that activate S phase are present in the cytoplasm of S phase cells.

Result: Regardless of the phase of the cell, the nucleus of the cell with which the M phase cell was fused immediately began the early stages of mitosis. This included condensation of the chromosomes, spindle formation, and breaking down of the nuclear envelope. For a cell in G_1 (shown in the diagram), the condensed chromosomes that appear have not replicated.

Conclusion: Taken together, the results showed that specific molecules in the cytoplasm direct the progression of cells from G_1 to S, and from G_2 to M in the cell cycle. Those molecules can move between the cytoplasm and the nucleus.

THINK LIKE A SCIENTIST Johnson and Rao also fused cells from other stages of the cell cycle. They observed that fused S and G_2 cells entered M phase earlier than did two fused S cells. What does this result indicate?

Source: R. T. Johnson and P. N. Rao. 1970. Mammalian cell fusion: induction of premature chromosome condensation in interphase nuclei. *Nature* 226:717–722.

The Cell Cycle Can Be Arrested at Specific Checkpoints

A cell has internal controls that monitor its progression through the cell cycle through the action of a particular set of control proteins called *cyclins.* As part of the internal controls, the cell cycle has three key **checkpoints** to prevent critical phases from beginning until the previous phases are completed correctly **(Figure 10.15):**

1. The G_1/S *checkpoint* is the main point in the cell cycle at which the mechanisms governing the cell cycle determine

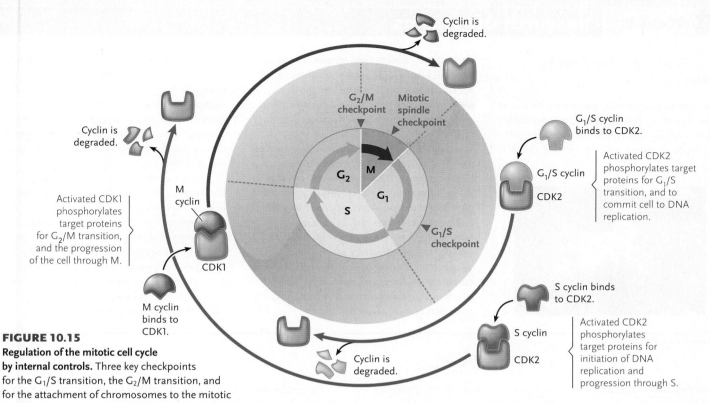

FIGURE 10.15

Regulation of the mitotic cell cycle by internal controls. Three key checkpoints for the G₁/S transition, the G₂/M transition, and for the attachment of chromosomes to the mitotic spindle monitor cell cycle events to prevent crucial phases of the cell cycle from starting until previous phases are completed correctly. Complexes of cyclins and cyclin-dependent kinases (Cdks) regulate the progression of the cell through the cell cycle. The three cyclin–Cdks present in all eukaryotes are shown. The Cdks are present throughout the cell cycle, but they are active only when complexed with a cyclin (shown by the broad arrows in the figure). Each cyclin is synthesized and degraded in a regulated way so that it is present only for a particular phase of the cell cycle. During that phase, the Cdk to which it is bound phosphorylates and, thereby, regulates the activity of target proteins in the cell that are involved in initiating or regulating key events of the cell cycle.

© Cengage Learning 2014

whether the cell will proceed through the rest of the cell cycle and divide. Once it passes this checkpoint, the cell is committed to continue the cell cycle through to cell division in M. The cell cycle arrests (the cell stops proceeding through the cell cycle) at the G₁/S checkpoint, for example, if the DNA is damaged by radiation or chemicals. The G₁/S checkpoint is also the primary point at which cells "read" extracellular signals for cell growth and division. Therefore, if a growth factor required for stimulating cell growth is absent, the cells are arrested at this checkpoint. (Extracellular signals and their effects on the cell cycle are discussed in more detail later.)

2. The *G₂/M checkpoint* is at the junction between the G₂ and M phases. Passage through this checkpoint commits a cell to mitosis. Cells are arrested at the G₂/M checkpoint if DNA was not replicated fully in S, or if the DNA has been damaged by radiation or chemicals. Complete DNA replication is essential for producing genetically identical daughter cells, highlighting the importance of this checkpoint.

3. The *mitotic spindle checkpoint* is within the M phase before metaphase. This checkpoint assesses whether chromosomes are attached properly to the mitotic spindle so that they are aligned correctly at the metaphase

plate. The checkpoint is essential for production of genetically identical daughter cells, which depends on separation of daughter chromosomes in anaphase, which, in turn, depends on the correct alignment of the chromosomes on the spindle in metaphase. Once the cell begins anaphase, it is irreversibly committed to completing M, underlining the importance of the mitotic spindle checkpoint.

The control systems that operate at the checkpoints are signals to stop; basically, they are brakes. This becomes evident when a checkpoint is inactivated by mutation or chemical treatment. The consequence of inactivation of the checkpoints is that the cell cycle proceeds, even if DNA is damaged, DNA replication is incomplete, or the spindle did not assemble completely.

Cyclins and Cyclin-Dependent Kinases Are the Internal Controls That Directly Regulate Cell Division

The internal control system that acts at checkpoints causes cells to arrest in the cycle should proceeding through the cycle be harmful. The direct regulation of the cell cycle itself

The Yeast *Saccharomyces cerevisiae*

Saccharomyces cerevisiae, commonly known as baker's yeast or brewer's yeast, was probably the first microorganism to have been grown and kept in cultures—a beer-brewing vessel is

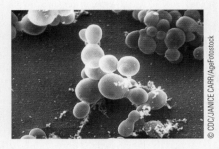

© CDC/JANICE CARR/AgeFotostock

basically a *Saccharomyces* culture. The yeast has also been widely used in scientific research; its microscopic size and relatively short generation time make it easy and inexpensive to culture in large numbers in the laboratory.

Genetic studies with *Saccharomyces* led to the discovery of some of the genes that control the eukaryotic cell cycle. Many of these genes, after their first discovery in yeast cells, were found to have counterparts in animals and plants, evidence of a very ancient evolutionary origin for these genes. Mutated versions of the genes often contribute to the development of cancer in mammals. The complete DNA sequence of *S. cerevisiae,* which includes more than 12 million base pairs that encode about 6,000 genes, was the

first eukaryotic genome to be obtained. Analysis of the genome sequence revealed that yeast has many genes similar in sequence to counterpart genes in animals, including mammals, making this relatively simple microorganism an excellent subject for research that can be applied to the more complex species of interest, especially humans.

The genes of yeast can also be manipulated easily using genetic engineering techniques. This has made it possible for researchers to alter essentially any of the yeast genes experimentally to test their functions and to introduce genes or DNA segments from other organisms for testing or cloning. *Saccharomyces* has been so important to genetic studies in eukaryotes that it is often called the eukaryotic *E. coli.*

involves an internal control system consisting of proteins called **cyclins,** and enzymes called **cyclin-dependent kinases (Cdks)** (see Figure 10.15). A Cdk is a *protein kinase,* which phosphorylates and thereby regulates the activity of target proteins. Cdk enzymes are "cyclin-dependent" because they are active *only* when bound to a cyclin molecule. Cyclins are named because their concentrations change as the cell cycle progresses. R. Timothy Hunt, Imperial Cancer Research Fund, London, UK, received a Nobel Prize in 2001 for discovering cyclins. The basic control of the cell cycle by Cdks and cyclins is the same in all eukaryotes, but there are differences in the number and types of the molecules. We will focus on cell cycle regulation in vertebrates to explain how these proteins work.

The concentrations of the various Cdks remain constant throughout the cell cycle, while the concentrations of cyclins change as they are synthesized and degraded at specific stages of the cell cycle. Thus, a specific Cdk becomes active when the cell synthesizes the cyclin that binds to it and remains active until the cyclin is degraded. Each active Cdk phosphorylates particular target proteins, which play roles in initiating or regulating key events of the cell cycle. The phosphorylation regulates the activities of those proteins, and keeps the cycle operating in an orderly way. Those key events are DNA replication, mitosis, and cytokinesis. A succession of cyclin–Cdk complexes, each of which has specific regulatory effects, ensures that these stages follow in sequence somewhat like a clock passing through the sequence of hours. Regulation of the activity of cyclin–Cdk complexes is integrated with the regulatory events at the cell cycle checkpoints to ensure that daughter cells with damaged DNA or abnormal amounts of DNA are not generated.

Three classes of cyclins, each named for the stage of the cell cycle at which they bind and activate Cdks, operate in all eukaryotes (see Figure 10.15):

1. G_1/S cyclin binds to Cdk2 near the end of G_1 forming a complex required for the cell to make the transition from G_1 to S, and to commit the cell to DNA replication.
2. S cyclin binds to Cdk2 in the S phase forming a complex required for the initiation of DNA replication and the progression of the cell through S.
3. M cyclin binds to Cdk1 in G_2 forming a complex required for the transition from G_2 and M, and the progression of the cell through mitosis.

In most cells, a fourth class of cyclins, G_1 cyclin, binds to two additional Cdks, Cdk4 and Cdk6, before step 1 (the G_1/S transition) to form two cyclin–Cdk complexes. These complexes are needed to move the cell through the G_1 checkpoint stimulating it then to proceed from G_1 to S.

The M cyclin–Cdk1 complex is also called **M phase-promoting factor (MPF).** In addition to initiating mitosis, the M cyclin–Cdk1 complex (MPF) also orchestrates some of its key events. When all chromosomes are correctly attached to the mitotic spindle near the end of metaphase, the M cyclin–Cdk1 complex activates another enzyme complex, the **anaphase-promoting complex (APC).** Activated APC degrades an inhibitor of anaphase, and this leads to the separation of sister chromatids and the onset of daughter chromosome separation in anaphase. Later in anaphase, APC directs the degradation of the M cyclin, causing Cdk1 to lose its activity. The loss of Cdk1 activity then allows the separated chromosomes to become extended again, the nuclear envelope to reform around the two clusters of daughter chromosomes in telophase, and the cytoplasm then to divide in cytokinesis.

External Controls Coordinate the Mitotic Cell Cycle of Individual Cells with the Overall Activities of the Organism

The internal controls that regulate the cell cycle are modified by signal molecules that originate from outside the dividing cells. In animals, these signal molecules include the peptide hormones and similar proteins called *growth factors*.

The hormones and growth factors act on the cell by the reception–transduction–response pattern that applies to cell communication in general (see Chapter 7). The external factors bind to receptors at the cell surface, which respond by triggering reactions inside the cell. These reactions often include steps that add inhibiting or stimulating phosphate groups to the cyclin–Cdk complexes, particularly to the Cdks. The reactions triggered by the activated receptor may also directly affect the same proteins regulated by the cyclin–Cdk complexes. The overall effect is to speed, slow, or stop the progress of cell division, depending on the particular hormone or growth factor and the internal pathway that is stimulated.

Cell-surface receptors in animals also recognize contact with other cells or with molecules of the extracellular matrix (see Section 5.5). The contact triggers internal reaction pathways that inhibit division by arresting the cell cycle, usually in the G_1 phase. The response, called **contact inhibition,** stabilizes cell growth in fully developed organs and tissues. As long as the cells of most tissues are in contact with one another or the extracellular matrix, they are shunted into the G_0 phase and prevented from dividing. If the contacts are broken, the freed cells often enter rounds of division.

Contact inhibition is easily observed in cultured mammalian cells grown on a glass or plastic surface. In such cultures, division proceeds until all the cells are in contact with their neighbors in a continuous, unbroken, single layer. At this point, division stops. If a researcher then scrapes some of the cells from the surface, cells at the edges of the "wound" are released from inhibition and divide until they form a continuous layer and all the cells are again in contact with their neighbors.

Cell Cycle Controls Are Lost in Cancer

Cancer occurs when cells lose the normal controls that determine when and how often they will divide. Cancer cells divide continuously and uncontrollably, producing a rapidly growing mass called a *tumor*. Cancer cells also typically lose their adhesions to other cells and often become actively mobile. As a result, in a process called *metastasis,* they break loose from an original tumor, spread throughout the body, and grow into new tumors in other body regions. Metastasis is promoted by changes that block contact inhibition and alter the cell-surface molecules that link cells together or to the extracellular matrix.

Growing tumors damage surrounding normal tissues by compressing them and interfering with blood supply and nerve function. Tumors may also break through barriers such as the outer skin, internal cell layers, or the gut wall. The breakthroughs cause bleeding, open the body to infection by microorganisms, and destroy the separation of body compartments necessary for normal functioning. Both compression and breakthroughs can cause pain that, in advanced cases, may become extreme. As tumors increase in mass, the actively growing and dividing cancer cells may deprive normal cells of their required nutrients, leading to generally impaired body functions, muscular weakness, fatigue, and weight loss.

Cancer cells typically have a number of mutated genes of different types. The altered functions of those mutated genes in some way promote uncontrolled cell division or metastasis. In their normal (non-mutated) form, many of these genes code for components of the cyclin/Cdk system that regulates cell division; others encode proteins that regulate gene expression, form cell surface receptors, or make up elements of the systems controlled by the receptors. The mutated form of the genes, called **oncogenes** (*oncos* = bulk or mass), encode altered versions of these products.

For example, a mutation in a gene that codes for a surface receptor might result in a protein that is constantly active, even without binding an extracellular signal molecule. As a result, the internal reaction pathways triggered by the receptor, which induce cell division, are continually stimulated. Another mutation, this time in a cyclin gene, could decrease the cyclin–Cdk binding that triggers DNA replication and the rest of the cell cycle. (*Insights from the Molecular Revolution* describes an experiment testing the effects of a viral system that induces cancer by overriding normal controls of the cyclin/CDK system.) Cancer, oncogenes, and the alterations that convert normal genes to oncogenes are discussed in further detail in Chapter 16.

The overview of the mitotic cell cycle and its regulation presented in this chapter only hints at the complexity of cell growth and division. The likelihood of any given cell dividing is determined by the interplay of various internal signals in the context of external cues from the environment. If a cell is destined to divide, then the problem of accurately replicating and partitioning its DNA requires a highly regulated, intricately interrelated series of mechanisms. It is a challenging operation even for male Australian Jack Jumper ants (*Myrmecia pilosula*), which have only two chromosomes ($2n = 2$); and think of the challenges faced by the fern species, *Ophioglossum pycnostichum*, which has 1,260 chromosomes in each cell.

STUDY BREAK 10.4

1. Why is a Cdk not active throughout the entire cell cycle?
2. How do cyclin–Cdk complexes typically trigger transitions in the cell cycle?
3. What is an oncogene? How might an oncogene affect the cell cycle?
4. What is metastasis?

Herpesviruses and Uncontrolled Cell Division: How does herpesvirus 8 transform normal cells into cancer cells?

Almost all of us harbor one or more herpesviruses as more or less permanent residents in our cells. Fortunately, most of the herpesviruses are relatively benign—one group is responsible for the bothersome but nonlethal oral and genital ulcers known commonly as cold sores or "herpes." But another virus, *herpesvirus 8* **(Figure),** is a DNA tumor virus that causes two kinds of cancer: Kaposi's sar-

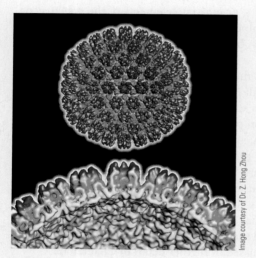

FIGURE Human herpesvirus 8.

Image courtesy of Dr. Z. Hong Zhou

coma and lymphomas of the body cavity. Kaposi's sarcoma is one of the most common cancers that develops in AIDS patients, and is the fourth most common cancer caused by infection worldwide.

Cancers are characterized by uncontrolled cell division. In normal cells, G_1 cyclin combined with either Cdk4 or Cdk6 contributes to the G_1/S transition and thus stimulates cell division. One way that normal cells control cell division is to use regulatory proteins that bind to and inhibit G_1 cyclin–Cdk complexes, thereby arresting cells in G_1 and preventing them from becoming transformed into cancer cells. The proteins that have this regulatory ability are called *tumor suppressor proteins.*

Research Question

How does herpesvirus 8 cause the uncontrolled cell division characteristic of malignant tumors?

Experiments

To answer this question, investigators in London and at the Friedrich-Alexander University in Germany examined the effects of herpesvirus 8 on the primary transition point that leads to cell division, the change from G_1 to S. The investigators focused on how the virus might interfere with regulatory mechanisms that control the rate of cell division.

The investigators knew that the DNA of herpesvirus 8 encodes a protein that acts as a cyclin. Could this viral cyclin, *K cyclin,* be the means by which the herpesvirus bypasses normal controls and triggers the rapid cell di-

vision characteristic of cancer? To answer this question, researchers first inserted the DNA coding for K cyclin into a benign virus. When they infected cultured human cells with this virus, the virus produced K cyclin, which bound to the human Cdk6. These K cyclin–Cdk6 complexes stimulated the initiation of the S phase much faster than do the cell's normal G_1 cyclin–Cdk6 complexes. In addition, the tumor suppressor proteins that normally regulate cell division by binding to G_1 cyclin–Cdk6 complexes were unable to bind to the K cyclin–Cdk6 complexes, resulting in uncontrolled division.

Conclusion

Herpesvirus 8 has evolved a mechanism that overrides normal cellular controls and triggers cell division. At some point in its evolution, the virus may have picked up a copy of a cyclin gene, which through mutation and selection became the K cyclin that is unaffected by the inhibitors.

THINK LIKE A SCIENTIST The researchers also created a cell line in which a K cyclin gene could be turned on specifically upon the addition of an inducer molecule. They treated these cells so that they entered G_0. When the inducer was added to these cells, a significant proportion of the cells subsequently entered the S phase. What does this result indicate?

Source: C. Swanton et al. Herpes viral cyclin/Cdk6 complexes evade inhibition by CDK inhibitor proteins. 1997. *Nature* 390:184–187.

THINK OUTSIDE THE BOOK

Collaboratively or on your own, summarize an experiment demonstrating that higher-than-normal levels of cyclin E (a G_1/S cyclin) can initiate breast cancer in humans.

10.5 Cell Division in Prokaryotes

Prokaryotes undergo a cycle of cytoplasmic growth, DNA replication, and cell division, producing two daughter cells from an original parent cell. The entire mechanism of prokaryotic cell division is called **binary fission**—that is, splitting or dividing into two parts.

Replication Occupies Most of the Cell Cycle in Rapidly Dividing Prokaryotic Cells

All prokaryotes use DNA as their genetic material. The vast majority of prokaryotes have a single, circular DNA molecule known as the **prokaryotic chromosome,** more specifically the **bacterial chromosome** for bacteria, and the **archaeal chromosome** for archaeans. When prokaryotic cells divide at the maximum rate, DNA replication occupies most of the period between cytoplasmic divisions. As soon as replication is complete, the cytoplasm divides to complete the cell cycle. For example, in *E. coli* cells, which are capable of dividing every 20 minutes, DNA replication occupies 19 minutes of the entire 20-minute division cycle.

Replicated Chromosomes Are Distributed Actively to the Halves of the Prokaryotic Cell

In the 1960s, François Jacob of The Pasteur Institute, Paris, France, proposed a model for the segregation of bacterial chromosomes to the daughter cells in which the two chromosomes attach to the plasma membrane near the middle of the cell and separate as a new plasma membrane is added between the two sites during cell elongation. The essence of this model is that chromosome separation is passive. However, current research indicates that bacterial chromosomes rapidly separate in an active way that is linked to DNA replication events and that is independent of cell elongation. The new model is shown in **Figure 10.16**.

Mitosis Evolved from Binary Fission

The prokaryotic mechanism works effectively because most prokaryotic cells have only a single chromosome. Thus, if a daughter cell receives at least one copy of the chromosome, its genetic information is complete. By contrast, in most cases the genetic information of eukaryotes is distributed among several chromosomes, with each chromosome containing a much greater length of DNA than a bacterial chromosome. If a daughter cell fails to receive a copy of even one chromosome, the effects may be lethal. The evolution of mitosis solved the mechanical problems associated with distributing long DNA molecules without breakage. Mitosis provided the level of precision required to ensure that each daughter cell receives a complete complement of the chromosomes the parent cell had.

Scientists believe that the ancestral division process was binary fission and that mitosis evolved from that process. Variations in the mitotic apparatus in modern-day organisms illuminate possible intermediates in this evolutionary pathway. For example, in many primitive eukaryotes, such as dinoflagellates (a type of single-celled protist [see Chapter 27]), the nuclear envelope remains intact during mitosis, and the chromosomes bind to the inner membrane of the nuclear membrane. When the nucleus divides, the chromosomes are segregated.

A more advanced form of the mitotic apparatus is seen in yeasts and diatoms (diatoms are another type of single-celled protist). In these organisms, the mitotic spindle forms and chromosomes segregate to daughter nuclei without the disassembly and reassembly of the nuclear envelope. Currently, scientists think that the types of mitosis seen in yeasts and diatoms, as well as in animals and higher plants, evolved separately from a common ancestral type.

Mitotic cell division, the subject of this chapter, produces two cells that have the same genetic information as the parental cell entering division. In the next chapter, you will learn about

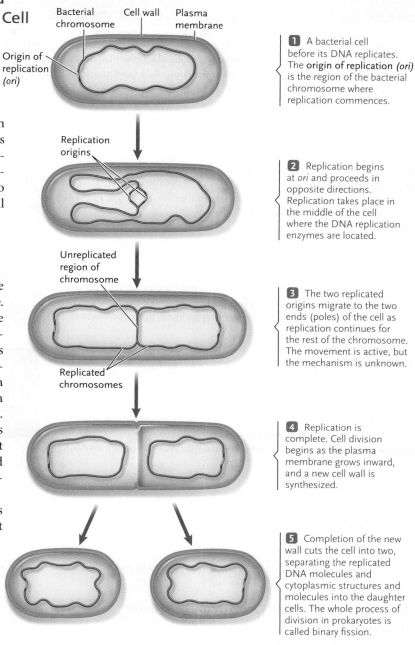

1 A bacterial cell before its DNA replicates. The **origin of replication** *(ori)* is the region of the bacterial chromosome where replication commences.

2 Replication begins at *ori* and proceeds in opposite directions. Replication takes place in the middle of the cell where the DNA replication enzymes are located.

3 The two replicated origins migrate to the two ends (poles) of the cell as replication continues for the rest of the chromosome. The movement is active, but the mechanism is unknown.

4 Replication is complete. Cell division begins as the plasma membrane grows inward, and a new cell wall is synthesized.

5 Completion of the new wall cuts the cell into two, separating the replicated DNA molecules and cytoplasmic structures and molecules into the daughter cells. The whole process of division in prokaryotes is called binary fission.

FIGURE 10.16 Model for the segregation of replicated bacterial chromosomes to daughter cells.
© Cengage Learning 2014

meiosis, a specialized form of cell division that produces gametes, which have half the number of chromosomes as that present in diploid cells.

STUDY BREAK 10.5

1. How do prokaryotes divide?
2. What processes involved in eukaryotic cell division are absent from prokaryotic cell division?

Disrupted or defective control of cell growth and division can lead to diseases such as cancer. Complex, interacting molecular networks within the cell fine-tune the division of each cell in both unicellular and multicellular organisms. Identifying the genes and proteins involved in these networks is crucial both for a complete understanding of cell growth and division, and for developing models for diseases caused by cell cycle defects. Many researchers worldwide are working in this area of research.

How are transitions between phases of the cell cycle regulated?

Research in many labs has shown that transitions are important control points for progression through the cell cycle. If a cell in G_1 phase has damaged DNA, for instance, the cell pauses to repair the DNA before entering S phase, to ensure that any mutations are not passed on to progeny cells. My lab is studying how cells execute two transitions in the cell division program: the transition from G_1 phase to S phase and the transition from mitosis to G_1 phase. Most of our studies are focused on the budding yeast *Saccharomyces cerevisiae* because this organism is easy to manipulate and uses many of the same proteins as human cells do to carry out cell division.

The G_1-to-S and mitosis-to-G_1 transitions involve turnover—breakdown—of proteins that serve to maintain cells in the pre-transition state. I focus here on the G_1-to-S transition. In the G_1 phase, the cyclin-dependent kinase (CDK) inhibitor Sic1 blocks the activity of the S phase cyclin–CDK complexes that promote DNA replication, and thereby delays the onset of S phase. Enzymes rapidly eliminate Sic1 at the end of the G_1 phase, thereby unmasking the activity of the S phase cyclin–CDK, which causes the rapid initiation of chromosome duplication. Although we now know much about how this transition works, a key unanswered question that is of interest to my laboratory is: how do the enzymes that eliminate Sic1 do their job? For example, we are trying to unravel the precise mechanism of action of the ubiquitin ligase enzyme that attaches ubiquitin to Sic1, which serves as a signal that activates Sic1 turnover. (Ubiquitin is a small protein added to proteins to designate those proteins for destruction; see Chapter 16. Ubiquitination of a particular protein is catalyzed by one of a family of ubiquitin ligases in the cell.) Another key question is how the timing of Sic1 turnover is controlled.

From our studies on both the G_1-to-S and mitosis-to-G_1 transitions, it is apparent that many of the key players have already been discovered. However, we still do not know how all of the proteins are organized and communicate with each other to bring about the transition in the cell division program at the right time, and we also do not know how the proteins operate as nano-machines to carry out a specific biological task.

THINK LIKE A SCIENTIST Why would researchers choose *Saccharomyces cerevisiae* as a model organism to address mitotic cell cycle defects over *Escherichia coli* or mammalian cells? How similar would you expect *Saccharomyces* ubiquitin to be to human ubiquitin?

Raymond Deshaies is professor of biology at CalTech and an investigator of Howard Hughes Medical Institute. The focus of his lab is investigation of the cellular machinery that mediates protein degradation by the ubiquitin-proteasome system, and how this machinery regulates cell division. Learn more about his work at http://www.its.caltech.edu/~rjdlab.

REVIEW KEY CONCEPTS

To access the course materials and companion resources for this text, please visit www.cengagebrain.com.

10.1 The Cycle of Cell Growth and Division: An Overview

- In mitotic cell division, DNA replication is followed by the equal separation—that is, segregation—of the replicated DNA molecules and their delivery to daughter cells. The process ensures that the two cell products of a division end up with the same genetic information as the parent cell entering division.

- Mitosis is the basis for growth and maintenance of body mass in multicelled eukaryotes, and for the reproduction of many single-celled eukaryotes.

- The DNA of eukaryotic cells is divided among individual, linear chromosomes located in the cell nucleus.

- DNA replication and duplication of chromosomal proteins converts each chromosome into two exact copies known as sister chromatids.

10.2 The Mitotic Cell Cycle

- Mitosis and interphase constitute the mitotic cell cycle. Mitosis occurs in five stages. In prophase (stage 1), the chromosomes condense into short rods and the spindle forms in the cytoplasm (Figures 10.3 and 10.4).

- In prometaphase (stage 2), the nuclear envelope breaks down, the spindle enters the former nuclear area, and the sister chromatids of each chromosome make connections to opposite spindle poles. Each chromatid has a kinetochore that attaches to spindle microtubules (Figures 10.3, 10.4, and 10.6).

- In metaphase (stage 3), the spindle is fully formed and the chromosomes, moved by the spindle microtubules, become aligned at the metaphase plate (Figures 10.3 and 10.4).

- In anaphase (stage 4), the spindle separates the sister chromatids and moves them to opposite spindle poles. At this point, chromosome segregation is complete (Figures 10.3 and 10.4).

- In telophase (stage 5), the chromosomes decondense and return to the extended state typical of interphase and a new nuclear envelope forms around the chromosomes (Figures 10.3 and 10.4).

Courtesy of Raymond Deshaies

- Cytokinesis, the division of the cytoplasm, completes cell division by producing two daughter cells, each containing a daughter nucleus produced by mitosis (Figures 10.3 and 10.4).
- Cytokinesis in animal cells proceeds by furrowing, in which a band of microfilaments just under the plasma membrane contracts, gradually separating the cytoplasm into two parts (Figure 10.8).
- In plant cytokinesis, cell wall material is deposited along the plane of the former spindle midpoint; the deposition continues until a continuous new wall, the cell plate, separates the daughter cells (Figure 10.9).

Animation: Mitosis

Animation: Cytoplasmic division

Animation: Phases of the Cell Cycle

10.3 Formation and Action of the Mitotic Spindle

- In animal cells, the centrosome divides and the two parts move apart. As they do so, the microtubules of the spindle form between them. In plant cells, which do not have a centrosome, the spindle microtubules assemble around the nucleus (Figure 10.10).
- In the spindle, kinetochore microtubules run from the poles to the kinetochores of the chromosomes, and nonkinetochore microtubules run from the poles to a zone of overlap at the spindle midpoint without connecting to the chromosomes (Figure 10.4).
- During anaphase, chromosomes can move when motor proteins attached to kinetochores walk along the kinetochore microtubules, moving the chromosomes to the poles (Figures 10.12 and 10.13), or when motor proteins at the spindle poles pull kinetochore microtubules toward the poles. Both mechanisms are used, with their relative contributions being species-specific and cell type-specific.
- Also during anaphase, nonkinetochore microtubules reduce their overlap at the spindle midpoint and lengthen simultaneously, pushing the poles farther apart thereby lengthening the cell.

Animation: Mitosis step-by-step

10.4 Cell Cycle Regulation

- The internal controls that monitor progression through the cell cycle include checkpoints at key points to ensure that critical phases do not commence before previous phases are completed correctly (Figure 10.15).
- The internal control system that directly regulates cell division involves complexes of a cyclin and a cyclin-dependent protein kinase (Cdk). Cdk is activated when combined with a cyclin and then phosphorylates target proteins, regulating their activities. The altered target proteins then initiate or regulate key events of the cell cycle. Four classes of cyclins—G_1, G_1/S, S, and M—are distinguished by the stage of the cell cycle at which they activate Cdks (Figure 10.15).
- External controls are based primarily on surface receptors that recognize and bind signals such as peptide hormones and growth factors, surface groups on other cells, or molecules of the extracellular matrix. The binding triggers internal reactions that speed, slow, or stop cell division.
- In cancer, control of cell division is lost, and cells divide continuously and uncontrollably, forming a rapidly growing mass of cells that interferes with body functions. Cancer cells can also break loose from their original tumor (metastasize) to form additional tumors in other parts of the body.

10.5 Cell Division in Prokaryotes

- Replication begins at the origin of replication of the bacterial chromosome in reactions catalyzed by enzymes located in the middle of the cell. Once the origin of replication is duplicated, the two origins migrate to the two ends of the cells. Division of the cytoplasm then occurs through a partition of cell wall material that grows inward until the cell is separated into two parts (Figure 10.16).

UNDERSTAND & APPLY

Test Your Knowledge

1. During the cell cycle, the DNA mass of a cell:
   a. decreases during G_1.
   b. decreases during metaphase.
   c. increases during the S phase.
   d. increases during G_2.
   e. decreases during interphase.

2. A tumor suppressor protein, p21, inhibits Cdk1. The earliest effect of p21 on the cell cycle would be to stop the cell cycle at:
   a. early G_1.
   b. late G_1.
   c. the S phase.
   d. G_2.
   e. the mitotic prophase.

3. A major difference between hereditary information in eukaryotes and prokaryotes is:
   a. in prokaryotes, the hereditary information is distributed among individual, linear DNA molecules in the nucleus.
   b. in eukaryotes, the hereditary information is encoded in a single, circular DNA molecule.
   c. in prokaryotes, the hereditary information is usually distributed among multiple circular DNA molecules in the cytoplasm.
   d. in eukaryotes, the hereditary information is distributed among individual, linear DNA molecules in the cytoplasm.
   e. in eukaryotes, the hereditary information is distributed among individual, linear DNA molecules in the nucleus.

4. The major microtubule organizing center of the animal cell is:
   a. chromosomes, composed of chromatids.
   b. the centrosome, composed of centrioles.
   c. the chromatin, composed of chromatids.
   d. chromosomes, composed of centromere.
   e. centrioles, composed of centrosome.

5. The chromatids separate into chromosomes:
   a. during prophase.
   b. going from prophase to metaphase.
   c. going from anaphase to telophase.
   d. going from metaphase to anaphase.
   e. going from telophase to interphase.

6. Which of the following statements about mitosis is *incorrect*?
   a. Microtubules from the spindle poles attach to the kinetochores on the chromosomes.
   b. In anaphase, the spindle separates sister chromatids and pulls them apart.
   c. In metaphase, spindle microtubules align the chromosomes at the spindle midpoint.
   d. Cytokinesis describes the movement of chromosomes.
   e. Both the animal cell furrow and the plant cell plate form at their former spindle midpoint.

7. Mitomycin C is an anticancer drug that stops cell division by inserting itself into the strands of DNA and binding them together. This action is thought to have its major effect at:
   a. late G_1, early S phases.
   b. late G_2.
   c. prophase.
   d. metaphase.
   e. anaphase.

8. Which of the following statements about cell cycle regulators is *incorrect*?
   a. The concentrations of cyclins change throughout the cell cycle.
   b. Cyclins are present in all stages of the cell cycle except S.
   c. Cyclin–Cdk complexes phosphorylate target proteins.
   d. Cdks combine with cyclin to move the cycle into mitosis.
   e. During anaphase of mitosis, cyclin is degraded, allowing mitosis to end.

9. Which of the following is *not* a characteristic of cancer cells?
   a. less cytoplasmic volume than normal cells
   b. an absence of cyclin
   c. loss of adhesion to other cells
   d. loss of control of cell division
   e. loss of normal control of G_1/S phase transition

10. In bacteria:
    a. several chromosomes undergo mitosis.
    b. binary fission produces four daughter cells.
    c. replication begins at the origin *(ori),* and the DNA strands separate.
    d. the plasma membrane plays an important role in separating the duplicated chromosomes into the two daughter cells.
    e. the daughter cells receive different genetic information from the parent cell.

Discuss the Concepts

1. You have a means of measuring the amount of DNA in a single cell. You first measure the amount of DNA during G_1. At what points during the remainder of the cell cycle would you expect the amount of DNA per cell to change?

2. A cell has 38 chromosomes. After mitosis and cell division, one daughter cell has 39 chromosomes and the other has 37. What might have caused these abnormal chromosome numbers? What effects do you suppose this might have on cell function? Why?

3. Paclitaxel (Taxol), a substance isolated from Pacific yew *(Taxus brevifolia),* is effective in the treatment of breast and ovarian cancers. It works by stabilizing microtubules, thereby preventing them from disassembling. Why would this activity slow or stop the growth of cancer cells?

4. A cell has 24 chromosomes at G_1 of interphase. How many chromosomes would you expect it to have at G_2 of interphase? At metaphase of mitosis? At telophase of mitosis?

Design an Experiment

Many chemicals in the food we eat have the potential to affect the growth of cancer cells. Chocolate, for example, contains a number of flavonoid compounds, which act as natural antioxidants. Design an experiment to determine whether any of the flavonoids in chocolate inhibit the cell cycle of breast cancer cells that are growing in culture.

Interpret the Data

Biologists have long been interested in the effects of radiation on cells. In one experiment, researchers examined the effect of radium on mitosis of chick embryo cells growing in culture. A population of experimental cells was examined under the microscope for the number of cells in telophase (as a measure of mitosis occurring) before, during, and after exposure to radium. The results are shown in the **Figure.**

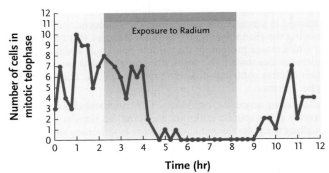

R. G. Canti and M. Donaldson. 1926. The effect of radium on mitosis *in vitro. Proceedings of the Royal Society of London, Series B, Containing Papers of a Biological Character* 100:413–419.

1. What is the effect of radium exposure on mitosis?

2. Was the effect of radium exposure permanent?

Source: R. G. Canti and M. Donaldson. 1926. The effect of radium on mitosis *in vitro. Proceedings of the Royal Society of London, Series B, Containing Papers of a Biological Character* 100:413–419.

Apply Evolutionary Thinking

The genes and proteins involved in cell cycle regulation in prokaryotes and eukaryotes are very different. However, both types of organisms use similar molecular regulatory reactions to coordinate DNA synthesis with cell division. What does this observation mean from an evolutionary perspective?

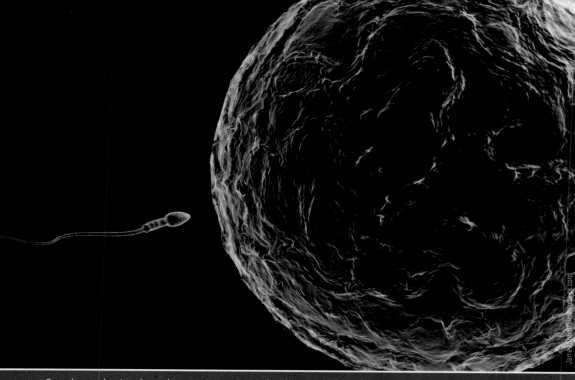

Sexual reproduction depends on meiosis, a specialized process of cell division that produces gametes such as eggs and sperm. Here a sperm is swimming toward an egg.

James Seddon/Shutterstock.com

Meiosis: The Cellular Basis of Sexual Reproduction

Why it matters . . . A couple clearly shows mutual interest. First, he caresses her with one arm, then another—then another, another, and another. She reciprocates. This interaction goes on for hours; a hug here, a squeeze there. At the climactic moment, the male reaches deftly under his mantle and removes a packet of sperm, which he inserts under the mantle of the female. For every one of his sperm that successfully performs its function, a fertilized egg can develop into a new octopus.

The octopuses are engaged in a form of **sexual reproduction,** the production of offspring through union of male and female **gametes**—for example, egg and sperm cells in animals. Sexual reproduction depends on **meiosis,** a specialized process of cell division that, in animals, produces gametes. Meiosis reduces the number of chromosomes, producing gametes with half the number of chromosomes present in the **somatic cells** (body cells) of a species. The derivation of the word *meiosis (meioun* = to diminish) reflects this reduction. At **fertilization,** the nuclei of an egg and sperm cell fuse, producing a cell called the **zygote,** in which the chromosome number typical of the species is restored. Without the halving of chromosome number by the meiotic divisions, fertilization would double the number of chromosomes in each subsequent generation.

Both meiosis and fertilization also mix genetic information into new combinations allowing for genetic diversity in the offspring of a mating pair. By contrast, asexual reproduction generates genetically identical offspring because they are the products of mitotic divisions (asexual reproduction is discussed in Chapter 10). Sexual reproduction allows the genetic variability, which originally arose from mutations, to be shuffled and reshuffled throughout generations, resulting in a wide array of genetic combinations.

The halving of the chromosome number and shuffling of genetic information into new combinations—both by meiosis—and the restoration of the chromosome number by fertilization are the biological foundations of sexual reproduction. Intermingled tentacles in octopuses and the courting and complex mating rituals of humans have evolved as a means for promoting fertilization.

11.1 The Mechanisms of Meiosis

In humans, and other animals, meiosis takes place in the primary reproductive organs, the **gonads.** Meiosis in mature gonads of the male, the **testes,** produces **spermatozoa (sperm),** the gametes of the male. Meiosis in mature gonads of the female, the **ovaries,** produces **ova (eggs),** the gametes of the female. The cellular mechanisms of gamete formation—**gametogenesis**—are described in Chapter 49.

Meiosis Is Based on the Interactions and Distribution of Homologous Chromosome Pairs

To follow the steps of meiosis, you must understand the significance of the chromosome pairs in diploid organisms. As discussed in Section 10.1, the two representatives of each chromosome in a diploid cell constitute a *homologous pair*—they have the same genes, arranged in the same order in the DNA of the chromosomes. One chromosome of each homologous pair, the **paternal chromosome,** comes from the male parent of the organism, and the other chromosome, the **maternal chromosome,** comes from its female parent.

Although the genes of the two chromosomes of a homologous pair are arranged in the same order, the versions of each gene, called **alleles,** present in the members of the pair may be the same or different. For a gene that encodes a protein, the different alleles might encode different versions of the same protein, which have different structures, molecular properties, or both, or perhaps an allele may not encode a protein at all.

Recall from Section 10.1 that humans normally have 46 chromosomes in their diploid cells, which consist of 22 homologous pairs and a pair of sex chromosomes (see Figure 10.7). However, each individual (except for identical twins, identical triplets, and so forth) has a unique combination of the alleles in the two chromosomes of each homologous pair. The distinct set of alleles, arising from the mixing mechanisms of meiosis and fertilization, gives each individual his or her unique combination of inherited traits, including such attributes as height, hair and eye color, susceptibility to certain diseases, and even aspects of personality and intelligence.

Meiosis separates homologous pairs, thereby reducing the diploid or *2n* number of chromosomes to the **haploid** or *n* number **(Figure 11.1).** Each gamete produced by meiosis receives only one member of each homologous pair. For example, a human egg produced in an ovary, or a sperm cell produced in a testis, contains 23 chromosomes, one of each pair. When the egg and sperm combine in sexual reproduction to produce the *zygote*—the first cell of the new individual—the diploid number of 46 chromosomes (23 pairs) is regenerated. The processes of DNA replication and mitotic cell division ensure that this diploid number is maintained in the body cells as the zygote develops.

The Meiotic Cell Cycle Produces Four Genetically Different Daughter Cells with Half the Parental Number of Chromosomes

Meiosis follows a premeiotic interphase in which DNA replicates and the chromosomal proteins are duplicated. (This interphase passes through G_1, S, and G_2 stages as does a premitotic interphase.) Meiosis consists of two cell divisions—**meiosis I** and **meiosis II**—in sexually reproducing organisms in which duplicated chromosomes in the parental cell are distributed to four daughter cells, each of which, therefore, has half the number of chromosomes as does the parental cell (outlined in **Figure 11.2**). By contrast, in mitosis, each chromosome duplication is followed by a division. Consequently, the chromosome number remains constant from one cell generation to the next.

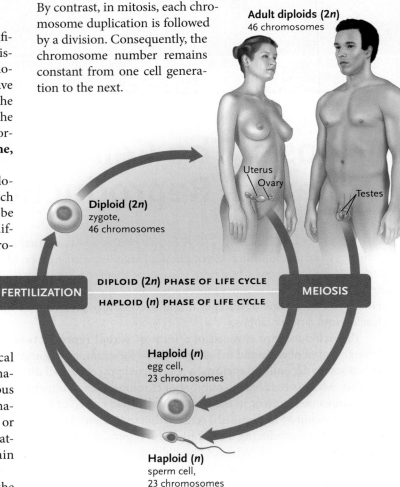

FIGURE 11.1 The cycle of meiosis and fertilization in animals, with humans as an example. Meiosis in animals produces gametes, spermatozoa (sperm) in the testes of the male, and ova (eggs) in the ovaries of the female. Meiosis reduces the chromosome number from the diploid level of two representatives of each chromosome to the haploid level of one representative of each chromosome. Fertilization restores the chromosome number to the diploid level.

© Cengage Learning 2014

FIGURE 11.2 Production of four haploid nuclei by the two meiotic divisions. For simplicity, just one pair of homologous chromosomes is followed through the divisions. The maternal chromosome is red, and the paternal chromosome is blue, a convention that is followed in other meiosis figures.
© Cengage Learning 2014

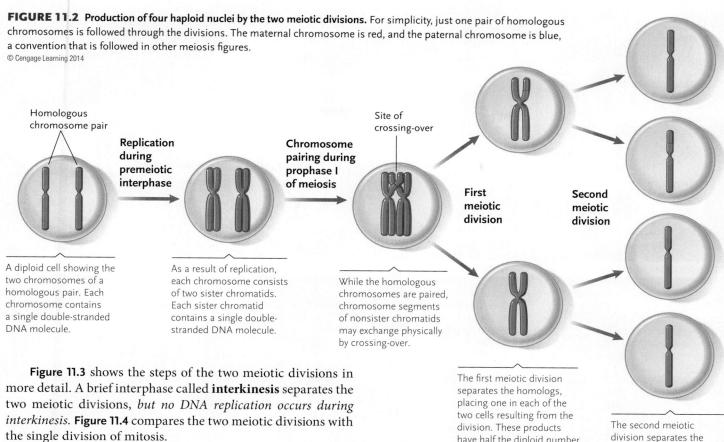

Homologous chromosome pair

Replication during premeiotic interphase

Chromosome pairing during prophase I of meiosis

Site of crossing-over

First meiotic division

Second meiotic division

A diploid cell showing the two chromosomes of a homologous pair. Each chromosome contains a single double-stranded DNA molecule.

As a result of replication, each chromosome consists of two sister chromatids. Each sister chromatid contains a single double-stranded DNA molecule.

While the homologous chromosomes are paired, chromosome segments of nonsister chromatids may exchange physically by crossing-over.

The first meiotic division separates the homologs, placing one in each of the two cells resulting from the division. These products have half the diploid number of chromosomes, but each chromosome still consists of two chromatids. Each of those chromatids contains a single double-stranded DNA molecule.

The second meiotic division separates the sister chromatids, and one of each of the now-daughter chromosomes is placed in each cell resulting from the division. Each daughter chromosome contains a single double-stranded DNA molecule.

Figure 11.3 shows the steps of the two meiotic divisions in more detail. A brief interphase called **interkinesis** separates the two meiotic divisions, *but no DNA replication occurs during interkinesis*. **Figure 11.4** compares the two meiotic divisions with the single division of mitosis.

CHROMOSOME SEGREGATION FAILURE Rarely, chromosome segregation fails. That is, both chromosomes of a homologous pair may connect to a kinetochore microtubule from the same spindle pole in meiosis I. The result is **nondisjunction** in which the spindle fails to separate the homologous chromosomes. As a result, one pole receives both chromosomes of the homologous pair, whereas the other pole has no copies of that chromosome. Nondisjunction can also occur in meiosis II. In this case, both chromatids of a sister-chromatid pair connect to a kinetochore microtubule from the same spindle pole. Nondisjunction in this case produces a similar result: one pole receives both sister chromatids (as daughter chromosomes), whereas the other pole receives no copies of that chromosome. Zygotes that receive an extra chromosome because of nondisjunction have three copies of one chromosome instead of two. In humans, most zygotes of this kind do not result in live births. One exception is zygotes that have three copies of chromosome 21 instead of the normal two copies, which develop into individuals with Down syndrome. Chapter 13 discusses nondisjunction and Down syndrome in more detail.

SEX CHROMOSOMES IN MEIOSIS As you learned in Section 10.1, in many eukaryotes, including most animals, one or more pairs of chromosomes, called the **sex chromosomes,** are different in male and female individuals of the same species. For example, in humans, the cells of females contain a pair of sex chromosomes called the *XX* pair (the sex chromosomes are visible in Figure 10.7). Male humans contain a pair of sex chromosomes

that consist of one X chromosome and a smaller chromosome called the *Y chromosome*. Developing into a male, in fact, is directly determined by the presence of the Y chromosome because of a gene it contains (see Chapter 50). In the absence of a Y chromosome, as in an XX individual, a female is produced.

The two X chromosomes in females are fully homologous. In mammals, the X and Y chromosomes in males are homologous through a short region. This means that an X chromosome from the mother is able to pair up with either an X or a Y from the father and follow the same pathways through the meiotic divisions as the other chromosome pairs.

As a result of meiosis, a gamete formed in a female (an egg) may receive either member of the X pair. A gamete formed in a male (a sperm) receives either an X or a Y chromosome.

The sequence of steps in the two meiotic divisions accomplishes the major outcomes of meiosis: the reduction of chromosome number and the generation of genetic variability. The latter is the subject of the next section. (Figure 11.4 reviews the two meiotic divisions and compares them with the single division of mitosis.)

Prophase I

Plasma membrane | Duplicated centrioles | Nuclear envelope

Tetrad

Crossover

Condensation of chromosomes

1. • At the beginning of prophase I, the replicated chromosomes, each consisting of two identical sister chromatids, begin to condense into threadlike structures.
   • As in mitosis, each pair of sister chromatids is held together tightly by sister chromatid cohesion, in which cohesin proteins encircle the sister chromatids along their length.

Synapsis

2. • The two chromosomes of each homologous pair come together and line up side-by-side in a zipperlike way in a process called **synapsis** or **pairing**. The tight association is facilitated by a protein framework called the **synaptonemal complex** (inset below).
   • When fully paired, the homologs are called tetrads because each consists of four chromatids. No equivalent of chromosome pairing exists in mitosis.

Crossing-Over

3. • While they are paired, the chromatids of homologous chromosomes exchange segments by **crossing-over**. In crossing-over, enzymes break and rejoin DNA molecules of chromatids with great precision.
   • The sites where crossing-over has occurred become visible under the light microscope when the chromosomes condense further as prophase I proceeds. The sites, called **crossovers** or **chiasmata** (singular, *chiasma* = crosspiece) (see enlarged circle).
   • Once crossing-over is complete toward the end of prophase I, the synaptonemal complex disassembles and disappears.

Prometaphase I

4. • By the end of prophase I, a spindle has formed in the cytoplasm by the same mechanisms as described in Section 10.3. At the start of prometaphase I, the nuclear envelope breaks down, and the spindle moves into the former nuclear area.
   • Kinetochore microtubules connect to the chromosomes—kinetochore microtubules from one pole attach to both sister kineto-chores of one duplicated chromosome, and kinetochore microtubules from the other pole attach to both sister kinetochores of the other duplicated chromosome. That is, both sister chromatids of one homolog attach to microtubules leading to one spindle pole, whereas both sister chromatids of the other homolog attach to microtubules leading to the opposite pole.
   • Nonkinetochore microtubules from the two poles overlap in the middle of the cell but do not attach to chromosomes.

Synaptonemal complex

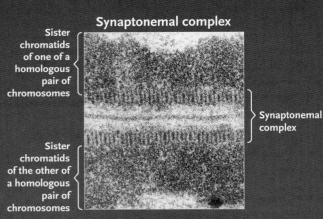

Sister chromatids of one of a homologous pair of chromosomes

Synaptonemal complex

Sister chromatids of the other of a homologous pair of chromosomes

INSET The synaptonemal complex as seen in a meiotic cell of the fungus *Neotiella*.
Courtesy of Diter Von Wettstein

Second meiotic division

Prophase II

8. • The chromosomes condense and a spindle forms.

Prometaphase II

9. • The nuclear envelope breaks down, and the spindle enters the former nuclear area.
   • Kinetochore microtubules from the opposite spindle poles attach to the kinetochores of each chromosome.

FIGURE 11.3 **The meiotic divisions.** Two homologous pairs of chromosomes are shown. Maternal chromosomes are red; paternal chromosomes are blue.

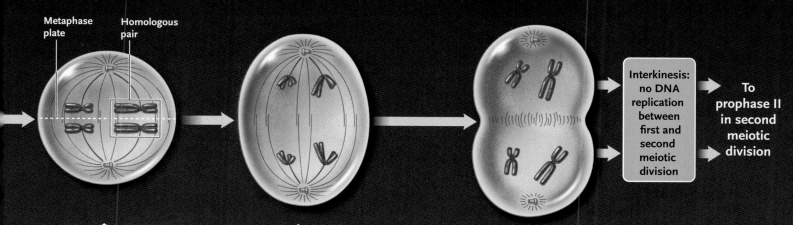

Metaphase I

5 • Movements of the kinetochore microtubules align the tetrads in the equatorial plane—metaphase plate—between the two spindle poles.

Anaphase I

6 • Anaphase I is triggered when the enzyme separase (see Section 10.2) cleaves the cohesin rings just along the arms of the sister chromatids, leaving sister chromatid cohesion intact at the centromere region.
• The two chromosomes of each homologous pair segregate and move to opposite spindle poles. The movement delivers one-half the diploid number of chromosomes to each pole of the spindle. However, all the chromosomes at the poles are still double structures composed of two sister chromatids.

Telophase I

7 • Telophase I is a brief, transitory stage in which there is little or no change in the chromosomes except for limited decondensation or unfolding in some species.
• New nuclear envelopes form in some species but not in others.
• Telophase I is followed by an interkinesis in which the single spindle of the first meiotic division disassembles and the microtubules reassemble into two new spindles for the second division.

Interkinesis: no DNA replication between first and second meiotic division

To prophase II in second meiotic division

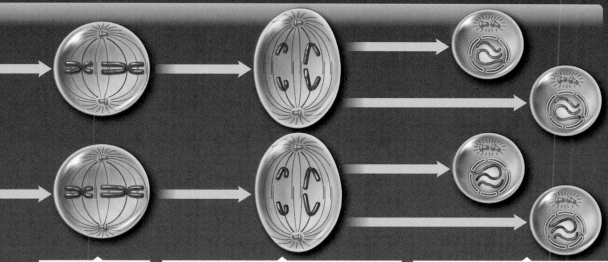

Metaphase II

10 • Movements of the spindle microtubules align the chromosomes on the metaphase plate.

Anaphase II

11 • Separase cleaves the remaining cohesin proteins that are holding the pairs of sister chromatids together in their centromere regions.
• Kinetochore microtubules separate the two chromatids of each chromosome and move them toward opposite spindle poles.
• At the completion of anaphase II, the chromatids—now called chromosomes—have been segregated to the two poles.

Telophase II

12 • The chromosomes begin decondensing, eventually reaching the extended interphase state.
• The spindles disassemble.
• New nuclear envelopes form around the masses of chromatin.
• Cytokinesis typically follows. The result is four haploid cells, each with a nucleus containing half the number of chromosomes present in a G1 nucleus of the same species.

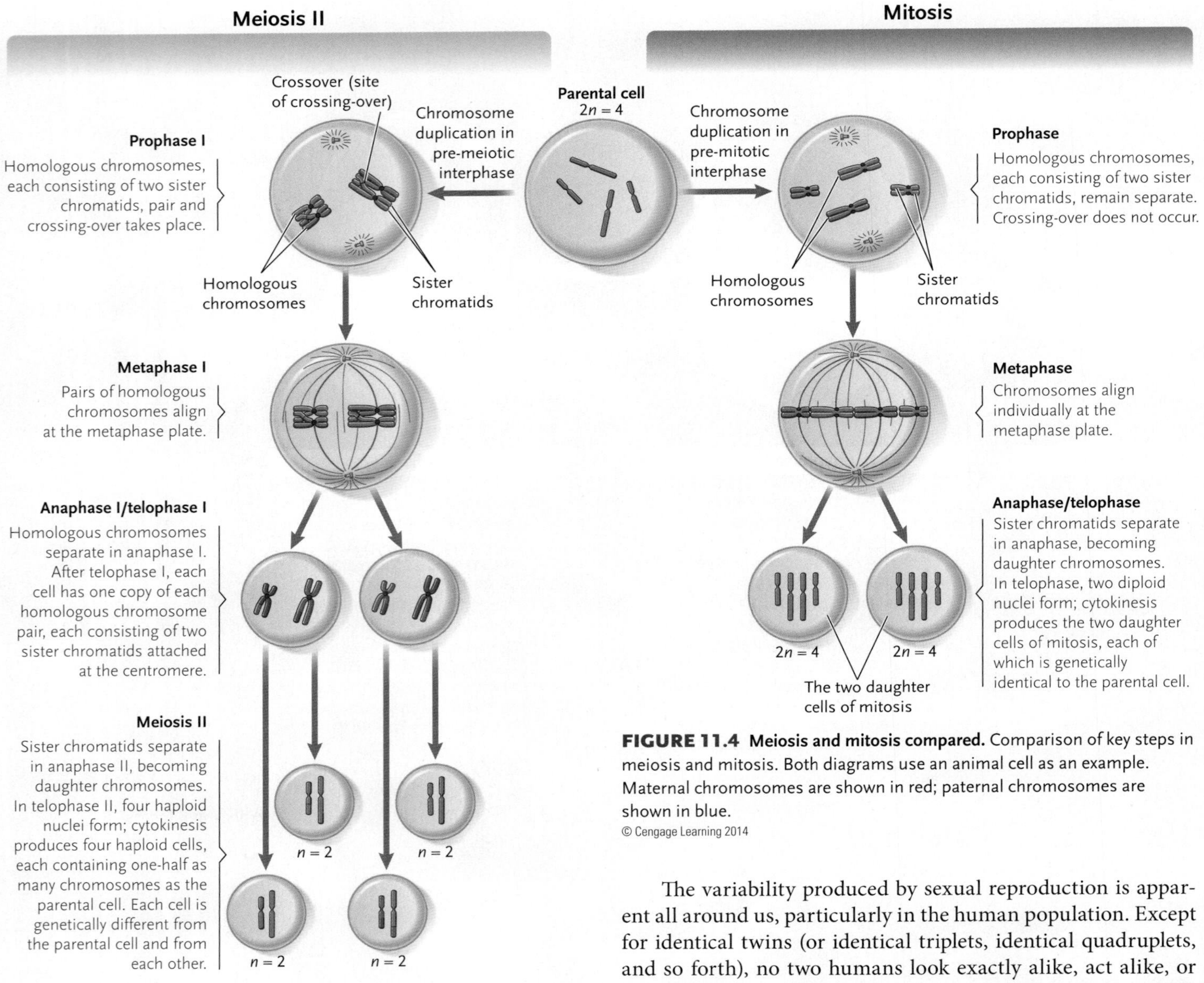

Prophase I

Homologous chromosomes, each consisting of two sister chromatids, pair and crossing-over takes place.

Crossover (site of crossing-over)

Chromosome duplication in pre-meiotic interphase

Parental cell
$2n = 4$

Chromosome duplication in pre-mitotic interphase

Prophase

Homologous chromosomes, each consisting of two sister chromatids, remain separate. Crossing-over does not occur.

Homologous chromosomes   Sister chromatids

Homologous chromosomes   Sister chromatids

Metaphase I

Pairs of homologous chromosomes align at the metaphase plate.

Metaphase

Chromosomes align individually at the metaphase plate.

Anaphase I/telophase I

Homologous chromosomes separate in anaphase I. After telophase I, each cell has one copy of each homologous chromosome pair, each consisting of two sister chromatids attached at the centromere.

Anaphase/telophase

Sister chromatids separate in anaphase, becoming daughter chromosomes. In telophase, two diploid nuclei form; cytokinesis produces the two daughter cells of mitosis, each of which is genetically identical to the parental cell.

$2n = 4$ $2n = 4$

The two daughter cells of mitosis

Meiosis II

Sister chromatids separate in anaphase II, becoming daughter chromosomes. In telophase II, four haploid nuclei form; cytokinesis produces four haploid cells, each containing one-half as many chromosomes as the parental cell. Each cell is genetically different from the parental cell and from each other.

$n = 2$ $n = 2$

$n = 2$ $n = 2$

FIGURE 11.4 Meiosis and mitosis compared. Comparison of key steps in meiosis and mitosis. Both diagrams use an animal cell as an example. Maternal chromosomes are shown in red; paternal chromosomes are shown in blue.
© Cengage Learning 2014

STUDY BREAK 11.1

1. How does the outcome of meiosis differ from that of mitosis?
2. What is recombination, and in what stage of meiosis does it occur?
3. Which of the two meiotic divisions is similar to a mitotic division?

11.2 Mechanisms That Generate Genetic Variability

The genetic variability due to the shuffling of alleles is a prime evolutionary advantage of sexual reproduction. The resulting variability increases the chance that at least some offspring will be successful in surviving and reproducing in changing environments.

The variability produced by sexual reproduction is apparent all around us, particularly in the human population. Except for identical twins (or identical triplets, identical quadruplets, and so forth), no two humans look exactly alike, act alike, or have identical biochemical and physiological characteristics, even if they are members of the same immediate family. Other species that reproduce sexually show the same type of variability arising from meiosis.

During meiosis and fertilization, genetic variability arises from three sources:

1. Crossing-over between paired homologous chromosomes during prophase I, which recombines alleles of genes on paired homologous chromosomes.
2. The independent assortment of chromosomes segregated to the poles during anaphase I, which recombines alleles of genes on nonhomologous chromosomes.
3. The particular sets of male and female gametes that unite in fertilization, which recombines the alleles of genes in the offspring of two parents.

Each of these sources of variability is discussed in further detail in the following sections.

Meiosis and Mammalian Gamete Formation: What determines whether an egg or a sperm will form?

The sex of a mammal is determined genetically. That is, the presence of a Y chromosome and the specific male-determining gene it contains, directs the development of a male. In the absence of a Y chromosome, the default development of a female occurs. The gonads contain germ cells—cells that eventually turn into eggs or sperm. Interestingly, germ cells can become eggs or sperm, regardless of the genetic sex of the individual. Whether a germ cell develops into an egg or a sperm depends on the time at which meiosis begins. In females, meiosis commences in the fetus before birth and, as a result, eggs are produced in the ovaries. In males, meiosis begins after birth and sperm is produced in the testes. The working model was that germ cells in fetuses are programmed to enter meiosis and produce eggs unless prevented from doing so by a meiosis-inhibiting factor, in which case sperm is produced.

Research Question

What molecular signal determines germ cell fate in mammals?

Experiment

The research group of Peter Koopman at the Institute for Molecular Bioscience, University of Queensland, Brisbane, Australia, set out to answer the question using mouse as the model mammal. The researchers used molecular tools to search for genes expressed in a sex-related fashion during gonad formation. They considered one gene, *Cyp26b1*, to be a strong candidate for the meiosis-inhibiting factor

gene. *Cyp26b1* codes for an enzyme that breaks down retinoic acid, which is a natural derivative of vitamin A. Retinoic acid has many regulatory functions, one of which is to cause germ cells in female embryos to begin meiosis.

The researchers obtained these results:

- The *Cyp26b1* gene is expressed in embryonic mouse testes but not in embryonic ovaries **(Figure)**. This result argued that the delay in meiosis in males was caused by the *Cyp26b1*-encoded enzyme degrading the retinoic acid.
- In *Cyp26b1*-knockout male mice, that is, mice in which the gene had been deleted, germ cells enter meiosis much earlier (see Figure).

The interpretation of the data is that, in the absence of the enzyme to degrade retinoic acid, meiosis was stimulated to occur at the same time as it does in the females. These two lines of evidence support the hypothesis that the product of the *Cyp26b1* gene is the meiosis-inhibiting factor in males.

Conclusion

The timing of the initiation of meiosis in mice (and probably in other mammals) is determined by the regulation of retinoic acid presence by the *Cyp26b1*-encoded enzyme. In females, the gene is not active in embryonic ovaries, so the retinoic acid present in the fetus stimulates the initiation of meiosis before birth leading to the production of eggs. In males, the gene is active in embryonic testes, and the resulting enzyme degrades the retinoic acid. As a result, meiosis is delayed, initiating after birth, leading to the production of sperm.

Although potentially answering a very important biological question, the study does not provide all of the answers for how germ cell fate is regulated. Other questions remain, including how does early meiosis favor egg formation over sperm formation.

THINK LIKE A SCIENTIST Sequence analysis has shown that the CYP26B1 protein is closely related in human, rat, mouse, and zebrafish. What do you think is the significance of this result with respect to germ cell development?

Source: J. Bowles et al. 2006. Retinoid signaling determines germ cell fate in mice. *Science* 312:596–600.

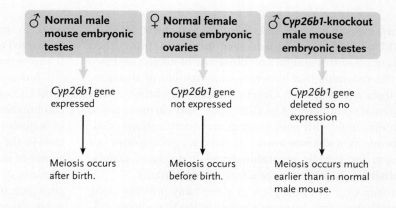

Shuffling of Alleles on the Same Chromosome Depends on Chromosome Pairing and Crossing-Over Events between Homologous Chromosomes

You learned earlier that crossing-over, the physical exchange of chromosome segments at corresponding positions along homologous chromosomes, occurs during prophase I of meiosis.

If there are genetic differences between the homologs, crossing-over can produce new allele combinations in a chromatid **(Figure 11.5)**. Consider two genes on a homologous pair of chromosomes: the alleles for the two genes are *A* and *B* on

the chromosome from one parent, and *a* and *b* on the chromosome from the other parent. After chromosome duplication the homologous chromosomes, each consisting of two sister chromatids, pair in prophase I. Crossing-over exchanges segments of nonsister chromatids, producing new combinations of alleles. In our example, crossing-over occurred between the two genes, exchanging chromosome segments to produce two nonparental arrangements of alleles on two of the chromatids. Specifically, one sister chromatid of the left homologous chromosome now has an *a* allele and a *B* allele, and one sister chromatid of the right homologous chromosome now has an *A* allele and a *b* allele.

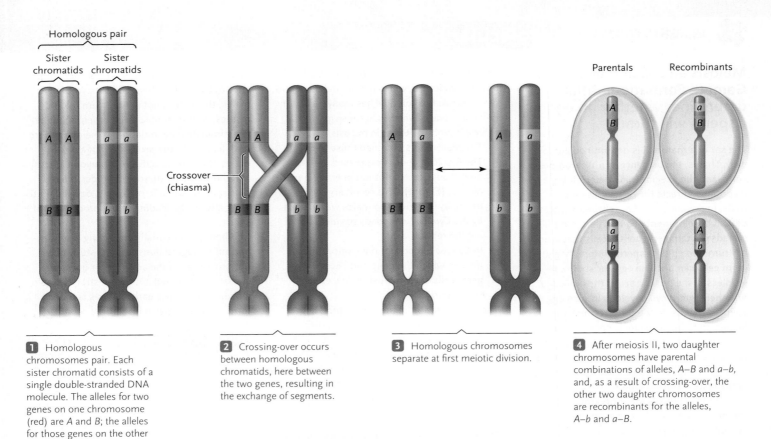

1 Homologous chromosomes pair. Each sister chromatid consists of a single double-stranded DNA molecule. The alleles for two genes on one chromosome (red) are *A* and *B*; the alleles for those genes on the other chromosome (blue) are *a* and *b*.

2 Crossing-over occurs between homologous chromatids, here between the two genes, resulting in the exchange of segments.

3 Homologous chromosomes separate at first meiotic division.

4 After meiosis II, two daughter chromosomes have parental combinations of alleles, *A–B* and *a–b*, and, as a result of crossing-over, the other two daughter chromosomes are recombinants for the alleles, *A–b* and *a–B*.

FIGURE 11.5 **Recombination by crossing-over.**
© Cengage Learning 2014

When meiosis is completed, there are four nuclei (see Figure 11.5). Two nuclei receive unchanged chromatids (now called chromosomes) with parental combinations of alleles, and two receive chromosomes that have new combinations of alleles resulting from crossing-over. Geneticists call the chromosomes with parental combinations of alleles **parental chromosomes,** and chromosomes with new, nonparental combinations of alleles **recombinant chromosomes.** Therefore, crossing-over is a mechanism for **genetic recombination**—it produces **genetic recombinants,** also called more simply **recombinants.**

Crossing-over can take place at almost any position along the chromosome arms, between any two of the four chromatids of a homologous pair. One or more additional crossing-over events may occur in the same chromosome pair and involve the same or different chromatids that exchanged segments in any single event. In most species, crossing-over occurs at two or three sites in each set of paired chromosomes.

Independent Assortment of Maternal and Paternal Chromosomes Is the Second Major Source of Genetic Variability in Meiosis

Recall that a diploid individual inherits one set of chromosomes from the mother and one from the father. The maternal and paternal members of a chromosome pair are homologous chromosomes. Before meiosis, each homologous chromosome dupli-

cates to produce a pair of sister chromatids that remain connected at the centromere until they are separated in meiosis II and become daughter chromosomes.

Independent assortment of paired homologous chromosomes accounts for the second major mechanism of genetic recombination in meiosis. Recall that prometaphase I is the stage of meiosis in which the homologous pairs of chromosomes attach to the spindle poles. The maternal and paternal chromosomes of each homologous pair typically carry different alleles of many of the genes on that chromosome. For each homologous pair, one chromosome (at this stage consisting of a pair of sister chromatids) makes spindle connections leading to one pole and the other chromosome (also a pair of sister chromatids) connects to the opposite pole. In making these connections, the orientation of one chromosome pair—which member of a pair faces one pole and which faces the other—has no influence on the orientation of any other chromosome pair. How chromosomes orient themselves when aligning along the metaphase plate is entirely random. As a result, any combination of chromosomes of maternal and paternal origin may be segregated to the spindle poles **(Figure 11.6),** a phenomenon called **independent assortment** of chromosomes. The second meiotic division separates the chromatids containing these random combinations to gamete nuclei.

The number of combinations possible due to independent assortment depends on the number of chromosome pairs in a

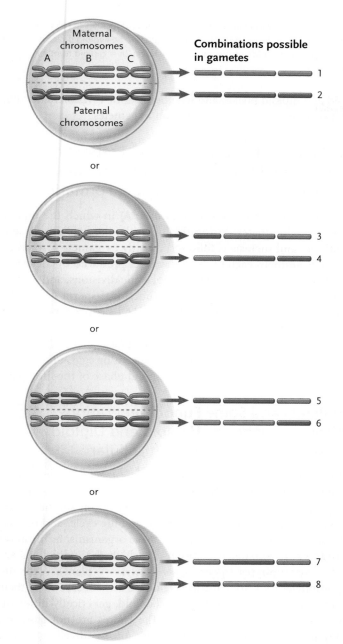

Combinations possible in gametes

Maternal chromosomes

A   B   C

Paternal chromosomes

or

or

or

FIGURE 11.6 Possible outcomes of the random spindle connections of three pairs of chromosomes at metaphase I of meiosis. The three types of chromosomes are labeled A, B, and C, and for simplification, crossing-over is not considered. Maternal chromosomes are red; paternal chromosomes are blue. There are four possible patterns of connections, giving eight possible combinations of maternal and paternal chromosomes in gametes (labeled 1–8).
© Cengage Learning 2014

species. For example, the 23 chromosome pairs of humans allow 2^{23} different combinations of maternal and paternal chromosomes to be delivered to the poles, producing potentially 8,388,608 genetically different gametes from independent assortment alone. When independent assortment is combined with the rearrangement of alleles accomplished by crossing over, the number of possible combinations is innumerable.

Random Joining of Male and Female Gametes in Fertilization Contributes to Additional Variability among Individuals

The male and female gametes produced by meiosis typically are genetically diverse due to crossing-over and independent assortment. Which two gametes join in fertilization is a matter of chance. This chance union of gametes amplifies the variability resulting from sexual reproduction. Considering just the variability from independent assortment of chromosomes and that from fertilization, the possibility that two children of the same parents could receive the same combination of chromosomes is 1 chance out of $(2^{23})^2$ or 1 in 70,368,744,000,000 ($\sim$70 trillion), a number that far exceeds the number of people in the entire human population. The further variability introduced by recombination makes it practically impossible for humans and many other sexually reproducing organisms to produce genetically identical gametes or offspring. A common exception in humans is identical twins (or identical triplets, identical quadruplets, and so forth), which arise not from the combination of identical gametes during fertilization but from mitotic division of a single fertilized egg into separate cells that give rise to genetically identical individuals.

STUDY BREAK 11.2

1. What are the three ways in which sexual reproduction enhances the degree of genetic variability among individuals?
2. Consider an animal with six pairs of chromosomes; one set of six chromosomes is from this animal's male parent, and the homologous set of six chromosomes is from this animal's female parent. How many combinations of chromosomes are possible in the gametes of an individual of this species if we look only at independent assortment of chromosomes, disregarding the effect of crossing over?

THINK OUTSIDE THE BOOK >

Mutations in several organisms have been identified that affect meiosis. Individually or collaboratively, use the Internet or research literature to find two examples of mutations that affect meiosis and outline: (1) how meiosis differs in the mutant cells of individuals compared with non-mutant cells of individuals; and (2) what we have learned about the molecular mechanisms of meiosis by studying these mutations.

11.3 The Time and Place of Meiosis in Organismal Life Cycles

The time and place at which meiosis occurs follow one of three major patterns in the life cycles of eukaryotes **(Figure 11.7)**. The differences reflect the portions of the life cycle spent in the hap-

A. Animal life cycles

Diploid phase dominates the life cycle, the haploid phase is reduced, and meiosis is followed directly by gamete formation.

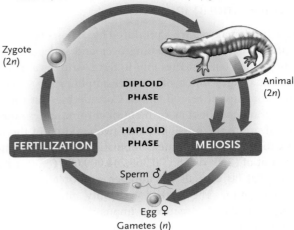

Zygote (2n)

DIPLOID PHASE

Animal (2n)

FERTILIZATION

HAPLOID PHASE

MEIOSIS

Sperm ♂

Egg ♀

Gametes (n)

B. All plants and some fungi and algae (fern shown; relative length of the two phases varies widely in plants)

Generations alternate between haploid and diploid phases, each of which is multicellular.

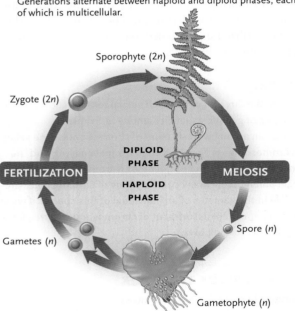

Sporophyte (2n)

Zygote (2n)

DIPLOID PHASE

FERTILIZATION

HAPLOID PHASE

MEIOSIS

Gametes (n)

Spore (n)

Gametophyte (n)

C. Other fungi and algae

Haploid phase is dominant and the diploid phase is reduced to a single cell.

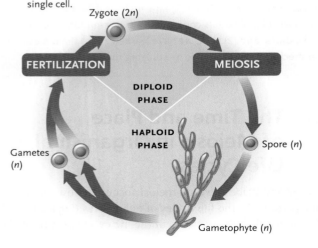

Zygote (2n)

FERTILIZATION

DIPLOID PHASE

MEIOSIS

HAPLOID PHASE

Gametes (n)

Spore (n)

Gametophyte (n)

FIGURE 11.7 Variations in the time and place of meiosis in eukaryotes. The diploid phase of the life cycles is shaded in green; the haploid phase is shaded in yellow. *n* = haploid number of chromosomes; 2*n* = diploid number.
© Cengage Learning 2014

loid and diploid phases and whether mitotic divisions intervene between meiosis and the formation of gametes.

In Animals, the Diploid Phase Predominates, the Haploid Phase Is Reduced, and Meiosis Is Followed Directly by Gamete Formation

Animals follow the pattern **(Figure 11.7A)** in which the diploid phase predominates during the life cycle, the haploid phase is reduced, and meiosis is followed directly by gamete formation. In male animals, each of the four nuclei produced by meiosis is enclosed in a separate cell by cytoplasmic divisions, and each of the four cells differentiates into a functional sperm cell. In female animals, only one of the four nuclei becomes functional as an egg cell nucleus.

Fertilization restores the diploid phase of the life cycle. Thus, animals are haploids only as sperm or eggs, and no mitotic divisions occur during the haploid phase of the life cycle.

In Plants and Some Fungi, Generations Alternate between Haploid and Diploid Phases That Are Both Multicellular

Plants and some algae and fungi follow the life cycle pattern shown in **Figure 11.7B.** These organisms alternate between haploid and diploid generations in which, depending on the organism, either generation may dominate the life cycle, and mitotic divisions occur in both phases. In these organisms, fertilization produces the diploid generation, in which the individuals are called **sporophytes** (*spora* = seed; *phyta* = plant). After the sporophytes grow to maturity by mitotic divisions, some of their cells undergo meiosis, producing haploid, genetically different, reproductive cells called **spores.** The spores are not gametes; they germinate and grow directly by mitotic divisions into a generation of haploid individuals called **gametophytes** (*gameta* = gamete). At maturity, the nuclei of some cells in gametophytes develop into egg or sperm nuclei. All the egg or sperm nuclei produced by a particular gametophyte are genetically identical because they arise through mitosis; meiosis does not occur in gametophytes. Fusion of a haploid egg and sperm nucleus produces a diploid zygote nucleus that divides by mitosis to produce the diploid sporophyte generation again.

In many plants, including most bushes, shrubs, trees, and flowers, the diploid sporophyte generation is the most visible part of the plant. The gametophyte generation is reduced to a mostly microscopic stage that develops in the reproductive parts of the sporophytes—for flowering plants, in the structures of the flower. The female gametophyte remains in the flower; the male gametophyte is often released from flowers as microscopic pollen grains. When pollen contacts the stigma (sticky receptacle on the female portion of a flower) of the same species, it generates a pollen tube that penetrates the ovule and ultimately releases a haploid nucleus that fertilizes a haploid egg cell of a female gametophyte in the flower. The resulting cell reproduces by mitosis to form a sporophyte.

In Some Fungi and Other Organisms, the Haploid Phase Is Dominant and the Diploid Phase Is Reduced to a Single Cell

The life cycle of some fungi and algae follows a third life cycle pattern **(Figure 11.7C)**. In these organisms, the diploid phase is limited to a single cell, the zygote, produced by fertilization. Immediately after fertilization, the diploid zygote undergoes meiosis to produce the haploid phase. Mitotic divisions occur only in the haploid phase.

During fertilization, two haploid gametes fuse to form a diploid zygote nucleus. This nucleus immediately enters meiosis, producing four haploid cells. These cells develop directly or after one or more mitotic divisions into haploid spores. These spores germinate to produce haploid individuals, the gameto-

phytes, which grow or increase in number by mitotic divisions. Eventually plus (+) and minus (−) mating types are formed in these individuals by differentiation of some of the cells produced by the mitotic divisions. Because the gametes ultimately are produced by mitosis from a single ancestral haploid cell, all the gametes of an individual are genetically identical.

In this chapter, we have seen that meiosis has three outcomes that are vital to sexual reproduction. Meiosis reduces the chromosomes to the haploid number so that the chromosome number does not double at fertilization. Through crossing-over and independent assortment of chromosomes, meiosis produces genetic variability; further variability is provided by the random combination of gametes in fertilization. The next chapter shows how the outcomes of meiosis and fertilization underlie the inheritance of traits in sexually reproducing organisms.

STUDY BREAK 11.3

How does the place of meiosis differ in the life cycles of animals and most plants?

UNANSWERED QUESTIONS

Whereas scientists have detailed the physical events that occur during meiosis, many of the molecular mechanisms and regulatory pathways that operate in meiosis are still poorly understood. Nonetheless, technological advances in high-resolution microscopy and the establishment of a wide range of genetically tractable model organisms have allowed researchers all over the world to make tremendous progress in understanding meiosis at the molecular level.

How do homologous chromosomes pair as the cell enters meiosis?

Early in meiosis I, homologous chromosomes comprising the newly replicated genome need to find each other within the nucleus and pair. Recent research with the fission yeast *Schizosaccharomyces pombe*, the budding yeast *Saccharomyces cerevisiae* (see *Focus on Model Research Organisms* in Chapter 10), and the worm *Caenorhabditis elegans* (see *Focus on Model Research Organisms* in Chapter 30) has shown that a specific family of nuclear envelope proteins connects cytoskeletal networks in the cytoplasm to chromosome ends in the nucleus. Different motors (dynein microtubules in the case of fission yeast and worms, and actin filaments in the case of budding yeast) drive the movement of chromosomes tethered to these nuclear envelope proteins. By means of these movements the matching sequences of homologous chromosomes are able to come together and become aligned.

How are the pairing interactions between homologous chromosomes stabilized?

Once formed, the pairing interactions between homologous chromosomes must be stabilized. A critical component for stabilizing the pairing interactions of homologous chromosomes is the formation of the synaptonemal complex (see Figure 11.5). While this structure, first observed over 50 years ago, is ubiquitously present during meiosis from yeast to humans, its components, organization, and function are only now starting

to be understood. For example, recent work in budding yeast, worms, mice, and the fruit fly *Drosophila melanogaster* (see *Focus on Model Research Organisms* in Chapter 13) has demonstrated that the formation of the synaptonemal complex is critical for the completion of crossover events between homologs. Some important questions remain unanswered, however. For example, how is the assembly and disassembly of the synaptonemal complex regulated and how does this structure interface with proteins involved in other processes occurring during meiosis?

How is the formation of crossovers ensured during meiosis?

Once the interaction between homologous chromosomes is stabilized, crossing-over must occur in order to link homologs physically so that there will be sufficient mechanical tension between them as they align later at the metaphase I plate and are separated from each other by microtubules during the metaphase I to anaphase I transition. A highly conserved meiosis-specific protein known as Spo11 has been shown to be important in the formation of crossovers. That is, Spo11 causes the generation of physical breaks in the DNA along the paired homologous chromosomes. Repair of the breaks is initiated, which, at a subset of sites along the chromosomes, results in the formation of crossovers between nonsister chromatids. Here, an unanswered question is: how are both the frequency and distribution of crossing-over events regulated?

How do environmental chemicals affect meiosis?

Synthetic chemicals are everywhere in our environment, from those present in plastics to those in household cleaning products. One of the lines of research in our lab involves identifying which chemicals impair meiosis and understanding their mechanism of action. We are addressing this in the worm, *C. elegans*, which is amenable to various genetic, molecular, cytological, and biochemical approaches. As a proof-of-concept we demonstrated that key events in meiosis, namely the formation of the synaptonemal complex and the progression of meiotic recombination, were impaired following exposure to a commonly used plasticizer called Bisphenol A (BPA). This result is of particular importance given that BPA has been associated with increased risk of miscarriages in humans. Moreover, we demonstrated that BPA exposure in worms, at doses that are physiologically relevant to humans, altered the germline expression of a subset of DNA break repair genes also present in humans. This simple biological system can therefore be extremely useful for the identification of the mechanism of action of these chemicals. We are currently designing various molecular approaches that will allow us to use the worm to study the effects on reproductive biology of hundreds of chemicals. The results from such experiments will provide key insights into the impact of environmental toxicants on human reproductive health.

THINK LIKE A SCIENTIST If homologous chromosomes need to pair and synapse on entrance into meiosis, what happens when one of the homologs has undergone rearrangements resulting in large regions being either duplicated or inverted?

Courtesy of Monica Colaiácovo

Monica Colaiácovo is an Associate Professor in the Department of Genetics at Harvard Medical School. Her main research interests include understanding the mechanisms underlying germline maintenance and accurate chromosome inheritance during meiosis. To learn more about Dr. Colaiácovo's research go to http://genepath.med.harvard.edu/colaiacovo/index.php.

REVIEW KEY CONCEPTS

To access the course materials and companion resources for this text, please visit www.cengagebrain.com.

11.1 The Mechanisms of Meiosis

- The major cellular processes that underlie sexual reproduction are the halving of chromosome number by meiosis and restoration of the number by fertilization. Meiosis and fertilization also produce new combinations of genetic information (Figure 11.1).

- Meiosis occurs in eukaryotes that reproduce sexually and typically in organisms that are at least diploid—that is, organisms that have at least two representatives of each chromosome.

- DNA replicates and the chromosomal proteins are duplicated during the premeiotic interphase, producing two copies, the sister chromatids, of each chromosome.

- During prophase I of the first meiotic division (meiosis I), the replicated chromosomes condense and come together and pair. While they are paired, chromatids of homologous chromosomes exchange segments by crossing-over. While these events are in progress, the spindle forms in the cytoplasm. The crossing-over events become visible later as chiasmata (Figures 11.2–11.4).

- During prometaphase I, the nuclear envelope breaks down, the spindle enters the former nuclear area, and kinetochore microtubules leading to opposite spindle poles attach to one kinetochore of each set of sister chromatids of homologous chromosomes (Figure 11.3).

- At metaphase I, spindle microtubule movements have aligned the tetrads on the metaphase plate, the equatorial plane between the two spindle poles. The connections of kinetochore microtubules to opposite poles ensure that the homologous chromosomes of each pair segregate and move to opposite spindle poles during anaphase I, reducing the chromosome number to the haploid value. Each chromosome at the poles still contains two chromatids.

- Telophase I and interkinesis are brief and transitory stages; no DNA replication occurs during interkinesis. During these stages, the single spindle of the first meiotic division disassembles.

- During prophase II, the chromosomes condense and a spindle forms. During prometaphase II, the nuclear envelope breaks down, the spindle enters the former nuclear area, and spindle microtubules leading to opposite spindle poles attach to the two kinetochores of each chromosome. At metaphase II, the chromosomes align on the metaphase plate. The connections of kinetochore microtubules to opposite spindle poles ensure that during anaphase II, the chromatids of each chromosome are separated and migrate to those opposite spindle poles.

- During telophase II, the chromosomes decondense to their extended interphase state, the spindles disassemble, and new nuclear envelopes form. The result is four haploid cells, each containing half the number of chromosomes present in a G_1 nucleus of the same species.

  Animation: Meiosis

  Animation: Generalized life cycles

  Animation: Mitosis and Meiosis compared

  Animation: Haploid and Diploid Cells

  Animation: Meiosis and Mitosis Drag and Drop

11.2 Mechanisms That Enhance Genetic Variability

- Crossing-over recombines genes on paired homologous chromosomes during meiosis (Figure 11.5). Chromatids acquire new combinations of alleles—recombinants—by physically exchanging segments in crossing-over. The exchange process involves precise breakage and exchange of DNA molecules through a complex mechanism. It is catalyzed by enzymes and occurs while the homologous chromosomes are held together tightly by the synaptonemal complex.

- The chiasmata visible between the chromosomes at late prophase I reflect the exchange of chromatid segments that occurred during the molecular steps of crossing-over.

- The independent assortment of homologous chromosomes is another mechanism through which meiosis produces genetic variability. The homologous pairs segregate at anaphase I of meiosis, the orientation and segregation of one pair having no influence on the orientation and segregation of any other pair (Figure 11.6).

- Random union of male and female gametes at fertilization is a third mechanism for enhancing genetic variability.

  Animation: Crossover review

  Animation: Independent assortment

11.3 The Time and Place of Meiosis in Organismal Life Cycles

- The time and place of meiosis follow one of three major pathways in the life cycles of eukaryotes, which reflect the portions of the life cycle spent in the haploid and diploid phases and whether mitotic divisions intervene between meiosis and the formation of gametes (Figure 11.7).

- In animals, the diploid phase predominates during the life cycle; mitotic divisions occur only in this phase. Meiosis in the diploid phase gives rise to products that develop directly into egg and sperm cells without undergoing mitosis (Figure 11.7A).
- In all plants and some fungi, the life cycle alternates between haploid and diploid generations that both grow by mitotic divisions. Fertilization produces the diploid sporophyte generation; after growth by mitotic divisions, some cells of the sporophyte undergo meiosis and produce haploid spores. The spores germinate and grow by mitotic divisions into the gametophyte genera-

tion. After growth of the gametophyte, cells develop directly into egg or sperm nuclei, which fuse in fertilization to produce the diploid sporophyte generation again (Figure 11.7B).
- In some fungi and protists, meiosis occurs immediately after fertilization, producing a haploid phase, which predominates during the life cycle; mitosis occurs only in the haploid phase. At some point in the life cycle, haploid cells differentiate directly into gametes, which fuse to produce the brief diploid phase (Figure 11.7C).

UNDERSTAND & APPLY

Test Your Knowledge

1. The chromosome constitution number of this individual is $2n = 6$.

© Cengage Learning 2014

   This drawing represents
   a. mitotic metaphase.
   b. meiotic metaphase I.
   c. meiotic metaphase II.
   d. a gamete.
   e. six nonhomologous chromosomes.

2. Which of the following is *not* associated with sperm production?
   a. daughter cells identical to the parent cell
   b. variety in resulting cells
   c. chromosome number halved in resulting cells
   d. four daughter cells arising from one parent cell
   e. 23 chromosomes in the human sperm

3. Chiasmata:
   a. form during metaphase II of meiosis.
   b. occur between two nonhomologous chromosomes.
   c. represent chromosomes independently assorting.
   d. are sites of DNA exchange between homologous chromatids.
   e. ensure the resulting cells are identical to the parent cell.

4. If $2n$ is four, the number of possible combinations of chromosomes in the resulting gametes, excluding any crossing-over is:
   a. 1.
   b. 2.
   c. 4.
   d. 8.
   e. 16.

5. In meiosis:
   a. homologous chromosomes pair at prophase II.
   b. chromosomes segregate from their homologous partners at anaphase I.
   c. the centromeres split at anaphase I.
   d. a female gamete has two X chromosomes.
   e. reduction of chromosome number occurs in meiosis II.

6. The DNA content in a diploid cell in G_2 is X. If that cell goes into meiosis at metaphase II, the DNA content will be:
   a. 0.1X.
   b. 0.5X.
   c. X.
   d. 2X.
   e. 4X.

7. Metaphase in mitosis is similar to what stage in meiosis?
   a. prophase I
   b. prophase II
   c. metaphase I
   d. metaphase II
   e. crossing-over

8. In the human sperm:
   a. there must be one chromosome of each type, except for the sex chromosomes, where either an X or a Y chromosome is present.
   b. a chromosome must be represented from each parent.
   c. there must be an unequal number of chromosomes from both parents.
   d. there must be representation of chromosomes from only one parent.
   e. there is the possibility of 246 different combinations of maternal and paternal chromosomes.

9. In plants, the adult diploid individuals are called:
   a. spores.
   b. sporophytes.
   c. gametes.
   d. gametophytes.
   e. zygotes.

10. Which of the following sequences of events describes the general life cycle of an animal?
    a. $2n \rightarrow$ meiosis $\rightarrow 2n \rightarrow$ fertilization $\rightarrow 1n$
    b. $1n \rightarrow$ meiosis $\rightarrow 2n \rightarrow$ fertilization $\rightarrow 1n$
    c. $2n \rightarrow$ meiosis $\rightarrow 1n \rightarrow$ fertilization $\rightarrow 2n$
    d. $2n \rightarrow$ mitosis $\rightarrow 1n \rightarrow$ fertilization $\rightarrow 2n$
    e. $2n \rightarrow$ mitosis $\rightarrow 1n \rightarrow$ fertilization $\rightarrow 1n$

Discuss the Concepts

1. You have a technique that allows you to measure the amount of DNA in a cell nucleus. You establish the amount of DNA in a sperm cell of an organism as your baseline. Which multiple of this amount would you expect to find in a nucleus of this organism at G_2 of premeiotic interphase? At telophase I of meiosis? During interkinesis? At telophase II of meiosis?

2. One of the human chromosome pairs carries a gene that influences eye color. In an individual human, one chromosome of this pair has an allele of this gene that contributes to the formation of blue eyes. The other chromosome of the pair has an allele that contributes to brown eye color (other genes also influence eye color in humans). After meiosis in the cells of this individual, what fraction of the nuclei will carry the allele that contributes to blue eyes? To brown eyes?

3. Mutations are changes in DNA sequence that can create new alleles. In which cells of an individual, somatic or meiotic cells, would mutations be of greatest significance to the survival of that individual? What about to the species to which the individual belongs?

Design an Experiment

Design experiments to determine whether a new pesticide on the market adversely affects egg production and fertilization in frogs.

Interpret the Data

A wingless female aphid can generate up to 100 almost-identical female offspring who themselves reproduce similarly, thereby creating a clone of offspring similar to the original female. In these cases, no eggs are laid, and the young are born as juveniles. However, if day length (photoperiod) and temperature change significantly (for example, in fall in temperate climates), the wingless female can switch and produce a combination of winged males, winged females that reproduce without laying eggs, or egg-laying females with wings. Depending on circumstances and aphid species, different proportions of each of these are produced.

The **Table** represents some data from an early study looking at the effect of temperature and day length (or night length) on the number of males produced by individual females of the species *Megoura viciae*.

Temperature	Photoperiod (hr light/24 hr)	Number of mothers	Number of males per mother	
			Range	Mean
25°C	12	9	0–0	0
	16	9	0–0	0
20°C	12	27	5–21	13.7
	16	44	0–21	8.0
15°C	12	37	3–17	11.6
	16	25	1–20	10.4
11°C	12	19	0–5	2.5
	16	32	0–9	2.9

Reprinted from A. D. Lees. 1959. The role of photoperiod and temperature in the determination of parthenogenetic and sexual forms in the aphid *Megoura viciae* Buckton—I: The influence of these factors on apterous virginoparae and their progeny. *Journal of Insect Physiology* 3:92–117, with permission from Elsevier.

1. At the temperatures used females produce about 100 offspring each, typically a mix of males and females. Assuming this number of offspring, what was the highest and lowest percentage of male offspring per female parent at 20°C? What temperature range appears to be optimal for male production? Which temperature(s) correspond(s) with the production of no males? With less than 10% males?

2. The effect of photoperiod on male production was also studied. Various day lengths were simulated in the laboratory using artificial light. Citing examples from the data, what effect does photoperiod have on male production? How does it compare to the effect of temperature?

Source: A. D. Lees. 1959. The role of photoperiod and temperature in the determination of parthenogenetic and sexual forms in the aphid *Megoura viciae* Buckton—I: The influence of these factors on apterous virginoparae and their progeny. *Journal of Insect Physiology* 3:92–117.

Apply Evolutionary Thinking

Explain aspects of the processes of mitosis and meiosis that would lead you to conclude that they are evolutionarily related processes. Do you think that mitosis evolved from meiosis, or did the opposite occur? Explain your conclusion.

Rabbits, showing genetic variation in coat color.

12

Mendel, Genes, and Inheritance

Why it matters . . . Parties and champagne were among the last things on Ernest Irons' mind on New Year's Eve, 1904. Irons, a medical intern, was examining a blood specimen from a new patient and was sketching what he saw through his microscope—peculiarly elongated red blood cells **(Figure 12.1)**. He and his supervisor, James Herrick, had never seen anything like them. The shape of the cells was reminiscent of a sickle, a cutting tool with a crescent-shaped blade.

The patient had complained of weakness, dizziness, shortness of breath, and pain. His father and two sisters had died from mysterious ailments that had damaged their lungs or kidneys. Did those deceased family members also have sickle-shaped red cells in their blood? Was there a connection between the abnormal cells and the ailments? How did the cells become sickled?

The medical problems that baffled Irons and Herrick killed their patient when he was only 32 years old. The patient's symptoms were characteristic of a genetic disorder now called *sickle-cell anemia*. This disease develops when a person has received two mutated copies of a gene (one from each parent) that codes for a subunit of hemoglobin, the oxygen-transporting protein in red blood cells. The mutated gene codes for a slightly altered form of the hemoglobin subunit. When oxygen supplies are low, the altered hemoglobin forms long, fibrous, helical structures that distend

A. A normal red blood cell

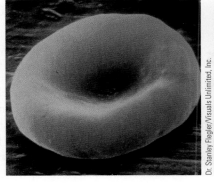

B. A sickled red blood cell

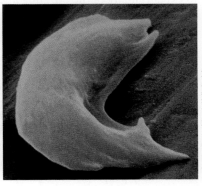

FIGURE 12.1 Red blood cell shape in sickle-cell anemia.

FIGURE 12.2
Gregor Mendel (1822–1884), the founder of modern genetics.

M. Hofer/National Library of Medicine

red blood cells into the sickle shape. The mutant protein differs from the nonmutant protein by just a single amino acid.

The sickled red blood cells are too elongated and inflexible to pass freely through the capillaries, the smallest vessels in the circulatory system. As a result, the cells tend to increase in number and cluster at capillary openings, obstructing the flow of blood, when oxygen levels are low. Tissues become starved for oxygen and saturated with metabolic wastes, causing the symptoms experienced by Irons and Herrick's patient. The problem worsens as oxygen concentration falls in tissues and more red blood cells are pushed into the sickled form. (You will learn more about sickle-cell anemia in this chapter and in Chapter 13.)

Researchers have studied sickle-cell anemia in great detail at both the molecular and the clinical levels. Interestingly, though, our understanding of sickle-cell anemia—and other inherited traits—actually began with studies of pea plants in a monastery garden.

Almost fifty years before Ernest Irons sketched sickled red blood cells, a scholarly monk named Gregor Mendel **(Figure 12.2)** used garden peas to study patterns of inheritance. To test his hypotheses about inheritance, Mendel bred generation after generation of pea plants and carefully observed the patterns by which parents transmit traits to their offspring. Through his experiments and observations, Mendel discovered the fundamental principles that govern inheritance. His discoveries and conclusions founded the science of genetics and still have the power to explain many of the puzzling and sometimes devastating aspects of inheritance.

12.1 The Beginnings of Genetics: Mendel's Garden Peas

Until about 1900, scientists and the general public believed in the **blending theory of inheritance,** which suggested that hereditary traits blend evenly in offspring through mixing of the parents' blood, much like the effect of mixing coffee and cream. Even today, many people assume that parental characteristics such as

skin color, body size, and facial features blend evenly in their offspring, with the traits of the children appearing about halfway between those of their parents. Yet if blending takes place, extremes, such as very tall and very short individuals, should gradually disappear over generations as repeated blending takes place, yet they do not. Also, why do children with blue eyes keep turning up among the offspring of brown-eyed parents?

Gregor Mendel's experiments with garden peas, performed in the 1860s, were not done to address directly the blending theory of inheritance, but their results provided answers to these questions and many more. Mendel was an Augustinian monk who lived in a monastery in Brno, now part of the Czech Republic. But he had an unusual education for a monk of his time. He had studied mathematics, chemistry, zoology, and botany at the University of Vienna under some of the foremost scientists of his day. He had also been reared on a farm and was well aware of agricultural principles and their application. He kept abreast of breeding experiments published in scientific journals. Mendel also won several awards for developing improved varieties of fruits and vegetables.

In his work with peas, Mendel studied a variety of *characters.* A **character** is a specific heritable attribute or property of an organism. It may be a visible attribute, or it may be one that can only be detected by biochemical or molecular analysis. The characters Mendel studied included seed shape, seed color, and flower color. Mendel studied plants that had alternative forms of these characters, known as **character differences** or **traits.** For example, for the flower color character, the two traits studied—the alternative forms—were purple and white. Mendel established that characters are passed to offspring in the form of discrete hereditary factors, which now are known as genes. Mendel observed that, rather than blending evenly, many parental traits appear unchanged in offspring, whereas others disappear in one generation to reappear unchanged in subsequent generations produced by particular breeding experiments. Although Mendel did not know it, the inheritance patterns he observed are the result of the segregation of chromosomes, on which the genes are located, to gametes in meiosis (see Chapter 11). Mendel's methods illustrate, perhaps as well as any experiments in the history of science, how rigorous scientific work is conducted: through observation, making hypotheses, and testing the hypotheses with experiments. And, although others had studied inheritance patterns before him, Mendel's most important innovation was his quantitative approach to science, specifically his rigor and statistical analysis in an era when qualitative, descriptive science was the accepted practice.

Mendel Chose True-Breeding Garden Peas for His Experiments

Mendel chose the garden pea *(Pisum sativum)* for his genetics experiments because the plant could be grown easily in the monastery garden, without elaborate equipment. **Figure 12.3** shows the structure of a pea flower. Normally, pea plants **self-fertilize** (also known as **self-pollinate,** or more simply, *self*): sperm nuclei in pollen produced by anthers fertilize egg cells

FIGURE 12.3 | **Research Method**

Making a Genetic Cross between Two Pea Plants

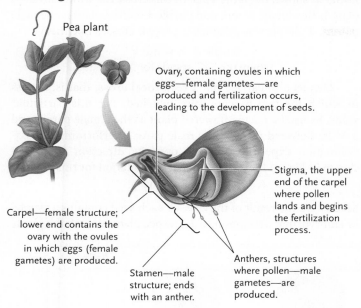

Pea plant

Ovary, containing ovules in which eggs—female gametes—are produced and fertilization occurs, leading to the development of seeds.

Stigma, the upper end of the carpel where pollen lands and begins the fertilization process.

Carpel—female structure; lower end contains the ovary with the ovules in which eggs (female gametes) are produced.

Anthers, structures where pollen—male gametes—are produced.

Stamen—male structure; ends with an anther.

Purpose: Mendel used the garden pea, *Pisum sativum,* for his genetic experiments. The goal of the experiments was to test various hypotheses about the patterns of inheritance by cross-breeding plants with easily observable characters, such as flower color and seed shape. He could then analyze whether the characters he observed and counted in the offspring supported the predictions made by a particular hypothesis.

In cross-breeding, the sperm and the egg must come from different plants. However, this type of flowering plant has both male and female structures within the same flower and is capable of self-fertilization, also called "selfing." The figure to the left shows a pea flower sectioned to show the location of the reproductive structures. (Details of plant fertilization are presented in Chapter 35).

The figure below shows how Mendel designed his experiments to prevent selfing and perform his crosses.

Protocol:

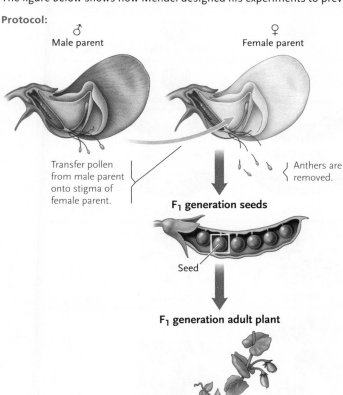

♂
Male parent

♀
Female parent

Transfer pollen from male parent onto stigma of female parent.

Anthers are removed.

F₁ generation seeds

Seed

F₁ generation adult plant

1. Remove the anthers from one of the parents (the white-flowered plant) to prevent self-fertilization. Transfer pollen from the male parent (the purple-flowered plant) onto the stigma of the white flower (the female parent). This results in cross-fertilization, the fertilization of one plant with pollen from another.

2. The cross-fertilized plant produces seeds. Seeds may be scored for seed traits, such as round vs. wrinkled shape.

3. Seeds are grown into adult plants. Plants may be scored for adult traits, such as purple vs. white flower color.

housed in the ovule of the same flower. (Reproduction in flowering plants is described in Chapter 36.) However, for his experiments, Mendel prevented self-fertilization simply by removing the anthers (see Figure 12.3). Pollen to fertilize these flowers had to come from a different plant. This technique is called **cross-fertilization** or **cross-pollination,** or more simply and more generally, a *cross*. Using this method, Mendel tested the effects of mating pea plants of different parental types.

To begin his experiments, Mendel chose pea plants that were known to be **true-breeding** (also called *pure-breeding*); that is, when self-fertilized—*selfed*—they passed traits without change from one generation to the next.

Mendel First Worked with Crosses of Plants Differing in One Character

Mendel selected seven characters for study **(Figure 12.4).** Flower color was among the characters: one true-breeding variety of

peas had purple flowers, and the other true-breeding variety had white flowers.

Mendel fertilized a white flower with pollen from a purple flower as shown in Figure 12.3. In this cross, the white-flowered plant is the female parent and purple-flowered plant is the male parent. A simple way geneticists present crosses like this is:

purple ♂ × white ♀
where the "×" stands for "cross."

Mendel also carried out a **reciprocal cross,** that is, a cross in which the two parents were switched. For this particular study, he used a purple-flowered plant as the female parent and a white-flowered plant as the male parent, performing a cross-pollination experiment in the opposite direction from that shown in Figure 12.3. The simple representation for this cross is:

white ♂ × purple ♀

Seeds were the result of the crosses; each seed contains a zygote, or embryo, that develops into a new pea plant. We call the plants

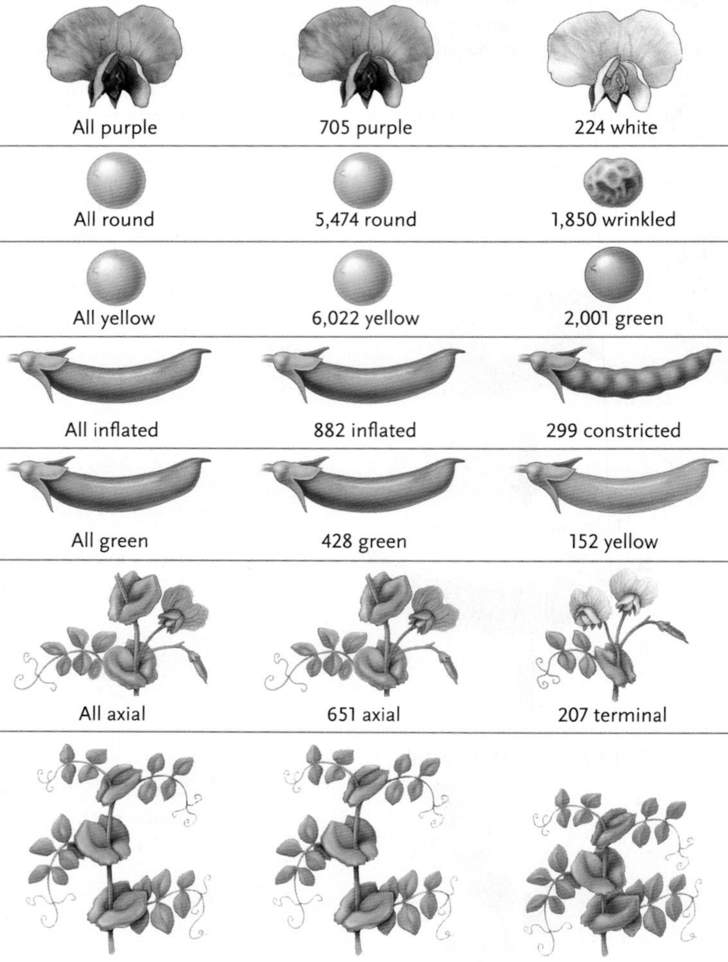

Character	Traits crossed	F_1	F_2		Ratio
Flower color	purple × white	All purple	705 purple	224 white	3.15 : 1
Seed shape	round × wrinkled	All round	5,474 round	1,850 wrinkled	2.96 : 1
Seed color	yellow × green	All yellow	6,022 yellow	2,001 green	3.01 : 1
Pod shape	inflated × constricted	All inflated	882 inflated	299 constricted	2.95 : 1
Pod color	green × yellow	All green	428 green	152 yellow	2.82 : 1
Flower position	axial (along stems) × terminal (at tips)	All axial	651 axial	207 terminal	3.14 : 1
Stem length	tall × dwarf	All tall	787 tall	277 dwarf	2.84 : 1

FIGURE 12.4 Mendel's crosses with seven characters in peas, including his results and the calculated ratios of offspring.
© Cengage Learning 2014

used in an initial cross between two true-breeding parents the parental or **P generation.** We call the first generation of off-spring from a cross between two true-breeding parents the **F₁ generation** (F stands for *filial; filius* = son).

From the purple ♂ × white ♀ cross, the plants that grew from the F₁ seeds all formed purple flowers, as if the trait for white flowers had disappeared. The flowers showed no evidence of blending. Exactly the same result was obtained for the reciprocal cross of white ♂ × purple ♀, showing that the sex of the parent did not affect the inheritance pattern of the traits.

Mendel then allowed purple-flowered F₁ plants to self, producing seeds that represented the **F₂ generation.** When he planted the F₂ seeds produced by this cross, plants with the white-flowered trait reappeared along with purple-flowered plants. Mendel counted 705 plants with purple flowers and 224 with white flowers, in a ratio that he noted was close to 3 purple:1 white.

Mendel made similar crosses that involved six other characters with pairs of traits; for example, the character of seed color has the traits yellow and green (see Figure 12.4). In all cases, he observed a uniform F₁ generation, in which only one of the two traits was present. In the F₂ generation, the missing trait reappeared, and both traits were present among the off-spring. Moreover, Mendel noted that, for each character, the ratio of the two traits was also close to 3:1.

Single-Character Crosses Led Mendel to Propose the Principle of Segregation

Mendel made three important conclusions to explain the results of his crosses of plants differing in one character. The first two conclusions led to the third and most important conclusion, the *principle of segregation.*

Conclusion 1: *The adult plants carry a* pair *of factors that govern the inheritance of each trait.* Mendel correctly deduced that for each trait, an organism inherits one factor from each parent.

In modern terminology, Mendel's factors are *genes,* which are located on chromosomes; the different versions of a gene, producing different traits of a character, are **alleles** of the gene (see Section 11.1). Although Mendel did not use the terms *genes* and *alleles,* we use them in this chapter in our description of Mendel's work. Thus, there are two alleles of the gene that governs flower color in garden peas: one allele for purple flower color and another allele for white flower color. Organisms with two copies of each gene are known as diploids (see Section 11.1); the two alleles of a gene in a diploid individual may be identical or different.

Conclusion 2: Mendel's second conclusion (in modern language) was: *If an individual's pair of genes consists of different alleles, one allele is dominant over the other, which is recessive.* Mendel had to explain why one of the traits, such as white flowers, disappears in the F₁ generation and then reappears in the F₂ generation. Mendel deduced that the trait that "disappeared" in the F₁ generation was actually present but was masked by the

"stronger" allele. Mendel called the masking effect **dominance.** When a dominant allele of a gene is paired with a recessive allele of that gene, the dominant allele determines the trait that appears. By contrast, the trait determined by the recessive allele appears only when two copies of that allele are present. For example, the allele for purple flowers is dominant and the allele for white flowers is recessive.

Conclusion 3: Mendel's third conclusion was: *The pairs of alleles that control a character segregate (separate) as gametes are formed; half the gametes carry one allele, and the other half carry the other allele.* This conclusion is now known as Mendel's **Principle of Segregation.** During fertilization, fusion of the haploid maternal and paternal gametes produces a diploid nucleus called the *zygote nucleus.* The zygote nucleus receives one allele for the character from the male gamete and one allele for the same character from the female gamete, reuniting the pairs.

Figure 12.5 steps through the purple × white pea cross to illustrate the principle of segregation. Some important genetic terms are used in the figure:

- **Homozygote** (*homo* = same): A true-breeding individual with both alleles of a gene the same. A homozygote produces only one type of gamete: It contains one copy of that allele.
- **Homozygous:** An individual that is a homozygote is said to be homozygous for the particular allele of the gene.
- **Heterozygote** (*hetero* = different): An individual with two different alleles of a gene. A heterozygote produces two types of gametes: one type has a copy of one allele, the other type has a copy of the other, different, allele.
- **Heterozygous:** An individual that is a heterozygote is said to be heterozygous for the pair of different alleles of a gene.
- **Monohybrid** (*mono* = one; *hybrid* = an offspring of parents with different traits): An F₁ heterozygote produced from a cross that involves a single character.
- **Monohybrid cross:** A cross between two individuals that are each heterozygous for the same pair of alleles.

Mendel's results explain how individuals may differ genetically but still look the same. The *PP* and *Pp* plants, although genetically different, both have purple flowers (see Figure 12.5). In modern terminology, **genotype** refers to the *genetic constitution of an organism in terms of genes and alleles,* and **phenotype** (Greek *phainein* = to show) refers to its *appearance.* In this case, the two different *genotypes PP* and *Pp* produce the same purple-flower *phenotype.*

How does the genotype relate to the phenotype for the flower-color trait? Conceptually, the *P* allele of the gene encodes a product that is needed for the synthesis of the purple pigment in the flower. Therefore, a *PP* plant has purple flowers. The *p* allele is a mutant form of the gene; the product of this allele is inactive or mostly inactive and therefore, in *pp* plants, purple pigment cannot be made. In the absence of purple pigment, the flower is white. The *Pp* heterozygote is purple, rather than intermediate between purple and white, because the amount and activity of the product of the *P* allele is sufficient to enable a normal amount of purple pigment to be synthesized.

FIGURE 12.5 | **Experimental Research**

The Principle of Segregation: Inheritance of Flower Color in Garden Peas

Question: How is flower color in garden peas inherited?

Experiment: Mendel crossed a true-breeding purple-flowered plant with a true-breeding white-flowered plant and analyzed the progeny through the F_1 and F_2 generations. We explain this cross here in modern terms.

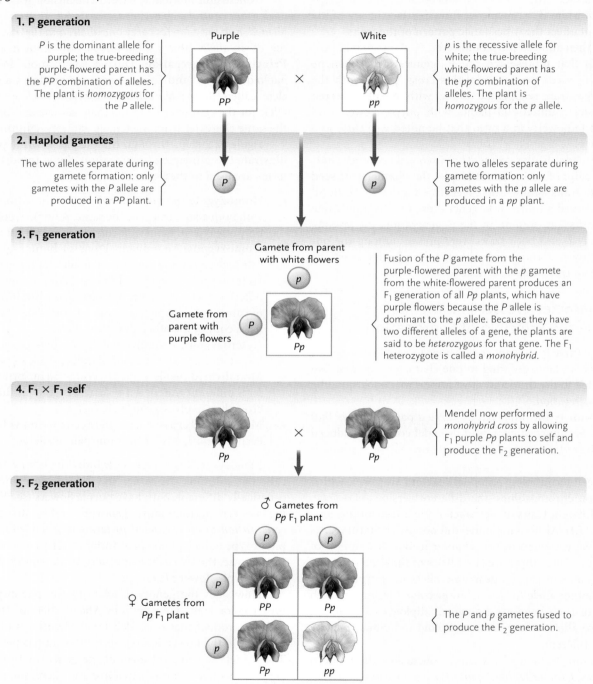

1. P generation

P is the dominant allele for purple; the true-breeding purple-flowered parent has the *PP* combination of alleles. The plant is *homozygous* for the *P* allele.

Purple — *PP* × White — *pp*

p is the recessive allele for white; the true-breeding white-flowered parent has the *pp* combination of alleles. The plant is *homozygous* for the *p* allele.

2. Haploid gametes

The two alleles separate during gamete formation: only gametes with the *P* allele are produced in a *PP* plant.

P   *p*

The two alleles separate during gamete formation: only gametes with the *p* allele are produced in a *pp* plant.

3. F_1 generation

Gamete from parent with white flowers — *p*

Gamete from parent with purple flowers — *P*

Pp

Fusion of the *P* gamete from the purple-flowered parent with the *p* gamete from the white-flowered parent produces an F_1 generation of all *Pp* plants, which have purple flowers because the *P* allele is dominant to the *p* allele. Because they have two different alleles of a gene, the plants are said to be *heterozygous* for that gene. The F_1 heterozygote is called a *monohybrid*.

4. F_1 × F_1 self

Pp   ×   *Pp*

Mendel now performed a *monohybrid cross* by allowing F_1 purple *Pp* plants to self and produce the F_2 generation.

5. F_2 generation

♂ Gametes from *Pp* F_1 plant

	P	*p*
P	*PP*	*Pp*
p	*Pp*	*pp*

♀ Gametes from *Pp* F_1 plant

The *P* and *p* gametes fused to produce the F_2 generation.

Results: Mendel's selfing of the F_1 purple-flowered plants produced an F_2 generation consisting of 3/4 purple-flowered and 1/4 white-flowered plants. White flowers were inherited as a recessive trait, disappearing in the F_1 and reappearing in the F_2.

Conclusion: The results supported Mendel's Principle of Segregation hypothesis that the pairs of alleles that control a character segregate as gametes are formed, with half of the gametes carrying one allele, and the other half carrying the other allele.

THINK LIKE A SCIENTIST Suppose you pick at random one of the F_2 purple-flowered plants and allow it to self. What is the chance that the progeny will include both purple-flowered and white-flowered plants?

Mendel Could Predict Both Classes and Proportions of Offspring from the Crosses He Made

Mendel could predict both which traits would appear in the offspring of a cross and their proportions. To understand this, let us review the mathematical rules that govern **probability**—that is, the possibility that an outcome will occur if it is a matter of chance, as in the random fertilization of an egg by a sperm cell that contains one allele or another.

In the mathematics of probability, we predict the likelihood of an outcome on a scale of 0 to 1. An outcome that is certain has a probability of 1, and an impossible outcome has a probability of 0. If two different outcomes are equally likely, as in getting heads or tails in tossing a coin, the probability of one of the outcomes is calculated by dividing that outcome by the total number of possible outcomes. The probability of obtaining a head in tossing a coin is 1 divided by 2, or 1/2. The probability of obtaining a tail is also 1 divided by 2, or 1/2. The probabilities of all the possible outcomes, when added together, must equal 1. Thus, a coin toss has only two possible outcomes, heads or tails, each with a probability of 1/2; the sum of these probabilities is: 1/2 + 1/2 = 1.

THE PRODUCT RULE IN PROBABILITY What is the chance of tossing two heads in succession? Because the outcome of one toss has no effect on the next one, the two successive tosses are independent. When two or more events are independent, the probability that they will occur in succession is calculated using the **product rule**—their individual probabilities are multiplied. That is, the probability that events A and B *both* will occur equals the probability of event A *multiplied* by the probability of event B. For example, the probability of getting heads on the first toss is 1/2; the probability of heads on the second toss is also 1/2 **(Figure 12.6)**. Because the events are independent, the probability of getting two heads in a row is $1/2 \times 1/2 = 1/4$. Applying the same principles, the probability of getting two tails is also $1/2 \times 1/2 = 1/4$ (see Figure 12.6). Similarly, because the sex of one child has no effect on the sex of the next child in a family, the probability of having four girls in a row is the product of their individual probabilities (very close to 1/2 for each birth): $1/2 \times 1/2 \times 1/2 \times 1/2 = 1/16$.

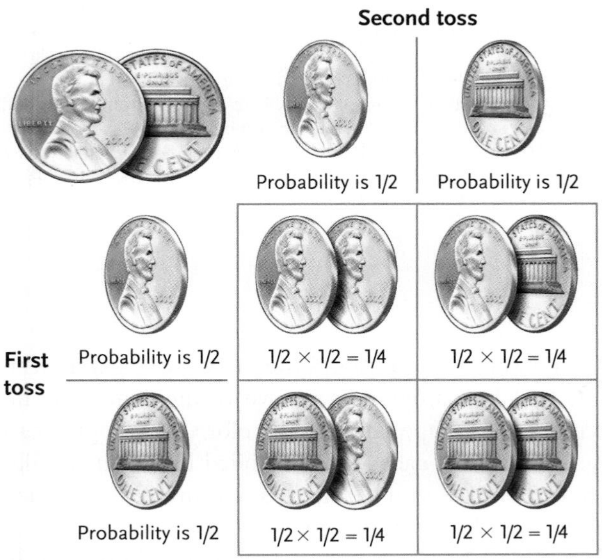

THE SUM RULE IN PROBABILITY We apply a different relationship, the **sum rule,** when there are two or more different ways of obtaining the same outcome. Returning to the coin toss example, we can determine the probability of getting a head and a tail in two tosses. We could toss the coin twice and get a head, then a tail. The probability that this will occur is 1/2 for the head $\times$ 1/2 for the tail = 1/4 (see Figure 12.6). However, we could toss the coin twice and get first a tail, then a head. The probability that this will occur is 1/2 for the tail $\times$ 1/2 for the head, which also = 1/4 (see Figure 12.6). Therefore, for the probability of tossing a head and a tail, we sum the individual probabilities to get the final probability: here, 1/4 + 1/4 = 1/2. (The other two probabilities for two coin tosses are 1/4 for two heads, and 1/4 for two tails.)

PROBABILITY IN MENDEL'S CROSSES The same rules of probability just discussed apply to Mendel's crosses. For example, **Figure 12.7** shows the rules of probability applied to calculating the proportion of *PP*, *Pp*, and *pp* F_2 plants from a cross of two F_1 *Pp* purple-flowered plants.

What if we want to know the probability of obtaining purple flowers in the cross *Pp* $\times$ *Pp*? (As you have learned, in peas this cross occurs by self-fertilization of heterozygous F_1 plants.) In this case, the rule of addition applies, because there are two ways to get purple flowers: genotypes *PP* and *Pp*. Adding the individual probabilities of these combinations shown in Figure 12.7, 1/4 *PP* + 1/2 *Pp*, gives a total of 3/4, indicating that three-fourths of the F_2 offspring are expected to have purple flowers.

What is shown in Figure 12.7 is the **Punnett square** method for determining the genotypes of offspring and their expected proportions. The method was named for its originator, Reginald Punnett, a British geneticist who worked in the early part of the twentieth century. To use the Punnett square, write the probability of obtaining gametes with each type of allele from one parent at the top of the diagram and write the chance of obtaining each type of allele from the other parent on the left side. Then fill in the cells by combining the alleles from the top and from the left and multiply their individual probabilities.

FIGURE 12.6 Rules of probability. For each coin toss, the probability of a head is 1/2; the probability of a tail is also 1/2. Because the outcome of the first toss is independent of the outcome of the second, the combined probabilities of the outcomes of successive tosses are calculated by multiplying their individual probabilities according to the product rule.

© Cengage Learning 2014

FIGURE 12.7 Punnett square method for predicting offspring and their ratios in genetic crosses. The example is the $F_1 \times F_1$ self of purple-flowered plants from Figure 12.5 to produce the F_2 generation. Each cell shows the genotype and proportion of one type of F_2 plant.
© Cengage Learning 2014

Mendel Used a Testcross to Check the Validity of His Conclusions

Mendel realized that he could assess the validity of his conclusions by determining whether they could be used successfully to *predict* the outcome of a cross of a different type than he had tried so far. Accordingly, he crossed an F_1 plant with purple flowers, assumed to have the heterozygous genotype *Pp*, with a true-breeding white-flowered plant, with the homozygous genotype *pp* **(Figure 12.8, Experiment 1).** There are two expected classes of offspring, *Pp* and *pp*, both with a probability of 1/2. Thus, the phenotypes of the offspring are expected to be 1 purple-flowered : 1 white-flowered. Mendel's actual results closely approach that ratio. If you performed the same type of cross with all the other traits used in his study, including those traits affecting seed shape, seed color, and plant height, you would find the same 1 : 1 ratio.

A cross between an individual with the dominant phenotype and a homozygous recessive individual, such as the one just described, is called a **testcross.** Geneticists use a testcross to determine whether an individual with a dominant trait is a heterozygote or a homozygote, because their phenotypes are identical. If the offspring of the testcross are of two types, with half displaying the dominant trait and half the recessive trait, then the individual in question must be a heterozygote (see Figure 12.8, Experiment 1). If all the offspring display the dominant trait, the individual in question must be a homozygote. For example, the cross $PP \times pp$ gives all *Pp* progeny, which show the dominant purple flower phenotype **(Figure 12.8, Experiment 2).**

Obviously, the testcross method cannot be used for humans. However, it can be used in reverse, by noting the traits present in families over several generations and working backward to deduce whether a parent must have been a homozygote or a heterozygote (see also Chapter 13).

Crosses Involving Two Characters Led Mendel To Propose the Principle of Independent Assortment

Mendel next asked what happens in crosses when more than one character is involved. Would the alleles of the genes controlling the different characters be inherited independently, or would they interact to alter their expected proportions in offspring?

Gametes from F_1 purple *Pp* plant

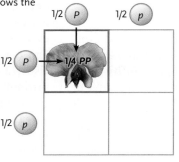

Gametes from F_1 purple *Pp* plant

A. To produce an F_2 plant with the *PP* genotype, two *P* gametes must combine. The probability of selecting a *P* gamete from one F_1 parent is 1/2, and the probability of selecting a *P* gamete from the other F_1 parent is also 1/2. Using the product rule, the probability of producing a purple-flowered *PP* plant from a $Pp \times Pp$ cross is $1/2 \times 1/2 = 1/4$.

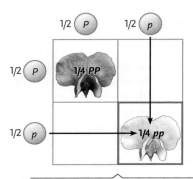

B. To produce an F_2 plant with the *pp* genotype, two *p* gametes must combine. The probability of selecting a *p* gamete from one F_1 parent is 1/2, and the probability of selecting a *p* gamete from the other F_1 parent is also 1/2. Using the product rule, the probability of producing a white-flowered *pp* plant from a $Pp \times Pp$ cross is $1/2 \times 1/2 = 1/4$.

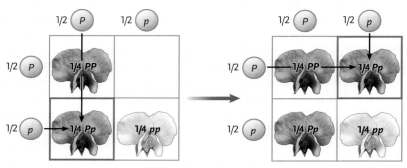

C. To produce an F_2 plant with the *Pp* genotype, a *P* gamete must combine with a *p* gamete. The cross $Pp \times Pp$ can produce *Pp* offspring in two different ways: (1) a *P* gamete from the first parent can combine with a *p* gamete from the second parent; or (2) a *p* gamete from the first parent can combine with a *P* gamete from the second parent. We apply the sum rule to obtain the combined probability: each of the ways to get *Pp* has an individual probability of 1/4, so the probability of *Pp*, purple-flowered offspring is $1/4 + 1/4 = 1/2$.

To answer these questions, Mendel crossed parental strains of peas that had differences in two of the characters he was studying: seed shape and seed color. His single-character crosses had shown each was controlled by a pair of alleles of one gene. For seed shape, round is dominant to wrinkled: the homozygous *RR* or heterozygous *Rr* genotypes produce round seeds and the homozygous *rr* genotype produces wrinkled seeds. For seed color, yellow is dominant to green: the *YY* or *Yy* genotypes produce yellow seeds; the homozygous *yy* genotype produces green seeds. **Figure 12.9** shows the cross of a true-breeding plant with round and yellow seeds (*RR YY*) with a true-breeding plant with wrinkled and green seeds (*rr yy*) through to the F_2 generation.

FIGURE 12.8 **Experimental Research**

Testing the Predicted Outcomes of Genetic Crosses

Question: How can it be determined whether a plant with the dominant phenotype is a heterozygote or a homozygote?

Experiment 1: Mendel crossed an F_1 plant with purple flowers, predicted to have a Pp genotype, with a true-breeding white-flowered plant and analyzed the flower color phenotypes in the offspring.

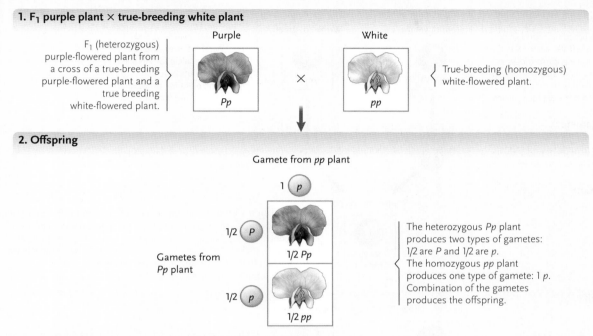

1. F_1 purple plant × true-breeding white plant

F_1 (heterozygous) purple-flowered plant from a cross of a true-breeding purple-flowered plant and a true breeding white-flowered plant.

Purple

Pp

×

White

pp

True-breeding (homozygous) white-flowered plant.

2. Offspring

Gamete from pp plant

1 p

Gametes from Pp plant

1/2 P

1/2 p

1/2 Pp

1/2 pp

The heterozygous Pp plant produces two types of gametes: 1/2 are P and 1/2 are p. The homozygous pp plant produces one type of gamete: 1 p. Combination of the gametes produces the offspring.

Results: Predicted progeny from a cross of a purple-flowered heterozygote with a true-breeding white-flowered plant is 1 Pp purple-flowered : 1 pp white-flowered. Mendel observed 85 purple-flowered and 81 white-flowered plants, close to the prediction.

Experiment 2: Mendel crossed a true-breeding plant with purple flowers, predicted to have a PP genotype, with a true-breeding white-flowered plant and analyzed the flower color phenotypes in the offspring.

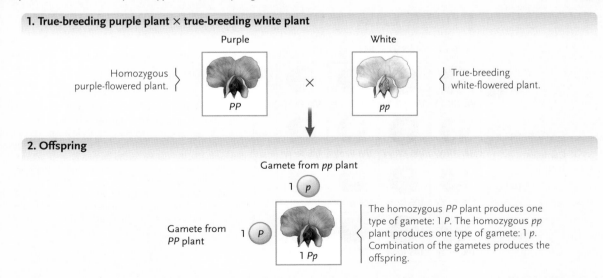

1. True-breeding purple plant × true-breeding white plant

Homozygous purple-flowered plant.

Purple

PP

×

White

pp

True-breeding white-flowered plant.

2. Offspring

Gamete from pp plant

1 p

Gamete from PP plant

1 P

1 Pp

The homozygous PP plant produces one type of gamete: 1 P. The homozygous pp plant produces one type of gamete: 1 p. Combination of the gametes produces the offspring.

Results: The outcome of crossing a purple-flowered homozygote with a true-breeding white-flowered plant is all Pp plants, which have purple flowers.

Conclusion: The outcome of a cross between (here) a plant with a dominant phenotype and a plant with a recessive phenotype—a testcross—gives a different result depending on whether the plant with the dominant phenotype is a homozygote or a heterozygote. Therefore, a testcross is a useful way to determine the genotype for an individual with a dominant phenotype.

THINK LIKE A SCIENTIST In rabbits, white fat beneath the skin is dominant to yellow fat. A white-fat rabbit is bred to a yellow-fat rabbit for a season, resulting in 13 white-fat and 12 yellow-fat offspring. What are the genotypes of the parents and their offspring?

The Principle of Independent Assortment

Question: Do alleles of genes for two different characters in garden peas assort independently in a cross?

Experiment: Mendel crossed a true-breeding plant with round and yellow seeds with a true-breeding plant with wrinkled and green seeds and analyzed the progeny through the F_1 and F_2 generations. We explain this cross here in modern terms.

1. P generation

The genotype of the true-breeding round, yellow parent is *RR YY*, where *R* is the dominant allele for round, and *Y* is the dominant allele for yellow. The plant is homozygous for both the *R* and *Y* alleles.

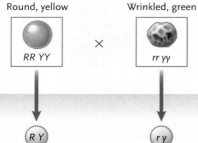

Round, yellow — *RR YY*   ×   Wrinkled, green — *rr yy*

The genotype of the true-breeding wrinkled and green parent is *rr yy*, where *r* is the recessive allele for wrinkled, and *y* is the recessive allele for green. The plant is homozygous for both the *r* and *y* alleles.

2. Haploid gametes

Only gametes with the *R* and *Y* alleles are produced in an *RR YY* plant.

R Y   *r y*

Only gametes with the *r* and *y* alleles are produced in an *rr yy* plant.

3. F₁ generation

Round and yellow

Rr Yy

Fusion of an *R Y* gamete from the round, yellow parent with an *r y* gamete from the wrinkled, green parent produces an F_1 generation all of which have the genotype *Rr Yy*, phenotype round, yellow seeds. The doubly heterozygous individual is called a *dihybrid*. The seeds are round because the *R* allele is dominant to the *r* allele, and yellow because the *Y* allele is dominant to the *y* allele.

4. F₁ × F₁ self

Round, yellow — *Rr Yy*   ×   Round, yellow — *Rr Yy*

Mendel then planted the F_1 seeds, grew the plants to maturity, and selfed them; that is, he crossed the F_1 to themselves. A cross such as this of two double heterozygotes is called a *dihybrid cross*.

5. F₂ generation

♂ Gametes

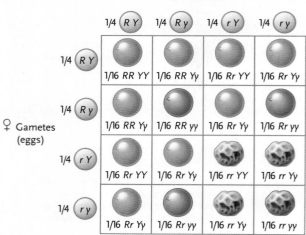

	1/4 *R Y*	1/4 *R y*	1/4 *r Y*	1/4 *r y*
1/4 *R Y*	1/16 *RR YY*	1/16 *RR Yy*	1/16 *Rr YY*	1/16 *Rr Yy*
1/4 *R y*	1/16 *RR Yy*	1/16 *RR yy*	1/16 *Rr Yy*	1/16 *Rr yy*
1/4 *r Y*	1/16 *Rr YY*	1/16 *Rr Yy*	1/16 *rr YY*	1/16 *rr Yy*
1/4 *r y*	1/16 *Rr Yy*	1/16 *Rr yy*	1/16 *rr Yy*	1/16 *rr yy*

♀ Gametes (eggs)

If the alleles that control seed shape and seed color assort independently, each F_1 plant grown from the seeds would produce four types of gametes: the *R* allele for seed shape would go to a gamete with either the *Y* or *y* allele for seed color, and similarly, the *r* allele would go to a gamete with either the *Y* or *y* allele. Thus, independent assortment of genes from the *Rr Yy* parents is expected to produce four types of gametes with equal probability: 1/4 *R Y*, 1/4 *R y*, 1/4 *r Y*, and 1/4 *r y*. Random fusion of the four different male gametes with the four different female gametes produces the F_2 generation.

Results: Filling in the cells of the Punnett square gives 16 combinations, each with an equal probability of 1 in every 16 offspring if the alleles of the two genes assort independently. The 16 combinations resulting from independent assortment give an expected F_2 phenotypic ratio of 9 round yellow : 3 round green : 3 wrinkled yellow : 1 wrinkled green. Mendel's selfing of the F_1 *Rr Yy* round yellow plants produced an F_2 generation with 315 round yellow : 108 round green : 101 wrinkled yellow : 32 wrinkled green, which is close to a 9 : 3 : 3 : 1 ratio (3 : 1 for round : wrinkled, and 3 : 1 for yellow : green).

Conclusion: The results indicate that the alleles of the genes for the two characters assort independently during the formation of gametes.

THINK LIKE A SCIENTIST Suppose that, instead of the $F_1 \times F_1$ cross (self-fertilization) shown in the figure, you cross an F_1 plant with a round, green plant of genotype *Rr yy*. What progeny would you get and in what ratio would they occur?

New genetic terms are used in the figure:

Dihybrid (*di* = two): An F$_1$ that is produced from a cross that involves two characters and is heterozygous for each of the pairs of alleles of the two genes involved. An example would be *Aa Bb*, where genes *A* and *B* control different characters, and the upper- and lowercase letters for the two genes are dominant and recessive alleles, respectively.

Dihybrid cross: A cross between two individuals that are each heterozygous for the pairs of alleles of two genes.

This 9:3:3:1 ratio observed in the F$_2$ generation of the cross was consistent with Mendel's previous findings if he added one further conclusion: *the alleles of the genes that govern the two characters assort independently during formation of gametes.* That is, the allele for seed shape that the gamete receives (*R* or *r*) has no influence on which allele for seed color it receives (*Y* or *y*) and vice versa. The two events are completely independent, a property known as **independent assortment;** it is now known as Mendel's **Principle of Independent Assortment.**

Filling in the cells of the diagram (see Figure 12.9) gives 16 combinations of alleles, all with an equal probability of 1 in every 16 offspring. Of these, the genotypes *RR YY, RR Yy, Rr YY,* and *Rr Yy* all have the same phenotype: round yellow seeds. These combinations occur in 9 of the 16 cells in the diagram, giving a total probability of 9/16. The genotypes *rr YY* and *rr Yy,* which produce the wrinkled yellow seeds, are found in three cells, giving a probability of 3/16 for this phenotype. Similarly, the genotypes *RR yy* and *Rr yy,* which yield round green seeds, occur in three cells, giving a probability of 3/16. Finally, the genotype *rr yy,* which produces wrinkled green seeds, is found in only one cell and therefore has a probability of 1/16.

The actual results obtained by Mendel closely approximated the expected 9:3:3:1 ratio of round yellow:round green:wrinkled yellow seeds:wrinkled green seeds, as developed in Figure 12.9. Thus, Mendel's first three conclusions, with the added hypothesis of independent assortment, explain the observed results of his dihybrid cross. Mendel's testcrosses completely confirmed his independent assortment conclusion; for example, the testcross *Rr Yy* × *rr yy* produced 55 round yellow seeds, 51 round green seeds, 49 wrinkled yellow seeds, and 53 wrinkled green seeds. This distribution corresponds well with the expected 1:1:1:1 ratio in the offspring. (Try to set up a Punnett square for this cross and predict the expected classes of offspring and their frequencies.)

What is the molecular basis for the seed-shape character differences? The normal *R* allele encodes an enzyme, starch-branching enzyme 1 (SBEI), which is required to produce a branched form of starch called amylopectin. The *r* allele is a mutant form of the gene, resulting in an inactive form of the enzyme and, therefore, no amylopectin production. Round seeds (*RR* and *Rr* genotypes) contain amylopectin, but wrinkled seeds (*rr* genotype) do not. The presence or absence of amylopectin is responsible for the seed shape. During development of *RR* or *Rr* seeds, the amylopectin limits the amount of water that accumulates in the seed. When the seeds dry as they

mature, they lose water and shrink uniformly, staying round. The *rr* seeds without amylopectin accumulate excessive water. When they dry, they lose more water and the seeds collapse, giving them a wrinkled appearance.

Mendel's first three conclusions provided a coherent explanation of the pattern of inheritance for alternate traits of the same character, such as purple and white for flower color. His fourth conclusion, independent assortment, addressed the inheritance of traits for different characters, such as seed shape, seed color, and flower color, and showed that, instead of being inherited together, the traits of different characters were distributed independently to offspring.

Mendel's Research Founded the Field of Genetics

Mendel's findings anticipated in detail the patterns by which genes and chromosomes determine inheritance. Yet, when Mendel first reported his findings, during the nineteenth century, the structure and function of chromosomes and the patterns by which they are separated and distributed to gametes were unknown; meiosis remained to be discovered. In addition, his use of mathematical analysis was a new and radical departure from the usual biological techniques of his day.

Mendel reported his results to a small group of fellow intellectuals in Brünn and presented his results in 1866 in a natural history journal published in the city. But Mendel's scientific conclusions were not immediately appreciated. His article received little notice outside of Brno, and those who read it were unable to appreciate the significance of his findings. His work was overlooked until the early 1900s, when three investigators—Hugo de Vries in Holland, Carl Correns in Germany, and Erich von Tschermak in Austria—independently performed a series of breeding experiments similar to Mendel's and reached the same conclusions. These investigators, in searching through previously published scientific articles, discovered to their surprise Mendel's article about his experiments conducted 34 years earlier. Each gave credit to Mendel's discoveries, and the quality and far-reaching implications of his work were at last realized. Mendel died in 1884, 16 years before the rediscovery of his experiments and conclusions, and thus he never received the recognition that he so richly deserved during his lifetime. Mendel was also unable to relate the behavior of his "factors" (genes) to chromosomes because the critical information he required was not obtained until later, through the discovery of meiosis during the 1890s.

Sutton's Chromosome Theory of Inheritance Related Mendel's Genes to Chromosomes

By the time Mendel's results were rediscovered in the early 1900s, critical information from studies of meiosis was available. It was not long before Walter Sutton, a genetics graduate student at Columbia University in New York, recognized the similarities between the inheritance of the genes discovered by

Mendel and the behavior of chromosomes in meiosis and fertilization (Figure 12.10).

In a historic article published in 1903, Sutton drew all the necessary parallels between genes and chromosomes:

- Chromosomes occur in pairs in sexually reproducing, diploid organisms, as do the alleles of each gene.
- The chromosomes of each pair are separated and delivered singly to gametes, as are the alleles of a gene.
- The separation of any pair of chromosomes in meiosis and gamete formation is independent of the separation of other pairs (see Figure 12.10), as in the independent assortment of the alleles of different genes in Mendel's dihybrid crosses.

- One member of each chromosome pair is derived in fertilization from the male parent, and the other member is derived from the female parent, in an exact parallel with the two alleles of a gene.

From this total coincidence in behavior, Sutton correctly concluded that genes and their alleles are carried on the chromosomes, a conclusion known today as the **chromosome theory of inheritance.**

The exact parallel between the principles set forth by Mendel and the behavior of chromosomes and genes during meiosis is shown in Figure 12.10 for an *Rr Yy* diploid. For a dihybrid cross of *Rr Yy* × *Rr Yy,* when the gametes produced as

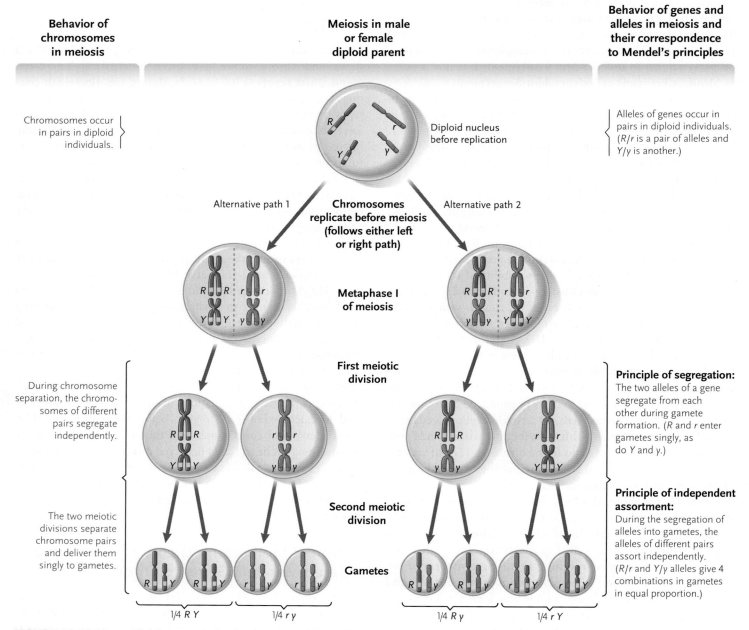

FIGURE 12.10 The parallels between the behavior of chromosomes and genes and alleles in meiosis. The gametes show four different combinations of alleles produced by independent segregation of chromosome pairs.
© Cengage Learning 2014

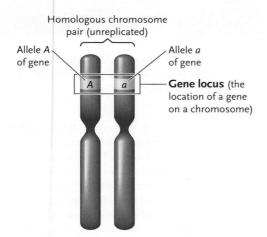

FIGURE 12.11 A locus, the site occupied by a gene on a pair of homologous chromosomes. Two alleles, *A* and *a*, of the gene are present at this locus in the homologous pair. These alleles have differences in the DNA sequence of the gene.
© Cengage Learning 2014

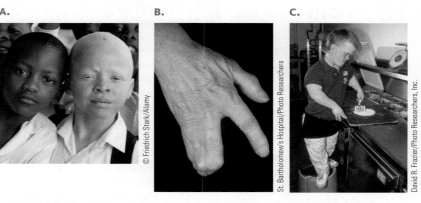

FIGURE 12.12 Human traits showing inheritance patterns that follow Mendelian principles. **(A)** Lack of normal skin color (albinism), a recessive trait. **(B)** Webbed fingers, a dominant trait. **(C)** Achondroplasia, or short-limbed dwarfism, a dominant trait.

in Figure 12.10 use randomly, the progeny will show a phenotypic ratio of 9:3:3:1. This mechanism explains the same ratio of gametes and progeny as the *Rr Yy × Ry Yy* cross in Figure 12.9.

The particular site on a chromosome at which a gene is located is called the **locus** (plural, *loci*) of the gene. The locus is a particular DNA sequence that encodes (typically) a protein responsible for the phenotype controlled by the gene. A locus for a gene with two alleles, *A* and *a,* on a homologous pair of chromosomes is shown in **Figure 12.11**. At the molecular level, different alleles of a gene have different DNA sequences which may result in functional gene have differences in the protein encoded by the gene. These differences are detected as distinct phenotypes in the offspring of a cross. *Insights from the Molecular Revolution* describes a molecular study that uncovered the mechanisms that control height in pea plants, one of the seven characteristics originally examined by Mendel.

All the genetics research conducted since the early 1900s has confirmed Mendel's basic conclusions about inheritance. Moreover, Mendel's conclusions apply to all types of organisms, from yeast and fruit flies to humans. In humans, a number of easily seen traits show inheritance patterns that follow Mendelian principles **(Figure 12.12)**; for example, *albinism,* the lack of normal skin color, is recessive to normal skin color, and normally separated fingers are recessive to fingers with webs between them. Similarly, *achondroplasia,* the most frequent form of short-limb dwarfism, is a dominant trait that involves abnormal bone growth. Many human disorders that cannot be seen easily also show simple inheritance patterns. For instance, *cystic fibrosis,* in which a defect in the membrane transport of chloride ions leads to pulmonary and digestive dysfunctions and eventually death, is a recessive trait.

Not all patterns of inheritance follow Mendel's principles discussed in this section. In the next section, we discuss patterns of inheritance that, in some circumstances, require modifications or additions to Mendel's principles.

12.2 Later Modifications and Additions to Mendel's Principles

The rediscovery of Mendel's research in the early 1900s produced an immediate burst of interest in genetics, and the research that followed greatly expanded our understanding of genes and their inheritance. That research supported Mendel's conclusions fully, but also revealed many variations on the principles he had outlined. Some of those variations are described in the following sections.

In Incomplete Dominance, Dominant Alleles Do Not Completely Mask Recessive Alleles

When one allele of a gene is not completely dominant over another allele of the same gene, it is said to show **incomplete dominance.** With incomplete dominance, the phenotype of the heterozygote is somewhere between the phenotypes of individuals that are homozygous for either of the alleles. Flower color in snapdragons shows incomplete dominance. One gene controls the flower color character, with one allele for red and another allele for white. Because one allele is not completely dominant to the other in incomplete dominance, we use a different genetic symbolism. That is, we use an italic letter that relates to the character, with superscripts for the different alleles. In this case, *C* signifies flower color and the superscript R is for red,

Mendel's Dwarf Pea Plants: How does a gene defect produce dwarfing?

One of the seven characters Mendel studied was stem length. The stem length gene *Le* controls the length of the stem between the leaf branches of the pea plant. Plants homozygous or heterozygous for the dominant *Le* allele have a normal stem length, resulting in tall plants, whereas plants homozygous for the recessive allele *le* have a much reduced stem length, resulting in dwarf plants.

Research Question

What is the function of Mendel's *Le* gene in controlling stem length?

Experiment

Two independent research teams worked out the molecular basis for stem length in garden peas. The investigators, including Diane Lester and her colleagues at the University of Tasmania in Australia, David Martin and his coworkers at Oregon State University, and Peter Hedden at the University of Bristol, England, were interested in learning the molecular differences between the *Le* and *le* alleles of the stem length gene.

Lester's team discovered that the gene codes for an enzyme that carries out the final step in the synthesis of the plant hormone gibberellic acid, which, among other effects, causes the stems of plants to elongate (see Chapter 37). Martin's group cloned the gene, determined its complete DNA sequence, and analyzed its function. (Cloning techniques and DNA sequencing are described in Chapters 18 and 19, respectively.)

Results

DNA sequencing showed that the *Le* and *le* alleles of the gene encode two versions of the enzyme that catalyzes gibberellic acid synthesis, which differ by only a single amino acid **(Figure)**. Lester's group found that the faulty enzyme encoded by the *le* allele carries out its step (addition of a hydroxyl group to a precursor) much more slowly than the enzyme encoded by the normal *Le* allele. As a result, plants with the *le* allele have only about 5% as much gibberellic acid in their stems as *Le* plants. The reduced gibberellic acid levels limit stem elongation, resulting in dwarf plants.

Conclusion

The methods of molecular biology allowed contemporary researchers to study a gene first studied genetically in the mid-nineteenth century. The findings leave little doubt that the gene codes for an enzyme that catalyzes formation of a plant hormone responsible for causing plant stems to elongate. Moreover, the recessive allele of the gene has a less active hormone due to a change in a single amino acid in the enzyme, and this change leads to the dwarf phenotype Mendel observed in his monastery garden.

THINK LIKE A SCIENTIST The *le*-encoded enzyme is much less active than the normal *Le*-encoded enzyme. Think back to what you learned about enzymes in Chapter 4. Explain how a single amino acid change might reduce the activity of an enzyme.

Sources: D. R. Lester, J. J. Ross, P. J. Davies, and J. B. Reid. 1997. Mendel's stem length gene *(Le)* encodes a gibberellin 3β-hydroxylase. *Plant Cell* 9:1435–1443; D. N. Martin, W. M. Proebsting, and P. Hedden. 1997. Mendel's dwarfing gene: cDNAs from the *Le* alleles and function of the expressed proteins. *Proceedings of the National Academy of Sciences USA* 94:8907–8911.

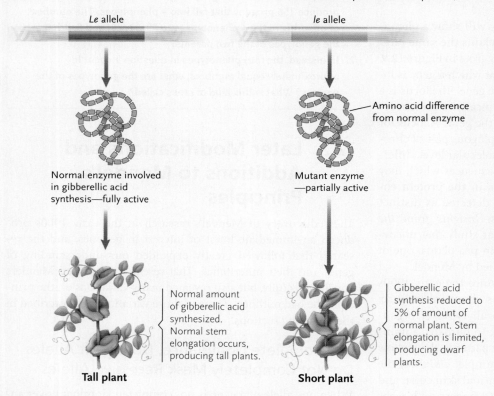

Le allele — Normal enzyme involved in gibberellic acid synthesis—fully active — Normal amount of gibberellic acid synthesized. Normal stem elongation occurs, producing tall plants. — **Tall plant**

le allele — Amino acid difference from normal enzyme — Mutant enzyme —partially active — Gibberellic acid synthesis reduced to 5% of amount of normal plant. Stem elongation is limited, producing dwarf plants. — **Short plant**

© Cengage Learning 2014

and the superscript W is for white. That is, C^R is the allele for red color and C^W is the allele for white color. We use these symbols in **Figure 12.13,** which follows a cross of a true-breeding red-flowered snapdragon with a true-breeding, white-flowered snapdragon through to the F_2 generation. The key differences from a cross in which complete dominance is the case are:

1. The F_1 phenotype is intermediate between the phenotypes of the two parents.

2. The phenotypes of F_2 individuals are directly determined by the different genotypes of those individuals, giving a 1:2:1 ratio rather than a 3:1 ratio as is characteristic for complete dominance.

We can explain the flower colors as follows: the C^R allele encodes an enzyme that produces a red pigment, but two alleles ($C^R C^R$) are necessary to produce enough of the active form of the enzyme to produce fully red flowers. The enzyme is com-

FIGURE 12.13 **Experimental Research**

Experiment Showing Incomplete Dominance of a Trait

Question: How is flower color in snapdragons inherited?

Experiment: Cross a true-breeding red-flowered snapdragon with a true-breeding white-flowered snapdragon and analyze the progeny through the F_1 and F_2 generations.

1. P generation

The red-flowered snapdragon is homozygous for the C^R allele.

Homozygous red parent   Red $C^R C^R$

×

White $C^W C^W$ Homozygous white parent

The white-flowered snapdragon is homozygous for the C^W allele.

2. F_1 generation

F_1 offspring all pink   Pink $C^R C^W$

Fusion of C^R gametes from the red-flowered plant and C^W gametes from the white-flowered plant produces $C^R C^W$ heterozygotes in the F_1. These plants have pink flowers, an intermediate phenotype between red and white. This phenotype is not that expected if one of the alleles shows complete dominance to the other allele. This phenotype is, however, consistent with incomplete dominance.

3. $F_1 \times F_1$ cross

Pink $C^R C^W$ ×   Pink $C^R C^W$

F_1 pink-flowered plants are crossed to produce the F_2 generation.

4. F_2 generation

Gametes from one $C^R C^W$ F_1 pink-flowered plant

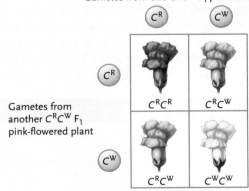

C^R C^W

C^R

Gametes from another $C^R C^W$ F_1 pink-flowered plant

$C^R C^R$ $C^R C^W$

C^W

$C^R C^W$ $C^W C^W$

Each parent plant produces two types of gametes, C^R and C^W. Random fusion of the gametes from the two parents produces the F_2 generation.

Results: The F_2 phenotypic ratio is 1 red : 2 pink : 1 white. Each phenotype results from a distinct genotype, $C^R C^R$ for red flowers, $C^R C^W$ for pink flowers, and $C^W C^W$ for white flowers. The 1 : 2 : 1 phenotypic ratio is consistent with incomplete dominance.

Conclusion: In incomplete dominance, each genotype has a distinct phenotype. From a cross of two heterozygotes, the outcome is a phenotypic ratio of 1 : 2 : 1 rather than the 3 : 1 ratio characteristic of complete dominance.

THINK LIKE A SCIENTIST What progeny would you get if you testcrossed an F_1 pink plant to a red parent? To a white parent?

pletely inactive in $C^W C^W$ plants, which produce colorless flowers that appear white because of the scattering of light by cell walls and other structures. With their single C^R allele, the $C^R C^W$ heterozygotes of the F_1 generation can produce only enough pigment to give the flowers a pink color. When pink $C^R C^W$ F_1 plants are crossed, the fully red and white colors reappear, together with the pink color, in exactly the same ratio as the ratio of genotypes produced from a cross of two heterozygotes in Mendel's experiments (for example, see Figure 12.7).

Some human disorders show incomplete dominance. For example, sickle-cell anemia (see *Why It Matters*) is characterized by an alteration in the hemoglobin molecule that changes the shape of red blood cells when oxygen levels are low. An individual with sickle-cell anemia is homozygous for a recessive allele that encodes a defective form of one of the polypeptides of the hemoglobin molecule. Individuals heterozygous for that recessive allele and the normal allele have a condition known as *sickle-cell trait,* which is a milder form of the disease because the individuals produce some normal polypeptides from the normal allele (Sickle-cell anemia and sickle-cell trait are two types of sickle-cell disease.).

In Codominance, the Effects of Different Alleles Are Equally Detectable in Heterozygotes

Codominance occurs when alleles have approximately equal effects in individuals, making the alleles equally detectable in heterozygotes. The inheritance of the human MN blood group presents an example of codominance. The L^M and L^N alleles of the MN blood group gene that control this character encode different forms of a glycoprotein molecule located on the surface of red blood cells. If the genotype is $L^M L^M$, only the M form of the glycoprotein is present and the blood type is M; if it is $L^N L^N$, only the N form is present and the blood type is N. In heterozygotes with the $L^M L^N$ genotype, both glycoprotein types are present and can be detected, producing the blood type MN. Because each genotype has a different phenotype, the inheritance pattern for the MN blood group alleles is generally the same as for incompletely dominant alleles. The MN blood types do not affect blood transfusions and have relatively little medical importance.

In Multiple Alleles, More Than Two Alleles of a Gene Are Present in a Population

One of Mendel's major and most fundamental assumptions was that alleles (his factors) occur in pairs in individuals; in the pairs, the alleles may be the same or different. After the rediscovery of Mendel's principles, it soon became apparent that although alleles do indeed occur in pairs in individuals, **multiple alleles** (more than two different alleles of a gene) may be present if all the individuals of a population are considered. For example, for a gene B, there could be the normal allele, B, and several alleles with alterations in this gene, for example, b_1, b_2, b_3, and so on. Some individuals in a population may have the B and b_1 al-

B allele	5'...ATGCAGATACCGATTACAGACCATAGG...3' 3'...TACGTCTATGGCTAATGTCTGGTATCC...5'
b_1 allele	5'...ATGCAGAGACCGATTACAGACCATAGG...3' 3'...TACGTCTCTGGCTAATGTCTGGTATCC...5'
b_2 allele	5'...ATGCAGATACCGACTACAGACCATAGG...3' 3'...TACGTCTATGGCTGATGTCTGGTATCC...5'
b_3 allele	5'...ATGCAGATACCGATTACAGTCCATAGG...3' 3'...TACGTCTATGGCTAATGTCAGGTATCC...5'

FIGURE 12.14 Multiple alleles. Multiple alleles consist of differences in the DNA sequence of a gene at one or more points, which result in detectable differences in the structure of the protein encoded by the gene. The differences shown here are single base-pair changes. The B allele is the normal allele, which encodes a protein with normal function. The three b alleles each have alterations of the normal protein-coding DNA sequence that may adversely affect the function of that protein.
© Cengage Learning 2014

leles of a gene; others, the b_2 and b_3 alleles; still others, the b_3 and b_5 alleles; and so on, for all possible combinations. That is, although *any one individual can have only two alleles of the gene,* there are more than two alleles in the population as a whole. Genes may certainly occur in many more than the four alleles of the example; for instance, one of the genes that plays a part in the acceptance or rejection of organ transplants in humans has more than 200 different alleles.

The multiple alleles of a gene each contain differences at one or more points in their DNA sequences **(Figure 12.14),** which cause detectable alterations in the structure and function of proteins encoded by the alleles. Multiple alleles present no real difficulty in genetic analysis because each diploid individual still has only two of the alleles, allowing gametes to be predicted and traced through crosses by the usual methods.

HUMAN ABO BLOOD GROUP The human *ABO* blood group provides an example of multiple alleles, in a system that also exhibits both dominance and codominance. The ABO blood group was discovered in 1901 by Karl Landsteiner, an Austrian biochemist who was investigating the sometimes fatal outcome of attempts to transfer blood from one person to another. Landsteiner found that only certain combinations of four blood types, designated A, B, AB, and O, can be mixed safely in transfusions **(Table 12.1).**

TABLE 12.1	Blood Types of the Human ABO Blood Group		
Blood Type	Antigens	Antibodies	Blood Types Accepted in a Transfusion
A	A	Anti-B	A or O
B	B	Anti-A	B or O
AB	A and B	None	A, B, AB, or O
O	None	Anti-A, anti-B	O

© Cengage Learning 2014

FIGURE 12.15 Inheritance of the blood types of the human ABO blood group.
© Cengage Learning 2014

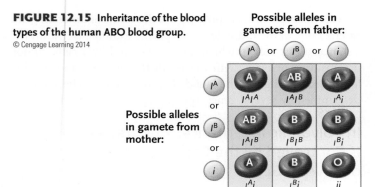

Possible alleles in gametes from father: I^A or I^B or i

Possible alleles in gamete from mother: I^A or I^B or i

	A I^AI^A	AB I^AI^B	A I^Ai
AB I^AI^B	B I^BI^B	B I^Bi
A I^Ai	B I^Bi	O ii

FIGURE 12.16 An example of epistasis: the inheritance of coat color in Labrador retrievers.

A. Black labrador

Erik Lam/Shutterstock.com

B. Chocolate brown labrador

c.byatt-norman/Shutterstock.com

C. Yellow labrador

c.byatt-norman/Shutterstock.com

D. Black × yellow labrador cross

Homozygous parents:

Black *BB EE* × Yellow *bb ee*

F₁ puppies:

Black *Bb Ee*

F₂ offspring from cross of two F₁ *Bb Ee* dogs:

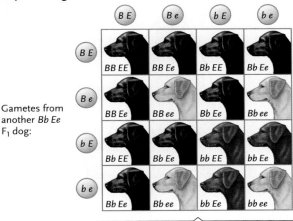

Gametes from one *Bb Ee* F₁ dog:

	B E	B e	b E	b e
B E	BB EE	BB Ee	Bb EE	Bb Ee
B e	BB Ee	BB ee	Bb Ee	Bb ee
b E	Bb EE	Bb Ee	bb EE	bb Ee
b e	Bb Ee	Bb ee	bb Ee	bb ee

Gametes from another *Bb Ee* F₁ dog:

© Cengage Learning 2014

F₂ phenotypic ratio is 9 black : 3 chocolate : 4 yellow

Landsteiner determined that, in the wrong combinations, red blood cells from one blood type are agglutinated (clumped) by an agent in the serum of another type (the serum is the fluid in which the blood cells are suspended). The clumping was later found to depend on the action of an antibody in the blood serum. (Antibodies, protein molecules that interact with specific substances called antigens, are discussed in Chapter 45.)

The antigens responsible for the blood types of the ABO blood group are the carbohydrate parts of glycoproteins located on the surfaces of red blood cells (unrelated to the glycoprotein carbohydrates responsible for the blood types of the MN blood group). For example, people with type A blood have *antigen A* on their red blood cells, and anti-B antibodies in their blood. If a person with type A blood receives a transfusion of type B blood, his or her anti-B antibodies will cause the blood to clump. Table 12.1 shows how the four blood types of the human ABO blood group determine compatibility in transfusions.

The four blood types—A, B, AB, and O—are produced by different combinations of multiple (three) alleles of a single gene *I* designated I^A, I^B, and *i* **(Figure 12.15)**. I^A and I^B are codominant alleles that are each dominant to the recessive *i* allele.

In Epistasis, Genes Interact, with the Activity of One Gene Influencing the Activity of Another Gene

The genetic characters discussed so far in this chapter, such as flower color, seed shape, and the blood types of the ABO group, all involve examples of genetic variation influenced by the alleles of single genes. This is not the case for every gene. In **epistasis** (*epi* = on or over; *stasis* = standing or stopping), two (or more) genes affect the same phenotype; the genes interact at the level of their products, with the phenotypic expression of one or more alleles of a gene at one locus inhibiting or masking the effects of the phenotypic expression of one or more alleles of a gene at a different locus. The result of epistasis is predictable ratios that can be explained by allele segregation patterns of the two (or more) genes involved.

Labrador retrievers (Labs) may have black, chocolate brown, or yellow fur **(Figure 12.16A–C)**. The different colors result from variations in the amount and distribution in hairs of a brownish black pigment called melanin. One gene, coding for an enzyme involved in melanin production, determines how much melanin is produced. The dominant *B* allele of this gene produces black fur color in *BB* or *Bb* Labs; less pigment is produced in *bb* dogs, which are chocolate brown. Another gene at a different locus determines whether the black or chocolate color appears at all, by controlling the deposition of pigment in hairs. A dominant allele *E* of this second gene permits pigment deposition, so that the black color in *BB* or *Bb* individuals, or the chocolate color in *bb* individuals, actually appears in the fur. Pigment deposition is almost completely blocked in homozygous recessive *ee* individuals, so the fur lacks melanin and has a yellow color whether the genotype for the *B* gene is *BB, Bb,* or *bb.* Thus, the *E* gene is epistatic to the *B* gene.

Epistasis between the *B* and *E* gene combines two expected phenotypic classes into one in the progeny of crosses among Labs **(Figure 12.16D).** Rather than two separate classes, as would be expected from a dihybrid cross without epistasis, the *BB ee, Bb ee,* and *bb ee* genotypes produce a single yellow phenotype. Therefore if we cross a true-breeding black Labrador with a true-breeding yellow Labrador of genotype *bb ee,* the F₁ puppies are *Bb Ee* black heterozygotes (see Figure 12.16). F₂ progeny produced by crossing F₁ dogs have the distribution: 9/16 black, 3/16 chocolate, and 4/16 yellow because of epistasis. That is, the ratio is 9:3:4 instead of the expected 9:3:3:1 ratio.

As you can see in this example, epistasis occurs when phenotypic expression of one or more alleles of a gene at one locus inhibits or masks the phenotypic expression of one or more alleles of a gene at a different locus. The result of epistasis is predictable ratios that can be explained by allele segregation patterns of the two (or more) genes involved. As in the Labrador coat color example, many other dihybrid crosses that involve epistatic interactions produce phenotypic distributions that differ from the expected 9:3:3:1 ratio.

In human biology, gene interactions and epistasis are common, and epistasis is an important factor in determining an individual's susceptibility to common human diseases. That is, different degrees of susceptibility are the result of different gene interactions in the individuals. A specific example is insulin resistance, a disorder in which muscle, fat, and liver cells do not use insulin correctly, with the result that glucose and insulin levels become high in the blood. This disorder is believed to be determined by several genes interacting with one another.

In Polygenic Inheritance, a Character Is Controlled by the Common Effects of Several Genes

Some characters follow a pattern of inheritance in which there is a more or less even gradation of types, forming a continuous distribution, rather than "on" or "off" (discontinuous) effects such as the production of purple or white flowers in pea plants. For example, in the human population, people range from short to tall, in a continuous distribution of gradations in height between limits of about 4 and 7 feet. Typically, a continuous distribution of this type is the result of **polygenic inheritance,** in which several to many different genes contribute to the *same* character. Other characters that exhibit a similar continuous distribution include skin color and body weight in humans, ear length in corn, seed color in wheat, and color spotting in mice. These characters are also known as **quantitative traits.** The individual genes that contribute to a quantitative trait are known as **quantitative trait loci** or QTLs.

A. Students at Brigham Young University, arranged according to height

Daniel Fairbanks

B. Actual distribution of individuals in the photo according to height

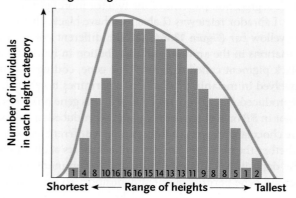

1 4 8 10 16 16 16 15 14 13 13 11 9 8 8 5 1 2

Shortest ← Range of heights → Tallest

C. Idealized bell-shaped curve for a population that displays continuous variation in a trait

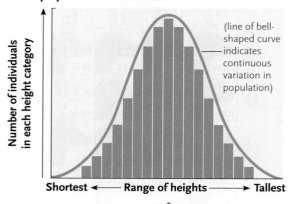

(line of bell-shaped curve indicates continuous variation in population)

Shortest ← Range of heights → Tallest

FIGURE 12.17 Continuous variation in height due to polygenic inheritance.
© Cengage Learning 2014

If the sample in the photo included more individuals, the distribution would more closely approach this ideal.

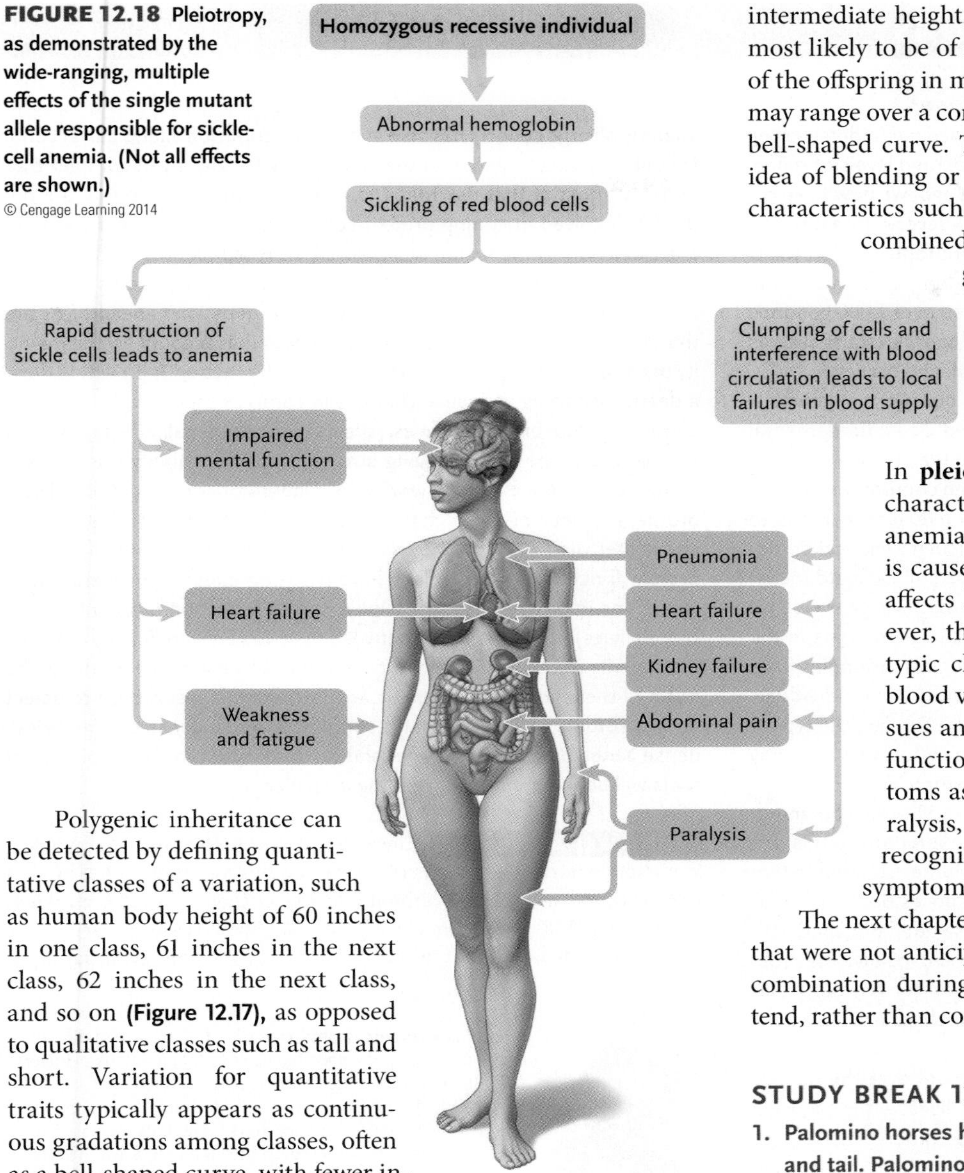

FIGURE 12.18 Pleiotropy, as demonstrated by the wide-ranging, multiple effects of the single mutant allele responsible for sickle-cell anemia. (Not all effects are shown.)
© Cengage Learning 2014

Homozygous recessive individual

↓

Abnormal hemoglobin

↓

Sickling of red blood cells

Rapid destruction of sickle cells leads to anemia

Clumping of cells and interference with blood circulation leads to local failures in blood supply

Impaired mental function

Heart failure

Weakness and fatigue

Pneumonia

Heart failure

Kidney failure

Abdominal pain

Paralysis

Polygenic inheritance can be detected by defining quantitative classes of a variation, such as human body height of 60 inches in one class, 61 inches in the next class, 62 inches in the next class, and so on **(Figure 12.17)**, as opposed to qualitative classes such as tall and short. Variation for quantitative traits typically appears as continuous gradations among classes, often as a bell-shaped curve, with fewer individuals at the extremes and the greatest numbers clustered around the midpoint; this is a good indication that the trait is quantitative.

Polygenic inheritance is often modified by the environment. For example, height in humans is not the result of genetics alone. Poor nutrition during infancy and childhood is one environmental factor that can limit growth and prevent individuals from reaching the height expected from genetic inheritance; good nutrition can have the opposite effect. Thus, the average young adult in Japan today is several inches taller than the average adult in the 1930s, when nutrition was poorer. Similarly, individuals who live in cloudy, northern or southern climates usually have lighter skin color than individuals with the same genotype who live in sunny climates.

At first glance, the effects of polygenic inheritance might appear to support the idea that characteristics of parents are blended in their offspring. Commonly, people believe that the children in a family with one tall and one short parent will be of intermediate height. Although the children of such parents are most likely to be of intermediate height, careful genetic analysis of the offspring in many such families shows that their offspring may range over a continuum from short to tall, forming a typical bell-shaped curve. Thus, genetic analysis does not support the idea of blending or even mixing of parental traits in polygenic characteristics such as body size or skin color; it is due to the combined effect of Mendelian inheritance for multiple genes influencing the same character.

In Pleiotropy, Two or More Characters Are Affected by a Single Gene

In **pleiotropy,** single genes affect more than one character of an organism. For example, sickle-cell anemia (see earlier discussion and *Why It Matters*) is caused by a recessive allele of a single gene that affects hemoglobin structure and function. However, the altered hemoglobin, the primary phenotypic change of the sickle-cell mutation, leads to blood vessel blockage, which can damage many tissues and organs in the body and affect many body functions, producing such wide-ranging symptoms as fatigue, abdominal pain, heart failure, paralysis, and pneumonia **(Figure 12.18)**. Physicians recognize these wide-ranging pleiotropic effects as symptoms of sickle-cell anemia.

The next chapter describes additional patterns of inheritance that were not anticipated by Mendel, including the effects of recombination during meiosis. These additional patterns also extend, rather than contradict, Mendel's fundamental principles.

STUDY BREAK 12.2

1. Palomino horses have a golden coat color, with a white mane and tail. Palominos do not breed true. Instead, there is a 50% chance that a foal with two Palomino parents will be a Palomino. What is the explanation?

2. A true-breeding rabbit with *agouti* (mottled, grayish brown) fur crossed with a true-breeding rabbit with *chinchilla* (silver) fur produces all agouti offspring. A true-breeding *chinchilla* rabbit crossed with a true-breeding *Himalayan* rabbit (white fur with pigmented nose, ears, tail, and legs) produces all *chinchilla* offspring. A true-breeding *Himalayan* rabbit crossed with a true-breeding *albino* rabbit produces all *Himalayan* offspring. Explain the inheritance of the fur colors.

THINK OUTSIDE THE BOOK

Individually or collaboratively explore the Internet or research literature to outline what is known about the genetics of human eye color, and about how certain individuals can have irises of different colors.

What is next for the genetics of human disease?

Technological and conceptual advances have propelled our understanding of the genetic basis of human disease to unprecedented levels of refinement. Comparing the *status quo* to a short decade ago, we now have detailed genetic and genomic information for humans, several primates, and a host of other organisms across all phyla, and our progress is set to accelerate. We have also witnessed the rapid identification of mutations that cause rare diseases, and we have recently discovered over 1,000 genomic segments that are associated with susceptibility to complex traits such as type I and type II diabetes, obesity, high cholesterol, and others.

Despite these advances, the most fundamental questions in genetics, raised a century ago, have yet to be fully addressed. Most fundamental of all is the question of how the genotype relates to the phenotype. Once a genetic disorder has been diagnosed, we can often confirm the diagnosis by mutation analysis. However, our ability to use such analysis to predict whether or not a patient will develop a particular genetic disorder is still extremely limited. This is in part because we still evaluate the effects of genetic and genomic variations in isolation. We have not yet developed the tools to understand the effects of such variations in the context of an entire genome (as well as the environment). Examples from a wide variety of single-gene disorders have shown us that individuals can be genotypically affected but clinically normal. Likewise, the typical clinical experience is that patients with the same mutation can vary greatly in the presence and severity of particular symptoms.

One of the challenges ahead of us is to understand the effect of mutations in the context of an individual's total genomic variation; that is, the sum total of likely pathogenic mutations in the genome. Coupled to that goal, the availability of the total genome sequence poses both an opportunity and a problem. Given that each human has several thousand variants predicted to affect gene/protein function, we need to solve the problems of how to assign the effect of genetic variation for any particular phenotype and how to predict the frequency with which each allele results in disease symptoms.

An overlapping question is the century-old debate over the relative contribution of rare and common alleles to genetic disease (the so-called common-allele common disease versus rare-allele common disease argument). The premise that complex traits such as schizophrenia, hypertension, and diabetes must be caused by common mutations in the general population (otherwise they would not be so common) has also proven true of some disorders, such as age-related macular degeneration, the most frequent cause of blindness in the elderly. However, exhaustive

analysis of large cohorts has also shown that common alleles can account for only a modest fraction of the genetic risk for any given complex trait and that most of that risk maps to regions of the genome that are not directly involved in coding protein sequences. For example, a common allele in a noncoding region of a gene called *Fused Toes (FTO)* that shows exceptionally significant association with obesity and diabetes accounts for only about 3 kg of increased weight. The questions ahead of us are therefore: (1) where are the genetic risk alleles that account for the majority of complex traits (the so-called genetic "dark matter"); (2) why is there a dearth of coding sequence changes in complex traits; and (3) how do common, probably mild, alleles (alleles that have small effects) interact with possibly rare alleles having strong effects to magnify the risk of a complex trait? For example, work in our laboratory on Bardet–Biedl syndrome, a genetic disorder characterized by obesity and learning defects, and related disorders has shown that some alleles that essentially abolish protein function (and therefore have a strong effect) can interact with mutations in other genes that have a modest effect on protein function (mild alleles). The interaction between the two can significantly enhance the severity of the clinical phenotype, but until we are able to model the effect of these mutations, genetic analysis alone is insufficient to detect such disorders. It is likely that a combination of extensive sequencing, dense genotyping, and a new generation of computational and biological tools will be required to address these challenges.

THINK LIKE A SCIENTIST Given significant resources, what would you choose to do as a means of understanding variation in the human genome and its involvement in disease: sequence 1,000 humans or sequence 1,000 other species? (*Hint:* One would give you access to a lot of human variation, the other would help you with evolutionary arguments.)

Courtesy of Nicholas Katsanis, PhD

Nicholas Katsanis is professor of Cell Biology and director of the Center for Human Disease Modeling at Duke University. His research interests focus on the genetic basis of Bardet–Biedl syndrome, where his laboratory is engaged in the identification of causative genes. The Katsanis lab also pursues questions centered on the signaling roles of vertebrate cilia, the translation of signaling pathway defects on the causality of ciliary disorders, and the dissection of second-site modification phenomena as a consequence of genetic load in a functional system. To learn more, go to http://www.cellbio.duke.edu/faculty/research/Katsanis.html.

REVIEW KEY CONCEPTS

To access the course materials and companion resources for this text, please visit www.cengagebrain.com.

12.1 The Beginnings of Genetics: Mendel's Garden Peas

- Mendel was successful in his research because of his good choice of experimental organism, which had clearly defined characters, such as flower color or seed shape, and because he analyzed his results quantitatively (Figures 12.3 and 12.4).

- Mendel showed that traits are passed from parents to offspring as hereditary factors (now called genes and alleles) in predictable ratios and combinations, disproving the notion of blended inheritance (Figure 12.5).

- From the results of crosses that involve single characters (monohybrid crosses), Mendel made three conclusions: (1) the genes that govern genetic characters occur in pairs in individuals; (2) if different alleles of a gene are present in the pair of an individual, one allele is dominant over the other; and (3) the two alleles of a gene segregate and enter gametes singly (Figures 12.5 and 12.7).

- Mendel confirmed his conclusions by a testcross between an F₁ heterozygote with a homozygous recessive parent. This type of testcross is still used to determine whether an individual is homozygous or heterozygous for a dominant allele (Figure 12.8).

- To explain the results of his crosses with individuals showing differences in two characters—dihybrid crosses—Mendel made a fourth conclusion: the alleles of the genes that govern the two characters segregate independently during formation of gametes (Figure 12.9).

- Walter Sutton was the first person to note the similarities between the inheritance of genes and the behavior of chromosomes in meiosis and fertilization. These parallels made it obvious that genes and alleles are carried on the chromosomes. Sutton's parallels are called the chromosome theory of inheritance (Figure 12.10).

- A locus is the site occupied by a gene on a chromosome (Figure 12.11).

Animation: Sickle-cell anemia

Animation: Crossing garden pea plants

Animation: Dihybrid cross

Animation: Genetic terms

Animation: Monohybrid cross

Animation: Test Cross

12.2 Later Modifications and Additions to Mendel's Principles

- In incomplete dominance, some or all alleles of a gene are neither completely dominant nor recessive. In such cases, the phenotype of heterozygotes with different alleles of the gene can be distinguished from that of either homozygote (Figure 12.13).

- In codominance, different alleles of a gene have approximately equal effects in heterozygotes, also allowing heterozygotes to be distinguished from either homozygote.

- Many genes may have multiple alleles if all the individuals in a population are taken into account. However, any diploid individual in a population has only two alleles of these genes, which are inherited and passed on according to Mendel's principles (Figures 12.14 and 12.15).

- In epistasis, two (or more) genes affect the same phenotype; the genes interact at the level of their products, with the phenotypic expression of one or more alleles of one locus inhibiting or masking the phenotypic expression of one or more alleles at a different locus. The result is that some phenotypic classes among progeny may be combined into one (Figure 12.16).

- In polygenic inheritance, genes at several to many different loci control the same character, producing a more or less continuous variation in the character from one extreme to another. Plotting the distribution of such characters among individuals often produces a bell-shaped curve (Figure 12.17).

- In pleiotropy, one gene affects more than one character of an organism (Figure 12.18).

Animation: Incomplete dominance

Animation: Dog Color

Animation: Height Graph

Animation: Transfusion

UNDERSTAND & APPLY

Test Your Knowledge

1. The dominant *C* allele of a gene that controls color in corn produces kernels with color; plants homozygous for a recessive *c* allele of this gene have colorless or white kernels. What kinds of gametes, and in what proportions, would be produced by the plants in the following crosses? What seed color, and in what proportions, would be expected in the offspring of the crosses?
   a. *CC* × *Cc*
   b. *Cc* × *cc*
   c. *Cc* × *Cc*

2. In peas, the allele *Le* produces tall plants and the allele *le* produces dwarf plants. The *Le* allele is dominant to *le*. If a tall plant is crossed with a dwarf, the offspring are distributed about equally between tall and dwarf plants. What are the genotypes of the parents?

3. The ability of humans to taste the bitter chemical phenylthiocarbamide (PTC) is a genetic trait. People with at least one copy of the normal, dominant allele of the *PTC* gene can taste PTC; those who are homozygous for a mutant, recessive allele cannot taste it. Could two parents able to taste PTC have a nontaster child? Could nontaster parents have a child able to taste PTC? A pair of taster parents, both of whom had one parent able to taste PTC and one nontaster parent, are expecting their first child. What is the chance that the child will be able to taste PTC? Unable to taste PTC? Suppose the first child is a nontaster. What is the chance that their second child will also be unable to taste PTC?

4. One gene has the alleles *A* and *a*; another gene has the alleles *B* and *b*. For each of the following genotypes, what types of gametes will be produced, and in what proportions, if the two gene pairs assort independently?
   a. *AA BB*
   b. *Aa BB*
   c. *Aa bb*
   d. *Aa Bb*

5. What genotypes, and in what frequencies, will be present in the offspring from the following matings?
   a. *AA BB* × *aa BB*
   b. *Aa Bb* × *aa bb*
   c. *Aa BB* × *AA Bb*
   d. *Aa Bb* × *Aa Bb*

6. In addition to the two genes in problem 4, assume you now study a third independently assorting gene that has the alleles *C* and *c*. For each of the following genotypes, indicate what types of gametes will be produced:
   a. *AA BB CC*
   b. *Aa BB Cc*
   c. *Aa BB cc*
   d. *Aa Bb Cc*

7. A man is homozygous dominant for alleles at 10 different genes that assort independently. How many genotypically different types of sperm cells can he produce? A woman is homozygous recessive for the alleles of 8 of these 10 genes, but she is heterozygous for the other 2 genes. How many genotypically different types of eggs can she produce? What hypothesis can you suggest to describe the relationship between the number of different possible gametes and the number of heterozygous and homozygous genes that are present?

8. In guinea pigs, an allele for rough fur *(R)* is dominant over an allele for smooth fur *(r)*; an allele for black coat *(B)* is dominant over that for white *(b)*. You have an animal with rough, black fur. What cross would you use to determine whether the animal is homozygous for these traits? What phenotype would you expect in the offspring if the animal is homozygous?

9. You cross a lima bean plant from a variety that breeds true for green pods with another lima bean from a variety that breeds true for yellow pods. You note that all the F₁ plants have green pods. These green-pod F₁ plants, when crossed, yield 675 plants with green pods and 217 with yellow pods. How many genes probably control pod color in this experiment? Give the alleles letter designations. Which is dominant?

10. Some recessive alleles have such a detrimental effect that they are lethal when present in both chromosomes of a pair. Homozygous recessives cannot survive and die at some point during embryonic development. Suppose that the allele r is lethal in the homozygous rr condition. What genotypic ratios would you expect among the living offspring of the following crosses?
    a. $RR \times Rr$
    b. $Rr \times Rr$

11. In garden peas, the genotypes GG or Gg produce green pods and gg produces yellow pods; $LeLe$ or $Lele$ plants are tall and $lele$ plants are dwarfed; RR or Rr produce round seeds and rr produces wrinkled seeds. If a plant of a true-breeding, tall variety with green pods and round seeds is crossed with a plant of a true-breeding, dwarf variety with yellow pods and wrinkled seeds, what phenotypes are expected, and in what ratios, in the F_1 generation? What phenotypes, and in what ratios, are expected if F_1 individuals are crossed (allowed to self-fertilize)?

12. In chickens, feathered legs are produced by a dominant allele F. Another allele f of the same gene produces featherless legs. The dominant allele P of a gene at a different locus produces pea combs; a recessive allele p of this gene causes single combs. A breeder makes the following crosses with birds 1, 2, 3, and 4; all parents have both feathered legs and pea combs:

Cross	Offspring
1×2	All feathered, pea comb
1×3	3/4 feathered; 1/4 featherless, all pea comb
1×4	9/16 feathered, pea comb; 3/16 featherless, pea comb; 3/16 feathered, single comb; 1/16 featherless, single comb

    What are the genotypes of the four birds?

13. A mix-up in a hospital ward caused a mother with O and MN blood types to think that a baby given to her really belonged to someone else. Tests in the hospital showed that the doubting mother was able to taste PTC (see problem 3). The baby given to her had O and MN blood types and had no reaction when the bitter PTC chemical was placed on its tongue. The mother had four other children with the following blood types and tasting abilities for PTC:
    a. Type A and MN blood, taster
    b. Type B and N blood, nontaster
    c. Type A and M blood, taster
    d. Type A and N blood, taster

    Without knowing the father's blood types and tasting ability, can you determine whether the child is really hers? (Assume that all her children have the same father.)

14. In cats, the genotype AA produces tabby fur color; Aa is also a tabby, and aa is black. Another gene at a different locus is epistatic to the gene for fur color. When present in its dominant W form (WW or Ww), this gene blocks the formation of fur color and all the offspring are white; ww individuals develop normal fur color. What fur colors, and in what proportions, would you expect from the cross $Aa\ Ww \times Aa\ Ww$?

15. Having malformed hands with shortened fingers is a dominant trait controlled by a single gene; people who are homozygous for the recessive allele have normal hands and fingers. Having woolly hair is a dominant trait controlled by a different gene; homozygous recessive individuals have normal, nonwoolly hair. Suppose a woman with normal hands and nonwoolly hair marries a man who has malformed hands and woolly hair. Their first child has normal hands and nonwoolly hair. What are the genotypes of the mother, the father, and the child? If this couple has a second child, what is the probability that it will have normal hands and woolly hair?

Discuss the Concepts

1. Explain how individuals of an organism that are phenotypically alike can produce different ratios of progeny phenotypes.

2. ABO blood type tests can be used to exclude paternity. Suppose a defendant who is the alleged father of a child takes a blood type test and the results do not exclude him as the father. Do the results indicate that he is the father? What arguments could a lawyer make based on the test results to exclude the defendant from being the father? (Assume the tests were performed correctly.)

Design an Experiment

Imagine that you are a breeder of Labrador retriever dogs. Labs can be black, chocolate brown, or yellow. Suppose that a yellow Lab is donated to you and you need to know its genotype. You have a range of dogs with known genotypes. What cross would you make to determine the genotype of the donated dog? Explain how the resulting puppies show you the Lab's genotype.

Interpret the Data

Half of the world's population eats rice at least twice a day. Much of this rice is grown in flooded conditions, and different strains of rice are tolerant (survive) or intolerant (die) under these conditions. Rice breeders used genetic crosses to test whether tolerance to flooding is a dominant trait. Researchers used three true-breeding flood-tolerant strains, FR143, BKNFR, and Kurk, and two true-breeding flood-intolerant strains, IR42 and NB, in the crosses. Results were obtained from three sets of crosses and are reported in the **Table** below:

1. F_2 results: Intolerant and tolerant strains were crossed and the resulting F_1 were interbred to produce the F_2.

2. Results of cross of F_1 to intolerant parent: F_1 plants were crossed with the intolerant parent of the cross.

3. Results of cross of F_1 to tolerant parent: F_1 plants were crossed with the tolerant parent of the cross.

Progeny Analyzed from	Number of Plants		
Intolerant × Tolerant Cross	Alive	Dead	Total
1. F_2 results of cross:			
IR42 × FR13A	187	77	264
IR42 × BKNFR	192	73	265
NB × Kurk	142	52	195
2. Results of cross of F_1 to intolerant parent:			
(F_1 of IR42 × FR13A) × IR42	14	17	31
(F_1 of IR42 × BKNFR) × IR42	15	10	25
(F_1 of NB × Kurk) × NB	21	35	56
3. Results of cross of F_1 to tolerant parent:			
(F_1 of IR42 × FR13A) × FR13A	31	0	31
(F_1 of IR42 × BKNFR) × BKNFR	28	0	28
(F_1 of NB × Kurk) × Kurk	40	0	40

Do the data support the hypothesis that the tolerance trait is dominant? Justify your conclusion by explaining the results from each of the three sets of crosses in terms of genotypes and phenotypic ratios.

Source: T. Setter et al. 1997. Physiology and genetics of submergence tolerance in rice. *Annals of Botany* 79:67–77.

Apply Evolutionary Thinking

How could an epistatic interaction shelter a harmful allele from the action of natural selection?

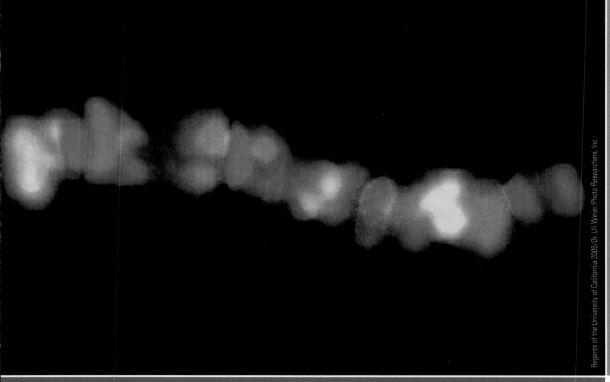

Fluorescent probes bound to specific sequences along human chromosome 10 (light micrograph). New ways of mapping chromosome structure yield insights into the inheritance of normal and abnormal traits.

Regents of the University of California 2005/Dr. Uli Weier/Photo Researchers, Inc.

13

Genes, Chromosomes, and Human Genetics

Why it matters . . . Imagine being 10 years old and trapped in a body that each day becomes more shriveled, frail, and old. You weigh less than 35 pounds, already you are bald, and you probably have only a few more years to live. But if you are like Mickey Hayes or Fransie Geringer **(Figure 13.1)**, you still have not lost your courage or your childlike curiosity about life. Like them, you still play, laugh, and celebrate birthdays.

Progeria, the premature aging that afflicts Mickey and Fransie, is caused by a genetic error that is present in only 1 of every 4 to 8 million human births. The error is in the gene for lamin A, one of the lamin proteins that reinforces the inner surface of the nuclear envelope in animal cells. In some way not yet understood, the defective lamin A makes the nucleus unstable, leading to the premature aging and reduced life expectancy characteristic of progeria.

Progeria affects both sexes equally, and all races and ethnic groups. Usually, symptoms begin to appear between 18 and 24 months of age. The rate of body growth declines to abnormally low levels. Skin becomes thinner, muscles become flaccid, and limb bones start to degenerate. Children with progeria die from a stroke or heart attack brought on by hardening of the arteries, a condition typical of advanced age. Death occurs at an average age of 13, with a range of about 8 to 21 years.

The plight of Mickey and Fransie provides a telling and tragic example of the dramatic effects that gene defects can have on living organisms. We are the products of our genes, and the characteristics of each individual, from humans to pine trees to protozoa, depend on the combination of genes, alleles, and chromosomes inherited from its parents, as well as on environmental effects. This chapter delves deeply into genes and the role of chromosomes in inheritance.

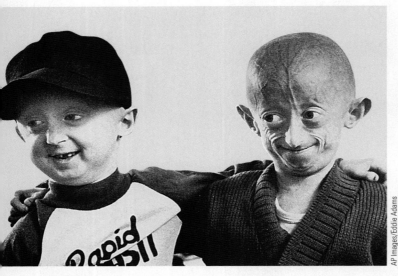

FIGURE 13.1 Two boys, both younger than 10, who have progeria, a genetic disorder characterized by accelerated aging and extremely reduced life expectancy.

13.1 Genetic Linkage and Recombination

In his historic experiments, Gregor Mendel found that each of the seven genes he studied assorted independently of the others in the formation of gametes. If Mendel had extended his study to numerous characters, he would have found exceptions to this principle. This should not be surprising, because an organism has far more genes than chromosomes. Chromosomes contain many genes, with each gene at a particular locus. Genes located on different chromosomes assort independently in gamete formation because the two chromosomes behave independently of one another during meiosis (see Chapter 11). Genes located on the same chromosome may be inherited together in genetic crosses—that is, *not* assort independently—because the chromosome is inherited as a single physical entity in meiosis. Genes near each other on the same chromosome are known as **linked genes,** and the phenomenon is called **linkage.**

The Principles of Linkage and Recombination Were Determined with *Drosophila*

In the early part of the twentieth century, Thomas Hunt Morgan and his coworkers at Columbia University were using the fruit fly, *Drosophila melanogaster,* as a model organism to investigate Mendel's principles in animals. (*Focus on Model Research Organisms* describes the development and use of *Drosophila* in research.) In 1911, Morgan crossed a true-breeding fruit fly with normal red eyes and normal wing length, genotype $pr^+pr^+ vg^+vg^+$, with a true-breeding fly with the recessive traits of purple eyes and vestigial (that is, short and crumpled) wings, genotype *prpr vgvg,* to analyze the segregation of the two traits.

This gene symbolism is new to us. In this system, the superscript plus (+) symbol associated with a letter or letters indicates a wild-type—normal—allele of a gene. Typically, but not always, a *wild-type* allele is the most common allele found in a population. In most instances, the wild-type allele is dominant to mutant alleles, but there are exceptions. The letters for the gene are based on the phenotype of the organism that expresses the *mutant* allele, for example, *pr* for *purple* eyes. Thus, we refer to the gene as the *purple* or *pr* gene; the dominant wild-type allele of the gene, pr^+, gives the wild-type red eye color.

Figure 13.2 steps through Morgan's cross of the two parents and his testcross of F_1 flies. Based on Mendel's principle of independent assortment (see Section 12.1), there should be four classes of phenotypes in the testcross offspring, in a 1:1:1:1 ratio of red eyes, normal wings:purple, vestigial:red, vestigial:purple, normal. But Morgan did not observe this result (step 4); instead, of the 2,839 progeny flies, 1,339 were red, normal and 1,195 were purple, vestigial. These phenotypes are identical to the two original P generation flies and are called **parental** phenotypes. The remaining progeny flies consisted of 151 red, vestigial and 154 purple, normal. These phenotypes have different combinations of traits from those of the P generation flies and are called **recombinant** phenotypes. If the genes had shown independent assortment, there would have been 25% of each of the four classes, or 50% parental and 50% recombinant phenotypes. In numbers, there would have been 710 (approximately) of each of the 4 classes.

How could the low frequency of recombinant phenotypes be explained? Morgan hypothesized that the two genes are linked genetically—physically close to each other on the same chromosome. That is, *pr* and *vg* are linked genes. He further hypothesized that the behavior of these linked genes in the testcross is explained by what he called *chromosome recombination,* a process in which two homologous chromosomes exchange segments with each other by crossing-over during meiosis (see Figure 11.5). Furthermore, he proposed that the frequency of this recombination is a function of the distance between linked genes. The nearer two genes are, the greater the chance they will be inherited together (resulting in parental phenotypes) and the lower the chance that recombinant phenotypes will be produced. The farther apart two genes are, the lower the chance that they will be inherited together and the greater the chance that recombinant phenotypes will be produced. These brilliant and far-reaching hypotheses were typical of Morgan, who founded genetics research in the United States, developed *Drosophila* as a research organism, and made discoveries that were almost as significant to the development of genetics as those of Mendel.

We show how Morgan's thinking applies to the purple-vestigial cross in **Figure 13.3.** Cartoons of the chromosomes themselves allow us to follow pictorially the consequences of crossing-over during meiosis in the production of gametes, and then the fusion of parental and recombinant gametes in the female with the gamete from the male testcross parent. We can

FIGURE 13.2 | Experimental Research

Evidence for Gene Linkage

Question: Do the purple-eye and vestigial-wing genes of *Drosophila* assort independently?

Experiment: Morgan crossed true-breeding wild-type flies with red eyes and normal wings with true-breeding purple-eyed, vestigial-winged flies. He then testcrossed the F_1 flies, which were wild type in phenotype, and analyzed the distribution of phenotypes in the progeny.

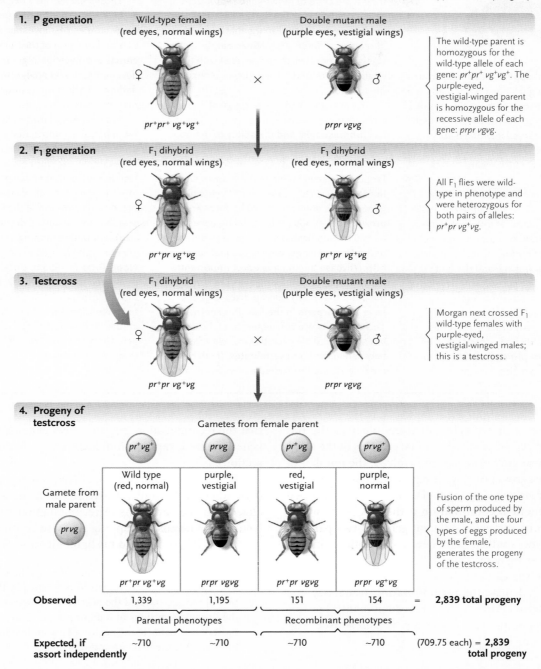

1. P generation

Wild-type female (red eyes, normal wings)

$pr^+pr^+\ vg^+vg^+$

×

Double mutant male (purple eyes, vestigial wings)

$prpr\ vgvg$

The wild-type parent is homozygous for the wild-type allele of each gene: $pr^+pr^+\ vg^+vg^+$. The purple-eyed, vestigial-winged parent is homozygous for the recessive allele of each gene: $prpr\ vgvg$.

2. F_1 generation

F_1 dihybrid (red eyes, normal wings)

$pr^+pr\ vg^+vg$

F_1 dihybrid (red eyes, normal wings)

$pr^+pr\ vg^+vg$

All F_1 flies were wild-type in phenotype and were heterozygous for both pairs of alleles: $pr^+pr\ vg^+vg$.

3. Testcross

F_1 dihybrid (red eyes, normal wings)

$pr^+pr\ vg^+vg$

×

Double mutant male (purple eyes, vestigial wings)

$prpr\ vgvg$

Morgan next crossed F_1 wild-type females with purple-eyed, vestigial-winged males; this is a testcross.

4. Progeny of testcross

Gametes from female parent

pr^+vg^+ $prvg$ pr^+vg $prvg^+$

Gamete from male parent

$prvg$

Wild type (red, normal)	purple, vestigial	red, vestigial	purple, normal
$pr^+pr\ vg^+vg$	$prpr\ vgvg$	$pr^+pr\ vgvg$	$prpr\ vg^+vg$

Fusion of the one type of sperm produced by the male, and the four types of eggs produced by the female, generates the progeny of the testcross.

Observed	1,339	1,195	151	154	**= 2,839 total progeny**
	Parental phenotypes		Recombinant phenotypes	
Expected, if assort independently	~710	~710	~710	~710	(709.75 each) = **2,839 total progeny**

Results: 2,534 of the testcross progeny flies had parental phenotypes, wild-type and purple, vestigial, whereas 305 of the progeny had recombinant phenotypes of red, vestigial and purple, normal. If the genes assorted independently, the expectation is a 1:1:1:1 ratio for testcross progeny: approximately 1,420 of both parental and recombinant progeny.

Conclusion: The purple-eye and vestigial-wing genes do not assort independently. The simplest alternative hypothesis is that the two genes are linked on the same chromosome. The small number of recombinant phenotypes is explained by crossing-over.

THINK LIKE A SCIENTIST Suppose that instead of the cross shown, the P generation cross was between a true-breeding red-eyed, vestigial-winged female and a true-breeding purple-eyed, normal-winged male. What would be the two largest classes in the progeny of the testcross? What would be the two smallest classes?

Source: C. B. Bridges. 1919. The genetics of purple eye color in *Drosophila. The Journal of Experimental Zoology* 28:265–305.

The Marvelous Fruit Fly, *Drosophila melanogaster*

Herman Eisenbeiss/
Photo Researchers, Inc.

The little fruit fly that appears seemingly from nowhere when rotting fruit or a fermented beverage is around is one of the mainstays of genetic research. It was first described in 1830 by C. F. Fallén, who named it *Drosophila,* meaning "dew lover." The species identifier became *melanogaster,* which means "black belly."

The great geneticist Thomas Hunt Morgan began to culture *D. melanogaster* in 1909 in the famous "Fly Room" at Columbia University. Many important discoveries in genetics were made in the Fly Room, including sex-linked genes and sex linkage and the first chromosome map. The subsequent development of methods to induce mutations in *Drosophila* led, through studies of the mutants produced, to many other discoveries that collectively established or confirmed essentially all the major principles and conclusions of eukaryotic genetics.

Among the many reasons for the success of *D. melanogaster* as a subject for genetics research are the ease of culturing these flies, their rapid cycle of growth and reproduction, easy identification of males and females, and the many types of mutations that cause morphological differences, such as purple eyes or vestigial wings, which can be seen with the unaided eye or under a low-power binocular microscope.

The availability of a wide range of mutants, comprehensive linkage maps of each of its chromosomes, and the ability to manipulate genes readily by molecular techniques made the fruit fly one of the model organisms for genome sequencing in the Human Genome Project. The sequence of *Drosophila's* genome was published in 2000; there are approximately 13,600 protein-coding genes in its 120 million base-pair genome. (A database of the *Drosophila* genome is available at http://flybase.bio.indiana.edu.) Importantly, many fruit fly and human genes are similar, to the point that many human disease genes have counterparts in the fruit fly genome. This similarity is a consequence of the distant but still detectable evolutionary relationship between insects and vertebrates. It enables the fly genes to be studied as models of hu-

man disease genes in efforts to understand better the functions of those genes and how alterations in them lead to disease.

Drosophila has also become established as an excellent experimental model for neurobiology studies (the investigation of the structure and function of the nervous system). Scientists are investigating, for example, neural development, and analyzing behavior. Again, findings are helping us understand the human nervous system also.

The analysis of fruit fly embryonic development has also contributed significantly to the understanding of development in humans. For example, experiments on mutants that affect fly development have provided insights into the genetic basis of many human birth defects. Recognition of the importance of research with *Drosophila* developmental genetics to our understanding of development in general came in the form of the award of the Nobel Prize in 1995 to three scientists who pioneered the fruit fly work: Edward Lewis of the California Institute of Technology, Christiane Nusslein-Volhard of the Max Planck Institute for Developmental Biology in Tübingen, Germany, and Eric Wieschaus of Princeton University.

---

see that the parental or recombinant phenotypes of the offspring directly reflect the genotypes of the gametes of the dihybrid parent. Because linked genes are involved, the number of offspring in the testcross progeny with parental phenotypes exceeds the number with recombinant phenotypes. As we learned in Chapter 11, **genetic recombination** is the process by which the combinations of alleles for different genes in two parental individuals become shuffled into new combinations in offspring as we are seeing here.

To determine the distance between the two genes on the chromosome, we calculate the **recombination frequency,** the percentage of testcross progeny that are recombinants. For this testcross, the recombination frequency is 10.7% (see Figure 13.3).

Recombination Frequency Can Be Used to Map Chromosomes

The recombination frequency of 10.7% for the *pr* and *vg* genes of *Drosophila* means that 10.7% of the gametes originating from the $pr^+pr\,vg^+vg$ parent contained recombined chromosomes. That recombination frequency is characteristic for those two genes. In other crosses that involve linked genes, Morgan found

that the recombination frequency was characteristic of the two genes involved, varying from less than 1% to 50% (see the next section).

From these observations, Alfred Sturtevant, then an undergraduate at Columbia University working with Morgan, realized that the variations in recombination frequencies could be used as a means of mapping genes on chromosomes. Sturtevant himself later recalled his lightbulb moment:

> I suddenly realized that the variations in the strength of linkage already attributed by Morgan to difference in the spatial separation of the gene offered the possibility of determining sequence in the linear dimensions of a chromosome. I went home and spent most of the night (to the neglect of my undergraduate homework) in producing the first chromosome map.

Sturtevant's revelation was that the recombination frequency observed between any two linked genes reflects the genetic distance between them on their chromosome. The greater this distance, the greater the chance that a crossover can form between the genes and the greater the recombination frequency.

Therefore, recombination frequencies can be used to make a **linkage map** of a chromosome showing the relative locations

FIGURE 13.3 Recombination between the purple-eye gene and the vestigial-wing gene, resulting from crossing-over between homologous chromosomes. The testcross of Figure 13.2 is redrawn here showing the two linked genes on chromosomes. Chromosomes or chromosome segments with wild-type alleles are red, whereas chromosomes or segments with mutant alleles are blue. The parental phenotypes in the testcross progeny are generated by segregation of the parental chromosomes, whereas the recombinant phenotypes are generated by crossing-over between the two linked genes.
© Cengage Learning 2014

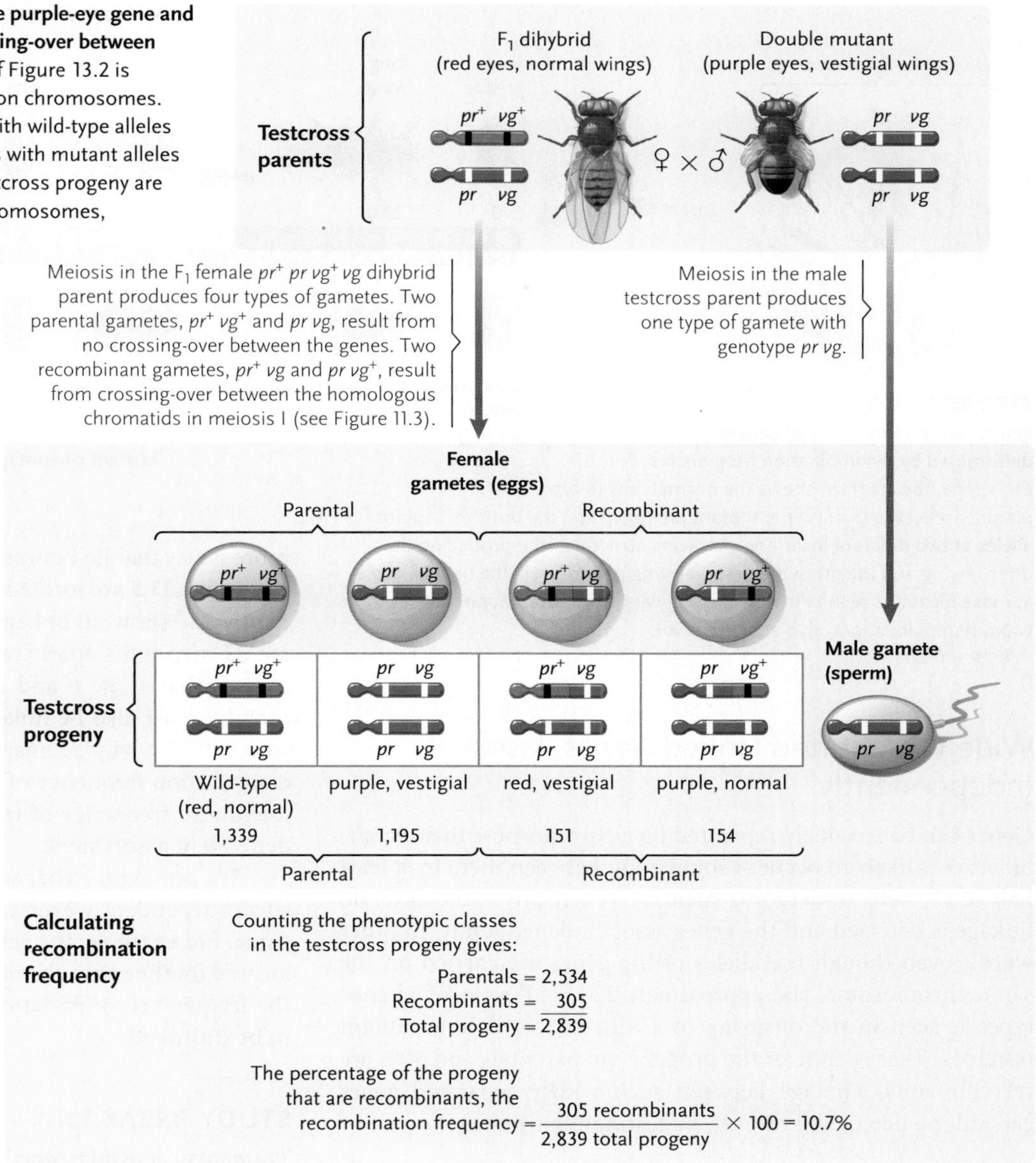

of genes. For example, assume that the three genes *a, b,* and *c* are carried together on the same chromosome. Crosses reveal a 9.6% recombination frequency for *a* and *b,* an 8% recombination frequency for *a* and *c,* and a 2% recombination frequency for *b* and *c.* These frequencies allow the genes to be arranged in only one sequence on the chromosomes as follows:

You will note that the *a–b* recombination frequency does not exactly equal the sum of the *a–c* and *c–b* recombination frequencies. This is because genes farther apart on a chromosome are more likely to have more than one crossover occur between them. Whereas a single crossover between two genes gives recombinants, a double crossover (two single crossovers occurring in the same meiosis) between two genes gives parentals. You can see this simply by drawing single and double crossovers between two genes on a piece of paper. In our example, double crossovers that occur between *a* and *b* have slightly decreased the recombination frequency between these two genes.

Using this method, Sturtevant created the first linkage map showing the arrangement of six genes on the *Drosophila* X chromosome. (A partial linkage map of a *Drosophila* chromosome is shown in **Figure 13.4.**)

Since the time of Morgan, many *Drosophila* genes and those of other eukaryotic organisms widely used for genetic research, including *Neurospora* (a fungus), yeast, maize (corn), and the mouse, have been mapped using the same approach. Recombination frequencies, together with the results of other techniques, have also been used to create linkage maps of the locations of genes in the DNA of prokaryotes such as the human intestinal bacterium *Escherichia coli* (see Chapter 17).

The unit of a linkage map, called a **map unit** (abbreviated mu), is equivalent to a recombination frequency of 1%. The map unit is also called the **centimorgan** (cM) in honor of Morgan's discoveries of linkage and recombination. Map units are not absolute physical distances in micrometers or nanometers; rather, they are *relative,* showing the positions of genes with respect to each other. One of the reasons that the units are relative and not absolute distances is that the frequency of crossing-over varies to some extent from one position to another on chromosomes.

In recent years, the linkage maps of a number of species have been supplemented by DNA sequencing of whole genomes, which shows the precise physical locations of genes in the chromosomes.

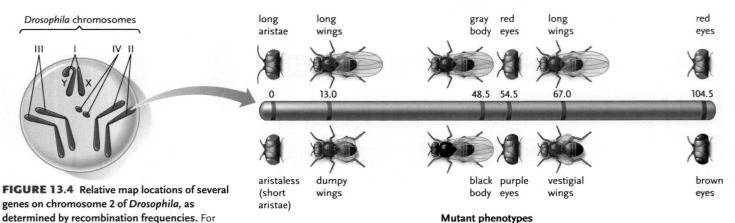

Wild-type phenotypes

long aristae | long wings | gray body | red eyes | long wings | red eyes

0   13.0   48.5   54.5   67.0   104.5

aristaless (short aristae) | dumpy wings | black body | purple eyes | vestigial wings | brown eyes

Mutant phenotypes

FIGURE 13.4 Relative map locations of several genes on chromosome 2 of *Drosophila*, as determined by recombination frequencies. For each gene, the diagram shows the normal or wild-type phenotype on the top and the mutant phenotype on the bottom. Mutant alleles at two different locations alter wing structure, one producing the dumpy wing and the other the vestigial wing phenotypes; the normal allele at these locations results in normal long-wing structure. Mutant alleles at two different locations also alter eye color.
© Cengage Learning 2014

Widely Separated Linked Genes Assort Independently

Genes can be so widely separated on a chromosome that recombination is likely to occur at some point between them in at least half of the cells undergoing meiosis. When this is the case, no linkage is detected and the genes assort independently. In other words, even though the alleles of the genes are carried on the same chromosome, the approximate 1:1:1:1 ratio of phenotypes is seen in the offspring of a dihybrid × double mutant testcross. That is, 50% of the progeny are parentals and 50% are recombinants. Linkage between such widely separated genes can still be detected, however, by testing their linkage to one or

more genes that lie between them. For example, the genes *a* and *c* in **Figure 13.5** are located so far apart that they assort independently and show no linkage. However, crosses show that *a* and *b* are 23 map units apart (recombination frequency of 23%), and crosses that show *b* and *c* are 34 map units apart. Therefore, *a* and *c* must also be linked and carried on the same chromosome at 23 + 34 = 57 map units apart. We could not see a recombination frequency of 57% in testcross progeny because the maximum frequency of recombinants is 50%, which equals independent assortment.

We now know that some of the genes Mendel studied assort independently even though they are on the same chromosome. For example, the genes for flower color and seed color are located on the same chromosome, but they are so far apart that the frequent recombination between them makes them appear to be unlinked.

STUDY BREAK 13.1

You want to determine whether genes *a* and *b* are linked. What cross would you use and why? How would this cross tell you if they are linked?

13.2 Sex-Linked Genes

In many organisms, one or more pairs of chromosomes are different in males and females (see Section 11.1). Genes located on these chromosomes, the *sex chromosomes,* are called **sex-linked genes;** they are inherited differently in males and females. (Note that the word *linked* in *sex-linked gene* means that the gene is on a sex chromosome, whereas the use of the term *linked* when considering two or more genes means that the genes are on the same chromosome, not necessarily a sex chromosome.) Chromosomes other than the sex chromosomes are called **autosomes;** genes on these chromosomes have the same patterns of inheritance in both sexes. In humans, chromosomes 1 to 22 are the autosomes.

Genes *a* and *c* are located so far apart that a crossover almost always occurs between them. Their linkage therefore cannot be detected.

23 mu

57 mu

34 mu

Gene *a* and *b*, and *b* and *c*, however, are close enough to show linkage; *a* and *c* must therefore also be linked.

FIGURE 13.5 Genes far apart on the same chromosome. Genes *a* and *c* are far apart and will not show linkage, suggesting they are on different chromosomes. However, linkage between such genes can be established by noting their linkage to another gene or genes located between them—in this case, gene *b*.

© Cengage Learning 2014

Females Are XX and Males Are XY in Both Humans and Fruit Flies

In most species with sex chromosomes, females have two copies of a chromosome known as the **X chromosome,** forming a homologous XX pair, whereas males have only one X chromosome. Another chromosome, the **Y chromosome,** occurs in males but not in females, giving males an XY combination. The XX–XY human chromosome complement is shown in Figure 10.7.

Because an XX female produces only one type of gamete with respect to the sex chromosomes, she is called the **homogametic sex.** That is, each normal gamete produced by an XX female carries an X chromosome. Because the XY male produces two types of gametes with respect to the sex chromosomes, one with an X and one with a Y, the male is called the **heterogametic sex.** That is, half the gametes produced by an XY male carry an X chromosome and half carry a Y. When a sperm cell carrying an X chromosome fertilizes an X-bearing egg cell, the new individual develops into an XX female. Conversely, when a sperm cell carrying a Y chromosome fertilizes an X-bearing egg cell, the combination produces an XY male **(Figure 13.6).** The Punnett square shows that fertilization is ex-

pected to produce females and males with an equal frequency of 1/2. This expectation is closely matched in human and *Drosophila* populations.

Other sex chromosome arrangements occur, as in some insects with XX females and XO males (the O means there is no Y chromosome). In some birds, butterflies, and some reptiles, the situation is reversed: males are the homogametic sex with a homologous pair of sex chromosomes (termed ZZ instead of XX), and females are the heterogametic sex with ZW sex chromosomes, equivalent to an XY combination. Researchers have compared the genes on sex chromosomes and have determined that the X and Y chromosomes of mammals are quite different from the Z and W chromosomes of birds. That is, mammalian X and Y chromosome genes typically are on bird autosomes, whereas bird Z and W chromosome genes are on mammalian autosomes. The interpretation is that mammalian and bird sex chromosomes have evolved from different autosomal pairs.

In bees and wasps, and certain other arthropods, sex is determined not by sex chromosomes but whether the individual is haploid or diploid. Essentially this means that sex depends on the number of sets of chromosomes. In this system, an individual produced by fusion of an egg and a sperm is diploid and develops into a female, whereas an unfertilized egg, which is haploid, develops into a male.

A number of eukaryotic microorganisms do not have sex chromosomes but have a "sex" system specified by simple alleles of a gene. For example, budding yeast, *Saccharomyces cerevisiae* (see *Focus on Research Organisms,* p. 218), is a haploid eukaryote with two *mating types* or sexes, designated **a** and α. The mating types are identical in appearance, but matings will only occur between two individuals of opposite type.

Human Sex Determination Depends on the Y Chromosome

The human X chromosome carries about 1,670 genes. Although some of these genes are associated with sexual traits, such as differing distributions of body fat in males and females, most are concerned with nonsexual traits such as the ability to perceive color, metabolize certain sugars, or form blood clots when tissues are injured. Human sex determination depends on the Y chromosome, which contains the *SRY* gene (for *sex-determining region* of the *Y*) that switches development toward maleness at an early point in embryonic development.

For the first month or so of embryonic development in humans, the rudimentary structures that give rise to reproductive organs and tissues are the same in XX or XY embryos. After 6 to 8 weeks, the *SRY* gene becomes active in XY embryos, producing a protein that regulates the expression of other genes, thereby stimulating part of these structures to develop as testes. As a part of stimulation by hormones secreted in the developing testes and elsewhere, tissues degenerate that would otherwise develop into female structures such as the vagina and oviducts. The remaining structures develop into the penis and scrotum. In XX embryos, which do not have a copy of the *SRY*

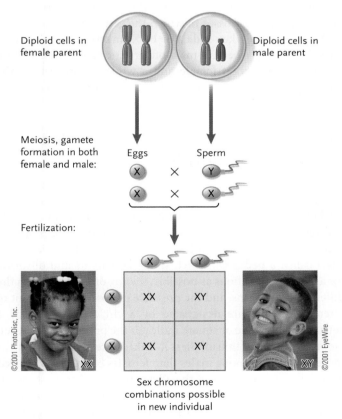

FIGURE 13.6 Sex chromosomes and the chromosomal basis of sex determination in humans. Females have two X chromosomes and produce gametes (eggs), all of which have the X sex chromosome. Males have one X and one Y chromosome and produce gametes, half with an X and half with a Y chromosome. Males transmit their Y chromosome to their sons, but not to their daughters. Males receive their X chromosome only from their mother.
© Cengage Learning 2014

gene, development proceeds toward female reproductive structures. The rudimentary male structures degenerate in XX embryos because the hormones released by the developing testes in XY embryos are not present. Further details of the *SRY* gene and its role in human sex determination are presented in Chapter 50 (specifically, see *Insights from the Molecular Revolution* in that chapter).

Sex-Linked Genes Were Discovered First in *Drosophila*

The different sets of sex chromosomes in males and females affect the inheritance of the alleles on these chromosomes in a distinct pattern known as *sex linkage*. Two features of the XX–XY arrangement cause sex linkage. One is that alleles carried on the X chromosome occur in two copies in females but in only one copy in males. The second feature is that alleles carried on the Y chromosome are present in males but not females; those alleles do not correspond to alleles on the X chromosome.

Morgan discovered sex-linked genes and their pattern of sex linkage in 1910. It started when he found a male fly in his stocks with white eyes instead of the normal red eyes **(Figure 13.7)**. To determine how the white-eye allele is inherited, Morgan first crossed a true-breeding female with red eyes with the white-eyed male, and then interbred the F_1 generation to produce the F_2 generation. The F_1 flies of both sexes had red eyes, indicating that the white-eye trait is recessive. The F_2 flies showed a phenotypic ratio of 3 red-eyed:1 white-eyed. But Morgan observed that, unexpectedly, the phenotypic ratio was not the same in males and females: all F_2 female flies had red eyes, whereas 1/2 the F_2 males had red eyes and the other 1/2 had white eyes.

Morgan hypothesized that the alleles segregating in the cross were of a gene located on the X chromosome and that the unexpected distribution of phenotypes in males and females in the F_2 generation could be accounted for by the inheritance

pattern of X and Y chromosomes. A gene on a sex chromosome is a *sex-linked gene,* as we discussed earlier. A gene on the X chromosome more precisely is called an **X-linked gene;** the pattern of inheritance of an X-linked gene is called **X-linked inheritance.** To designate X-linked genes and their alleles, we use a symbolism using the upper-case letter X for the X chromosome and superscript letter(s) for the gene. In this case, the white mutant allele is X^w, a white-eyed female is $X^w X^w$ and a white-eyed male is $X^w Y$. The wild-type allele of the white gene is X^{w^+}; X^w is recessive to this allele.

We can follow the alleles in Morgan's cross of a true-breeding red-eyed female ($X^{w^+} X^{w^+}$) with the white-eyed male ($X^w Y$) through to the F_2 generation in **Figure 13.8A.** The transmission of the white-eye allele shown in this cross—from a male parent to a female offspring ("child") to a male "grandchild" is called **crisscross inheritance.**

Morgan also performed a **reciprocal cross,** meaning that he switched the phenotypes of the parents. The reciprocal cross here was a true-breeding white-eyed female ($X^w X^w$) with a red-eyed male ($X^{w^+} Y$); we can follow the alleles in this cross through to the F_2 generation in **Figure 13.8B.**

The results of the reciprocal crosses differed markedly in both the F_1 and F_2 generations. Morgan's experiments had shown that there is a distinctive pattern in the phenotypic ratios for reciprocal crosses in which the gene involved is on the X chromosome. A key indicator of X-linked inheritance of a recessive trait is when all male offspring of a cross between a true-breeding mutant female and a wild-type male have the mutant phenotype. As we have seen, this occurs because a male receives his X chromosome from his female parent.

X-Linked Genes in Humans Are Inherited as They Are in *Drosophila*

For obvious reasons, experimental genetic crosses cannot be conducted with humans. However, a similar analysis can be made by interviewing and testing living members of a family and reconstructing the genotypes and phenotypes of past generations from family records. The results are summarized in a chart called a **pedigree,** which shows all parents and offspring for as many generations as possible, the sex of individuals in the different generations, and the presence or absence of the trait of interest. Females are designated by a circle and males by a square; a solid circle or square indicates the presence of the trait.

In humans, as in fruit flies, X-linked recessive traits appear more frequently among males than females because males need to receive only one copy of the allele on the X chromosome inherited from their mothers to develop the trait. Females must receive two copies of the recessive allele, one from each parent, to develop the trait. Two examples of human X-linked traits are red–green color blindness, a recessive trait in which the affected individual is unable to distinguish between the colors red and green because of a defect in light-sensing cells in the retina, and hemophilia, a recessive trait in which affected individuals have a defect in blood clotting.

A. Normal, red wild-type eye color

B. Mutant white eye color caused by recessive allele of a sex-linked gene on the X chromosome

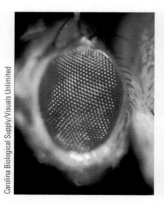
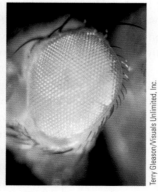

FIGURE 13.7 Eye color phenotypes in *Drosophila*. (A) Normal, red wild-type eye color. **(B)** Mutant white eye color caused by a recessive allele of a sex-linked gene carried on the X chromosome.

FIGURE 13.8 | **Experimental Research**

Evidence for Sex-Linked Genes

Question: How is the white-eye gene of *Drosophila* inherited?

Experiment: Morgan crossed a white-eyed male *Drosophila* with a true-breeding female with red eyes and then interbred the F₁ flies to produce the F₂ generation. He also performed the reciprocal cross in which the phenotypes were switched in the parental flies—true-breeding white-eyed female × red-eyed male.

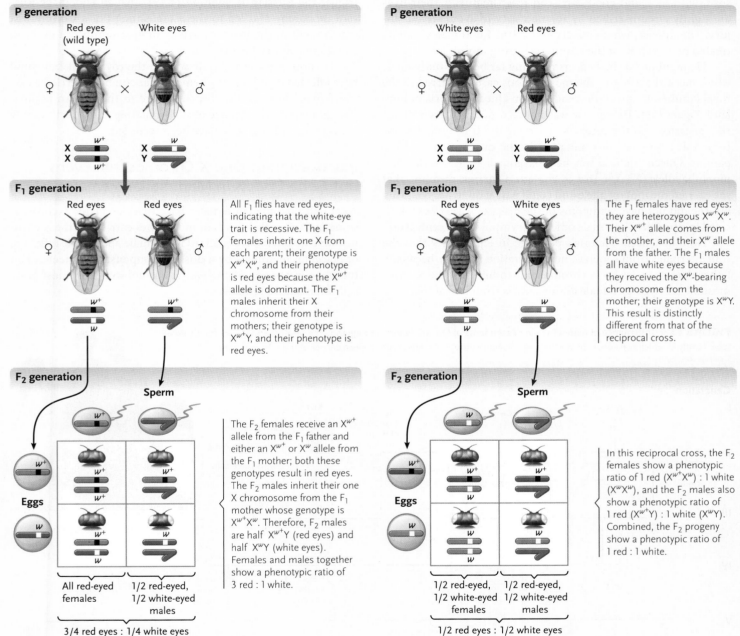

A. True-breeding red-eyed female × white-eyed male

P generation

Red eyes (wild type) ♀ × White eyes ♂

F₁ generation

Red eyes ♀   Red eyes ♂

All F₁ flies have red eyes, indicating that the white-eye trait is recessive. The F₁ females inherit one X from each parent; their genotype is X^{w+}X^{w}, and their phenotype is red eyes because the X^{w+} allele is dominant. The F₁ males inherit their X chromosome from their mothers; their genotype is X^{w+}Y, and their phenotype is red eyes.

F₂ generation

Sperm / Eggs

The F₂ females receive an X^{w+} allele from the F₁ father and either an X^{w+} or X^{w} allele from the F₁ mother; both these genotypes result in red eyes. The F₂ males inherit their one X chromosome from the F₁ mother whose genotype is X^{w+}X^{w}. Therefore, F₂ males are half X^{w+}Y (red eyes) and half X^{w}Y (white eyes). Females and males together show a phenotypic ratio of 3 red : 1 white.

All red-eyed females   1/2 red-eyed, 1/2 white-eyed males

3/4 red eyes : 1/4 white eyes

B. White-eyed female × red-eyed male

P generation

White eyes ♀ × Red eyes ♂

F₁ generation

Red eyes ♀   White eyes ♂

The F₁ females have red eyes: they are heterozygous X^{w+}X^{w}. Their X^{w+} allele comes from the mother, and their X^{w} allele from the father. The F₁ males all have white eyes because they received the X^{w}-bearing chromosome from the mother; their genotype is X^{w}Y. This result is distinctly different from that of the reciprocal cross.

F₂ generation

Sperm / Eggs

In this reciprocal cross, the F₂ females show a phenotypic ratio of 1 red (X^{w+}X^{w}) : 1 white (X^{w}X^{w}), and the F₂ males also show a phenotypic ratio of 1 red (X^{w+}Y) : 1 white (X^{w}Y). Combined, the F₂ progeny show a phenotypic ratio of 1 red : 1 white.

1/2 red-eyed, 1/2 white-eyed females   1/2 red-eyed, 1/2 white-eyed males

1/2 red eyes : 1/2 white eyes

Results: Differences were seen in both the F₁ and F₂ generations for the red ♀ × white ♂ and white ♀ × red ♂ reciprocal crosses.

Conclusion: The segregation pattern for the white-eye trait showed that the white-eye gene is a sex-linked gene located on the X chromosome.

THINK LIKE A SCIENTIST In humans, red-green color blindness (*c*) is recessive and X linked. The normal vision allele is c^+. A woman with normal vision whose father was color blind marries a man with normal vision whose father was also color blind. What possible children could they have with respect to color blindness, and in what proportion would they occur?

Hemophiliacs—people with hemophilia—are "bleeders"; that is, if they are injured, they bleed much more than usual and are susceptible to severe bruising because a protein required for forming blood clots is not produced in functional form. Males are bleeders if they receive an X chromosome that carries the recessive allele, X^h. The disease also develops in females with the recessive allele on both of their X chromosomes, genotype $X^h X^h$—a rare combination. Although affected persons, with luck and good care, can reach maturity, their lives are tightly circumscribed by the necessity to avoid serious injury. The disease, which affects about 1 in 7,000 males, can be treated by injection of the required clotting molecules.

Hemophilia has had effects reaching far beyond individuals who inherit the disease. The most famous cases occurred in the royal families of Europe descended from Queen Victoria of England (**Figure 13.9**). The disease was not recorded in Queen Victoria's ancestors, so the recessive allele for the trait probably appeared as a spontaneous mutation in the queen or one of her parents. Queen Victoria was heterozygous for the recessive hemophilia allele ($X^{h+} X^h$); that is, she was a **carrier,** meaning that she carried the mutant allele and could pass it on to her offspring but she did not have symptoms of the disease. A carrier is indicated in a pedigree by a male or female symbol with a central dot.

Note in Queen Victoria's pedigree in Figure 13.9 that the trait alternates from generation to generation in males because a father does not pass his X chromosome to his sons; the X chromosome received by a male always comes from his mother.

At one time, 18 of Queen Victoria's 69 descendants were affected males or female carriers. Because so many sons of European royalty were affected, the trait influenced the course of history. In Russia, Crown Prince Alexis (highlighted in Figure 13.9) was one of Victoria's hemophiliac descendants. The hypnotic monk Rasputin manipulated the Czar Nicholas II and Czarina Alexandra to his advantage by convincing them that only he could control their son's bleeding. The situation helped trigger the Russian Revolution of 1917, which ended the Russian monarchy and led to the establishment of a Communist government in the former Soviet Union, a significant event in twentieth century history.

Hemophilia affected only sons in the royal lines but could have affected daughters if a hemophiliac son had married a carrier female. Because the disease is rare in the human population as a whole, the chance of such a mating is so low that only a few unions of this type have been recorded.

Inactivation of One X Chromosome Evens out Gene Effects in Mammalian Females

Although mammalian females have twice as many X chromosomes as males, the effects of most genes carried on the X chromosome in females is equalized in the male and female offspring of placental mammals by a **dosage compensation mechanism** that inactivates one of the two X chromosomes in most body cells of female mammals.

FIGURE 13.9 **Inheritance of hemophilia in descendants of Queen Victoria of England.** The photograph shows the Russian royal family in which the son, Crown Prince Alexis, had hemophilia. His mother was a carrier of the mutated gene.
© Cengage Learning 2014

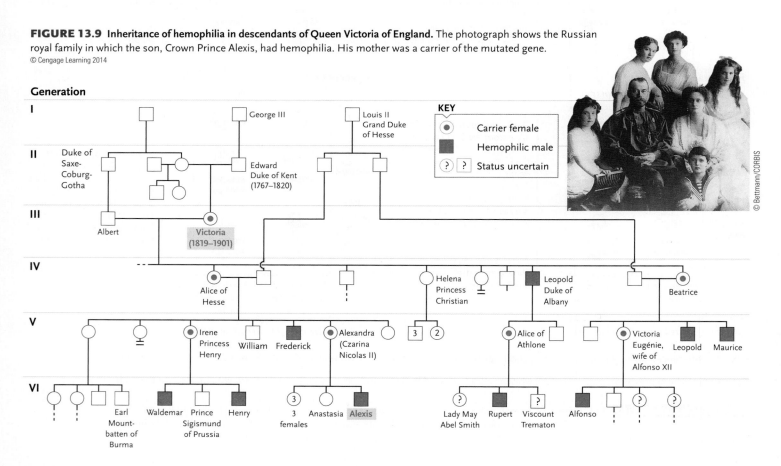

As a result of the equalizing mechanism, the activity of most genes carried on the X chromosome is essentially the same in males and females. The inactivation occurs by a condensation process that folds and packs the chromatin of one of the two X chromosomes into a tightly condensed state. The inactive, condensed X chromosome can be seen at one side of the nucleus in cells of females as a dense mass of chromatin called the **Barr body.**

The inactivation occurs during early embryonic development. Which of the two X chromosomes becomes inactive in a particular embryonic cell line is a random event. But once one of the X chromosomes is inactivated in a cell, that same X is inactivated in all descendants of the cell. Thus, within one female, one of the X chromosomes is active in particular cells and inactive in others and vice versa.

If the two X chromosomes carry different alleles of a gene, one allele will be active in cell lines in which one X chromosome is active, and the other allele will be active in cell lines in which the other X chromosome is active. For many sex-linked alleles, such as the recessive allele that causes hemophilia, random inactivation of either X chromosome has little overall whole-body effect in heterozygous females because the dominant allele is active in enough of the critical cells to produce a normal phenotype. However, for some genes, the inactivation of either X chromosome in heterozygotes produces recognizably different effects in distinct regions of the body.

For example, the orange and black patches of fur in calico cats result from inactivation of one of the two X chromosomes in regions of the skin of heterozygous females **(Figure 13.10).** Males, which get only one of the two alleles, normally have either black or orange fur.

An X-linked trait in humans has a similar, but less visible phenotype. Called anhidrotic ectodermal dysplasia, the trait is characterized by the absence of sweat glands. Females heterozygous for the mutation that causes the trait may have a patchy distribution of skin areas with and without the glands.

As we have seen, the discovery of genetic linkage, recombination, and sex-linked genes led to the elaboration and expansion of Mendel's principles of inheritance. Next, we examine what happens when patterns of inheritance are modified by changes in the chromosomes.

STUDY BREAK 13.2

You have a true-breeding strain of miniature-winged fruit flies, where this wing trait is recessive to the normal long wings. How would you show whether the miniature wing trait is sex-linked or autosomal?

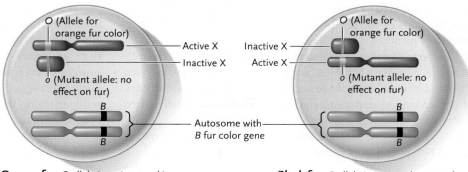

Orange fur: O allele is active, masking phenotypic expression of the B gene (an example of epistasis; see Section 12.2).

Black fur: O allele is inactive because the X chromosome it is on is inactivated; the mutant o allele on the active X chromosome does not mask the phenotypic expression of the B gene.

White patches result from interactions with a different, autosomal gene that blocks pigment deposition in the fur completely.

FIGURE 13.10 A female cat with the calico color pattern in which patches of orange and black fur are produced by random inactivation of one of the two X chromosomes. Two genes control the black and orange colors: the O allele on the X chromosome is for orange fur color, and the mutated o allele has no effect on color. The B gene on an autosome is for black fur color. A calico cat has the genotype Oo BB or Oo Bb; the former genotype is illustrated in the figure. In tortoiseshell cats, the same orange–black patching occurs as in calico cats but the gene for the white patching is not active.
© Cengage Learning 2014

THINK OUTSIDE THE BOOK

While *Drosophila* has X and Y chromosomes with XX flies being female and XY males being Y, this species does not have an *SRY* gene. Individually or collaboratively, explore the Internet or research papers to determine how sex determination occurs in *Drosophila*.

13.3 Chromosomal Mutations That Affect Inheritance

Chromosomal mutations are variations from the normal condition in chromosome structure or chromosome number. Changes in chromosome structure occur when the DNA breaks, which can be generated by agents such as radiation or certain chemicals or by enzymes encoded in some infecting viruses. The broken chromosome fragments may be lost, or they may reattach to the same or different chromosomes. Chromosomal structure changes may have genetic consequences if alleles are eliminated, mixed in new combinations, duplicated, or placed in new locations by the alterations in cell lines that lead to the formation of gametes.

Genetic changes may also occur through changes in chromosome number, including addition or loss of one or more chromosomes or even entire sets of chromosomes. Both forms of chromosomal mutation, changes in chromosome structure and changes in chromosome number, can be a source of disease and disability, and are also processes that have played a role in the evolution of genomes.

Deletions, Duplications, Translocations, and Inversions Are the Most Common Chromosomal Mutations Affecting Chromosome Structure

Chromosomal mutations after breakages occur in four major forms **(Figure 13.11)**:

- A **deletion** occurs if a broken segment is lost from a chromosome.
- A **duplication** occurs if a segment is transferred from one chromosome and inserted into its homolog. In the receiving homolog, the alleles in the inserted fragment are added to the ones already there.
- A **translocation** occurs if a broken segment is attached to a different, nonhomologous chromosome.
- An **inversion** occurs if a broken segment reattaches to the same chromosome from which it was lost, but in reversed orientation, so that the order of genes is reversed.

To be inherited, chromosomal alterations must occur or be included in cells of the germ line leading to development of eggs or sperm.

DELETIONS AND DUPLICATIONS A deletion (see Figure 13.11A) may cause severe problems if the missing segment contains genes that are essential for normal development or cellular functions. For example, an individual heterozygous for a deletion of part of human chromosome 5 typically has severe mental retardation, a variety of physical abnormalities and a malformed larynx. The cries of an affected infant sound more like a meow than a human cry, hence the name of the disorder, *cri-du-chat* (meaning "cry of the cat").

A duplication (see Figure 13.11B) may have effects that vary from harmful to beneficial, depending on the genes and alleles contained in the duplicated region. Although most duplications are likely to be detrimental, some have been important sources of evolutionary change. That is, because there are duplicate genes, one copy can mutate into new forms without seriously affecting the basic functions of the organism. For example, mammals have genes that encode several types of hemoglobin that are not present in vertebrates, such as sharks, which evolved earlier; the additional hemoglobin genes of mammals arose through evolutionary time through duplications, followed by mutations in the duplicates that created new and beneficial forms of hemoglobin as further evolution took place. Duplications sometimes arise if crossing-over occurs unequally during meiosis, so that a segment is deleted from one chromosome of a homologous pair and inserted in the other (see Section 19.4).

TRANSLOCATIONS AND INVERSIONS In a translocation, a segment breaks from one chromosome and attaches to another, nonhomologous chromosome. In many cases, a translocation is reciprocal, meaning that two nonhomologous chromosomes exchange segments (see Figure 13.11C). Reciprocal translocations resemble genetic recombination, except that the two chromosomes involved in the exchange do not contain the same genes.

In an inversion, a chromosome segment breaks and then reattaches to the same chromosome, but in reverse order (see Figure 13.11D). Inversions have essentially the same effects as translocations—genes may be broken internally by the inversion, with loss of function, or they may be transferred intact to a new location within the same chromosome, producing effects that range from beneficial to harmful.

Many cancers have chromosomal mutations, and the most common type of chromosomal mutation involved is a translocation. For example, 90% of patients with chronic myelogenous leukemia (CML) have a chromosomal mutation called the Philadelphia chromosome. CML is a type of cancer of the blood involving the uncontrolled division of stem cells for white blood cells. The Philadelphia chromosome arises from a reciprocal translocation event involving chromosomes 9 and 22 **(Figure 13.12)**. The movement of the segment of chromosome 9 to chromosome 22 fuses together on the resulting Philadelphia chromosome, the *ABL*

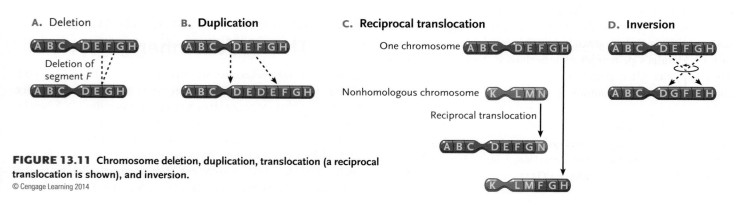

FIGURE 13.11 Chromosome deletion, duplication, translocation (a reciprocal translocation is shown), and inversion.
© Cengage Learning 2014

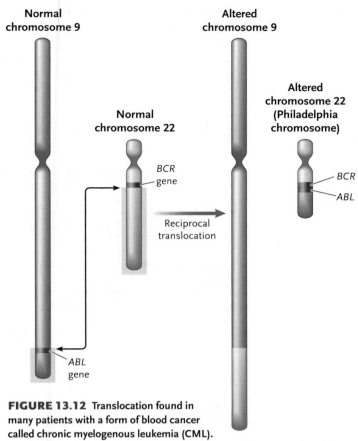

Normal chromosome 9

Altered chromosome 9

Normal chromosome 22

Altered chromosome 22 (Philadelphia chromosome)

BCR gene

BCR
ABL

Reciprocal translocation

ABL gene

FIGURE 13.12 **Translocation found in many patients with a form of blood cancer called chronic myelogenous leukemia (CML).** A reciprocal translocation involving chromosomes 9 and 22 produces a short chromosome named the Philadelphia. On this chromosome, the chromosome 9 *ABL* gene has become fused to the chromosome 22 *BCR*. The resulting overactivity of the *ABL* gene, which normally helps control cell division, causes the cell to convert to a cancer cell.
© Cengage Learning 2014

gene from 9 with the *BCR* gene on 22. The *ABL* gene is one of many genes that control cell growth and division. (Cell division control is described in Chapter 11.) Its product is a tyrosine kinase, an enzyme that adds phosphate to tyrosine amino acids in target proteins. We learned about tyrosine kinases in Section 7.3. In its new location on the Philadelphia chromosome, the *ABL* gene becomes much more active because of its fusion with the *BCR* gene and much more than normal of its tyrosine kinase product is made. As a result of its overactivity, normal cell cycle control breaks down and the cells are stimulated to growth and divide uncontrollably, becoming cancer cells. The drug Gleevec® is used to treat CML patients. It works by inhibiting the tyrosine kinase enzyme so that the body stops, or at least reduces, the production of too many white blood cells.

Inversions and translocations have been important factors in the evolution of the genomes of plants and some animals, including insects and primates. For example, nine of the chromosome pairs of humans show evidence of translocations and inversions that differ between humans and chimpanzees, and therefore must have occurred after the ancestral lineages leading to chimpanzees and humans split.

Some Chromosome Mutations Involve Changes in the Number of Entire Chromosomes

At times, whole, single chromosomes are lost or gained from cells entering or undergoing meiosis, resulting in a change of chromosome number. Most often, these changes occur through **nondisjunction**—the failure of homologous pairs to separate during the first meiotic division or of chromatids to separate during the second meiotic division **(Figure 13.13)**. As a result, gametes are produced that lack one or more chromosomes or contain extra copies of the chromosomes. Fertilization by these gametes produces an individual with extra or missing chromosomes. Such individuals are called **aneuploids,** whereas individuals with a normal set of chromosomes are called **euploids.**

Changes in chromosome number can also occur through duplication or loss of entire sets, meaning individuals may receive fewer or more than the normal number of the entire haploid complement of chromosomes. Individuals with one set of chromosomes instead of the normal two are **monoploids;** individuals with more than the two sets of chromosomes are called **polyploids.** *Triploids* have three copies of each chromosome instead of two; *tetraploids* have four copies of each chromosome; *hexaploids* have six copies of each chromosome. Multiples higher than hexaploids also occur.

ANEUPLOIDS The effects of addition or loss of whole chromosomes vary depending on the chromosome and the species. In animals, aneuploidy of autosomes usually produces debilitating or lethal developmental abnormalities. In humans, addition or loss of an autosomal chromosome causes embryos to develop so abnormally that generally they are aborted naturally. For reasons that are not understood, aneuploidy is as much as 10 times more frequent in humans than in other mammals. Of human fetuses that have been miscarried and examined, about 70% are aneuploids.

In some cases, autosomal aneuploids survive. This is the case with humans who receive an extra copy of chromosome 21—the smallest human chromosomes **(Figure 13.14A).** Many of these individuals survive until young adulthood. The condition produced by the extra chromosome, called *Down syndrome* or *trisomy 21* (for "three chromosome 21s"), is characterized by short stature and moderate to severe mental retardation. About 40% of individuals with Down syndrome have heart defects, and skeletal development is slower than normal. Most do not mature sexually and remain infertile. However, with attentive care and special training, individuals with Down syndrome can participate with reasonable success in many activities.

Down syndrome arises from nondisjunction of chromosome 21 during the meiotic divisions, primarily in women (about 5% of nondisjunctions that lead to Down syndrome occur in men). The nondisjunction occurs more frequently as women age, increasing the chance that a child may be born with the syndrome **(Figure 13.14B).** In the United States, 1 in every 1,000 children is born with Down syndrome, making it one of the most common serious human genetic disorders.

A. Nondisjunction during first meiotic division

Nondisjunction

Meiosis I   Meiosis II   Gametes

Extra chromosome ($n + 1$)

Extra chromosome ($n + 1$)

Missing chromosome ($n - 1$)

Missing chromosome ($n - 1$)

Nondisjunction during the first meiotic division causes both chromosomes of one pair to be delivered to the same pole of the spindle. The nondisjunction produces two gametes with an extra chromosome and two with a missing chromosome.

B. Nondisjunction during second meiotic division

Nondisjunction

Meiosis I   Meiosis II   Gametes

Extra chromosome ($n + 1$)

Missing chromosome ($n - 1$)

Normal (n)

Normal (n)

Nondisjunction during the second meiotic division produces two normal gametes, one gamete with an extra chromosome and one gamete with a missing chromosome.

FIGURE 13.13 Nondisjunction during (A) the first meiotic division and (B) the second meiotic division.
© Cengage Learning 2014

A. The chromosomes of a human female with Down syndrome showing three copies of chromosome 21 (circled in red).

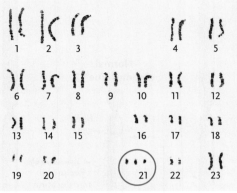

1997, Hironao Numabe, M.D., Tokyo Medical University

B. The increase in the incidence of Down syndrome with increasing age of the mother, from a study conducted in Victoria, Australia, between 1942 and 1957.

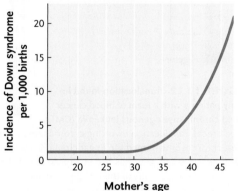

C. Person with Down syndrome.

R. Gino Santa Maria/Shutterstock.com

FIGURE 13.14 Down syndrome.
© Cengage Learning 2014

Trisomy of chromosomes 13 and 18 also is seen in live births of humans. Trisomy 13 produces Patau syndrome, with characteristics including cleft lip and palate, small eyes, extra fingers and toes, mental and developmental retardation, and cardiac anomalies. Most Patau syndrome individuals die be- fore the age of 3 months. Trisomy 18 produces Edwards syn- drome, with characteristics including small size at birth and multiple congenital malformations affecting almost every organ of the body. Most Edwards syndrome individuals die within the first 6 months after birth.

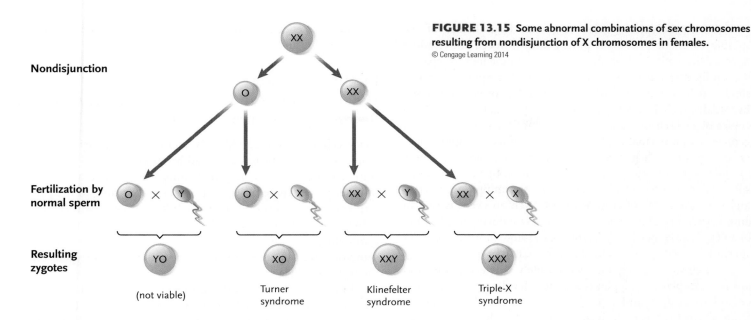

Nondisjunction

Fertilization by normal sperm

Resulting zygotes

YO (not viable)

XO Turner syndrome

XXY Klinefelter syndrome

XXX Triple-X syndrome

FIGURE 13.15 Some abnormal combinations of sex chromosomes resulting from nondisjunction of X chromosomes in females.
© Cengage Learning 2014

Aneuploidy of sex chromosomes can also arise by nondisjunction during meiosis (**Figure 13.15** and **Table 13.1**). Unlike autosomal aneuploidy, which usually has drastic effects on survival, altered numbers of X and Y chromosomes are often tolerated, producing individuals who progress through embryonic development and grow to adulthood. This is because, in the case of multiple X chromosomes, the X-chromosome inactivation mechanism converts all but one of the X chromosomes to a Barr body, so the dosage of active X-chromosome genes is the same as in normal XX females and XY males. Thus, XO (Turner syndrome) females have no Barr bodies, XXY (Klinefelter syndrome) males have one Barr body, and XXX (triple-X syndrome) females have two Barr bodies (see Table 13.1). However, X chromosomes are not inactivated until about 15 to 16 days after fertilization. Expression of the extra X chromosome genes early in development results in any deleterious effects associated with a particular sex chromosome aneuploidy.

Because sexual development in humans is pushed toward male or female reproductive organs primarily by the presence or absence of the *SRY* gene on the Y chromosome, people with a Y chromosome are externally malelike, no matter how many X chromosomes are present. If no Y chromosome is present, X chromosomes in various numbers give rise to femalelike individuals. (Table 13.1 lists the effects of some alterations in sex chromosome number.) Similar abnormal combinations of sex chromosomes also occur in other animals with varying effects on viability.

MONOPLOIDS A monoploid has only one set of chromosomes instead of the normal two. Monoploidy is lethal in most animal species, but is tolerated more in plants. Certain animal species produce monoploid organisms as a normal part of their life cycle. For instance, as mentioned earlier, some male wasps, ants, and bees are monoploid individuals that have developed from unfertilized eggs.

TABLE 13.1	Effects of Unusual Combinations of Sex Chromosomes in Humans		
Combination of Sex Chromosomes	Approximate Frequency	Barr Bodies	Effects
XO	1 in 5,000 births	0	Turner syndrome: females with underdeveloped ovaries; sterile; intelligence and external genitalia are normal; typically, individuals are short in stature with underdeveloped breasts
XXY	1 in 2,000 births	1	Klinefelter syndrome: male external genitalia with very small and underdeveloped testes; sterile; intelligence usually normal; sparse body hair and some development of the breasts; similar characteristics in XXXY and XXXXY individuals
XYY	1 in 1,000 births	0	XYY syndrome: apparently normal males but often taller than average
XXX	1 in 1,000 births	2	Triple-X syndrome: apparently normal female with normal or slightly retarded mental function

© Cengage Learning 2014

POLYPLOIDS Polyploidy often originates from failure of the spindle to function normally during mitosis in cell lines leading to germ-line cells. In these divisions, the spindle fails to separate the duplicated chromosomes, which are incorporated into a single nucleus with twice the usual number of chromosomes. Eventually, meiosis takes place and produces gametes with two copies of each chromosome instead of one. Fusion of one such gamete with a normal haploid gamete produces a triploid, and fusion of two such gametes produces a tetraploid. Polyploidy may also occur when a single egg is fertilized by more than one sperm. For example, fertilization by two sperm produces a triploid with three sets of chromosomes. Such an event is rare in animals because of mechanisms that operate during fertilization of an egg with a sperm to prevent the subsequent fusion of other sperm with that egg (discussed more in Chapter 49).

The effects of polyploidy vary widely between plants and animals. In plants, polyploids are often hardier and more successful in growth and reproduction than the diploid plants from which they were derived. As a result, polyploidy is common and has been an important process in the evolution of plant genomes. About half of all flowering plant species are polyploids, including important crop plants such as wheat and other cereals, cotton, strawberries, and bananas. Commercial bread wheat, for instance, is hexaploid, and cultivated bananas are triploid.

By contrast, among animals, polyploidy is uncommon because it usually has lethal effects during embryonic development. For example, in humans, all but about 1% of polyploids die before birth, and the few who are born usually die within a month. The lethality is probably caused by disturbance of animal developmental pathways, which are typically much more complex than those of plants.

We now turn to a description of the effects of altered alleles on human health and development.

STUDY BREAK 13.3

What mechanisms are responsible for: (a) duplication of a chromosome segment; (b) generation of a Down syndrome individual; (c) a chromosome translocation; and (d) polyploidy?

>

THINK OUTSIDE THE BOOK

Diagram the various ways a normal XX female and normal XY male could produce an XXY zygote.

13.4 Human Genetics and Genetic Counseling

We have already noted a number of human genetic traits and conditions caused by mutant alleles or chromosomal alterations. All these traits are of interest as examples of patterns of inheritance that amplify and extend Mendel's basic principles. Those with harmful effects are also important because of their impact on human life and society.

In Autosomal Recessive Inheritance, Heterozygotes Are Carriers and Homozygous Recessives Are Affected by the Trait

Sickle-cell anemia, cystic fibrosis, and phenylketonuria are examples of human disorders caused by recessive alleles on autosomes. Many other human genetic traits follow a similar pattern of inheritance. These traits are passed on according to the pattern known as **autosomal recessive inheritance,** in which individuals who are homozygous for the dominant allele are free of symptoms and are not carriers; heterozygotes are usually symptom-free but are carriers. People who are homozygous for the recessive allele show the trait.

In sickle-cell anemia, the amino acid change in hemoglobin causes red blood cells to assume a sickle shape (see Figure 12.1, the Chapter 12 *Why It Matters,* and Section 12.2). The problems the sickled red blood cells have in passing through capillaries cause the serious symptoms of sickle-cell anemia. Between 10% and 15% of African Americans in the United States are carriers for this disorder—they have sickle-cell trait (see Section 12.2). Although carriers make enough normal hemoglobin through the activity of the dominant allele to be essentially unaffected, the mutant, sickle-cell form of the hemoglobin molecule is also present in their red blood cells. Carriers can be identified by a molecular test for the mutant hemoglobin. In countries where malaria is common, including several countries in Africa, carriers are less susceptible to malaria, which helps explain the increased proportions of the mutant allele in people whose ancestors originated in areas of the world where the malarial parasite is common.

Cystic fibrosis (CF), one of the most common genetic disorders among persons of Northern European descent, is another autosomal recessive trait (**Figure 13.16**) (see Chapter 6 *Why It Matters*). About 1 in every 25 people from this line of descent is an unaffected carrier with one copy of the recessive allele.

FIGURE 13.16 A child affected by cystic fibrosis. Daily chest thumps, back thumps, and repositioning dislodge thick mucus that collects in airways to the lungs.

Approximately 1 per 4,000 children born in the United States have CF. They have a mutated form of the transport protein called cystic fibrosis transmembrane conductance regulator (CFTR; see Figure 6.1). CFTR is embedded in the plasma membrane of epithelial cells, such as those of the passageways and ducts of the lungs, pancreas, and digestive tract. The mutant CFTR is deficient in the transport of Cl⁻ (chloride ions) out of the cells into the extracellular fluids. This alteration in chloride transport causes thick, sticky mucus to collect in airways of the lungs, in the ducts of glands such as the pancreas, and in the digestive tract. The accumulated mucus impairs body functions and, in the lungs, promotes pneumonia and other infections. With current management procedures, the life expectancy for a person with cystic fibrosis is about 40 years.

Another autosomal recessive disease, *phenylketonuria* (PKU), appears in about 1 of every 15,000 births. Affected individuals cannot produce an enzyme that converts the amino acid phenylalanine to another amino acid, tyrosine. As a result, phenylalanine builds up in the blood and is converted in the body into other products, including phenylpyruvate. Elevations in both phenylalanine and phenylpyruvate damage brain tissue and can lead to mental retardation. If diagnosed early enough, an affected infant can be placed on a phenylalanine-restricted diet, which can prevent the PKU symptoms. The diet must maintain a level of phenylalanine in the blood high enough to allow normal development of the nervous system, but be low enough to prevent mental retardation. Treatment must begin within the first one or two months after birth, or the brain will be damaged and treatment will be less effective. By U.S. law, all infants born in the country are tested for PKU.

You may have seen warnings on certain foods and drinks for phenylketonuriacs (individuals with PKU) not to use them. This is because they contain the artificial sweetener aspartame (trade name NutraSweet®). Aspartame is a small molecule consisting of the amino acids aspartic acid and phenylalanine joined together. Aspartame binds to taste receptors in the mouth signaling that the substance is sweet. However, it has essentially no calories. Once ingested, aspartame is broken down to its component amino acids. Phenylalanine released in this way could be harmful to an individual with PKU, hence the warnings.

In Autosomal Dominant Inheritance, Only Homozygous Recessives Are Unaffected

Some human traits follow a pattern of **autosomal dominant inheritance.** In this case, the mutant allele that causes the trait is dominant, and people who are either homozygous or heterozygous for the dominant allele are affected. Individuals homozygous for the recessive nonmutant allele are unaffected.

Achondroplasia (see Figure 12.12C), a type of dwarfing that occurs in about 1 in 25,000 births worldwide, is caused by an autosomal dominant allele of a gene on chromosome 4. Of individuals with the dominant allele, only heterozygotes survive embryonic development; homozygous dominants are usually stillborn. When limb bones develop in heterozygous children, cartilage formation is defective, leading to disproportionately short arms and legs. They also have a relatively large head, but the trunk and torso are of normal size. Affected adults are usually not much more than 4 feet tall. Achondroplastic dwarfs are of normal intelligence, are fertile, and can have children. The gene responsible for this trait has been identified and is described in *Insights from the Molecular Revolution*. Progeria, the condition described in *Why It Matters* in this chapter, is also an autosomal dominant trait.

X-Linked Recessive Traits Affect Males More Than Females

Red–green color blindness and hemophilia have already been presented as examples of human traits that demonstrate **X-linked recessive inheritance,** that is, traits resulting from inheritance of recessive alleles carried on the X chromosome. Another X-linked recessive human disease trait is Duchenne muscular dystrophy (DMD) **(Figure 13.17).** In affected individuals, muscle tissue begins to degenerate late in childhood; by the onset of puberty, most individuals with this disease are unable to walk. Muscular weakness progresses, with later involvement of the heart muscle; the average life expectancy for individuals with DMD is 25 years. The nonmutant form of the gene that causes DMD encodes the protein dystrophin, which anchors a particular glycoprotein complex in the plasma membrane of a muscle fiber to the cytoskeleton in the cytosol. In patients with DMD, a mutation in the dystrophin gene results in a nonfunctional protein. As a result, the plasma membrane of muscle fibers is susceptible to tearing during contraction, which leads to muscle destruction.

FIGURE 13.17 Individual with Duchenne muscular dystrophy (DMD), an X-linked recessive trait.

Achondroplasia: What is the gene defect that is responsible for the trait?

Achondroplasia (also called achondroplastic dwarfing; see Figure 12.12C) is an autosomal dominant trait that is the most common form of short-limb dwarfism in humans. Over 80% of the cases of achondroplasia result from a new mutation; that is, the individual is born to parents each of normal height. Research has shown that the new gene mutation occurs during sperm formation in the male parent and is inherited from that parent.

Using genetic analysis, researchers found that the gene responsible for achondroplastic dwarfing is on chromosome 4. Mapping to the same region of the chromosome is *FGFR3* (*fibroblast growth factor receptor 3*), a gene encoding one of a family of cell membrane receptors that bind *fibroblast growth factor (FGF)*, a hormone that stimulates a wide range of cells to grow and divide. (Receptors are described in Chapter 7; this particular receptor is a receptor tyrosine kinase.) The *FGFR3* gene is active in chondrocytes—cells that form cartilage and bone.

Research Question

Is achondroplasia caused by a mutation in the *FGFR3* gene? In other words, are the achondroplasia and *FGFR3* genes one and the same?

Experiments

To answer the questions, Arnold Munnich and his colleagues at the Hospital of Children's Diseases in Paris, France, and John Wasmuth and his colleagues at the University of California, Irvine, independently analyzed the DNA sequence of the *FGFR3* gene in people with achondroplasia.

Results

Both research groups identified mutations in the *FGFR3* gene in DNA from all individuals they examined with achondroplasia, and no such mutations in DNA from normal people. The two groups found that every mutation in the gene affected the same guanine–cytosine (G–C) nucleotide pair, which had mutated to either an adenine–thymine (A–T) or a cytosine-guanine (C–G) base pair. Both mutations result in the substitution in the protein of arginine for glycine, which has very different chemical properties. The substitution occurs in a segment of the protein that spans the membrane connecting the growth factor-binding site with a site inside the cell that triggers the internal response.

Conclusions

The correlations between mutations in the *FGFR3* gene and achondroplasia supported the hypothesis that a mutation in the gene for the fibroblast growth factor receptor 3 protein is responsible for achondroplastic dwarfing.

How does the single amino acid substitution cause dwarfing? In individuals homozygous for the normal *FGFR3* gene, the growth factor receptor participates in pathways to regulate bone growth by switching between active and inactive forms bringing about the appropriate amount of chondrocyte proliferation for normal development **(Figure).** In heterozygotes with the mutation in *FGFR3,* the receptor is active all the time, resulting in a long-term negative effect on bone growth and, hence, the dwarfing symptoms of achondroplasia (see Figure).

The conclusions have been confirmed with a mouse model of achondroplasia. David Givol and colleagues at the Weizmann Institute of Science, Rehovot, Israel, and the Agricultural Research Organization, Bet Dagan, Israel, genetically engineered mice heterozygous for the same *FGFR3* mutation found in humans. These mice have small size and other similar characteristics of achondroplasia. The experiments showed that the dominant mutant allele of *FGFR3* leads to inhibition of chondrocyte proliferation, producing the symptoms of the disease.

THINK LIKE A SCIENTIST The mutations in the *FGFR3* discussed in the box resulted in the substitution of arginine for glycine in the amino acid sequence of the encoded protein. What are the chemical differences between the two amino acids? Why might these differences affect the function of the protein? Speculate whether you think a change from glycine to alanine would have had a similar effect on function.

Source: F. Rousseau et al. 1994. Mutations in the gene encoding fibroblast growth factor receptor-3 in achondroplasia. *Nature* 371:252–254; R. Shiang et al. 1994. Mutations in the transmembrane domain of FGFR3 cause the most common genetic form of dwarfism, achondroplasia. *Cell* 78:335–342; Y. Wang et al. 1999. A mouse model for achondroplasia produced by targeting fibroblast growth factor receptor 3. *Proceedings of the National Academy of Sciences USA* 96:4455–4460.

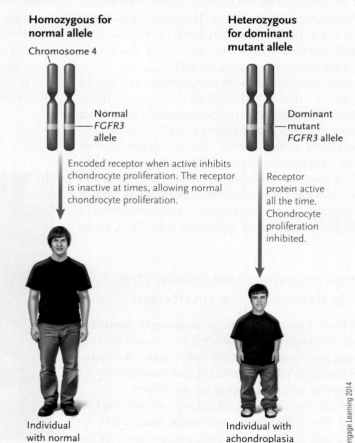

Homozygous for normal allele

Chromosome 4

Normal *FGFR3* allele

Encoded receptor when active inhibits chondrocyte proliferation. The receptor is inactive at times, allowing normal chondrocyte proliferation.

Individual with normal size develops.

Heterozygous for dominant mutant allele

Dominant mutant *FGFR3* allele

Receptor protein active all the time. Chondrocyte proliferation inhibited.

Individual with achondroplasia develops.

FIGURE Molecular basis for achondroplasia.

© Cengage Learning 2014

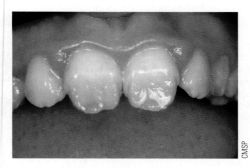

FIGURE 13.18 Teeth showing hereditary faulty enamel and dental discoloration, an X-linked dominant trait.

X-Linked Dominant Traits Are Rare

Only a few X-linked dominant traits have been identified in humans. One example is hereditary faulty enamel and dental discoloration, technical name *hereditary enamel hypoplasia* **(Figure 13.18)**. Another is a severe bleeding anomaly called constitutional thrombopathy.

Human Genetic Disorders Can Be Predicted, and Many Can Be Treated

Of all newborns, between 1% and 3% are homozygous for mutant alleles that encode defective forms of proteins required for normal functions. Possibly 1% have pronounced difficulties resulting from a chromosomal rearrangement or other aberration. Of all patients in children's hospitals, 10% to 25% are treated for problems arising from inherited disorders. Several approaches, which include genetic counseling, prenatal diagnosis, and genetic screening, can reduce the number of children born with genetic diseases.

Genetic counseling allows prospective parents to assess the possibility that they might have an affected child. For example, parents may seek counseling if they, a close relative, or one of their existing children has a genetic disorder. Genetic counseling begins with identification of parental genotypes through family pedigrees and often direct testing for an altered protein or DNA sequence. With this information in hand, counselors can often predict with a high degree of accuracy the chances of having a child with the trait in question. Couples can then make an informed decision about whether to have a child.

Genetic counseling is often combined with techniques of **prenatal diagnosis,** in which cells derived from a developing embryo or its surrounding tissues or fluids are tested for the presence of mutant alleles (by DNA testing or biochemical analysis) or for chromosomal mutations. In **amniocentesis,** cells are obtained from the amniotic fluid—the watery fluid surrounding the embryo or fetus in the mother's uterus **(Figure 13.19)**. In **chorionic villus sampling,** cells are obtained from portions of the placenta that develop from tissues of the embryo. More than 100 genetic disorders can now be detected by these tests. If prenatal diagnosis detects a serious genetic disorder, the prospective parents can reach an informed decision about whether to continue the pregnancy,

including religious and moral considerations, as well as genetic and medical advice.

Once a child is born, certain inherited disorders are identified by **genetic screening,** in which biochemical or molecular tests for disorders are routinely applied to children and adults or to newborn infants in hospitals. The tests can detect inherited disorders early enough to start any available preventive measures before symptoms develop. We have noted that most hospitals in the United States now test all newborns for PKU, making it possible to use dietary restrictions to prevent symptoms of the disorder from developing. As a result, it is becoming less common to see individuals debilitated by PKU.

In addition to the characters and traits described so far in this chapter, some patterns of inheritance depend on genes located not in the cell nucleus, but in mitochondria or chloroplasts in the cytoplasm, as discussed in the following section.

STUDY BREAK 13.4

1. A man has Simpson syndrome, an addiction to a certain television show. His wife does not have this syndrome. This couple has four children, two boys and two girls. One of the boys and one of the girls has this syndrome; the other children are normal. Can Simpson syndrome be an autosomal recessive trait? A sex-linked recessive trait?
2. In another family, a female child has wiggly ears, whereas her brother does not. Both parents are normal. Can the wiggly ear trait be an autosomal recessive trait? A sex-linked recessive trait?

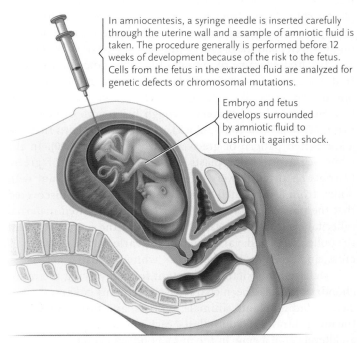

In amniocentesis, a syringe needle is inserted carefully through the uterine wall and a sample of amniotic fluid is taken. The procedure generally is performed before 12 weeks of development because of the risk to the fetus. Cells from the fetus in the extracted fluid are analyzed for genetic defects or chromosomal mutations.

Embryo and fetus develops surrounded by amniotic fluid to cushion it against shock.

FIGURE 13.19 Amniocentesis, a procedure used for prenatal diagnosis of genetic defects. The procedure is complicated and costly and, therefore, it is used primarily in high-risk cases.
© Cengage Learning 2014

13.5 Non-Mendelian Patterns of Inheritance

We consider two examples of patterns of inheritance in this section that do not follow the principles of Mendelian inheritance we have developed in this chapter and the previous one. In **cytoplasmic inheritance,** the pattern of inheritance follows that of genes in the genomes of mitochondria or chloroplasts. In **genomic imprinting,** the expression of a nuclear gene is based on whether an individual organism inherits the gene from the male or female parent.

Cytoplasmic Inheritance Follows the Pattern of Inheritance of Mitochondria or Chloroplasts

As noted in Chapter 5, not all DNA is contained in the nucleus; both chloroplasts and mitochondria also contain DNA. Like nuclear genes, chloroplast and mitochondrial genes are subject to mutation. However, the inheritance pattern of these mutant genes—called **cytoplasmic inheritance**—is fundamentally different from that of mutant genes in the nucleus. First, these genes do not segregate by meiosis, so the ratios of mutated and parental genes typical of Mendelian segregation are absent. Second, the genes usually show uniparental inheritance from generation to generation. In **uniparental inheritance,** all progeny (both males and females) have the phenotype of only one of the parents. For most multicellular eukaryotes, offspring inherit only the mother's phenotype, a phenomenon called **maternal inheritance.** In sexual reproduction, both the male and female gamete provide nuclear DNA, but the female provides most of the cytoplasm in the fertilized cell. Maternal inheritance occurs because, in animals, a zygote receives most of its cytoplasm, including mitochondria and (in plants) chloroplasts, from the female parent and little from the male parent. In plants, cytoplasmic inheritance varies depending on the species and the organelle (mitochondrion or chloroplast).

The first example of cytoplasmic inheritance of a mutant trait was found in 1909 by the German scientist Carl Correns, one of the geneticists who rediscovered Mendel's principles. Correns made his discovery through his genetic studies of a plant, *Mirabilis* (the four-o'clock), using mutant plants that had a variegated pattern of green and white **(Figure 13.20).** In the white segments, chloroplasts are colorless instead of green. Correns fertilized flowers in a green region of the plant with pollen from a variegated region and vice versa. He discovered that the phenotype of the progeny seedlings showed maternal inheritance. That is, it was always that of the female segment the pollen fertilized: variegated for the variegated ♀ × green ♂ cross, and green for the green ♀ × white ♂ cross.

Maternal inheritance of mutant traits involving the mitochondria have also been characterized in many eukaryotic species, including plants, animals, protists, and fungi. Similar to the mutant traits of chloroplasts, each mutant trait results from an alteration of a gene in the mitochondrial genome.

In humans, several inherited diseases have been traced to mutations in mitochondrial genes. Recall that the mitochondrion plays a critical role in synthesizing ATP, the energy

Fertilization of a flower in a green region of a plant with pollen from a white area results in green seedlings.

Fertilization of a flower in a white region of a plant with pollen from a green area results in white seedlings.

FIGURE 13.20 A four-o'clock *(Mirabilis)* plant with a variegated (patchy) distribution of green and white segments.
© Cengage Learning 2014

source for many cellular reactions. Several maternally inherited diseases in humans involve mutations in mitochondrial genes that encode components of the ATP-generating system of the organelle. The resulting mitochondrial defects are especially destructive to the organ systems most dependent on mitochondrial reactions for energy: the central nervous system, skeletal and cardiac muscle, the liver, and the kidneys.

For example, *Leber's hereditary optic neuropathy* (LHON) is a maternally inherited human disease that affects midlife adults and is characterized by complete or partial blindness caused by optic nerve degeneration. Mutations in any one of the genes for eight electron transfer system proteins (see Chapter 8) all can lead to LHON. The electron transfer system is responsible for ATP synthesis by oxidative phosphorylation (see Chapter 8). Death of the optic nerve is a common result of defects in oxidative phosphorylation which, in LHON, are caused by the inhibition of the electron transfer system.

Another example of a maternally inherited human disease is *myoclonic epilepsy and ragged-red fiber (MERRF) disease.* The "ragged-red fibers" in the name refers to the abnormal appearance of tissues under the microscope. Symptoms of the disease include jerking spasms of the limbs or the whole body, a defect in coordinating movement, and the accumulation of lactic acid in the blood. MERRF disease is caused by a mutation in a transfer RNA (tRNA) gene in the mitochondrial genome. Transfer RNA molecules play important roles in protein synthesis. The mutated tRNA adversely affects protein synthesis in the mitochondria which, in some way, causes the various phenotypes of the disease.

In Genomic Imprinting, the Allele Inherited from One of the Parents Is Expressed Whereas the Other Allele Is Silent

Throughout our discussions of Mendelian inheritance, we have assumed that a particular allele has the same effect in an individual whether it was inherited from the mother or father. For the vast majority of genes, the assumption is correct. However, in mammals, researchers have identified 30 or so genes whose effects do, in fact, depend on whether an allele is inherited from the mother or the father. For some of these genes, only the paternal, sperm-derived, allele is expressed; for others, only the maternal, egg-derived, allele is expressed. The phenomenon in which the expression of an allele of a gene depends on the parent that contributed it is called **genomic imprinting.** The silent allele—the inherited allele that is not expressed—is called the *imprinted allele.*

The first imprinted gene identified was *Igf2* in the mouse. *Igf2* encodes insulin-like growth factor 2, a protein that stimulates cell growth and division. The growth factor is needed for early embryos to develop normally. Researchers studying mice heterozygous for a deletion of the entire *Igf2* gene from the genome observed that if mice inherited the mutated chromosome from the father they were small, but if they had inherited the mutated chromosome from the mother they were normal size **(Figure 13.21A).** It appeared that only the paternally inherited gene had an effect on size: the maternally inherited gene, whether normal or not, had no effect. The scientists reasoned

that, in normal mice homozygous for *Igf2,* the active form of the gene is the copy on the paternal chromosome, whereas the maternal copy of the gene is imprinted (silent) **(Figure 13.21B).** If a heterozygote inherits the deletion from the father, that copy of the gene is inactive. Even if the maternal copy is nonmutant, it is not expressed because it is imprinted. As a result, normal development does not occur and the adult mouse produced is small.

The mechanism of genomic imprinting involves the modification of the DNA in the region that controls the expression of an allele by the addition of methyl ($-CH_3$) groups to cytosine (C) nucleotides. The methylation of the control region of a gene prevents it from being expressed. (You will learn more about the regulation of gene expression by methylation of DNA in Chapter 16.) Genomic imprinting occurs in the germ cells that develop into gametes. In those germ cells, the allele destined to be inactive in the new embryo after fertilization is methylated. That is, in the production of sperm, alleles for paternally imprinted genes are methylated, and in the production of eggs, alleles for maternally imprinted genes are methylated. That methylated (silenced) state of the gene is passed on cell generation to cell generation as the cells grow and divide to produce the somatic (body) cells of the organism.

Inherited imprints must first be erased in the germ cells before new imprinting occurs in the production of gametes. Consider a gene that is maternally imprinted, for example. In the adult, the maternal chromosome has an imprinted allele of that gene whereas the paternal chromosome has an active allele. When that adult produces gametes, it needs to imprint all alleles in a way appropriate to its sex—for example, if it is male, it must erase the imprint from the maternally inherited gene. In the diploid cells that go through meiosis to produce the gametes, all imprints are first erased providing a clean slate for individuals to imprint alleles appropriately to their gender.

We must be clear that, although we are talking about the expression of alleles inherited from one or the other parent, genomic imprinting is a completely different phenomenon than sex linkage. For sex-linked genes, alleles inherited from the mother or father are both expressed. The phenotypic ratios are different than those for autosomal genes, but that is because of the difference in sex chromosome composition in males and females. And, in fact, most known imprinted genes are autosomal genes.

In this chapter, you have learned about genes and the role of chromosomes in inheritance. In the next chapter, you will learn about the molecular structure and function of the genetic material and about the molecular mechanism by which DNA is replicated.

A. Phenotypes of mice heterozygous for a deletion of gene *Igf2*.

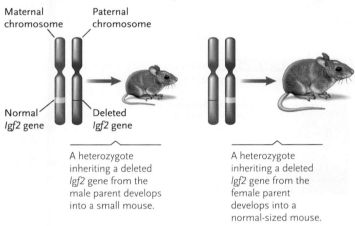

Maternal chromosome   Paternal chromosome

Normal *Igf2* gene   Deleted *Igf2* gene

A heterozygote inheriting a deleted *Igf2* gene from the male parent develops into a small mouse.

A heterozygote inheriting a deleted *Igf2* gene from the female parent develops into a normal-sized mouse.

B. Phenotype of mice homozygous for the normal allele of *Igf2*.

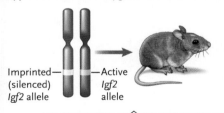

Imprinted (silenced) *Igf2* allele   Active *Igf2* allele

In a mouse heterozygous for the normal allele of *Igf2*, the paternal allele is active, and the maternal allele is imprinted (silenced). As long as a normal allele is inherited from the male parent, the mouse develops into a normal-sized adult.

FIGURE 13.21 Imprinting of the mouse *Igf2* (insulin-like growth factor 2) gene.
© Cengage Learning 2014

STUDY BREAK 13.5

What key feature or features would suggest to you that a mutant trait shows cytoplasmic inheritance?

How do deletions of chromosome 5 cause myeloid neoplasms?

In this chapter, you learned that deletions are a type of chromosomal mutation that occurs if a broken segment is lost from a chromosome. Deletions may cause serious problems if the missing segment contains genes that are critical for normal cellular functions. Patients who have received chemotherapy and/or radiation for a primary cancer may subsequently develop acute myeloid leukemia (AML) or myelodysplastic syndrome (MDS), that are classified as therapy-related *myeloid neoplasms* (t-MNs). (A neoplasm is an abnormal growth of tissue.) A t-MN following chemotherapy with an alkylating agent typically develops after 3 to 7 years, and has characteristic loss or deletion of chromosomes 5 and/or 7. Cytogeneticists (geneticists who specialize in the study of chromosomes) refer to these patients as del(5q) and −7/del(7q), since either the entire chromosome is lost (−) or portions of the long or "q" arms of the chromosomes are deleted (del). The short or "petit" (p) arm of these chromosomes is typically intact in these patients.

Cytogenetic analysis of many cases of t-MN revealed that complex karyotypes are associated with abnormalities of chromosome 5, rather than 7. A karyotype is the number and appearance of chromosomes in the nucleus (see Figure 10.17). A complex karyotype means that 3 chromosomal aberrations, including deletions, duplications, translocations and/or inversions, are detected. Recurring chromosomal abnormalities observed at a high frequency in patients with del(5q) included trisomy 8 and loss of portions of chromosomes: 13q, 16q, 17p, 18 and 20q. Our research has focused on the consequences of chromosome 5 deletions.

We and other groups of investigators mapped the deletions on chromosome 5 in t-MN patients with a del(5q) and defined a commonly deleted segment (CDS) that was predicted to contain one or more myeloid tumor suppressor genes. (Tumor suppressor genes are part of an elaborate regulatory system that controls cell division. As their name implies, they normally help suppress the growth of tumors or cancer. When they are not expressed normally, this can contribute to the development of cancer. They are discussed in Section 16.5 *The Genetics of Cancer.*) The CDS contains 19 genes, which are deleted from one copy of chromosome 5 in every patient diagnosed with a del(5q). Additional genes on 5q, adjacent to the CDS, are also deleted frequently (up to 95% of cases), and are also likely to be involved in disease pathogenesis. The function of these genes is diverse and includes the regulation of mitosis, transcriptional control and translational regulation.

Many genes require two copies to maintain normal expression levels and normal levels of protein. If one copy is lacking due to a chromosomal deletion and only one allele remains, there can be profound effects on cell growth and an abnormal phenotype may result. This phenomenon of a lowered gene dosage is sometimes referred to as "haploinsufficiency". In t-MN, a number of genes located on 5q, including *RPS14*, *EGR1*, *APC*, *CTNNA1*, *HSPA9*, and *DIAPH1*, have been implicated in the development of myeloid disorders due to a gene dosage effect. Current studies support a model, in which loss of a single allele of *more than one* gene on 5q contributes to the pathogenesis of t-MN with a del(5q). Scientists are still investigating if there are additional genes on del(5q) that may be involved, and trying to determine which combination of "haploinsufficient" genes on chromosome 5 may be working together to disrupt normal growth and cause cancer.

Our lab is using mouse models to identify the combination of genes on 5q that is critical, and which cellular pathways may be affected. We use mice that express only a single copy of one or more of the del(5q) genes to mimic what happens in the patients. In one approach, these mice are treated with a particular alkylating agent to mimic the effects of the chemotherapy administered to patients. These chemotherapy agents may induce mutations in the DNA that may not be harmful on their own. However, if these mutations are present in a cell that also expresses one or more 5q genes at a lowered dose, this may result in abnormal growth and the development of cancer. Our mouse models can help us determine which of the deleted 5q genes are critical for the development of myeloid neoplasms and whether mutations on other chromosomes (due to chemotherapy or other environmental exposures) may also be involved. Currently, it is very difficult to treat t-MN patients with a del(5q), and the patients' prognosis is poor. Development of mouse models of t-MN will be of utmost importance in the development of more effective therapies.

THINK LIKE A SCIENTIST In t-MN patients with a del(5q), are the genes in the deleted 5q segment completely inactive?

Courtesy of Michelle Le Beau

Courtesy of Angela Stoddart

Michelle Le Beau and **Angela Stoddart** are researchers in the Section of Hematology/Oncology at the University of Chicago. Dr. Le Beau has had a long-standing interest in correlating specific chromosomal abnormalities with clinical features of the neoplastic disease. Dr. Stoddart joined Dr. Le Beau's group in 2004, and together they are characterizing the genes involved in therapy-related myeloid neoplasms with abnormalities of chromosome 5.

REVIEW KEY CONCEPTS

To access the course materials and companion resources for this text, please visit www.cengagebrain.com.

13.1 Genetic Linkage and Recombination

- Genes, consisting of sequences of nucleotides in DNA, are arranged linearly in chromosomes.

- Genes near each other on the same chromosome are linked together in their transmission from parent to offspring. Linked genes are inherited in patterns similar to those of single genes, except for changes in the linkage caused by recombination (Figures 13.2 and 13.3).

- In genetic recombination, alleles linked on the same chromosome are mixed into new combinations by exchange of segments between the chromosomes of a homologous pair. The exchanges occur by the process of crossing-over while homologous chromosomes are paired during prophase I of meiosis.

- The amount of recombination between any two genes located on the same chromosome pair reflects the distance between them on the chromosome. The greater this distance, the greater the chance that chromatids will exchange segments at points between the genes and the greater the recombination frequency.

- The relationship between separation and recombination frequencies is used to produce chromosome maps in which genes are assigned relative locations with respect to each other (Figure 13.4).

13.2 Sex-Linked Genes

- Sex linkage is a pattern of inheritance produced by genes carried on sex chromosomes: chromosomes that differ in males and females. In humans and fruit flies, which have XX females and XY males, most sex-linked genes are carried on the X chromosome.

- Because males have only one X chromosome, they need to receive only one copy of a recessive allele from their mothers to develop the trait. Females must receive two copies of the recessive allele, one from each parent, to develop the trait (Figures 13.6–13.8).

- In mammals, inactivation of one of the two X chromosomes in cells of the female makes the dosage of X-linked genes the same in males and females (Figure 13.10).

Animation: X-chromosome inactivation

Animation: X-linked inheritance

13.3 Chromosomal Mutations That Affect Inheritance

- Inheritance is influenced by processes that delete, duplicate, or invert segments within chromosomes, or translocate segments between chromosomes (Figures 13.11 and 13.12).

- Chromosomes also change in number by addition or removal of individual chromosomes or entire sets. Changes in single chromosomes usually occur through nondisjunction, in which homologous pairs fail to separate during meiosis I, or sister chromatids fail to separate during meiosis II. As a result, one set of gametes receives an extra copy of a chromosome and the other set is deprived of the chromosome (Figures 13.13–13.15).

- Monoploids have only one set of chromosomes; monoploidy is lethal in most animal species. Polyploids have three or more copies of the entire chromosome set. Polyploids usually arise when the spindle fails to function during mitosis in cell lines leading to gamete formation, producing gametes that contain double the number of chromosomes typical for the species.

Animation: Nondisjunction

13.4 Human Genetics and Genetic Counseling

- Three modes of inheritance are most significant in human heredity: autosomal recessive, autosomal dominant, and X-linked recessive inheritance. X-linked dominant traits are rare.

- In autosomal recessive inheritance, males or females carry a recessive allele on an autosome. Heterozygotes are carriers that are usually unaffected, but homozygous individuals show symptoms of the trait (Figure 13.16).

- In autosomal dominant inheritance, a dominant gene is carried on an autosome. Individuals that are homozygous or heterozygous for the trait show symptoms of the trait; homozygous recessives are normal.

- In X-linked recessive inheritance, a recessive allele for the trait is carried on the X chromosome. Male individuals with the recessive allele on their X chromosome or female individuals with the recessive allele on both X chromosomes show symptoms of the trait. Heterozygous females are carriers but usually show no symptoms of the trait (Figure 13.17).

- In X-linked dominant inheritance, a dominant allele for the trait is carried on the X chromosome. The trait is expressed in males and females who receive such an X chromosome (Figure 13.18).

- Genetic counseling, based on identification of parental genotypes by constructing family pedigrees and prenatal diagnosis, allow prospective parents to reach an informed decision about whether to have a child or continue a pregnancy (Figure 13.19).

Animation: Amniocentesis

13.5 Non-Mendelian Patterns of Inheritance

- Cytoplasmic inheritance depends on genes carried on DNA in mitochondria or chloroplasts. Cytoplasmic inheritance follows the maternal line: it parallels the inheritance of the cytoplasm in fertilization, in which most or all of the cytoplasm of the zygote originates from the egg cell (Figure 13.20).

- Genomic imprinting is a phenomenon in which the expression of an allele of a gene is determined by the parent that contributed it. In some cases, the allele inherited from the father is expressed; in others, the allele from the mother is expressed. Commonly, the silencing of the other allele is the result of methylation of the region adjacent to the gene that is responsible for controlling the expression of that gene (Figure 13.21).

Test Your Knowledge

1. In humans, red–green color blindness is an X-linked recessive trait. If a man with normal vision and a color-blind woman have a son, what is the chance that the son will be color-blind? What is the chance that a daughter will be color-blind?

2. The following pedigree shows the pattern of inheritance of red–green color blindness in a family. Females are shown as circles and males as squares; the squares or circles of individuals affected by the trait are filled in black.

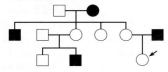

What is the chance that a son of the third-generation female indicated by the arrow will be color-blind if the father is a normal man? If the father is color-blind?

3. Individuals affected by a condition known as polydactyly have extra fingers or toes. The following pedigree shows the pattern of inheritance of this trait in one family:

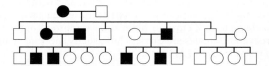

From the pedigree, can you tell if polydactyly comes from a dominant or recessive allele? Is the trait sex-linked? As far as you can determine, what is the genotype of each person in the pedigree with respect to the trait?

4. A number of genes carried on the same chromosome are tested and show the following crossover frequencies. What is their sequence in the map of the chromosome?

Genes	Crossover Frequencies between Them
C and A	7%
B and D	3%
B and A	4%
C and D	6%
C and B	3%

5. In *Drosophila,* two genes, one for body color and one for eye color, are carried on the same chromosome. The wild-type gray body color is dominant to black body color, and wild-type red eyes are dominant to purple eyes. You make a cross between a fly with gray body and red eyes and a fly with black body and purple eyes. Among the offspring, about half have gray bodies and red eyes and half have black bodies and purple eyes. A small percentage have (a) black bodies and red eyes or (b) gray bodies and purple eyes. What alleles are carried together on the chromosomes in each of the flies used in the cross? What alleles are carried together on the chromosomes of the F_1 flies with black bodies and red eyes, and those with gray bodies and purple eyes?

6. Another gene in *Drosophila* determines wing length. The dominant wild-type allele of this gene produces long wings; a recessive allele produces vestigial (short) wings. A female that is true-breeding for red eyes and long wings is mated with a male that has purple eyes and vestigial wings. F_1 females are then crossed with purple-eyed, vestigial-winged males. From this second cross, a total of 600 offspring are obtained with the following combinations of traits:

    252 with red eyes and long wings

    276 with purple eyes and vestigial wings

     42 with red eyes and vestigial wings

     30 with purple eyes and long wings

Are the genes linked, unlinked, or sex-linked? If they are linked, how many map units separate them on the chromosome?

7. In *Drosophila,* white eyes is an X-linked recessive trait. A white-eyed female is crossed with a male with normal red eyes. An F_1 female from this cross is mated with her father, and an F_1 male is mated with his mother. What will be the phenotypic ratios for eye color in the two sexes of the offspring of the last two crosses?

8. You conduct a cross in *Drosophila* that produces only half as many male as female offspring. What might you suspect as a cause?

Discuss the Concepts

1. Can a linkage map be made for a haploid organism that reproduces sexually?

2. Crossing-over does not occur between any pair of homologous chromosomes during meiosis in male *Drosophila.* From what you have learned about meiosis and crossing-over, propose one hypothesis for why this might be the case.

3. Even though X inactivation occurs in XXY (Klinefelter syndrome) humans, they do not have the same phenotype as normal XY males. Similarly, even though X inactivation occurs in XX individuals, they do not have the same phenotype as XO (Turner syndrome) humans. Why might this be the case?

4. All mammals have evolved from a common ancestor. However, the chromosome number varies among mammals. By what mechanism might this have occurred?

Design an Experiment

Assume that genes *a, b, c, d, e,* and *f* are linked. Explain how you would construct a linkage map that shows the order of these six genes and the map units between them.

Interpret the Data

Exposure to tobacco, by chewing it, smoking cigars, cigarettes, or pipes, and passive exposure to smoke has been linked to cancer of the mouth and throat. Chemical compounds released during tobacco use form covalent bonds with the DNA in the oral cavity cells to generate structures called "adducts." These oral cavity cells replicate frequently, and the presence of DNA adducts is suspected to contribute to mutations, chromosomal alterations, and hence cellular defects passed onto offspring cells. On the next page is a graph showing the incidence (attomol = 10^{-18} mol) of adduct formation in samples of *healthy* tissue from nonsmokers, smokers of various frequencies, and ex-smokers, all of whom required surgery to remove oral cancerous growths.

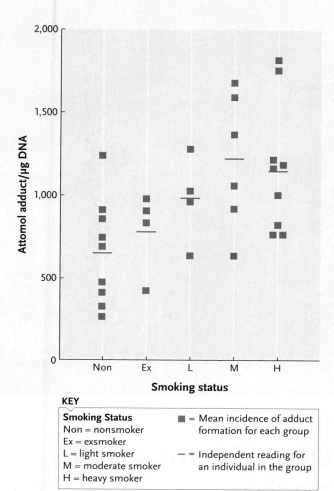

KEY

Smoking Status
Non = nonsmoker
Ex = exsmoker
L = light smoker
M = moderate smoker
H = heavy smoker

■ = Mean incidence of adduct
formation for each group

— = Independent reading for
an individual in the group

N. J. Jones, A. D. McGregor, and R. Waters. 1993. Detection of DNA adducts in human oral tissue: Correlation of adduct levels with tobacco smoking and differential enhancement of adducts using the butanol extraction and nuclease P1 versions of ^{32}P postlabeling. *Cancer Research* 53:1522–1528. Reprinted by permission of the American Association for Cancer Research.

1. On the *y* axis, why is it necessary to express the data in terms of the quantity of adduct per μg of DNA, rather than simply noting the total quantity of adduct?

2. What do these data suggest about the relation between smoking and the formation of adducts?

3. What impact does smoking cessation have on the frequency of adduct formation?

4. Which data point(s) could suggest a relation between the potential to develop oral cancer and passive exposure to tobacco smoke?

Source: N. J. Jones, A. D. McGregor, and R. Waters. 1993. Detection of DNA adducts in human oral tissue: Correlation of adduct levels with tobacco smoking and differential enhancement of adducts using the butanol extraction and nuclease P1 versions of ^{32}P postlabeling. *Cancer Research* 53:1522–1528.

Apply Evolutionary Thinking

How would the effects of natural selection differ on alleles that cause diseases fatal in childhood (such as progeria) and those that cause diseases that shorten life expectancy to 40 or 50 years (such as cystic fibrosis)?

14

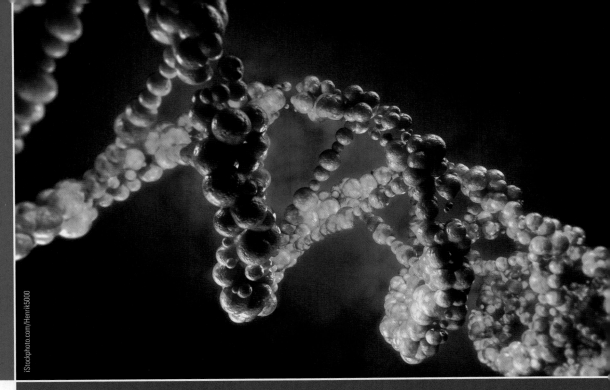

iStockphoto.com/Henrik5000

A digital model of DNA (based on data generated by X-ray crystallography).

DNA Structure, Replication, and Organization

Why it matters . . . In the spring of 1868, Johann Friedrich Miescher, a Swiss physician and physiological chemist, was collecting pus from discarded bandages. Miescher wanted to study the chemical composition of the nuclei of the white blood cells found in the pus. From the nuclei of these cells, Miescher extracted large quantities of an acidic substance with a high phosphorus content. He called the unusual substance "nuclein." His discovery is at the root of the development of our molecular understanding of life: nuclein is now known by its modern name, **deoxyribonucleic acid,** or **DNA,** the molecule that is the genetic material of all living organisms.

At the time of Miescher's discovery, scientists knew nothing about the molecular basis of heredity and very little about genetics. Although Mendel had already published the results of his genetic experiments with garden peas, the significance of his findings was not widely known or appreciated. It was not known which chemical substance in cells actually carries the instructions for reproducing parental traits in offspring. Not until 1952, more than 80 years after Miescher's discovery, did scientists fully recognize that the hereditary molecule was DNA.

After DNA was established as the hereditary molecule, the focus of research changed to the three-dimensional structure of DNA. Among the scientists striving to work out the structure were James D. Watson, a young American postdoctoral student at Cambridge University in England, and the British scientist Francis H. C. Crick, then a graduate student at Cambridge University. Using chemical and physical information about DNA, in particular Rosalind Franklin's analysis of the arrangement of atoms in DNA, the two investigators assembled molecular models from pieces of cardboard and bits of wire. Eventually they constructed a model for DNA that fit all the known data **(Figure 14.1).** Their discovery was of momentous importance in biology. The model immediately made it apparent how genetic information is stored and how it could be rep-

FIGURE 14.1 James D. Watson and Francis H. C. Crick demonstrating their 1953 model for DNA structure, which revolutionized the biological sciences.

licated faithfully. Unquestionably, the discovery launched a molecular revolution within biology, making it possible for the first time to relate the genetic traits of living organisms to a universal molecular code present in the DNA of every cell. In addition, Watson and Crick's discovery opened the way for numerous advances in fields such as medicine, forensics, pharmacology, and agriculture, and eventually gave rise to the current rapid growth of the biotechnology industry.

14.1 Establishing DNA as the Hereditary Molecule

In the first half of the twentieth century, many scientists believed that proteins were the most likely candidates for the hereditary molecules because they appeared to offer greater opportunities for information coding than did nucleic acids. That is, proteins contain 20 different amino acids, whereas nucleic acids have only four different nitrogenous bases available for coding. Other scientists believed that nucleic acids were the hereditary molecules. In this section, we describe the experiments showing that DNA, and not protein, is the genetic material.

Experiments Began When Griffith Found a Substance That Could Transform Pneumonia Bacteria Genetically

In 1928, Frederick Griffith, a British medical officer, observed an interesting phenomenon in his experiments with the bacterium *Streptococcus pneumoniae,* the species that causes a severe form of pneumonia in mammals. He studied two strains of the bacterium. The smooth strain—*S*—has a polysaccharide capsule sur-

rounding each cell and forms colonies that appear smooth and shiny when grown on a culture plate. The *S* strain is virulent (highly infective, or pathogenic), causing pneumonia and killing the mice in a day or two. The rough strain—*R*—does not have a polysaccharide capsule and forms colonies with a rough, nonshiny appearance. The *R* strain is nonvirulent (not infective, or nonpathogenic), meaning that it does not affect mice.

Griffith's experiments are shown in **Figure 14.2**. His critical experimental result was that the mice died if he injected them with a mixture of living *R* bacteria and heat-killed *S* bacteria. He was able to isolate living *S* bacteria with polysaccharide capsules from the infected mice. In some way, living *R* bacteria had acquired the ability to make the polysaccharide capsule from the dead *S* bacteria, and they had changed—transformed—into virulent *S* cells. The smooth, virulence trait was stably inherited by the descendants of the transformed bacteria. Griffith called the conversion of *R* bacteria to *S* bacteria *transformation* and called the agent responsible the *transforming principle*. What was the nature of the molecule responsible for the transformation? The most likely candidates were proteins or nucleic acids.

Avery and His Coworkers Identified DNA as the Molecule That Transforms Nonvirulent *Streptococcus* Bacteria to the Virulent Form

In the 1940s, Oswald Avery, a physician and medical researcher at the Hospital at Rockefeller Institute for Medical Research, and his coworkers Colin MacLeod and Maclyn McCarty performed an experiment designed to identify the chemical nature of the transforming principle that can change *R* nonvirulent *Streptococcus* bacteria into the *S* virulent form. Rather than working with mice, they reproduced the transformation using bacteria growing in culture tubes. They used heat to kill virulent *S* bacteria and then treated the macromolecules extracted from the *S* bacterial cells in turn with enzymes that break down each of the three main candidate molecules for the hereditary material—protein; DNA; and the other nucleic acid, RNA. When they destroyed proteins or RNA, the researchers saw no effect; the extract of *S* bacteria still transformed nonvirulent *R* bacteria into virulent *S* bacteria—the cells had polysaccharide capsules and produced smooth colonies on culture plates. When they destroyed DNA, however, no transformation occurred and no smooth colonies were seen on culture plates.

In 1944, Avery and his colleagues published their discovery that the transforming principle was DNA. At that time, many biologists were convinced that the genetic material was protein. So, although their findings were clearly revolutionary, Avery and his colleagues presented their conclusions in the paper cautiously, offering several interpretations of their results. Some biologists accepted their results almost immediately. However, those who believed that the genetic material was protein argued that it was possible that not all protein was destroyed by the enzyme treatments and, as contaminants in their DNA transformation reaction, those remaining proteins were responsible for the transformation.

FIGURE 14.2 | **Experimental Research**

Griffith's Experiment with Virulent and Nonvirulent Strains of *Streptococcus pneumoniae*

Question: What is the nature of the genetic material?

Experiment: Frederick Griffith studied the conversion of a nonvirulent (noninfective) *R* form of the bacterium *Streptococcus pneumoniae* to a virulent (infective) *S* form. The *S* form has a capsule surrounding the cell, giving colonies of it on a laboratory dish a smooth, shiny appearance. The *R* form has no capsule, so the colonies have a rough, nonshiny appearance. Griffith injected the bacteria into mice and determined how the mice were infected.

1. Mice injected with live *S* cells (control to show effect of *S* cells)

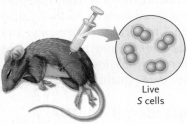

Live
S cells

Result: Mice die. Live *S* cells in their blood; shows that *S* cells are virulent.

2. Mice injected with live *R* cells (control to show effect of *R* cells)

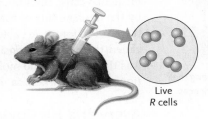

Live
R cells

Result: Mice live. No live *R* cells in their blood; shows that *R* cells are nonvirulent. Evidently the capsule is responsible for virulence of the *S* strain.

3. Mice injected with heat-killed *S* cells (control to show effect of dead *S* cells)

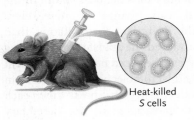

Heat-killed
S cells

Result: Mice live. No live *S* cells in their blood; shows that live *S* cells are necessary to be virulent to mice.

4. Mice injected with heat-killed *S* cells plus live *R* cells

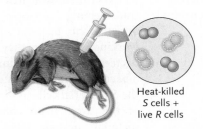

Heat-killed
S cells +
live *R* cells

Result: Mice die. Live *S* cells in their blood; shows that living *R* cells can be converted to virulent *S* cells with some factor from dead *S* cells.

Conclusion: Griffith concluded that some molecules released when *S* cells were killed could change living nonvirulent *R* cells genetically to the virulent *S* form. He called the molecule the *transforming principle* and the process of genetic change *transformation*.

THINK LIKE A SCIENTIST Theoretically could the existence of the transforming principle have been shown with the opposite set up; that is, by injecting mice with heat-killed *R* cells plus live *S* cells?

Source: F. Griffith. 1928. The significance of pneumococcal types. *Journal of Hygiene* (*London*) 27:113–159.

Hershey and Chase Found the Final Experimental Evidence Establishing DNA as the Hereditary Molecule

A final series of experiments conducted in 1952 by bacteriologist Alfred D. Hershey and his laboratory assistant Martha Chase at the Cold Spring Harbor Laboratory removed any remaining doubts that DNA is the hereditary molecule. Hershey and Chase studied the infection of the bacterium *Escherichia coli* by bacteriophage T2. *E. coli* is a bacterium normally found in the intestines of mammals. **Bacteriophages** (or simply **phages;** see Chapter 17) are viruses that infect bacteria. A **virus** is an infectious agent that is made of either DNA or RNA surrounded by a protein coat. Viruses can reproduce only in a host cell. When a virus infects a cell, it can use the cell's resources to produce more virus particles.

The phage life cycle begins when a phage attaches to the surface of a bacterium and infects it. Phages such as T2 quickly stop the infected cell from producing its own molecules and instead use the cell's resources for making progeny phages. After about 100 to 200 phages are assembled inside the bacterial cell, a viral enzyme breaks down the cell wall, killing the cell and releasing the new phages. The whole life cycle takes approximately 90 minutes.

The T2 phage that Hershey and Chase studied consists of *only* a core of DNA surrounded by proteins. Therefore, one of these molecules must be the genetic material that enters the bacterial cell and directs the infective cycle within. But which one? Hershey and Chase's definitive experiment to answer that question is presented in **Figure 14.3.** Their experimental approach was to label just the DNA or just the protein radioactively and then to use the label as a tag to follow the molecule through the phage life cycle. More specifically, because DNA contains phosphorus but not sulfur, they used radioactive phosphorus (^{32}P) to label only DNA, and because protein contains sulfur but not phosphorus, they used radioactive sulfur (^{35}S) to label only protein. Their results showed that labeled DNA, but not labeled protein, entered the cell and appeared in progeny phages. Unequivocally they had shown that the genetic material of phages is DNA, not protein.

When considered together, the experiments of Griffith, Avery and his coworkers, and Hershey and Chase established that DNA, not proteins, carries genetic information. The research also established the term *transformation,* which is still used in molecular biology.

FIGURE 14.3 **Experimental Research**

The Hershey and Chase Experiment Demonstrating That DNA Is the Hereditary Molecule

Question: Is DNA or protein the genetic material?

Experiment: Hershey and Chase performed a definitive experiment to show whether DNA or protein is the genetic material. They used phage T2 for their experiment; it consists only of DNA and protein. Because DNA contains phosphorus and not sulfur, they could label DNA selectively with radioactive ^{32}P. And, because protein contains sulfur and not phosphorus, they could label protein selectively with radioactive ^{35}S.

1. They infected *E. coli* growing in the presence of radioactive ^{32}P or ^{35}S with phage T2. The progeny phages were either labeled in their protein with ^{35}S (top), or in their DNA with ^{32}P (bottom).

2. Separate cultures of *E. coli* were infected with the radioactively labeled phages.

3. After a short period of time to allow the genetic material to enter the bacterial cell, the bacteria were mixed in a blender. The blending sheared from the cell surface the phage coats that did not enter the bacteria. The components were analyzed for radioactivity.

4. Progeny phages analyzed for radioactivity.

Progeny phages from *E. coli* growing in ^{35}S

^{35}S-labeled protein

E. coli

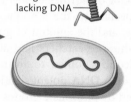
Phage coat lacking DNA

Result: No radioactivity within cell; ^{35}S in phage coat

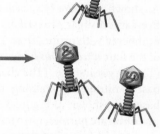
Phage coat lacking DNA

Result: No radioactivity in progeny phages

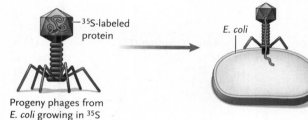
Progeny phages from *E. coli* growing in ^{32}P

^{32}P-labeled DNA

E. coli

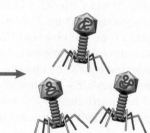

Result: ^{32}P within cell; not in phage coat

Result: ^{32}P in progeny phages

Conclusion: ^{32}P, the isotope used to label DNA, was found within phage-infected cells and in progeny phages, indicating that DNA is the genetic material. ^{35}S, the radioisotope used to label proteins, was found in phage coats after infection, but was not found in the infected cell or in progeny phages, showing that protein is not the genetic material.

THINK LIKE A SCIENTIST What isotope distribution would you expect to see if the phages used to infect *E. coli* were labeled with ^{14}C, the radioactive isotope of carbon?

Source: A. D. Hershey and M. Chase. 1952. Independent functions of viral protein and nucleic acid in growth of bacteriophage. *Journal of General Physiology* 36:39–56.

Transformation is the alteration of a cell's hereditary type by the uptake of DNA released by the breakdown of another cell, as in the Griffith and Avery experiments. Having identified DNA as the hereditary molecule, scientists turned next to determining its structure.

STUDY BREAK 14.1

Imagine that ^{35}S labeled *both* protein and DNA, whereas ^{32}P labeled only DNA. How would Hershey and Chase's results have been different?

14.2 DNA Structure

The experiments that established DNA as the hereditary molecule were followed by a highly competitive scientific race to discover the structure of DNA. The race ended in 1953, when Watson and Crick elucidated the structure of DNA, ushering in a new era of molecular biology.

Watson and Crick Brought Together Information from Several Sources to Develop a Model for DNA Structure

Before Watson and Crick began their research, other investigators had established that DNA contains four different nucleotides. Nucleotides were first described in Section 3.5 and are shown in detail in Figure 3.27. Recall that the nucleotide in DNA is called a deoxyribonucleotide because the sugar it contains is deoxyribose. As shown in **Figure 14.4,** each deoxyribonucleotide consists of the five-carbon sugar *deoxyribose* (carbon atoms on deoxyribose are numbered with primes from 1′ to 5′), a phosphate group, and one of the four *nitrogenous bases*—adenine (A), guanine (G), thymine (T), or cytosine (C). (The chemical structures of the nitrogenous bases—nitrogen-containing molecules with the property of a base—are shown in Figure 3.29.) Two of the bases, **adenine** and **guanine,** are *purines,* nitrogenous bases built from a pair of fused rings of carbon and nitrogen atoms. The other two bases, **thymine** and **cytosine,** are *pyrimidines,* built from a single carbon–nitrogen ring. Erwin Chargaff, a biochemist, measured the amounts of nitrogenous bases in DNA and discovered that they occur in definite ratios. He observed that the amount of purines equals the amount of pyrimidines, but more specifically, the amount of adenine equals the amount of thymine, and the amount of guanine equals the amount of cytosine; these researchers had also determined that DNA contains nucleotides joined to form a *polynucleotide chain* (see Figure 14.4). In a DNA polynucleotide chain, the deoxyribose sugars are linked by phosphate groups in an alternating sugar–phosphate–sugar–phosphate pattern, forming a **sugar–phosphate backbone** (highlighted in gray in Figure 14.4). Each phosphate group is a "bridge" between the 3′ carbon of one sugar and the 5′ carbon of the next sugar; the entire linkage, including the bridging phosphate group, is called a *phosphodiester bond,* also shown in Figure 14.4.

The polynucleotide chain of DNA has polarity—directionality. That is, the two ends of the chain are not the same: at one end, a phosphate group is bound to the 5′ carbon of a deoxyribose sugar, whereas at the other end, a hydroxyl group is bonded to the 3′ carbon of a deoxyribose sugar (see Figure 14.4). Consequently, the two ends are called the **5′ end** and **3′ end,** respectively.

Those were the known facts when Watson and Crick began their collaboration in the early 1950s. However, the number of polynucleotide chains in a DNA molecule and the manner in which they fold or twist in DNA were unknown. Watson and Crick themselves did not conduct experiments to study the struc-

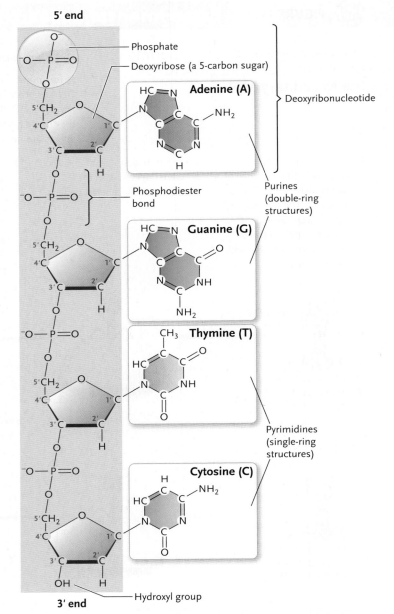

FIGURE 14.4 The four deoxyribonucleotide subunits of DNA, linked into a polynucleotide chain. The sugar–phosphate backbone of the chain is highlighted in gray. The connection between adjacent deoxyribose sugars is a phosphodiester bond. The polynucleotide chain has polarity; at one end, the 5′ end, a phosphate group is bound to the 5′ carbon of a deoxyribose sugar, whereas at the other end, the 3′ end, a hydroxyl group is bound to the 3′ carbon of a deoxyribose sugar.
© Cengage Learning 2014

ture of DNA. Instead, they used the research data of others for their analysis, relying heavily on data gathered by physicist Maurice H. F. Wilkins and his research associate Rosalind Franklin at King's College, London. These researchers were using X-ray diffraction to study the structure of DNA **(Figure 14.5A).** In **X-ray diffraction,** an X-ray beam is directed at a molecule in the form of a regular solid, ideally in the form of a crystal. Within the crystal, regularly arranged rows and banks of atoms bend and reflect

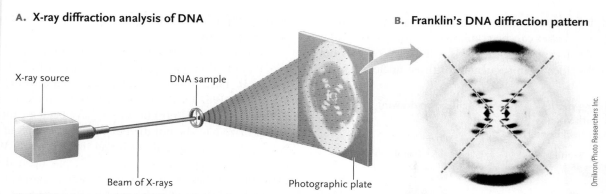

A. X-ray diffraction analysis of DNA

X-ray source

DNA sample

Beam of X-rays

Photographic plate

B. Franklin's DNA diffraction pattern

Omikron/Photo Researchers Inc.

FIGURE 14.5 X-ray diffraction analysis of DNA. (A) The X-ray diffraction method to study DNA. **(B)** The diffraction pattern Rosalind Franklin obtained. The X-shaped pattern of spots (dashed lines) was correctly interpreted by Franklin to indicate that DNA has a helical structure similar to a spiral staircase.
© Cengage Learning 2014

the X-rays into smaller beams that exit the crystal at definite angles determined by the arrangement of atoms in the crystal. If an X-ray film is placed behind the crystal, the exiting beams produce a pattern of exposed spots. From that pattern, researchers can deduce the positions of the atoms in the crystal.

Wilkins and Franklin did not have DNA crystals, but they were able to obtain X-ray diffraction patterns from DNA molecules that had been pulled out into a fiber **(Figure 14.5B).** The patterns indicated that the DNA molecules within the fiber were cylindrical and about 2 nm in diameter. Separations between the spots showed that major patterns of atoms repeat at intervals of 0.34 and 3.4 nm within the DNA. Franklin interpreted an X-shaped distribution of spots in the diffraction pattern (see dashed lines in Figure 14.5B) to mean that DNA has a helical structure.

The New Model Proposed That Two Polynucleotide Chains Wind into a DNA Double Helix

Watson and Crick constructed scale models of the four DNA nucleotides and fitted them together in different ways until they arrived at an arrangement that satisfied both Wilkins' and Franklin's X-ray data and Chargaff's chemical analysis. Watson and Crick's trials led them to the **double-helix model** for DNA, a double-stranded model for DNA structure in which two polynucleotide chains twist around each other in a right-handed way, like a double-spiral staircase **(Figure 14.6).**

In the double-helix model, the two sugar–phosphate backbones are separated from each other by the same distance— 2 nm—throughout the length of the double helix. The bases extend into and fill this central space. A purine and a pyrimidine, if paired together, are exactly wide enough to fill the space between the backbone chains in the double helix. However, two purines are too wide to fit the space exactly, and two pyrimidines are too narrow. The purine–pyrimidine base pairs in Watson and Crick's model are A–T and G–C pairs.

That is, wherever an A occurs in one strand, a T must be opposite it in the other strand; wherever a G occurs in one strand, a C must be opposite it. This feature of DNA is called **complementary base pairing,** and one strand is said to be *complementary* to the other. Complementary base pairing involving A–T and G–C base pairs fits Chargaff's rules. The base pairs, which fit together like pieces of a jigsaw puzzle, are stabilized by hydrogen bonds—two between A and T and three between G and C (see Figures 14.6 and 3.31; hydrogen bonds are discussed in Section 2.3. Note that A does not pair with C, and G does not pair with T because of the hydrogen bonding requirements.) The hydrogen bonds between the paired bases, repeated along the double helix, hold the two strands together in the helix.

The base pairs lie in flat planes almost perpendicular to the long axis of the DNA molecule. In this state, each base pair occupies a length of 0.34 nm along the long axis of the double helix (see Figure 14.6). This spacing accounts for the repeating 0.34 nm pattern noted in the X-ray diffraction patterns. The larger 3.4 nm repeating pattern was interpreted to mean that each full turn of the double helix takes up 3.4 nm along the length of the molecule and therefore 10 base pairs are packed into a full turn.

Watson and Crick also realized that the two strands of a double helix fit together in a stable chemical way only if they are **antiparallel** (have opposite polarity), meaning that they run in opposite directions (see Figure 14.6C, arrows). In other words, the *3′ end* of one strand is opposite the *5′ end* of the other strand. This antiparallel arrangement is highly significant for the process of replication, which is discussed in the next section.

As hereditary material, DNA must faithfully store and transmit genetic information for the entire life cycle of an organism. Watson and Crick recognized that this information is coded into the DNA by the particular sequence of the four nucleotides. Although only four different kinds of nucleotides exist, combining them in groups allows an essentially infinite number of different sequences to be "written," just as the 26 letters of the alphabet can be combined in groups to create a virtually unlimited number of words. Chapter 15 shows how the four nucleotides form sequences of three nucleotides that form enough "words" to "spell out" the structure of any conceivable protein.

Watson and Crick announced their model for DNA structure in a brief but monumental paper published in the journal *Nature* in 1953. Watson and Crick shared a Nobel Prize with Wilkins in 1962 for their discovery of the molecular structure of DNA. Rosalind Franklin might have been a candidate for a Nobel Prize had she not died of cancer at age 38 in 1958. (The Nobel Prize is given only to living investigators.) Watson and Crick's discovery of the structure of DNA opened the way to molecular studies of genetics and heredity, leading to our modern understanding of gene structure and action at the molecular level.

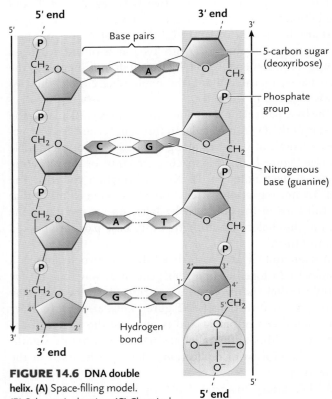

A. Space-filling model overlaid with sugar–phosphate backbones

B. Schematic drawing

2 nm

Each full twist of the DNA double helix = 3.4 nm

Distance between each pair of bases = 0.34 nm

C. Chemical structure drawing

Base pairs

5-carbon sugar (deoxyribose)

Phosphate group

Nitrogenous base (guanine)

Hydrogen bond

FIGURE 14.6 DNA double helix. (A) Space-filling model. **(B)** Schematic drawing. **(C)** Chemical structure drawing. Arrows and labeling of the ends show that the two polynucleotide chains of the double helix are antiparallel—that is, they have opposite polarity in that they run in opposite directions. In the space-filling model at the top, the spaces occupied by atoms are indicated by spheres. There are 10 base pairs per turn of the helix; only 8 base pairs are visible because the other 2 are obscured where the backbones pass over each other.
© Cengage Learning 2014

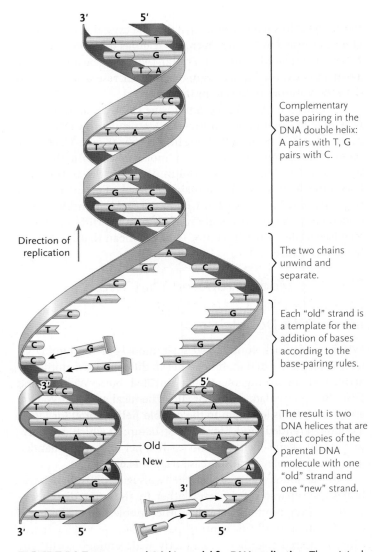

Complementary base pairing in the DNA double helix: A pairs with T, G pairs with C.

Direction of replication

The two chains unwind and separate.

Each "old" strand is a template for the addition of bases according to the base-pairing rules.

Old

New

The result is two DNA helices that are exact copies of the parental DNA molecule with one "old" strand and one "new" strand.

FIGURE 14.7 Watson and Crick's model for DNA replication. The original DNA molecule is shown in gray. A new polynucleotide chain (red) is assembled on each original chain as the two chains unwind. The template and complementary copy chains remain wound together when replication is complete, producing molecules that are half old and half new. The model is known as the semiconservative model for DNA replication.
© Cengage Learning 2014

STUDY BREAK 14.2

1. Which bases in DNA are purines? Which are pyrimidines?
2. What bonds form between complementary base pairs? Between a base and the deoxyribose sugar?
3. Which features of the DNA molecule did Watson and Crick describe?
4. The percentage of A in a double-stranded DNA molecule is 20. What is the percentage of C in that DNA molecule?

14.3 DNA Replication

Once they had discovered the structure of DNA, Watson and Crick realized immediately that complementary base pairing between the two strands could explain how DNA replicates **(Figure 14.7)**. They imagined that, for replication, the hydrogen bonds between the two strands break, and the two strands unwind and separate. Each strand then acts as a template for the synthesis of its partner strand. When replication is complete, there are two double helices, each of which has one strand derived from the parental DNA molecule base paired with a newly synthesized strand. Most important, each of the two new double

helices has a base-pair sequence identical to that of the parental DNA molecule.

The model of replication Watson and Crick proposed is termed **semiconservative replication (Figure 14.8A)**. Other scientists proposed two other models for replication. In the *conservative replication model,* the two strands of the original molecule serve as templates for the two strands of a new DNA molecule, then rewind into an all "old" molecule **(Figure 14.8B)**. After the two complementary copies separate from their templates, they wind together into an all "new" molecule. In the *dispersive replication model,* neither parental strand is conserved and both chains of each replicated molecule contain old and new segments **(Figure 14.8C)**.

The Meselson and Stahl Experiment Showed That DNA Replication Is Semiconservative

A definitive experiment published in 1958 by Matthew Meselson and Franklin Stahl of the California Institute of Technology demonstrated that DNA replication is semiconservative **(Figure 14.9)**. In their experiment, Meselson and Stahl had to be able to distinguish parental DNA strands from newly synthesized DNA. To do

A. Semiconservative replication   **B. Conservative replication**   **C. Dispersive replication**

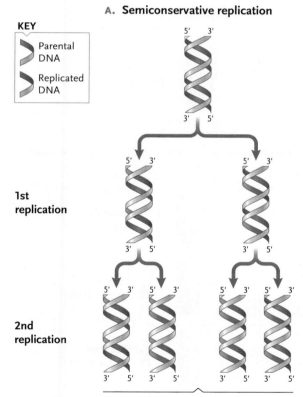
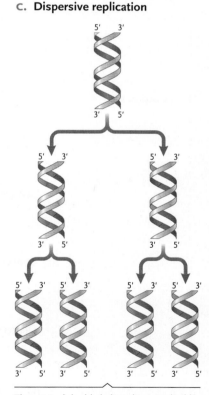

KEY

Parental DNA

Replicated DNA

1st replication

2nd replication

The two parental strands of DNA unwind, and each is a template for synthesis of a new strand. After replication has occurred, each double helix has one old strand paired with one new strand. This model was the one proposed by Watson and Crick themselves.

The parental strands of DNA unwind, and each is a template for synthesis of a new strand. After replication has occurred, the parental strands pair up again. Therefore, the two resulting double helices consist of one with two old strands and the other with two new strands.

The original double helix splits into double-stranded segments onto which new double-stranded segments form. These newly formed sections somehow assemble into two double helices, both of which are a mixture of the original double-stranded DNA interspersed with new double-stranded DNA.

FIGURE 14.8 Semiconservative (A), conservative (B), and dispersive (C) models for DNA replication.
© Cengage Learning 2014

FIGURE 14.9 | **Experimental Research**

The Meselson and Stahl Experiment Demonstrating the Semiconservative Model for DNA Replication to Be Correct

Question: Does DNA replicate semiconservatively?

Experiment: Matthew Meselson and Franklin Stahl proved that the semiconservative model of DNA replication is correct and that the conservative and dispersive models are incorrect.

1. Bacteria grown in ^{15}N (heavy) medium. The heavy isotope is incorporated into the bases of DNA, resulting in all the DNA being heavy, that is, labeled with ^{15}N.

2. Bacteria transferred to ^{14}N (light) medium and allowed to grow and divide for several generations. All new DNA is light.

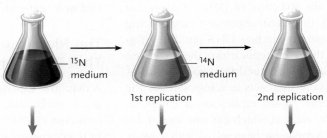

3. DNA extracted from bacteria cultured in ^{15}N medium and after each generation in ^{14}N medium. Extracted DNA was centrifuged in a special solution to separate DNA of different densities.

Results: Meselson and Stahl obtained the following results:

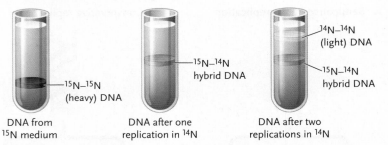

Conclusion: The predicted DNA banding patterns for the three DNA replication models shown in Figure 14.8 were:

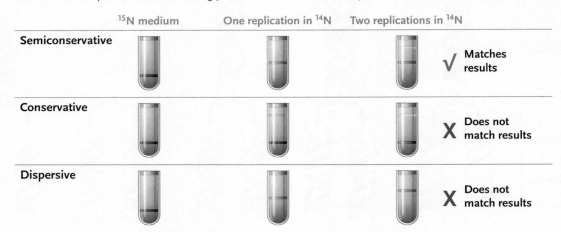

	^{15}N medium	One replication in ^{14}N	Two replications in ^{14}N	
Semiconservative				√ Matches results
Conservative				X Does not match results
Dispersive				X Does not match results

The results support the semiconservative model.

THINK LIKE A SCIENTIST For the semiconservative replication model, what proportion of ^{15}N–^{15}N, ^{15}N–^{14}N, and ^{14}N–^{14}N molecules would you expect after four and five replications in ^{14}N?

Source: M. Meselson and F. W. Stahl. 1958. The replication of DNA in *Escherichia coli. Proceedings of the National Academy of Sciences USA* 44:671–682.

this they used a nonradioactive "heavy" nitrogen isotope to tag the parental DNA strands. The heavy isotope, ^{15}N, has one more neutron in its nucleus than the normal ^{14}N isotope. Molecules containing ^{15}N are measurably heavier (denser) than molecules of the same type containing ^{14}N. DNA molecules with different densities were distinguished by a special type of centrifugation.

DNA Polymerases Are the Primary Enzymes of DNA Replication

During replication, complementary polynucleotide chains are assembled from individual deoxyribonucleotides by enzymes known as DNA polymerases. More than one kind of DNA polymerase is required for DNA replication in bacteria, archaea, and eukaryotes. *Deoxyribonucleoside triphosphates* are the substrates for the polymerization reaction catalyzed by DNA polymerases **(Figure 14.10)**. A nucleoside triphosphate is a nitrogenous base linked to a sugar, which is linked, in turn, to a chain of three phosphate groups (see Figure 3.27). You have encountered a nucleoside triphosphate before, namely the ATP produced in cellular respiration (see Chapter 8). In that case, the sugar is ribose, making ATP a *ribonucleoside triphosphate*. The deoxyribonucleoside triphosphates used in DNA replication have the sugar *deoxyribose* rather than the sugar *ribose*. Because four different bases are found in DNA—adenine (A), guanine (G), cytosine (C), and thymine (T)—four different deoxyribonucleoside triphosphates are used for DNA replication. In keeping with the ATP naming convention, the deoxyribonucleoside triphosphates for DNA replication are given the short names dATP, dGTP, dCTP, and dTTP, where the "d" stands for "deoxyribose." The abbreviation dNTP will be used for *deoxyribonucleoside triphosphate* in some of our discussions where the "N" refers to a purine or pyrimidine base without specifying which one.

Figure 14.10 presents a section of a DNA polynucleotide chain being replicated, and shows how DNA polymerase catalyzes the assembly of a new DNA strand that is complementary to the template strand. To understand Figure 14.10, remember that the carbons in the deoxyriboses of nucleotides are numbered with primes. Each DNA strand has two distinct ends: the 5′ end has an exposed phos-

phate group attached to the 5′ carbon of the sugar, and the 3′ end has an exposed hydroxyl group attached to the 3′ carbon of the sugar. As you learned earlier, because of the antiparallel nature of the DNA double helix, the 5′ end of one strand is opposite the 3′ end of the other.

DNA polymerase can add a nucleotide *only to the 3′* end of an existing nucleotide chain. As a new DNA strand is assembled, a 3′–OH group is always exposed at its "newest" end; the "oldest" end of the new chain has an exposed 5′ triphosphate. DNA polymerases are therefore said to assemble nucleotide chains in the 5′→3′ direction. Because of the antiparallel nature of DNA, DNA polymerase "reads" the template strand in the 3′→5′ direction for this new synthesis.

DNA polymerases of bacteria, archaea, and eukaryotes all consist of several polypeptide subunits arranged to form different domains (see Figure 3.26 and Section 3.4). The polymerases share a shape that is said to resemble a partially-closed human right

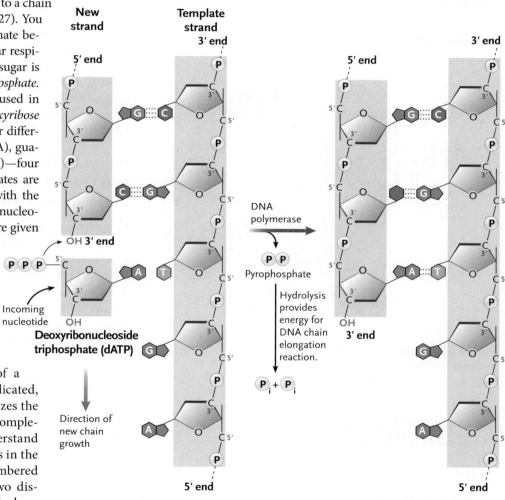

1 DNA polymerase forms a complementary base pair between a deoxyribonucleoside triphosphate with an A base (dATP) from the surrounding solution with the next, T, nucleotide of the template strand.

2 DNA polymerase catalyzes the formation of a phosphodiester bond involving the 3′–OH group at the end of the new chain and the innermost of the three phosphate groups of the dATP. The other two phosphates are released as a pyrophosphate molecule. The new chain has been lengthened by one nucleotide. The process continues, with DNA polymerase adding complementary nucleotides one by one to the growing DNA chain.

FIGURE 14.10 Reaction assembling a complementary DNA chain in the 5′→3′ direction on a template DNA strand, showing the phosphodiester bond formed when the DNA polymerase enzyme adds each deoxyribonucleotide to the chain.

© Cengage Learning 2014

hand in which the template DNA lies over the "palm" in a groove formed by the "fingers" and "thumb" **(Figure 14.11A).** The palm domain is evolutionarily related among the polymerases of bacteria, archaea, and eukaryotes, while the finger and thumb domains are different sequences in each of those three types of organisms. The template strand does not pass through the tunnel formed by the thumb and finger domains, however. Instead, the template strand and the 3′–OH of the new strand meet at the active site for the polymerization reaction of DNA synthesis, located in the palm domain. A nucleotide is added to the new strand when an incoming dNTP enters the active site carrying a base complementary to the template strand base positioned in the active site. By moving along the template strand, one nucleotide at a time, DNA polymerase extends the new DNA strand as we saw in Figure 14.10.

Figure 14.11B shows the representation of DNA polymerase used in the following DNA replication figures, and it also shows a *sliding DNA clamp.* The **sliding DNA clamp** is a protein that encircles the DNA and binds to the rear of the DNA polymerase in terms of the enzyme's forward movement during replication. The sliding DNA clamp tethers the DNA polymerase to the template strand. This tethering makes replication more efficient because without it, the enzyme detaches from the template after only a few tens of polymerizations. With the clamp, many tens of thousands of polymerizations occur before the enzyme detaches. Overall, the rate of DNA synthesis is much faster because of the sliding DNA clamp.

In sum, the key molecular events of DNA replication are as follows:

1. The two strands of the DNA molecule unwind for replication to occur.
2. DNA polymerase adds nucleotides to an existing chain.
3. The overall direction of new synthesis is in the 5′→3′ direction, which is a direction antiparallel to that of the template strand.
4. Nucleotides enter into a newly synthesized chain according to the A–T and G–C complementary base-pairing rules.

The following sections describe how enzymes and other proteins conduct these molecular events. Our focus is on the well-characterized replication system of *E. coli.* Replication in archaea and eukaryotes is highly similar, although there are differences in the replication machinery. The replication machinery of archaea is strikingly similar to that of eukaryotes and is clearly different from that of bacteria.

Helicases Unwind DNA for New DNA Synthesis and Other Proteins Stabilize the DNA at the Replication Fork

In semiconservative replication, the two strands of the parental DNA molecule unwind and separate to expose the template strands for new DNA synthesis **(Figure 14.12).** Unwinding of the DNA for replication occurs at a small, specific region in the bacterial chromosome known as an **origin of replication (*ori*).** Specific proteins recognize an *ori* and recruit **DNA helicase,** which unwinds the DNA strands. The unwinding produces a Y-shaped structure called a **replication fork,** which consists of the two unwound template strands transitioning to double-helical DNA. **Single-stranded binding proteins (SSBs)** coat the exposed single-stranded DNA segments, stabilizing the DNA and keeping the two strands from pairing (see Figure 14.12). The SSBs are displaced as the replication enzymes make the new polynucleotide chain on the template strands.

For circular chromosomes, such as the genomes of most bacteria, unwinding the DNA will eventually cause the still-wound DNA ahead of the unwinding to become highly twisted. You can visualize this phenomenon with some string. Take two equal lengths of string and twist them around each other. Now tie the two ends of each string together. You have created a model of a circular DNA double helix. Pick anywhere in the circle and pull apart the two pieces of string. The more you pull, the more the region where the two strings are still together becomes highly twisted. In the cell, the twisting of DNA during

A. Bacterial DNA polymerase

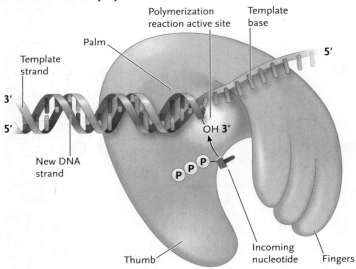

B. How a DNA polymerase and sliding clamp is shown in the book

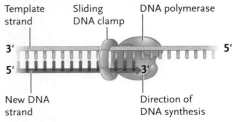

FIGURE 14.11 DNA polymerase structure. (A) Stylized drawing of a bacterial DNA polymerase. The enzyme viewed from the side resembles a human right hand. The polymerization reaction site lies on the palm. When the incoming nucleotide is added, the thumb and fingers close over the site to facilitate the reaction. **(B)** How DNA polymerase is shown in subsequent figures of DNA replication. The figure also shows a sliding DNA clamp tethering the DNA polymerase to the template strand.

© Cengage Learning 2014

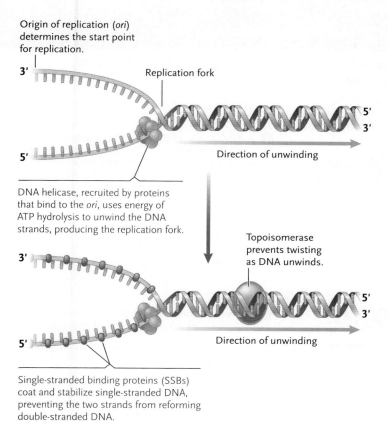

Origin of replication (*ori*) determines the start point for replication.

3′

Replication fork

5′
3′

Direction of unwinding

DNA helicase, recruited by proteins that bind to the *ori*, uses energy of ATP hydrolysis to unwind the DNA strands, producing the replication fork.

3′

Topoisomerase prevents twisting as DNA unwinds.

5′
3′

5′

Direction of unwinding

Single-stranded binding proteins (SSBs) coat and stabilize single-stranded DNA, preventing the two strands from reforming double-stranded DNA.

FIGURE 14.12 The roles of DNA helicase, single-stranded binding proteins (SSBs), and topoisomerase in DNA replication.
© Cengage Learning 2014

replication is prevented by **topoisomerase,** which cuts the DNA ahead of the replication fork, turns the DNA on one side of the break in the opposite direction of the twisting force, and rejoins the two strands again (see Figure 14.12).

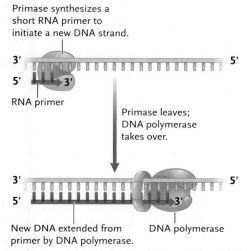

Primase synthesizes a short RNA primer to initiate a new DNA strand.

3′
5′
3′
5′

RNA primer

Primase leaves; DNA polymerase takes over.

3′
5′
5′
3′

New DNA extended from primer by DNA polymerase.

DNA polymerase

FIGURE 14.13 Initiation of a new DNA strand by synthesis of a short RNA primer by primase, and the extension of the primer as DNA by DNA polymerase.
© Cengage Learning 2014

RNA Primers Provide the Starting Point for DNA Polymerase to Begin Synthesizing a New DNA Chain

DNA polymerases can add nucleotides only to the 3′ end of an existing strand. How can a new strand begin when there is no existing strand in place? The answer lies in a short chain a few nucleotides long called a **primer** that is made of RNA instead of DNA **(Figure 14.13).** The primer is synthesized by the enzyme **primase.** Primase then leaves the template, and DNA polymerase takes over, extending the RNA primer with DNA nucleotides as it synthesizes the new DNA chain. RNA primers are removed and replaced with DNA later in replication.

One New DNA Strand Is Synthesized Continuously; the Other, Discontinuously

DNA polymerases synthesize a new DNA strand on a template strand in the 5′→3′ direction. Because the two strands of a DNA molecule are antiparallel, only one of the template strands runs in a direction that allows DNA polymerase to make a 5′→3′ complementary copy in the direction of unwinding. That is, on this template strand—top strand in **Figure 14.14**—the new DNA strand is synthesized continuously in the direction of unwinding of the double helix. However, the other template strand—bottom strand in Figure 14.14—runs in the opposite direction; this means DNA polymerase has to copy it in the direction opposite to the unwinding direction.

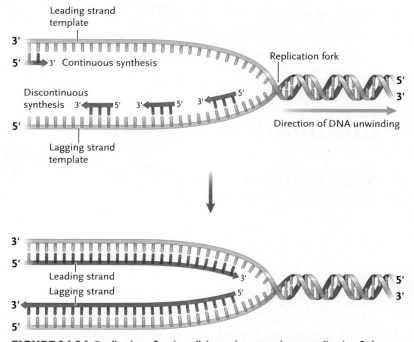

Leading strand template

3′
5′
3′ Continuous synthesis

Replication fork

5′
3′

Discontinuous synthesis   3′ 5′   3′ 5′   3′ 5′

Direction of DNA unwinding

5′

Lagging strand template

3′
5′
Leading strand
Lagging strand
3′
5′

5′
3′

FIGURE 14.14 Replication of antiparallel template strands at a replication fork. Synthesis of the new DNA strand on the top template strand is continuous. Synthesis on the new DNA strand on the bottom template strand is discontinuous—short lengths of DNA are made, which are then joined into a continuous chain. The overall effect is synthesis of both strands in the direction of replication fork movement.
© Cengage Learning 2014

The drawings simplify the process. In reality, the enzymes assemble at the fork, replicating both strands from that position as the template strands fold and pass through the assembly.

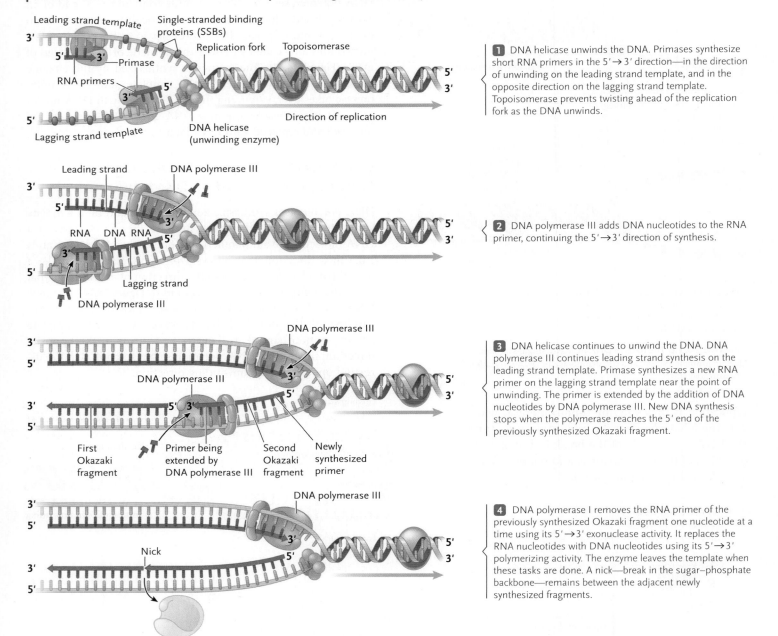

1 DNA helicase unwinds the DNA. Primases synthesize short RNA primers in the 5′→3′ direction—in the direction of unwinding on the leading strand template, and in the opposite direction on the lagging strand template. Topoisomerase prevents twisting ahead of the replication fork as the DNA unwinds.

2 DNA polymerase III adds DNA nucleotides to the RNA primer, continuing the 5′→3′ direction of synthesis.

3 DNA helicase continues to unwind the DNA. DNA polymerase III continues leading strand synthesis on the leading strand template. Primase synthesizes a new RNA primer on the lagging strand template near the point of unwinding. The primer is extended by the addition of DNA nucleotides by DNA polymerase III. New DNA synthesis stops when the polymerase reaches the 5′ end of the previously synthesized Okazaki fragment.

4 DNA polymerase I removes the RNA primer of the previously synthesized Okazaki fragment one nucleotide at a time using its 5′→3′ exonuclease activity. It replaces the RNA nucleotides with DNA nucleotides using its 5′→3′ polymerizing activity. The enzyme leaves the template when these tasks are done. A nick—break in the sugar–phosphate backbone—remains between the adjacent newly synthesized fragments.

© Cengage Learning 2014

How is the new DNA strand made in the opposite direction to the unwinding? The polymerases make this strand in short lengths that are synthesized in the direction opposite to that of DNA unwinding (see Figure 14.14). The short lengths produced by this **discontinuous replication** are then covalently linked into a continuous polynucleotide chain. The short lengths are called **Okazaki fragments,** after Reiji Okazaki, the scientist who first detected them. The new DNA strand synthesized in the direction of DNA unwinding is called the **leading strand** of DNA replication; the template strand for that strand is the **leading strand template.** The strand synthesized discontinuously in the opposite direction is called the **lagging strand;** the template strand for that strand is the **lagging strand template.**

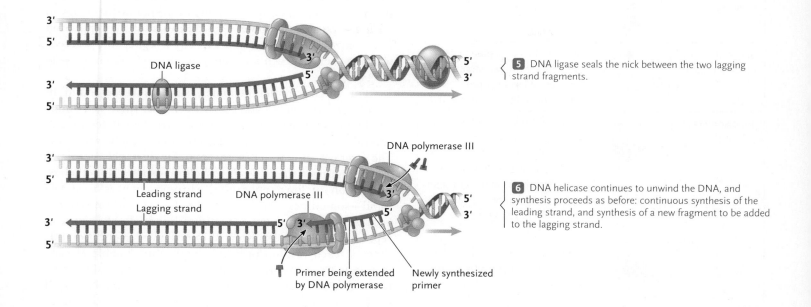

5 DNA ligase seals the nick between the two lagging strand fragments.

6 DNA helicase continues to unwind the DNA, and synthesis proceeds as before: continuous synthesis of the leading strand, and synthesis of a new fragment to be added to the lagging strand.

Leading strand
Lagging strand
DNA polymerase III
DNA polymerase III
DNA ligase

Primer being extended by DNA polymerase
Newly synthesized primer

SUMMARY As the replication fork moves, DNA synthesis is continuous on the leading strand template, and discontinuous on the lagging strand template. Synthesis of the leading strand involves the synthesis of the primer, and then the addition of DNA nucleotides by DNA polymerase III. Synthesis of the lagging strand involves the cyclic series of events of primer synthesis (primase), addition of DNA nucleotides by DNA polymerase III to create an Okazaki fragment, the simultaneous removal of the RNA primer at the 5′ end of the adjacent Okazaki fragment and replacement with DNA nucleotides by DNA polymerase I, and joining of adjacent Okazaki fragments by DNA ligase. Unwinding of the DNA for replication is done by helicase, and topoisomerase prevents twisting of the DNA ahead of the replication fork.

THINK LIKE A SCIENTIST Suppose DNA ligase was defective. How would that affect DNA replication?

Multiple Enzymes Coordinate Their Activities in DNA Replication

Figure 14.15 shows how the enzymes and proteins we have introduced act in a coordinated way to replicate DNA. Primase initiates all new strands by synthesizing an RNA primer. **DNA polymerase III,** the main polymerase, extends the primer by adding DNA nucleotides. For the lagging strand, **DNA polymerase I** removes the RNA primer at the 5′ end of the previous newly synthesized Okazaki fragment, replacing the RNA nucleotides one by one with DNA nucleotides. RNA nucleotide removal uses the 5′→3′ exonuclease activity of the enzyme. (An exonuclease removes nucleotides from the end of a molecule.) DNA synthesis uses the 5′→3′ polymerization activity. DNA ligase (*ligare* = to tie) seals the nick left between the two fragments. The replication process continues in the same way until the entire DNA molecule is copied. **Table 14.1** summarizes the activities of the major enzymes replicating DNA.

Replication advances at a rate of about 500 to 1,000 nucleotides per second in *E. coli* and other bacteria, and at a rate of about 50 to 100 per second in eukaryotes. The entire process is so rapid that the RNA primers and nicks left by discontinuous synthesis persist for only seconds or fractions of a second. A short distance behind the fork, the new DNA chains are fully continuous and wound into complete DNA double helices. Each helix consists of one "old" and one "new" polynucleotide chain.

Researchers identified the enzymes that replicate DNA through experiments with a variety of bacteria and eukaryotes and with viruses that infect both types of cells. Experiments with the bacterium *E. coli* have provided the most complete information about DNA replication, particularly in the laboratory of Arthur Kornberg at Stanford University. Kornberg received a Nobel Prize in 1959 for his discovery of the mechanism for DNA synthesis.

TABLE 14.1 — Major Enzymes of DNA Replication

Enzyme	Symbol	Function
Helicase		Unwinds DNA helix
Single-stranded binding proteins		Stabilize single-stranded DNA and prevent the two strands at the replication fork from reforming double-stranded DNA
Topoisomerase		Avoids twisting of the DNA ahead of replication fork (in circular DNA) by cutting the DNA, turning the DNA on one side of the break in the direction opposite to that of the twisting force, and rejoining the two strands again
Primase		Synthesizes RNA primer in the 5′→3′ direction to initiate a new DNA strand
DNA polymerase III		Main replication enzyme in E. coli. Extends the RNA primer by adding DNA nucleotides to it.
DNA polymerase I		E. coli enzyme that uses its 5′→3′ exonuclease activity to remove the RNA of the previously synthesized Okazaki fragment, and uses its 5′→3′ polymerization activity to replace the RNA nucleotides with DNA nucleotides.
Sliding DNA clamp		Tethers DNA polymerase III to the DNA template, making replication more efficient.
DNA ligase		Seals nick left between adjacent fragments after RNA primers replaced with DNA

© Cengage Learning 2014

Multiple Replication Origins Enable Rapid Replication of Large Chromosomes

Unwinding at an *ori* within a DNA molecule actually produces two replication forks: two Ys joined together at their tops to form a **replication bubble.** Typically, each of the replication forks moves away from the *ori* as DNA replication proceeds with the events at each fork mirroring those in the other **(Figure 14.16).**

For small circular genomes, such as those found in *E. coli,* and in many bacteria and archaea, there is a single *ori.* Eukaryotic genomes, by contrast, are distributed among several linear chromosomes, each of which can be very long. The average human chromosome, for instance, is about 25 times longer than the *E. coli* chromosome. Nonetheless, replication of long, eukaryotic chromosomes is relatively rapid—sometimes faster than the *E. coli* chromosome—because there are many, sometimes hundreds of origins of replications along eukaryotic chromosomes. Replication initiates at each origin, forming a replication bubble at each **(Figure 14.17).** Movement of the two forks in opposite directions from each origin extends the replication bubbles until the forks eventually meet along the chromosomes to produce fully replicated DNA molecules.

Normally, a replication origin is activated only once during the S phase of a eukaryotic cell cycle, so no portion of the DNA is replicated more than once. *Insights from the Molecular Revolution* describes the production of an abnormal number of copies of a segment of DNA that underlies a common cause of mental retardation in humans.

Telomerases Solve a Special Replication Problem at the Ends of Linear DNA Molecules in Eukaryotes

The RNA primer synthesized in DNA replication (see Figures 14.13 and 14.15) produces a problem for replicating the linear chromosomes of eukaryotes. Think about the end of a linear DNA molecule. New DNA synthesis on the 3′→5′ template strand starts with an RNA primer. That primer will subsequently be removed, leaving a single-stranded region at the 5′ end of the

FIGURE 14.16 Synthesis of leading and lagging strands in the two replication forks of a replication bubble formed at an origin of replication.
© Cengage Learning 2014

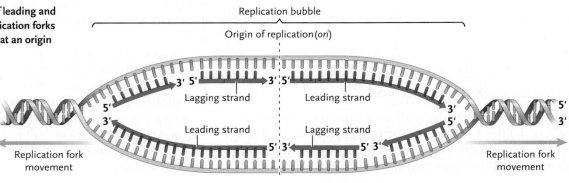

A Fragile Connection between DNA Replication and Mental Retardation: What is the molecular basis for fragile X syndrome?

The second most common cause of mental retardation in humans after Down syndrome results from breaks that occur in a narrow,

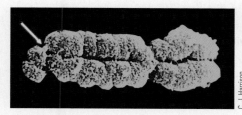

FIGURE 1 The constricted region *(arrow)* in the human X chromosome associated with fragile X syndrome. The chromosome is double because it has been duplicated in preparation for cell division.

constricted region near one end of the X chromosome **(Figure 1).** Because the region breaks easily when cultured cells divide, the associated disabilities are called *fragile X syndrome.*

Like most other X-linked traits, the disorder affects males more frequently than females: about 1 in 4,000 males and 1 in 6,000 to 8,000 females worldwide have fragile X syndrome. However, the syndrome has an unusual inheritance pattern in that it can be passed from an unaffected grandfather through his unaffected daughter to his grandchildren, in whom abnormal X chromosomes and the symptoms of the disease are seen.

Research Question

What is the molecular basis for fragile X syndrome?

Experiments

Researchers in the laboratories of Grant R. Sutherland of Adelaide Children's Hospital in Australia and others examined DNA from individuals with fragile X syndrome using

"probes"—short, artificially synthesized DNA sequences that are complementary to, and can pair with, DNA sequences of interest. They found that the fragile X region contains few-to-many repeats of the three-nucleotide sequence 5′-CGG-3′, the number of copies varying with the individual. The repeated sequences are in the *FMR1* gene. The fragility of the X chromosome correlated with higher numbers of the repeated sequence.

Conclusion

The researchers concluded that the molecular basis for the fragility of the X chromosome in fragile X syndrome is an abnormally high number of the CGG repeated sequence in the *FMR1* gene. The normal allele of the gene contains between 6 and 54 copies of the repeat; no symptoms are associated with this allele. Individuals with 55 to 200 copies are said to have a premutation allele because the gene is abnormal, yet there are few or no symptoms of fragile X syndrome. Individuals with more than 200 copies (sometimes thousands of copies) have the fully mutant allele; all males with this allele have all the symptoms of fragile X syndrome whereas females have somewhat milder symptoms. **Figure 2** shows more closely the grandfather-daughter-grandson inheritance pattern mentioned in the beginning.

If a mother passes a premutation X chromosome to her daughter, the daughter would likely have the premutation allele, perhaps with some increase in the number of repeats. In a few instances, the daughter will undergo expansion of the repeats to produce the full mutation and develop fragile X syndrome with the milder symptoms characteristic of females.

THINK LIKE A SCIENTIST Why do X-linked recessive traits affect males more frequently than females?

Source: E. J. Kremer, et al. 1991. Mapping of DNA instability at the fragile X to a trinucleotide repeat sequence p(CCG)*n. Science* 252:1711–1714.

A. Grandfather's X chromosome

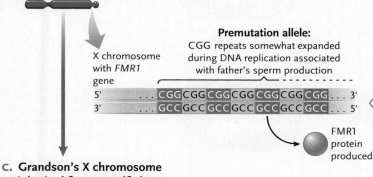

Premutation allele:
55–200 repeats of CGG

FMR1 Gene

5′ . . . CGG CGG CGG CGG CGG . . . 3′
3′ . . . GCC GCC GCC GCC GCC . . . 5′

Grandfather's single X chromosome has premutation allele near beginning of *FMR1* gene.

FMR1 protein produced

B. Daughter's paternal X chromosome

Premutation allele:
CGG repeats somewhat expanded during DNA replication associated with father's sperm production

X chromosome with *FMR1* gene

5′ . . . CGG CGG CGG CGG CGG CGG CGG . . . 3′
3′ . . . GCC GCC GCC GCC GCC GCC GCC . . . 5′

One of daughter's X chromosomes has premutation allele. The daughter has no symptoms.

FMR1 protein produced

C. Grandson's X chromosome inherited from grandfather

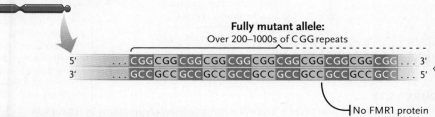

Fully mutant allele:
Over 200–1000s of C GG repeats

5′ . . . CGG CGG CGG CGG CGG CGG CGG CGG CGG CGG CGG . . . 3′
3′ . . . GCC GCC GCC GCC GCC GCC GCC GCC GCC GCC GCC . . . 5′

A grandson who inherits this allele develops full mutation by further expansion of the repeats. The lack of the *FMR1*-encoded protein alters control of neuron connection strength and mental retardation results.

No FMR1 protein produced

FIGURE 2 Inheritance pattern in fragile X syndrome.
© Cengage Learning 2014

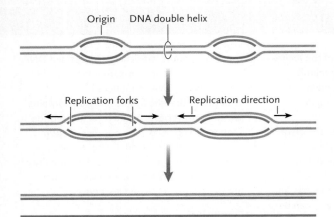

Origin   DNA double helix

Replication forks   Replication direction

FIGURE 14.17 Replication from multiple origins in the linear chromosomes of eukaryotes.
© Cengage Learning 2014

new DNA strand **(Figure 14.18).** But, because there is no existing nucleotide chain that can be used, DNA polymerase cannot fill in that region with DNA nucleotides. In a similar way, a single-stranded region is produced at the 5′ end of the new strand made starting at the other end of the chromosome. When these new, now shortened DNA strands are used as a template for the next round of DNA replication, the new chromosome will be shorter. Indeed, when most somatic cells go through the cell cycle, the chromosomes shorten with each division. Deletion of genes by such shortening can eventually have lethal consequences for the cell.

In most eukaryotic chromosomes the genes near the ends of chromosomes are protected by a buffer of noncoding DNA. The region of noncoding DNA is called the *telomere* (*telo* = end, *mere* = segment). A telomere consists of short sequences repeated hundreds to thousands of times. In humans, the repeated sequence, the *telomere repeat*, is 5′-TTAGGG-3′ on the template strand (the top strand in Figure 14.18). With each replication, a fraction of the telomere repeats is lost but the genes are unaffected. The buffering fails only when the entire telomere is lost.

The enzyme **telomerase** can stop the shortening of the telomeres by adding telomere repeats to the chromosome ends. Telomerase consists of proteins and an RNA molecule. The RNA of telomerase is the template for the addition of telomere repeats. Telomerase binds to the DNA template strand by complementary base pairing between the telomerase RNA and the DNA. It then adds telomere repeats to the DNA

using the RNA as a template (see Figure 14.18; the figure shows the addition of one repeat). Now when the top strand is used as a template for replication by primase and DNA polymerase and the RNA primer is removed, there will be a single-stranded region at the end of the chromosome as before. However, the chromosome has not shortened because of the extra telomere repeats added by the telomerase.

In humans, telomerase is active only in the rapidly dividing cells of the early embryo and in germ cells (the precursors to sperm and eggs). However, telomerase is inactive in somatic cells, meaning telomeres shorten when such cells divide. As a result, somatic cells are capable of only a certain number of mitotic divisions before they stop dividing and die.

Telomerase explains how cancer cells can divide indefinitely and not be limited to a certain number of divisions as a result of telomere shortening. For many cancers, as normal cells develop into cancer cells, their telomerases are reactivated, preserving chromosome length during the rapid divisions characteristic of cancer. A positive side of this discovery is that it may lead to an effective cancer treatment, if a means can be found to switch off the telomerases in tumor cells. The chromosomes in the rapidly dividing cancer cells would then eventually shorten to the length at which they break down, leading to cell death and elimination of the tumor.

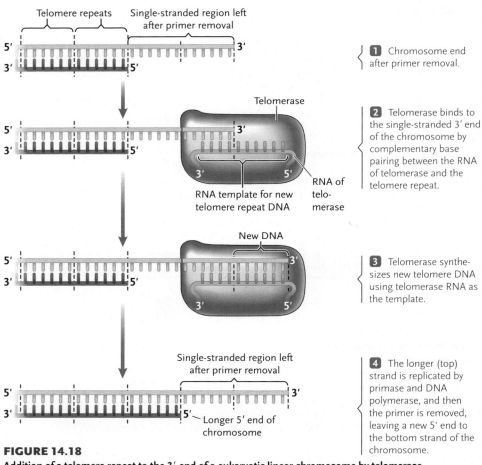

1 Chromosome end after primer removal.

2 Telomerase binds to the single-stranded 3′ end of the chromosome by complementary base pairing between the RNA of telomerase and the telomere repeat.

3 Telomerase synthesizes new telomere DNA using telomerase RNA as the template.

4 The longer (top) strand is replicated by primase and DNA polymerase, and then the primer is removed, leaving a new 5′ end to the bottom strand of the chromosome.

FIGURE 14.18
Addition of a telomere repeat to the 3′ end of a eukaryotic linear chromosome by telomerase.
© Cengage Learning 2014

Elizabeth Blackburn, Carol Greider, and Jack Szostak were awarded a Nobel Prize in 2009 for their discovery of how chromosomes are protected by telomeres and the enzyme telomerase.

STUDY BREAK 14.3

1. What is the importance of complementary base pairing to DNA replication?
2. Why is a primer needed for DNA replication? How is the primer made?
3. DNA polymerase III and DNA polymerase I are used in DNA replication in *E. coli*. What are their roles?
4. Why are telomeres important?

THINK OUTSIDE THE BOOK

We introduced progeria in *Why It Matters* in Chapter 13, and pictured two individuals with the disease in Figure 13.1. Progeria is a rare premature aging genetic disorder; most patients die in their early teens. A number of research studies have shown that telomere length (that is, the number of telomere repeats) decreases with age in humans (and in many other organisms). Collaboratively, or on your own, explore the research literature to determine if a decrease in telomere length is involved in progeria. If so, how does the mutation in progeria cause the decrease?

14.4 Mechanisms That Correct Replication Errors

DNA polymerases make very few errors as they assemble new nucleotide chains. Most of the mistakes that do occur, called **base-pair mismatches,** are corrected, either by a proofreading mechanism carried out during replication by the DNA polymerases themselves or by a DNA repair mechanism that corrects mismatched base pairs after replication is complete.

Proofreading Depends on the Ability of DNA Polymerases to Reverse and Remove Mismatched Bases

The **proofreading mechanism,** first proposed in 1972 by Arthur Kornberg and Douglas L. Brutlag of Stanford University, depends on the ability of DNA polymerases to back up and remove mispaired nucleotides from a DNA strand. For most of the polymerization reactions, DNA polymerase adds the correct nucleotide to the growing chain (**Figure 14.19,** step 1). But the polymerase also has a small window of time for proofreading. If a newly added nucleotide is mismatched (step 2), the DNA polymerase reverses, using a built-in 3′→5′ exonuclease activity to remove the newly added incorrect nucleotide (step 3). The enzyme then resumes forward synthesis, now inserting the correct nucleotide (step 4).

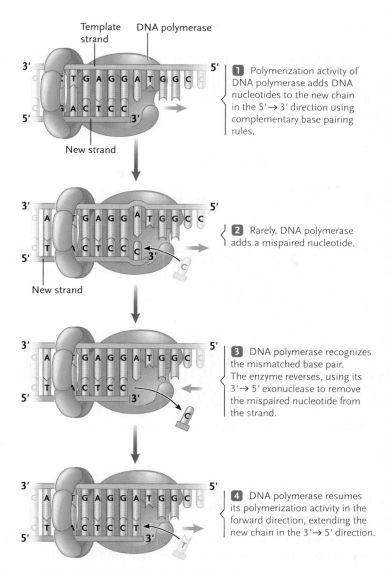

1. Polymerization activity of DNA polymerase adds DNA nucleotides to the new chain in the 5′→3′ direction using complementary base pairing rules.

2. Rarely, DNA polymerase adds a mispaired nucleotide.

3. DNA polymerase recognizes the mismatched base pair. The enzyme reverses, using its 3′→5′ exonuclease to remove the mispaired nucleotide from the strand.

4. DNA polymerase resumes its polymerization activity in the forward direction, extending the new chain in the 3′→5′ direction.

FIGURE 14.19 Proofreading by a DNA polymerase.
© Cengage Learning 2014

Several experiments have confirmed that the major DNA polymerases of replication can actually proofread their work. For example, when the *E. coli* DNA polymerase III is fully functional, its overall error rate is astonishingly low with only about 1 mispair surviving in the DNA for every 1 million nucleotides polymerized in the test tube. If the proofreading activity of the enzyme is experimentally inhibited, the error rate increases to about 1 mistake for every 1,000 to 10,000 nucleotides polymerized. Experiments with eukaryotes have yielded similar results.

DNA Repair Corrects Errors That Escape Proofreading

Any base-pair mismatches that remain after proofreading face still another round of correction by **DNA repair mechanisms.** These **mismatch repair** mechanisms increase the accuracy of

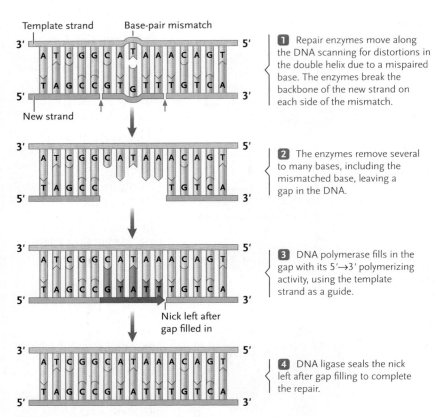

1 Repair enzymes move along the DNA scanning for distortions in the double helix due to a mispaired base. The enzymes break the backbone of the new strand on each side of the mismatch.

2 The enzymes remove several to many bases, including the mismatched base, leaving a gap in the DNA.

3 DNA polymerase fills in the gap with its 5'→3' polymerizing activity, using the template strand as a guide.

4 DNA ligase seals the nick left after gap filling to complete the repair.

FIGURE 14.20 Repair of mismatched bases in replicated DNA.
© Cengage Learning 2014

Very few replication errors remain in DNA after proofreading and DNA repair. The errors that persist, although extremely rare, are a primary source of **mutations,** differences in DNA sequence that appear and remain in the replicated copies. When a mutation occurs in a gene, it can alter the property of the protein encoded by the gene, which, in turn, may alter how the organism functions. Hence, mutations are highly important to the evolutionary process because they are the ultimate source of the variability in offspring acted on by natural selection.

We now turn from DNA replication and error correction to the arrangements of DNA in eukaryotic and prokaryotic cells. These arrangements organize superstructures that fit the long DNA molecules into the microscopic dimensions of cells and also contribute to the regulation of DNA activity.

STUDY BREAK 14.4

Why is a proofreading mechanism important for DNA replication, and what are the mechanisms that correct errors?

DNA replication well beyond the one-in-a million errors that persist after proofreading. The mechanisms operate similarly in all organisms.

As noted earlier, the "correct" A–T and G–C base pairs fit together like pieces of a jigsaw puzzle, and their dimensions separate the sugar–phosphate backbone chains by a constant distance. Mispaired bases are too large or small to maintain the correct separation, and they cannot form the hydrogen bonds characteristic of the normal base pairs. As a result, base mismatches distort the structure of the DNA helix. These distortions provide recognition sites for the enzymes catalyzing mismatch repair.

The repair enzymes detect the mispaired base, cut the new DNA strand on each side of the mismatch, and remove a portion of the chain **(Figure 14.20).** DNA polymerase fills in the gap with new DNA. The repair is completed by DNA ligase, which seals the nucleotide chain into a continuous DNA molecule.

Similar repair systems also detect and correct alterations in DNA caused by the damaging effects of chemicals and radiation, including the ultraviolet light in sunlight. Some idea of the importance of the repair mechanisms comes from the unfortunate plight of individuals with *xeroderma pigmentosum,* a hereditary disorder in which the repair mechanism is faulty. Because of the effects of unrepaired alterations in their DNA, skin cancer can develop quickly in these individuals if they are exposed to sunlight.

14.5 DNA Organization in Eukaryotes and Prokaryotes

Enzymatic proteins are the essential catalysts of every step in DNA replication. In addition, numerous proteins of other types organize the DNA in both eukaryotes and prokaryotes and control its function.

In eukaryotes, two major types of proteins, the histone and nonhistone proteins, are associated with DNA structure and regulation in the nucleus. These proteins are known collectively as the **chromosomal proteins** of eukaryotes. The complex of DNA and its associated proteins, termed **chromatin,** is the structural building block of a chromosome.

By comparison, the single chromosome of a prokaryotic cell is more simply organized and has fewer associated proteins. However, prokaryotic DNA is still associated with two classes of proteins with functions similar to those of the eukaryotic histones and nonhistones: one class that organizes the DNA structurally and one that regulates gene activity. We begin this section with the major DNA-associated proteins of eukaryotes.

Histones Pack Eukaryotic DNA at Successive Levels of Organization

The **histones** are a class of small, positively charged (basic) proteins that are complexed with DNA in the chromosomes of eukaryotes. (Most other cellular proteins are larger and are neutral or negatively charged.) The histones bind to DNA by an attraction between their positive charges and the negatively charged phosphate groups of the DNA.

Five types of histones exist in most eukaryotic cells: H1, H2A, H2B, H3, and H4. The amino acid sequences of these proteins are highly similar among eukaryotes, suggesting that they perform the same functions in all eukaryotic organisms.

One function of histones is to pack DNA molecules into the narrow confines of the cell nucleus. For example, each human cell nucleus contains 2 meters of DNA. Combination with the histones compacts this length so much that it fits into nuclei that are only about 10 μm in diameter. Another function is the regulation of DNA activity.

HISTONES AND DNA PACKING The histones pack DNA at several levels of chromatin structure. In the most fundamental structure, called a **nucleosome,** two molecules each of H2A, H2B, H3, and H4 combine to form a beadlike, eight-protein **nucleosome core particle** around which DNA winds for almost two turns **(Figure 14.21).** A short segment of DNA, the **linker,** extends between one nucleosome and the next. Under the electron microscope, this structure looks like beads on a string. The diameter of the beads (the nucleosomes) gives this structure its name—the **10-nm chromatin fiber** (see Figure 14.21).

Each nucleosome and linker includes about 200 base pairs of DNA. Nucleosomes compact DNA by a factor of about 7; that is, a length of DNA becomes about 7 times shorter when it is wrapped into nucleosomes.

HISTONES AND CHROMATIN FIBERS The fifth histone, H1, brings about the next level of chromatin packing. One H1 molecule binds both to the nucleosome (at the point where the DNA enters and leaves the core particle) and to the linker DNA. This binding causes the nucleosomes to package into a coiled structure 30 nm in diameter, called the **30-nm chromatin fiber.** One possible model for the 30-nm fiber is the **solenoid model,** with the nucleosomes spiraling helically with about six nucleosomes per turn (see Figure 14.21).

The arrangement of DNA in nucleosomes and the 30-nm fiber compacts the DNA and probably also protects it from chemical and mechanical damage. In the test tube, DNA wound into nucleosomes and chromatin fibers is much more resistant to attack by deoxyribonuclease (a DNA-digesting enzyme) than when it is not bound to histone proteins. It is also less accessible to the proteins and enzymes required for gene expression. Therefore, the association of the DNA with histones must loosen in order for a gene to become active (see Section 16.2).

PACKING OF EUKARYOTIC CHROMOSOMES AT STILL HIGHER LEVELS: EUCHROMATIN AND HETEROCHROMATIN In interphase nuclei, chromatin fibers are loosely packed in some regions and densely packed in others. The loosely packed regions are known as euchromatin (*emu* = true, regular, or typical), and the densely packed regions are called heterochromatin (*hetero* = different). Chromatin fibers also fold and pack into the thick, rodlike chromosomes visible during mitosis and meiosis. Experiments indicate that links formed between H1 histone molecules contribute to the packing of chromatin fibers, both into heterochromatin and into the chromosomes visible during nuclear division (see discussion in Section 10.2). However, the exact mechanism for the more complex folding and packing is not known.

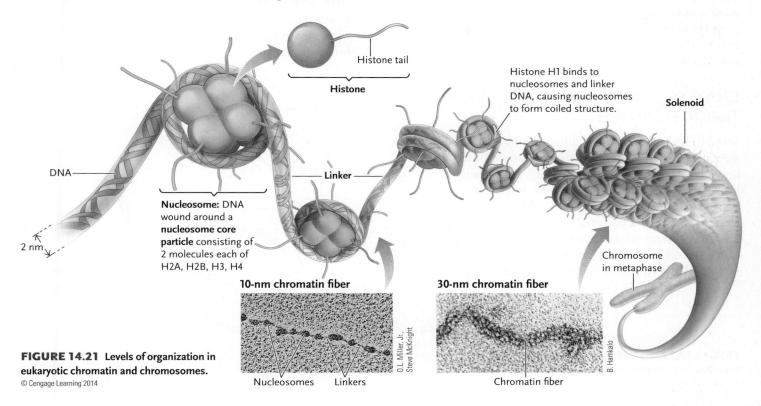

FIGURE 14.21 Levels of organization in eukaryotic chromatin and chromosomes.
© Cengage Learning 2014

Several experiments indicate that some regions of heterochromatin include genes that have been turned off and placed in a compact storage form. For example, recall the process of X-chromosome inactivation in mammalian females (see Section 13.2). As one of the two X chromosomes becomes inactive in cells early in development, it packs down into a block of heterochromatin called the *Barr body,* which is large enough to see under the light microscope. These findings support the idea that, in addition to organizing nuclear DNA, histones play a role in regulating gene activity.

Many Nonhistone Proteins Have Key Roles in the Regulation of Gene Expression in Eukaryotes

Nonhistone proteins are loosely defined as all the proteins associated with DNA that are not histones. Nonhistones vary widely in structure; most are negatively charged or neutral, but some are positively charged. They range in size from polypeptides smaller than histones to some of the largest cellular proteins.

Many nonhistone proteins help control the expression of individual genes. (The regulation of gene expression is the subject of Chapter 16.) For example, expression of a gene requires that the enzymes and proteins for that process be able to access the gene in the chromatin. If a gene is packed into heterochromatin, it is unavailable for activation. If the gene is in the form of the more-extended euchromatin, it is more accessible. Many nonhistone proteins affect gene accessibility by modifying histones to change how the histones associate with DNA in chromatin, either loosening or tightening the association. Other nonhistone proteins are regulatory proteins that activate or repress the expression of a gene. Yet others are components of the enzyme–protein complexes that are needed for the expression of any gene.

DNA Is Organized More Simply in Bacteria Than in Eukaryotes

Several features of DNA organization in bacteria differ fundamentally from eukaryotic DNA organization. In contrast to the linear DNA in eukaryotes, the primary DNA molecule of most bacteria is circular, with only one copy per cell. In parallel with eukaryotic terminology, the DNA molecule is called a **bacterial chromosome.** The chromosome of the best-known bacterium, *E. coli,* includes about 1,460 μm of DNA, which is equivalent to 4.6 million base pairs. There are exceptions: some bacteria have two or more different chromosomes in the cell, and some bacterial chromosomes are linear.

Replication begins from a single origin in the DNA circle, forming two forks that travel around the circle in opposite directions. Eventually, the forks meet at the opposite side from the origin to complete replication **(Figure 14.22).**

Inside bacterial cells, the DNA circle is packed and folded into an irregularly shaped mass called the **nucleoid** (shown in Figure 5.7). The DNA of the nucleoid is suspended directly in the cytoplasm with no surrounding membrane.

Many bacterial cells also contain other DNA molecules, called **plasmids,** in addition to the main chromosome of the nucleoid. Most plasmids are circular, although some are linear. Plasmids have replication origins and are duplicated and distributed to daughter cells together with the bacterial chromosome during cell division.

Although bacterial DNA is not organized into nucleosomes, certain positively charged proteins do combine with bacterial DNA. Some of these proteins help organize the DNA into loops, thereby providing some compaction of the molecule. Bacterial DNA also combines with many types of genetic regulatory proteins that have functions similar to those of the nonhistone proteins of eukaryotes (see Chapter 16).

With this description of bacterial DNA organization, our survey of DNA structure and its replication and organization is complete. The next chapter revisits the same structures and discusses how they function in the expression of information encoded in the DNA.

STUDY BREAK 14.5

1. What is the structure of the nucleosome?
2. What is the role of histone H1 in eukaryotic chromosome structure?

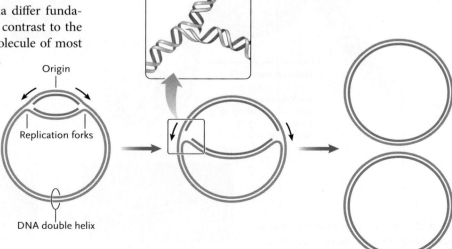

FIGURE 14.22 Replication from a single origin of replication in a circular bacterial chromosome.
© Cengage Learning 2014

Does size matter?

In this chapter, you learned that the addition of DNA onto telomeres by the enzyme telomerase can counteract the shortening of chromosomes that is predicted by the end replication problem. In humans, telomerase is active in cells destined to become sperm and egg, and in stem cells (cells capable of cell division to replenish tissues such as bone marrow) and other highly proliferative cells. It is also active in most cancers and immortalized cell lines (cells that can grow and divide indefinitely in a culture dish). The presence of telomerase does not result in ever-growing telomeres, however. Cancer cells maintain a specific average telomere length, which can be shorter than the average telomere length for normal somatic cells. Furthermore, some species express high amounts of telomerase activity in all tissues, yet maintain species-specific average telomere lengths.

How do cells measure the length of the telomeric DNA tract?

Several proteins that assemble at the telomere have been identified, including three that bind specifically to the telomeric repeat DNA sequence, and are thought to regulate telomere length. Exactly how this is accomplished is an active area of research. The function of these proteins can be tested by manipulating the protein in question with contemporary genetic approaches and observing effects on the length of the telomeric DNA. For example, if a protein's role is to prevent telomerase access to the telomere, preventing the protein from being synthesized should result in telomere elongation; if it functions to recruit telomerase, telomeres should shorten. Proteins that bind to the double-stranded telomeric DNA sequences can somehow affect access of telomerase to the single-stranded 3′ end. Biochemical experiments and studies of the three-dimensional structure of these proteins are being used to dissect how information about the length of the double-stranded telomere region is communicated to the very terminus where telomerase acts.

Shortened telomeres, whether due to cell division in the absence of telomerase or experimental manipulations of telomere proteins, resemble broken chromosomes and can prevent the cell from progressing through the cell cycle. When normal human cells lacking telomerase are cultured in a dish, they undergo only a finite number of cell divisions. What kind of evidence would rigorously link this behavior specifically to shortened telomeres? A more complicated question is whether short telomeres are involved in organismal aging. Model organisms lacking the telomerase gene are an important tool to address this question, as are humans who have rare defects in telomerase function. Such cases display physiological signs of premature aging.

Courtesy of Janis Shampay

THINK LIKE A SCIENTIST Contradictory results have been reported regarding the effect of exercise on telomerase and/or telomere length in renewable tissues like blood. In some instances, telomerase levels have increased; in others, no change in telomerase is detectable, but telomere length increases. How might telomere length change if telomerase levels remain modest?

Janis Shampay is a professor of biology at Reed College. Her research interests include the regulation of telomere metabolism and conservation of telomere protein function in non-mammalian model systems. Learn more about her work at http://academic.reed.edu/biology/professors/jshampay/index.html.

REVIEW KEY CONCEPTS

To access the course materials and companion resources for this text, please visit www.cengagebrain.com.

14.1 Establishing DNA as the Hereditary Molecule

- Griffith found that a substance derived from killed virulent *Streptococcus pneumoniae* bacteria could transform nonvirulent living *S. pneumoniae* bacteria to the virulent type (Figure 14.2).
- Avery and his coworkers showed that DNA, and not protein or RNA, was the molecule responsible for transforming *S. pneumoniae* bacteria into the virulent form.
- Hershey and Chase showed that the DNA of a phage, not the protein, enters bacterial cells to direct the life cycle of the virus. Taken together, the experiments of Griffith, Avery and his coworkers, and Hershey and Chase established that DNA is the hereditary molecule (Figure 14.3).

  Animation: Hershey-Chase experiments

  Animation: Griffith's experiment

14.2 DNA Structure

- Watson and Crick discovered that a DNA molecule consists of two polynucleotide chains twisted around each other into a right-handed double helix. Each nucleotide of the chains consists of deoxyribose, a phosphate group, and either adenine, thymine, guanine, or cytosine. The deoxyribose sugars are linked by phosphate groups to form an alternating sugar–phosphate backbone. The two strands are held together by adenine–thymine (A–T) and guanine–cytosine (G–C) base pairs. Each full turn of the double helix involves 10 base pairs (Figures 14.4 and 14.6).
- The two strands of the DNA double helix are antiparallel.

  Animation: Structure of DNA

  Animation: Semi Conservative Replication

14.3 DNA Replication

- DNA is duplicated by semiconservative replication, in which the two strands of a parental DNA molecule unwind and each serves as a template for the synthesis of a complementary copy (Figures 14.7–14.9).

- DNA replication is catalyzed by several enzymes. Helicase unwinds the DNA; primase synthesizes an RNA primer used as a starting point for nucleotide assembly by DNA polymerases. DNA polymerases assemble nucleotides into a chain one at a time, in a sequence complementary to the sequence of bases in the template strand. After a DNA polymerase removes the primers and fills in the resulting gaps, DNA ligase closes the remaining single-strand nicks (Figures 14.10–14.13 and 14.15).

- As the DNA helix unwinds, only one template strand runs in a direction allowing the new DNA strand to be made continuously in the direction of unwinding. The other template strand is copied in short lengths that run in the direction opposite to unwinding. The short lengths produced by this discontinuous replication are then linked into a continuous strand (Figures 14.14 and 14.15).

- DNA synthesis begins at sites that act as replication origins and proceeds from the origins as two replication forks moving in opposite directions (Figures 14.16 and 14.17).

- The ends of eukaryotic chromosomes consist of telomeres, short sequences repeated hundreds to thousands of times. These repeats provide a buffer against chromosome shortening during replication. Although most somatic cells show this chromosome shortening, some cell types do not because they have a telomerase enzyme that adds telomere repeats to the chromosome ends (Figure 14.18).

Animation: DNA Replication

14.4 Mechanisms That Correct Replication Errors

- In proofreading, the DNA polymerase reverses and removes the most recently added base if it is mispaired as a result of a replication error. The enzyme then resumes DNA synthesis in the forward direction (Figure 14.19).

- In DNA mismatch repair, enzymes recognize distorted regions caused by mispaired base pairs and remove a section of DNA that includes the mispaired base from the newly synthesized nucleotide chain. A DNA polymerase then resynthesizes the section correctly, using the original template chain as a guide (Figure 14.20).

Animation: Base-pair substitution

14.5 DNA Organization in Eukaryotes and Prokaryotes

- Eukaryotic chromosomes consist of DNA complexed with histone and nonhistone proteins.

- In eukaryotic chromosomes, DNA is wrapped around a core consisting of two molecules each of histones H2A, H2B, H3, and H4 to produce a nucleosome. Linker DNA connects adjacent nucleosomes. The chromosome structure in this form is the 10-nm chromatin fiber. The binding of histone H1 causes the nucleosomes to package into a coiled structure called the 30-nm chromatin fiber (Figure 14.21).

- Chromatin is distributed between euchromatin, a loosely packed region in which genes are active in RNA transcription, and heterochromatin, densely packed masses in which genes, if present, are inactive. Chromatin also folds and packs to form thick, rodlike chromosomes during nuclear division.

- Nonhistone proteins help control the expression of individual genes.

- The bacterial chromosome is a closed, circular molecule of DNA; it is packed into the nucleoid region of the cell. Replication begins from a single origin and proceeds in both directions. Many bacteria also contain plasmids, which replicate independently of the host chromosome (Figure 14.22).

- Bacterial DNA is organized into loops through interaction with proteins. Other proteins similar to eukaryotic nonhistones regulate gene activity in prokaryotes.

Animation: Chromosome structural organization

UNDERSTAND & APPLY

Test Your Knowledge

1. Working on the Amazon River, a biologist isolated DNA from two unknown organisms, P and Q. He discovered that the adenine content of P was 15% and the cytosine content of Q was 42%. This means that:
   a. the amount of guanine in P is 15%.
   b. the amount of guanine and cytosine combined in P is 70%.
   c. the amount of adenine in Q is 42%.
   d. the amount of thymine in Q is 21%.
   e. it takes more energy to unwind the DNA of P than the DNA of Q.

2. The Hershey and Chase experiment showed that phage:
   a. ^{35}S entered bacterial cells.
   b. ^{32}P remained outside of bacterial cells.
   c. protein entered bacterial cells.
   d. DNA entered bacterial cells.
   e. DNA mutated in bacterial cells.

3. Pyrimidines built from a single carbon ring are:
   a. cytosine and thymine.
   b. adenine, cytosine, and guanine.
   c. adenine and thymine.
   d. cytosine and guanine.
   e. adenine and guanine.

4. Which of the following statements about DNA replication is *false*?
   a. Synthesis of the new DNA strand is from 3' to 5'.
   b. Synthesis of the new DNA strand is from 5' to 3'.
   c. DNA unwinds, primase adds RNA primer, and DNA polymerases synthesize the new strand and remove the RNA primer.
   d. Many initiation points exist in each eukaryotic chromosome.
   e. Okazaki fragments are synthesized in the opposite direction from the direction in which the replication fork moves.

5. Which of the following statements about DNA is *false*?
   a. Phosphate is linked to the 5' and 3' carbons of adjacent deoxyribose molecules.
   b. DNA is bidirectional in its synthesis.
   c. Each side of the helix is antiparallel to the other.
   d. The binding of adenine to thymine is through three hydrogen bonds.
   e. Avery identified DNA as the transforming factor in crosses between smooth and rough bacteria.

6. In the Meselson and Stahl experiment, the DNA in the parental generation was all $^{15}N^{15}N$, and after one round of replication, the DNA was all $^{15}N^{14}N$. What DNAs were seen after three rounds of replication, and in what ratio were they found?
   a. one $^{15}N^{14}N$; one $^{14}N^{14}N$
   b. one $^{15}N^{14}N$; two $^{14}N^{14}N$
   c. one $^{15}N^{14}N$; three $^{14}N^{14}N$
   d. one $^{15}N^{14}N$; four $^{14}N^{14}N$
   e. one $^{15}N^{14}N$; seven $^{14}N^{14}N$

7. During replication, DNA is synthesized in a $5'{\rightarrow}3'$ direction. This implies that:
   a. the template is read in a $5'{\rightarrow}3'$ direction.
   b. successive nucleotides are added to the $3'$–OH end of the newly forming chain.
   c. because both strands are replicated nearly simultaneously, replication must be continuous on both.
   d. ligase unwinds DNA in a $5'{\rightarrow}3'$ direction.
   e. primase acts on the $3'$ end of the replicating strand.

8. Telomerase:
   a. is active in many cancer cells.
   b. is more active in adult than embryonic cells.
   c. complexes with the ribosome to form telomeres.
   d. acts on unique genes called telomeres.
   e. shortens the ends of chromosomes.

9. Mismatch repair is the ability:
   a. to seal Okazaki fragments with ligase into a continual DNA strand.
   b. of primase to remove the RNA primer and replace it with the correct DNA.
   c. of some enzymes to sense the insertion of an incorrect nucleotide, remove it, and use a DNA polymerase to insert the correct one.
   d. to correct mispaired chromosomes in prophase I of meiosis.
   e. to remove worn-out DNA by telomerase and replace it with newly synthesized nucleotides.

10. Which of the following does not accurately characterize bacterial DNA?
   a. The bacterial chromosome is a closed, circular molecule of DNA.
   b. Bacterial DNA is organized into nucleosomes.
   c. The primary DNA molecule of most bacteria is circular.
   d. DNA is organized more simply in bacteria than in eukaryotes.
   e. Positively charged proteins can combine with bacterial DNA.

Discuss the Concepts

1. Chargaff's data suggested that adenine pairs with thymine and guanine pairs with cytosine. What other data available to Watson and Crick suggested that adenine–guanine and cytosine–thymine pairs normally do not form?

2. Eukaryotic chromosomes can be labeled by exposing cells to radioactive thymidine during the S phase of interphase. If cells are exposed to radioactive thymidine during the S phase, would you expect both or only one of the sister chromatids of a duplicated chromosome to be labeled at metaphase of the following mitosis (see Section 10.2)?

3. If the cells in question 2 finish division and then enter another round of DNA replication in a medium that has been washed free of radioactive label, would you expect both or only one of the sister chromatids of a duplicated chromosome to be labeled at metaphase of the following mitosis?

4. During replication, an error uncorrected by proofreading or mismatch repair produces a DNA molecule with a base mismatch at the indicated position:

$$\text{AATTCCGACTCCTATGG}$$
$$\text{TTAAGGTTGAGGATACC}$$
$$\uparrow$$

The mismatch results in a mutation. This DNA molecule is received by one of the two daughter cells produced by mitosis. In the next round of replication and division, the mutation appears in only one of the two daughter cells. Develop a hypothesis to explain this observation.

5. Strains of bacteria that are resistant to an antibiotic sometimes appear spontaneously among other bacteria of the same type that are killed by the antibiotic. In view of the information in this chapter about DNA replication, what might account for the appearance of this resistance?

Design an Experiment

Design an experiment using radioactive isotopes to show that the process of bacterial transformation involves DNA and not protein.

Interpret the Data

Some cancer treatments target rapidly dividing cells while leaving nonproliferating cells undisturbed. The chemicals 5-fluorouracil (5-FU) and cisplatin (CDDP) are drugs that work in this way. 5-FU inhibits DNA replication, while CDDP binds to DNA causing changes that cannot be corrected by DNA repair enzymes so that programmed cell death (see Chapter 7) is triggered. Researchers suspected that these drugs might be useful in treating human gastric cancer and tested their effectiveness in gastric cancer cells growing in culture. They added 5-FU alone, or 5-FU and CDDP in various timed combinations (schedules) to cultured gastric cancer cells and measured the inhibitory effects of the drugs on cell proliferation compared with untreated cells. The results in the **Figure** show for each schedule the % cell proliferation, meaning the proliferation of treated cells/proliferation of control, untreated cells $\times$ 100%.

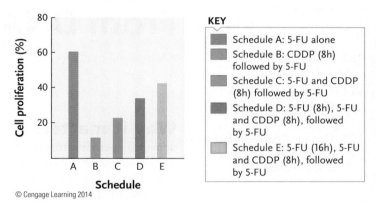

KEY
Schedule A: 5-FU alone
Schedule B: CDDP (8h) followed by 5-FU
Schedule C: 5-FU and CDDP (8h) followed by 5-FU
Schedule D: 5-FU (8h), 5-FU and CDDP (8h), followed by 5-FU
Schedule E: 5-FU (16h), 5-FU and CDDP (8h), followed by 5-FU

© Cengage Learning 2014

1. Which drug schedule was the most effective?

2. How did the drug schedule in the most effective treatment differ from the schedules in all the other treatments?

Source: H. Cho, *et al.* 2002. In-vitro effect of a combination of 5-fluorouracil (5-FU) and cisplatin (CDDP) on human gastric cancer cell lines: timing of cisplatin treatment. *Gastric Cancer* 5:43–46.

Apply Evolutionary Thinking

The amino acid sequences of the DNA polymerases found in bacteria show little similarity to those of the DNA polymerases found in eukaryotes and in archaea. By contrast, the amino acid sequences of the DNA polymerases of eukaryotes and archaea show a high degree of similarity. Interpret these observations from an evolutionary point of view.

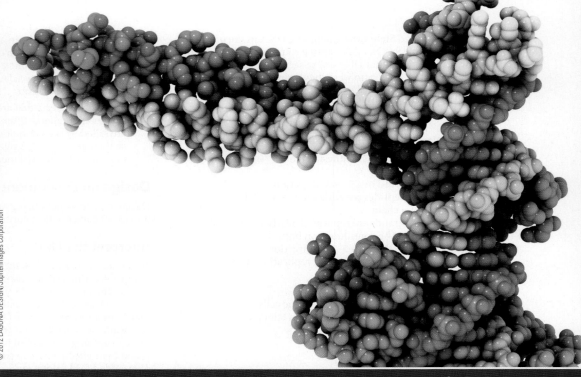

© 2012 LAGUNA DESIGN/Jupiterimages Corporation

Two transcription factors bound to DNA. In eukaryotes, transcription factors bound to the promoter of a gene recruit RNA polymerase to transcribe the gene.

15

From DNA to Protein

Why it matters . . . The marine mussel *Mytilus* **(Figure 15.1)** lives in one of the most demanding environments on Earth—it clings permanently to rocks pounded by surf day in and day out, constantly in danger of being dashed to pieces or torn loose by foraging predators. The mussel is remarkably resistant to disturbance, however; if you try to pry one loose you will find how difficult it is to tear the tough, flexible fibers that hold it fast, or even to cut them with a knife.

The fibers holding mussels to the rocks are a complex of proteins secreted by the muscular foot of the animal. The proteins, which include *keratin* (an intermediate filament protein; discussed in Section 5.3) form a tough, adhesive material called *byssus*.

Byssus is a premier underwater adhesive. It interests biochemists, adhesive manufacturers, dentists, and surgeons looking for better ways to hold repaired body parts together. Genetic engineers have inserted segments of mussel DNA into yeast cells, which reproduce in large numbers and serve as "factories" translating the mussel genes into byssus and other proteins. With byssus produced in this way, investigators are learning how to use or imitate the mussel glue for human needs. This work, like the mussel's own byssus-building, starts with one of life's universal truths: *every protein is assembled on ribosomes according to instructions that are copied from DNA.*

In this chapter, we trace the reactions by which proteins are made, beginning with the instructions encoded in DNA and leading through RNA to the sequence of amino acids in a protein. Many enzymes and other proteins are players as well as products in this story, as are several kinds of RNA and the

FIGURE 15.1 The marine mussel *Mytilus.*
WildPictures/Alamy

310

cell's protein-making molecular machines, the ribosomes. The same basic steps produce the proteins of all organisms. Our discussion begins with an overview of the entire process, starting with DNA and ending with a finished protein.

15.1 The Connection between DNA, RNA, and Protein

Genes that code for proteins are called *protein-coding genes.* In this section you will learn how such genes encode—specify the amino acid sequence of—proteins. This section also presents an overview of the molecular steps from gene to protein: transcription and translation.

Proteins Are Specified by Genes

How do scientists know that proteins are specified by genes? Two key pieces of research involving defects in metabolism proved this connection unequivocally. The first began in 1896 with Archibald Garrod, an English physician. He studied *alkaptonuria,* a human disease that does little harm but is easily detected: the patient's urine turns black in air. Garrod and an English geneticist, William Bateson, studied families of patients with the disease and concluded that it is an inherited trait. Garrod also found that people with alkaptonuria excrete a particular chemical in their urine. It is this chemical that turns black in air. Garrod deduced that normal people can metabolize the chemical, whereas people with alkaptonuria cannot. In 1908 Garrod concluded that the disease was an *inborn error of metabolism.* He did not know it at the time, but alkaptonuria results from a change in a gene that encodes an enzyme that metabolizes a key chemical. The altered gene causes a defect in the function of the enzyme, which leads to the disease phenotype. Garrod's work was the first evidence of a specific relationship between genes and metabolism.

In the second piece of research, George Beadle and Edward Tatum, working at Stanford University in the 1940s with the orange bread mold *Neurospora crassa,* obtained results showing a direct relationship between genes and enzymes. Beadle and Tatum chose *Neurospora* for their work because it is a haploid fungus, with simple nutritional needs. That is, wild-type *Neurospora*—the form of the mold found in nature—grows readily on a minimal medium (MM) consisting of a number of inorganic salts, sucrose, and vitamins. The researchers reasoned that the fungus uses the simple chemicals in MM to synthesize all of the more complex molecules needed for growth and reproduction, including amino acids for proteins and nucleotides for DNA and RNA.

Beadle and Tatum exposed spores of wild-type *Neurospora* to X-rays. An X-ray is a *mutagen,* an agent that causes mutations. They found that some of the treated spores would not germinate and grow on MM unless they supplemented the medium with additional nutrients, such as amino acids or vitamins. Mutant strains that require a nutrient supplement in the

MM to grow are called **auxotrophs** (*auxo* = increased; *troph* = eater) or *nutritional mutants.* Beadle and Tatum hypothesized that each auxotrophic strain had a defect in a gene that codes for an enzyme needed to synthesize a particular nutrient. The wild-type strain could make the nutrient for itself from raw materials in the MM, but the mutant strain could grow only if the researchers supplied the nutrient. By testing each mutant strain on MM with a single added nutrient, they discovered what specific nutrient the strain needed to grow and, therefore, generally what gene defect it had. For example, a mutant that requires the addition of the amino acid arginine to grow has a defect in a gene for an enzyme involved in the synthesis of arginine. Such arginine auxotrophs are known as *arg* mutants.

The synthesis of arginine in the cell from raw materials is a multistep "assembly-line" process with a different enzyme catalyzing each step. Beadle and Tatum studied four *arg* mutants—*argE, argF, argG,* and *argH*—to determine the metabolic defect each had; that is, where in the arginine synthesis pathway each was blocked. Their experimental approach was to test whether the *arg* mutants, all of which could grow on MM + arginine but not on MM, could also grow on MM supplemented with compounds known to be involved in arginine synthesis **(Figure 15.2).** Their analysis of *arg* auxotrophs and of auxotrophs of other kinds demonstrated a direct relationship between genes and enzymes, which they put forward as the **one gene–one enzyme hypothesis.** Their experiment was a keystone in the development of molecular biology. As a result of their work, they were awarded a Nobel Prize in 1958.

As you learned in Chapter 3, enzymes are just one form of proteins, the amino acid-containing macromolecules that carry out many vital functions in living organisms. A functional protein consists of one or more subunits, called *polypeptides.* The protein hemoglobin, for instance, is made up of four polypeptides, two each of an α subunit and a β subunit. Hemoglobin's ability to transport oxygen is a functional property belonging only to the complete protein, and not to any of the polypeptides individually. A different gene encodes each distinct polypeptide, meaning that two different genes are needed to specify the hemoglobin protein: one for the α polypeptide and one for the β polypeptide. Since some proteins consist of more than one polypeptide, and not all proteins are enzymes, Beadle and Tatum's hypothesis was updated to the **one gene–one polypeptide hypothesis.** It is important to keep in mind the distinction between a protein, the functional molecule, and a polypeptide, the molecule specified by a gene, as we discuss transcription and translation in the rest of this chapter.

The Pathway from Gene to Polypeptide Involves Transcription and Translation

The pathway from gene to polypeptide has two major steps, *transcription* and *translation.* **Transcription** is the mechanism by which the information encoded in DNA is made into a complementary RNA copy. It is called transcription because the information in one nucleic acid type is transferred to another nucleic acid

FIGURE 15.2 **Experimental Research**

Relationship between Genes and Enzymes

Question: Are enzymes specified by genes?

Experiment: Beadle and Tatum isolated auxotrophic mutants of the orange bread mold *Neurospora crassa*. Auxotrophic mutants require a nutritional supplement added to MM (minimal medium) to grow. They analyzed arginine auxotrophs—*arg* mutants—to determine the relationship between genes and enzymes. The wild type grows on MM, whereas *arg* mutants cannot. All *arg* mutants grow if arginine is added to the medium. Beadle and Tatum tested whether the *arg* mutants could also grow on MM supplemented with ornithine, citrulline, or argininosuccinate, three compounds known to be involved in the synthesis of arginine.

Results:

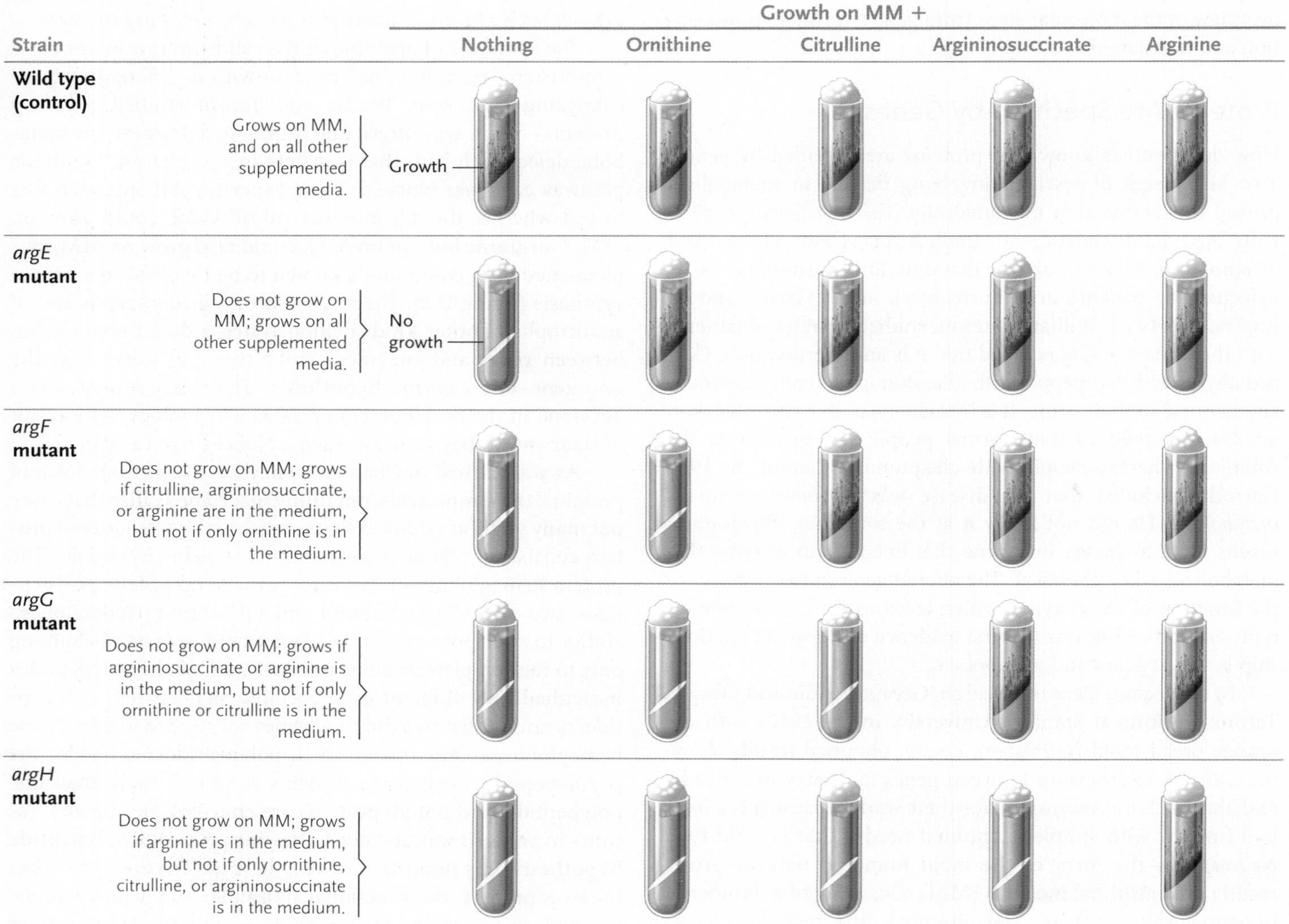

Strain		Growth on MM +				
		Nothing	Ornithine	Citrulline	Argininosuccinate	Arginine
Wild type (control)	Grows on MM, and on all other supplemented media.	Growth —				
argE **mutant**	Does not grow on MM; grows on all other supplemented media.	No growth —				
argF **mutant**	Does not grow on MM; grows if citrulline, argininosuccinate, or arginine are in the medium, but not if only ornithine is in the medium.					
argG **mutant**	Does not grow on MM; grows if argininosuccinate or arginine is in the medium, but not if only ornithine or citrulline is in the medium.					
argH **mutant**	Does not grow on MM; grows if arginine is in the medium, but not if only ornithine, citrulline, or argininosuccinate is in the medium.					

type. **Translation** is the use of the information encoded in the RNA to assemble amino acids into a polypeptide. It is called translation because the information in a nucleic acid, in the form of nucleotides, is converted into a different kind of molecule—amino acids. In 1956, Francis Crick gave the name **central dogma** to the flow of information from DNA → RNA → protein.

In transcription, the enzyme RNA polymerase copies the DNA sequence of a gene into an RNA sequence. The process is similar to DNA replication, except that only one of the two DNA strands—the **template strand**—is copied into an RNA strand, and only part of the DNA sequence of the genome is copied in any cell at any given time. A gene encoding a poly-

Conclusion: Each of the *arg* mutants shows a different pattern of growth on the supplemented MM. Beadle and Tatum deduced that the biosynthesis of arginine occurs in a number of steps, with each step controlled by a gene that encodes the enzyme for the step.

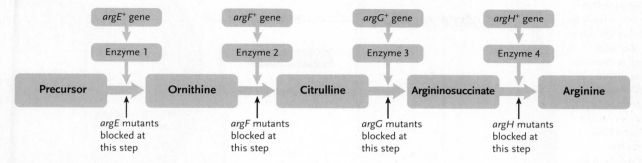

The logic is as follows, working from the end of the pathway back to its beginning:

- The *argH* mutant grows on MM + arginine, but not on MM + any of the other three compounds; this means that the mutant is blocked at the last step in the pathway that produces arginine.
- The *argG* mutant grows on MM + arginine or argininosuccinate, but not on MM + any of the other supplements; this means that *argG* is blocked in the pathway before argininosuccinate is made.
- Similarly, the *argF* mutant's growth pattern shows that it is blocked in the pathway before citrulline is made, and the *argE* mutant's growth pattern shows that it is blocked in the pathway before ornithine is made.

In this way Beadle and Tatum worked out the genetic control of the arginine biosynthesis pathway. With this and similar studies of other types of auxotrophs, they showed that there is a direct relationship between genes and enzymes.

THINK LIKE A SCIENTIST Consider two couples in which each partner has the recessive trait of albinism (lack of normal skin color; see Section 12.1). Each couple has four children. The children of the first couple all have albinism, whereas the children of the second couple all have normal skin color. How can you explain these results?

Source: G. W. Beadle and E. L. Tatum. 1942. Genetic control of biochemical reactions in *Neurospora. Proceedings of the National Academy of Sciences USA* 27:499–506.

---

peptide is a **protein-coding gene,** and the RNA transcribed from it is called **messenger RNA (mRNA).**

In translation, an mRNA associates with a *ribosome,* a particle on which amino acids are linked into polypeptide chains. As the ribosome moves along the mRNA, the amino acids specified by the mRNA are joined one by one to form the polypeptide encoded by the gene.

Transcription and translation occur in all organisms. Both processes are similar in prokaryotes and eukaryotes but there are differences **(Figure 15.3).** One key difference is that in eukaryotes, transcription in the nucleus produces a precursor-mRNA (pre-mRNA) that must be altered to generate the functional mRNA. Specifically, each end of the pre-mRNA is modified, and then extra segments within its sequence are removed by RNA processing. The result is the functional mRNA that exits the nucleus and is translated in the cytoplasm. In prokaryotes, transcription in the cytoplasm produces a functional mRNA directly, with no modifications.

It is important to understand that not all genes encode polypeptides; some encode various RNA molecules that play roles in transcription and translation, and in some other processes in the cell. You will learn about some of these important RNA molecules later in the chapter.

The Genetic Code Is Written in Three-Letter Words Using a Four-Letter Alphabet

Conceptually, the transcription of DNA into RNA is straightforward. The DNA "alphabet" consists of the four letters A, T, G, and C, representing the four bases of DNA nucleotides: adenine, thymine, guanine, and cytosine. The RNA "alphabet" consists of the four letters A, U, G, and C, representing the four RNA bases: adenine, uracil, guanine, and cytosine. In other words, the nucleic acids share three of the four bases but differ in the other one; T in DNA is equivalent to U in RNA. Translation of mRNA to form a polypeptide is more complex because, although there are four RNA bases, there are 20 amino acids. How is nucleotide information in an mRNA translated into the amino acid sequence of a polypeptide?

BREAKING THE GENETIC CODE The nucleotide information that specifies the amino acid sequence of a polypeptide is called the **genetic code.** Scientists realized that the four bases in an mRNA (A, U, G, and C) would have to be used in combinations of at least three to provide the capacity to code for 20 different amino acids. One- and two-letter words were eliminated because if the code used one-letter words, only four different

FIGURE 15.3

Transcription and translation in: (A) prokaryotes; and **(B)** eukaryotes. In prokaryotes, RNA polymerase synthesizes an mRNA molecule that is ready for translation on ribosomes. In eukaryotes, RNA polymerase synthesizes a precursor-mRNA (pre-mRNA molecule) that is processed to produce a translatable mRNA. That mRNA exits the nucleus through a nuclear pore and is translated on ribosomes in the cytoplasm.

© Cengage Learning 2014

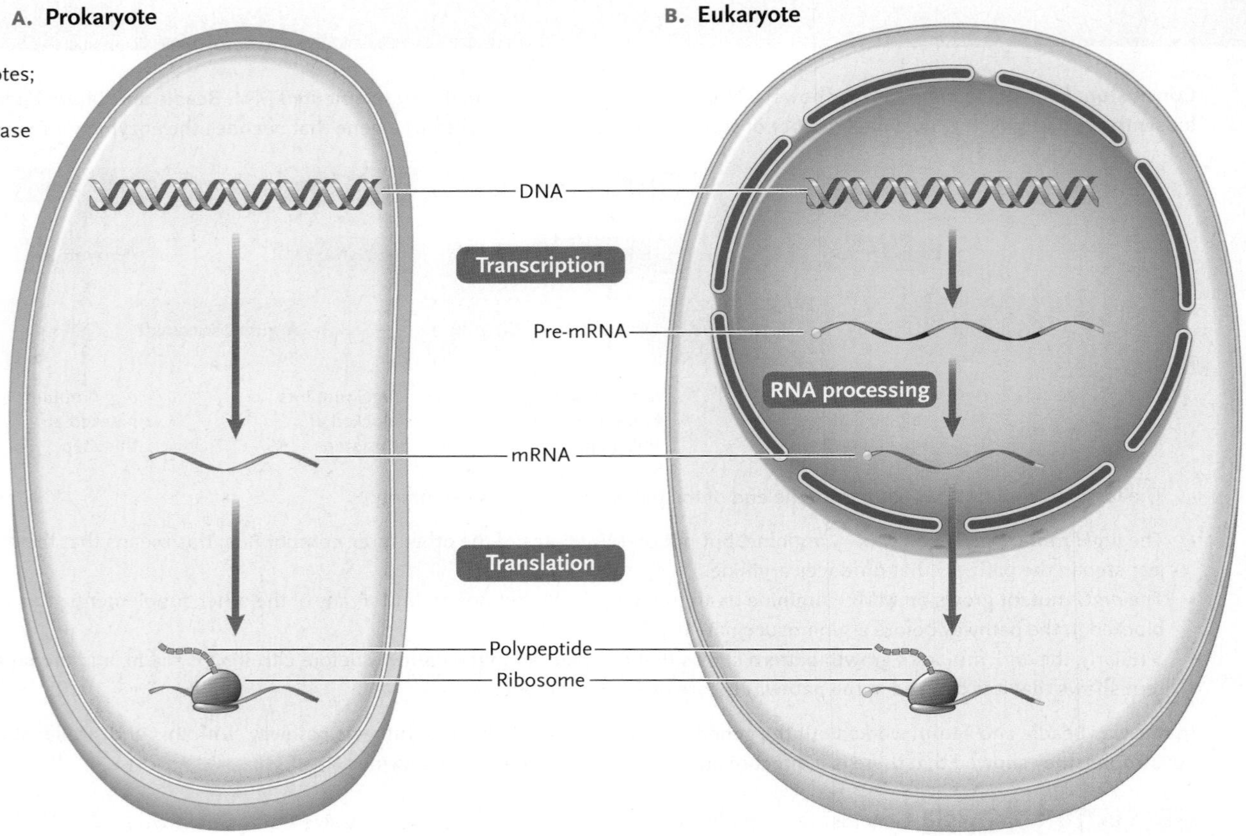

A. Prokaryote

B. Eukaryote

DNA

Transcription

Pre-mRNA

RNA processing

mRNA

Translation

Polypeptide

Ribosome

amino acids could be specified (that is, 4^1); if two-letter words were used, only 16 different amino acids could be specified (that is, 4^2). But if the code used three-letter words, 64 different amino acids could be specified (that is, 4^3), more than enough to specify 20 amino acids. Experimental research showed that the genetic code is a three-letter code; each three-letter word (triplet) of the code is called a **codon. Figure 15.4** illustrates the relationship between codons in a gene, codons in an mRNA, and the amino acid sequence of a polypeptide. The three-letter codons in DNA are first transcribed into complementary three-letter RNA codons. The process is similar to DNA replication except that in mRNA, the complement to adenine (A) in the template strand is uracil (U) instead of thymine (T) as in DNA replication.

How do the RNA codons correspond to the amino acids? The identity of most of the codons was established in 1964 by Marshall Nirenberg and Philip Leder of the National Institutes of Health (NIH). These researchers found that short, artificial mRNAs of codon length—three nucleotides—could bind to ribosomes in a test tube and cause a single transfer RNA (tRNA), with its linked amino acid, to bind to the ribosome. (As you will learn in Section 15.4, tRNAs are a special class of RNA molecules that bring amino acids to the ribosome for assembly into the polypeptide chain.) Nirenberg and Leder then made 64 of the short mRNAs, each consisting of a different, single codon. They added the mRNAs, one at a time, to a test tube containing ribosomes and all the different tRNAs, each linked

to its own amino acid. The idea was that each single-codon mRNA would link to the tRNA in the mixture that carried the amino acid corresponding to the codon. The experiment worked for 50 of the 64 codons, allowing those codons to be assigned to amino acids definitively.

Another approach, carried out in 1966 by H. Gobind Khorana and his coworkers at Massachusetts Institute of Technology, used long, artificial mRNA molecules containing only one nucleotide repeated continuously, or different nucleotides in repeating patterns. The researchers added each artificial mRNA to ribosomes in a test tube, and analyzed the sequence of amino acids in the polypeptide chain made by the ribosomes. For example, an artificial mRNA containing only uracil nucleotides in the sequence UUUUUU… resulted in a polypeptide containing only the amino acid phenylalanine: UUU must be a codon for phenylalanine. Khorana's approach, combined with the results of Nirenberg and Leder's experiments, identified the coding assignments of all the codons. Nirenberg and Khorana received a Nobel Prize in 1968 for their research in solving the nucleic acid code.

FEATURES OF THE GENETIC CODE **Figure 15.5** shows the genetic code of the 64 possible codons. By convention, scientists write the codons in the $5' \rightarrow 3'$ direction, as they appear in mRNAs, in which U substitutes for the T of DNA. Of the 64 codons, 61 specify amino acids. These are known as **sense codons.** One sense codon, AUG, specifying the amino acid methionine, is the first

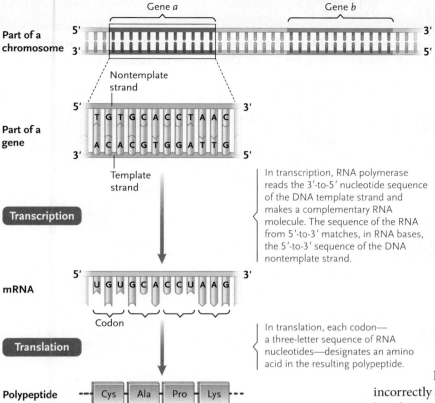

Part of a chromosome

Gene *a*   Gene *b*

5′ ... 3′
3′ ... 5′

Nontemplate strand

5′ T G T G C A C C T A A C 3′

Part of a gene

3′ A C A C G T G G A T T G 5′

Template strand

Transcription

In transcription, RNA polymerase reads the 3′-to-5′ nucleotide sequence of the DNA template strand and makes a complementary RNA molecule. The sequence of the RNA from 5′-to-3′ matches, in RNA bases, the 5′-to-3′ sequence of the DNA nontemplate strand.

mRNA

5′ U G U G C A C C U A A G 3′

Codon

Translation

In translation, each codon—a three-letter sequence of RNA nucleotides—designates an amino acid in the resulting polypeptide.

Polypeptide --- Cys — Ala — Pro — Lys ---

Amino acid

KEY

Cys = cysteine   Pro = proline
Ala = alanine   Lys = lysine

FIGURE 15.4 Relationship between a gene, codons in an mRNA, and the amino acid sequence of a polypeptide.
© Cengage Learning 2014

Only two amino acids, methionine and tryptophan, are specified by a single codon. All the rest are each represented by more than one codon, some by as many as six. In other words, there are many *synonyms* in the nucleic acid code, a feature known as **degeneracy** (also called *redundancy*). For example, UGU and UGC both specify cysteine, and CCU, CCC, CCA, and CCG all specify proline.

The genetic code is also **commaless;** that is, the words of the nucleic acid code are sequential, with no indicators such as commas or spaces to mark the end of one codon and the beginning of the next. The code can be read correctly only by starting at the right place—at the first base of the first three-letter codon at the beginning of a coded message—and reading three nucleotides at a time from this beginning codon. In other words, there is only one correct *reading frame* for each mRNA. By analogy, if you read the message SADMOMHASMOPCUTOFFBOYTOT three letters at a time, starting with the first letter of the first "codon," you would find that a mother reluctantly had her small child's hair cut. However, if you start incorrectly at the second letter of the first codon, you read the gibberish message ADM OMH ASM OPC UTO FFB OYT OT.

The code is **universal.** With a few exceptions, the same codons specify the same amino acids in all living organisms, and also in viruses. In other words, the eukaryotic translation machinery can read a prokaryotic mRNA to make the same polypeptide as in the prokaryote, and vice versa. The universality of the nucleic acid code indicates that it was established in its present form very early in the evolution of life and has remained virtually unchanged since then. (The evolution of life

codon read in an mRNA in translation in both prokaryotes and eukaryotes. In that position, AUG is called a **start codon** or **initiator codon.** The three codons that do not specify amino acids—UAA, UAG, and UGA—are **stop codons** (also called **nonsense codons** and **termination codons**) that act as "periods" indicating the end of a polypeptide-encoding "sentence." When a ribosome reaches one of the stop codons, polypeptide synthesis stops and the new polypeptide chain is released from the ribosome.

FIGURE 15.5 The genetic code, written in the form in which the codons appear in mRNA. The AUG initiator codon, which codes for methionine, is shown in green; the three terminator codons are boxed in red.
© Cengage Learning 2014

Second base of codon

First base of codon	U	C	A	G	Third base of codon
U	UUU UUC } Phe UUA UUG } Leu	UCU UCC UCA UCG } Ser	UAU UAC } Tyr UAA UAG	UGU UGC } Cys UGA UGG Trp	U C A G
C	CUU CUC CUA CUG } Leu	CCU CCC CCA CCG } Pro	CAU CAC } His CAA CAG } Gln	CGU CGC CGA CGG } Arg	U C A G
A	AUU AUC } Ile AUA AUG Met	ACU ACC ACA ACG } Thr	AAU AAC } Asn AAA AAG } Lys	AGU AGC } Ser AGA AGG } Arg	U C A G
G	GUU GUC GUA GUG } Val	GCU GCC GCA GCG } Ala	GAU GAC } Asp GAA GAG } Glu	GGU GGC GGA GGG } Gly	U C A G

KEY

Ala = alanine
Arg = arginine
Asn = asparagine
Asp = aspartic acid
Cys = cysteine
Gln = glutamine
Glu = glutamic acid
Gly = glycine
His = histidine
Ile = isoleucine
Leu = leucine
Lys = lysine
Met = methionine
Phe = phenylalanine
Pro = proline
Ser = serine
Thr = threonine
Trp = tryptophan
Tyr = tyrosine
Val = valine

Transcription has three stages: initiation, elongation, and termination. RNA polymerase moves along the gene, separating the two DNA strands to allow RNA synthesis in the 5′→3′ direction using the 3′→5′ DNA strand as template.

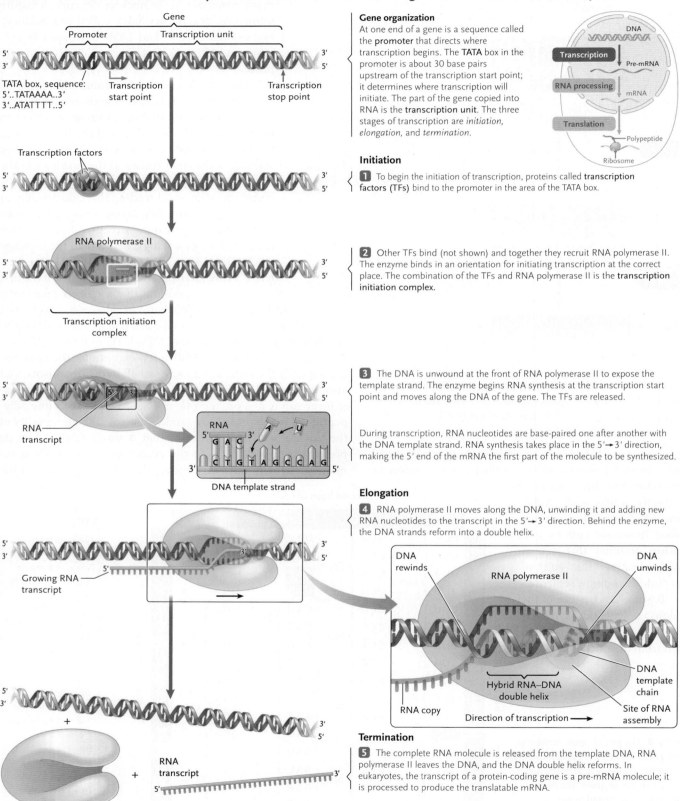

Gene organization

At one end of a gene is a sequence called the **promoter** that directs where transcription begins. The **TATA** box in the promoter is about 30 base pairs upstream of the transcription start point; it determines where transcription will initiate. The part of the gene copied into RNA is the **transcription unit**. The three stages of transcription are *initiation*, *elongation*, and *termination*.

Initiation

1 To begin the initiation of transcription, proteins called **transcription factors (TFs)** bind to the promoter in the area of the TATA box.

2 Other TFs bind (not shown) and together they recruit RNA polymerase II. The enzyme binds in an orientation for initiating transcription at the correct place. The combination of the TFs and RNA polymerase II is the **transcription initiation complex**.

3 The DNA is unwound at the front of RNA polymerase II to expose the template strand. The enzyme begins RNA synthesis at the transcription start point and moves along the DNA of the gene. The TFs are released.

During transcription, RNA nucleotides are base-paired one after another with the DNA template strand. RNA synthesis takes place in the 5′→3′ direction, making the 5′ end of the mRNA the first part of the molecule to be synthesized.

Elongation

4 RNA polymerase II moves along the DNA, unwinding it and adding new RNA nucleotides to the transcript in the 5′→3′ direction. Behind the enzyme, the DNA strands reform into a double helix.

Termination

5 The complete RNA molecule is released from the template DNA, RNA polymerase II leaves the DNA, and the DNA double helix reforms. In eukaryotes, the transcript of a protein-coding gene is a pre-mRNA molecule; it is processed to produce the translatable mRNA.

and the genetic code are discussed further in Chapter 25.) Minor exceptions to the universality of the genetic code have been found in a few organisms including a yeast, some protozoans, a prokaryote, and in the genetic systems of mitochondria and chloroplasts.

STUDY BREAK 15.1

1. **On the basis of their work with auxotrophic mutants of the fungus *Neurospora crassa*, Beadle and Tatum proposed the one gene–one enzyme hypothesis. Why is it now known as the one gene–one polypeptide hypothesis?**

2. **If the codon were five bases long, how many different codons would exist in the genetic code?**

THINK OUTSIDE THE BOOK

The section you have just read states that the genetic code is not completely universal. Use the Internet or research literature to determine what variants of the genetic code exist and in which organisms they occur.

15.2 Transcription: DNA-Directed RNA Synthesis

An organism's genome contains a large number of genes. For example, the human genome sequence has about 20,500 protein-coding genes. Transcription is the process of transferring the information coded in the DNA sequences of particular genes to complementary RNA copies. As you have learned, some of those genes are protein-coding genes that encode mRNAs that are translated; others are non-protein-coding genes that encode RNAs that are not translated, such as ribosomal RNAs (rRNAs) and transfer RNAs (tRNAs). Much of our initial understanding of transcription came from experimental research done with the model organism, *Escherichia coli*. We now know that transcription is generally similar, but not identi-

cal, in prokaryotes and eukaryotes. Throughout this section, we will point out the important differences between bacterial and eukaryotic processes.

Transcription Proceeds in Three Stages

Figure 15.6 illustrates the general organization of a eukaryotic protein-coding gene and shows how it is transcribed. The gene consists of two main parts, a **promoter,** which is a control sequence for transcription, and a **transcription unit,** the section of the gene that is copied into an RNA molecule.

Transcription takes place in three stages:

1. In **initiation,** the molecular machinery that carries out transcription assembles at the promoter and begins synthesizing an RNA copy of the gene (Figure 15.6, steps 1 and 2). The molecular machinery includes particular **transcription factors (TFs),** proteins that bind to the promoter in the area of a special sequence known as the **TATA box,** and an **RNA polymerase,** an enzyme that catalyzes the assembly of RNA nucleotides into an RNA strand. In eukaryotes, RNA polymerase II is the particular enzyme that transcribes protein-coding genes. Unlike DNA polymerases, RNA polymerases do not require a primer to start a chain.

    In the initiation process, the DNA is unwound in the front of the RNA polymerase to expose the template strand. RNA polymerase II then begins RNA synthesis at the transcription start point (Figure 15.6, stage 3); the TFs are then released. As shown in Figure 15.4, only one of the two DNA strands is copied into an mRNA strand during transcription. The RNA strand is made in the $5' \rightarrow 3'$ direction using the $3' \rightarrow 5'$ DNA strand as template. Therefore, we refer to the beginning of the RNA strand as the *5' end,* and the other end as the *3' end.* You will recall that synthesis of new DNA strands also occurs in the $5' \rightarrow 3'$ direction.

    The sequence of the mRNA strand is determined by the DNA template strand and proceeds in a manner similar to DNA replication with new RNA nucleotides being added according to complementary base pairing rules. The one exception is that, when adenine appears in the DNA template strand, a uracil is paired with it in the RNA transcript (see Figure 15.4). Uracil has base-pairing properties similar to thymine.

2. In **elongation,** in which the RNA polymerase II moves along the gene extending the RNA chain, with the DNA continuing to unwind ahead of the enzyme (Figure 15.6, step 4).

3. In **termination,** transcription ends and the RNA molecule—the transcript—and the RNA polymerase II are released from the DNA template (Figure 15.6, step 5).

Roger Kornberg of Stanford University received a Nobel Prize in 2006 for describing the molecular structure of the eukaryotic transcription apparatus and how it acts in transcription.

Similarities and differences in transcription of eukaryotic and bacterial protein-coding genes are as follows:

- Gene organization is the same, although the specific sequences in the promoter where the transcription apparatus assembles differ.
- In eukaryotes, RNA polymerase II, the enzyme that transcribes protein-coding genes, cannot bind directly to DNA; it is recruited to the promoter once proteins called **transcription factors** have first bound. In bacteria, RNA polymerase binds directly to DNA at bacterial-specific promoter sequences; the enzyme is directed to the promoter by a protein factor that is then released once transcription begins.
- Elongation is essentially identical in the two types of organisms.
- In bacteria, specific DNA sequences called **terminators** signal the end of transcription of the gene. A specific protein binds to the terminator, triggering the termination of transcription and the release of the RNA and RNA polymerase from the template. Eukaryotic DNA has no equivalent sequences. Instead, the 3′ end of the mRNA is specified by a very different process, which is discussed in the next section.

Once an RNA polymerase molecule has started transcription and progressed past the beginning of a gene, another molecule of RNA polymerase may start transcribing the same gene as soon as there is room at the promoter. In most genes, this process continues until there are many RNA polymerase molecules spaced closely along a gene, each making an RNA transcript. In this way, a large number of RNA transcripts can be produced from one gene.

Overall, transcription is similar to DNA replication. The main differences are that: (1) only one of the two DNA strands acts as a template for synthesis of the RNA transcript, instead of both for replication; and (2) only a relatively small part of a DNA molecule—the RNA-coding sequence of a gene—serves as a template, rather than all of both strands as in DNA replication.

Transcription of Non-Protein-Coding Genes Occurs in a Similar Way

Non-protein-coding genes include those for tRNAs and rRNAs (ribosomal RNAs, the RNA components of ribosomes). Genes encoding RNAs that are not translated are called **noncoding RNA genes.** In eukaryotes, whereas RNA polymerase II transcribes protein-coding genes, RNA polymerase III transcribes tRNA genes and the gene for one of the four rRNAs, and RNA polymerase I transcribes the genes for the three other rRNAs. The promoters for noncoding RNA genes differ from those of protein-coding genes, being specialized for the assembly of the transcription machinery that involves the correct RNA polymerase type. In bacteria, a single RNA polymerase transcribes all types of genes. The promoters for bacterial noncoding RNA genes are essentially the same as those of protein-coding genes.

STUDY BREAK 15.2

1. For the DNA template below, what would be the sequence of an RNA transcribed from it?

   3′-CAAATTGGCTTATTACCGGATG-5′

2. What is the role of the promoter in transcription?

15.3 Production of mRNAs in Eukaryotes

Both prokaryotic and eukaryotic mRNAs contain regions that code for proteins, along with noncoding regions that play key roles in the process of protein synthesis. In prokaryotic mRNAs, the coding region is flanked by untranslated ends, the **5′ untranslated region (5′ UTR)** and the **3′ untranslated region (3′ UTR)**. The same elements are present in eukaryotic mRNAs along with additional noncoding elements. The synthesis of mRNA in eukaryotes is the focus of this section.

Eukaryotic Protein-Coding Genes Are Transcribed into Precursor-mRNAs That Are Modified in the Nucleus

A eukaryotic protein-coding gene is typically transcribed into a **precursor-mRNA (pre-mRNA)** that must be processed in the nucleus to produce the translatable mRNA (**Figure 15.7;** and see Figures 15.3 and 15.6, step 5). The mRNA exits the nucleus and is translated in the cytoplasm.

MODIFICATIONS OF PRE-mRNA AND mRNA ENDS At the 5′ end of the pre-mRNA is the **5′ cap,** consisting of a guanine-containing nucleotide that is reversed so that its 3′-OH group faces the beginning rather than the end of the molecule. A *capping enzyme* adds the 5′ cap to the pre-mRNA soon after RNA polymerase II begins transcription. The cap, which is connected to the rest of the chain by three phosphate groups, remains when pre-mRNA is processed to mRNA. The cap is the site where ribosomes attach to mRNAs at the start of translation.

Transcription of a eukaryotic protein-coding gene is terminated differently from that of a prokaryotic gene. The eukaryotic gene has no terminator sequence that signals RNA polymerase to stop transcription. Instead, near the 3′ end of the gene is a sequence called a **polyadenylation signal** that is transcribed into the pre-mRNA. Proteins bind to this *sequence*, and cleave the pre-mRNA just downstream of that point. Then, the enzyme **poly(A) polymerase** adds a chain of 50 to 250 adenine nucleotides, one nucleotide at a time, to that 3′ end of the pre-mRNA. This string of A nucleotides, called the **poly(A) tail,** enables the mRNA produced from the pre-mRNA to be translated efficiently, and protects it from attack by RNA-digesting enzymes in the cytoplasm. If the poly(A) tail of an mRNA is removed experimentally, the mRNA is quickly degraded inside cells.

SEQUENCES INTERRUPTING THE RNA-CODING SEQUENCE

The transcription unit of a eukaryotic protein-coding gene—the RNA-coding sequence—also contains one or more non-protein-coding sequences called **introns** that interrupt the protein-coding sequence (shown in Figure 15.7). The segments of the RNA-coding sequence interrupted by the introns are called **exons.** The introns are transcribed into pre-mRNA, but removed from pre-mRNA during processing in the nucleus, so that the amino acid-coding sequence in the finished mRNA is read continuously, without interruptions. As Figure 15.7 shows, the exons in finished mRNAs contain the protein-coding sequence of the gene. The exons at the two ends of the mRNA also contain the 5′ UTR and 3′ UTR sequences.

Introns were discovered by several methods, including direct comparisons between the nucleotide sequences of mature mRNAs and either pre-mRNAs or the genes encoding them. The majority of known eukaryotic genes contain at least one intron; some contain more than 60. The original discoverers of introns, Richard Roberts of New England Biolabs and Phillip Sharp of Massachusetts Institute of Technology, received a Nobel Prize in 1993 for their findings.

Introns Are Removed During Pre-mRNA Processing to Produce the Translatable mRNA

A process called **mRNA splicing,** which occurs in the nucleus, removes introns from pre-mRNAs and joins exons together. As an illustration of mRNA splicing, **Figure 15.8** shows the processing of a pre-mRNA with a single intron to produce a mature mRNA. mRNA splicing takes place in a **spliceosome,** a complex formed between the pre-mRNA and a handful of **small ribonucleoprotein particles.** A *ribonucleoprotein particle* is a complex of RNA and proteins. The small ribonucleoprotein particles involved in mRNA splicing are located in the nucleus; each consists of a relatively short *small nuclear RNA* (snRNA) bound to a number of proteins. The particles are therefore known as snRNPs, pronounced "snurps." The snRNPs bind in a particular order to an intron in the pre-mRNA and form the active spliceosome. The spliceosome cleaves the pre-mRNA to release the intron, and joins the flanking exons.

The cutting and splicing are so exact that not a single base of an intron is retained in the finished mRNA, nor is a single base removed from the exons. Without this precision, removing introns would change the reading frame of the coding portion of the mRNA, producing gibberish from the point of a mistake onward.

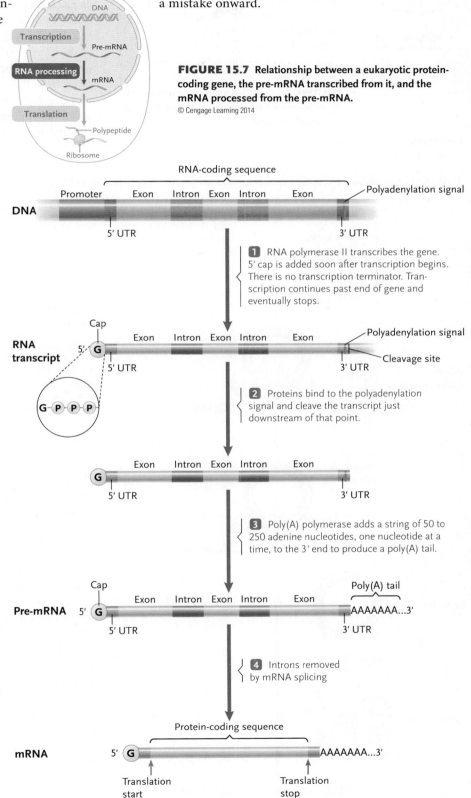

FIGURE 15.7 Relationship between a eukaryotic protein-coding gene, the pre-mRNA transcribed from it, and the mRNA processed from the pre-mRNA.
© Cengage Learning 2014

Introns Contribute to Protein Variability

Introns require elaborate cellular machinery to remove them during pre-mRNA processing. Why are they present in mRNA-encoding genes? Among a number of possibilities, introns may provide advantages by increasing the coding capacity of existing genes through a process called *alternative splicing* and by generating new proteins through a process called *exon shuffling*.

FIGURE 15.8 mRNA splicing—the removal from pre-mRNA of introns and joining of exons in the spliceosome.

© Cengage Learning 2014

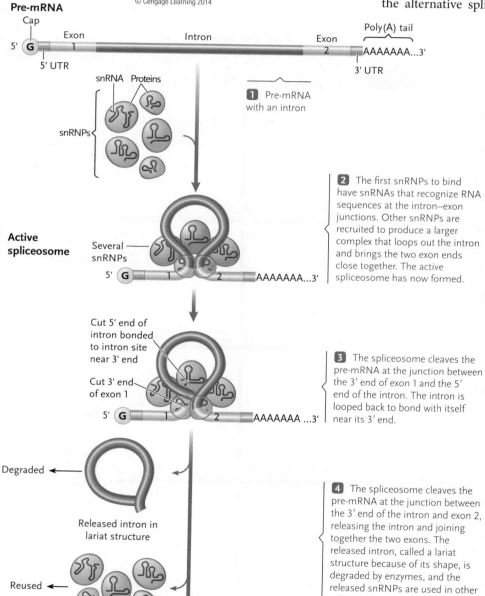

Pre-mRNA

1 Pre-mRNA with an intron

2 The first snRNPs to bind have snRNAs that recognize RNA sequences at the intron–exon junctions. Other snRNPs are recruited to produce a larger complex that loops out the intron and brings the two exon ends close together. The active spliceosome has now formed.

3 The spliceosome cleaves the pre-mRNA at the junction between the 3′ end of exon 1 and the 5′ end of the intron. The intron is looped back to bond with itself near its 3′ end.

4 The spliceosome cleaves the pre-mRNA at the junction between the 3′ end of the intron and exon 2, releasing the intron and joining together the two exons. The released intron, called a lariat structure because of its shape, is degraded by enzymes, and the released snRNPs are used in other mRNA splicing reactions.

ALTERNATIVE SPLICING Many pre-mRNAs are processed by reactions that join exons in different combinations to produce different mRNAs from a single gene. The mechanism, called **alternative splicing,** greatly increases the number and variety of proteins encoded in the cell nucleus without increasing the size of the genome. For example, geneticists estimate that three-quarters of all human pre-mRNAs are subjected to alternative splicing. In each case, the different mRNAs produced from the "parent" pre-mRNA are translated to produce a family of related proteins with various combinations of amino acid sequences derived from the exons. Each protein in the family, then, will vary to a degree in its function. Alternative splicing helps us understand why humans have only about 20,500 genes. As a result of the alternative splicing process, the number of proteins produced far exceeds the number of genes, and it is proteins that mostly direct an organism's functions.

As an example, the pre-mRNA transcript of the mammalian α-tropomyosin gene, which encodes a protein component of muscle fibers, is alternatively spliced in smooth muscle (for example, muscles of the intestine and bladder), skeletal muscle (for example, biceps, glutes), fibroblast (connective tissue cell that makes collagen), liver, and brain. The result of alternative splicing is different forms of the α-tropomyosin protein. **Figure 15.9** shows the alternative splicing of the α-tropomyosin pre-mRNA to the mRNAs found in smooth muscle and striated muscle. Exons 2 and 12 are exclusive to the smooth muscle mRNA, whereas exons 3, 10, and 11 are exclusive to the striated muscle mRNA.

The polypeptides made from the two mRNAs have some identical stretches of amino acids, along with others that differ. As we learned in Section 3.4, the primary structure of a protein—its amino acid structure—directs the folding of the chain into its three-dimensional shape. Therefore, the two forms of α-tropomyosin fold into related, but different shapes. In its role in muscle contraction in smooth muscles and striated muscles, α-tropomyosin interacts with other proteins. The interactions depend on the specific structural form of the α-tropomyosin and, as you might expect, the two forms participate in different types of muscle action; typically smooth muscles perform squeezing actions in blood vessels and internal organs, whereas skeletal muscles pull on the bones of the skeleton to move body parts.

FIGURE 15.9 Alternative splicing of the α-tropomyosin pre-mRNA to distinct mRNA forms found in smooth muscle and skeletal muscle. All of the introns are removed in both mRNA splicing pathways, but exons 3, 10, and 11 are also removed to produce the smooth muscle mRNA, and exons 2 and 12 are also removed to produce the skeletal muscle mRNA.
© Cengage Learning 2014

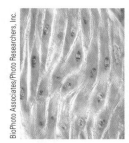

Smooth muscle
Found in walls of tubes and cavities of the body, including blood vessels, the stomach and intestine, the bladder, and the uterus. Contraction of smooth muscles typically produces a squeezing motion.

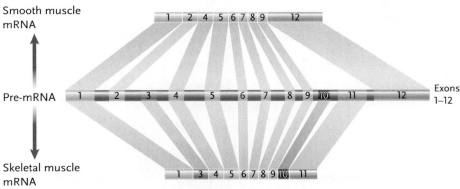

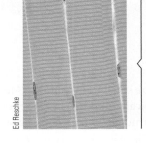

Skeletal muscle
Most muscles of this type are attached by tendons to the skeleton. Their function is locomotion and movement of body parts. The human body has more than 600 skeletal muscles, ranging in size from the small muscles that move the eyeballs, to the large muscles that move the legs.

Alternative splicing forces us to reconsider the one gene–one polypeptide hypothesis introduced earlier in the chapter. We must now accept the fact that for some genes at least, one gene may specify a number of polypeptides each of which has a related function.

EXON SHUFFLING Another advantage provided by introns may come from the fact that intron–exon junctions often fall at points dividing major functional regions in encoded proteins, as they do in the genes for antibody proteins, hemoglobin blood proteins, and the peptide hormone insulin. The functional divisions may have allowed new proteins to evolve by **exon shuffling,** a process by which existing protein regions or domains, already selected for their functions by the evolutionary process, are mixed into novel combinations to create new proteins. Evolution of new proteins by this mechanism would produce changes much more quickly and efficiently than by alterations in individual amino acids at random points. (Exon shuffling as a mechanism for the evolution of new proteins is discussed in Section 19.4.) The process resembles automobile design, in which new models are produced by combining proven parts and substructures of previous models, rather than starting with an entirely new design each year.

STUDY BREAK 15.3

1. What are the similarities and differences between pre-mRNAs and mRNAs?

2. What is the role of snRNPs in mRNA splicing?

THINK OUTSIDE THE BOOK

Explore the Internet or research literature to find a molecular model that explains how an exon and its flanking introns are removed by alternative splicing.

15.4 Translation: mRNA-Directed Polypeptide Synthesis

Translation is the reading of an mRNA to assemble amino acids into a polypeptide. In prokaryotes, translation takes place throughout the cell, whereas in eukaryotes it takes place mostly in the cytoplasm, although, as you will see, a few specialized genes are transcribed and translated in mitochondria and chloroplasts.

Figure 15.10 summarizes the translation process. For prokaryotes, the mRNA produced by transcription is immediately available for translation. For eukaryotes, the mRNA produced by splicing of the pre-mRNA first exits the nucleus, and then is translated in the cytoplasm. In translation, the mRNA associates with a ribosome, and tRNAs, another type of RNA, bring amino acids to the complex to be joined one by one into the polypeptide chain. The sequence of amino acids in the polypeptide chain is determined by the sequence of codons in the mRNA, whereas the ribosome is simply a facilitator of the translation process. The mRNA is read from the 5′ end to the 3′ end, and the polypeptide is assembled from the N-terminal end to the C-terminal end.

In this section, we start by discussing the key players in the process, the tRNAs and ribosomes, and then walk through the translation process from a start codon to a stop codon.

tRNAs Are Small, Highly Specialized RNAs That Bring Amino Acids to the Ribosome

A **transfer RNA (tRNA)** brings an amino acid to the ribosome for addition to the polypeptide chain.

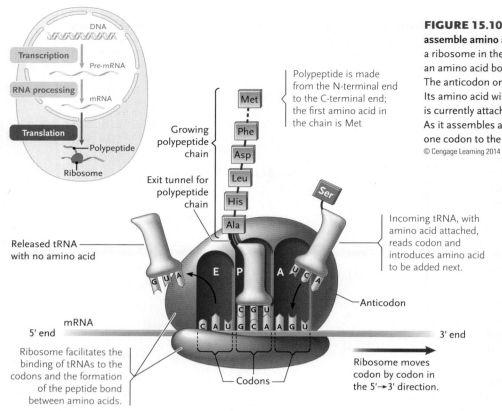

FIGURE 15.10 An overview of translation, in which ribosomes assemble amino acids into a polypeptide chain. The figure shows a ribosome in the process of translation. A tRNA molecule with an amino acid bound to it is entering the ribosome on the right. The anticodon on the tRNA will pair with the codon in the mRNA. Its amino acid will then be added to the growing polypeptide that is currently attached to the tRNA in the middle of the ribosome. As it assembles a polypeptide chain, the ribosome moves from one codon to the next along the mRNA in the 5′→3′ direction.
© Cengage Learning 2014

double-helical segments, forming in two dimensions what is known as the *cloverleaf* pattern. At the tip of one of the double-helical segments is the **anticodon,** the three-nucleotide segment that base pairs with a codon in mRNAs. Opposite the anticodon, at the other end of the cloverleaf, is a double-helical segment that links to the amino acid corresponding to the anticodon. For example, a tRNA that base pairs with the codon 5′-AGU-3′ has serine (Ser) linked to it (see Figure 15.10). The anticodon of the tRNA that pairs with this codon is 3′-UCA-5′. (The anticodon and codon pair in an antiparallel manner, as do the strands in DNA. We will write anticodons in the 3′→5′ direction to make it easy to see how they pair with codons.)

The tRNA cloverleaf folds in three dimensions into the structure shown in Figure 15.11B—it is generally referred to as

tRNA STRUCTURE tRNAs are small RNAs, about 75 to 90 nucleotides long (mRNAs are typically hundreds of nucleotides long), with a highly distinctive structure that accomplishes their role in translation **(Figure 15.11).** All tRNAs can wind into four

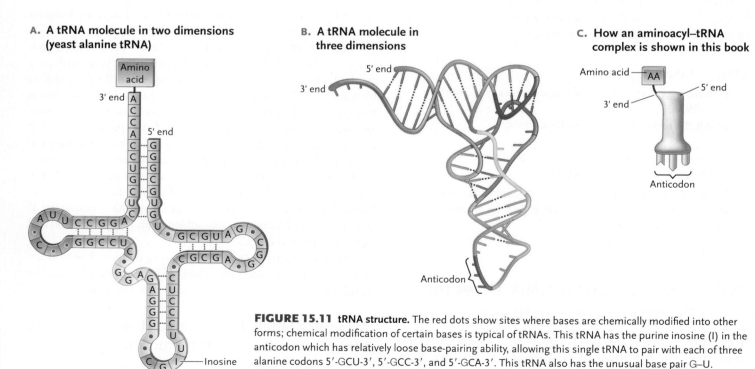

A. A tRNA molecule in two dimensions (yeast alanine tRNA)

B. A tRNA molecule in three dimensions

C. How an aminoacyl–tRNA complex is shown in this book

FIGURE 15.11 tRNA structure. The red dots show sites where bases are chemically modified into other forms; chemical modification of certain bases is typical of tRNAs. This tRNA has the purine inosine (I) in the anticodon which has relatively loose base-pairing ability, allowing this single tRNA to pair with each of three alanine codons 5′-GCU-3′, 5′-GCC-3′, and 5′-GCA-3′. This tRNA also has the unusual base pair G–U. Unusual base pairs, allowed by the greater flexibility of short RNA chains, are common in tRNAs.
© Cengage Learning 2014

an upside-down L. The anticodon and the segment binding the amino acid are located at the opposite tips of the structure.

We learned earlier that 61 of the 64 codons of the genetic code specify an amino acid. Does this mean that 61 different tRNAs read the sense codons? The answer is no. Francis Crick's **wobble hypothesis** states that the complete set of 61 sense codons can be read by fewer than 61 distinct tRNAs because of particular pairing properties of the bases in the anticodons. That is, the pairing of the anticodon with the first two nucleotides of the codon is always precise, but the anticodon has more flexibility in pairing with the third nucleotide of the codon. In many cases, the same tRNA anticodon can read codons that have either U or C in the third position; for example, a tRNA carrying phenylalanine can read both codons 5′-UUU-3′ and 5′-UUC-3′. Similarly the same tRNA anticodon can read two codons that have A or G in the third position; for example, a tRNA carrying glutamine can read

both codons 5′-CAA-3′ and 5′-CAG-3′. The special inosine purine in the alanine tRNA shown in Figure 15.11A allows even more extensive wobble by allowing the tRNA to pair with codons that have either U, C, or A in the third position.

ADDITION OF AMINO ACIDS TO THEIR CORRESPONDING tRNAs The correct amino acid must be present on a tRNA if translation is to be accurate. The process of adding an amino acid to a tRNA is called **aminoacylation** (literally, the addition of an amino acid) or **charging** (because the process adds free energy as the amino acid–tRNA combinations are formed). The finished product of charging, a tRNA linked to its "correct" amino acid, is called an **aminoacyl–tRNA.** Twenty different enzymes called **aminoacyl–tRNA synthetases**—one synthetase for each of the 20 amino acids—catalyze aminoacylation in the four steps shown in **Figure 15.12.**

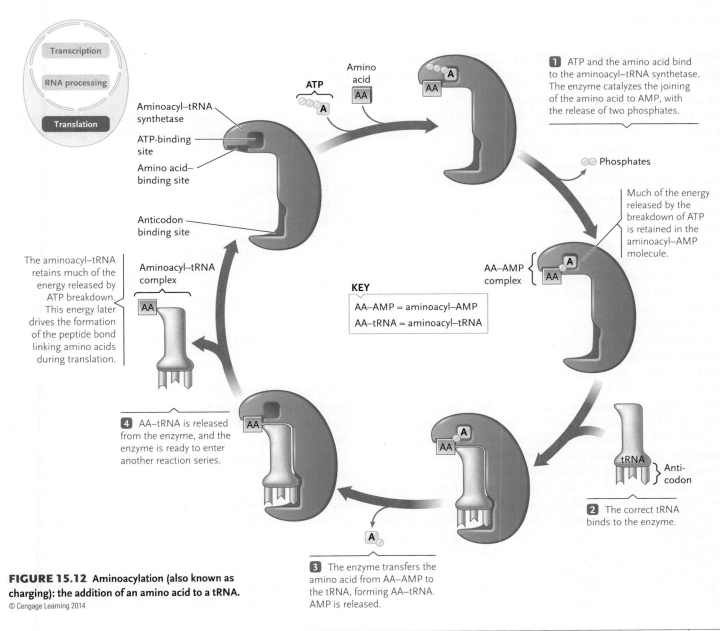

FIGURE 15.12 Aminoacylation (also known as charging): the addition of an amino acid to a tRNA.
© Cengage Learning 2014

FIGURE 15.13 Ribosome structure. **(A)** Computer model of a ribosome in the process of translation. **(B)** The ribosome as we will show it during translation.
© Cengage Learning 2014

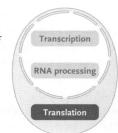

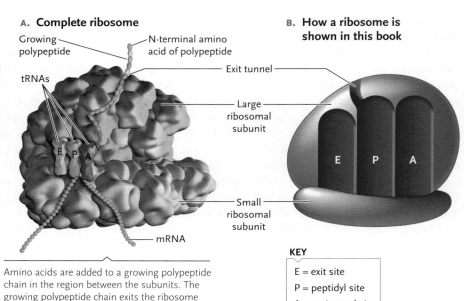

A. Complete ribosome

Growing polypeptide

N-terminal amino acid of polypeptide

tRNAs

Exit tunnel

Large ribosomal subunit

E P A

Small ribosomal subunit

mRNA

Amino acids are added to a growing polypeptide chain in the region between the subunits. The growing polypeptide chain exits the ribosome through the exit tunnel in the large subunit.

B. How a ribosome is shown in this book

E P A

KEY

E = exit site
P = peptidyl site
A = aminoacyl site

With the tRNAs attached to their corresponding amino acids, our attention moves to the ribosome, where the amino acids are removed from their tRNAs and linked into polypeptide chains.

Ribosomes Are rRNA–Protein Complexes That Work as Automated Protein Assembly Machines

Ribosomes are ribonucleoprotein particles that carry out protein synthesis by translating mRNA into chains of amino acids. A ribosome reads the codons on an mRNA and joins the appropriate amino acids to make a polypeptide chain.

In prokaryotes, ribosomes carry out their assembly functions throughout the cell. In eukaryotes, ribosomes function in the cytoplasm, either suspended freely in the cytoplasmic solution, or attached to the membranes of the endoplasmic reticulum (ER), the system of tubular or flattened sacs in the cytoplasm (discussed in Section 5.3).

A finished ribosome is made up of two parts of dissimilar size, called the *large* and *small ribosomal subunits* **(Figure 15.13)**. Each subunit is a combination of **ribosomal RNA (rRNA)** and ribosomal proteins.

Prokaryotic and eukaryotic ribosomes are similar in function, and quite similar in structure. However, certain differences in their molecular structure, particularly in the ribosomal proteins, give them distinguishable properties. For example, the antibiotics streptomycin and erythromycin are effective antibacterial agents because they inhibit the function of the bacterial ribosome, but not the eukaryotic ribosome.

In translation, the mRNA follows a bent path through a groove in the ribosome. The ribosome also has binding sites where tRNAs interact with the mRNA (see Figure 15.13 and refer also to Figure 15.10). The **A site** (aminoacyl site) is where the incoming aminoacyl–tRNA carrying the next amino acid to be added to the polypeptide chain binds to the RNA. The **P site** (peptidyl site) is where the tRNA carrying the growing polypeptide chain is bound. The **E site** (exit site) is where a tRNA, now without an attached amino acid, binds before exiting the ribosome. You will learn more about these functional sites as we discuss the stages of translation.

Translation Initiation Brings the Ribosomal Subunits, an mRNA, and the First Aminoacyl–tRNA Together

Translation is similar in bacteria and eukaryotes. We will present translation from a eukaryotic perspective and indicate how it differs in bacteria.

There are three major stages of translation:

1. During **initiation,** the translation components assemble on the start codon of the mRNA.
2. In **elongation,** the assembled complex reads the string of codons in the mRNA one at a time while joining the specified amino acids into the polypeptide.
3. **Termination** completes the translation process when the complex disassembles after the last amino acid of the polypeptide specified by the mRNA has been added to the polypeptide.

The energy of GTP hydrolysis to GDP + P_i fuels each of the three stages.

TRANSLATION INITIATION **Figure 15.14** illustrates the steps of translation initiation in eukaryotes. Each initiation step is aided by proteins called **initiation factors (IFs).** The IFs are released when initiation is complete in step 3.

In bacteria, translation initiation is similar in using a special initiator Met–tRNA, GTP, and IFs, but the way in which the ribosome assembles at the start codon is different. Rather than scanning from the 5′ end of the mRNA, the small ribosomal subunit, the initiator Met–tRNA, GTP, and IFs bind directly to the region of the mRNA with the AUG start codon. A **ribosome binding site**—a short, specific RNA sequence—just upstream of the start codon directs the small ribosomal subunit in this initiation step. The large ribosomal subunit then binds to the small subunit to complete the ribosome. GTP hydrolysis then releases the IFs.

After the initiator tRNA pairs with the AUG initiator codon, the subsequent stages of translation simply read the codons one at a time on the mRNA. The initiator tRNA–AUG pairing thus establishes the correct **reading frame**—the series of codons for the polypeptide encoded by the mRNA.

Polypeptide Chains Grow during the Elongation Stage of Translation

The central reactions of translation take place in the elongation stage, which adds amino acids one at a time to a growing polypeptide chain. The individual steps of elongation depend on the binding properties of the P, A, and E sites of the ribosome, as we will see.

Figure 15.15 shows the elongation cycle of translation. The cycle begins at the point when an initiator tRNA with its attached methionine is bound to the P site, and the A site is empty (top of figure). Four steps occur during the cycle:

1. The appropriate aminoacyl–tRNA binds to the codon in the A site of the ribosome (step 1). This binding is facilitated by a protein **elongation factor (EF)** that is bound to the aminoacyl–tRNA and that is released once the tRNA binds to the codon.

2. A peptide bond is formed between the C-terminal end of the growing polypeptide on the P site tRNA and the amino acid on the A site tRNA (step 2). **Peptidyl transferase,** part of the large ribosomal subunit, catalyzes this reaction. When peptide bond formation is complete, there is an empty tRNA (a tRNA without a bound amino acid) in the P site and the tRNA in the A site has two amino acids—the growing polypeptide chain—attached to it. The P site binds that is, A tRNA linked to a growing polypeptide chain containing two or more amino acids is called a **peptidyl–tRNA.**

3. The ribosome translocates; that is, it moves to the next codon along the mRNA, while the tRNAs remain bound to the mRNA in their same positions (step 3). An EF is used for

this step and then is released. After translocation, the tRNA that was in the P site is now in the E site, the peptidyl–tRNA that was in the A site is now in the P site, and the A site is now empty.

4. The empty tRNA is released from the E site and the ribosome is ready to begin the next round of the elongation cycle.

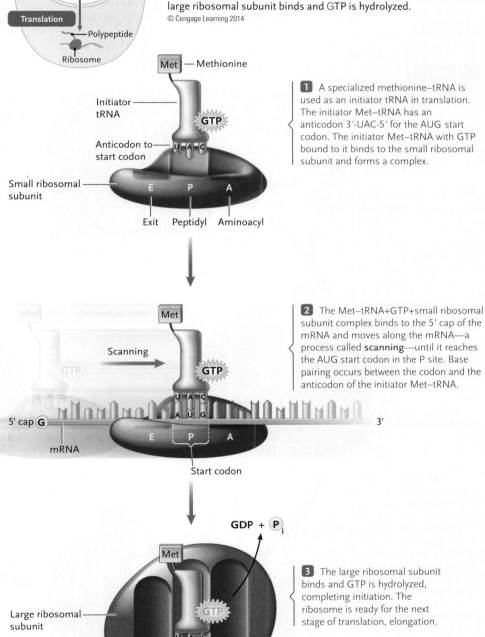

FIGURE 15.14 Translation initiation in eukaryotes. Protein initiation factors (IFs) participate in the event but, for simplicity, they are not shown in the figure. The IFs are released when the large ribosomal subunit binds and GTP is hydrolyzed.

© Cengage Learning 2014

1 A specialized methionine–tRNA is used as an initiator tRNA in translation. The initiator Met–tRNA has an anticodon 3'-UAC-5' for the AUG start codon. The initiator Met–tRNA with GTP bound to it binds to the small ribosomal subunit and forms a complex.

2 The Met–tRNA+GTP+small ribosomal subunit complex binds to the 5' cap of the mRNA and moves along the mRNA—a process called **scanning**—until it reaches the AUG start codon in the P site. Base pairing occurs between the codon and the anticodon of the initiator Met–tRNA.

3 The large ribosomal subunit binds and GTP is hydrolyzed, completing initiation. The ribosome is ready for the next stage of translation, elongation.

Elongation is highly similar in prokaryotes and eukaryotes, with no significant conceptual differences. The elongation cycle turns at the rate of about one to three times per second in eukaryotes and 15 to 20 times per second in bacteria. Once it is long enough, the growing polypeptide chain extends from the ribosome through the exit tunnel (see Figure 15.13) as elongation continues.

Researchers were surprised to discover that peptidyl transferase, the enzyme that forms peptide bonds in the elongation cycle, is not a protein but a part of an rRNA of the large ribosomal subunit. An RNA molecule that catalyzes a reaction like a protein enzyme does is called a *catalytic RNA* or a **ribozyme** (*ribo*nucleic acid en*zyme*). (*Insights from the Molecular Revolution* describes the experimental evidence showing that peptidyl transferase is a ribozyme.)

FIGURE 15.15 **Translation elongation.** A protein elongation factor (EF) complexes with the aminoacyl–tRNA to bring it to the ribosome, and another EF is needed for ribosome translocation. For simplicity, the EFs are not shown in the figure.
© Cengage Learning 2014

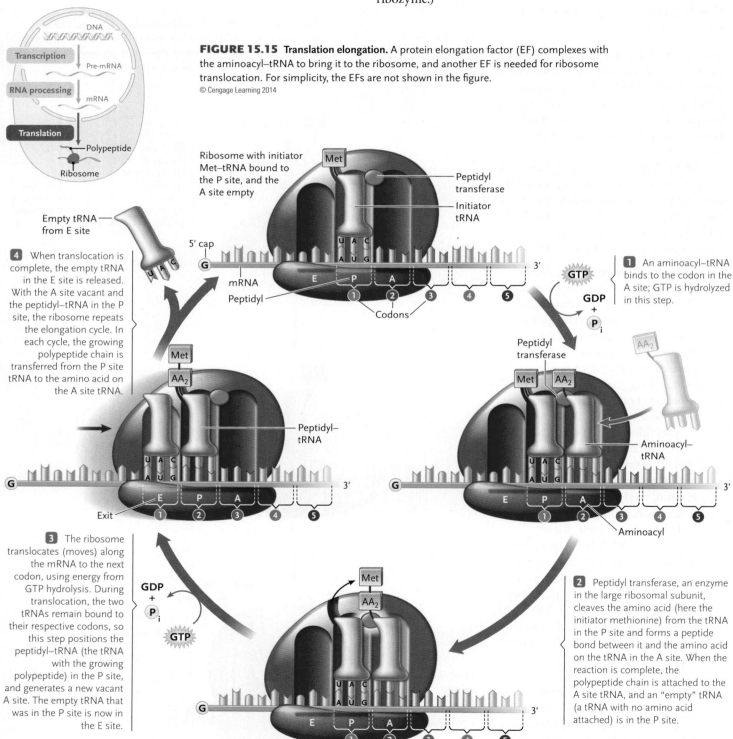

4 When translocation is complete, the empty tRNA in the E site is released. With the A site vacant and the peptidyl–tRNA in the P site, the ribosome repeats the elongation cycle. In each cycle, the growing polypeptide chain is transferred from the P site tRNA to the amino acid on the A site tRNA.

1 An aminoacyl–tRNA binds to the codon in the A site; GTP is hydrolyzed in this step.

2 Peptidyl transferase, an enzyme in the large ribosomal subunit, cleaves the amino acid (here the initiator methionine) from the tRNA in the P site and forms a peptide bond between it and the amino acid on the tRNA in the A site. When the reaction is complete, the polypeptide chain is attached to the A site tRNA, and an "empty" tRNA (a tRNA with no amino acid attached) is in the P site.

3 The ribosome translocates (moves) along the mRNA to the next codon, using energy from GTP hydrolysis. During translocation, the two tRNAs remain bound to their respective codons, so this step positions the peptidyl–tRNA (the tRNA with the growing polypeptide) in the P site, and generates a new vacant A site. The empty tRNA that was in the P site is now in the E site.

Peptidyl Transferase: Protein or RNA?

A key event in polypeptide synthesis is the formation of a peptide bond when the amino acid or growing polypeptide on the tRNA in the P site of the ribosome is transferred to the amino acid on the tRNA in the A site. Peptide bond formation is catalyzed by a peptidyl transferase enzyme in the large ribosomal subunit. In the large 50S subunit of bacteria, regions of the 23S rRNA and several ribosomal proteins are in the peptidyl transferase area, indicating that they may participate in its enzymatic activity.

Research Question

Is the peptidyl transferase activity of the bacterial ribosome a function of a ribosomal protein or an RNA molecule?

Experiment

To answer this question, Harry F. Noller and his coworkers at the University of California, Santa Cruz, set up a *fragment reaction,* a simplified peptidyl transferase reaction in a test tube **(Figure).**

In the fragment reaction, 5′-CAACCA–*Met is a short RNA segment with *Met (radioactively labeled methionine) attached; it mimics the initiator tRNA. Puromycin is a molecule that has the shape of a tRNA and can enter the A site of the 50S ribosomal sub-

5′-CAACCA–*Met + Puromycin

↓ 50S ribosomal subunits

↓

*Met–puromycin + 5′-CAACCA-3′

FIGURE Fragment reaction.

units that are part of the reaction. Once in the A site, the *Met in the P site can be transferred to the puromycin by peptide bond formation, producing *Met–puromycin. Biochemical separation techniques can distinguish the *Met–puromycin from *Met not bonded to puromycin; the amount of radioactivity in *Met–puromycin is a measure of peptidyl transferase activity.

Results

Once they had demonstrated that peptide bond formation occurred in the fragment reaction, Noller's group tested the effects of various treatments on peptide bond formation with the results shown in the **Table**.

Conclusion

The researchers concluded that the peptidyl transferase activity of the 50S ribosomal subunit is a function of the 23S rRNA component of that subunit, rather than of ribosomal proteins. That is, the peptidyl transferase activity remained when proteins were removed or degraded, but was lost when RNA was degraded. The investigators allowed that their

protein removal techniques may have left fragments of ribosomal proteins that potentially may be part of the peptidyl transferase region of the 50S subunit and that definitive proof of their conclusion would require a demonstration that a completely protein-free preparation of the rRNA can carry out the peptidyl transferase reaction *in vitro*. To date, however, peptidyl transferase activity by rRNA alone has not been demonstrated.

THINK LIKE A SCIENTIST Other experiments have shown that a region of 23S rRNA in the vicinity of the site of peptide bond formation has a sequence in which a number of bases are absolutely conserved in all known large ribosomal subunit rRNA sequences, including those from bacteria, archaea, eukaryotes, mitochondria, and chloroplasts. Does this observation support or refute the conclusions of the study?

Source: H. F. Noller, V. Hoffarth, and L. Zimniak. 1992. Unusual resistance of peptidyl transferase to protein extraction procedures. *Science* 256:1416–1419.

TABLE	Effects of Treatments on Peptide Bond Formation	
Treatment	Result	Interpretation
None	Met–puromycin forms	Peptidyl transferase activity is present in 50S subunits
Sodium dodecyl sulfate (SDS) or phenol to remove proteins from subunits	Met–puromycin forms	Peptidyl transferase activity remains after proteins removed from 50S subunits
Proteinase K to digest proteins of subunits	Met–puromycin forms	Peptidyl transferase activity remains after proteins of 50S subunits digested away
Ribonuclease to degrade RNA of subunits	No Met–puromycin forms	Peptidyl transferase activity is lost when RNA of subunits is degraded

© Cengage Learning 2014

Termination Releases a Completed Polypeptide from the Ribosome

Translation termination takes place when the A site of a ribosome arrives at one of the stop codons on the mRNA, UAG, UAA, or UGA **(Figure 15.16).** The stop codon is read by a protein **release factor** (**RF;** also called a **termination factor**). Termination is highly similar in prokaryotes and eukaryotes.

Multiple Ribosomes Translate a Single mRNA Simultaneously

Once the first ribosome has begun translation, another one can assemble with an initiator tRNA as soon as there is room on the mRNA. Ribosomes continue to attach as translation continues and become spaced along the mRNA like beads on a string. The entire structure of an mRNA molecule and the multiple ribo-

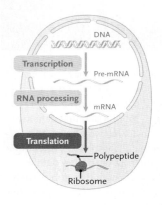

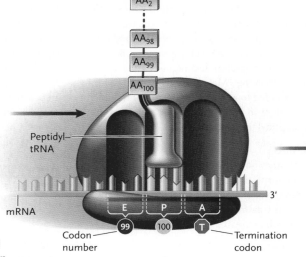

FIGURE 15.16
Translation termination.
© Cengage Learning 2014

Peptidyl–tRNA

mRNA

Codon number

Termination codon

1 The ribosome reaches a stop codon, UAG, UAA, or UGA.

2 No tRNA has an anticodon that can pair with a stop codon. Instead, a release factor (RF) binds to the stop codon in the A site. The shape of the release factor mimics that of a tRNA, including regions that read the stop codons.

somes attached to it is known as a **polysome** (a contraction of *polyribosome;* **Figure 15.17**). The multiple ribosomes greatly increase the overall rate of polypeptide synthesis from a single mRNA. The total number of ribosomes in a polysome depends on the length of the coding region of its mRNA molecule, ranging from a minimum of one or two ribosomes on the smallest mRNAs to as many as 100 on the longest mRNAs.

In prokaryotes, because of the absence of a nuclear envelope, transcription and translation typically are coupled. That is, as soon as the 5′ end of a new mRNA emerges from the RNA polymerase, ribosomal subunits attach and initiate translation **(Figure 15.18)**. In essence, the polysome forms while the mRNA is still being made. By the time the mRNA is completely transcribed, it is covered with ribosomes from end to end, each assembling a copy of the encoded polypeptide.

Newly Synthesized Polypeptides Are Processed and Folded into Finished Form

Most eukaryotic proteins are in an inactive, unfinished form when ribosomes release them. Processing reactions that convert the new proteins into finished form include the removal of amino acids from the ends or interior of the polypeptide chain and the addition of larger organic groups, including carbohydrate or lipid structures.

Proteins fold into their final three-dimensional shapes as the processing reactions take place. For many proteins, helper proteins called **chaperones** or **chaperonins** assist the folding process by combining with the folding protein, promoting "correct" three-dimensional structures, and inhibiting incorrect ones (see Section 3.4 and Figure 3.25).

In some cases, the same initial polypeptide may be processed by alternative pathways that produce different mature polypeptides, usually by removing different, long stretches of amino acids from the interior of the polypeptide chain. Alternative processing of a pre-mRNA is an-

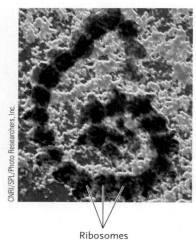

Ribosomes

FIGURE 15.17 Polysomes, consisting of a series of ribosomes reading the same mRNA.
© Cengage Learning 2014

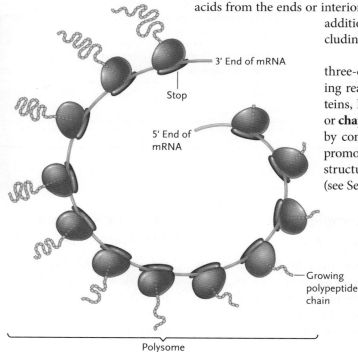

3′ End of mRNA

Stop

5′ End of mRNA

Growing polypeptide chain

Polysome

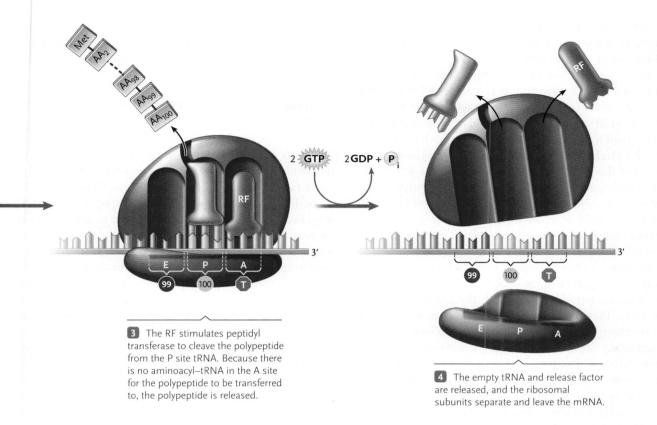

3 The RF stimulates peptidyl transferase to cleave the polypeptide from the P site tRNA. Because there is no aminoacyl–tRNA in the A site for the polypeptide to be transferred to, the polypeptide is released.

4 The empty tRNA and release factor are released, and the ribosomal subunits separate and leave the mRNA.

other mechanism that increases the number of polypeptides encoded by a single gene.

Other proteins are processed into an initial, inactive form that is later activated at a particular time or location by removal of a covering segment of the amino acid chain. The digestive enzyme pepsin, for example, is made by processing reactions within cells lining the stomach into an inactive form called *pepsinogen.* When the cells secrete pepsinogen into the stomach, the high acidity of that organ triggers removal of a segment of amino acids from one end of the protein's amino acid chain. The amino acid removal converts the enzyme into the

active form in which it rapidly breaks proteins in food particles into shorter pieces. The initial production of the protein as inactive pepsinogen protects the cells that make it from having their proteins degraded by the enzyme.

Finished Proteins Are Sorted to the Cellular Locations Where They Function

In a eukaryotic cell, every protein that is made must be sorted to the compartment where it performs a necessary function. Without a sorting system, cells would wind up as a jumble of proteins floating about in the cytoplasm, with none of the spatial organization that makes cellular life possible.

As proteins are sorted, they are channeled into one of three compartments in the cell: (1) the cytosol; (2) the endomembrane system, which includes the endoplasmic reticulum (ER), Golgi complex, lysosomes, secretory vesicles, the nuclear envelope, and the plasma membrane (see Section 5.3); and (3) other membrane-bound organelles distinct from the endomembrane system, including mitochondria, chloroplasts, microbodies (for example, peroxisomes), and the nucleus (see Section 5.3).

PROTEIN SORTING TO THE CYTOSOL Proteins that function in the cytosol are synthesized on *free ribosomes* in the cytosol. The polypeptides are simply released from the ribosomes once translation is completed. Examples of proteins that function in the cytoplasm are microtubule proteins and the enzymes that carry out glycolysis (see Section 8.2).

mRNAs with attached ribosomes

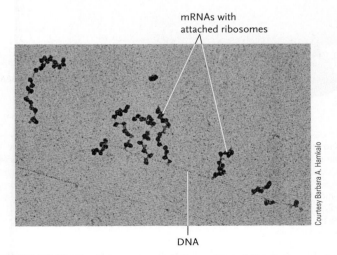

Courtesy Barbara A. Hamkalo

DNA

FIGURE 15.18 Simultaneous transcription and translation in progress in an electron microscope preparation extracted from *E. coli.* ×57,000.

PROTEIN SORTING TO THE ENDOMEMBRANE SYSTEM The endomembrane system is a major traffic network for proteins. Polypeptides that sort to the endomembrane system begin their synthesis on free ribosomes in the cytosol. Specific to these polypeptides is a short segment of amino acids called a **signal sequence** (also called a **signal peptide**) near their N-terminal ends. As **Figure 15.19** shows, the signal sequence initiates a series of steps that result in the polypeptide entering the lumen of the rough ER. This mechanism is called **cotranslational import** because import of the polypeptide into the ER occurs simultaneously with translation of the mRNA encoding the polypeptide. The signal sequence was discovered in 1975 by Günter Blobel, B. Dobberstein, and colleagues at Rockefeller University, when they observed that proteins sorted through the endomembrane system initially contain extra amino acids at their N-terminal ends. Blobel received a Nobel Prize in 1999 for his work with the mechanism of sorting proteins in cells.

Once in the lumen of the rough ER, the proteins fold into their final form. They also have or attain a tag—a "zip code" if you will—that targets each protein for sorting to its final destination. Depending on the protein and its destination, the tag may be an amino acid sequence already in the protein, or a functional group or short sugar chain added in the lumen. Some proteins remain in the ER, whereas others are transported to the Golgi complex where they may be modified further. From the Golgi complex, proteins are packaged into vesicles, which may deliver them to lysosomes, secrete them from

the cell (digestive enzymes, for example), or deposit them in the plasma membrane (cell surface receptors, for instance). Vesicle traffic in the cytoplasm involving the rough ER and Golgi complex is illustrated in Figure 5.14.

PROTEIN SORTING TO MITOCHONDRIA, CHLOROPLASTS, MICROBODIES, AND NUCLEUS Proteins are sorted to mitochondria, chloroplasts, microbodies, and the nucleus after they have been made on free ribosomes in the cytosol. This mechanism of sorting is called **posttranslational import.** Proteins destined for the mitochondria, chloroplasts, and microbodies have short amino acid sequences called **transit sequences** at their N-terminal ends that target them to the appropriate organelle. The protein is taken up into the correct organelle by interactions between its transit sequences and membrane transport complexes specific to the appropriate organelle. A transit peptidase enzyme within the organelle then removes the transit sequence. Proteins sorted to the nucleus, such as the enzymes for DNA replication, and RNA transcription, have short amino acid sequences called **nuclear localization signals.** A cytosolic protein binds to the signal and moves it to the nuclear pore complex (see Section 5.3 and Figures 5.10 and 5.11) where it is then transported into the nucleus through the pore. For these proteins, the nuclear localization signal remains because the proteins need to enter the nucleus each time the nuclear envelope breaks down and reforms during the cell division cycle.

The same basic system of sorting signals distributes proteins in prokaryotic cells, indicating that this mechanism probably evolved with the first cells. In prokaryotes, signals similar to the ER-directing signals of eukaryotes direct newly synthesized bacterial proteins to the plasma membrane (bac-

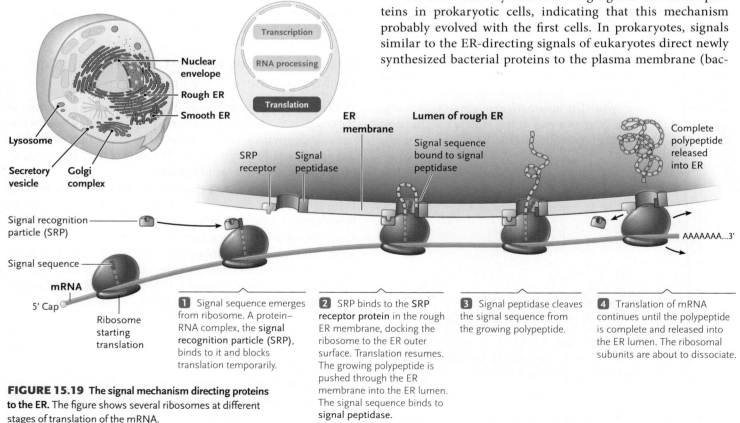

FIGURE 15.19 The signal mechanism directing proteins to the ER. The figure shows several ribosomes at different stages of translation of the mRNA.
© Cengage Learning 2014

1. Signal sequence emerges from ribosome. A protein–RNA complex, the **signal recognition particle (SRP)**, binds to it and blocks translation temporarily.

2. SRP binds to the **SRP receptor protein** in the rough ER membrane, docking the ribosome to the ER outer surface. Translation resumes. The growing polypeptide is pushed through the ER membrane into the ER lumen. The signal sequence binds to signal peptidase.

3. Signal peptidase cleaves the signal sequence from the growing polypeptide.

4. Translation of mRNA continues until the polypeptide is complete and released into the ER lumen. The ribosomal subunits are about to dissociate.

teria do not have ER membranes); further information built into the proteins keeps them in the plasma membrane or allows them to enter the cell wall or to be secreted outside the cell. Proteins without sorting signals remain in the cytoplasmic solution.

STUDY BREAK 15.4

1. How does translation initiation occur in eukaryotes versus prokaryotes?
2. Distinguish between the P, A, and E sites of the ribosome.
3. How are proteins directed to different compartments of a eukaryotic cell?

15.5 Genetic Changes That Affect Protein Structure and Function

We learned in Chapters 12 and 13 that a mutant allele of a gene can alter the phenotype controlled by the gene. In this section, we discuss two types of genetic change and how they can alter protein structure and function and, therefore, produce an altered phenotype. One type is mutation of a base pair in the DNA, that is, a change from one base pair to another. The other is when certain genetic elements known as transposable elements (TEs) move from one location to another in the genome.

Base-Pair Mutations Can Alter Protein Structure and Function

Mutations, in general, are changes to the genetic material. Base-pair substitution mutations are particular mutations involving changes to individual base pairs in the genetic material. (Mutations can involve more than one base pair, but they are not considered in this discussion.) If a base-pair substitution mutation occurs in the protein-coding portion of a gene, it can change a base in a codon in the mRNA and thereby affect the structure and function of the encoded protein. More generally, mutations affecting the functions of genes are known as gene mutations.

Consider a theoretical stretch of normal (unmutated) DNA encoding a string of amino acids in a polypeptide **(Figure 15.20A).** Four types of base-pair substitution mutations affecting a protein-coding gene are:

1. **Missense mutation (Figure 15.20B):** A sense codon is changed to a different sense codon

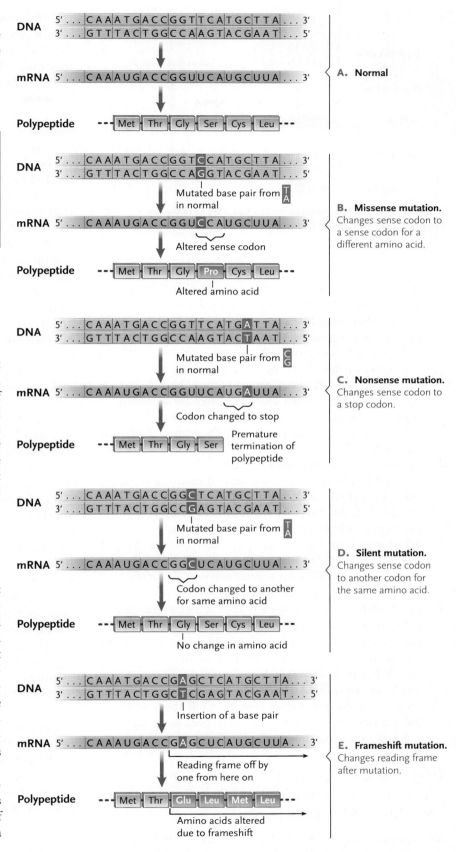

FIGURE 15.20 Effects of base-pair mutations in protein-coding genes on the amino acid sequence of the encoded polypeptide.

© Cengage Learning 2014

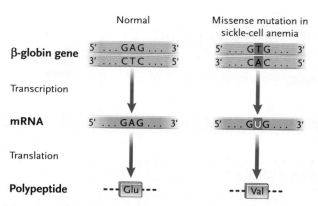

β-globin gene

Normal
5′ ...GAG... 3′
3′ ...CTC... 5′

Missense mutation in
sickle-cell anemia
5′ ...GTG... 3′
3′ ...CAC... 5′

Transcription

mRNA
5′ ...GAG... 3′   5′ ...GUG... 3′

Translation

Polypeptide   --- Glu ---   --- Val ---

FIGURE 15.21 Missense mutation in a gene for one of the two polypeptides of hemoglobin that is the cause of sickle-cell anemia.
© Cengage Learning 2014

that specifies a different amino acid. Whether the function of a polypeptide is altered significantly depends on the amino acid change that occurs. Individuals homozygous for a missense mutation in the gene for one of the two polypeptide types found in the oxygen-carrying protein hemoglobin **(Figure 15.21)** have the genetic disease sickle-cell anemia, described in Chapter 12 (pp. 239–240; 254; 257). Many other human genetic diseases are caused by missense mutations, including albinism, hemophilia, and achondroplasia (see Chapter 13 *Insights from the Molecular Revolution*).

2. **Nonsense mutation (Figure 15.20C):** A sense codon is changed to a nonsense (stop) codon. Translation of an mRNA containing a nonsense mutation results in a shorter-than-normal polypeptide and, in many cases, this polypeptide will be only partially functional at best.

3. **Silent mutation (Figure 15.20D):** A sense codon is changed to a different sense codon, but that codon specifies the same amino acid as in the normal polypeptide, so the function of the polypeptide is unchanged.

4. **Frameshift mutation (Figure 15.20E):** A single base-pair deletion or insertion in the coding region of a gene alters the reading frame of the resulting mRNA (the figure shows a single base-pair insertion). After the point of the mutation, the ribosome reads codons that are not the same as for the normal mRNA, producing a different amino acid sequence in the polypeptide from then on. The resulting polypeptide typically is nonfunctional because of the significantly altered amino acid sequence.

Transposable Elements Move from One Location to Another in the Genome and May Affect Gene Function

All organisms contain particular segments of DNA that can move from one place to another within a cell's genome. The movable sequences are called *transposable genetic elements,* or more simply, **transposable elements (TEs).**

The movement of TEs, called **transposition,** involves a type of genetic recombination process. However, the location in the DNA where the TE moves to—the **target site**—is not homologous with the TE. In this respect, transposition differs from genetic recombination in meiosis in eukaryotes, and in the processes that produce recombinants in bacteria (see Chapter 17), which involve crossing-over between homologous DNA molecules.

Transposition of a TE occurs at a low frequency. Depending on the TE, transposition occurs in one of two ways: (1) a cut-and-paste process, in which the TE leaves its original location and transposes to a new location **(Figure 15.22A);** or (2) a copy-and-paste process, in which a copy of a TE transposes to a new location, leaving the original TE behind **(Figure 15.22B).** For most TEs, transposition starts with contact between the TE and the target site. This also means that TEs do not exist free of the DNA in which they are integrated.

TEs are important because of the genetic changes they cause. For example, they produce mutations by transposing into and thereby disrupting the coding sequences of genes, knocking out their functions. If they transpose instead into regulatory sequences of genes, they may increase or decrease the level of gene expression of those genes. As such, TEs are an important source of genetic variability and in genome evolution (see Section 19.4).

BACTERIAL TRANSPOSABLE ELEMENTS Bacterial TEs were discovered in the 1960s. They move from site to site within the bacterial chromosome, between the bacterial chromosome and plasmids, and between plasmids.

The two major types of bacterial TEs are *insertion sequences (IS)* and *transposons* **(Figure 15.23). Insertion sequences** are the simplest TEs. They are relatively small and contain only the gene for **transposase,** an enzyme that catalyzes the reactions inserting or removing the TE from the DNA (see Figure 15.23A). At the two ends of an IS is a short *inverted repeat* sequence—the same DNA sequence running in opposite directions (shown by directional arrows in the figure). The inverted repeat sequences enable the transposase enzyme to identify the ends of the TE when it catalyzes transposition.

The second type of bacterial TE, called a **transposon,** has an inverted repeat sequence at each end enclosing a central region with one or more genes (Figure 15.23B). In a number of bacterial transposons, the inverted repeat sequences are insertion sequences, one of which provides the transposase for movement of the element. Bacterial transposons without IS ends have short inverted repeat end sequences, and a transposase gene is within the central region. Additional gene(s) in the central region of both types of transposons typically are for antibiotic resistance; they originated from the main bacterial DNA circle or from plasmids. The additional genes are carried along as the TEs move from place to place within and between species.

Many antibiotics, such as penicillin, erythromycin, tetracycline, ampicillin, and streptomycin, that were once successful in curing bacterial infections have lost much of their effec-

A. Cut-and-paste transposition. The TE leaves one location in the DNA and moves to a new location.

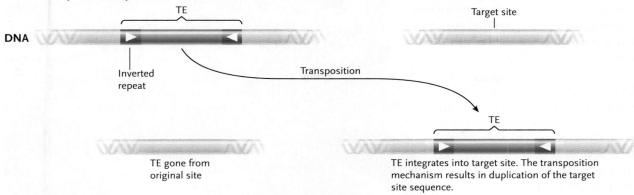

B. Copy-and-paste transposition. A copy of the TE moves to a new location, leaving the original TE behind.

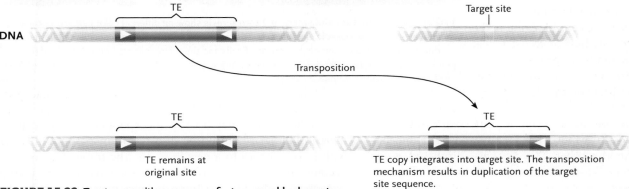

FIGURE 15.22 Two transposition processes for transposable elements.
© Cengage Learning 2014

tiveness because of resistance genes carried in transposons. Movements of the transposons, particularly to plasmids that have been transferred between bacteria within the same species and between different species, greatly increase the spread of genes providing antibiotic resistance. Resistance genes have made many bacterial diseases difficult or impossible to treat with standard antibiotics. (Chapter 26 discusses bacterial antibiotic resistance further.)

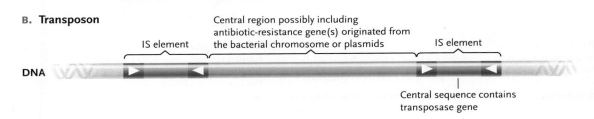

FIGURE 15.23 Types of bacterial transposable elements. **(A)** Insertion sequence. **(B)** Transposon.
© Cengage Learning 2014

TRANSPOSABLE ELEMENTS IN EUKARYOTES TEs actually were first discovered in a eukaryote, maize (corn), in the 1940s by Barbara McClintock, a geneticist working at the Cold Spring Harbor Laboratory in New York. McClintock noted that some mutations affecting kernel and leaf color appeared and disappeared rapidly under certain conditions. Mapping the alleles by linkage studies produced a surprising result—the map positions changed frequently, indicating that the alleles could move from place to place in the corn chromosomes. Some of the movements were so frequent that changes in their effects could be noticed at different times in a single developing kernel **(Figure 15.24)**.

When McClintock first reported her results, her findings were regarded as an isolated curiosity, possibly applying only to corn. This was because the then-prevailing opinion among geneticists was that genes are fixed in the chromosomes and do not move to other locations. Her conclusions were widely accepted only after TEs were detected and characterized in bacteria in the 1960s. By the 1970s, further examples of TEs were discovered in other eukaryotes, including yeast and mammals. We now know that TEs are probably universally distributed among both prokaryotes and eukaryotes. McClintock was awarded a Nobel Prize in 1983 for her discovery of mobile genetic elements.

Eukaryotic TEs fall into two major classes, *transposons* and *retrotransposons,* distinguished by the way the TE sequence moves from place to place in the DNA. Researchers detect both classes of eukaryotic TEs through DNA sequencing or through their effects on genes at or near their sites of insertion. Unlike prokaryotes, eukaryotes have no TEs resembling insertion sequences.

Eukaryotic transposons are similar to bacterial transposons in their general structure and in the way they transpose. A gene for transposase is in the central region of the transposon, and most have inverted repeat sequences at their ends. Depending on the transposon, transposition is by the cut-and-paste or copy-and-paste mechanism (see Figure 15.22).

FIGURE 15.24 **Barbara McClintock and corn kernels showing different color patterns because of the movement of transposable elements.** As TEs move into or out of genes controlling pigment production in developing kernels, the ability of cells and their descendants to produce the dark pigment is destroyed or restored. The result is random patterns of pigmented and colorless (yellow) segments in individual kernels.

Nik Kleinberg

Members of the other class of eukaryotic TEs, the **retrotransposons,** transpose by a copy-and-paste mechanism but, unlike the other TEs we have discussed, their transposition occurs via an intermediate RNA copy of the TE **(Figure 15.25).** Some retrotransposons are bounded by sequences that are directly repeated rather than in inverted form; others have no repeated sequences at their ends.

Cellular genes may become incorporated into the central region of either a transposon or a retrotransposon and travel with it as it moves to a new location. The trapped genes may become continuously active through the effects of regulatory sequences in the TE. The trapped genes may also become abnormally active if moved in a TE to the vicinity of a regulatory region or promoter of an intensely transcribed cellular gene.

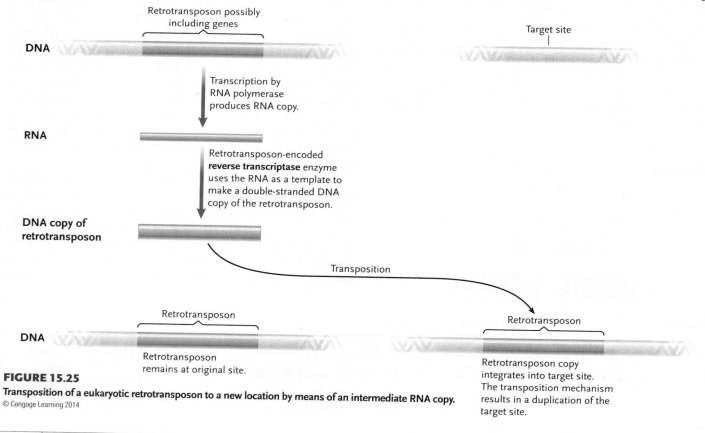

FIGURE 15.25
Transposition of a eukaryotic retrotransposon to a new location by means of an intermediate RNA copy.
© Cengage Learning 2014

Certain forms of cancer have been linked to the TE-instigated abnormal activation of genes that regulate cell division.

Once TEs are inserted in the chromosomes, they become more or less permanent residents, duplicated and passed on during cell division along with the rest of the DNA. TEs inserted into the DNA of reproductive cells that produce gametes may be inherited, thereby becoming a permanent part of the genetic material of a species.

Long-standing TEs are subject to mutation along with other sequences in the genome. Such mutations may accumulate in a TE, gradually altering it into a nonmobile, residual sequence in the DNA. The genomes of many eukaryotes, including humans, contain many nonfunctional TEs likely created in this way.

In this chapter we have learned how gene expression occurs by the processes of transcription and translation. We also learned how gene expression may be changed by mutation or by the actions of transposable elements. In the next chapter, you will see how organisms and cells exert control over how their genes are expressed.

STUDY BREAK 15.5

1. How does a missense mutation differ from a silent mutation?
2. How do genetic recombination and TE transposition differ?
3. In what ways are bacterial IS elements and transposons alike?
4. How do eukaryotic transposons and retrotransposons differ?

 UNANSWERED QUESTIONS

How does the ribosome work?

When they were first discovered, ribosomes were thought to be passive "workbenches" on which proteins were synthesized, presumably by enzymes. But in the late 1960s, it was found that the ribosome itself catalyzes formation of peptide bonds. Binding and lining up the mRNA and the correct tRNAs were other things that appeared to be jobs carried out by the ribosome. But the ribosome is a huge macromolecular complex, containing more than 50 ribosomal proteins and large ribosomal RNAs containing many thousands of nucleotides.

In my laboratory, we wanted to identify the parts of the ribosome that carried out these important functions, as a step toward understanding *how* they manage to do it. Following the common wisdom that it is the proteins that carry out the biological functions of the cell, we assumed that ribosomal proteins were responsible for ribosome functions, but which ones? We decided to use *chemical modification* to approach this question.

Our plan was to inactivate the functions of the ribosome by attacking the ribosome with chemical reagents that were known to modify the reactive groups in proteins; we would then figure out which proteins were inactivated by rescuing the inactivated proteins with individual active ones, using *in vitro* reconstitution of a functional ribosome as the measure for a successful result. To our great surprise, it was very difficult to inactivate the ribosome using reagents that react with proteins. This disappointed me, because I was eager to use my skills as a protein chemist to understand how the ribosome works. As a sort of control experiment, we tried using a reagent that attacks RNA, and for the first time we observed rapid inactivation. Furthermore, ribosome activity could be rescued by reconstituting with active RNA, proving that inactivation of the ribosome was caused by modification of the ribosomal RNA. In 1972 we published a paper proposing that ribosomal RNA plays a functional role in protein synthesis, and followed it with evidence from many other kinds of studies in the following years, but no one was ready to believe such a "crackpot idea" (as one prominent scientist was heard to describe our findings).

Twenty years after our original proposal, I was frustrated that people were still resistant to the idea, although less so because of the discovery of ribozymes (catalytic RNAs) in the 1980s. So in 1992 we carried out another set of experiments in which we subjected ribosomes to relatively brutal extraction procedures that are used to remove proteins from complexes with RNA and DNA (see *Insights from the Molecular Revolution* for this chapter). We washed the ribosomes with a strong detergent, treated them with a protease that is known to chew up proteins into small fragments and vigorously extracted them with phenol, which is known to separate protein from RNA. At the end of this ordeal, the ribosomes were completely active in catalyzing formation of peptide bonds. Although we had not completely removed all of the protein from the ribosome, the idea that the biological activity of ribosomes might be based on its RNA, rather than its proteins, now became accepted widely by researchers. It was finally demonstrated convincingly by high-resolution crystallography of the ribosome that the site of peptide bond formation contains no protein, only RNA.

This experience provides insight into how science works; the scientific community is resistant to changes in fundamental paradigms ("Only proteins can carry out biological functions"), and acceptance of new ideas often depends on proposing them at "the right time." Also typical of science, the new insight leaves many questions still unanswered. Even after 50 years of research by hundreds of investigators, we are still trying to figure out the ribosome's fundamental mechanisms of action. And an even more daunting question is, how did the ribosome evolve?

THINK LIKE A SCIENTIST

1. Ribosomes contain both RNA and protein. Because ribosomes are responsible for making proteins, how can you explain the evolution of the first ribosome?
2. If you think you can answer that question, how about the following: the chance that you could make a functional protein from mRNA containing a random RNA sequence is essentially zero. Why, then, would you evolve a ribosome in the first place?

Harry Noller is Robert L. Sinsheimer Professor of Molecular Biology at the University of California, Santa Cruz, where he studies the structure and function of the ribosome, using such diverse approaches as biochemistry, molecular genetics, fluorescence spectroscopy, and X-ray crystallography. To learn more about his research, go to http://rna.ucsc.edu/rnacenter/noller_lab.html.

Courtesy of Harry Noller

REVIEW KEY CONCEPTS

To access the course materials and companion resources for this text, please visit www.cengagebrain.com.

15.1 The Connection between DNA, RNA, and Protein

- In their genetic experiments with *Neurospora crassa*, Beadle and Tatum found a direct correspondence between gene mutations and alterations of enzymes. Their one gene–one enzyme hypothesis was updated as the one gene–one polypeptide hypothesis (Figure 15.2).

- The pathway from genes to proteins involves transcription then translation. In transcription, a sequence of nucleotides in DNA is copied into a complementary sequence in an RNA molecule. In translation, the sequence of nucleotides in an mRNA molecule specifies an amino acid sequence in a polypeptide (Figure 15.3).

- The genetic code is a triplet code. AUG at the beginning of a coded message establishes a reading frame for reading the codons three nucleotides at a time until a stop codon is reached. The code is degenerate: most of the amino acids are specified by more than one codon (Figures 15.4 and 15.5).

- The genetic code is essentially universal.

Animation: Transcription—A molecular view

Practice: The major differences between prokaryotic and eukaryotic protein synthesis

15.2 Transcription: DNA-Directed RNA Synthesis

- Transcription, the process by which information coded in DNA is transferred to a complementary RNA copy, begins when an RNA polymerase binds to a promoter sequence in the DNA and starts synthesizing an RNA molecule. The enzyme then adds RNA nucleotides in sequence according to the DNA template. At the end of the transcribed sequence, the enzyme and the completed RNA transcript release from the DNA template.

- Transcription occurs in three stages: initiation, elongation, and termination. Each stage is similar in eukaryotes and bacteria. For transcription of eukaryotic protein-coding genes, transcription factors bind to the promoter and recruit RNA polymerase II which then begins transcription of the mRNA. Elongation of the mRNA occurs as the polymerase reads the coding sequence of the gene. The 3′ end of the mRNA is specified differently in prokaryotes and eukaryotes; termination of transcription in bacteria is determined by a terminator sequence in the gene (Figure 15.6).

- Transcription of non-protein-coding genes occurs in a similar way to transcription of protein-coding genes. In eukaryotes, special RNA polymerases are used to transcribe tRNA and rRNA genes, whereas in bacteria the same RNA polymerase that transcribes protein-coding genes transcribes those genes.

15.3 Production of mRNAs in Eukaryotes

- A gene encoding an mRNA molecule includes the promoter, which is recognized by the regulatory proteins and transcription factors that promote DNA unwinding and the initiation of transcription by an RNA polymerase. Transcription in eukaryotes produces a pre-mRNA molecule that consists of a 5′ cap, the 5′ untranslated region, interspersed exons (amino acid-coding segments) and introns, the 3′ untranslated region, and the 3′ poly(A) tail. The 5′ cap and 3′ poly(A) tail are not encoded in the DNA. The 5′ cap is added by a capping enzyme, and the 3′ poly(A) tail is added by

poly(A) polymerase once a 3′ end of the pre-mRNA is generated by cleavage downstream of a polyadenylation signal in the transcript (Figure 15.7).

- Introns in pre-mRNAs are removed to produce functional mRNAs by splicing. snRNPs bind to the introns, loop them out of the pre-mRNA, clip the intron at each exon boundary, and join the adjacent exons together (Figure 15.8).

- Many pre-mRNAs are subjected to alternative splicing, a process that joins exons in different combinations to produce different mRNAs encoded by the same gene. Translation of each mRNA produced in this way generates a protein with different function (Figure 15.9).

Animation: Transcription—Intron and exons

15.4 Translation: mRNA-Directed Polypeptide Synthesis

- Translation is the assembly of amino acids into polypeptides. Translation occurs on ribosomes. The P, A, and E sites of the ribosome are used for the stepwise addition of amino acids to the polypeptide as directed by the mRNA (Figures 15.10 and 15.13).

- Amino acids are brought to the ribosome attached to specific tRNAs. Amino acids are linked to their corresponding tRNAs by aminoacyl–tRNA synthetases. By matching amino acids with tRNAs, the reactions also provide the ultimate basis for the accuracy of translation (Figures 15.11 and 15.12).

- Translation proceeds through the stages of initiation, elongation, and termination. In initiation, a ribosome assembles with an mRNA molecule and an initiator methionine-tRNA. In elongation, amino acids linked to tRNAs add one at a time to the growing polypeptide chain. In termination, the new polypeptide is released from the ribosome, and the ribosomal subunits separate from the mRNA (Figures 15.14–15.16).

- Multiple ribosomes translate the same mRNA simultaneously, forming a polysome (Figures 15.17 and 15.18).

- After they are synthesized on ribosomes, polypeptides are converted into finished form by processing reactions, which include removal of one or more amino acids from the protein chains, addition of organic groups, and folding guided by chaperones.

- Finished proteins are sorted to the cellular locations where they function. Proteins that function in the cytosol are synthesized on free ribosomes. Proteins are sorted to the endomembrane system by cotranslational import and proteins are sorted to the mitochondria, chloroplast, microbodies, and nucleus by posttranslational import. In each of these two cases, specific amino acid sequences in the polypeptides direct them to their destinations (Figure 15.19).

Animation: The 4 steps of translation

Animation: Structure of a ribosome

15.5 Genetic Changes That Affect Protein Structure and Function

- Base-pair substitution mutations alter the mRNA and can lead to changes in the amino acid sequence of the encoded polypeptide. A missense mutation changes one sense codon to one that specifies a different amino acid, a nonsense mutation changes a sense codon to a stop codon, and a silent mutation changes one sense codon to another sense codon that specifies the same amino acid. A base-pair insertion or deletion is a frameshift mutation that alters the reading frame beyond the point of the mutation, leading to a different amino acid sequence from then on in the polypeptide (Figures 15.20 and 15.21).

- Both prokaryotes and eukaryotes contain TEs (transposable elements)—DNA sequences that can move from place to place in the DNA. The TEs may move from one location in the DNA to another, or generate duplicated copies that insert in new locations while leaving the "parent" copy in its original location (Figure 15.22).

- Genes of the host cell DNA may become incorporated into a TE and may be carried with it to a new location. There, the genes may become abnormally active when placed near sequences that control the activity of genes within the TE, or near the control elements of active host genes.

- Bacterial TEs occur as insertion sequences and transposons. Both contain a gene for the transposase enzyme needed for transposition. The transposon may also contain genes, such as for antibiotic resistance, which originate in host DNA (Figure 15.23).

- Eukaryotic TEs occur as transposons, which release from one location in the DNA and insert at a different site, or as retrotransposons, which move by making an RNA copy, which is then replicated into a DNA copy that is inserted at a new location. The original copy remains at the original location (Figures 15.24 and 15.25).

- TE-instigated abnormal activation of genes regulating cell division has been linked to the development of some forms of cancer in humans and other complex animals.

Animation: Mutations and translation

Animation: Frameshift mutation

UNDERSTAND & APPLY

Test Your Knowledge

1. Which statement about the following pathway is false?

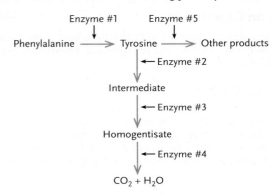

Enzyme #1   Enzyme #5

Phenylalanine ⟶ Tyrosine ⟶ Other products

⟵ Enzyme #2

Intermediate

⟵ Enzyme #3

Homogentisate

⟵ Enzyme #4

$CO_2 + H_2O$

   a. A mutation for enzyme #1 causes phenylalanine to build up.
   b. A mutation for enzyme #2 prevents tyrosine from being synthesized.
   c. A mutation at enzyme #3 prevents homogentistate from being synthesized.
   d. A mutation for enzyme #2 could hide a mutation in enzyme #4.
   e. Each step in a pathway such as this is catalyzed by an enzyme, which is coded by a gene.

2. Eukaryotic mRNA:
   a. uses snRNPs to cut out introns and seal together translatable exons.
   b. uses a spliceosome mechanism made of DNA to recognize consensus regions to cut and splice.
   c. has a guanine cap on its 3′ end and a poly(A) tail on its 5′ end.
   d. is composed of adenine, thymine, guanine, and cytosine.
   e. codes the guanine cap and poly(A) tail from the DNA template.

3. A segment of a strand of DNA has a base sequence of 5′-GCATTAGAC-3′. What would be the sequence of an RNA molecule complementary to that sequence?
   a. 5′-GUCTAATGC-3′
   b. 5′-GCAUUAGAC-3′
   c. 5′-CGTAATCTG-3′
   d. 5′-GUCUAAUGC-3′
   e. 5′-CGUAAUCUG-3′

4. Which of the following statements about the initiation phase of translation is false?
   a. An initiation factor allows 5′ mRNA to attach to the small ribosomal subunit.
   b. Initiation factors complex with GTP to help Met–tRNA and AUG pair.
   c. mRNA attaches first to the small ribosomal subunit.
   d. GTP is synthesized.
   e. 3′-UAC-5′ on the tRNA binds 5′-AUG-3′ on mRNA.

5. Which of the following statements about aminoacylation is false?
   a. It precedes translation.
   b. It occurs in the ribosome.
   c. It requires ATP to bind an aminoacyl–tRNA synthetase.
   d. It joins the correct amino acid to a specific tRNA based on the tRNA's anticodon.
   e. It uses three binding sites on aminoacyl–tRNA synthetase.

6. Which of the following statements is false?
   a. GTP is an energy source during various stages of translation.
   b. In the ribosome, peptidyl transferase catalyses peptide bond formation between amino acids.
   c. When the mRNA code UAA reaches the ribosome, there is no tRNA to bind to it.
   d. A long polypeptide is cut off the tRNA in the A site so its Met amino acid links to the amino acid in the P site.
   e. Forty-two amino acids of a protein are encoded by 126 nucleotides of the mRNA.

7. Which item binds to SRP receptor and to the signal sequence to guide a newly synthesized protein to be secreted to its proper "channel"?
   a. ribosome
   b. signal recognition particle
   c. endoplasmic reticulum
   d. signal peptidase
   e. receptor protein

8. A part of an mRNA molecule with the sequence 5'-UGC GCA-3' is being translated by a ribosome. The following activated tRNA molecules are available. Two of them can correctly bind the mRNA so that a dipeptide can form.

tRNA Anticodon	Amino Acid
3'-GGC-5'	Proline
3'-CGU-5'	Alanine
3'-UGC-5'	Threonine
3'-CCG-5'	Glycine
3'-ACG-5'	Cysteine
3'-CGG-5'	Alanine

   a. cysteine–alanine
   b. proline–cysteine
   c. glycine–cysteine
   d. alanine–alanine
   e. threonine–glycine

9. A missense mutation cannot be:
   a. the code for the sickle-cell gene.
   b. caused by a frameshift.
   c. the deletion of a base in a coding sequence.
   d. the addition of two bases in a coding sequence.
   e. the same as a silent mutation.

10. Which of the following is *not* correct about transposable elements?
   a. They can be recognized by their ends of inverted transposable elements.
   b. They have an internal portion that can be transcribed.
   c. They encode a transposase enzyme.
   d. They have no harmful effects on cell function.
   e. They move by a cut-and-paste or copy-and-paste mechanism.

Discuss the Concepts

1. Which do you think are more important to the accuracy by which amino acids are linked into proteins: nucleic acids or enzymatic proteins? Why?

2. A mutation occurs that alters an anticodon in a tRNA from 3'-AAU-5' to 3'-AUU-5'. What effect will this mutation have on protein synthesis?

3. The normal form of a gene contains the nucleotide sequence:

    5'-ATGCCCGCCTTTGCTACTTGGTAG-3'
    3'-TACGGGCGGAAACGATGAACCATC-5'

When this gene is transcribed, the result is the following mRNA molecule:

    5'-AUGCCCGCCUUUGCUACUUGGUAG-3'

In a mutated form of the gene, two extra base pairs (underlined) are inserted:

    5'-ATGCCCGCCT<u>AA</u>TTGCTACTTGGTAG-3'
    3'-TACGGGCGGA<u>TT</u>AACGATGAACCATC-5'

What effect will this particular mutation have on the structure of the protein encoded in the gene?

4. A geneticist is attempting to isolate mutations in the genes for four enzymes acting in a metabolic pathway in the bacterium *Escherichia coli*. The end product *E* of the pathway is absolutely essential for life:

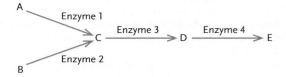

The geneticist has been able to isolate mutations in the genes for enzymes 1 and 2, but not for enzymes 3 and 4. Develop a hypothesis to explain why.

5. Experimental systems have been developed in which transposable elements can be induced to move under the control of a researcher. Following the induced transposition of a yeast TE element, two mutants were identified with altered activities of enzyme X. One of the mutants lacked enzyme activity completely, whereas the other had five times as much enzyme activity as normal cells did. Both mutants were found to have the TE inserted into the gene for enzyme X. Propose hypotheses for how the two different mutant phenotypes were produced.

Design an Experiment

How could you show experimentally that the genetic code is universal; namely, that it is the same in bacteria as it is in eukaryotes such as fungi, plants, and animals?

Interpret the Data

The **Figure** below shows amino acid changes that occurred at a particular position in a polypeptide as a result of mutations in the gene encoding the polypeptide. The amino acids connected by a line are specified by codons that differ in a single base. Using the genetic code shown in Figure 15.5, deduce the codons that specify the amino acids.

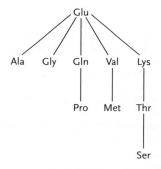

Apply Evolutionary Thinking

How might the process of alternative splicing and exon shuffling affect the rate at which new proteins evolve?

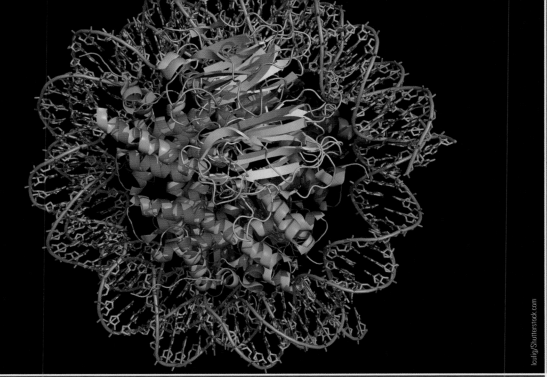

DNA wrapped around a core of eight histone proteins to form a nucleosome. A nucleosome is the basic structural unit of eukaryotic chromosomes.

Regulation of Gene Expression

Why it matters . . . A human egg cell is almost completely metabolically inactive when it is released from the ovary. It remains quiescent as it begins its journey down the oviduct leading from the ovary to the uterus, carried along by movements of cilia lining the walls of the tube **(Figure 16.1)**. It is here, in the oviduct, that egg and sperm cells meet and embryonic development begins. Within seconds after the cells unite, the fertilized egg breaks its quiescent state and begins a series of divisions that continues as it moves through the fallopian tube and enters the uterus. Subsequent divisions produce specialized cells that *differentiate* into the distinct types tailored for specific functions in the body, such as muscle cells and cells of the nervous system.

All the nucleated cells in both the developing embryo and the adult retain the same set of genes. The structural and functional differences in the cell types are determined not by the presence or absence of certain genes but rather through differences in patterns of *gene expression* that result in cell type-specific sets of proteins. Some genes, known as *housekeeping genes,* are expressed in almost all cell types. *Regulated genes,* by contrast, are expressed in a controlled way, meaning that they may or may not be expressed at any given time or in a given cell type. That is, each differentiated cell is characterized by genes that are active in only that cell type. For example, all mammalian cells carry the genes for hemoglobin, but these genes are active only in cells that give rise to red blood cells. Specific regulatory events activate the hemoglobin genes only in the precursor cells to red blood cells and not in other cell types.

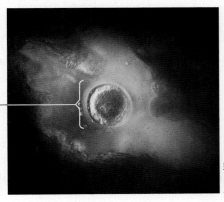

Egg

FIGURE 16.1 A human egg at the time of its release from the ovary. The outer layer appearing light blue in color is a coat of polysaccharides and glycoproteins that surrounds the egg. Within the egg, genes and regulatory proteins are poised to enter the pathways initiating embryonic development.

The fundamental mechanisms that control gene expression are common to all multicellular eukaryotes. Even single-celled eukaryotes and prokaryotes have systems that regulate gene expression. With few exceptions, however, prokaryotic systems are limited almost exclusively to short-term responses to environmental changes; eukaryotic cells exhibit both short-term responses and long-term differentiation.

The regulation of gene expression occurs at several levels. **Transcriptional regulation,** the fundamental level of control, determines which genes are transcribed into mRNA. Additional controls fine-tune regulation by affecting the processing of mRNA *(posttranscriptional regulation),* its translation into proteins *(translational regulation),* and the life span and activity of the proteins themselves *(posttranslational regulation).*

These levels of regulation ultimately affect more than proteins, because among the proteins are enzymes that determine the types and kinds of all other molecules made in the developing cell. So, effectively, these regulatory mechanisms tailor the production of all cellular molecules. The entire spectrum of controls constitutes an exquisitely sensitive mechanism regulating when, where, and what kinds and numbers of cellular molecules are produced.

In this chapter we examine the regulation of gene expression in prokaryotes and eukaryotes. Our discussion begins with bacterial systems, where researchers first discovered a mechanism for transcriptional regulation, and then moves to eukaryotic systems where regulation of gene activity is more complex. We then discuss the mechanisms by which genes regulate embryonic development, and finally how cancers develop when regulatory controls are lost.

16.1 Regulation of Gene Expression in Prokaryotes

Transcription and translation are closely regulated in prokaryotes in ways that reflect prokaryotic life histories. Prokaryotes are relatively simple, single-celled organisms with generations that take a matter of minutes. Rather than the complex patterns of long-term cell differentiation and development typical of multicellular eukaryotes, prokaryotic cells typically undergo rapid and reversible alterations in biochemical pathways that allow them to adapt quickly to changes in their environment. These changes are the outcomes of regulatory events that control gene expression.

The human intestinal bacterium *Escherichia coli,* for example, can catabolize a number of sugars and other molecules to provide carbon and energy for the cell. One of those sugars is lactose (milk sugar). Lactose is not essential for *E. coli* growth but, when lactose is present, the bacterium makes three proteins for catabolizing the sugar. That is, under these conditions, the three regulated genes for those proteins are expressed. In the absence of lactose, however, it does not make those proteins. That is, under these conditions, the three regulated genes for lactose catabolism are not expressed.

More generally, when the environment in which a bacterium lives changes, some metabolic processes are stopped and others are started. Typically, this involves turning off the genes for the metabolic processes not needed and turning on the genes for the new metabolic processes. Each metabolic process involves a few to many genes, and the regulation of those genes must be coordinated. Overall, the regulation of genes in this way conserves energy for the bacterium because gene products are made only when they are needed. In the following subsections we learn about some examples of coordinated regulation of gene expression in prokaryotes.

Some Regulated Genes Occur in Clusters Called Operons

With the exception of housekeeping genes, genes in the prokaryotic genome are regulated. Some of those genes occur singly, meaning that the gene is transcribed by RNA polymerase to produce an mRNA molecule that is translated to produce a single polypeptide. A useful term to introduce here is the **transcription unit,** which means the segment of DNA from the initiation point of transcription to the termination point of transcription. The transcription unit for the gene that occurs singly in a genome corresponds to the mRNA-coding sequence of that gene.

Other regulated genes occur in tightly associated clusters. Each cluster constitutes one transcription unit, meaning that the set of genes in the cluster is transcribed into a single mRNA molecule. Translation of the mRNA produces polypeptides corresponding to each of the genes in the transcription unit. The organization of genes in a cluster provides a means for efficient coordinate regulation of those genes, called the *operon model* for the control of gene expression. The operon model was proposed in 1961 by François Jacob and Jacques Monod of the Pasteur Institute in Paris to explain the control of the expression of genes for lactose catabolism in *E. coli* (described in the next two subsections). Subsequently, research has shown that the operon model is applicable to the regulation of expression of many genes in prokaryotes and their viruses. Jacob and Monod received the Nobel Prize in 1965 for their discovery of bacterial operons and their regulation by repressors. As stated earlier, though, not all regulated genes in prokaryotes are organized into operons.

Formally, an **operon** is a cluster of prokaryotic genes organized into a single transcription unit and its associated **regulatory sequences.** Regulatory sequences are DNA sequences involved in the regulation of a gene or genes. Proteins bind to these sequences to control the transcription of the gene or genes. A protein that binds to a specific DNA sequence is a *DNA-binding protein.* A particular DNA-binding protein binds to a specific DNA sequence by interactions between particular amino acid regions in the three-dimensional shape of the protein and specific base pairs of the DNA in a hand-in-glove manner. One regulatory DNA sequence is the **promoter,** which is the site to which RNA polymerase binds to begin transcription. The other regulatory DNA sequence in an operon is the **operator,** a short segment to which a specific protein binds to affect the expression of the operon. The protein is an example of a **regulatory protein,** a DNA-binding protein that binds to a regulatory sequence and affects the expression of an associated gene or genes. For operons, the regulatory protein is encoded by a gene that is separate from the operon the protein controls. Some operons are controlled by a regulatory protein termed a **repressor,** which, when active, prevents the operon genes from being expressed. Other operons are controlled by a regulatory protein termed an **activator,** which, when active, turns on the expression of the genes.

In the rest of this section, we discuss two operons in *E. coli.*

The *lac* Operon for Lactose Metabolism Is Transcribed When an Inducer Inactivates a Repressor

Lactose is a sugar that, when catabolized, provides energy for the cell. Jacob and Monod used genetic and biochemical approaches to study the genetic control of lactose metabolism in *E. coli.* Their genetic studies showed that the protein products of three genes, *lacZ, lacY,* and *lacA,* are involved in lactose catabolism **(Figure 16.2).** The three genes are adjacent to one another on the chromosome in the order *Z–Y–A* and constitute a single transcription unit. The single promoter is upstream of *lacZ,* and the genes are transcribed as a unit into a single mRNA starting with the *lacZ* gene. The *lacZ* gene encodes the enzyme β-galactosidase, which has two catalytic activities. One activity is the hydrolysis of the disaccharide sugar, lactose, into the monosaccharide sugars, glucose and galactose. These sugars are then catabolized by other enzymes, producing energy for the cell. The other activity is converting lactose to *allolactose,* an isomer of lactose. (Isomers are discussed in Section 3.1.) As we will see, allolactose plays an important role in regulating the expres-

sion of the genes of the *lac* operon. The *lacY* gene encodes a permease (not an enzyme, despite its name), a protein that actively transports lactose into the cell. The *lacA* gene encodes a transacetylase enzyme; its function is unclear.

Jacob and Monod called the cluster of genes and adjacent sequences that control their expression the *lac* operon (see Figure 16.2). They coined the name *operon* from a key DNA sequence they discovered between the promoter and the *lacZ* gene that, through binding a protein, regulates transcription of the operon—the *operator.*

The *lac* operon is a negatively regulated system controlled by a regulatory protein termed the **Lac repressor.** The Lac repressor is encoded by the regulatory gene *lacI,* which is nearby but separate from the *lac* operon, and is synthesized in *active* form (see Figure 16.2). In general, the term **structural gene** is used for a gene that encodes a protein that has a function other than gene regulation; examples are the three *lac* operon genes. The term **regulatory gene** is used for a gene that encodes a protein that regulates the expression of structural genes.

Figure 16.3A shows how the Lac repressor inhibits transcription when lactose is absent from the medium. Here, the three-dimensional shape of the active Lac repressor positions amino acid regions that can recognize and bind specifically to base pairs in the operator of the *lac* operon. When lactose is added to the medium, expression of the *lac* operon increases about 100-fold **(Figure 16.3B).** The molecular switch for this change is set when some lactose molecules that enter the cell are converted by β-galactosidase in the cell to **allolactose.** Allolactose is the **inducer** for the *lac* operon, so-called because it causes—induces—the transcription of the operon's structural genes. Allolactose works by inactivating the Lac repressor (see Figure 16.3B). It does so by binding to a specific site on the repressor protein, causing it to undergo an allosteric shift (change in shape). In the altered protein, the DNA-binding amino acid regions are no longer positioned appropriately to recognize the DNA sequence of the operator and, hence, the repressor cannot bind to the operator. Because an inducer molecule increases its expression, the *lac* operon is called an **inducible operon.**

When the lactose is used up from the medium, the regulatory system again switches the *lac* operon off. That is, in the absence of lactose, no allolactose inducer molecules are pro-

FIGURE 16.2 The *E. coli lac* operon. The *lacZ, lacY,* and *lacA* genes encode the proteins taking part in lactose metabolism. The separate regulatory gene, *lacI,* encodes the Lac repressor, which plays a pivotal role in the control of the operon. The promoter binds RNA polymerase, and the operator binds activated Lac repressor. The transcription unit, which extends from the transcription initiation site to the transcription termination site, contains the structural genes.

© Cengage Learning 2014

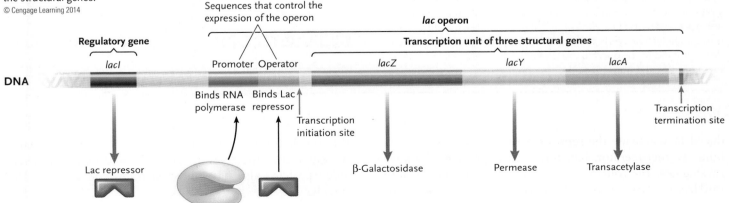

A. Lactose absent from medium: structural genes not transcribed

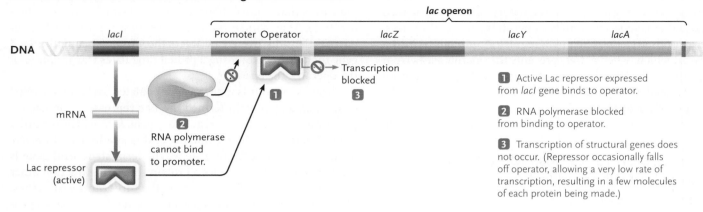

1. Active Lac repressor expressed from *lacI* gene binds to operator.

2. RNA polymerase blocked from binding to operator.

3. Transcription of structural genes does not occur. (Repressor occasionally falls off operator, allowing a very low rate of transcription, resulting in a few molecules of each protein being made.)

B. Lactose present in medium: structural genes transcribed

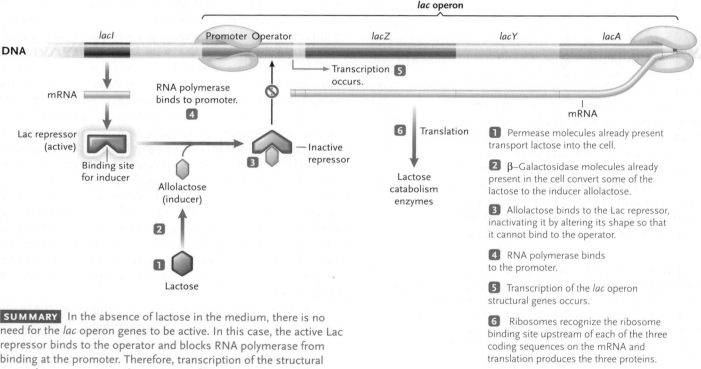

1. Permease molecules already present transport lactose into the cell.

2. β–Galactosidase molecules already present in the cell convert some of the lactose to the inducer allolactose.

3. Allolactose binds to the Lac repressor, inactivating it by altering its shape so that it cannot bind to the operator.

4. RNA polymerase binds to the promoter.

5. Transcription of the *lac* operon structural genes occurs.

6. Ribosomes recognize the ribosome binding site upstream of each of the three coding sequences on the mRNA and translation produces the three proteins.

SUMMARY In the absence of lactose in the medium, there is no need for the *lac* operon genes to be active. In this case, the active Lac repressor binds to the operator and blocks RNA polymerase from binding at the promoter. Therefore, transcription of the structural genes does not occur. In the presence of lactose, the *lac* operon genes are expressed so that catabolism of lactose can occur. In this case, the inducer allolactose binds to the Lac repressor inactivating it so that it does not bind to the operator. As a result, RNA polymerase binds to the promoter and transcribes the structural genes.

THINK LIKE A SCIENTIST If there was a mutation in the *lacI* gene that results in a Lac repressor that could not bind to the operator, what effect would that have on the regulation of the *lac* operon?

© Cengage Learning 2014

duced to inactivate the repressor. The now-active repressor binds to the operator, blocking transcription of the structural genes. The controls are aided by the fact that bacterial mRNAs are very short-lived, about 3 minutes on the average.

This quick turnover permits the cytoplasm to be cleared quickly of the mRNAs transcribed from an operon. The encoded proteins also have short lifetimes and are degraded quickly.

Transcription of the *lac* Operon Is Also Controlled by a Positive Regulatory System

A *positive gene regulation* system also regulates the *lac* operon. This system ensures that the *lac* operon is transcribed efficiently if lactose is provided as an energy source, but not if glucose is present in addition to lactose. This is because glucose is a more efficient source of energy than lactose. Glucose can be used directly in the glycolysis pathway to produce energy for the cell (see Chapter 8). Lactose, on the other hand, must first be con-verted into glucose and galactose, and the galactose then converted into glucose. These conversions require energy from the cell. Thus the cell gains more net energy by catabolizing glucose than by catabolizing lactose, or for that matter, any other sugar.

Figure 16.4 shows the positive gene regulation system under the two conditions of lactose present + glucose low or absent (efficient transcription of *lac* operon genes), and lactose present + glucose present (very low level of transcription of *lac* operon genes). In essence, we are adding to the model shown in Figure 16.3B. The key regulatory molecule involved in positive

A. Lactose present and glucose low or absent: structural genes expressed at high levels

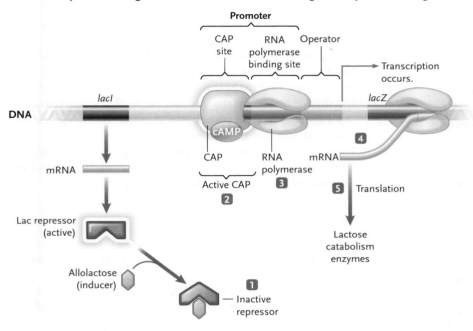

1 Lactose converted to the inducer, allolactose, which inactivates Lac repressor.

2 Active adenylyl cyclase synthesizes cAMP to high levels. cAMP binds to activator CAP, activating it. Activated CAP binds to CAP site in the promoter.

3 RNA polymerase binds efficiently to the promoter.

4 Genes of operon transcribed to high levels.

5 Translation produces high amounts of proteins.

B. Lactose present and glucose present: structural genes expressed at very low levels

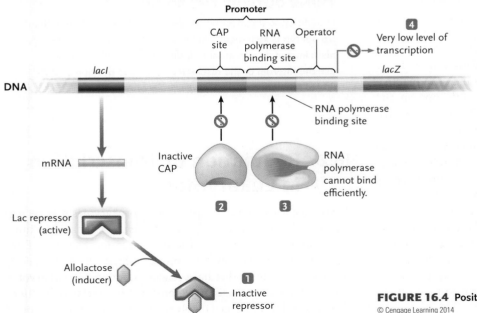

1 Lactose converted to the inducer, allolactose, which inactivates Lac repressor.

2 Catabolism of incoming glucose leads to inactivation of adenylyl cyclase, which causes the amount of cAMP in the cell to drop to a level too low to activate CAP. Inactive CAP cannot bind to the CAP site.

3 RNA polymerase is unable to bind to the promoter efficiently.

4 Transcription occurs at a very low level: Because the Lac repressor is not present to block RNA polymerase from binding to the promoter, the level of transcription is higher than when lactose is absent, but far lower than when lactose is present and glucose is absent.

FIGURE 16.4 Positive regulation of the *lac* operon by the CAP activator.
© Cengage Learning 2014

gene regulation of the *lac* operon is the DNA-binding protein, **CAP (catabolite activator protein)**. CAP is an **activator,** a regulatory protein that stimulates gene expression. It is synthesized in *inactive* form. When activated by cAMP (cyclic AMP, a nucleotide that plays a role in regulating cellular processes in both prokaryotes in eukaryotes; see Section 7.4), CAP binds to the **CAP site** in the promoter for the *lac* operon and enables RNA polymerase to bind efficiently and transcribe the operon's genes. If glucose is absent, cAMP levels are high, resulting in active CAP. If glucose is present, cAMP levels are low, resulting in inactive CAP, which cannot bind to the CAP site.

The same positive gene regulation system using CAP and cAMP regulates a large number of other operons that control the catabolism of many sugars. In each case, the system functions so that glucose, if it is present in the growth medium, is catabolized first.

Transcription of the *trp* Operon Genes for Tryptophan Biosynthesis Is Repressed When Tryptophan Activates a Repressor

Tryptophan is an amino acid that is used in the synthesis of proteins. As such, tryptophan is a critical component for cell growth and survival because protein synthesis is the main activity that the cell performs. Therefore, if tryptophan is absent from the growth medium, *E. coli* must synthesize tryptophan so that it can make its proteins. If tryptophan is present in the medium, the cell will use that source of the amino acid rather than synthesizing its own.

Tryptophan biosynthesis also involves an operon, the *trp* operon **(Figure 16.5)**. The five structural genes in this operon, *trpE–trpA,* encode the enzymes for the steps in the tryptophan biosynthesis pathway. Upstream of the *trpE* gene are the operon's promoter and operator sequences. Expression of the *trp* operon is controlled by the Trp repressor, a regulatory protein encoded by the regulatory gene, *trpR,* which is located elsewhere in the genome. In contrast to the Lac repressor, the Trp repressor is synthesized in an inactive form that cannot bind to the operator. That is, an inactive Trp repressor has a three-dimensional shape that cannot recognize and bind to the DNA sequence of the operator for which it is specific. The inactive state of the repressor leads to the default state of the operon, the expression of the *trp* operon structural genes (see Figure 16.5A). The repressor is activated when tryptophan levels are high, such as when tryptophan is present in the medium (see Figure 16.5B). Activation occurs when tryptophan binds to a specific site on the Trp repressor and causes the DNA-binding protein to undergo an allosteric shift (change in shape) to a three-dimensional shape with amino acid regions that now can recognize and bind to the DNA sequence of the operator. Because the presence of tryptophan represses the expression of the tryptophan biosynthesis genes, this operon is an example of a **repressible operon.** Here, tryptophan acts as a **corepressor,** a regulatory molecule that activates the repressor to turn off expression of the operon.

To compare and contrast the two operons we have discussed:

- *lac* **operon:** the repressor is synthesized in an active form. The inducer allolactose inactivates the repressor. The structural genes are then transcribed.
- *trp* **operon:** the repressor is synthesized in an inactive form so the structural genes are transcribed. The corepressor (tryptophan from the growth medium) activates the repressor. Active repressor blocks transcription of the operon.

Inducible and repressible operons illustrate two types of *negative gene regulation* because both are regulated by a repressor that turns off gene expression when it is in active form. Genes are expressed only when the repressor is in inactive form.

In sum, regulation of gene expression in prokaryotes occurs primarily at the transcription level. There are also some examples of regulation at the translation level. For example, some proteins can bind to the mRNAs that produce them and modulate their translation. This serves as a feedback mechanism to fine-tune the amounts of the proteins in the cell. In the next section, we discuss the regulation of gene expression in eukaryotes. You will see that regulation occurs at several points between the gene and the protein, and that regulatory mechanisms of eukaryotes are more complex than those in prokaryotes.

STUDY BREAK 16.1

1. Suppose the *lacI* gene is mutated so that the Lac repressor is not made. How does this mutation affect the regulation of the *lac* operon?
2. Answer the equivalent question for the *trp* operon: how would a mutation that prevents the Trp repressor from being made affect the regulation of the *trp* operon?

THINK OUTSIDE THE BOOK

In addition to the mechanism described, transcription of the *trp* operon is also regulated by another regulatory mechanism known as attenuation. Individually or collaboratively, use the Internet or the research literature to develop an outline of regulation of the *trp* operon by attenuation.

16.2 Regulation of Transcription in Eukaryotes

As you just learned, gene expression in prokaryotes may be regulated at the transcription level with genes organized in functional units called operons. The molecular mechanisms of operon regulation provide a simple means of coordinating synthesis of proteins that have related functions. In eukaryotes, the coordinated synthesis of proteins that have related functions also occurs, but the genes involved are usually scattered around

FIGURE 16.5 Regulation of the repressible *trp* operon by the Trp repressor in the absence (A) and presence (B) of tryptophan.

A. Tryptophan absent from medium: tryptophan must be made by the cell—structural genes transcribed

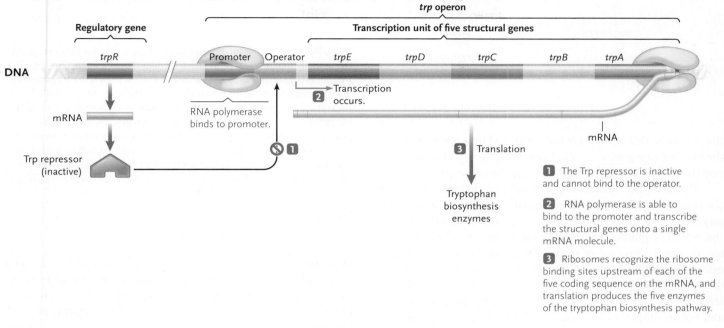

1. The Trp repressor is inactive and cannot bind to the operator.

2. RNA polymerase is able to bind to the promoter and transcribe the structural genes onto a single mRNA molecule.

3. Ribosomes recognize the ribosome binding sites upstream of each of the five coding sequence on the mRNA, and translation produces the five enzymes of the tryptophan biosynthesis pathway.

B. Tryptophan present in medium: cell uses tryptophan in medium rather than synthesizing it—structural genes not transcribed

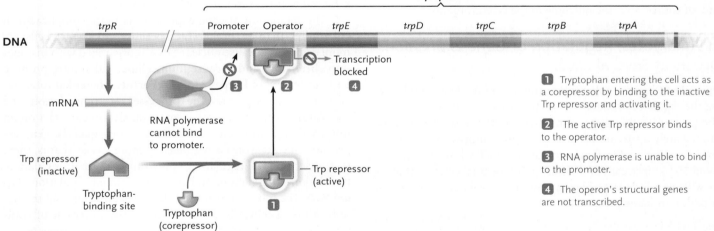

1. Tryptophan entering the cell acts as a corepressor by binding to the inactive Trp repressor and activating it.

2. The active Trp repressor binds to the operator.

3. RNA polymerase is unable to bind to the promoter.

4. The operon's structural genes are not transcribed.

SUMMARY In the absence of tryptophan in the medium, tryptophan must be made. In this scenario, the Trp repressor is inactive, which enables RNA polymerase to bind to the promoter and transcribe the structural genes for the tryptophan biosynthesis enzymes. In the presence of tryptophan in the medium, that tryptophan can be used for protein synthesis, so there is no need for the cell to make tryptophan. Tryptophan binds to the Trp repressor, activating it. The active repressor binds to the operator, thereby blocking RNA polymerase from binding to the promoter. Transcription of the structural genes does not occur.

THINK LIKE A SCIENTIST If there was a mutation in the *trpR* gene that results in a Trp repressor that cannot bind tryptophan, what effect would that have on the regulation of the *trp* operon.

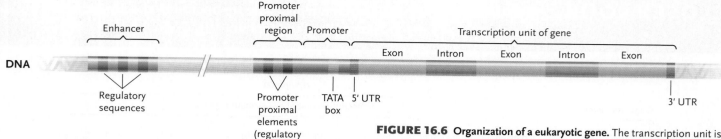

FIGURE 16.6 **Organization of a eukaryotic gene.** The transcription unit is the segment that is transcribed into the pre-mRNA; it contains the 5′ UTR (untranslated region), exons, introns, and 3′ UTR. Immediately upstream of the transcription unit is the promoter, which often contains the TATA box. Adjacent to the promoter and further upstream of the transcription unit is the promoter proximal region, which contains regulatory sequences called promoter proximal elements. More distant from the gene in some cases is the enhancer, which contains regulatory sequences that control the rate of transcription of the gene. Transcription of the gene produces a pre-mRNA molecule with a 5′ cap and 3′ poly(A) tail; processing of the pre-mRNA to remove introns generates the functional mRNA (see Chapter 15).
© Cengage Learning 2014

the genomes; that is, they are not organized into operons. Nonetheless, like operons, individual eukaryotic genes also consist of protein-coding sequences and their regulatory sequences.

There are two general types of eukaryotic gene regulation. Short-term regulation involves regulatory events in which gene sets are quickly turned on or off in response to changes in environmental or physiological conditions in the cell's or organism's environment. This type of regulation is most similar to prokaryotic gene regulation. Long-term gene regulation involves regulatory events required for an organism to develop and differentiate. Long-term gene regulation occurs in multicellular eukaryotes and not in simpler, unicellular eukaryotes. The mechanisms we discuss in this and the next section are applicable to both short-term and long-term regulation. The specific molecules and genes involved in short-term and long-term regulation are different and, of course, so is the outcome to the cell or organism.

In Eukaryotes, Regulation of Gene Expression Occurs at Several Levels

Figure 16.6 shows a eukaryotic protein-coding gene, emphasizing the regulatory sequences involved in its expression. Eukaryotic protein-coding genes consist of single transcription units. Immediately upstream of the transcription unit is the promoter. The promoter in the figure contains a TATA box, a sequence about 25 bp upstream of the start point for transcription that, as we will see a little later, plays an important role in transcription initiation in many promoters.

The TATA box has the 7-bp consensus sequence $\begin{matrix} 5'\text{-TATAAAA-}3' \\ 3'\text{-ATATTTT-}5' \end{matrix}$. Promoters without TATA boxes have other sequence elements that play a similar role. The following discussions involve a TATA box–containing promoter.

RNA polymerase II itself cannot recognize the promoter sequence. Instead, particular **transcription factors**—proteins required for RNA polymerase to initiate transcription or that regulate that process—recognize and bind to the TATA box and then recruit the polymerase. Once the RNA polymerase II–transcription factor complex forms, the polymerase unwinds the DNA and transcription begins. Adjacent to the promoter, further upstream, is the **promoter proximal region,** which contains regulatory sequences called **promoter proximal**

elements. Regulatory proteins (types of transcription factors) that bind to promoter proximal elements may stimulate or inhibit the rate of transcription initiation. More distant from the beginning of some protein-coding genes is the **enhancer.** Regulatory proteins (types of transcription factors) binding to regulatory sequences within an enhancer stimulate or inhibit the rate of transcription initiation, fine-tuning the regulation achieved at the promoter proximal elements.

The regulation of gene expression is more complicated in eukaryotes than in prokaryotes because eukaryotic cells are more complex, because nuclear DNA is organized with histones into chromatin, and because multicellular eukaryotes produce large numbers and types of cells. Further, the eukaryotic nuclear envelope separates the processes of transcription and translation, whereas in prokaryotes translation can start on an mRNA that is still being synthesized. Consequently, gene expression in eukaryotes is regulated at more levels. That is, there is transcriptional regulation, posttranscriptional regulation, translational regulation, and posttranslational regulation **(Figure 16.7).** The most important of these is transcriptional regulation. More specifically, for most genes the initiation of transcription is the major control point and involves changes in the structure of the chromatin at the promoter, and regulatory events at a gene's promoter and regulatory sequences. In the following two subsections, we first discuss regulatory events at a gene's promoter so as to understand the mechanisms operating at the DNA level, and then we discuss the chromatin changes.

Regulation of Transcription Initiation Involves the Effects of Transcription Factors Binding to a Gene's Promoter and Regulatory Sites

Transcription initiation is the most important level at which regulation of gene expression takes place.

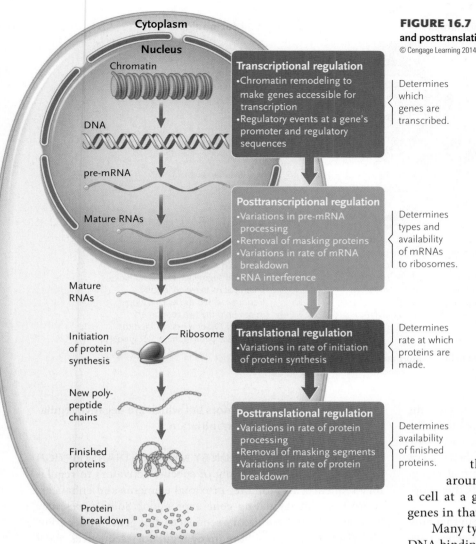

FIGURE 16.7 Steps in transcriptional, posttranscriptional, translational, and posttranslational regulation of gene expression in eukaryotes.
© Cengage Learning 2014

Transcriptional regulation
• Chromatin remodeling to make genes accessible for transcription
• Regulatory events at a gene's promoter and regulatory sequences

Determines which genes are transcribed.

Posttranscriptional regulation
• Variations in pre-mRNA processing
• Removal of masking proteins
• Variations in rate of mRNA breakdown
• RNA interference

Determines types and availability of mRNAs to ribosomes.

Translational regulation
• Variations in rate of initiation of protein synthesis

Determines rate at which proteins are made.

Posttranslational regulation
• Variations in rate of protein processing
• Removal of masking segments
• Variations in rate of protein breakdown

Determines availability of finished proteins.

polymerase II and orient the enzyme to start transcription at the correct place. The combination of general transcription factors with RNA polymerase II is the **transcription initiation complex.** On its own, this complex brings about only a low rate of transcription initiation, which leads to just a few mRNA transcripts.

Activators, a type of regulatory protein, are transcription factors that stimulate transcription initiation. Activators that bind to the promoter proximal elements interact directly with the general transcription factors at the promoter to stimulate transcription initiation, so many more transcripts are synthesized in a given time. Housekeeping genes—genes that are expressed in all cell types for basic cellular functions such as glucose metabolism—have promoter proximal elements that are recognized by activators present in all cell types. By contrast, genes expressed only in particular cell types or at particular times have promoter proximal elements that are recognized by activators found only in those cell types, or at those times when transcription of these genes needs to be activated. To turn this around, the particular set of activators present within a cell at a given time is responsible for determining which genes in that cell are expressed to a significant level.

Many types of activators are found in eukaryotic cells. The DNA binding and activation functions of activators are properties of two distinct domains in the proteins. (Protein domains were introduced in Section 3.4.) The three-dimensional arrangement of amino acid chains in the domains produces highly specialized regions called **motifs.** The motifs give the protein its specialized function. Motifs found in the DNA-binding do-

ACTIVATION OF TRANSCRIPTION To initiate transcription, proteins called **general transcription factors** (also called *basal transcription factors*) bind to the promoter in the area of the TATA box **(Figure 16.8).** These factors recruit the enzyme RNA

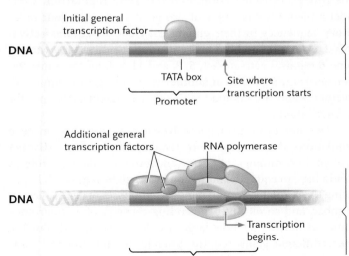

1 The first general transcription factor recognizes and binds to the TATA box of a protein-coding gene's promoter.

2 Additional general transcription factors and then RNA polymerase adds to the complex. A general transcription factor unwinds the promoter DNA, and then transcription begins.

FIGURE 16.8 Formation of the transcription complex on the promoter of a protein-coding gene by the combination of general transcription factors with RNA polymerase. The general transcription factors are needed for RNA polymerase to bind and initiate transcription at the correct place.
© Cengage Learning 2014

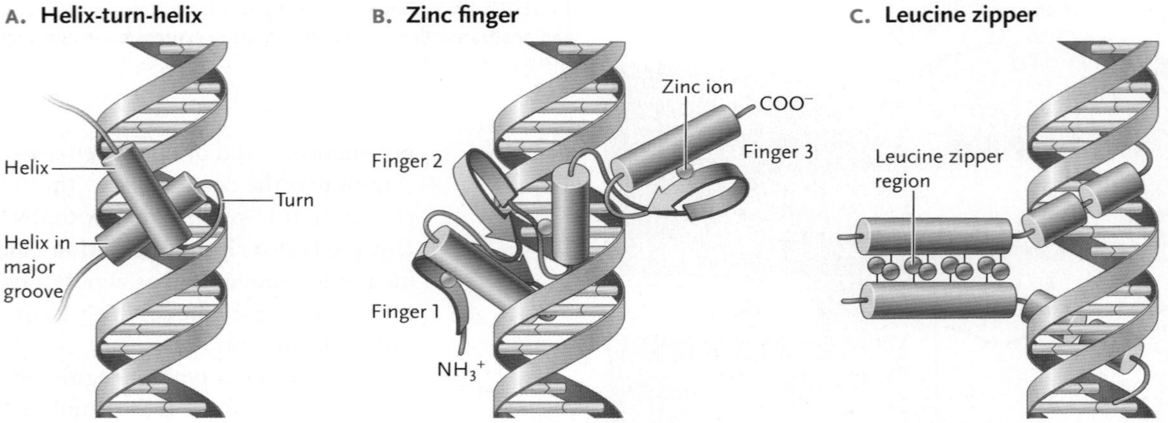

A. Helix-turn-helix

Helix

Turn

Helix in major groove

A helix-turn-helix motif part of a protein bound to DNA. One of the α-helices binds to base pairs in the major groove of the DNA. A looped region of the protein—the turn—connects to a second α-helix that helps hold the first helix in place.

B. Zinc finger

Zinc ion

COO⁻

Finger 2

Finger 3

Finger 1

NH_3^+

Zinc finger motifs are parts of proteins named for their resemblance to fingers and the presence of a bound zinc atom. Zinc fingers bind to specific base pairs in the grooves of DNA.

C. Leucine zipper

Leucine zipper region

Leucine zipper proteins are dimers, with each monomer consisting of α-helical segments. Hydrophobic interactions between leucine residues within the leucine zipper motif hold the two monomers together. Other α-helices bind to DNA base pairs in the major groove.

FIGURE 16.9 Three DNA-binding motifs found in activators and other regulatory proteins.
© Cengage Learning 2014

mains of regulatory proteins, such as activators, include the helix-turn-helix, zinc finger, and leucine zipper **(Figure 16.9)**. For each type, an amino acid sequence binds to a sequence of base pairs accessed in the grooves of the DNA double helix. Each DNA-binding protein has an amino acid sequence that recognizes a specific base-pair sequence in DNA and it is this property which gives the protein its specificity of action.

Some genes have enhancers associated with them that bind activators **(Figure 16.10)**. The enhancers of different genes have specific sets of regulatory sequences, which bind particular activators. A **coactivator** (also called a *mediator*), a large multiprotein complex, forms a bridge between the activators at the enhancer and the proteins at the promoter and promoter proximal region. As a result, the concentration of activators at the promoter is increased, which stimulates transcription up to its maximal rate.

REPRESSION OF TRANSCRIPTION In some genes, repressors—transcription factors that are inhibitory to transcription initiation—oppose the effect of activators, thereby reducing or blocking the rate of transcription. The final rate of transcription then depends on the "battle" between the activation signal and the repression signal.

Different kinds of repressors in eukaryotes work in different ways. Some repressors bind to the same regulatory sequence to which activators bind (often in the enhancer), thereby preventing activators from binding to that site. Other repressors bind to their own specific site in the DNA near where the activator binds and interact with the activator so that it cannot interact with the coactivator. Yet other repressors bind to specific sites in the DNA and recruit **corepressors,** multiprotein com-

plexes analogous to coactivators but which are negative regulators, inhibiting transcription initiation.

CONTROL OF TRANSCRIPTION BY MULTIPLE TRANSCRIPTION FACTORS How is the binding of specific activators to regulatory sequences in promoter proximal elements and enhancers coordinated in regulating gene expression? Some genes may have one to a few regulatory sequences, but genes under complex regulatory control have many regulatory sequences. Each regulatory sequence binds a specific transcription factor. A relatively small number of transcription factors (activators and repressors) control transcription of all protein-coding genes. By combining a few transcription factors in particular ways, the transcription of a wide array of genes can be controlled. The process is called **combinatorial gene regulation.** Consider a theoretical example of two genes, each with four regulatory sequences in their enhancers. Gene *A* requires activators 2, 5, 7, and 8 to bind for transcription activation, whereas gene *B* requires activators 1, 5, 8, and 11 to bind for transcription activation. Looked at another way, both genes require activators 5 and 8 combined with other different activators for full activation.

This operating principle solves a basic dilemma in gene regulation—if each gene were regulated by a single, distinct protein, the number of genes encoding regulatory proteins would have to equal the number of genes to be regulated. Regulating the regulators would require another set of genes of equal number, and so on until the coding capacity of any chromosome set, no matter how large, would be exhausted. But because different genes require different combinations of transcription factors, the number of genes encoding transcription

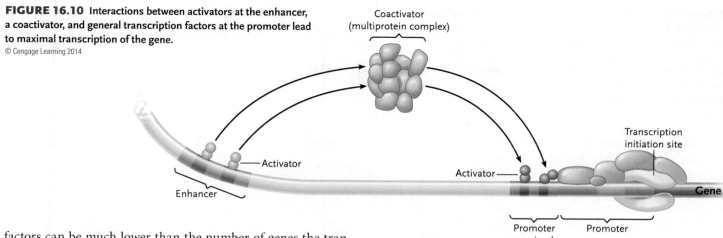

FIGURE 16.10 Interactions between activators at the enhancer, a coactivator, and general transcription factors at the promoter lead to maximal transcription of the gene.
© Cengage Learning 2014

factors can be much lower than the number of genes the transcription factors control.

The involvement of multiple transcription factors in controlling many eukaryotic genes helps us explain cell-specific gene expression. Because multiple activators are required to activate the transcription of a particular gene, the gene will be transcribed only in cells which contain those activators. **Figure 16.11** illustrates this for the human β-interferon gene, the product of which functions as part of the immune system to combat virus infections (see Chapter 45). In a normal cell that is not infected with a virus, no activators are present to activate the gene so transcription does not occur **(Figure 16.11A)**. In a virus-infected cell, activators J, I, and N are produced (the letters are shortened forms of their actual names) and bind to their specific regulatory sequences in the gene's enhancer, activating transcription of the gene **(Figure 16.11B)**. You can see from this example how cell type-specific gene expression can be achieved in a similar way. That is, transcription of a gene will be activated only if the appropriate array of activators is present for that gene in a cell. We can also use this example to illustrate combinatorial gene regulation. The N activator, for example, regulates many other genes including polypeptides that form part of immunoglobulin proteins (see Chapter 45).

COORDINATED REGULATION OF TRANSCRIPTION OF GENES WITH RELATED FUNCTIONS In the discussion of prokaryotic operons, you learned that genes with related function are often clustered in a single transcription unit, providing an effective way to control the synthesis of the products of those genes in a coordinate manner. There are no operons in eukaryotes, yet the transcription of genes with related function is coordinately controlled. How is this accomplished?

A. **Human β-interferon gene in a normal cell**

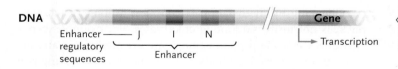

In a normal cell, no activators are bound to the enhancer regulatory sequences so the gene is not transcribed.

B. **Human β-interferon gene in a cell infected by a virus**

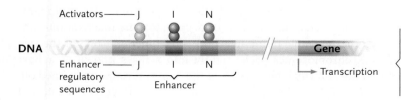

In a virus-infected cell, three different activators bind to the three enhancer regulatory sequences and transcription of the gene occurs.

FIGURE 16.11 Cell-specific regulation of transcription by activators.
© Cengage Learning 2014

FIGURE 16.12 Steroid hormone regulation of gene expression. A steroid hormone enters the cell and forms a complex in the cytoplasm with a steroid hormone receptor that is specific to the hormone. Steroid hormone–receptor complexes migrate to the nucleus, bind to the steroid hormone response element next to each gene they control (one such gene is shown in the figure), and affect transcription of those genes.
© Cengage Learning 2014

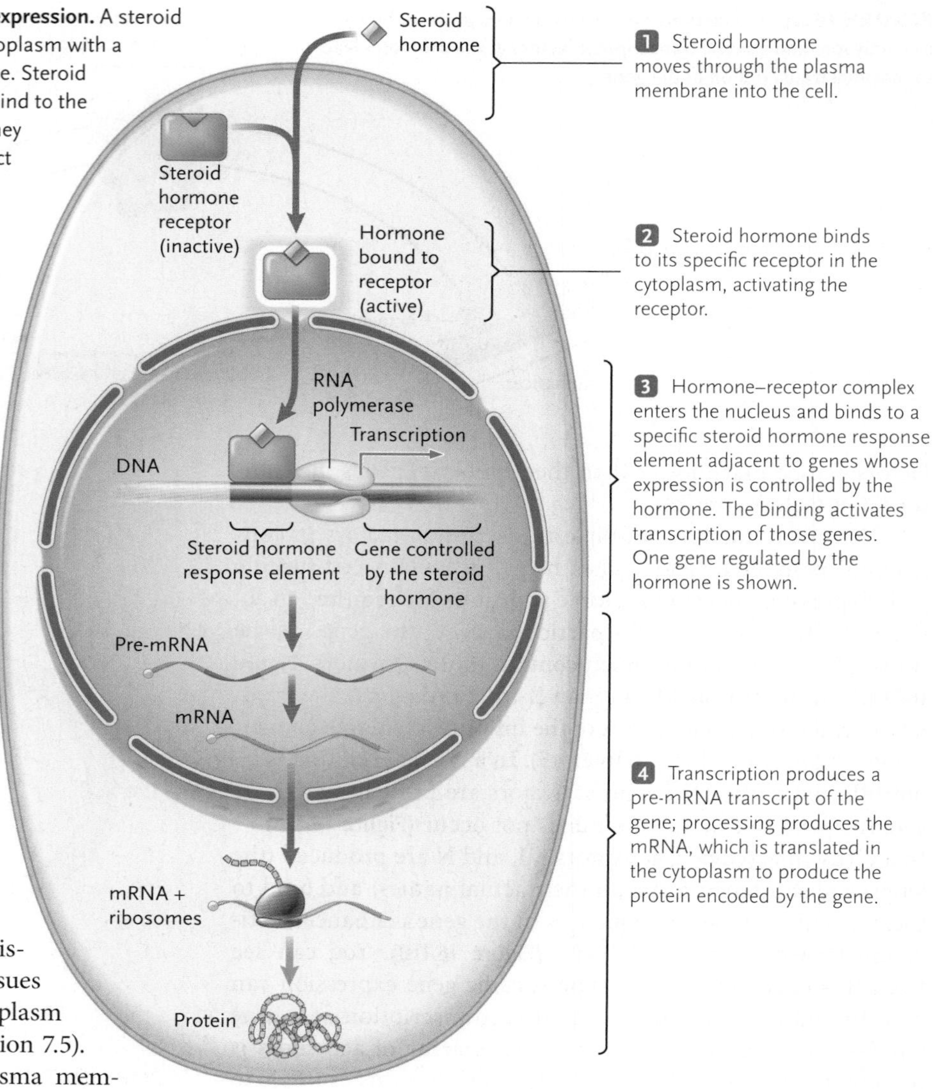

1 Steroid hormone moves through the plasma membrane into the cell.

2 Steroid hormone binds to its specific receptor in the cytoplasm, activating the receptor.

3 Hormone–receptor complex enters the nucleus and binds to a specific steroid hormone response element adjacent to genes whose expression is controlled by the hormone. The binding activates transcription of those genes. One gene regulated by the hormone is shown.

4 Transcription produces a pre-mRNA transcript of the gene; processing produces the mRNA, which is translated in the cytoplasm to produce the protein encoded by the gene.

The answer is that all genes that are coordinately regulated have the same regulatory sequences associated with them. Therefore, with one signal, the transcription of all of the genes can be controlled simultaneously. Consider the control of gene expression by steroid hormones in mammals. A **hormone** is a molecule produced by one tissue and transported via the bloodstream to a target tissue or tissues to alter physiological activity. A **steroid** is a type of lipid derived from cholesterol (see Section 3.3). Examples of steroid hormones are testosterone and glucocorticoid. Testosterone regulates the expression of a large number of genes associated with the maintenance of primary and secondary male characteristics. Glucocorticoid, among other actions, regulates the expression of genes involved in the maintenance of the concentration of glucose and other fuel molecules in the blood.

A steroid hormone acts on specific target tissues in the body because only cells in those tissues have **steroid hormone receptors** in their cytoplasm that recognize and bind the hormone (see Section 7.5). The steroid hormone moves through the plasma membrane into the cytoplasm and binds to its specific receptor, activating it **(Figure 16.12)**. The hormone–receptor complex then enters the nucleus and binds to specific regulatory sequences that are adjacent to the genes whose expression is controlled by the hormone. This binding activates transcription of those genes, and proteins encoded by the genes are synthesized rapidly.

A single steroid hormone can regulate many different genes because all of the genes have an identical DNA sequence—a **steroid hormone response element**—to which the hormone–receptor complex binds. For example, all genes controlled by glucocorticoid have a glucocorticoid response element associated with them. Therefore, the release of glucocorticoid into the bloodstream coordinately activates the transcription of genes with that response element.

Methylation of DNA Can Control Gene Transcription

In **DNA methylation,** enzymes add a methyl group ($—CH_3$) to cytosine bases in the DNA. Methylated cytosines in promoter regions can regulate transcription through a process called **silencing,** in which transcription of genes controlled by those promoters is turned off. This is an example of **epigenetics,** a phenomenon in which a change in gene expression does not involve a change in the DNA sequence of the gene or of the genome.

Silencing by methylation is common among vertebrates, but it is not universal among eukaryotes. For example, genes encoding the blood protein hemoglobin are methylated and inactive in most vertebrate body cells. In the cell lines giving rise to red blood cells, enzymes remove the methyl groups from the hemoglobin genes, which are then transcribed.

DNA methylation in some cases silences large blocks of genes, or even chromosomes. Recall from Section 13.2 that a dosage compensation mechanism inactivates one of the two X chromosomes in most body cells of female placental mammals, including humans. In X chromosome inactivation—another example of an epigenetic phenomenon—one of the two X chromosomes packs tightly into a mass known as a Barr body, in which most genes of the X chromosome are turned off. The inactivation occurs during embryonic development, and which X chromosome is inactivated in a particular embryonic cell line is a random event. As part of X chromosome inactivation, cytosines in the DNA become methylated.

DNA methylation underlies **genomic imprinting,** an epigenetic phenomenon in which the expression of an allele is determined by the parent that contributed it (see Section 13.5). In genomic imprinting, methylation permanently silences transcription of either the inherited maternal or paternal allele of a particular gene. The methylation occurs during gametogenesis in a parent. An inherited methylated allele, the *imprinted allele,* is not expressed—it is silenced. The expression of the gene involved therefore depends on expression of the nonimprinted allele inherited from the other parent. The methylation of the parental allele is maintained as the DNA is replicated, so that the silenced allele remains inactive in progeny cells. Figure 13.21 shows an example of genomic imprinting.

Chromatin Structure Plays an Important Role in Whether a Gene Is Active or Inactive

Eukaryotic DNA is organized into chromatin by combination with histone proteins (discussed in Section 14.5). Recall that DNA is wrapped around a core of two molecules each of histones H2A, H2B, H3, and H4, forming the nucleosome (see Figure 14.21 and Chapter Opener). Higher levels of chromatin organization occur when histone H1 links adjacent nucleosomes.

A eukaryotic promoter can exist in two states. In the inactive state, which is the normal state in eukaryotic cells, the nucleosomes in normal chromatin prevent general transcription factors and RNA polymerase II from binding so transcription does not occur. However, regulatory transcription factors can bind to the DNA and lead to a change in chromatin to make it active so transcription can occur. In the active state, general transcription factors and RNA polymerase II bind to the promoter and, controlled by the molecular events already discussed, transcription regulation can occur. A key regulatory event for regulating transcription initiation, then, is controlling the transition between the inactive and active states of chromatin in the region of a promoter.

Acetylation of histone tails (see Figure 14.21) is one mechanism that plays an important role in determining whether chromatin is inactive or active. In inactive chromatin, the his-

tone tails are not acetylated and, in this form, the tails form a tight association with the DNA wrapped around the histone octamer of a nucleosome **(Figure 16.13).** When a regulatory transcription factor binds to a regulatory sequence associated with a gene, it can recruit protein complexes that include *histone acetyltransferase,* an enzyme that acetylates (adds acetyl groups; $CH_3CO—$) to specific amino acids of the histone tails. Acetylation changes the charge of the histone tails and results in a loosening of the association of the histones with the DNA (see Figure 16.13). Usually acetylation of histones is not enough to make the chromatin completely active. Typically large multiprotein complexes bind to displace the acetylated nucleosomes in the promoter region from the DNA, or move them along the DNA away from the promoter. This type of change in chromatin structure is called *chromatin remodeling.* Then, general transcription factors and RNA polymerase II are free to bind and initiate transcription.

Inactivation of an active gene involves essentially the opposite of this process. With respect to the histones, the enzyme *histone deacetylase* catalyzes the removal of acetyl groups from the histone tails restoring the inactive state of the chromatin in that region (see Figure 16.13).

The tails of histones can also be modified at specific positions by the enzyme-catalyzed covalent addition of methyl groups or phosphate groups. These chemical modifications can also affect chromatin structure and gene expression. Histone methylation, for instance, is associated with gene inactivation. Like acetylation, methylation and phosphorylation of histone tails is reversible. Overall, the conclusion is that the patterns of modification of histone tails are important in determining chromatin structure and gene activity. This has led to the concept of the **histone code,** which is a regulatory mechanism for altering chromatin structure and, therefore, gene activity, based on signals in histone tails represented by chemical modification patterns.

Once mRNAs are transcribed from active genes, further regulation occurs at each of the major steps in the pathway from genes to proteins: during pre-mRNA processing and the movement of finished mRNAs to the cytoplasm (posttranscriptional regulation), during protein synthesis (translational regulation), and after translation is complete (posttranslational regulation). The next section takes up the regulatory mechanisms operating at each of these steps.

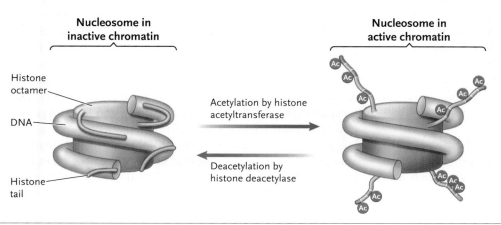

Nucleosome in inactive chromatin

Nucleosome in active chromatin

Histone octamer

DNA

Histone tail

Acetylation by histone acetyltransferase

Deacetylation by histone deacetylase

FIGURE 16.13 Conversion of inactive chromatin to active chromatin by acetylation of histone tails, and the reverse by deacetylation of histone tails.

© Cengage Learning 2014

STUDY BREAK 16.2
1. What are the roles of general transcription factors, activators, and coactivators in transcription of a protein-coding gene?
2. What is the role of histones in gene expression? How does acetylation of the histones affect gene expression?

16.3 Posttranscriptional, Translational, and Posttranslational Regulation

Transcriptional regulation determines which genes are copied into mRNAs. Once mRNAs are transcribed from active genes, further regulation occurs at each of the major steps in the pathway from genes to proteins: during pre-mRNA processing and the movement of finished mRNAs to the cytoplasm (posttranscriptional regulation), during protein synthesis (translational regulation), and after translation is complete (posttranslational regulation) (refer again to Figure 16.6).

Posttranscriptional Regulation Controls mRNA Availability

Posttranscriptional regulation regulates translation by controlling the availability of mRNAs to ribosomes. The controls work by several mechanisms, including changes in pre-mRNA processing and the rate at which mRNAs are degraded.

VARIATIONS IN PRE-mRNA PROCESSING In Chapter 15 you learned that eukaryotic mRNAs are transcribed initially as pre-mRNA molecules. These pre-mRNAs are processed to produce the finished mRNAs, which are then available for protein synthesis. Variations in pre-mRNA processing can regulate *which* proteins are made in cells. As described in Section 15.3, pre-mRNAs can be processed by *alternative splicing,* which produces different mRNAs from the same pre-mRNA by removing different combinations of exons (amino acid-coding segments) along with the introns (the noncoding spacers). The resulting mRNAs are translated to produce a family of related proteins with various combinations of amino acid sequences derived from the exons. Alternative splicing itself is under regulatory control. Regulatory proteins specific to the type of cell control which exons are removed from pre-mRNA molecules by binding to regulatory sequences within those molecules. The outcome of alternative splicing is that related, but structurally different proteins within a family are synthesized in different cell types or tissues, and have functions that are keyed to the activities of those cell types and tissues. Perhaps three-quarters of human genes are alternatively spliced at the pre-mRNA level.

VARIATIONS IN THE RATE OF THE mRNA BREAKDOWN The rate at which eukaryotic mRNAs break down can also be controlled posttranscriptionally. Regulatory molecules, such as a steroid hormone, directly or indirectly affect the mRNA break-

down steps, either slowing or increasing the rate of those steps. For example, in the mammary gland of the rat, the mRNA for casein (a milk protein) has a half-life of about 5 hours (meaning that it takes 5 hours for half of the mRNA present at a given time to break down). The half-life of casein mRNA changes to about 92 hours if the peptide hormone prolactin is present. Prolactin is synthesized in the brain and in other tissues, including the breast. The most important effect of prolactin is to stimulate the mammary glands to produce milk. During milk production, a large amount of casein must be synthesized, and this is accomplished in part by radically decreasing the rate of breakdown of the casein mRNA.

Modulation of the stability of an mRNA typically involves a specific sequence or sequences in its 3′ UTR (untranslated region at the 3′ end of an mRNA following the stop codon; see Section 15.3). When specific proteins or regulatory RNAs (see next subsection) recognize and bind to the bases of the sequence, a degradation pathway is triggered for breaking down that RNA.

REGULATION OF GENE EXPRESSION IN EUKARYOTES BY SMALL RNAs In 1998, Andrew Fire of Stanford University School of Medicine and Craig Mello of University of Massachusetts Medical School showed that RNA silenced the expression of a particular gene in the nematode worm, *Caenorhabditis elegans.* They called the phenomenon **RNA interference (RNAi).** Their discovery revolutionized the way scientists thought about and studied gene regulation in eukaryotes because it showed that posttranscriptional regulation may involve not only regulatory proteins, but also noncoding single-stranded RNAs, which can bind to mRNAs and affect their translation. We now know that RNAi is widespread among eukaryotes. Fire and Mello received a Nobel Prize in 2006 for their discovery of RNA interference.

Two major groups of small regulatory RNAs are involved in RNAi, **microRNAs (miRNAs)** and **small interfering RNAs (siRNAs). Figure 16.14** shows the transcription of an miRNA gene and the processing of its transcript to produce a functional miRNA molecule. The miRNA, in a protein complex called the **miRNA-induced silencing complex (miRISC)** binds to specific sequences in the 3′ UTRs of target mRNAs by base pairing. If the miRNA and mRNA pair imperfectly, the double-stranded segment formed between the miRNA and the mRNA blocks ribosomes from translating the mRNA (shown in Figure 16.14). In this case, expression of the target mRNA is silenced, but it is not destroyed. If the miRNA and mRNA pair perfectly, an enzyme in the protein complex cleaves the target mRNA where the miRNA is bound to it, destroying the mRNA which, of course, silences its expression. RNAi by imperfect pairing and translation inhibition is the most common mechanism in animals. RNAi by perfect pairing and RNA degradation is the most common mechanism in plants.

Many thousands of different miRNAs have been identified among eukaryotic organisms examined. MicroRNA genes have been found in all multicellular eukaryotes that have been examined, and also in some unicellular ones. MicroRNAs play

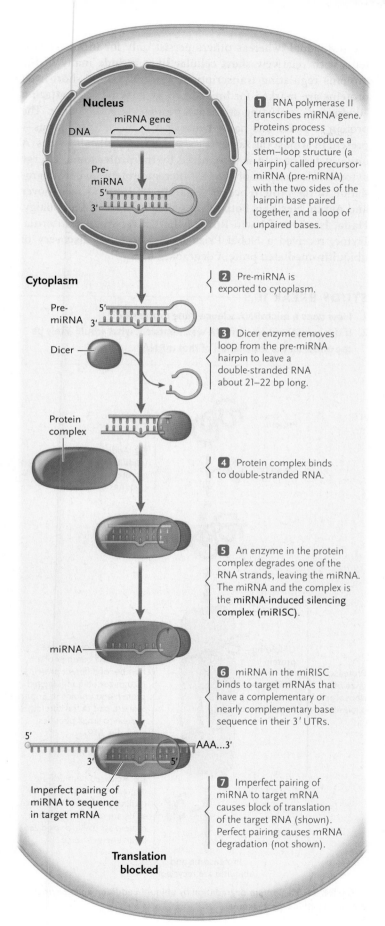

FIGURE 16.14 RNA interference—regulation of gene expression by microRNAs (miRNAs).
© Cengage Learning 2014

1 RNA polymerase II transcribes miRNA gene. Proteins process transcript to produce a stem–loop structure (a hairpin) called precursor-miRNA (pre-miRNA) with the two sides of the hairpin base paired together, and a loop of unpaired bases.

Nucleus
DNA
miRNA gene
Pre-miRNA
5'
3'

Cytoplasm

2 Pre-miRNA is exported to cytoplasm.

Pre-miRNA
5'
3'

Dicer

3 Dicer enzyme removes loop from the pre-miRNA hairpin to leave a double-stranded RNA about 21–22 bp long.

Protein complex

4 Protein complex binds to double-stranded RNA.

5 An enzyme in the protein complex degrades one of the RNA strands, leaving the miRNA. The miRNA and the complex is the miRNA-induced silencing complex (miRISC).

miRNA

6 miRNA in the miRISC binds to target mRNAs that have a complementary or nearly complementary base sequence in their 3′ UTRs.

5'
3'
AAA...3'
5'

Imperfect pairing of miRNA to sequence in target mRNA

7 Imperfect pairing of miRNA to target mRNA causes block of translation of the target RNA (shown). Perfect pairing causes mRNA degradation (not shown).

Translation blocked

central roles in controlling gene expression in a variety of cellular, physiological, and developmental processes in animals and plants. We will see some specific examples in the next two sections.

The other major type of small regulatory RNAs is the **small interfering RNA (siRNA).** Whereas miRNA is produced from RNA that is encoded in the cell's genome, siRNA is produced from double-stranded RNA that is *not* encoded by nuclear genes. For example, the life cycle and replication of many viruses with RNA genomes involves a double-stranded RNA stage. Cells attacked by such a virus can defend themselves using siRNA, which they produce from the virus' own RNA. The viral double-stranded RNA enters the cell's RNAi process in a way very similar to that described for miRNAs; double-stranded RNA is cut by Dicer (see Figure 16.14) into short double-stranded RNA molecules, and a single-stranded siRNA is generated that is complexed with proteins to produce an **siRNA-induced silencing complex (siRISC).** In the RNAi process, the siRNA in the siRISC acts similarly to miRNA in the miRISC; that is, single-stranded RNAs complementary to the siRNA are targeted and the target RNA is cleaved and the pieces are then degraded. In our viral example, the targeted RNAs would be viral mRNAs for proteins the virus uses to replicate itself, or a single-stranded RNA that is the viral genome itself, or that is produced from the viral genome during replication.

The expression of any gene can be knocked down to low levels or knocked out completely in experiments involving RNAi with siRNA. To silence a gene, researchers introduce into the cell a double-stranded RNA that can be processed into an siRNA complementary to the mRNA transcribed from that gene. Knocking down or knocking out the function of a gene is equivalent to creating a mutated version of that gene, but without changing the gene's DNA sequence. Researchers are using this experimental approach, for example, to study genes that have been detected by sequencing complete genomes, but whose functions are completely unknown. After an siRNA specific to a gene of interest is introduced into the cell, researchers look for a change in phenotype, such as properties relating to growth or metabolism. If such a change is seen, the researchers now have some insight into the gene's function.

Translational Regulation Controls the Rate of Protein Synthesis

At the next regulatory level, translational regulation controls the rate at which mRNAs are used in protein synthesis. Translational regulation occurs in essentially all cell types and species. For example, translational regulation is involved in cell cycle control in all eukaryotes and in many processes during development in multicellular eukaryotes, such as red blood cell differentiation in animals. Significantly, many viruses exploit translational regulation to control their infection of cells and to shut off the host cell's own genes.

Consider the general role of translational regulation in animal development. During early development of most animals, little transcription occurs. The changes in protein synthesis patterns seen in developing cell types and tissues instead are the result of the activation, repression, or degradation of maternal mRNAs, the mRNAs that were in the mother's egg before fertilization. One important mechanism for translational regulation involves adjusting the length of the poly(A) tail of the mRNA. (Recall from Section 15.3 that the poly(A) tail—a string of adenine-containing nucleotides—is added to the 3′ end of pre-mRNA and is retained on the mRNA produced from the pre-mRNA after introns are removed.) That is, enzymes can change the length of the poly(A) tail on an mRNA in the cytoplasm in either direction: by shortening it or lengthening it. Increases in poly(A) tail length result in increased translation; decreases in length result in decreased translation. For example, during embryogenesis (the formation of the embryo) of the fruit fly, *Drosophila*, key proteins are synthesized when the poly(A) tails on the mRNAs for those proteins are lengthened in a regulated way. Evidence for this came from experiments in which poly(A) tail lengthening was blocked; the result was that embryogenesis was inhibited. But although researchers know that the length of poly(A) tails is regulated in the cytoplasm, how this process occurs is not completely understood.

Posttranslational Regulation Controls the Availability of Functional Proteins

Posttranslational regulation controls the availability of functional proteins mainly in three ways: chemical modification, processing, and degradation. Chemical modification involves the addition or removal of chemical groups, which reversibly alters the activity of the protein. For example, you learned in Section 7.2 how the addition of phosphate groups to proteins involved in signal transduction pathways either stimulates or inhibits the activity of those proteins. Further, in Section 10.4 you learned how the addition of phosphate groups to target proteins plays a crucial role in regulating how a cell progresses through the cell division cycle. And in Section 16.2 you learned how acetylation of histones altered the properties of the nucleosome, loosening its association with DNA in chromatin.

In processing, a protein is synthesized as an inactive precursor that is activated under regulatory control by removal of a segment. For example, you learned in Section 15.4 that the digestive enzyme pepsin is synthesized as pepsinogen, an inactive precursor that activates by removal of a segment of amino acids. Similarly, the glucose-regulating hormone insulin is synthesized as a precursor called proinsulin; processing of the precursor removes a central segment but leaves the insulin molecule, which consists of two polypeptide chains linked by disulfide bridges.

The rate of degradation of proteins is also under regulatory control. Some proteins in eukaryotic cells last for the lifetime of the individual, whereas others persist only for minutes. Proteins with relatively short cellular lives include many of the proteins regulating transcription. Typically, these short-lived proteins are marked for breakdown by enzymes that attach a "doom tag" consisting of a small protein called *ubiquitin*. The protein is given this name because it is indeed ubiquitous—present in almost the same form in essentially all eukaryotes. A ubiquinated protein is recognized by the **proteasome,** a large cytoplasmic complex of several different proteins, where degradation of the protein occurs **(Figure 16.15).** Aaron Ciechanover and Avram Hershko, both of the Israel Institute of Technology, Haifa, Israel, and Irwin Rose of the University of California, Irvine, received a Nobel Prize in 2004 for the discovery of ubiquitin-mediated protein degradation.

STUDY BREAK 16.3

1. How does a microRNA silence gene expression?
2. If the poly(A) tail on a mRNA was removed, what would likely be the effect on the translation of that mRNA?

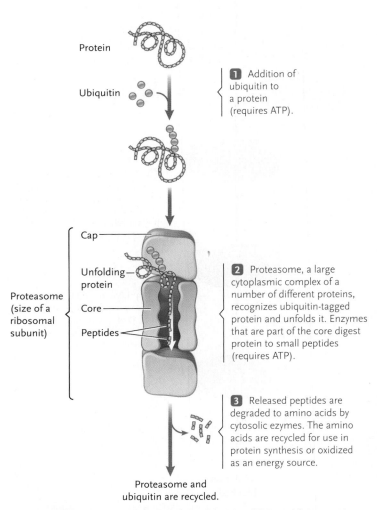

1. Addition of ubiquitin to a protein (requires ATP).

Protein

Ubiquitin

Cap

Unfolding protein

Proteasome (size of a ribosomal subunit)

Core

Peptides

2. Proteasome, a large cytoplasmic complex of a number of different proteins, recognizes ubiquitin-tagged protein and unfolds it. Enzymes that are part of the core digest protein to small peptides (requires ATP).

3. Released peptides are degraded to amino acids by cytosolic ezymes. The amino acids are recycled for use in protein synthesis or oxidized as an energy source.

Proteasome and ubiquitin are recycled.

FIGURE 16.15 Protein degradation by ubiquitin addition and enzymatic digestion within a proteasome.
© Cengage Learning 2014

16.4 Genetic and Molecular Regulation of Development

In the development of multicellular eukaryotes, a fertilized egg divides into many cells that are ultimately transformed into an adult, which is itself capable of reproduction. The process is orchestrated by a series of programmed changes encoded in the genome, although the development of a multicellular organism is also influenced to some extent by the environment. The programmed changes do not alter the genome itself, which remains identical in virtually all of the cells of the developing organism. (Thus, the processes involved in development are all examples of epigenetic phenomena.) Instead, controlled changes in gene expression direct sequential developments that are appropriate both in time—certain stages must unfold before others—and in place—new structures must arise in the correct location. An understanding of how genes are regulated therefore aids researchers in their genetic analysis of development. Developmental geneticists are very interested in identifying and characterizing the genes involved in development, and understanding how the products of the genes regulate and bring about the elaborate events that occur.

One productive research approach has been to isolate mutants that affect developmental processes. Researchers can then identify the genes involved, clone the genes (see Chapter 18), and analyze them in detail to build models for the molecular functions of the gene products in development. A number of model organisms are used for these studies because of the relative ease with which mutants can be made and studied, and the ease of performing molecular analyses. These organisms include the fruit fly (*Drosophila melanogaster*) (see *Focus on Model Research Organisms* in Chapter 13) and *Caenorhabditis elegans* (a nematode worm) (see *Focus on Model Research Organisms* in Chapter 31) among invertebrates; the zebrafish (*Danio rerio*) (see Figure 10.1), the house mouse (*Mus musculus*) (see *Focus on Model Research Organisms* in Chapter 45) among vertebrates; and thale cress (*Arabidopsis thaliana*) among plants (see *Focus on Basic Research* in Chapter 36).

In this section we discuss some of the genetic and molecular mechanisms that regulate development in animals to illustrate the principles involved. In Chapter 50 you will learn about the cellular and morphological events that occur as an adult animal develops from a fertilized egg. The gene regulation of development in plants is described in Chapter 36.

Development in Animals Is Accomplished by Several Genetically Regulated Mechanisms

Fertilization of an egg by a sperm cell produces a zygote. Development begins at this point. Development in all animals is accomplished by a number of mechanisms that are under genetic control but are also influenced to some extent by the environment. The mechanisms include mitotic cell divisions, movements of cells, *induction, determination,* and *differentiation.*

Mitotic divisions produce the cells that are the subjects of the gene regulatory processes that create the various cell types in the adults. Cell movements during development are part of the program to create specific tissues and organs.

Induction is the process in which one group of cells (the inducer cells) causes or influences another nearby group of cells (the responder cells) to follow a particular developmental pathway.

Determination is the process by which the developmental fate of a cell is set. Before determination, a cell is **totipotent** (has the potential to become any cell type of the adult) but after determination, the cell commits to becoming a particular cell type. Induction is the major process responsible for determination.

Differentiation follows determination, and involves the establishment of a cell-specific developmental program in cells. Differentiation results in cell types with clearly defined structures and functions; those features derive from specific patterns of gene expression in cells.

Developmental Information Is Located in Both the Nucleus and Cytoplasm of the Fertilized Egg

The information that directs the initiation of development is stored in two locations in the zygote. Part of the information is stored in the zygote nucleus, in the DNA derived from the egg and sperm nuclei. This information directs development as individual genes are activated or turned off in a highly ordered manner. The rest of the information is stored in the egg cytoplasm, in the form of messenger RNA (mRNA) and protein molecules.

Because the fertilizing sperm contributes essentially no cytoplasm, nearly all the cytoplasmic information of the fertilized egg (zygote) is maternal in origin. Key mRNA and proteins stored in the egg cytoplasm that direct the early stages of development are known as **cytoplasmic determinants.** Those cytoplasmic determinants are distributed asymmetrically in the cell, rather than being evenly distributed. Therefore, when the zygote divides, the cytoplasmic determinants are distributed asymmetrically to the daughter cells, reflecting their distribution in the zygote. As a result, the two daughter cells resulting from division differ in the signals they have, and this leads to different patterns of gene expression. Cytoplasmic determinants direct the first stages of animal development before genes of the zygote become active. Depending on the animal group, the control of early development by cytoplasmic determinants may be limited to the first few divisions of the zygote, as in mammals, or it may last until the actual tissues of the embryo are formed, as in most invertebrates.

Induction Is the Major Process Responsible for Determination

Induction is the major process responsible for determination, which is the process by which the developmental fate of a cell is set. Induction is a highly selective process; only certain responder cells can respond to the signal from the inducer cells. Many experiments have shown that induction occurs through the interaction of signal molecules of inducer cells with surface receptors on the responding cells. In some cases, a signal molecule released by the inducer cell interacts with a plasma membrane-embedded receptor on the surface of a responder cell. The binding activates the receptor, triggering a signal transduction pathway within the cell, which produces a developmental change, often involving a change in gene activity (the cellular response to the signal). (Surface receptors and their associated signal transduction pathways are discussed in Sections 7.3 and 7.4.) In other cases, induction occurs by direct cell-to-cell contact involving interaction between a membrane-embedded protein on the inducer cell and a membrane-embedded receptor protein on the surface of the responder cell. The receptor is activated by the interaction, and in the same way as for the diffusible signal molecule example, the cell undergoes a developmental change.

Differentiation Produces Specialized Cells without Loss of Genes

Differentiation is the process by which cells that have been committed to a particular developmental fate by the determination process now develop into specialized cell types with distinct structures and functions, such as skin cells and nerve cells. As part of differentiation, cells produce molecules characteristic of the specific types. For example, in the lens cells of the eye, 80% to 90% of the total protein synthesized is crystallin, the protein responsible for the transparency of the lens.

Research into differentiation confirmed that, as cells specialize, the DNA in the genome remains constant, matching that of the original zygote. In other words, differentiation generally occurs as a result of differential gene activity, and not by a process in which DNA is lost, so that each type of differentiated cell retains only those genes required for that cell type. (There are just a few examples in particular organisms where gene loss occurs during differentiation.)

The most compelling evidence showing that the DNA in the genome remains constant during differentiation has come from experiments in which animals and plants have been cloned. Animal clones have been made by taking an egg produced by one animal, and then replacing its nucleus with a nucleus taken from a somatic cell of a different adult animal, showing that the adult nucleus is still totipotent. Different species of frogs and many types of mammals have been cloned in this way. The first mammal cloned was Dolly, a sheep (the cloning experiment is described in Chapter 18, and shown in Figure 18.14). Similarly, plants can be cloned from single cells isolated from a mature plant.

Genes Control Determination and Differentiation

Both determination and differentiation involve specific, regulated changes in gene expression. One well-studied example of the genetic control of determination and differentiation is the production of skeletal muscle cells from somites in mammals **(Figure 16.16)**. Somites give rise to the vertebral column, the ribs, the repeating sets of skeletal muscles associated with the ribs and vertebral column, and the skeletal muscles of the limbs. Under the control of the master regulatory gene, *myoD*, particular cells of a somite differentiate into skeletal muscle cells. The product of *myoD* is the transcription factor MyoD.

Generally speaking, the molecular mechanisms involved in determination and differentiation depend on master regulatory genes that encode transcription factors. Expression of the master regulatory genes is controlled by induction in most cases. The genes controlled by the transcription factors encoded by the master regulatory genes may themselves be regulatory genes encoding transcription factors that, in turn, regulate the expression of other sets of genes.

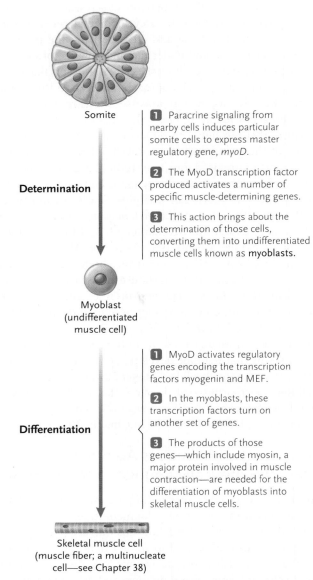

Somite

Determination

1 Paracrine signaling from nearby cells induces particular somite cells to express master regulatory gene, *myoD*.

2 The MyoD transcription factor produced activates a number of specific muscle-determining genes.

3 This action brings about the determination of those cells, converting them into undifferentiated muscle cells known as **myoblasts**.

Myoblast (undifferentiated muscle cell)

Differentiation

1 MyoD activates regulatory genes encoding the transcription factors myogenin and MEF.

2 In the myoblasts, these transcription factors turn on another set of genes.

3 The products of those genes—which include myosin, a major protein involved in muscle contraction—are needed for the differentiation of myoblasts into skeletal muscle cells.

Skeletal muscle cell (muscle fiber; a multinucleate cell—see Chapter 38)

FIGURE 16.16 The genetic control of determination and differentiation involved in mammalian skeletal muscle cell formation.
© Cengage Learning 2014

Genes Regulate Pattern Formation during Development

As a part of the signals guiding differentiation, cells receive positional information that tells them where they are in the embryo. The positional information is vital to **pattern formation:** the arrangement of organs and body structures in their proper three-dimensional relationships. Positional information is laid down primarily in the form of concentration gradients of regulatory molecules produced under genetic control. In most cases, gradients of several different regulatory molecules interact to tell a cell, or a cell nucleus, where it is in the embryo. Below, we describe in brief the results of studies of the genetic control of pattern formation during the development of the fruit fly, *Drosophila melanogaster*. The developmental principles discovered from these studies apply to many other animal species, including humans.

THE LIFE CYCLE OF *DROSOPHILA* The production of an adult fruit fly from a fertilized egg occurs in a sequence of genetically controlled development events. The *Drosophila* life cycle is shown in **Figure 16.17.** As is typical of most insects, the life cycle proceeds from fertilized egg to embryo within the egg. The embryo then hatches from the egg and the life cycle proceeds through three larval stages to a pupal stage and then to the adult stage. The stages of development from fertilized egg to hatching collectively are called **embryogenesis.** As illustrated by the color usage in Figure 16.17, the segments of the embryo can be mapped to the segments of the adult fly. The development of verte-brates, including mammals, occurs quite differently, as you will learn in Chapter 50.

GENETIC ANALYSIS OF *DROSOPHILA* DEVELOPMENT The study of developmental mutants by a large number of researchers has given us important information about *Drosophila* development. Three researchers performed key, pioneering research with developmental mutants: Edward B. Lewis of the California Institute of Technology, Christiane Nüsslein-Volhard of the Max Planck Institute for Developmental Biology in Tübingen, Germany, and Eric Wieschaus of Princeton University. The three shared a Nobel Prize in 1995 "for their discoveries concerning the genetic control of early embryonic development."

Nüsslein-Volhard and Wieschaus studied early embryogenesis. They searched for *every* gene required for early pattern formation in the embryo. They did this by looking for recessive *embryonic lethal* mutations. These mutations, when homozygous, result in the death of the embryo during development. By examining the stage of development at which an embryo died, and how development was disrupted, they gained insights into the role of the particular genes in embryogenesis.

Lewis studied mutants that changed the fates of cells in particular regions in the embryo, producing structures in the adult that normally were produced by other regions. His work was the foundation of research identifying master regulatory genes that control the development of body regions in a wide range of organisms.

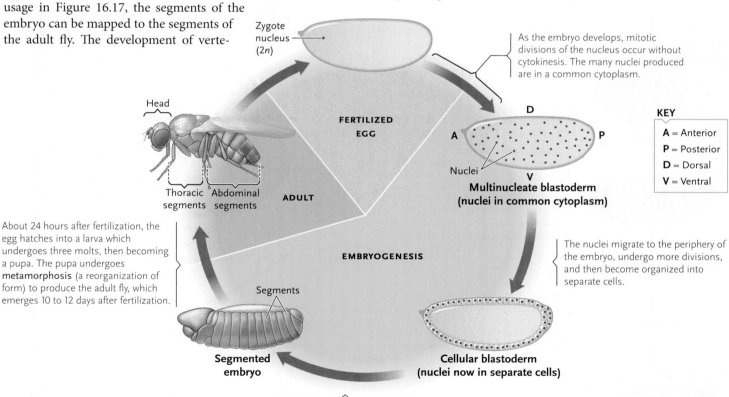

Zygote nucleus (2n)

As the embryo develops, mitotic divisions of the nucleus occur without cytokinesis. The many nuclei produced are in a common cytoplasm.

Head

Thoracic segments   Abdominal segments

FERTILIZED EGG

ADULT

EMBRYOGENESIS

About 24 hours after fertilization, the egg hatches into a larva which undergoes three molts, then becoming a pupa. The pupa undergoes **metamorphosis** (a reorganization of form) to produce the adult fly, which emerges 10 to 12 days after fertilization.

Segments

Segmented embryo

D

A   P

Nuclei

V

Multinucleate blastoderm (nuclei in common cytoplasm)

KEY

A = Anterior
P = Posterior
D = Dorsal
V = Ventral

The nuclei migrate to the periphery of the embryo, undergo more divisions, and then become organized into separate cells.

Cellular blastoderm (nuclei now in separate cells)

The cellular blastoderm develops into a segmented embryo. At this point, 10 hours have passed since the egg was fertilized.

FIGURE 16.17 The life cycle of *Drosophila* and the relationship between segments of the embryo and segments of the adult.
© Cengage Learning 2014

MATERNAL-EFFECT GENES AND SEGMENTATION GENES FOR ESTABLISHING THE BODY PLAN IN THE EMBRYO Two classes of genes—*maternal-effect genes* and *segmentation genes*—work sequentially to control the establishment of the embryo's body plan; that is, how the organism is laid out in its anterior-to-posterior, ventral-to-dorsal, and side-to-side axes **(Figure 16.18)**. These genes code for regulatory proteins which regulate the expression of other genes.

Many **maternal-effect genes** are expressed by the mother during oogenesis. These genes control the anterior-to-posterior polarity of the egg and, therefore, of the embryo. Some control

the formation of the anterior structures of the embryo, others control the formation of the posterior structures, and yet others control the formation of the terminal end.

The *bicoid* gene is the key maternal-effect gene responsible for head and thorax development. The *bicoid* gene is transcribed in the mother during oogenesis, and the resulting mRNAs are deposited in the egg, localizing near the anterior end **(Figure 16.19)**. After the egg is fertilized, translation of the mRNAs produces Bicoid protein, which diffuses through the egg to form a gradient with its highest concentration at the anterior end of the egg. The Bicoid protein is a transcription factor that activates some genes and represses others along the anterior–posterior axis of the embryo. Embryos with mutations in the *bicoid* gene have no thoracic structures, but have posterior structures at each end. Researchers concluded, therefore, that the *bicoid* gene in normal embryos is a master regulator gene controlling the expression of genes for the development of anterior structures (head and thorax).

A number of other maternal-effect genes, through the activities of their products in gradients in the embryo, are also involved in axis formation. The *nanos* gene, for instance, is the key maternal-effect gene for the posterior structures. When the *nanos* gene is mutated, embryos lack abdominal segments.

Maternal-effect genes

Maternal-effect genes are expressed by the mother during oogenesis. These genes control the polarity of the egg and, therefore, of the embryo.

Segmentation genes

Segmentation genes—gap genes, pair-rule genes, and segment polarity genes—subdivide the embryo into regions. **Gap genes** are activated based on their positions in the maternally directed anterior–posterior axis of the egg by reading the concentrations of Bicoid and other proteins. Gap gene products are transcription factors that activate pair-rule genes. Gap genes subdivide the embryo along the anterior–posterior axis into broad regions that later develop into several distinct segments.

Pair-rule genes encode transcription factors that regulate the expression of segment polarity genes. Pair-rule genes control the division of the embryo into units of two segments each.

Segment polarity genes encode transcription factors and signaling proteins that regulate other genes involved in laying down the pattern of the embryo. Segment polarity genes establish the boundaries and anterior–posterior axis of each segment, thereby determining the regions that become segments of larvae and adults.

KEY

A = Anterior
P = Posterior
D = Dorsal
V = Ventral

FIGURE 16.18 Maternal-effect genes and segmentation genes and their role in *Drosophila* embryogenesis.
© Cengage Learning 2014

Distribution of maternal *bicoid* mRNA (green)

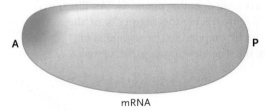

mRNA

Distribution of Bicoid protein (blue)

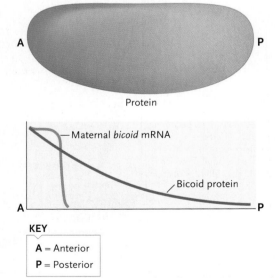

Protein

Maternal *bicoid* mRNA

Bicoid protein

KEY

A = Anterior
P = Posterior

FIGURE 16.19 Gradients of *bicoid* mRNA and Bicoid protein in the *Drosophila* egg.
© Cengage Learning 2014

Once the anterior–posterior axis of the embryo is set, the expression of at least 24 **segmentation genes** progressively subdivides the embryo into regions, determining the segments of the embryo and the adult (see Figure 16.18). Gradients of Bicoid and other proteins encoded by maternal-effect genes regulate expression of the embryo's segmentation genes differentially. That is, each segmentation gene is expressed at a particular time and in a particular location during embryogenesis. Three classes of segmentation genes act sequentially; their activities are regulated in a cascade of gene activations:

- **Gap genes,** through their activation of the next genes in the regulatory cascade, control the subdivision of the embryo along the anterior–posterior axis into several broad regions. Mutations in gap genes result in the loss of one or more body segments in the embryo.

- **Pair-rule genes,** through their activation of the next genes in the regulatory cascade, control the division of the embryo into units of two segments each. Mutations in pair-rule genes delete every other segment of the embryo.

- **Segment polarity genes** set the boundaries and anterior–posterior axis of each segment in the embryo. Mutations in segment polarity genes produce segments in which one part is missing and the other part is duplicated as a mirror image.

HOMEOTIC GENES FOR SPECIFYING THE DEVELOPMENTAL FATE OF EACH SEGMENT Once the segmentation pattern has been set, **homeotic** ("structure-determining") **genes** of the embryo specify what each segment will become after metamorphosis. In normal flies, homeotic genes are master regulatory genes that control the development of structures such as eyes, antennae, legs, and wings on particular segments (see Figure 16.17). Researchers discovered the role of homeotic genes from the study of mutations in these genes; such mutations alter the developmental fate of a segment in the embryo in a major way. For example, in flies with a mutation in the *Antennapedia* gene, legs develop in place of antennae **(Figure 16.20).**

How do homeotic genes regulate development? Homeotic genes encode transcription factors that regulate expression of genes responsible for the development of adult structures. Each homeotic gene has a common region called the **homeobox** that is key to its function. A homeobox corresponds to an amino acid section of the encoded transcription factor called the **homeodomain.** The homeodomain of each protein binds to regulatory sequences of the genes whose transcription it regulates.

Homeobox-containing genes are called *Hox* genes. There are eight *Hox* genes in *Drosophila* and, interestingly, they are organized along a chromosome in the same order as they are expressed along the anterior–posterior body axis **(Figure 16.21).**

Hox genes are present in all major animal phyla. In each case, the genes control the development of the segments/regions of the body and are arranged in order in the genome. The homeobox sequences in the *Hox* genes are highly conserved, indicating a common function in the wide range of animals in which they are found. For example, the homeobox sequences of mammals are the same or very similar to those of the fruit fly (see Figure 16.21).

Homeotic genes are also found in plants. For example, many homeotic mutations that affect flower development have been identified and analyzed in *Arabidopsis* (see Section 36.5).

MicroRNAs Play Critical Roles in Development

Earlier in the chapter you learned how microRNAs (miRNAs) can regulate gene expression at the translational level. Studies of miRNA gene mutants, and of organisms with defective miRNA synthesis, have shown that miRNAs have critical roles in development and differentiation, including embryogenesis, the formation of organs, and the development of the germline (the lineage of cells from which gametes are produced). For instance, in *Drosophila,* miRNAs are required for development of both somatic tissues and the germline, and in the maintenance of stem cells in the germline. (Stem cells are cells capable of differentiating into almost any adult cell type.) In zebrafish, miRNAs are essential for development; for example, a knock-

Normal   *Antennapedia* mutant

FIGURE 16.20 *Antennapedia,* a homeotic mutant of *Drosophila,* in which legs develop in place of antennae.

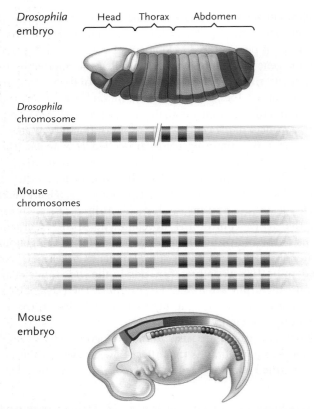

Drosophila embryo — Head Thorax Abdomen

Drosophila chromosome

Mouse chromosomes

Mouse embryo

FIGURE 16.21 The *Hox* genes of the fruit fly and the corresponding regions of the embryo they affect. The mouse has four sets of *Hox* genes on four different chromosomes. Their relationship to the fruit fly genes is shown by the colors.
© Cengage Learning 2014

out of Dicer results in a developmental arrest at 7 to 10 days after fertilization. miRNAs are also involved in brain formation, somitogenesis (generation of somites), and heart development. In mice (and, by extrapolation, other mammals), miRNAs are essential for development; for example, a knockout of Dicer dies at 7.5 days of gestation. Dicer is also required for embryonic stem cell differentiation *in vitro* and, therefore, probably *in vivo* also.

MicroRNAs also regulate plant development. An example involves the regulation of plant growth and development by auxins, a class of plant hormones exemplified by indoleacetic acid (IAA) (see Section 37.1). Auxin affects development by modulating the expression of a number of genes that control cell division and cell elongation in specific parts of the plant and at specific stages during a plant's life cycle. The effects of auxin on plant development are controlled by several families of transcription factors. Among other results, researchers have shown that genetically engineered plants that express a form of one of those transcription factors that cannot be cleaved as usual by an miRNA have significant growth defects. In another study, miRNA-directed cleavage of another of the transcription factors caused defects in auxin regulation of lateral root development. Further, *Arabidopsis thaliana* (thale cress) mutants

lacking a key, evolutionarily conserved, protein of the miRNA-induced silencing complex (miRISC; see Figure 16.14), exhibit severe developmental defects and are sterile. Many other experiments have shown the importance of miRNA regulation of mRNA expression in plant development.

In short, regulatory proteins play critical roles in development and differentiation; however, they are but one player, and not the only player. Researchers must now discover how miRNAs regulate the expression of protein-coding mRNAs to develop a more complete understanding of the regulatory circuits underlying development.

STUDY BREAK 16.4

1. What are determination and differentiation and, in general, how are they controlled?
2. How do the segmentation genes and homeotic genes of *Drosophila* differ in function?

16.5 The Genetics of Cancer

In Chapter 10 you learned that the cell division cycle in all eukaryotes is carefully regulated by genes (see Section 10.4 and Figure 10.15). Complex signaling systems involving extracellular and cellular molecules are used in cell division regulation. The extracellular signaling molecules include polypeptide hormones and steroid hormones made in one tissue that influence the growth and division of cells in other tissues. The specific effects of the signaling molecules depend on the presence of receptors for those molecules on or in the cells they target. When a signaling molecule binds to its receptor, the receptor is activated which triggers a signal transduction pathway in the target cell that brings about a cellular response. (Signaling and signal transduction pathways are discussed in Chapter 7.)

Figure 16.22 presents simplified illustrations of examples of the effects of two types of cell division-related signaling molecules on normal cells. **Figure 16.22A** shows a **growth factor**—a molecule that stimulates cell division of a target cell—binding to a surface receptor and triggering a signal transduction pathway. In the nucleus, the pathway results in a change in gene expression via its effect on transcription factors and a protein (or proteins) that stimulates cell division is produced. **Figure 16.22B** shows a **growth-inhibiting factor**—a molecule that inhibits cell division of a target cell by binding to a surface receptor and stimulating transcription of a gene for a protein that inhibits cell division. In this case the cellular response is the production of a protein (or proteins) that are inhibitory to cell division.

Cell division control depends on many stimulatory and inhibitory factors. For normal cells, the relationship between gene products that stimulate cell division and gene products that inhibit cell division govern whether the cell remains in a nondividing state or whether it grows and divides. Only when the balance shifts towards stimulatory signals does the cell grow and divide.

A. Stimulation of cell division by a growth factor

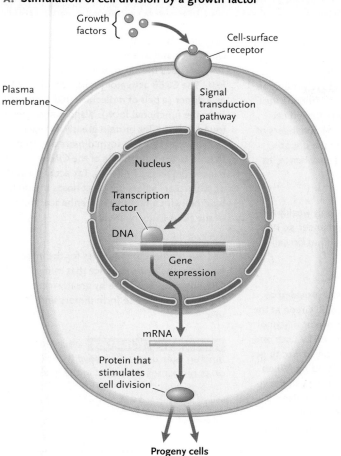

B. Inhibition of cell division by a growth-inhibiting factor

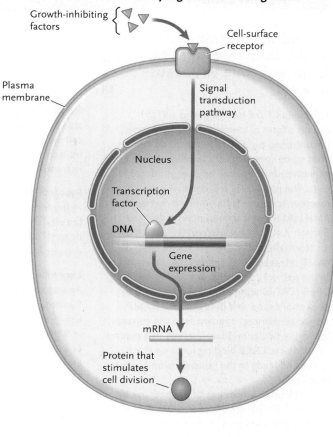

FIGURE 16.22 Stimulation of cell division by a growth factor (A) and inhibition of cell division by a growth-inhibiting factor (B) in a normal cell via signal transduction pathways.
© Cengage Learning 2014

Occasionally, differentiated cells of complex multicellular organisms deviate from their normal genetic program and begin to grow and divide, giving rise to tissue masses called **tumors.** Those cells have lost their normal regulatory controls and have reverted partially or completely to an embryonic developmental state, in a process called **dedifferentiation.** If the altered cells stay together in a single mass, the tumor is *benign.* Benign tumors usually are invasive, but not life threatening, and their surgical removal generally results in a complete cure.

If the cells of a tumor invade and disrupt surrounding tissues, the tumor is *malignant* and is called a **cancer.** Sometimes, cells from malignant tumors break off and move through the blood system or lymphatic system, forming new tumors at other locations in the body. The spreading of a malignant tumor is called **metastasis** (meaning "change of state"). Malignant tumors can result in debilitation and death in various ways, including damage to critical organs, metabolic problems, hemorrhage, and secondary malignancies. In some cases, malignant tumors can be eliminated from the body by surgery or be destroyed by chemicals *(chemotherapy)* or radiation.

Cancers Are Genetic Diseases

Experimental evidence of various kinds shows that cancers are genetic diseases:

1. Particular cancers can have a high incidence in some human families. Cancers that run in families are known as **familial (hereditary) cancers.** Cancers that are not inherited are known as **sporadic (nonhereditary) cancers.** Familial cancers are less frequent than sporadic cancers.

2. Descendants of cancer cells are all cancer cells. In fact, it is the cloned descendants of a cancer cell that form a tumor.

3. The incidence of cancers increases on exposure to mutagens, agents that cause mutations in DNA. Particular chemicals and certain kinds of radiation are effective mutagens.

4. Particular chromosomal mutations are associated with specific forms of cancer (see Section 13.3 and Figure 13.12). In these cases, chromosomal breakage affects the expression of genes associated with the regulation of cell division.

5. Some viruses can induce cancer. They do so by the expression of viral genes in the host, which disrupts normal cell cycle control.

A Viral Tax on Transcriptional Regulation: How does human T-cell leukemia virus cause cancer?

The *human T-cell leukemia virus (HTLV)* causes a form of cancer by triggering rapid, uncontrolled division of white blood cells. It does so by speeding up a pathway that triggers division at normal rates in uninfected cells. The pathway is a G-protein–coupled receptor-response pathway involving cyclic AMP (cAMP) as a second messenger (see Section 7.4).

Normally, the pathway is triggered by cAMP released when an infection takes place. In the pathway, a specific activator called CREB (CRE-binding protein) is activated by phosphorylation and binds to CRE (cAMP-response element) sequences in the enhancers for particular regulatory genes that control cell division. CREB binding turns on the genes and leads to the rapid division of white blood cells.

HTLV takes advantage of the pathway by means of a short sequence 5′-CGTCA-3′ in its DNA that mimics part of the human CRE en-hancer sequence 5′-TGACGTCA-3′. When the host cell's CREB is activated, it binds strongly to the enhancer sequence in the virus, turns on the viral genes, and leads to reproduction of the virus. Unfortunately for someone infected with HTLV, CREB in an infected cell also binds more strongly to the enhancers of cell division genes, leading to the uncontrolled division of white cells and, hence, to leukemia.

Research Question

In the test tube, CREB binds only weakly to the viral mimic of the CRE enhancer, so how does HTLV cause the effects it has on cell division?

Experiment

An answer to the question was provided by Susanne Wagner and Michael R. Green at the University of Massachusetts Medical School in Worcester. They found that HTLV uses one of its own encoded proteins, called *Tax*, to get around the problem of weak binding. When Tax is present, CREB binds strongly to the viral mimic of the CRE enhancer. Tax also greatly increases the ability of CREB to bind to the cell's normal CRE enhancers, leading to rapid and uncontrolled growth of infected white cells and to leukemia.

How does the Tax protein accomplish this feat? The CREB activator interacts with DNA as a dimer (a pair of molecules that together form the functional form). Wagner and Green found that the Tax protein greatly increases the ability of CREB to form dimers and thus to bind to the viral mimic of the CRE enhancer sequence. Evidently, Tax acts as a sort of molecular "safety pin" that holds the dimer together with either the viral enhancer mimic or the cell's CRE enhancers.

Conclusion

The Tax protein compensates for the imperfection of the viral sequence that mimics the CRE enhancer sequence by greatly increasing the ability of CREB to form dimers and bind to it.

THINK LIKE A SCIENTIST What three binding sites does CREB have that explain all of its properties described in the box?

Source: S. Wagner and M. R. Green. 1993. HTLV-1 Tax protein stimulation of DNA binding of bZIP proteins by enhancing dimerization. *Science* 262:395–399.

All the characteristics of cancer cells that have been mentioned—dedifferentiation, uncontrolled division, and metastasis—reflect changes in gene activity.

Three Main Classes of Genes Are Implicated in Cancer

Three major classes of genes are altered frequently in cancers: *proto-oncogenes, tumor suppressor genes,* and microRNA (miRNA) genes.

PROTO-ONCOGENES **Proto-oncogenes** (*onkos* = bulk or mass) are genes in normal cells that encode various kinds of proteins which stimulate cell division. Examples are growth factors (see Figure 16.22A), receptors on target cells that are activated by growth factors (see Chapter 7), components of cellular signal transduction pathways triggered by cell division stimulatory signals (see Chapter 7), and transcription factors that regulate the expression of the structural genes for progression through the cell cycle. In normal cells, the products of proto-oncogenes are balanced by inhibitory proteins encoded by tumor suppressor genes. In cancer cells, the proto-oncogenes are altered to become **oncogenes,** genes that stimulate the cell to progress to the cancerous state. Various alterations can convert a proto-oncogene into an oncogene, and only one allele needs to be altered for the cellular changes to occur:

- A mutation in a gene's promoter or other control sequences that results in the gene becoming more active than normal.
- A mutation in the coding segment of the gene may produce an altered form of the encoded protein that is more active than normal.
- Translocation, a process in which a segment of a chromosome breaks off and attaches to a different chromosome (discussed in Section 13.3), may move a gene that controls cell division to a new location near the promoter or enhancer sequence of a highly active gene, making the cell division gene overactive (see Figure 13.12).
- Infecting viruses may introduce genes that affect the expression of genes for cell cycle control, or alter regulatory proteins to turn genes on. (*Insights from the Molecular Revolution* describes a virus that causes a blood cancer by altering a transcription factor.)

TUMOR SUPPRESSOR GENES **Tumor suppressor genes** are genes in normal cells that encode growth-inhibiting factors—

A. Sporadic breast cancer. Two independent mutations of the *BRCA1* tumor suppressor.

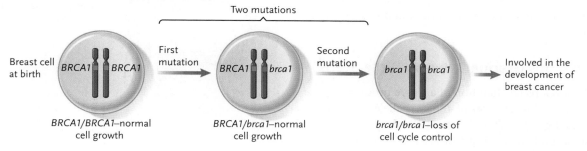

B. Familial breast cancer. An individual has a predisposition for breast cancer because of inheriting one mutated *brca1* allele; mutation of the other normal *BRCA1* allele then occurs.

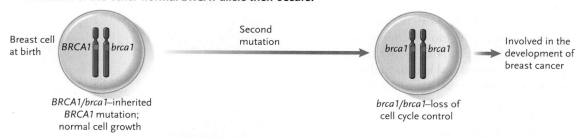

FIGURE 16.23 Mutational inactivation of tumor suppressor gene alleles in sporadic (A) and familial (B) cancers as exemplified by the *BRCA1* gene associated with breast cancer.
© Cengage Learning 2014

proteins that inhibit cell division (see Figure 16.22B). The best known tumor suppressor gene is *TP53,* so called because its encoded protein, p53, has a molecular weight of 53,000 daltons. The p53 protein is a transcription factor that turns on the expression of cell division inhibiting proteins. While p53 activity is important as part of the checks and balances involved in cell division of normal cells, it is also important if the cell has sustained DNA damage, that is, mutations. If a mutated cell undergoes DNA replication and divides, the mutations are propagated to progeny cells. This is relevant to cancer because mutations contribute to the development of cancer. However, cancer may be prevented if p53, along with other tumor-suppressor gene-encoded proteins, arrests the cell cycle to give the cell time to repair the damage, or, if the damage cannot be repaired, to trigger the cell to undergo programmed cell death (apoptosis; see Chapter 45). However, if the *TP53* gene is mutated so that the p53 protein is not produced or is produced in an inactive form, this may allow a damaged cell to continue through the cell cycle, passing its mutations on to progeny cells. The importance of p53 to cell division control is shown by the fact that inactive *TP53* genes are found in at least 50 percent of all cancers. In general, mutations of tumor-suppressor genes contribute to the onset of cancer because the mutations result in a decrease in or a loss of the inhibitory action of the cell cycle controlling proteins they encode.

Both alleles of a tumor suppressor gene must be inactivated for inhibitory activity to be lost in cancer cells. **Figure 16.23** illustrates inactivation of the tumor suppressor gene *BRCA1*

(*breast cancer 1*) in sporadic and familial forms of breast cancer. Inactivating both alleles of *BRCA1* is not by itself sufficient for the development of breast cancer, but is one of the gene changes typically involved. Since sporadic breast cancer requires the mutational inactivation of two normal alleles of *BRCA1,* this form of the disease typically occurs later in life than the familial form. For familial breast cancer and other familial cancers, we use the term *predisposition* for the cancer. This term relates to the inactivation mechanism just described. That is, an individual is predisposed to develop a particular cancer if they inherit one mutant allele of an associated tumor suppressor disease because then a mutation inactivating the other allele is all that is needed to lose the growth inhibitory properties of the tumor suppressor gene's product.

miRNA GENES You learned earlier in this chapter about the role of microRNAs (miRNAs) in regulating expression of target mRNAs. In human cancers, many miRNA genes show altered, cancer-specific expression patterns. Studying these miRNA genes has given scientists insights into the normal activities of their encoded miRNAs in cell cycle control. Some miRNAs regulate the expression of mRNAs that are the transcripts of tumor suppressor genes. If these miRNAs are overexpressed because of alterations of the genes encoding them, expression of the target mRNAs can be completely blocked, thereby removing or decreasing inhibitory signals for cell proliferation. Other miRNAs regulate the translation of mRNAs that are transcripts of particular proto-oncogenes. If these miRNA genes are inactivated, or expression of these genes is markedly reduced, expression of the proto-oncogenes is higher than normal and cell proliferation is stimulated.

Cancer Develops Gradually by Multiple Steps

Cancer rarely develops by alteration of a single proto-oncogene to an oncogene, or inactivation of the two alleles of a single tumor-suppressor gene. Rather, in almost all cancers, successive alterations in several to many genes gradually accumulate to transform normal cells to cancer cells. This gradual mechanism is called the *multistep progression of cancer*. **Figure 16.24** shows one example of the steps that can occur, in this case for a form of colorectal cancer.

The ravages of cancer, probably more than any other example, bring home the critical extent to which humans and all other multicellular organisms depend on the mechanisms controlling gene expression to develop and live normally. The most amazing thing about these control mechanisms is that, in spite of their complexity, they operate without failures throughout most of the lives of all eukaryotes.

In the next chapter, you will learn about the molecular genetics of bacteria and their phages, and about DNA sequences in prokaryotic and eukaryotic genomes that have the ability to move to different chromosomal locations.

STUDY BREAK 16.5

1. What is the normal function of a tumor-suppressor gene? How do mutations in tumor-suppressor genes contribute to the onset of cancer?
2. What is the normal function of a proto-oncogene? How can mutations in proto-oncogenes contribute to the onset of cancer?
3. How can changes in expression of miRNA genes contribute to the onset of cancer?

THINK OUTSIDE THE BOOK

Individually or collaboratively, use the Internet and/or research papers to find two examples of how an miRNA potentially may be an effective therapeutic molecule to treat cancer. For the examples, outline the natures of the cancers and the state of the research on the particular miRNAs under investigation.

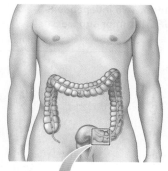

FIGURE 16.24 A multistep model for the development of a type of colorectal cancer.
© Cengage Learning 2014

Normal colon cells

Loss of the *APC* tumor suppressor gene activity, and other DNA changes

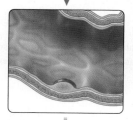

Small adenoma (benign growth)

ras oncogene activation; loss of *DCC* tumor suppressor gene

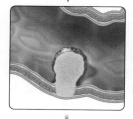

Large adenoma (benign growth)

Loss of *TP53* tumor suppressor gene activity and other mutations

Carcinoma (malignant tumor with metastasis)

Can RNA interference silence disease?

RNA interference (RNAi) is the process in which a single-stranded RNA such as small interfering RNA (siRNA) of about 21 nucleotides associated with a protein complex (RISC) inhibits gene expression by associating with an mRNA containing a totally or partially complementary sequence. The siRNA inhibits the mRNA by site-specific cleavage, enhanced degradation, or inducing a block to its translation. RNAi can be induced experimentally in cells by delivering either a double-stranded siRNA that is converted within the cell to mature siRNA, or a DNA transcriptional template, which is transcribed to produce short-hairpin RNAs that are then converted into the mature siRNAs. Each of these approaches has advantages and disadvantages. Physicians and scientists are working together to develop RNAi for many clinical applications. Let us consider some examples of RNAi therapies that are in the works.

What are examples of diseases that are amenable to RNAi based therapies?

Anti-viral treatment. Viruses are essentially exogenous genes that enter an organism and through the infection process can cause disease. Can RNAi treat viral infections? Our laboratory has had a long-standing interest in using RNAi to inhibit virus genes in order to short-circuit hepatitis virus infection. After we were able to establish that RNAi worked in whole mammals, we were successful in knocking down both hepatitis virus C gene sequences and hepatitis virus B replication in mice. Currently, scientists are using RNAi approaches to inactivate all types of viruses, including HIV, by directly attacking the viral genes and/or key cell proteins (for example, viral receptor) required for viral infection and spread. A number of different treatments for viral infections are in various phases of testing in animals and humans.

Macular degeneration treatment. Macular degeneration, the leading cause of blindness among those age 55 and older in the United States, is characterized by an overabundance of blood vessel growth in the retina of the eye due to the up-regulation of the VEGF (vascular endothelial growth factor) pathway. These blood vessels leak, leading to clouded and often complete loss of vision. Researchers are investigating whether RNAi could be an effective therapy for such diseases by targeting VEGF or its receptor. Although there was early enthusiasm, some of the medical improvement may have been the result of a side effect, due to the siRNA

eliciting the nonspecific release of proteins that indirectly interfere with the VEGF pathway. Such side effects can be reduced by modification of the siRNA and/or its delivery method.

Anti-cancer treatment. RNAi can also be used to turn down various genes that are misregulated in cancer. Again, promising results in animals have prompted early clinical studies. RNAi therapies in combination with more standard therapies are now being tried.

How long will it take before there are approved RNAi drugs?

Scientists believe that it may take some years because clinical trials proceed at a slower rate than preclinical animal studies. Moreover, clinical scientists will remind us that a mouse is not a man, and that while success in humans is likely there will likely be technical barriers that need to be worked out before the full benefits of this new therapeutic approach will be realized. Interestingly, scientists are finding all sorts of new classes of small and very large noncoding RNAs. In fact, scientists have found that over 90% of the human genome is transcribed into RNA. This field is sometimes referred to as the dark matter of the genome. What are the functions of these RNAs? The biological importance of these noncoding RNAs is an intense area of fundamental research. While discovery is ongoing, the next intriguing question is whether these RNAs can be harnessed or manipulated in a manner similar to RNAi for therapeutic benefit.

THINK LIKE A SCIENTIST Researchers proceed cautiously with trials of RNAi disease therapies in animals before jumping into human trials. If you were working in Dr. Kay's lab, what would be your strategy to test for dangerous side effects of the drugs?

Courtesy of Mark A. Kay, M.D., PhD

Mark A. Kay is the Dennis Farrey Family Professor in the Departments of Pediatrics and Genetics at the Stanford University School of Medicine. One major focus of his work is the role of small RNAs in mammalian gene regulation and their manipulation to produce new therapeutics. To learn more about his research, go to http://kaylab.stanford.edu.

REVIEW KEY CONCEPTS

To access the course materials and companion resources for this text, please visit www.cengagebrain.com.

16.1 Regulation of Gene Expression in Prokaryotes

- Transcriptional control in prokaryotes involves short-term changes that turn specific genes on or off in response to changes in environmental conditions. The changes in gene activity are controlled by regulatory proteins that recognize operators of operons (Figure 16.2).

- Regulatory proteins may be repressors, which slow the rate of transcription of operons, or activators, which increase the rate of transcription.

- Some repressors are made in an active form, in which they bind to the operator of an operon and inhibit its transcription. Combination with an inducer blocks the activity of the repressor and allows the operon to be transcribed (Figure 16.3).

- Other repressors are made in an inactive form, in which they are unable to inhibit transcription of an operon unless they combine with a corepressor (Figure 16.4).

- Activators typically are made in inactive form, in which they cannot bind to their binding site next to an operon. Combining with another molecule, often a nucleotide, converts the activator into the form in which it binds with its binding site and recruits RNA polymerase, thereby stimulating transcription of the operon (Figure 16.5).

Animation: Negative control of the lactose operon

Animation: The lactose operon

16.2 Regulation of Transcription in Eukaryotes

- Operons are not found in eukaryotes. Instead, genes that encode proteins with related functions typically are scattered through the genome, while being regulated in a coordinated manner.

- Two general types of gene regulation occur in eukaryotes. Short-term regulation involves relatively rapid changes in gene expression in response to changes in environmental or physiological conditions. Long-term regulation involves changes in gene expression that are associated with the development and differentiation of an organism.

- A eukaryotic protein-coding gene contains a single transcription unit. Regulatory sequences are located upstream of the promoter and, for some genes, at a more distant site (Figure 16.6).

- Gene expression in eukaryotes is regulated at the transcriptional level (where most regulation occurs) and at posttranscriptional, translational, and posttranslational levels (Figure 16.7).

- Regulatory events at a gene's promoter and regulatory sequences control transcription initiation and involve the binding of an array of proteins. At the promoter, general transcription factors bind and recruit RNA polymerase II, giving a very low level of transcription. Activator proteins bind to promoter proximal elements and increase the rate of transcription. Other activators bind to the enhancer if one is present and, through interaction with a coactivator, which also binds to the proteins at the promoter, greatly stimulate the rate of transcription (Figures 16.8–16.10).

- The overall control of transcription of a gene depends on the particular regulatory transcription factors that bind to the regulatory sequences of genes. The regulatory transcription factors are cell-type specific and may be activators or repressors. This gene regulation is achieved by a relatively low number of regulatory proteins, acting in various combinations (Figure 16.11).

- The coordinate expression of genes with related functions is achieved by each of the related genes having the same regulatory sequences associated with them.

- Sections of chromosomes or whole chromosomes can be inactivated by DNA methylation, a phenomenon called silencing. DNA methylation is also involved in genomic imprinting, in which transcription of either the inherited maternal or paternal allele of a gene is inhibited permanently.

- Transcriptionally active genes have a looser chromatin structure than transcriptionally inactive genes. The change in chromatin structure that accompanies the activation of transcription of a gene involves specific histone modifications, as well as chromatin remodeling, particularly in the region of a gene's promoter (Figure 16.13).

Animation: Controls of eukaryotic gene expression

16.3 Posttranscriptional, Translational, and Posttranslational Regulation

- Posttranscriptional, translational, and posttranslational controls operate primarily to regulate the quantities of proteins synthesized in cells (Figure 16.6).

- Posttranscriptional controls regulate pre-mRNA processing, mRNA availability for translation, and the rate at which mRNAs are degraded. In alternative splicing, different mRNAs are derived from the same pre-mRNA. In another process, small single-stranded RNAs complexed with proteins bind to mRNAs that have complementary sequences, and either the mRNA is cleaved or translation is blocked (Figure 16.13).

- Translational regulation controls the rate at which mRNAs are used by ribosomes in protein synthesis.

- Posttranslational controls regulate the availability of functional proteins. Mechanisms of regulation include the alteration of protein activity by chemical modification, protein activation by processing of inactive precursors, and affecting the rate of degradation of a protein.

16.4 Genetic and Molecular Regulation of Development

- Development proceeds as a result of cell division, cell movements, induction, determination, and differentiation.

- Developmental information is stored in both the nucleus and cytoplasm of the fertilized egg. The mRNA and protein molecules that direct the first stages of development are the cytoplasmic determinants.

- In determination, the developmental fate of a cell is set. The major process responsible for determination is induction, which results from the effects of signaling molecules of inducing cells on responding cells.

- In differentiation, cells change from embryonic form to specialized types with distinct structures and functions. Differentiation produces specialized cells without loss of genes from the genome.

- Determination and differentiation both involve regulated changes in gene expression (Figure 16.16).

- Pattern formation derives from the positions of cells in the embryo. Typically, positional information is detected by the cells in the form of concentration gradients of regulatory molecules encoded by genes (Figures 16.17–16.19).

- *Hox* genes are evolutionarily conserved regulatory genes that control the development of the segments or regions of the body (Figures 16.20 and 16.21).

16.5 The Genetics of Cancer

- Complex signaling systems are involved in cell division regulation. Growth factors bind to cellular receptors of target cells and result in the production of proteins that stimulate cell division. Growth-inhibiting factors bind to cellular receptors of target cells and result in the production of proteins that inhibit cell division (Figure 16.22)

- In cancer, cells partially or completely dedifferentiate, divide rapidly and uncontrollably, and may break loose to form additional tumors in other parts of the body.

- Proto-oncogenes, tumor suppressor genes, and miRNA genes typically are altered in cancer cells. Proto-oncogenes encode proteins that stimulate cell division. Their altered forms, oncogenes, are abnormally active. Tumor suppressor genes in their normal form encode proteins that inhibit cell division. Mutated forms of these genes lose this inhibitory activity (Figure 16.23). MicroRNA genes control the activity of mRNA transcripts of particular tumor suppressor genes and proto-oncogenes. Alteration of activity of such an miRNA gene can lead to a lower than normal activity of tumor suppressor gene products or a higher than normal activity of proto-oncogene products depending on the target of the miRNA. In either case, cell proliferation can be stimulated.

- Most cancers develop by multistep progression involving the successive alteration of several to many genes (Figure 16.24).

UNDERSTAND & APPLY

Test Your Knowledge

1. The control of the delivery of finished mRNAs to the cytoplasm is an example of:
   a. translational regulation.
   b. posttranslational regulation.
   c. transcriptional regulation.
   d. posttranscriptional regulation.
   e. deoxyribonucleic regulation.

2. For the *E. coli lac* operon, when lactose is present:
   a. and glucose is absent, cAMP binds and activates catabolic activator protein (CAP).
   b. and glucose is absent, the level of cAMP decreases.
   c. activated CAP binds the repressor protein to remove it from the operator gene.
   d. the cell prefers lactose over glucose.
   e. RNA polymerase cannot bind to the promoter.

3. For the *trp* operon:
   a. tryptophan is an inducer.
   b. when tryptophan binds to the Trp repressor, transcription is blocked.
   c. Trp repressor is synthesized in an active form.
   d. low levels of tryptophan bind to the *trp* operator and block transcription of the tryptophan biosynthesis genes.
   e. high levels of tryptophan activate RNA polymerase and induce transcription.

4. Transcriptional regulation is important because it:
   a. is the final and most important step in gene regulation.
   b. determines the availability of finished proteins.
   c. determines which genes are expressed.
   d. determines the rate at which proteins are made.
   e. removes masking proteins that block initiation of transcription.

5. Activation of a eukaryotic gene typically involves:
   a. release of transcription factors from the gene's promoter.
   b. acetylation of histone tails of nucleosomes in the region of the promoter.
   c. genomic imprinting.
   d. DNA methylation in the region of the promoter.
   e. binding of microRNAs to the DNA sequence of the promoter.

6. Which statement about activation of transcription is *not* correct?
   a. A transcription factor binds to the promoter in the area of the TATA box.
   b. A coactivator forms a bridge between the promoter and the gene to be transcribed.
   c. Transcription factors bind the promoter and RNA polymerase.
   d. Activators bind to the enhancer region on DNA.
   e. RNA is transcribed downstream from the promoter region.

7. Normal ears in a certain mammal are perky; mutants have droopy ears. In males of these mammals, the gene encoding perky ears is transcribed only from the female parent. This is because the gene from the male parent is silenced by methylation. If the maternal gene is mutated:
   a. male offspring have droopy ears.
   b. offspring have perky ears.
   c. male offspring have one droopy ear and one perky ear.
   d. the genetic mechanism is called alternative splicing.
   e. this is an example of posttranscriptional regulation.

8. Which of the following statements does *not* describe microRNA?
   a. MicroRNA is encoded by non-protein-coding genes.
   b. MicroRNA has a precursor that is folded and then cut by a Dicer enzyme.
   c. MicroRNA is an example of a molecule that induces RNA interference or gene silencing.
   d. MicroRNA is synthesized *in vitro* but probably not *in vivo*.
   e. MicroRNA has a similar function to that of small interfering RNAs.

9. In mammals, the nose is located at the anterior end of the embryo, and the heart at the center of the embryo. These positions are the result of activation of:
   a. *Hox* genes that are arranged along a number of chromosomes in the same order as they are expressed along the anterior–posterior body axis.
   b. maternal-effect genes, after somites differentiate into muscle.
   c. *Hox* genes that are scattered randomly among different chromosomes.
   d. a transcription factor called the homeobox.
   e. a homeodomain that binds ribosomes.

10. Which of the following is not a characteristic of cancer cells?
    a. proto-oncogenes altered to become oncogenes
    b. the mutation of a suppressor gene that results in normal genes becoming inactive
    c. the mutation of the *TP53* gene
    d. multistep progression
    e. amplification of growth factors and growth factor receptors

Discuss the Concepts

1. In a mutant strain of *E. coli,* the CAP protein is unable to combine with its target region of the *lac* operon. How would you expect the mutation to affect transcription when cells of this strain are subjected to the following conditions?

   lactose and glucose are both available
   lactose is available but glucose is not
   both lactose and glucose are unavailable

2. Duchenne muscular dystrophy, an inherited genetic disorder, affects boys almost exclusively. Early in childhood, muscle tissue begins to break down in affected individuals, who typically die in their teens or early twenties as a result of respiratory failure. Muscle samples from women who carry the mutation reveal some regions of degenerating muscle tissue adjacent to other regions that are normal. Develop a hypothesis explaining these observations.

3. Eukaryotic transcription is generally controlled by the binding of regulatory proteins to DNA sequences rather than by the modification of RNA polymerases. Develop a hypothesis explaining why this is so.

Design an Experiment

Design an experiment using rats as the model organism to test the hypothesis that human chorionic gonadotrophin (hCG), a hormone produced during pregnancy, leads to a significant protection against breast cancer.

Interpret the Data

Investigating a correlation between specific cancer-causing mutations and risk of mortality in humans is challenging, in part because each cancer patient is given the best treatment available at the time. There are no "untreated control" cancer patients, and the idea of what treatments are optimal changes quickly as new drugs become available and new discoveries are made.

The **Table** below shows the results of a study in which 442 women who had been diagnosed with breast cancer were checked for mutations of the *BRCA1* tumor suppressor gene and of a second tumor suppressor gene involved with breast cancer, *BRCA2*. The treatments and progress of the women were followed over several years. All of the women in the study had at least two affected close relatives, so their risk of developing breast cancer due to an inherited factor was estimated to be greater than that of the general population.

BRCA Mutations in Women Diagnosed with Breast Cancer[1]

	BRCA1 Mutations	*BRCA2* Mutations	No *BRCA* Mutations	Total
Total number of patients	89	35	318	442.0
Average age at diagnosis	43.9	46.2	50.4	
Preventive mastectomy[2]	6	3	14	23
Preventive oophorectomy[2]	38	7	22	67
Number of deaths	16	1	21	38
Percent died	18.0	2.8	6.9	8.6

[1]Results from a 2007 study investigating *BRCA* mutations in women diagnosed with breast cancer. All women in the study had a family history of breast cancer.

[2]Some of the women underwent preventive mastectomy (removal of the noncancerous breast) during their course of treatment. Others had preventive oophorectomy (surgical removal of the ovaries).

© Cengage Learning 2014

1. According to this study, what is the woman's risk of dying of cancer if two of her close relatives have breast cancer?

2. What is her risk of dying of cancer if she carries a mutated *BRCA1* gene?

3. Is a *BRCA1* or *BRCA2* mutation more dangerous in breast cancer cases?

4. What other data would you have to see in order to make a conclusion about the effectiveness of preventive surgeries?

Apply Evolutionary Thinking

Fruit flies homozygous for a mutation in the tumor suppressor gene *HIPPO* develop tumors in every organ. Expression of the human gene *MST2* in flies homozygous for *HIPPO* show greatly reduced or no tumors. What does this result suggest about the evolution of tumor suppressor genes in animals?

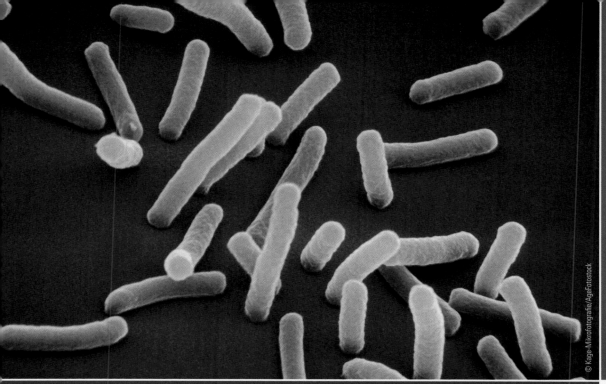

Escherichia coli, a model research organism for several types of biological studies, including bacterial genetics (colorized scanning electron micrograph).

Bacterial and Viral Genetics

Why it matters . . . In 1885, a German pediatrician, Theodor Escherich, identified a bacterium that caused severe diarrhea in infants. He named it *Bacterium coli.* Researchers were surprised to discover, however, that *B. coli* is also present in healthy infants, and is a normal inhabitant of the human intestine. It was also discovered that only certain strains of *B. coli* cause human diseases. Further, researchers in the twentieth century found that, if bacteria of different strains were mixed together, the organisms produced some progeny with a mixture of traits from more than one strain—evidence that bacteria could undergo genetic recombination.

As an organism that is readily available and easy to grow, this intestinal bacterium has been of central interest to scientists since its first discovery. Renamed *Escherichia coli* in honor of Escherich, the bacterium brings several distinct advantages to scientific investigation. It can be grown quickly in huge numbers in nutrient solutions that are simple to prepare. And it can be infected with a group of viruses called **bacteriophages** (**phages** for short) that have been as valuable to scientists as *E. coli* has because phages can be grown by the billions in cultures of the host bacterium. The rapid generation times and numerous offspring of *E. coli* and its phages make them especially valuable to geneticists. Geneticists have used them to analyze genetic crosses and their outcomes much more quickly than they can with eukaryotes. Researchers can also detect genetic events that occur only once within millions of offspring. The characteristics of these rare events helped scientists to work out the structure, activity, and recombination of genes at the molecular level.

Because *E. coli* can be cultured in completely defined chemical media—solutions in which the identity and amount of each chemical is known—it is particularly useful for biochemical investigations. What goes in, what comes out, and what biochemical changes occur inside the cells while growing in a particular medium can be detected and closely followed in normal and mutant bacteria, and in bacteria infected by phages. These biochemical studies have added immea-

Escherichia coli

We probably know more about *E. coli* than any other organism. For example, microbiologists have deciphered the complete DNA sequence of the genome of a standard laboratory strain of *E. coli*, including the sequence of its approximately 4,200 protein-coding genes. However, the functions of about one-third of these genes are still unidentified.

E. coli got its start in laboratory research because of the ease with which it can be grown in culture. *E. coli* cells divide about every 20 minutes under optimal conditions, producing a clone of 1 billion cells in a matter of hours, in only 10 mL of culture medium. The same amount of medium could accommodate as many as 10 billion cells before the growth rate of *E. coli* begins to slow. *E. coli*

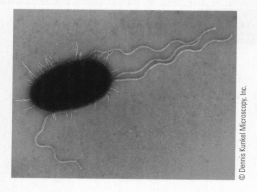

© Dennis Kunkel Microscopy, Inc.

can be grown with minimal equipment, requiring little more than culture vessels in an incubator that is held at 37°C.

Early on, however, a major advantage of *E. coli* for research was the ability to do genetics experiments with it. When Joshua Lederberg and Edward Tatum discovered that *E. coli* could conjugate, with genetic material passing from one bacterium to the other, they and other scientists realized they could carry out genetic crosses with the bacterium, producing genetic recombinants that could indicate the relative positions of genes on the chromosome. Knowing these relative positions, researchers were able to generate a genetic map of the *E. coli* chromosome. The map showed that genes with related functions are clustered together, a fact that had significant implications for the regulation of expression of those genes. For example, François Jacob and Jacques Monod's work with the genes for lactose metabolism led to the pioneering operon model (described in Section 16.1). In their work, they used conjugation to map the genes and generated partial diploids by genetic means to help understand the details of the regulation of transcription of those genes.

The development of *E. coli* as a model organism for studying gene organization and the regulation of gene expression led to the field of molecular genetics. The study of naturally occurring plasmids in *E. coli* and of enzymes that cut DNA at specific sequences eventually resulted in techniques for combining DNA from different sources, such as inserting a gene from an organism into a plasmid. Today *E. coli* is used for amplifying (cloning) plasmids that contain inserted genes or other sequences.

In essence, the biotechnology industry has its foundation in molecular genetics studies of *E. coli*, and large-scale *E. coli* cultures are widely used as "factories" for production of desired proteins. For example, the human insulin hormone, required for treatment of certain forms of diabetes, is produced by *E. coli* factories. (Chapter 18 explains more about cloning and other types of DNA manipulation.)

Laboratory strains of *E. coli* are harmless to humans. Similarly, the natural *E. coli* cells in the colon of humans and other mammals are usually harmless. There are pathogenic strains of *E. coli*, though; sometimes they make the news when humans who eat food that contains a pathogenic strain develop disease symptoms, notably colitis (inflammation of the colon) and bloody diarrhea. The genomes of several pathogenic *E. coli* strains have been sequenced; each has more genes than either the lab strain or the strain that normally lives in the human colon. The extra genes of the pathogenic strains include the genes that make the bacteria pathogenic.

© Cengage Learning 2014

---

surably to the definition of genes and their activities, and have identified many biochemical pathways and the enzymes catalyzing them. (*Focus on Model Research Organisms* tells more about *E. coli's* advantages as a model laboratory organism.)

Biologists successfully applied the same techniques used with bacteria and their phages to eukaryotes such as *Neurospora* and *Aspergillus*, fungi with short generation times that can also be grown and analyzed biochemically in large numbers. Molecular studies are now easy to carry out with a wide variety of eukaryotes, including the yeast *Saccharomyces cerevisiae*, the fruit fly *Drosophila melanogaster*, and the plant *Arabidopsis thaliana*. The results of this research showed that the molecular characteristics of genes discovered in prokaryotes apply to eukaryotes as well.

This chapter outlines the basic findings of molecular genetics in bacteria and their viruses. It also describes more broadly the structure and properties of viruses, viroids, and prions. We begin our discussion with genetic recombination mechanisms in bacteria.

17.1 Gene Transfer and Genetic Recombination in Bacteria

In the first half of the twentieth century, foundational genetic experiments with eukaryotes revealed the processes of genetic recombination during sexual reproduction, which led to the construction of genetic maps of chromosomes for a number of organisms (see Chapters 12 and 13). Bacteria became the subject of genetics research in the middle of the twentieth century. A key early question was whether gene transfer and genetic recombination can occur in bacteria even though these organisms do not undergo meiosis. For particular bacteria, the answer to the question was yes—genes can be transferred from one bacterium to another by several different mechanisms, and the newly introduced DNA can recombine with DNA already present. Such genetic recombination performs the same function as it does in eukaryotes: it generates genetic variability through the exchange of alleles between homologous regions of DNA molecules from two different individuals.

FIGURE 17.1 **Experimental Research**

Genetic Recombination in Bacteria

Question: Does genetic recombination occur in bacteria?

Experiment: To answer the question, Lederberg and Tatum used two mutant strains of *E. coli:* Mutant strain 1's genotype was *bio⁻ met⁻ leu⁺ thr⁺ thi⁺*, where the "+" means a normal allele and the "−" means a mutant allele. This strain required biotin and methionine to grow. Mutant strain 2's genotype was *bio⁺ met⁺ leu⁻ thr⁻ thi⁻*; it required leucine, threonine, and thiamine to grow.

Lederberg and Tatum plated about 100 million cells of a mixture of the two mutant strains on minimal medium, which lacked any of the nutrients the strains needed for growth. As controls, they also plated large numbers of the two mutant strains individually on minimal medium.

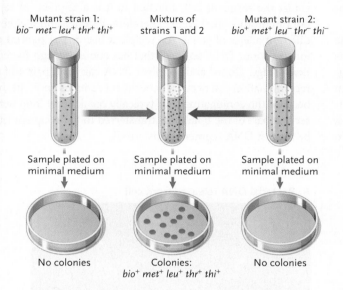

Results: No colonies grew on the control plates, meaning that the mutant alleles in the strains had not mutated back to normal alleles. However, for the mixture of mutant strain 1 and mutant strain 2, several hundred colonies grew on the minimal medium.

Conclusion: To grow on minimal medium, the bacteria must have been able to make biotin, methinine, leucine, threonine, and thiamine, meaning that they had the genotype: *bio⁺ met⁺ leu⁺ thr⁺ thi⁺*.

Lederberg and Tatum concluded that the colonies on the plate must have resulted from genetic recombination between mutant strains 1 and 2.

THINK LIKE A SCIENTIST Suppose mutant strain 1 had the genotype *met⁻ leu⁺* and mutant strain 2 had the genotype *met⁺ leu⁻*. Would these have been better strains to use in the experiment? Explain why or why not.

Source: J. Lederberg and E. Tatum. 1946. Gene recombination in *Escherichia coli. Nature* 158:558.

© Cengage Learning 2014

E. coli is one of the bacteria in which genetic recombination occurs. By the 1940s, geneticists knew that *E. coli* and many other bacteria could be grown in a **minimal medium** containing water, an organic carbon source such as glucose, and a selection of inorganic salts—including one that provides nitrogen—such as ammonium chloride. The growth medium can be in liquid form or in the form of a gel that is made by adding agar to the liquid medium. (Agar is a polysaccharide material, indigestible by most bacteria, that is extracted from algae.)

Since it is not practical to study a single bacterium for most experiments, researchers soon developed techniques for starting bacterial cultures from a single cell, generating cultures with a large number of genetically identical cells. Cultures of this type are called **clones.** To start bacterial clones, the scientist spreads a drop of a bacterial culture over a sterile agar gel in a culture dish. The culture has been diluted enough to ensure that cells will be widely separated on the agar surface. Each individual cell divides many times to produce a separate colony that is a clone of the initial cell. Cells can be removed from a clone and introduced into liquid cultures or spread on agar and grown in essentially any quantity.

Genetic Recombination Occurs in *E. coli*

In 1946, Joshua Lederberg and Edward L. Tatum of Yale University performed an experiment to determine if genetic recombination occurs in bacteria, using *E. coli* as their experimental organism **(Figure 17.1).** For this experiment, they used *auxotrophs,* mutant strains that could not grow on minimal medium (see Section 15.1). When they mixed together cells of two mutant strains, each of which had different requirements for growth on minimal medium, they observed that some cells were able to grow on minimal medium. They interpreted the result to mean that genetic recombination had occurred between the two strains.

Bacterial Conjugation Brings DNA of Two Cells Together, Allowing Genetic Recombination to Occur

Lederberg and Tatum's results led to a major question: how did DNA molecules with different alleles get together to undergo genetic recombination? Recombination in eukaryotes occurs in diploid cells undergoing meiosis by an exchange of segments between the chromatids of homologous chromosome pairs (see Section 13.1). Bacteria typically have a single, circular chromosome—they are haploid organisms (see Section 14.5). A different mechanism occurs in bacteria as you will learn in this section.

CONJUGATION AND THE F FACTOR Rather than fusing together to produce the prokaryotic equivalent of a diploid zygote, bacterial cells *conjugate:* they contact each other, initially becoming connected by a long tubular structure on the cell surface called a *sex pilus* **(Figure 17.2A)**, and then forming a cytoplasmic bridge that connects two cells **(Figure 17.2B)**. During **conjugation**, DNA of one cell, the *donor* (the bristly cell in Figure 17.2A), moves through the cytoplasmic bridge into the other cell, the *recipient*.

The ability to conjugate depends on the presence within a donor cell of a plasmid called the F factor (F = fertility). Plasmids are small circles of DNA that occur in bacteria in addition to the main circular chromosomal DNA molecule **(Figure 17.3)**. Plasmids contain several to many genes and a replication origin that permits them to be duplicated and passed on during bacterial division. Donor cells in conjugation are called **F⁺** cells because they contain the F factor. They are able to conjugate (mate) with recipient cells but not with other donor cells. Recipient cells, which lack the F factor, are unable to initiate conjugation; they are called **F⁻** cells.

The F factor contains genes that encode proteins of the **sex pilus,** also called the **F pilus** (plural, *pili*). The sex pilus allows an F⁺ donor cell to attach to an F⁻ recipient (see Figure 17.2A). Once attached, the donor and recipient cells form a cytoplasmic bridge and conjugate **(Figure 17.4A)**. The outcome of F⁺ × F⁻ conjugation is that the F⁻ cell receives a copy of the F factor and, therefore, becomes an F⁺ cell. No chromosomal DNA is transferred between cells in F⁺ × F⁻ conjugation, however, so no genetic recombination of bacterial genes occurs.

Hfr CELLS AND GENETIC RECOMBINATION How does genetic recombination of bacterial genes occur as a result of conjugation if no chromosomal DNA transfers when an F factor is transferred in conjugation? The answer is that in some F⁺ cells the F factor integrates into the bacterial chromosome producing a donor that can conjugate with and transfer genes on the bacterial chromosome to a recipient **(Figure 17.4B)**. These special donor cells are known as **Hfr cells** (Hfr = high frequency recombination). The outcome of Hfr × F⁻ conjugation is that genetic recombination of bacterial genes can occur because, as the F factor replicates and part of it enters the recipient cell, attached to it is a segment of replicated donor chromosomal DNA. Therefore, if donor and recipient have different alleles of genes, the recipient becomes a **partial diploid** for the donor DNA segment that has come through the conjugation bridge. Donor and recipient DNA can then pair and genetic recombination can occur. The recipient remains an F⁻ in this case, because the conjugating cells typically break apart long before the second part of the F factor is transferred to the recipient (it would be the last DNA segment transferred).

A. Attachment by sex pilus

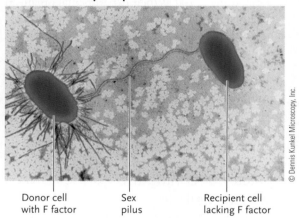

Donor cell with F factor

Sex pilus

Recipient cell lacking F factor

© Dennis Kunkel Microscopy, Inc.

B. Cytoplasmic bridge formed

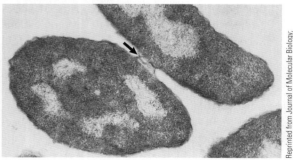

Reprinted from Journal of Molecular Biology, Vol 16/Issue 2; Julian D. Gross & Lucien G. Caro; DNA transfer in bacterial conjugation; Page 269; 1966; with permission from Elsevier

FIGURE 17.2 Conjugating *E. coli* cells. **(A)** Initial attachment of two cells by the sex pilus. **(B)** A cytoplasmic bridge (arrow) has formed between the cells, through which DNA moves from one cell to the other.

A. Bacterial DNA released from cell

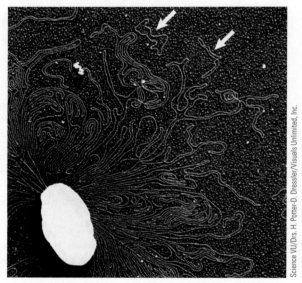

Science VU/Drs. H. Potter-D. Dressler/Visuals Unlimited, Inc.

B. Plasmid

Professor Stanley Cohen/Science Photo Library/Photo Researchers, Inc.

FIGURE 17.3 Electron micrographs of DNA released from a bacterial cell. **(A)** Plasmids (arrows) near the mass of chromosomal DNA. **(B)** A single plasmid at higher magnification (colorized).

A. Transfer of the F factor during conjugation between F⁺ and F⁻ cells.

B. Transfer of bacterial genes and production of recombinants during conjugation between Hfr and F⁻ cells.

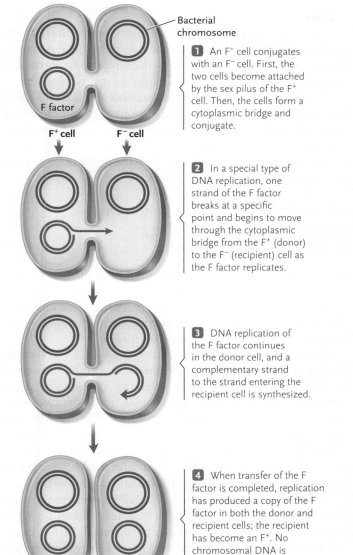

1 An F⁺ cell conjugates with an F⁻ cell. First, the two cells become attached by the sex pilus of the F⁺ cell. Then, the cells form a cytoplasmic bridge and conjugate.

2 In a special type of DNA replication, one strand of the F factor breaks at a specific point and begins to move through the cytoplasmic bridge from the F⁺ (donor) to the F⁻ (recipient) cell as the F factor replicates.

3 DNA replication of the F factor continues in the donor cell, and a complementary strand to the strand entering the recipient cell is synthesized.

4 When transfer of the F factor is completed, replication has produced a copy of the F factor in both the donor and recipient cells; the recipient has become an F⁺. No chromosomal DNA is transferred in this mating.

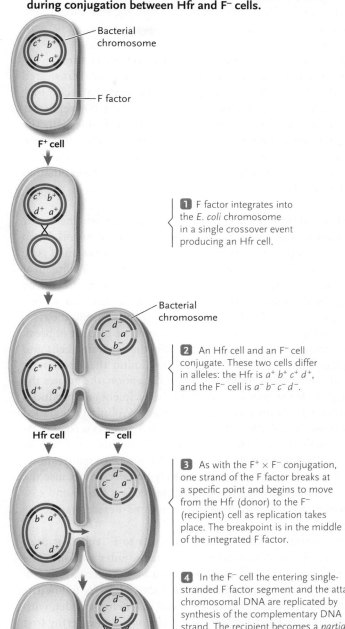
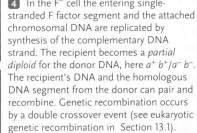
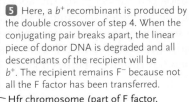

1 F factor integrates into the *E. coli* chromosome in a single crossover event producing an Hfr cell.

2 An Hfr cell and an F⁻ cell conjugate. These two cells differ in alleles: the Hfr is *a⁺ b⁺ c⁺ d⁺*, and the F⁻ cell is *a⁻ b⁻ c⁻ d⁻*.

3 As with the F⁺ × F⁻ conjugation, one strand of the F factor breaks at a specific point and begins to move from the Hfr (donor) to the F⁻ (recipient) cell as replication takes place. The breakpoint is in the middle of the integrated F factor.

4 In the F⁻ cell the entering single-stranded F factor segment and the attached chromosomal DNA are replicated by synthesis of the complementary DNA strand. The recipient becomes a *partial diploid* for the donor DNA, here *a⁺ b⁺/a⁻ b⁻*. The recipient's DNA and the homologous DNA segment from the donor can pair and recombine. Genetic recombination occurs by a double crossover event (see eukaryotic genetic recombination in Section 13.1).

5 Here, a *b⁺* recombinant is produced by the double crossover of step 4. When the conjugating pair breaks apart, the linear piece of donor DNA is degraded and all descendants of the recipient will be *b⁺*. The recipient remains F⁻ because not all the F factor has been transferred.

Hfr chromosome (part of F factor, followed by bacterial genes)

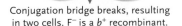

Conjugation bridge breaks, resulting in two cells. F⁻ is a *b⁺* recombinant.

SUMMARY The F factor plasmid is responsible for conjugation. F⁺ *E. coli* bacteria have an F factor free in the cytoplasm. An F⁺ donor cell can conjugate with an F⁻ recipient cell and transfer a copy of the F factor to the recipient, converting it to an F⁺ cell. No genetic recombination between donor and recipient genes can occur **(A).** Hfr bacteria have the F factor integrated into the chromosome. When an Hfr conjugates with an F⁻ cell, the transfer of DNA from donor to recipient includes a copy of some of the donor chromosomal genes, enabling genetic recombination to occur between donor and recipient genes in the recipient **(B).**

THINK LIKE A SCIENTIST Why does transfer of chromosomal genes not occur from recipient to donor?

In the particular example shown in Figure 17.4B, a b^+ recombinant is generated. In other pairs in the mating population, the a^+ gene could recombine with the homologous recipient gene, or both a^+ and b^+ genes could recombine. The genetic recombinants observed in Lederberg and Tatum's experiment (see Figure 17.1) were produced in this same general way.

Recombinants produced during conjugation can be detected only if the alleles of the genes in the DNA transferred from the donor differ from those in the recipient's chromosome. Following recombination, the bacterial DNA replicates and the cell divides normally, producing a cell line with the new gene combination. Any remnants of the DNA fragment that originally entered the cell are degraded as division proceeds and do not contribute further to genetic recombination or cell heredity.

MAPPING GENES BY CONJUGATION Genetic recombination by conjugation was discovered by two scientists, François Jacob (who also proposed the operon model for the regulation of gene expression in bacteria; see Section 16.1) and Elie L. Wollman, at the Pasteur Institute in Paris. They began their experiments by conjugating Hfr and F⁻ cells that differed in a number of alleles. At regular intervals after conjugation commenced, they removed some of the cells and agitated them in a blender to break apart attached cells. They then cultured the separated cells and analyzed them for recombinants. They found that the longer they allowed cells to conjugate before separation, the greater the number of donor genes that entered the recipient and produced recombinants. From this result, Jacob and Wollman concluded that during conjugation, the Hfr cell slowly injects a copy of its DNA into the F⁻ cell. Full transfer of an entire DNA molecule to an F⁻ cell would take about 90 to 100 minutes. In nature, however, the entire DNA molecule is rarely transferred because the cytoplasmic bridge between conjugating cells is fragile and easily broken by random molecular motions before transfer is complete.

The pattern of gene transfer from Hfr to F⁻ cells was used to map the *E. coli* chromosome. The F factor integrates into one of a few possible fixed positions around the circular *E. coli* DNA. As a result, the genes of the bacterial DNA follow the F factor segment into the recipient cell in a definite order, with the gene immediately behind the F factor segment entering first and the next genes following. In the theoretical example shown in Figure 17.4B, donor genes will enter in the order $a^+-b^+-c^+-d^+$. By breaking off conjugation at gradually increasing times, investigators allowed longer and longer pieces of DNA to enter the recipient cell, carrying more and more genes from the donor cell (detected by the appearance of recombinants). By noting the order and time at which genes were transferred, investigators were able to map and assign the relative positions of most genes in the *E. coli* chromosome. The resulting genetic map has distances between genes in units of minutes. To this day, the genetic map of *E. coli* shows map distances as minutes, reflecting the mapping of genes by conjugation.

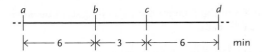

The genetic maps from *E. coli* conjugation experiments indicate that the genes are arranged in a circle, reflecting the circular form of the *E. coli* chromosome. Sequencing of the *E. coli* genome has confirmed the results obtained by genetic mapping.

In addition to the F plasmid, bacteria also contain other types of plasmids. **R plasmids,** for example, contain genes that provide resistance to unfavorable conditions, such as exposure to antibiotics. The competitive advantage provided by the genes in some plasmids may account for the wide distribution of plasmids of all kinds in prokaryotic cells.

In Transformation, DNA Taken Up by a Bacterium Is the Source of Genetic Recombination

DNA can transfer from one bacterial cell to another by two additional mechanisms, *transformation* and *transduction*. Like conjugation, these mechanisms transfer DNA in one direction, and create partial diploids in which recombination can occur between alleles in the homologous DNA regions.

In **transformation,** bacteria take up pieces of DNA that are released as other cells disintegrate. Frederick Griffith discovered this mechanism in 1928, when he found that a noninfective form of the bacterium *Streptococcus pneumoniae,* unable to cause pneumonia in mice, could be transformed to the infective form if it was exposed to heat-killed cells of an infective strain (see Section 14.1 and Figure 14.2). In 1944, Oswald Avery and his colleagues at New York University found that the substance capable of transforming noninfective bacteria to the infective form was DNA.

Subsequently, geneticists established that in transformation experiments the linear DNA fragments taken up from disrupted cells recombine with the chromosomal DNA of recipient cells, in much the same way as genetic recombination takes place in conjugation.

Approximately 1% of bacterial species can take up DNA from the surrounding medium by natural mechanisms. Transformation in those species is known as *natural transformation.* Such bacteria typically have a DNA-binding protein on the outer surface of the cell wall. When DNA from the cell's surroundings binds to the protein, a deoxyribonuclease enzyme breaks the DNA into short pieces that pass through the cell wall and plasma membrane into the cytosol. The entering DNA can then recombine with the recipient cell's chromosome if it contains homologous regions.

Natural transformation does not occur with *E. coli* cells. However, they can be induced to take up DNA by *artificial transformation.* One transformation technique is to expose *E. coli* cells to calcium ions and the DNA of interest, incubate them on ice, and then give them a quick heat shock. This treatment alters the plasma membrane so that DNA can penetrate and enter. The entering DNA undergoes recombination if it contains regions that are homologous to part of the recipient's chromosomal DNA. Another technique for artificial transformation, called *electroporation,* exposes cells briefly to rapid pulses of an electrical current. The electrical shock alters the plasma membrane so that DNA can enter. This method works well with many bacterial species, and also with many types of eukaryotic cells.

Artificial transformation is often used to insert plasmids containing DNA sequences of interest into *E. coli* cells as a part of DNA cloning techniques. After the cells are transformed, clones of the cells are grown in large numbers to increase the quantity of the inserted DNA to the amounts necessary for sequencing or genetic engineering. (DNA cloning and genetic engineering are discussed further in Chapter 18.)

In Transduction, DNA Introduced into a Bacterium by Phage Infection Is the Source of Genetic Recombination

In **transduction,** DNA is transferred to recipient bacterial cells by an infecting phage (see Section 14.1). When new phages assemble in an infected bacterial cell, they sometimes incorporate fragments of the host cell DNA instead of all or some of the viral DNA. After the phages are released from the host cell, they may attach to another cell and inject—*transduce*—the bacterial DNA (and the viral DNA if it is present) into that cell. The introduction of this DNA, as in conjugation and transformation, makes the recipient cell a partial diploid and creates the potential for recombination to take place. And, because the phage carrying the donor bacterium's DNA does not have a complete phage genome, it cannot kill the recipient cell, allowing recombinant colonies to form. The DNA-containing capacity of a phage is limited, so the amount of donor DNA that can be transferred to a recipient by this mechanism is far less than is the case for conjugation. Joshua Lederberg and his graduate student, Norton Zinder, then at the University of Wisconsin–Madison, discovered transduction in 1952 in experiments with the bacterium *Salmonella typhimurium* and phage P22. Lederberg received a Nobel Prize in 1958 for his discovery of conjugation and transduction in bacteria.

Replica Plating Allows Genetic Recombinants to Be Identified and Counted

How do researchers identify and count genetic recombinants in conjugation, transformation, or transduction experiments? Joshua Lederberg and Esther Lederberg developed a now widely applied technique for doing this called **replica plating (Figure 17.5).** In replica plating, researchers begin with bacterial colonies grown on a plate with a **complete medium,** that is, a medium containing a full complement of nutrients, including amino acids and other chemicals that normal strains can make for themselves. This master plate is pressed gently onto sterile velveteen, which transfers some of each colony to the velveteen in the same pattern as the colonies on the plate. The velveteen is then pressed onto new plates containing a minimal growth medium—the replica plates—thereby transferring some cells from each original colony to those plates. The composition of the minimal medium on the replica plates is adjusted to promote the growth of colonies with particular characteristics. The transfer inoculates each new plate with a "replica" of the original set of colonies on the starting plate, and the replica plates are incubated to allow new colonies to grow.

Figure 17.5 shows the identification of auxotrophic mutants of *E. coli* by replica plating. An investigator can determine the mutations a strain carries by comparing the original plate to the replica plates to identify missing colonies on the minimal-medium plate. Normal cells will grow in a minimal medium, but auxotrophic mutants will not, because they are unable to make one or more of the missing substances. Thus, the investigator takes for further study the colonies from the original plate that correspond to the missing colonies on the minimal-medium plates.

In an actual experiment, the compositions of the media are appropriate for the goals of the experiment. For example, to identify a met^+ recombinant in a conjugation experiment, the starting plate contains methionine and the colonies are replica plated to a plate lacking methionine. Comparison of the colony patterns on the two plates identifies met^+ recombinants because they grow on the plate lacking methionine, whereas met^- parentals do not.

Horizontal Gene Transfer Is an Important Process for Genome Evolution

Our examples of gene inheritance patterns in earlier chapters all involved the movement of genetic material from generation to generation, whether that is from cell to cell or from parent to offspring. This is movement of genetic material by descent and, at its root, it is a feature of both asexual and sexual reproduction. By contrast, **horizontal gene transfer** is the movement of genetic material between organisms other than by descent. Horizontal gene transfer can occur between different species, introducing genes from one species into another. It is most common in bacteria, but there is evidence that eukaryotes, including the ancestors of humans, obtained genes from viruses by horizontal transfer.

Conjugation, transformation, and transduction are three major mechanisms by which horizontal gene transfer occurs in bacterial species. Transduction occurs in archaea, and there is some evidence for natural transformation. It is not clear, however, whether conjugation occurs in archaea.

In conjugation, genetic material is transferred horizontally from donor to recipient under the control of a plasmid that facilitates cell–cell contact, and by the formation of a bridge

FIGURE 17.5 **Research Method**

Replica Plating

Purpose: Replica plating is used to identify different strains of bacteria with respect to their growth requirements in a heterogeneous mixture of strains.

Protocol:

1. Press sterile velveteen gently onto the master plate of solid growth medium with bacterial colonies on it. Some of each colony transfers to the velveteen in the same pattern as the colonies on the plate. In the example, a mixture of colonies of normal and auxotrophic strains is on a plate of complete medium.

2. Press the velveteen gently onto a sterile replica plate to transfer some of each strain. In the example, the replica plate contains minimal medium. Incubate to allow colonies to grow, and compare the pattern of colonies on the replica plate with that on the master plate.

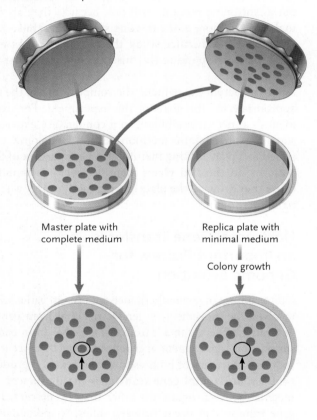

Master plate with complete medium

Replica plate with minimal medium

Colony growth

Interpreting the Results: A colony present on the master plate but not on the replica plate indicates that the strain requires some substance missing from the minimal medium in order to grow. In other words, the strain is an auxotroph. In actual experiments, the compositions of the master plate and replica plate media are chosen to be appropriate for the goals of the experiment.

Source: J. Lederberg and E. M. Lederberg. 1952. Replica plating and indirect selection of bacterial mutants. *Journal of Bacteriology* 63:399–406.

© Cengage Learning 2014

through which DNA is transferred. Relatively long segments of DNA can be transferred in this process. The particular donor segment of DNA transferred depends on where the plasmid responsible for conjugation is integrated into the donor bacteri-

um's chromosome. Our examples involved conjugation between donor and recipient cells of the same species, but the process can occur between bacteria that are not closely related.

In transformation, a DNA molecule in the environment surrounding a bacterial cell is taken up by that cell. In natural populations of bacteria, the bacteria taking up DNA typically are those capable of natural transformation. Any piece of DNA can be taken up in transformation, but usually only relatively short DNA molecules are involved.

In transduction, a phage transfers bacterial DNA from a donor to a recipient cell. The amount of DNA that can be transferred in this way is relatively small because it is limited by the capacity of the phage head in which the DNA must be packed. Horizontal gene transfer by transduction generally is limited to closely related bacteria because the phage carrying the donor DNA has to bind to specific receptors on a recipient bacterium in order to inject the genetic material. Unless the bacteria are closely related, the necessary receptor will be absent.

In our examples of conjugation, transformation, and transduction in the same species, genetic recombination between donor and recipient DNA occurs because the sequences match, allowing recombination to occur. When different species are involved, the transferred DNA may have little sequence similarity to the recipient's DNA. However, if some stretches of sequences match sufficiently, recombination can lead to the integration of a segment of the transferred donor DNA. In this way, a recipient species can acquire new genes. If those genes confer some type of selective advantage to the recipient, then the altered genome will be retained and potentially the old genome will be lost in the population. In this way, horizontal gene transfer has resulted in the evolution of genomes.

Now that the genomes of many bacterial and archaeal species have been sequenced completely, geneticists can compare whole genomes to see how much horizontal gene transfer has occurred. Such analysis shows that horizontal gene transfer has involved a significant fraction of prokaryotic genes, making it a major process in the evolution of prokaryotic genomes. For instance, 20% of the *E. coli* genome is thought to have derived from horizontal gene transfer. The genes acquired in this way have enabled *E. coli* to adapt to the various environments in which it lives, and have contributed to its effectiveness as a pathogen. As an example, the genome of *E. coli* strain O157:H7, the cause of a number of food-related deaths, has 1,387 genes not present in the genome of the standard lab strain of *E. coli.* Some of these extra genes, which represent about 25% of the organism's total number of genes, encode proteins responsible for the pathogenicity of the bacterium. All of the extra genes were acquired by horizontal gene transfer from other bacterial species.

Compared with prokaryotes, far fewer eukaryotic genomes are available to analyze for horizontal gene transfer. Nonetheless, such analysis reveals that horizontal gene transfer also appears to have had an important role in eukaryotic genome evolution. Too little information is available to give us a complete understanding of the phenomenon in eukaryotes, however. For instance, we know little about the mechanism of horizontal gene transfer in eukaryotes, or the factors that enhance or discourage it. The most common transfers seen are from prokaryotes, although the amount of prokaryotic DNA found in eukaryotes varies widely. Most of these transfers are to the nuclear genomes, with only a few to organelle genomes. Eukaryote–eukaryote transfers of nuclear genes are also known, as are eukaryote–prokaryote transfers. Both of these types of transfer are rarer than prokaryote–prokaryote transfers, with eukaryote–prokaryote being extremely rare.

In the next section, we turn to viruses. Whereas the phages mentioned earlier in this section infect only bacteria, viruses infect all living organisms. Many viruses have become important subjects for research into, among other things, the molecular nature of recombination and the genetic control of viral infection.

STUDY BREAK 17.1

1. What are the properties of F⁺, F⁻, and Hfr cells of *E. coli*?
2. How is horizontal gene transfer in bacteria distinctive from gene segregation in sexually reproducing organisms?

17.2 Viruses and Viral Genetics

As agents of transduction, the phages that infect bacteria are important tools in research on bacterial genetics. The same viruses are also important for studying *viral* recombination and genetics. Viruses can undergo genetic recombination when the DNA of two viruses, carrying different alleles of one or more genes, infect a single cell. Using phages, researchers study viral genetics with the same molecular and biochemical techniques used to investigate their bacterial hosts. In this section, we examine the structure and infectious properties of viruses.

Viruses in the Free Form Consist of a Nucleic Acid Core Surrounded by a Protein Coat

A **virus** (Latin for poison) is a biological particle that can infect the cells of a living organism. Viral infections usually have detrimental effects on their hosts. The study of viruses is called *virology,* and researchers studying viruses are known as *virologists.*

All viruses consist of a DNA or RNA genome surrounded by a protein coat—a layer of proteins—called the **capsid.** The complete viral particle is also called a **virion.** Most, but not all, vi-ruses are significantly smaller than bacteria. Among the few exceptions are viruses that are 300 to 500 nm long and 120 to 300 nm in diameter. The smallest bacteria are 200 to 300 nm long.

Are viruses living or nonliving? Many scientists consider viruses to be on the border of living and nonliving. That is, although they share some properties with living things, viruses are missing several important characteristics: (1) they cannot reproduce independently, but only within a host cell; (2) they are not made up of cells—they are merely pieces of DNA or RNA surrounded by a protein capsid and in mammals an additional membranous envelope; and (3) they do not grow, develop, or generate metabolic energy. They do, however, share their chemical constituents with living things, and indeed these constituents possess the same biological order; however, this order typically stops at the macromolecule level. Viruses can adapt very readily over time since they routinely mutate the glycoproteins in their capsids or their envelopes, which provide the lock-and-key fit for entry into their host cells. In summary, viruses cannot be defined as being alive at the cellular level; they do, however, share enough characteristics with organisms that they cannot be defined as nonliving, either, like a rock.

Viral Structure Is the Minimum Necessary to Transmit Nucleic Acid Molecules from One Host Cell to Another

The genetic material of all organisms is double-stranded DNA, but the nucleic acid genome of a virus may be double-stranded DNA, singled-stranded DNA, double-stranded RNA, or single-stranded RNA, depending on the viral type. Moreover, the genome of some viruses consists of one nucleic acid molecule, while the genomes of others are distributed among two or more nucleic acid molecules. For example, the genome of a herpes virus consists of a single molecule of double-stranded DNA, while the genome of influenza virus is distributed among eight (influenza A and B strains) or seven (influenza C strains) segments of single-stranded RNA.

The simplest viruses contain only a few genes, whereas those of the most complex viruses may contain a hundred or more. All viruses have genes encoding the proteins of their capsid. For some viruses, the capsid is assembled from protein molecules of a single type. More complex viruses have capsids assembled from several different proteins, including the recognition proteins that bind to host cells. All viruses also have a gene or genes encoding the proteins that make them pathogenic to their hosts. The particles of some viruses also contain the DNA or RNA polymerase enzymes required for viral genome replication and/or an enzyme that attacks cell walls or membranes.

Most viruses take one of two basic structural forms, helical or polyhedral. In **helical viruses,** the capsid proteins assemble in a rodlike spiral around the genome **(Figure 17.6A).** A number of viruses that infect plant cells are helical. In **polyhedral viruses,** the capsid proteins form triangular units that form an

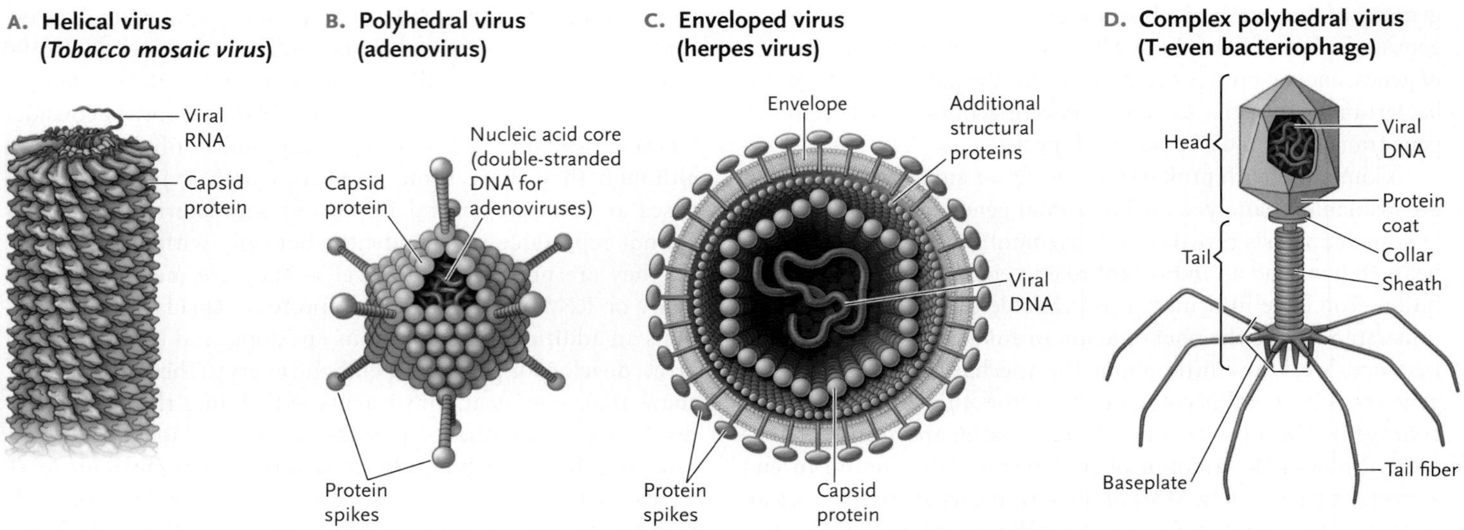

A. Helical virus (*Tobacco mosaic virus*)
- Viral RNA
- Capsid protein

B. Polyhedral virus (adenovirus)
- Nucleic acid core (double-stranded DNA for adenoviruses)
- Capsid protein
- Protein spikes

C. Enveloped virus (herpes virus)
- Envelope
- Additional structural proteins
- Viral DNA
- Protein spikes
- Capsid protein

D. Complex polyhedral virus (T-even bacteriophage)
- Head
- Viral DNA
- Protein coat
- Collar
- Sheath
- Tail
- Baseplate
- Tail fiber

FIGURE 17.6 **Basic viral structures.** The tobacco mosaic virus in **(A)** assembles from more than 2,000 identical protein subunits. Protein spikes contain recognition proteins that allow the viral particle to bind to the surface of a host cell.
© Cengage Learning 2014

icosahedral structure **(Figure 17.6B).** The polyhedral viruses include forms that infect animals, plants, and bacteria. In some polyhedral viruses, protein spikes that provide host cell recognition extend from the corners where the facets fit together. Some viruses, the **enveloped viruses,** are covered by a surface membrane derived from the plasma membrane of their host cells; both enveloped helical and enveloped polyhedral viruses are known **(Figure 17.6C).** For example, **HIV** (for *Human Immunodeficiency Virus*), the virus that causes AIDS, is an enveloped polyhedral virus. Protein spikes extend through the membrane, giving the particle its recognition and adhesion functions.

A number of bacteriophages with DNA genomes, such as T2 (see Section 14.1), have a **tail** attached at one side of a polyhedral head, forming what is known as a **complex virus (Figure 17.6D).** The genome is packed into the head; the tail is made up of proteins forming a collar, sheath, baseplate, and tail fibers. The tail has recognition proteins at its tip and, once attached to a host cell, functions as a sort of syringe that injects the DNA genome into the cell.

Viruses of Different Kinds Infect All Living Cells

Viruses are classified by the International Committee on Taxonomy of Viruses into orders, families, genera, and species using several criteria, including size and structure, type and number of nucleic acid molecules, method of replication of the nucleic acid molecules inside host cells, host range, and infective cycle. More than 4,000 species of viruses have been classified into more than 80 families according to these criteria.

One or more kinds of viruses probably infect all living organisms. Usually a virus infects only a single species or a few closely related species. A virus may even infect only one organ

system, or a single tissue or cell type in its host. However, some viruses are able to infect unrelated species, either naturally or after mutating. For example, some humans have contracted bird flu from being infected with the natural avian flu virus as a result of contact with virus-infected birds. Of the viral families, 21 include viruses that cause human diseases. Viruses also cause diseases of wild and domestic animals; plant viruses cause annual losses of millions of tons of crops, especially cereals, potatoes, sugar beets, and sugar cane. **Table 17.1** lists some virus families that infect animals, and **Table 17.2** lists some virus families that infect plants. Note that the dsRNA and ssRNA viruses replicate using an RNA-to-RNA mechanisms, whereas retroviruses, which have ssRNA genomes, replicate via a DNA intermediate (see later in the chapter).

The effects of viruses on the organisms they infect range from undetectable, through merely bothersome, to seriously debilitating or lethal diseases. For instance, some viral infections of humans, such as those causing cold sores, chicken pox, and the common cold, are usually little more than a nuisance to healthy adults. Others, including AIDS, encephalitis, yellow fever, and smallpox, are and have been among the most severe and deadly human diseases.

While most viruses have detrimental effects, some may be considered beneficial. One of the primary reasons why bacteria do not completely overrun the planet is that they are destroyed in incredibly huge numbers by bacteriophages. Viruses also provide a natural means to control some insect pests.

Viruses Infect Bacterial, Animal, and Plant Cells by Similar Pathways

Free viruses move by random molecular motions until they contact the surface of a host cell. For infection to occur, the virus or the viral genome must enter the cell. Inside the cell, typically the

TABLE 17.1 | Examples of Animal Viruses

Virus Family	Envelope	Diseases and Infections in Humans and Other Animals
Double-stranded DNA (dsDNA) viruses		
Adenoviridae	No	Respiratory infections, tumors
Baculoviridae	Yes	Gastrointestinal infections in arthropods
Hepadnaviridae	Yes	Hepatitis B
Herpesviridae	Yes	
Herpes simplex I		Oral herpes, cold sores
Herpes simplex II		Genital herpes
Varicella-zoster		Chicken pox, shingles
Papillomaviridae	No	Benign and malignant warts
Poxviridae	Yes	Smallpox, cowpox
Single-stranded DNA Viruses		
Parvoviridae	No	Fifth disease (humans), canine parvovirus disease (dogs)
Double-stranded RNA (dsRNA) Viruses		
Reoviridae	No	Upper respiratory infections, enteritis, diarrhea
Single-stranded RNA (ssRNA) Viruses		
Filoviridae	No	Ebola hemorrhagic fever, Marburg hemorrhagic fever
Flaviviridae	Yes	Yellow fever, dengue, hepatitis C, West Nile fever
Orthomyxoviridae	Yes	Influenza
Paramyxoviridae	Yes	Measles, mumps, pneumonia
Picornaviridae	No	
Enteroviruses		Polio, hemorrhagic eye disease, gastroenteritis
Rhinoviruses		Common cold
Hepatitis A virus		Hepatitis A
Aphthovirus		Foot-and-mouth disease (livestock)
Rhabdoviridae	Yes	Rabies, other animal diseases
Retroviruses (ssRNA genome that replicates via a DNA intermediate)		
Retroviridae	Yes	
FeLV		Feline leukemia
HTLV I, II		T-cell leukemia
HIV		AIDS

© Cengage Learning 2014

TABLE 17.2 | Examples of Plant Viruses

Virus Family	Examples of Viruses (the name indicates a plant infected and its symptoms)
Double-stranded DNA (dsDNA) Viruses	
Caulimoviridae	*Cauliflower mosaic virus, Blueberry red ringspot virus*
Single-stranded DNA (ssDNA) Viruses	
Geminiviridae	*Maize streak virus, Beet curly top virus*
Double-stranded RNA (dsRNA) Viruses	
Reoviridae	*Maize rough dwarf virus, Rice ragged stunt virus*
Single-stranded RNA (ssRNA) Viruses	
Bromoviridae	*Brome mosaic virus, Broad bean mottle virus*
Luteoviridae	*Barley yellow dwarf virus, Soybean dwarf virus*
Rhabdoviridae	*Broccoli necrotic yellow virus, Potato yellow dwarf virus*
Tobamoviruses	*Tobacco mosaic virus, Pepper mold mottle virus*
Tombusviridae	*Tomato bushy stunt virus, Carnation Italian ringspot virus*

© Cengage Learning 2014

INFECTION OF BACTERIAL CELLS Bacteriophages differ in how they infect and kill hosts cells; **virulent bacteriophages** kill their host cells during each cycle of infection, and **temperate bacteriophages** may enter an inactive phase in which the host cell replicates and passes on the bacteriophage DNA for generations before the phage becomes active and kills the host. Bacteriophages of both types that infect *E. coli* are widely used in genetic research, and we will use examples of each type to illustrate their life cycles.

Virulent Bacteriophages. Among the virulent bacteriophages infecting *E. coli,* the **T-even bacteriophages** T2, T4, and T6 have been most valuable in genetic studies (see Figure 17.6D). The coat of these phages is divided into a *head* and a *tail*. Packed into the head is a single linear molecule of double-stranded DNA. The tail, assembled from several different proteins, has recognition proteins at its tip that can bind to the surface of the host cell.

viral genes are expressed, leading to replication of the viral genome and assembly of progeny viruses. The viruses are then released from the host cell, a process that often ruptures the host cell, killing it.

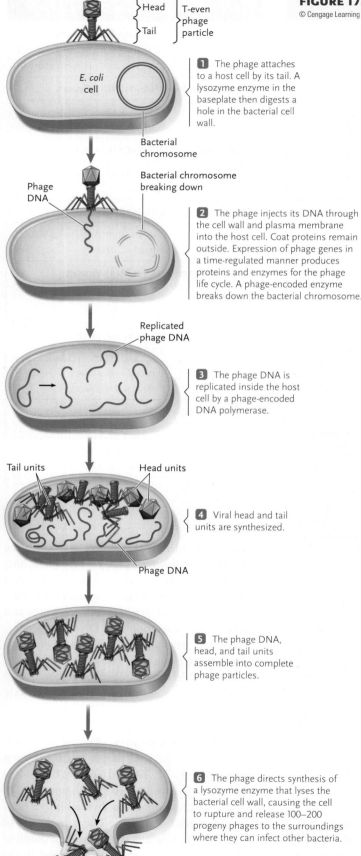

FIGURE 17.7 The infective cycle of a T-even bacteriophage, an example of a virulent phage.
© Cengage Learning 2014

Head }
Tail } T-even phage particle

E. coli cell

Bacterial chromosome

1 The phage attaches to a host cell by its tail. A lysozyme enzyme in the baseplate then digests a hole in the bacterial cell wall.

Phage DNA

Bacterial chromosome breaking down

2 The phage injects its DNA through the cell wall and plasma membrane into the host cell. Coat proteins remain outside. Expression of phage genes in a time-regulated manner produces proteins and enzymes for the phage life cycle. A phage-encoded enzyme breaks down the bacterial chromosome.

Replicated phage DNA

3 The phage DNA is replicated inside the host cell by a phage-encoded DNA polymerase.

Tail units   Head units

4 Viral head and tail units are synthesized.

Phage DNA

5 The phage DNA, head, and tail units assemble into complete phage particles.

6 The phage directs synthesis of a lysozyme enzyme that lyses the bacterial cell wall, causing the cell to rupture and release 100–200 progeny phages to the surroundings where they can infect other bacteria.

When a virulent DNA bacteriophage such as phage T2 infects *E. coli,* it enters the **lytic cycle (Figure 17.7),** in which the host cell is killed in each cycle of infection. The phage attaches to a host bacterial cell and injects its DNA genome into the cell; there, expression of phage genes directs the phage life cycle, leading to the production of progeny phages, which are released from the cell when a phage-encoded enzyme is synthesized that breaks open— *lyses*—the cell. Those phages can now infect other bacteria.

For some virulent phages (although not T-even phages), fragments of the host DNA may be included in the heads as the viral particles assemble, providing the basis for transduction of bacterial genes during the next cycle of infection. Because genes are randomly incorporated from essentially any DNA fragments, gene transfer by this mechanism is termed **generalized transduction.**

Temperate Bacteriophages. Temperate bacteriophages alternate between a lytic cycle and a **lysogenic cycle,** in which the viral DNA inserts into the host cell DNA and production of new viral particles is delayed. During the lysogenic cycle, the integrated viral DNA, known as the **prophage,** remains partially or completely inactive, but is replicated and passed on with the host DNA to all descendants of the infected cell. In response to certain environmental signals, the prophage loops out of the chromosome and the lytic cycle of the phage proceeds.

A much-researched temperate phage that infects *E. coli* is lambda (λ). Phage λ infects *E. coli* in much the same way as the T-even phages do **(Figure 17.8).** The phage injects its linear double-stranded DNA chromosome into the bacterium. Phage gene products then control whether the phage follows a lytic cycle (Figure 17.8, steps 7–9), or follows a lysogenic cycle (Figure 17.8, steps 3–5).

A key feature of the lysogenic cycle is that the phage DNA integrates into the host chromosome to produce a *prophage* (Figure 17.8, step 3). The DNA of a temperate phage typically inserts at one or possibly a few specific sites in the bacterial chromosome through the action of a phage-encoded enzyme that recognizes certain sequences in the host DNA. In the case of λ, there is one integration site in the *E. coli* chromosome. In the prophage state, the λ genes are mostly inactive and, therefore, no phage components are made. An exception is the gene encoding a repressor protein that acts to block expression of the λ genes that control the lytic cycle. As a consequence, the λ prophage does not affect its host cell and its descendants. The prophage is replicated and passed on in division along with the host cell DNA (Figure 17.8, steps 4 and 5).

In response to certain environmental signals, such as ultraviolet (UV) irradiation, the repressor protein is inactivated or destroyed and the lytic cycle genes of λ phage become active. The phage now enters the lytic cycle. The λ genome is excised

CLOSER LOOK

FIGURE 17.8 The infective cycle of lambda (λ), an example of a temperate phage, which can go through the lytic cycle or the lysogenic cycle.

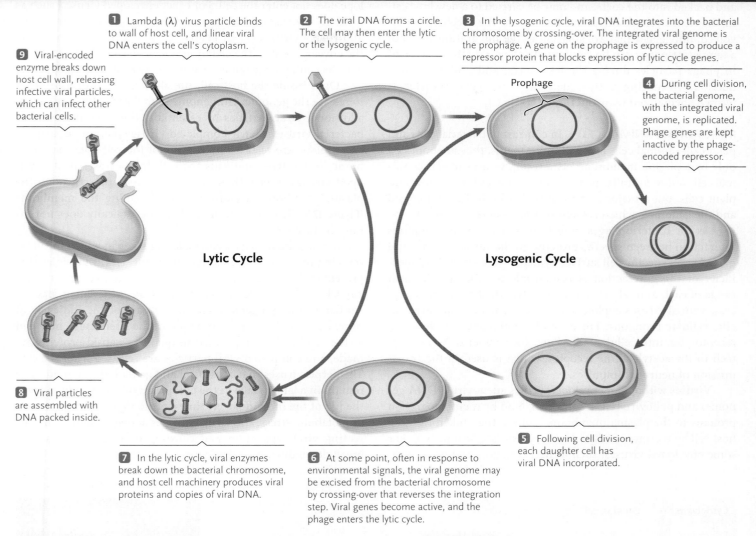

1 Lambda (λ) virus particle binds to wall of host cell, and linear viral DNA enters the cell's cytoplasm.

2 The viral DNA forms a circle. The cell may then enter the lytic or the lysogenic cycle.

3 In the lysogenic cycle, viral DNA integrates into the bacterial chromosome by crossing-over. The integrated viral genome is the prophage. A gene on the prophage is expressed to produce a repressor protein that blocks expression of lytic cycle genes.

9 Viral-encoded enzyme breaks down host cell wall, releasing infective viral particles, which can infect other bacterial cells.

Prophage

4 During cell division, the bacterial genome, with the integrated viral genome, is replicated. Phage genes are kept inactive by the phage-encoded repressor.

Lytic Cycle

Lysogenic Cycle

8 Viral particles are assembled with DNA packed inside.

5 Following cell division, each daughter cell has viral DNA incorporated.

7 In the lytic cycle, viral enzymes break down the bacterial chromosome, and host cell machinery produces viral proteins and copies of viral DNA.

6 At some point, often in response to environmental signals, the viral genome may be excised from the bacterial chromosome by crossing-over that reverses the integration step. Viral genes become active, and the phage enters the lytic cycle.

SUMMARY Lambda phage injects its linear chromosome into a cell (step 1). Inside the cell, the chromosome forms a circle (step 2), and then follows one of two paths. Viral gene products act as molecular switches to determine which path is followed at the time of infection. One path is the lytic cycle, which is like the lytic cycles of virulent phages. During the lytic cycle, progeny phages are produced which are then released from the cell (steps 7–9). The second and more common path, the lysogenic cycle, begins when the circular λ chromosome integrates into the host cell's chromosome by crossing-over, becoming a prophage (step 3). The prophage is replicated and passed on in division along with the host cell chromosome (steps 4–5).

THINK LIKE A SCIENTIST Suppose an Hfr strain of *E. coli* with an integrated λ prophage conjugates with an F⁻ recipient that does not have a λ prophage. Hypothesize what will happen when the prophage enters the recipient during the conjugation process. Would your answer be different if the recipient contained a λ prophage?

from the host chromosome by a recombination event that reverses the integration step, generating a circular λ chromosome (Figure 17.8, step 6). A special replication mechanism produces many copies of linear λ chromosomes from that circular chromosome. Expression of genes on those chromosomes generates coat proteins, which assemble with the chromosomes to produce the viral particles (Figure 17.8, steps 7 and 8). This active stage culminates in rupture of the host cell with the release of infective viral particles (Figure 17.8, step 9), and the beginning of a new cycle (Figure 17.8, step 1).

At times, excision of the λ chromosome from the *E. coli* DNA is not precise, resulting in the inclusion of one or more host cell genes. These genes are replicated with the viral DNA and packed into the coats, and may be carried to a new host cell in the next cycle of infection. Because of the mechanism involved, only genes that are adjacent to the integration site(s) of a temperate phage can be cut out with the viral DNA, included in phage particles during the lytic stage, and undergo transduction. Accordingly, this mechanism of gene transfer is termed **specialized transduction.**

INFECTION OF ANIMAL CELLS In contrast to bacterial cells and plant cells, animal cells lack cell walls; that is, they are bounded only by a plasma membrane. As a result, viruses infecting animal cells follow a similar pattern of infection as for bacterial and plant cells, but a major difference is that both the viral capsid and the genome—which is DNA or RNA—enter a host cell.

As for bacteriophages, infection of animal cells requires interaction between specific proteins on the surface of the viral particle and specific cell surface proteins—receptors. The specificity of this interaction is responsible for the specific host range of each animal virus. Importantly, for the infection process, animal viruses exploit cell surface proteins that have specific cellular functions. For example, poliovirus attaches to a receptor for intracellular adhesion, and influenza viruses attach to the acetylcholine receptor, which is used in the transmission of neuronal impulses.

Viruses without an envelope, such as adenovirus (DNA genome) and poliovirus (RNA genome), bind by their recognition proteins to the plasma membrane and are then taken into the host cell by receptor-mediated endocytosis (see Section 6.5). For some enveloped viruses, such as herpesviruses and poxviruses

(DNA genome), and HIV (RNA genome), the viral capsid and its contained genome enter the host cell by fusion of their envelope with the host cell plasma membrane. In this case, the envelope does not enter the cell. For other enveloped viruses, such as orthomyxoviruses (RNA genome; for example, influenza virus) and rhabdoviruses (RNA genome), the entire virus, including the envelope, enters the cell by endocytosis and the viral capsid with its contained genome then is released into the cytoplasm.

Once inside the host cell, the viral capsid is disassembled to release the genome. The genome then directs the synthesis of additional viral particles by essentially the same pathways as bacterial viruses. Newly completed viruses that do not acquire an envelope are released by rupture of the cell's plasma membrane, which lyses and kills the cell. In contrast, most enveloped viruses receive their envelope as they pass through the plasma membrane, usually without breaking the membrane **(Figure 17.9).** This pattern of viral release typically does not injure the host cell.

Some animal viruses enter a **latent stage** in which the virus remains in the cell in inactive form: the viral nucleic acid is present in the cytoplasm or nuclear DNA, but no complete viral particles or viral release can be detected. (The latent stage is similar to the lysogenic cycle that is part of the life cycle of some bacteriophages.) At some point, the latent stage may end as the viral DNA is replicated in quantity, capsid proteins are made, and completed viral particles are released from the cell. The herpesviruses that cause oral and genital ulcers in humans remain in a latent stage in the cytoplasm of some body cells for the life of the individual. At times, particularly during periods of metabolic stress, the virus becomes active in some cells, directing viral replication and causing ulcers to form as cells break down during viral release.

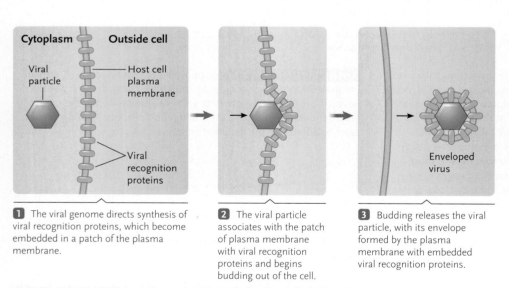

1 The viral genome directs synthesis of viral recognition proteins, which become embedded in a patch of the plasma membrane.

2 The viral particle associates with the patch of plasma membrane with viral recognition proteins and begins budding out of the cell.

3 Budding releases the viral particle, with its envelope formed by the plasma membrane with embedded viral recognition proteins.

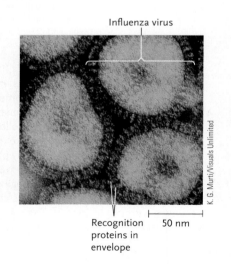

Influenza virus

Recognition proteins in envelope

50 nm

K. G. Murti/Visuals Unlimited

FIGURE 17.9 **How enveloped viruses acquire their envelope.** The micrograph shows the influenza virus with its envelope. Note the recognition proteins studding the envelope.
© Cengage Learning 2014

INFECTION OF PLANT CELLS Plant viruses may be rodlike or polyhedral (see Figure 17.6B); although most include RNA as their nucleic acid, some contain DNA. None of the known plant viruses has an envelope. Plant viruses enter cells through mechanical injuries to leaves and stems or through transmission from plant to plant by biting and feeding insects such as leaf hoppers and aphids, by nematode worms, and by pollen during fertilization. Plant viruses can also be transmitted from generation to generation in seeds. Once inside a cell, plant viruses replicate in the same patterns as animal viruses. Within plants, viral particles pass from infected to healthy cells through plasmodesmata, the openings in cell walls that directly connect plant cells (see Figure 5.26), and through the vascular system.

Plant viruses are generally named and classified by the type of plant they infect and their most visible effects. *Tomato bushy stunt virus,* for example, causes dwarfing and overgrowth of leaves and stems of tomato plants, and *Tobacco mosaic virus* causes a mosaic-like pattern of spots on leaves of tobacco plants. Most species of crop plants can be infected by at least one destructive virus.

Tobacco mosaic virus (see Figure 17.6A) was the first virus to be isolated, crystallized, disassembled, and reassembled in the test tube, and the first viral structure to be established in full molecular detail.

Viral Infections Are Typically Difficult to Treat

The vast majority of animal viral infections are asymptomatic. However, there are many types of pathogenic viruses, and they employ a variety of ways to cause disease. In some instances, release of progeny viruses from the cell leads to massive cell death, destroying vital tissues such as nervous tissue or white or red blood cells, or causing lesions such as ulcers in skin and mucous membranes. Some other viruses release cellular molecules when infected cells break down, which can induce fever and inflammation. Yet other viruses alter gene function when they insert into the host cell DNA, leading to cancer and other abnormalities.

Viral infections are unaffected by the antibiotics and other treatments used for bacterial infections. As a result, many viral infections are allowed to run their course, with treatment limited to relieving the symptoms while the immune defenses of the patient attack the virus. Some viruses, however, cause serious and sometimes deadly symptoms on infection and, consequently, researchers have spent considerable effort to develop antiviral drugs to treat them. Many of these drugs target a stage of the viral life cycle. For example, amantadine inhibits entry of hepatitis B and hepatitis C virus into cells, acyclovir (an analog of nucleosides; *analog* means it is chemically similar) inhibits replication of the genomes of herpesviruses, and zanamivir inhibits release of influenza virus particles from cells.

Let us consider influenza viruses in more detail. Influenza viruses (see Figure 17.9) infect a number of vertebrates, including birds (avian flu), pigs (swine flu), and humans and other mammals. In humans, influenza viruses are transmitted from person to person primarily by aerosols and droplets, and enter the host through the respiratory tract. The virus enters a host cell when the viral protein spike binds to a cell surface receptor and induces endocytosis. Release of progeny viruses lyses and kills the cell, which triggers inflammatory responses. The presence of viruses in the body also stimulates the immune system to make antibodies targeted at the virus. (Antibodies are highly specific protein molecules produced by the immune system that recognize and bind to foreign proteins originating from a pathogen; see Chapter 45.)

Virologists classify influenza viruses into three types, influenzavirus A, influenzavirus B, and influenzavirus C. Strains of each type can infect humans; the A and B types are highly contagious. Yearly epidemic outbreaks of flu are routine, even in developed countries, resulting in many deaths. However, each virus strain that causes an epidemic has a particular virulence because of differences in the molecular properties of the viruses, so the death rate varies with each epidemic. Moreover, influenza viruses have the potential to cause pandemics, meaning the spread of a new strain through many human populations in a country, region, or even worldwide. Some pandemics are relatively benign, whereas others result in high mortality rates. For example, the so-called Spanish Flu of 1918 to 1919 was caused by a particularly virulent strain of the influenza virus; 50 to 100 million people died worldwide after an estimated 500 million people, one-third of the human population at the time, were infected. Most who died were between 20 and 40 years old.

The influenza virus type that caused the 1918 to 1919 influenza pandemic was influenzavirus A/H1N1. The "H" and "N" numbers refer to the subtypes of two glycoproteins found in the envelope of the virus. The "H" glycoprotein is hemagglutinin (the protein spike that binds to cell surface receptors), and the "N" glycoprotein is neuraminidase. The H protein gives the virus the ability to attach to the host cell, and the N protein allows the virus to be released from the cell and spread infection. The H and N proteins are *antigens,* meaning that we recognize them as foreign and then generate antibodies to remove the virus from the system (for more details, see Chapter 45). The reason that we continue to get influenza, even though we have produced antibodies against an infecting influenza virus, is that the virus mutates and infectious viruses are produced with new combinations of H and N proteins. Infection by a mutated virus means, therefore, that we see new antigens for which we now have to make new antibodies. In the meantime, the virus attacks cells and the symptoms of influenza develop.

The virus type of the 1918 to 1919 influenza virus was determined experimentally in 2005, when researchers led by Jeffrey Taubenberger at the U.S. Armed Forces Institute of Pathology reconstructed the genome of the virus and produced infectious, pathogenic viruses in the laboratory. The team worked mainly with tissue from a 1918 flu victim found in permafrost in Alaska. Using modern DNA technology (see Chapter 18), they pieced together the sequences of the virus's 11 genes and characterized their protein products. They also transformed clones of the genes into animal cells and were able to produce complete viruses. These reconstructed 1918 viruses were about 50 times more virulent than most modern-day human influenza viruses; they killed a higher percentage of mice and killed them much more quickly, for instance. (All of these experiments were done with appropriate approval and under highly controlled experimental conditions.) By studying the 1918 virus genome and its pathogenicity, the researchers are learning how highly virulent viruses can be produced. So far, they have learned that the 1918 virus had mutations in polymerase genes for replicating the viral genome in host cells, likely making this strain capable of replicating more efficiently. An influenza pandemic starting in 2009 involving a new strain of influenzavirus A/H1N1 was not as deadly as the 1918 flu virus (at least 18,000 people died worldwide through May 2010), although the exact molecular properties responsible for their differences in virulence are not completely understood.

The influenza type A and B viruses have many unusual features that tend to keep them a step ahead of efforts to counteract their infections. One is the genome of the virus, which consists of eight segments of RNA. When two different influenza viruses infect the same individual, the pieces can assemble in random combinations derived from either parent virus. The new combinations can change the proteins of the capsid, making the virus unrecognizable to antibodies developed against either parent virus. The invisibility to antibodies means that new virus strains can infect and produce flu symptoms in people who have already had the flu or who have had flu shots that stimulate the formation of antibodies effective only against the earlier strains of the virus. Random mutations in the RNA genome of the virus add to the variations in the capsid proteins that make previously formed antibodies ineffective, and also can alter the virulence of the virus.

Retroviruses Are Viruses That Replicate Their RNA Genomes via a DNA Intermediate

Most viruses with RNA genomes directly replicate those genomes to produce progeny RNA genomes. However, a **retrovirus** is an enveloped virus with an RNA genome that replicates via a DNA intermediate. In that respect, retroviruses resemble transposable elements known as retrotransposons (see Section 15.5). When a retrovirus infects a host cell, a *reverse transcriptase* enzyme carried in the viral particle (and encoded by the viral genome) is released and copies the single-stranded RNA genome into a double-stranded DNA copy. The viral DNA is then inserted into the host DNA, where it is replicated and passed to progeny cells during cell division. Similar to the prophage of temperate bacteriophages, the inserted viral DNA is known as a **provirus (Figure 17.10).** *Insights from the Molecular Revolution* describes the discovery of reverse transcriptase.

PROPERTIES OF RETROVIRUSES Retroviruses are found in a wide range of organisms, with most so far identified in vertebrates. You, as well as most other humans and mammals, probably contain from 1 to as many as 100 or more retroviruses in your genome as proviruses. Many of these retroviruses never produce viral particles. However, they may sometimes cause genetic disturbances of various kinds, including alterations of gene activity or DNA rearrangements such as deletions and translocations, some of which may be harmful to the host. HIV, the causal agent of acquired immunodeficiency syndrome (AIDS), does produce viral particles. (More discussion of AIDS is in Chapter 45.)

Some retroviruses, such as *avian sarcoma virus,* have been linked to cancer (in this case in chickens). Many of the cancer-causing retroviruses have picked up a host gene that triggers the entry of cells into uncontrolled DNA replication and cell division. When included in a retrovirus, the host gene comes under the influence of the highly active retroviral promoter, which makes the gene continually active and leads to the uncontrolled cell division which is characteristic of cancer. In other words, the host gene has become an oncogene, a gene that promotes the development of cancer by stimulating cell division (see Section 16.5). Usually, the host gene replaces one or more retrovirus genes, making the virus unable to produce viral particles.

Retroviruses also may activate genes related to cell division by moving them to the vicinity of an active host cell promoter or enhancer, or by delivering an enhancer or active promoter to the vicinity of a host cell gene. In either case, the result may be uncontrolled cell division. Retroviruses that have been modified to make them harmless are used in genetic engineering to introduce genes into mammalian and other animal cells.

FIGURE 17.10 A mammalian retrovirus in the provirus form in which it is inserted into chromosomal DNA. The direct repeats at either end contain sequences capable of acting as enhancer, promoter, and termination signals for transcription. The central sequence contains genes coding for proteins, concentrated in the *gag, pol,* and *env* regions. The provirus of HIV, the virus that causes AIDS, takes this form.
© Cengage Learning 2014

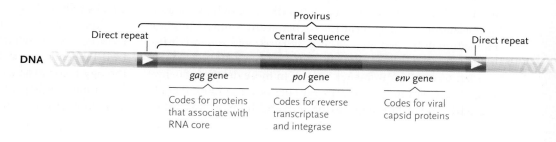

Reversing the Central Dogma: How do RNA tumor viruses replicate their genomes?

RNA tumor viruses, which represent some of the viruses with RNA genomes, convert the animal cells they infect to tumor cells. In 1964, Howard Temin of the University of Wisconsin–Madison hypothesized that the mechanism by which RNA tumor viruses convert normal cells to tumor cells involves an enzyme encoded by the virus that copies the RNA genome into DNA, called provirus DNA. He proposed that the provirus DNA integrated into the DNA of a host cell chromosome, enabling the viral genetic material to persist as the tumor cell reproduced. At the time, most virologists were highly doubtful that such an enzyme existed because it went against the central dogma of molecular biology, namely that DNA is transcribed to produce RNA, and RNA is translated into protein.

Research Question

How do RNA tumor viruses replicate their genomes?

Experiment

David Baltimore of MIT searched for the hypothesized enzyme using reaction mixtures that contained virions of Rauscher murine leukemia virus (R-MuLV, an RNA tumor virus that infects mice), and the four precursors for DNA, one of which was radioactive **(Figure)**. Only if the enzyme was present could the precursors be polymerized into a DNA molecule. To detect DNA synthesis, he took samples of the reaction mixtures and added an acid to precipitate macromolecules. Then he measured the radioactivity in the precipitated, so-called acid-insoluble material.

Results

1. Radioactivity was detected in the acid-insoluble material when a radioactive DNA precursor was used in the reaction. This result showed that the virions were capable of catalyzing the synthesis of DNA.
2. No radioactivity was detected in acid-insoluble material when a radioactive DNA precursor was used in the reaction if the virions were first treated with RNase. RNase is an enzyme that degrades RNA. This result showed that the template for DNA synthesis was RNA.
3. No radioactivity was detected in acid-insoluble material when the four RNA precursors, one of which was radioactive, were used in the reaction instead of DNA precursors. This result showed that the virions were not capable of catalyzing the synthesis of RNA.

Conclusion

Baltimore concluded that R-MuLV virions contain an enzyme that can synthesize DNA using an RNA template. In parallel research, he demonstrated that the same enzyme was present in virions of another RNA tumor virus, Rous sarcoma virus, a result confirmed independently by Howard Temin. In commenting on the work, the editors of *Nature* dubbed the enzyme reverse transcriptase because the synthesis direction was the opposite of that for transcription, namely RNA to DNA rather than DNA to RNA. The name stuck. RNA tumor viruses with this enzyme then became known as retroviruses. Baltimore and Temin received a Nobel Prize in 1975 "for their discoveries concerning the interaction between tumor viruses and the genetic material of the cell."

THINK LIKE A SCIENTIST When Baltimore treated the acid-insoluble radioactive product from his first experiment (Results #1) with DNase (deoxyribonuclease), an enzyme that degrades DNA, the radioactivity became acid-soluble. When he treated it with RNase (ribonuclease), an enzyme that degrades RNA, the radioactivity remained acid-insoluble. How do these results confirm Baltimore's conclusions?

Source: D. Baltimore. 1970. Viral RNA-dependent DNA polymerase: RNA-dependent DNA polymerase in virions of RNA tumour viruses. *Nature* 226:1209–1211.

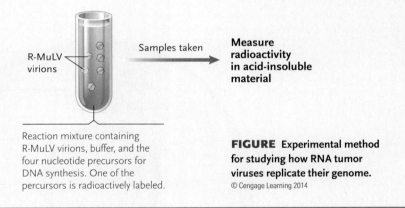

Reaction mixture containing R-MuLV virions, buffer, and the four nucleotide precursors for DNA synthesis. One of the percursors is radioactively labeled.

R-MuLV virions

Samples taken

Measure radioactivity in acid-insoluble material

FIGURE Experimental method for studying how RNA tumor viruses replicate their genome.
© Cengage Learning 2014

© Cengage Learning 2014

THE PROPERTIES AND LIFE CYCLE OF HIV HIV infects particular cells of the immune system. At the time of initial infection with HIV, many people suffer a mild fever and other symptoms that may be mistaken for the flu or the common cold. The symptoms disappear as antibodies against the viral proteins appear in the body, and the number of viral particles drops in the bloodstream. However, the genome of the virus is still present, integrated into the DNA of the host cells as a provirus, and the virus steadily spreads to infect other immune system cells. An infected person may remain apparently healthy for years, yet can transmit the virus to others. Both the transmitter and recipient of the virus may be unaware that the disease is present, making it difficult to control the spread of HIV infections. Ultimately, many important cells of the immune system are destroyed, wiping out the body's immune response and making the HIV-infected person susceptible to other infections. Steady debilitation may then occur, sometimes resulting in death.

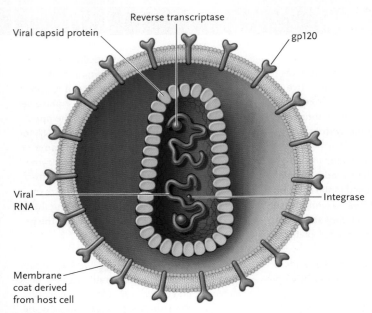

FIGURE 17.11 Structure of HIV, a retrovirus.
© Cengage Learning 2014

Like all retroviruses, HIV is an enveloped virus with two copies of a single-stranded RNA genome contained within a capsid **(Figure 17.11)**. When HIV first infects a cell, a *gp120* glyco-protein of the capsid attaches the virus to a particular type of receptor in the cell's plasma membrane. Then, another viral protein triggers fusion of the viral envelope with the host cell's plasma membrane, releasing the virus into the cell; the life cycle

then commences **(Figure 17.12)**. A key event of the life cycle is that a double-stranded DNA copy of the RNA genome made by the viral *reverse transcriptase* is integrated into the host chromosome by the viral *integrase* to produce the provirus, which, like an integrated prophage, is dormant. When the dormancy of the provirus is broken (for example, when the host cell in which it is located is stimulated to grow and divide), the genes of the retrovirus are expressed leading to the production of infective HIV particles. Those viral particles are released from the host cell by budding. The viral particles may infect more body cells, or another person. Transmission to another person occurs when an infected person's body fluids, especially blood or semen, enter the blood or tissue fluids of another person's body.

Viruses May Have Originated and Evolved from Fragments of Cellular DNA or RNA

All viruses are *obligate intracellular parasites,* which means that they rely on host cell machinery for their life cycles. Viruses are known to infect all types of cellular organisms, which suggests there could be commonality of virus sequences and functions that could lead to an understanding of their origin. Sequence analysis of viral genomes indicates that there is not a single origin for viruses; rather, it is likely that viruses originated multiple times over a long period of evolutionary time as cellular life was evolving. Researchers generally agree that many viruses originated as fragments of DNA or RNA, and then became independent entities when, in some way, the fragments became sur-

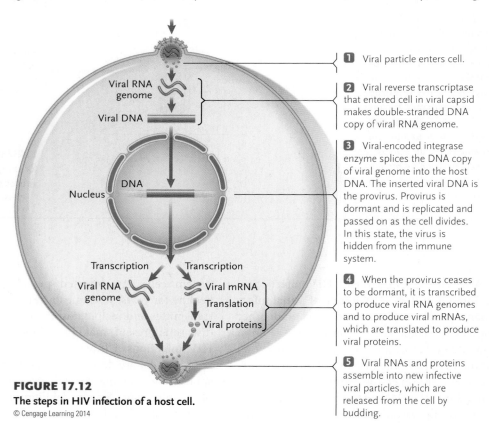

FIGURE 17.12
The steps in HIV infection of a host cell.
© Cengage Learning 2014

1 Viral particle enters cell.

2 Viral reverse transcriptase that entered cell in viral capsid makes double-stranded DNA copy of viral RNA genome.

3 Viral-encoded integrase enzyme splices the DNA copy of viral genome into the host DNA. The inserted viral DNA is the provirus. Provirus is dormant and is replicated and passed on as the cell divides. In this state, the virus is hidden from the immune system.

4 When the provirus ceases to be dormant, it is transcribed to produce viral RNA genomes and to produce viral mRNAs, which are translated to produce viral proteins.

5 Viral RNAs and proteins assemble into new infective viral particles, which are released from the cell by budding.

rounded by a protective layer of protein that allowed them to escape from their parent cells. Over evolutionary time, the DNA or RNA segments have combined and mutated extensively, producing the present-day virus genomes. The genome changes have been so extensive that it is not possible to determine definitively whether or not some viruses diverged anciently from common ancestry or arose independently. Certainly it is clear that there is not a single origin for viruses. Instead, there were multiple and varied ancient origins for viruses although those origins may not be understood entirely.

STUDY BREAK 17.2

1. What is the difference between a virulent phage and a temperate phage?
2. How does viral infection of an animal cell and a plant cell differ?
3. How are retroviruses distinctive among RNA viruses with respect to replication of their genome?

> **THINK OUTSIDE THE BOOK**

Use the Internet or research literature to determine which gene(s) of the influenza virus genome is(are) responsible for the pathogenicity of the virus, and to outline what is known about their molecular functions.

17.3 Viroids and Prions, Infectious Agents Lacking Protein Coats

Although viruses are so chemically simple that most scientists do not consider them to be living organisms, even simpler and smaller infective agents exist. Two infectious agents that are simpler than viruses are *viroids* and *prions*. Both of these agents lack capsids.

Viroids, first discovered in 1971, are plant pathogens that consist solely of single-stranded, circular RNA without a protein coat. Infection by viroids can rapidly destroy entire fields of citrus, potatoes, tomatoes, coconut palms, and other crop plants.

About 30 types of viroids are known. Their RNA genomes range from 246 to 401 nucleotides and in no case do they encode a protein. The genomes are about 10-fold smaller than the smallest known viral RNA genome. Depending on the viroid, the genomes replicate and accumulate either in the nucleus or in the chloroplast of the host cell. Replication is catalyzed by host nuclear or chloroplast RNA polymerases.

Somehow the tiny RNA genomes of viroids contain sufficient information to infect host plants and to induce the host to replicate the genomes. As a consequence of the replication, the viroids cause specific diseases. The manner in which viroids cause disease remains ill-defined. In fact, researchers believe that there is more than one mechanism. Some recent research has defined one pathway to disease in which viroid RNA acti-

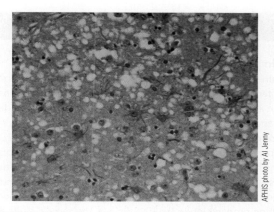

FIGURE 17.13 Bovine spongiform encephalopathy (BSE). The light-colored patches in this section from a brain damaged by BSE are areas where tissue has been destroyed.

vates a protein kinase (an enzyme that adds phosphate groups to proteins; see Sections 4.5 and 7.2) in plants. This process leads to a reduction in protein synthesis and protein activity, and disease symptoms result.

Prions, named in 1982 by Stanley Prusiner of the University of California, San Francisco, for *proteinaceous infection,* are the only known infectious agents that do not include a nucleic acid molecule.

Prions have been identified as the causal agents of certain diseases that degenerate the nervous system in mammals. One of these diseases is *scrapie,* a brain disease that causes sheep to rub against fences, rocks, or trees until they scrape off most of their wool. Another prion-based disease is bovine spongiform encephalopathy (BSE), also called *mad cow disease.* The disease produces spongy holes and deposits of proteinaceous material in brain tissue **(Figure 17.13).** In 1996, 150,000 cattle in Great Britain died from an outbreak of BSE, which was traced to cattle feed containing ground-up tissues of sheep that had died of scrapie. Humans are subject to a fatal prion infection called *Creutzfeldt-Jakob disease (CJD).* The symptoms of CJD include rapid mental deterioration, loss of vision and speech, and paralysis; autopsies show spongy holes and deposits in brain tissue similar to those of cattle with BSE. Classic CJD occurs as a result of the spontaneous transformation of normal proteins into prion proteins. Fewer than 300 cases a year occur in the United States. Variant CJD is a form of the disease caused by eating meat or meat products containing nervous system tissue from cattle with BSE. Another prion-based disease of humans, *kuru,* is originally thought to have spread in a tribe in New Guinea when relatives ritualistically ate a deceased individual, including the brain, as a way (in their view) of returning that person's life force to the tribe.

STUDY BREAK 17.3

What distinguishes viroids and prions from viruses?

How do prions spread?

The brain-wasting diseases caused by prions are poorly understood despite intensive research, and the mechanisms by which prions spread between animals are under investigation. In scrapie, a prion disease of sheep, and chronic wasting disease (CWD), which affects deer and elk, infected animals shed prions into the environment. Remarkably, infectivity can persist in the soil for years. Under natural circumstances, prions enter the host at a peripheral site (for example, oral), invade the central nervous system (CNS), and spread within the CNS. These events require that prions infect new cells and spread between distant locations in the body. Prion propagation in the CNS follows the same circuits that neurons use to communicate with each other.

To understand mechanisms of prion infection, scientists have investigated how prions move through the CNS. Research led by our group at the Rocky Mountain Laboratories (RML), and Marco Prado at the University of Minas Gerais, followed prion aggregates as they invaded cultured mouse neurons. Using prion preparations labeled with a red fluorescent dye, we tracked the fate of the prion particles in living cells using fluorescence microscopy. Interestingly, the prion aggregates moved through neuronal projections to points of contact with other cells. In rare cases, there seemed to be transfer between cells. We proposed that newly described structures, called tunneling nanotubes (TNTs), mediate the intercellular transfer of prion aggregates. TNTs are delicate intercellular projections that can transport proteins between cells. Using methodologies developed by our team, another group later reported that prion aggregates can be transported between cells through TNTs.

Infection requires that prion aggregates interact with the normal prion protein in host cells and induce its conversion to the disease-associated form. We were unable to observe this process because the normal prion protein lacked a fluorescent tag. To permit simultaneous visualization of the normal and disease-associated aggregates of prion proteins, we developed a technique to label the normal prion protein in live cells with a green fluorescent dye. Importantly, we found that the green normal prion protein could convert to the aggregated prion disease-associated form. This technique, called IDEAL-labeling, provides the tools for visualizing the entire prion infection process in living cells by tracking green and red fluorescent prion proteins simultaneously. The advances in understanding how prions move through the nervous system were heralded as a significant step toward developing therapies to stop the spread of brain-wasting diseases by blocking pathways of prion replication and intercellular spreading.

Some important outstanding questions in the field include:

What are the most relevant mechanisms of prion spreading in animals? Do they involve TNTs or perhaps prion-containing vesicles released from infected cells?

How are prions shed into the environment?

How does prion propagation cause the death of neurons? Does this involve corruption of the function of the normal prion protein?

THINK LIKE A SCIENTIST One direction of research on prions is to determine if conversion to the infectious form causes corruption of the function of the normal prion protein. What are some approaches to understand further this process?

Courtesy of Gerald Baron

Gerald Baron is an investigator and head of the TSE/Prion cell biology section at Rocky Mountain Laboratories in Montana. To learn more about Dr. Baron's research, go to http://www3.niaid.nih.gov/labs/aboutlabs/lpvd/TSEPrionCellBiologySection/.

This research was supported in part by the Intramural Research Program of the NIH, NIAID.

REVIEW KEY CONCEPTS

To access the course materials and companion resources for this text, please visit www.cengagebrain.com.

17.1 Gene Transfer and Genetic Recombination in Bacteria

- The rapid generation times and numerous offspring of bacteria and viruses make it possible to trace genetic crosses and their outcomes much more quickly than in eukaryotes. These characteristics make it possible to detect rare genetic events. The results of these crosses show that recombination may occur within the boundaries of a gene, as well as between genes.

- Recombination occurs in both bacteria and eukaryotes by exchange of segments between homologous DNA molecules. In bacteria, the DNA of the bacterial chromosome may recombine with DNA brought into the cell from outside.

- Three primary mechanisms bring DNA into bacterial cells from the outside: conjugation, transformation, and transduction.

- In conjugation, two bacterial cells form a cytoplasmic bridge, and part or all of the DNA of one cell moves into the other through the bridge. The donated DNA can then recombine with homologous sequences of the recipient cell's DNA (Figures 17.1, 17.2, and 17.4).

- *E. coli* bacteria that are able to act as DNA donors in conjugation have an F plasmid, making them F$^+$; recipients have no F plasmid and are F$^-$. In Hfr strains of *E. coli,* the F plasmid is integrated into the main chromosome. As a result, genes of the main chromosome are often transferred into F$^-$ cells along with a portion of the F plasmid DNA. Researchers have mapped genes on the *E. coli* chromosome by noting the order in which they are transferred from Hfr to F$^-$ cells during conjugation (Figure 17.4).

- In transformation, intact cells take up pieces of DNA released from cells that have disintegrated. The entering DNA fragments can recombine with the recipient cell's chromosomal DNA.

- In transduction, bacterial DNA is transferred from one cell to another by an infecting phage.

- Conjugation, transformation, and transduction do not move genes from generation to generation (that is, by descent) but move genes between organisms by horizontal gene transfer.

Animation: Prokaryotic conjugation

Animation: Distinguishing between the three major processes: conjugation, transformation, and transduction

17.2 Viruses and Viral Genetics

- A virus is a biological particle that can infect the cells of a living organism. A free viral particle consists of a nucleic acid core, either double-stranded or single-stranded DNA or double-stranded or single-stranded RNA, surrounded by a protein coat (the capsid). Some animal viruses also have a surrounding envelope derived from the host cell's plasma membrane (Figure 17.6). Many viruses have recognition proteins on the virus surface that enable the virus to attach to host cells.

- One or more kinds of viruses likely infect all living organisms. Viruses show specificity with respect to the host(s) they infect; typically, a particular virus infects only a single species or a few closely related species.

- The cycle of viral infection begins when the nucleic acid molecule of a virus is introduced into a host cell. The virus then directs the cellular machinery to assemble viral capsid proteins with the new viral genomes into new viral particles.

- Virulent phages kill a host bacterial cell during a lytic cycle by releasing an enzyme that ruptures the plasma membrane and cell wall and releases the new viral particles (Figure 17.7).

- Temperate phages do not always kill their host bacterial cell. They may enter the lytic cycle, in which the phage DNA becomes active, exits the host DNA, and begins replication, or a lysogenic cycle. In the lysogenic cycle, the phage's DNA is integrated into the host cell's DNA and may remain for many generations. At some point, the phage may enter the lytic cycle and begin replication. After production of viral capsids, the DNA is assembled into new progeny phages, which are released as the cell ruptures (Figure 17.8).

- During a cycle of viral infection with particular phages, one or more fragments of host cell DNA may be incorporated into phage particles. As an infected cell breaks down, it releases the viral particles containing host cell DNA. These phages, which form the basis of bacterial transduction, may infect a second cell and introduce the bacterial DNA segment into the new host, where it may recombine with the host DNA.

- The infection of animal cells by viruses depends on interaction between specific proteins on the viral particle surface and specific cell surface proteins. Viruses without envelopes enter the cell by receptor-mediated endocytosis. Enveloped viruses infect cells by endocytosis, or by release of the genome-containing capsid into the cell following fusion of the envelope with the plasma membrane.

- Progeny of nonenveloped animal viruses typically are released from a cell by rupture of the cell's plasma membrane, which lyses and kills the cell. Most enveloped viruses bud from the cell in a process that surrounds the protein capsid–nucleic acid genome with an envelope derived from the cell's plasma membrane and containing embedded viral encoded recognition proteins (Figure 17.9).

- Viruses are unaffected by antibiotics and most other treatment methods; hence, infections caused by them are difficult to treat.

- Retroviruses are viruses with single-stranded RNA genomes that are replicated via a DNA intermediate that integrates into a nuclear chromosome to produce a provirus (Figures 17.10–17.12).

Animation: HIV replication cycle

Animation: Lytic pathway

Animation: Body plans of viruses

Animation: Bacteriophage multiplication cycles

17.3 Viroids and Prions, Infectious Agents Lacking Protein Coats

- Viroids, which infect crop plants, consist only of a very small, single-stranded RNA molecule. Prions, which cause brain diseases in some animals, are infectious proteins with no associated nucleic acid. Prions are misfolded versions of normal cellular proteins that can induce other normal proteins to misfold (Figure 17.13).

UNDERSTAND & APPLY

Test Your Knowledge

1. When studying the differences in the genes of bacteria, researchers:
   a. do not grow bacteria on a minimal medium as the medium lacks needed nutrients.
   b. use a bacterial clone, which is a group of cells from different bacteria of varying genetic makeup.
   c. use bacteria diploid for their full genome because they can grow on minimal medium.
   d. can study only one genetic trait in a single recombinant event.
   e. can measure the passage of genes between cells during conjugation, transduction, and transformation.

2. If recombination occurred between two bacterial genomes as shown in the figure, the result would be:

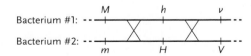

   a. *MHv* and *mhV.*
   b. *MHV* and *mhv.*
   c. *Mhv* and *mHv.*
   d. *MHV* and *mhV.*
   e. *mhv* and *MhV.*

3. In conjugation, when a bacterial F factor is transferred:
   a. the donor cell becomes F^-.
   b. the recipient cell becomes F^+.
   c. the recipient cell becomes F^-.
   d. the donor cell turns into a recipient cell.
   e. chromosomal DNA is transferred in the mating.

4. Which of the following is *not* correct for bacterial conjugation?
   a. Both Hfr and F^+ bacteria have the ability to code for a sex pilus.
   b. After an F^- cell has conjugated with an F^+, its plasmid holds the F^+ factor.
   c. The recipient cell is Hfr following conjugation.
   d. In an Hfr × F^- conjugation, DNA of the main chromosome moves to a recipient cell.
   e. Genes on the F factor encode proteins of the sex pilus.

5. Which of the following is *not* correct for bacterial transformation?
   a. Artificial transformation is used in cloning procedures.
   b. Avery was able to transform live noninfective bacteria with DNA from dead infective bacteria.
   c. The cell wall and plasma membrane must be penetrated for transformation to proceed.
   d. A virus is required for the process.
   e. Electroporation is a form of artificial transformation used to introduce DNA into cells.

6. Transduction:
   a. may allow recombination of newly introduced DNA with host cell DNA.
   b. is the movement of DNA from one bacterial cell to another by means of a plasmid.
   c. can cause the DNA of the donor to change but not the DNA of the recipient.
   d. is the movement of viral DNA, but not bacterial DNA, into a recipient bacterium.
   e. requires a physical contact between two bacteria.

7. Viruses:
   a. have a protein core.
   b. have a nucleic acid coat.
   c. that infect and kill bacteria during a cell cycle are called virulent bacteriophages.
   d. were probably the first forms of life on Earth.
   e. if they are temperate bacteriophages, kill host cells immediately.

8. When a virus enters the lysogenic stage:
   a. the viral DNA is replicated outside the host cell.
   b. it enters the host cell and kills it immediately.
   c. it enters the host cell, picks up host DNA, and leaves the cell unharmed.
   d. it sits on the host cell plasma membrane with which it covers itself and then leaves the cell.
   e. the viral DNA integrates into the host genome.

9. An infectious material is isolated from a nerve cell. It contains protein with amino acid sequences identical to the host protein but no nucleic acids. It belongs to the group:
   a. prions.
   b. Archaea.
   c. toxin producers.
   d. viroids.
   e. sporeformers.

10. Which is *not* correct about retroviruses?
    a. They are RNA viruses.
    b. They are believed to be the source of retrotransposons.
    c. They encode an enzyme for their insertion into host cell DNA.
    d. They encode single-stranded viral DNA from viral RNA.
    e. They encode a reverse transcriptase enzyme for RNA to DNA synthesis.

Discuss the Concepts

1. You set up an experiment like the one carried out by Lederberg and Tatum, mixing millions of *E. coli* of two strains with the following genetic constitutions:

$$bio^- \quad met^- \quad thr^+ \quad leu^+$$
Strain 1: --+——+——+——+--

$$bio^+ \quad met^+ \quad thr^- \quad leu^-$$
Strain 2: --+——+——+——+--

Among the bacteria obtained after mixing, you find some cells that do not require threonine, leucine, or biotin to grow, but still need methionine. How might you explain this result?

2. As a control for their experiments with bacterial recombination, Lederberg and Tatum placed cells of either "parental" strain 1 or 2 on the surface of a minimal medium. If you set up this control and a few scattered colonies showed up, what might you propose as an explanation? How could you test your explanation?

3. What rules would you suggest to prevent the spread of mad cow disease (BSE)?

Design an Experiment

You have a culture of Hfr *E. coli* cells that cannot make biotin for themselves. To this culture you add some wild-type *E. coli* cells that have been heat killed, and then subject the culture to electroporation. After the addition, you find some cells that can grow on minimal medium. How could you establish whether the wild-type bio^+ allele was inserted in a plasmid or the chromosomal DNA of the Hfr cells?

Interpret the Data

As indicated in the chapter, the F factor integrates into one of several positions around the circular *E. coli* chromosome, producing in each case a different Hfr strain. The various Hfr strains can be used to map the locations of genes around the bacterial chromosome. The **Table** presents the results of conjugation experiments with four different Hfr strains. For each Hfr strain, the order of transfer of the bacterial genes to the recipient is given. From the data, construct a map to show the arrangement of the genes on the circular *E. coli* chromosome.

Hfr strain	First					Last
1	e	i	f	u	j	b
2	w	k	c	r	b	j
3	r	c	k	w	h	p
4	w	h	p	y	e	i

Order of gene transfer

Apply Evolutionary Thinking

Are viruses evolutionarily derived from complex organisms that can reproduce themselves, or are they remnants of precellular "life"? Argue your case.

Researcher with a number of different genetically modified organisms (GMOs), here plants. GMOs are organisms whose genetic material has been altered using genetic engineering techniques.

DNA Technologies: Analyzing and Modifying Genes

Why it matters . . . In early October 1994, 32-year-old Shirley Duguay, a mother of five, disappeared from her home on Prince Edward Island, Canada. Within a few days, her car was found abandoned; bloodstains inside matched her blood type. Several months later the Royal Canadian Mounted Police (RCMP) found Duguay's body in a shallow grave. Among the chief suspects in the murder was her estranged common-law husband, Douglas Beamish, who was living nearby with his parents.

While searching for Duguay, the RCMP discovered a plastic bag containing a man's leather jacket with the victim's blood on it. Beamish's friends and family acknowledged that Beamish had a similar jacket, but none could or would positively identify it. In the lining of the jacket, investigators found 27 white hairs, which forensic scientists identified as cat hairs. The RCMP remembered that Beamish's parents had a white cat named Snowball. Could they prove that the cat hair in the jacket was Snowball's?

Two experts on cat genomes, Marilyn Menotti-Raymond and Stephen J. O'Brien of the Laboratory of Genome Diversity at the U.S. National Cancer Institute, analyzed DNA from the root of one of the cat hairs taken from the jacket and from a blood sample taken from Snowball. They then used a technique called the polymerase chain reaction (PCR) to amplify 10 specific regions of the cat genome, each of which varies among cats in the number of copies of a short (two-nucleotide) repeated sequence. They found that the hair and blood samples matched perfectly, providing strong evidence that the hair came from Snowball. Beamish was tried for the murder of Shirley Duguay. The evidence presented by Menotti-Raymond and O'Brien helped convict him, and he was sentenced to 18 years in prison.

The researchers' analysis of the cat hair and blood is an example of DNA fingerprinting. (Though human DNA fingerprinting evidence is now extensively used in court cases, the Beamish

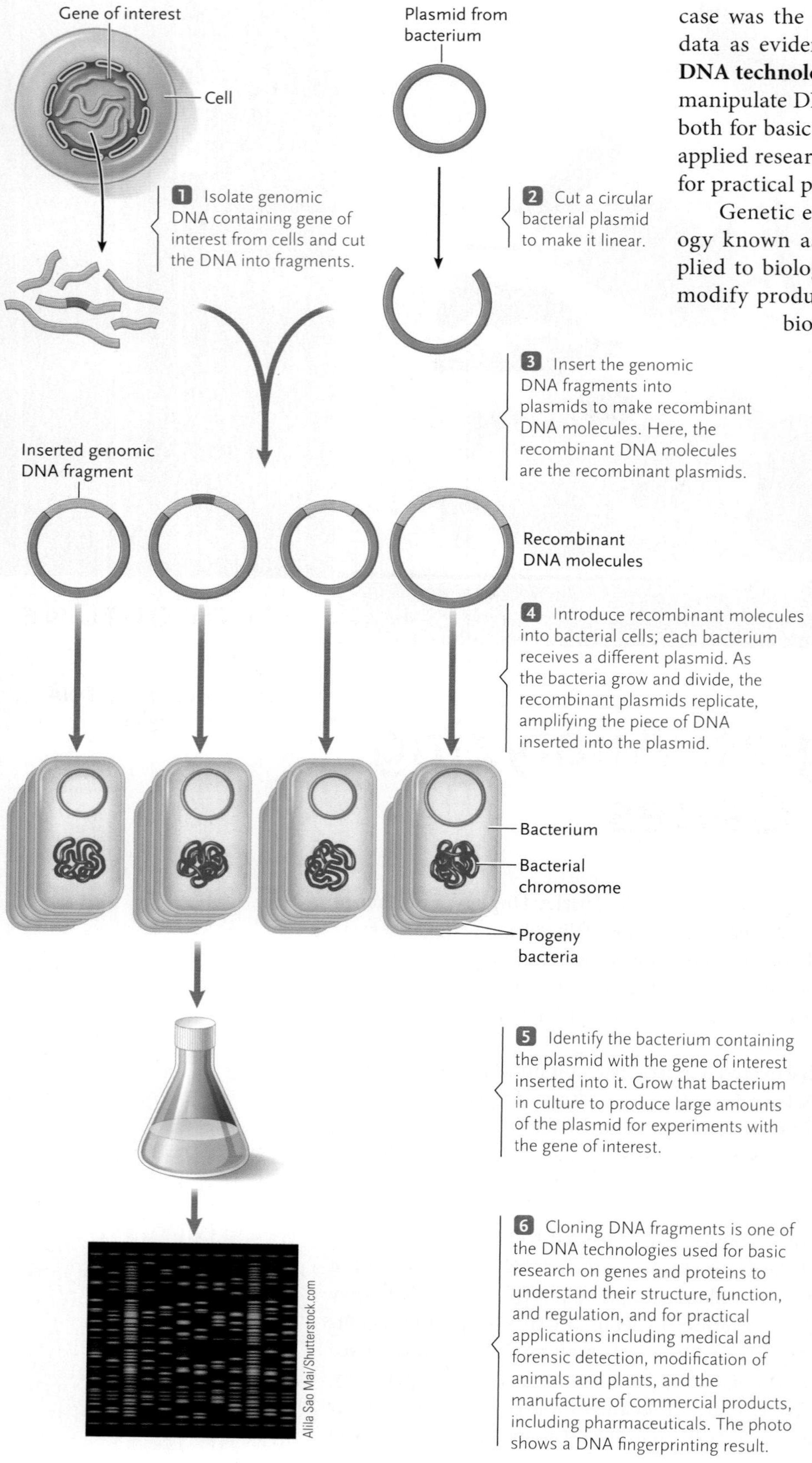

Gene of interest

Cell

1 Isolate genomic DNA containing gene of interest from cells and cut the DNA into fragments.

Plasmid from bacterium

2 Cut a circular bacterial plasmid to make it linear.

3 Insert the genomic DNA fragments into plasmids to make recombinant DNA molecules. Here, the recombinant DNA molecules are the recombinant plasmids.

Inserted genomic DNA fragment

Recombinant DNA molecules

4 Introduce recombinant molecules into bacterial cells; each bacterium receives a different plasmid. As the bacteria grow and divide, the recombinant plasmids replicate, amplifying the piece of DNA inserted into the plasmid.

Bacterium

Bacterial chromosome

Progeny bacteria

5 Identify the bacterium containing the plasmid with the gene of interest inserted into it. Grow that bacterium in culture to produce large amounts of the plasmid for experiments with the gene of interest.

6 Cloning DNA fragments is one of the DNA technologies used for basic research on genes and proteins to understand their structure, function, and regulation, and for practical applications including medical and forensic detection, modification of animals and plants, and the manufacture of commercial products, including pharmaceuticals. The photo shows a DNA fingerprinting result.

Alila Sao Mai/Shutterstock.com

FIGURE 18.1 Overview of cloning DNA fragments in a bacterial plasmid.
© Cengage Learning 2014

case was the first to admit nonhuman DNA fingerprinting data as evidence.) DNA fingerprinting is an application of **DNA technologies,** techniques to isolate, purify, analyze, and manipulate DNA sequences. Scientists use DNA technologies both for basic research into the biology of organisms and for applied research. The use of DNA technologies to alter genes for practical purposes is called **genetic engineering.**

Genetic engineering is used in the field of applied biology known as **biotechnology,** which is any technique applied to biological systems or living organisms to make or modify products or processes for a specific purpose. Thus, biotechnology includes manipulations that do not involve DNA technologies, such as the manipulation of the yeasts that brew beer and bake bread and the manipulation of the bacteria to make yogurt and cheese using standard genetics techniques.

In this chapter, you will learn about how biologists isolate genes and study and manipulate them for basic and applied research. You will learn about the basic DNA technologies and their applications to research in biology, and to genetic engineering.

We begin our discussion with a description of methods used to obtain genes in large quantities, an essential step for their analysis or manipulation.

18.1 DNA Cloning

Remember from Chapter 17 that a *clone* is a line of genetically identical cells or individuals derived from a single ancestor. **DNA cloning** is a method for producing many copies of a piece of DNA, such as a *gene of interest,* defined as a gene that a researcher wants to study or manipulate. DNA cloning is also used to clone important DNA sequences that are not genes. We will focus on genes in our discussion.

When DNA cloning involves a gene, it is called **gene cloning.** Having many copies of a gene generated by gene cloning opens the way for research on and manipulation of a gene that is not possible without cloning. Consider a typical diploid organism. Each cell contains only two copies of most genes, amounting to a tiny fraction of the total amount of DNA in a diploid cell. Therefore, in its natural state in the genome, a gene is difficult to study and manipulate.

An overview of one common method for cloning a gene of interest from a genome is shown in **Figure 18.1.** The method uses bacteria (commonly, *Escherichia coli*) and plasmids, small circular DNA molecules that replicate separately from the bacterial chromosome (see Section 17.1). The researcher extracts genomic DNA containing a gene of interest from cells and cuts the genomic DNA into fragments. The fragments are inserted into plasmids producing *recombinant DNA molecules*. **Recombinant DNA** is DNA fragments from two or more different sources that have been joined together to form a single molecule, in this case a recombinant plasmid. The recombinant plasmids are introduced into bacteria by artificial transformation (see Section 17.1). Each transformed bacterium receives a different plasmid and, as the bacterium continues to grow and divide, the plasmid replicates. Through the replication of the plasmid, amplification of the piece of DNA inserted into the plasmid occurs. The final step is to identify the bacteria containing the plasmid with the gene of interest and then isolate the plasmid for further study of the gene.

Cloned genes and other cloned DNA sequences from genomes are used in both basic research and applied research. In *basic research,* a researcher might want to study a cloned gene to learn about its structure, including its DNA sequence and sequences that regulate its expression. A researcher might also want to study the gene's function, including how its expression is regulated, and the nature of the gene's product. For protein-coding genes, for instance, the cloned gene could be used to produce the protein product in large quantities in a microorganism host to facilitate study of that protein's structure and function. As part of this research, the cloned gene could be manipulated in the laboratory to help dissect a gene's function.

In *applied research,* the interest in cloned genes or other cloned DNA sequences is not in the structure and function of a gene or sequence; that typically is understood, at least to a significant degree, at the beginning of research projects. Rather, cloned genes or cloned DNA sequences are used, for instance, for medical, forensic, agricultural, or commercial applications. Some examples are:

- Gene therapy to correct or treat genetic diseases.
- Diagnosis of genetic diseases, such as sickle-cell anemia.
- DNA fingerprinting in forensics.
- Production of pharmaceuticals, such as humulin, human insulin to treat diabetes, and tissue plasminogen activator to break down blood clots.
- Generation of genetically modified animals and plants, including animals that synthesize pharmaceuticals, and plants that are nutritionally enriched, insect-resistant, or herbicide-resistant.
- Modification of bacteria to use in cleanup of oil spills or toxic waste.

FIGURE 18.2 The restriction site for the restriction enzyme *Eco*RI, and the generation of a recombinant DNA molecule by complementary base pairing of DNA fragments that are produced by digestion with the same restriction enzyme.
© Cengage Learning 2014

Bacterial Enzymes Called Restriction Endonucleases Are the Basis of DNA Cloning

The key to DNA cloning is the joining of two DNA molecules from different sources, such as a genomic DNA fragment and a bacterial plasmid (see Figure 18.1). Bacterial enzymes called **restriction endonucleases** (also called **restriction enzymes**) can be used to generate DNA fragments that can be joined to produce recombinant DNA molecules. Restriction enzymes recognize *restriction sites,* specific DNA sequences that are typically 4 to 8 base pairs (bp) long, and cut the DNA at specific locations within those sites. The DNA fragments produced by a restriction enzyme are known as **restriction fragments.**

The "restriction" in the name of the enzymes refers to their normal role inside bacteria, in which the enzymes defend against viral attack by breaking down (restricting) the DNA molecules of infecting viruses. The bacterium protects the restriction sites in its own DNA from cutting by chemically modifying bases in those sites enzymatically, thereby preventing them from being recognized and cut by its restriction enzyme.

Hundreds of different restriction enzymes have been identified, each one cutting DNA at a specific restriction site. As illustrated by the restriction site of *Eco*RI **(Figure 18.2),** most re-

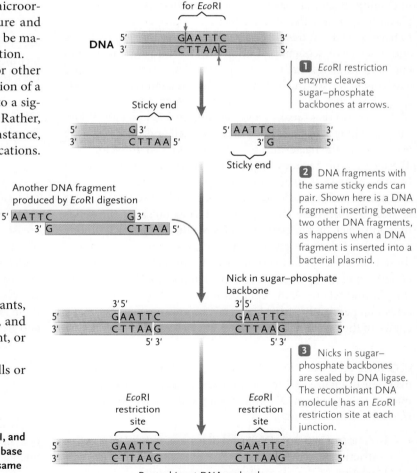

striction sites are symmetrical in that the sequence of nucleotides read in the 5′→3′ direction on one strand is same as the sequence read in the 5′→3′ direction on the complementary strand. The restriction enzymes most used in cloning—such as EcoRI—cleave the sugar–phosphate backbones of DNA to produce DNA fragments with single-stranded ends (step 1). These ends are called **sticky ends** because the short single-stranded regions can form hydrogen bonds with complementary sticky ends on any other DNA molecules cut with the same enzyme (step 2). The pairings leave nicks in the sugar–phosphate backbones of the DNA strands that are sealed by *DNA ligase,* an enzyme that has the same function in DNA replication (step 3; see Section 14.3). The process of sealing with DNA ligase is called **ligation.** The result is a recombinant DNA molecule. The ligation process restores the sequences of the two restriction sites producing a molecule that, in this case, has an *EcoRI* restriction site at each junction.

Bacterial Plasmids Illustrate the Use of Restriction Enzymes in Cloning

The bacterial plasmids used for cloning are examples of **cloning vectors**—DNA molecules into which a DNA fragment can be inserted to form a recombinant DNA molecule for the purpose of cloning. Bacterial plasmid cloning vectors are derivatives of plasmids naturally found in bacteria, and have been engineered to have special features that make them useful for cloning genes and other DNA sequences. Commonly, plasmid cloning vectors contain two genes that are useful in the final steps of a cloning experiment for sorting bacteria that have recombinant plasmids from those that do not: (1) The *amp*^R gene encodes an enzyme (β-lactamase) that breaks down the antibiotic ampicillin. When the plasmid is introduced into an *E. coli* cell and the *amp*^R gene is expressed, the bacterium becomes resistant to ampicillin. (2) The *lacZ*^+ gene encodes β-galactosidase (recall the *lac* operon from Section 16.1), which hydrolyzes the sugar lactose, as well as a number of synthetic substrates. A cluster of restriction sites is located within the *lacZ*^+ gene, but does not alter the gene's function. Each of the restriction sites within the cluster is unique to the plasmid. For a given cloning experiment, one or two of these restriction sites is chosen. The plasmid also has an *origin of replication (ori).* An origin of replication is the DNA region of the plasmid where DNA replication begins. The plasmid replicates independently of the bacterial chromosome so that many copies of the plasmid are produced in a cell.

CLONING A GENE OF INTEREST **Figure 18.3** expands on the overview of Figure 18.1 to show the steps used to clone a gene of interest using a plasmid cloning vector and restriction enzymes. In outline, the steps are:

- Isolation of genomic DNA and digestion of that DNA with a restriction enzyme (step 1).
- Digestion of the plasmid cloning vector with the same restriction enzyme (step 2).

- Ligation of cut genomic DNA fragments and cut plasmid DNA together using DNA ligase (step 3). This produces a mixture of recombinant plasmids (plasmids with DNA fragments inserted into the cloning vector), nonrecombinant plasmids (resealed cloning vectors with no DNA fragment inserted), and joined-together pieces of genomic DNA with no cloning vector involved.
- Transformation of the DNA into *E. coli* (step 4). Some bacteria will take up a plasmid whereas others will not.
- Spreading the bacterial cells on growth medium containing ampicillin and X-gal and incubate to allow colonies to grow (step 5).
- *Selection:* Bacteria containing plasmids are selected for because of the ampicillin in the growth medium. Within each cell of a colony, the plasmids replicate until approximately 100 are present.
- *Screening:* The X-gal in the medium distinguishes between bacteria that have been transformed with recombinant plasmids and nonrecombinant plasmids by *blue-white screening* (see Figure 18.3, Interpreting the Results). White colonies contain recombinant plasmids whereas blue colonies contain nonrecombinant plasmids. Among the white colonies is the one with a recombinant plasmid that contains the gene of interest. We will see a little later how we can identify that particular plasmid.

Three researchers, Paul Berg, Stanley N. Cohen, and Herbert Boyer, in 1973 pioneered the development of DNA cloning techniques using restriction enzymes and bacterial plasmids. Berg received a Nobel Prize in 1980 for his research.

STORAGE OF CLONED DNA FRAGMENTS AS DNA LIBRARIES By following the procedure shown in Figure 18.3, a researcher can generate a set of clones that collectively contains a copy of every DNA sequence in a genome. Such a collection is called a **genomic library.** A genomic library can be made using plasmid cloning vectors (as in Figure 18.3) or any other kind of cloning vector. The number of clones in a genomic library increases with the size of the genome. For example, a yeast genomic library of plasmid clones consists of hundreds of plasmids, whereas a human genomic library of plasmid clones consists of thousands of plasmids. Storage of a genomic library is accomplished by picking individual colonies and placing them in growth medium-containing wells of microwell plates and freezing the plates (see Figure 18.3).

A genomic library is a resource containing the entire DNA of an organism cut into pieces. That means all of the genes are represented, as well as the DNA between the genes. Similar to a book library, where you can search through the same set of books on various occasions to find different passages of interest, researchers can screen the same genomic library on various occasions to find and isolate different genes or other DNA sequences. We will illustrate one type of such screening shortly.

Researchers also commonly use another kind of DNA library that is made by starting with mRNA molecules isolated from a

FIGURE 18.3 **Research Method**

Cloning a Gene of Interest in a Plasmid Cloning Vector

Purpose: Cloning a gene produces many copies of a gene of interest that can be used, for example, to determine the DNA sequence of the gene, to manipulate the gene in basic research experiments, to understand its function, and to produce the protein encoded by the gene either for basic or applied research.

Protocol:

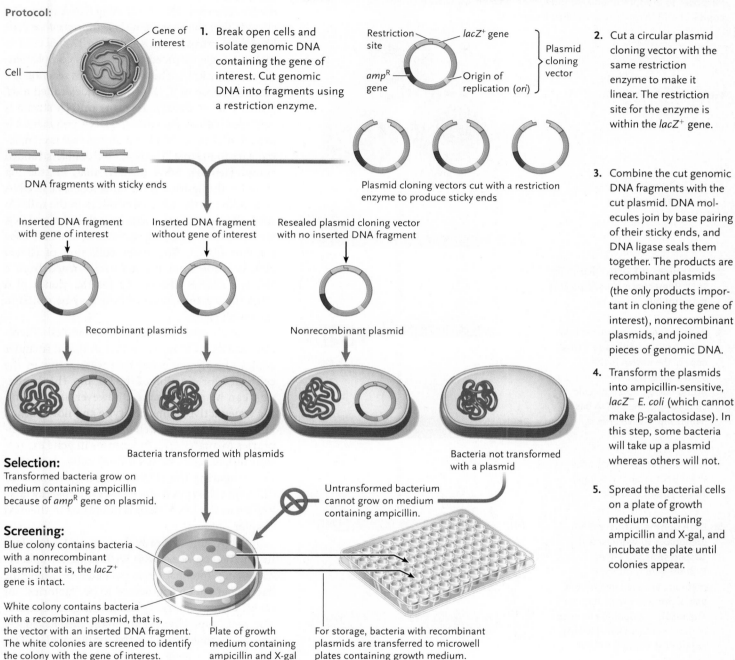

1. Break open cells and isolate genomic DNA containing the gene of interest. Cut genomic DNA into fragments using a restriction enzyme.

2. Cut a circular plasmid cloning vector with the same restriction enzyme to make it linear. The restriction site for the enzyme is within the *lacZ*⁺ gene.

3. Combine the cut genomic DNA fragments with the cut plasmid. DNA molecules join by base pairing of their sticky ends, and DNA ligase seals them together. The products are recombinant plasmids (the only products important in cloning the gene of interest), nonrecombinant plasmids, and joined pieces of genomic DNA.

4. Transform the plasmids into ampicillin-sensitive, *lacZ*⁻ *E. coli* (which cannot make β-galactosidase). In this step, some bacteria will take up a plasmid whereas others will not.

5. Spread the bacterial cells on a plate of growth medium containing ampicillin and X-gal, and incubate the plate until colonies appear.

Selection:
Transformed bacteria grow on medium containing ampicillin because of *amp*ᴿ gene on plasmid.

Screening:
Blue colony contains bacteria with a nonrecombinant plasmid; that is, the *lacZ*⁺ gene is intact.

White colony contains bacteria with a recombinant plasmid, that is, the vector with an inserted DNA fragment. The white colonies are screened to identify the colony with the gene of interest.

Plate of growth medium containing ampicillin and X-gal

For storage, bacteria with recombinant plasmids are transferred to microwell plates containing growth medium.

Interpreting the Results: All of the colonies on the plate contain plasmids because the bacteria that form the colonies are resistant to the ampicillin present in the growth medium. Blue-white screening distinguishes bacterial colonies with nonrecombinant plasmids from those with recombinant plasmids. Blue colonies have nonrecombinant plasmids. These plasmids have intact *lacZ*⁺ genes and produce β-galactosidase, which changes X-gal to a blue product. White colonies have recombinant plasmids. These plasmids have DNA fragments inserted into the *lacZ*⁺ gene, so they do not produce β-galactosidase. As a result, they cannot convert X-gal to the blue product and the colonies are white. Among the white colonies is a colony containing the plasmid with the gene of interest. Further screening is done to identify that particular white colony (see Figure 18.5). Once identified, the colony is cultured to produce large quantities of the recombinant plasmid for analysis or manipulation of the gene.

FIGURE 18.4 | **Research Method**

Synthesis of DNA from mRNA Using Reverse Transcriptase

Purpose: To produce double-stranded, complementary DNA (cDNA) copies of mRNA molecules isolated from cells.

Protocol:

1. Isolate mRNAs from cells. One mRNA is shown.

2. Add primer of T DNA nucleotides (dT). Primer base-pairs to poly(A) tail of mRNA.

3. Reverse transcriptase uses DNA precursors to synthesize a DNA copy of the mRNA in the 5′-to-3′ direction. The result is a hybrid nucleic acid molecule consisting of the mRNA base paired with a DNA strand.

4. An RNase enzyme degrades the mRNA strand, leaving a single strand of DNA.

5. DNA polymerase uses DNA precursors to synthesize the second strand of DNA. Experimentally different methods are available for the use of primers in this reaction. The result is a double-stranded complementary DNA (cDNA) copy of the starting mRNA.

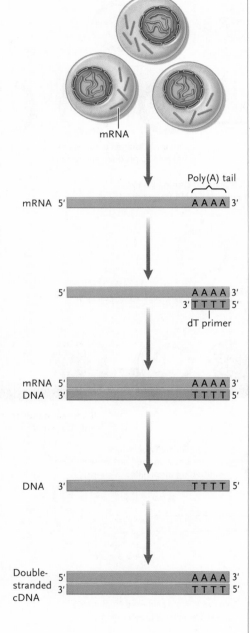

Outcome: The outcome is a population of double-stranded, cDNA molecules that have base-pair sequences corresponding to the base sequences of the mRNA molecules isolated from the cell.

© Cengage Learning 2014

cell (**Figure 18.4**). mRNA molecules are the transcripts of protein-coding genes in the cell (see Section 15.2), so the collection of mRNA molecules isolated represents a snapshot of the expression of protein-coding genes in the cell type used in the experiment. To convert single-stranded mRNA to double-stranded DNA for cloning (RNA cannot be cloned), first the enzyme *reverse transcriptase* (isolated from retroviruses; see Section 17.2) is used to make a single-stranded DNA that is complementary to the mRNA. The primer for this reaction is a short sequence of T DNA nucleotides called a *dT primer* ("d" = deoxy). Then the mRNA strand is degraded with an enzyme, and DNA polymerase is used to make a second DNA strand that is complementary to the first. The result is a double-stranded **complementary DNA (cDNA)** that, for the top strand in the figure, has the same sequence in DNA nucleotides as the RNA nucleotides in the mRNA. After adding restriction sites to each end, the cDNA is inserted into a cloning vector as described for the genomic library. The entire collection of cloned cDNAs made from the mRNAs isolated from a cell is a **cDNA library.** As for the clones in a DNA library, the cloned cDNAs may be stored in microwell plates.

Not all genes are active in every cell. Therefore, a cDNA library is limited in that it includes copies of only the genes that were active in the cells used for creation of the library. This limitation can be an advantage, however, in identifying genes active in one cell type and not another. For example, a researcher could use cDNA libraries to analyze the differences in gene expression in liver cells versus kidney cells. A method for comparing the cDNA libraries produced by different cell types using a DNA microarray (also known as the DNA chip) is described in the next chapter.

cDNA libraries provide a critical advantage to genetic engineers who wish to insert eukaryotic genes into bacteria, particularly when the bacteria are to be engineered to be "factories" for making the protein encoded by the gene. Eukaryotic genes typically contain many *introns,* spacer sequences that interrupt the amino acid-coding sequence of a gene (see Section 15.3). Bacterial genes do not contain introns and, consequently, bacteria have no system to remove introns from transcripts of eukaryotic genes. However, a eukaryotic mRNA does not contain introns and, therefore, neither does the cDNA copy. Therefore, bacteria can express the cDNA accurately to synthesize the encoded eukaryotic protein.

FIGURE 18.5 **Research Method**

DNA Hybridization to Identify a DNA Sequence of Interest

Purpose: Hybridization with a specific DNA probe allows researchers to detect a specific DNA sequence, such as a gene, within a population of DNA molecules. Here, DNA hybridization is used to screen a genomic library of plasmid clones to identify those containing a recombinant plasmid with a gene of interest.

Protocol:

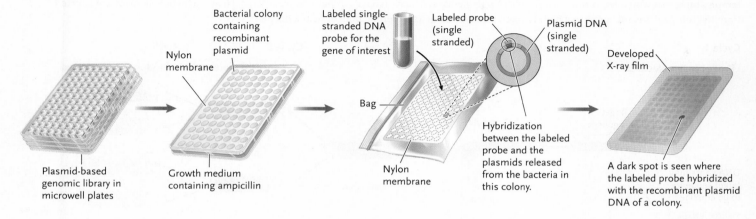

1. Replica plate from each genomic library microwell plate onto a plate of growth medium containing ampicillin with a nylon membrane on the surface. The pattern of colonies that grow on a membrane matches that of the clones in the microwell plate.

2. Remove the membrane and treat it to break open the cells and to denature the released DNA into single strands. The single-stranded DNA sticks to the filter in the same position as the colony from which it was derived. Place the filter in a plastic bag. Add a labeled single-stranded DNA probe (DNA or RNA) for the gene of interest and incubate. If a recombinant plasmid's inserted DNA fragment is complementary to the probe, the two will hybridize, that is, form base pairs. Wash off excess labeled probe.

3. Detect the hybridization by looking for the labeled tag on the probe. If the probe was radioactively labeled, place the filter against X-ray film. The decaying radioactive compound exposes the film, giving a dark spot when the film is developed. Correlate the position of any dark sport on the film with the original microwell plate. Bacteria from that well can then be used for further study of the gene clone.

Interpreting the Results: DNA hybridization with a labeled probe identifies a sequence of interest. If the probe is for a particular gene (as illustrated in this figure), it allows the specific identification of a bacterial clone with recombinant plasmids containing that gene. The method can be used similarly to identify a cDNA clone corresponding to a particular gene. The specificity of the method depends directly on the probe used. The same collection of bacterial clones can be used to search for recombinant plasmids carrying different genes or different plasmids of interest by changing the probe.

SCREENING A DNA LIBRARY FOR A GENE OF INTEREST One method used to screen a DNA library to identify a clone containing a gene of interest is based on the fact that a gene has a unique DNA sequence. In this technique, called **DNA hybridization,** a gene of interest is identified in the set of clones when it base-pairs with a short, single-stranded complementary DNA or RNA molecule called a *nucleic acid probe* **(Figure 18.5).** The figure illustrates screening for a gene of interest in a genomic library. Similarly, the approach can be used to screen a cDNA library for the cDNA corresponding to a gene of interest. The probe is labeled with a radioactive or a nonradioactive tag,

so investigators can detect it. In our example, if we know the sequence of part of a gene of interest, we can use that information to design and synthesize a nucleic acid probe. Or, we can take advantage of DNA sequence similarities of evolutionarily related organisms. For instance, we could make a probe for the human actin gene based on the sequence of the cloned mouse actin gene and expect that the two nucleic acids would hybridize because of the evolutionary conservation of that gene. Once a colony containing plasmids with a gene of interest has been identified, that colony can be used to produce large quantities of the cloned gene.

FIGURE 18.6 | **Research Method**

The Polymerase Chain Reaction (PCR)

Purpose: To amplify—produce large numbers of copies of—a target DNA sequence in the test tube without cloning.

Protocol: A polymerase chain reaction mixture has four key elements: (1) the DNA with the target sequence to be amplified; (2) a pair of DNA primers, one complementary to one end of the target sequence and the other complementary to the other end of the target sequence; (3) the four nucleoside triphosphate precursors for DNA synthesis (dATP, dTTP, dGTP, and dCTP); and (4) DNA polymerase. Since PCR uses high temperatures that would break down normal DNA polymerases, a heat-stable DNA polymerase is used. Heat-stable polymerases are isolated from microorganisms that grow in a high-temperature area such as a thermal pool or near a deep-sea vent.

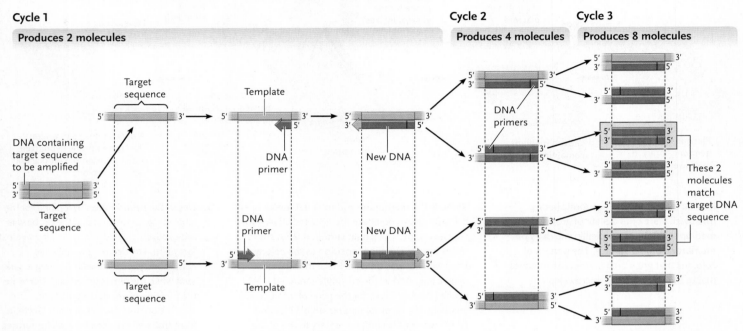

1. **Denaturation:** heat DNA containing target sequence to 95°C to denature it to single strands.

2. **Annealing:** cool the mixture to 55 to 65°C (depending on the primers) to allow the two primers to anneal their complementary sequences at the two ends of the target sequence.

3. **Extension:** heat to 72°C, the optimal temperature for DNA polymerase to extend the primers, using the four nucleoside triphosphate precursors to make complementary copies of the two template strands. This completes cycle 1 of PCR; the end result is two molecules.

4. **Repeat the same** steps of denaturation, annealing of primers, and extension in cycle 2, producing a total of four molecules.

5. **Repeat the same steps in** cycle 3, producing a total of eight molecules. Two of the eight match the exact length of the target DNA sequence (highlighted in yellow).

Interpreting the Results: After three cycles, PCR produces a pair of molecules matching the target sequence. Subsequent cycles amplify these molecules to the point where they outnumber all other molecules in the reaction by many orders of magnitude.

© Cengage Learning 2014

The Polymerase Chain Reaction (PCR) Amplifies DNA *in Vitro*

Producing multiple DNA copies by cloning in a living cell requires a series of techniques and considerable time. A much more rapid process, **polymerase chain reaction (PCR),** produces an extremely large number of copies of a specific DNA sequence—such as a gene—from a DNA mixture without cloning the sequence. The process is called *amplification* be-

cause it increases the amount of DNA to the point where it can be analyzed or manipulated easily. Developed in 1983 by Kary B. Mullis and Fred Faloona at Cetus Corporation (Emeryville, CA), PCR has become one of the most important tools in modern molecular biology, having a wide range of applications in all areas of biology. Mullis received a Nobel Prize in 1993 for his role in the development of PCR.

Figure 18.6 shows how PCR is performed. In essence, PCR is a special case of DNA replication in which a DNA polymerase replicates only a portion of a DNA molecule. PCR takes advantage of a characteristic common to all DNA polymerases, namely, that these enzymes add nucleotides only to the end of an existing chain called the *primer* (see Section 14.3). For replication to take place, a primer therefore must be in place, base-paired to the template chain at which replication is to begin. By cycling 20 to 30 times through a series of steps, PCR amplifies the target sequence, producing millions of copies—an amount that is sufficient for analysis and/or manipulation.

Two primers are used in PCR to bracket the sequence of interest. The cycles of PCR then replicate only this sequence from a mixture of essentially any DNA molecules. Thus PCR not only finds the "needle in the haystack" among all the sequences in a mixture, but also makes millions of copies of the "needle"—the DNA sequence of interest. Usually, no further purification of the amplified sequence is necessary. The limitation of PCR is that sequence information must be available to design the primers. Increasingly with genome sequences of many organisms becoming available, this is less of a problem than it used to be. Typically, researchers use commercial companies to make the primers using DNA synthesizers.

In basic research, for instance, as long as DNA sequence information is available to design the primers, PCR can be used as a method to amplify a gene of interest from genomic DNA for study or cloning. This method is far quicker than the plasmid cloning/DNA hybridization method just described. There are many other uses for PCR. For example, PCR is used in forensics to produce enough DNA for analysis from the root of a single human hair, or from a small amount of blood, semen, or saliva, such as the traces left at the scene of a crime. It is also used to extract and multiply DNA sequences from skeletal remains; ancient sources such as mammoths, Neanderthals, and Egyptian mummies; and, in rare cases, from amber-entombed fossils, fossil bones, and fossil plant remains.

A successful outcome of PCR is shown by analyzing a sample of the amplified DNA using **agarose gel electrophoresis** to see if the copies are the same length as the target DNA sequence **(Figure 18.7)**. Gel electrophoresis is a technique by which DNA, RNA, or protein molecules are separated in a gel that is subjected to an electric field. The type of gel and the conditions used vary with the experiment, but in each case the gel functions as a molecular sieve to separate the macromolecules based on their size, electrical charge, or other properties. For separating large DNA molecules, such as those typically produced by PCR or by cutting with a restriction enzyme, a gel made of agarose (a natural molecule isolated from seaweed) is used because of its large pore size.

For PCR experiments, the size of the amplified DNA is determined by comparing the position of the DNA band with the positions of bands of a DNA ladder, that is, DNA fragments of known size that are separated on the gel at the same time. PCR is successful, if the sample fragment size matches the predicted size for the target DNA fragment. In some cases, such as DNA from ancient sources, a size prediction may not be possible; here, agarose gel electrophoresis analysis simply indicates whether there was DNA in the sample that could be amplified.

The advantages of PCR have made it an extremely valuable technique for researchers, law enforcement agencies, and forensic specialists whose primary interest is in the amplification of specific DNA fragments up to a practical maximum of a few thousand base pairs. Cloning remains the technique of choice for amplification of longer fragments. As already mentioned, the major limitation of PCR relates to the primers. In order to design a primer for PCR, the researcher must first have sequence information about the target DNA. By contrast, cloning can be used to amplify DNA of unknown sequence.

Review of Some of the Materials, Concepts, and Techniques Introduced in this Section

More than any other chapter in the book, this chapter discusses a lot of research methods and therefore has contained a lot of new terms and techniques to learn. Here is a collection of a number of these terms and techniques and what they are or do:

- *Genetic engineering.* The use of DNA technologies to alter genes for practical purposes.
- *DNA cloning.* A method for producing many copies of a piece of DNA.
- *Gene cloning.* When DNA cloning involves a gene.
- *Recombinant DNA.* DNA fragments from two or more sources that have joined together.
- *Restriction enzyme (restriction endonuclease).* An enzyme that recognizes a specific DNA sequence and cuts the DNA within that sequence. Fragments produced by cutting DNA with a restriction enzyme are *restriction fragments.*
- *Ligation.* The process of joining two or more DNA fragments together to make one DNA molecule.
- *DNA ligase.* The enzyme that seals together DNA fragments generated by restriction enzyme digestion to produce a recombinant DNA molecule.
- *Cloning vector.* DNA molecules into which a DNA fragment can be inserted to form a recombinant DNA molecule that can be replicated in a host organism for the purpose of cloning the DNA fragment.
- *Genomic DNA library.* A set of clones that collectively contains a copy of every DNA sequence in a genome.
- *cDNA (complementary DNA).* A double-stranded DNA copy of a single-stranded mRNA molecule.
- *cDNA library.* A collection of cloned cDNAs made from the mRNAs isolated from a cell.
- *DNA hybridization.* A technique to identify a gene of interest in a set of clones using a nucleic acid probe that can base-pair with the DNA sequence of the gene.
- *Polymerase chain reaction (PCR).* A DNA replication-based technique for amplifying DNA sequences, including genes, without cloning.

FIGURE 18.7 | **Research Method**

Separation of DNA Fragments by Agarose Gel Electrophoresis

Purpose: Gel electrophoresis separates DNA molecules, RNA molecules, or proteins according to their sizes, electrical charges, or other properties through a gel in an electric field. Different gel types and conditions are used for different molecules and types of applications. A common gel for separating large DNA fragments is made of agarose.

Protocol:

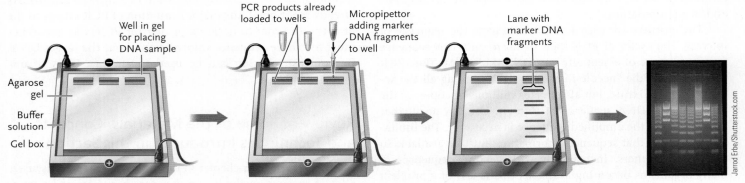

1. Prepare a gel consisting of a thin slab of agarose and place it in a gel box in between two electrodes. The gel has wells for placing the DNA samples to be analyzed. Add buffer to cover the gels.

2. Load DNA sample solutions, such as PCR products, into wells of the gel, alongside a well loaded with a DNA ladder (DNA fragments of known sizes). All samples have a dye added to help see the liquid when loading the wells. The dye migrates during electrophoresis, enabling the progress of electrophoresis to be followed.)

3. Apply an electric current to the gel; the negatively charged DNA fragments migrate to the positive pole. Shorter DNA fragments migrate faster than longer DNA fragments. At the completion of separation, DNA fragments of the same length have formed bands in the gel. At this point, the bands are invisible.

4. Stain the gel with a DNA-binding dye. The dye fluoresces under UV light, enabling the DNA bands to be seen and photographed. Shown is an actual gel photograph.

Interpreting the Results: Agarose gel electrophoresis separates DNA fragments according to their length. The lengths of the DNA fragments being analyzed are determined by measuring their migration distances and comparing those distances to a calibration curve of the migration distances of the bands of the DNA ladder, which have known lengths. For PCR, agarose gel electrophoresis shows whether DNA of the correct length was amplified. For restriction enzyme digests, this technique shows whether fragments are produced as expected.

- *Agarose gel electrophoresis.* A technique in which an electric field passing through an agarose gel is used to separate DNA or RNA molecules on the basis of size.

STUDY BREAK 18.1

1. What features do restriction enzymes have in common? How do they differ?

2. Plasmid cloning vectors are one type of cloning vector that can be used with *E. coli* as a host organism. What features of a plasmid cloning vector make it useful for constructing and cloning recombinant DNA molecules?

3. What is a cDNA library, and from what cellular material is it derived? How does a cDNA library differ from a genomic library?

4. What information and materials are needed to amplify a region of DNA using PCR?

18.2 Applications of DNA Technologies

The ability to clone pieces of DNA—genes, especially—and to amplify specific segments of DNA by PCR has revolutionized biology. These and other DNA technologies are now used for research in all areas of biology, including cloning genes to determine their structure, function, and regulation of expression; manipulating genes to determine how their products function in cellular or developmental processes; and identifying differences in DNA sequences among individuals in ecological studies. The same DNA technologies also have practical applications, including medical and forensic detection, modification of animals and plants, and the manufacture of commercial products. In this section, case studies provide examples of how the techniques are used to answer questions and solve problems.

DNA Technologies Are Used in Molecular Testing for Many Human Genetic Diseases

Many human genetic diseases are caused by defects in enzymes or other proteins that result from mutations at the DNA level. Once scientists have identified the specific mutations responsible for human genetic diseases, they can often use DNA technologies to develop molecular tests for those diseases. One example is sickle-cell anemia (see *Why It Matters* in Chapter 12 and Sections 12.2 and 13.4). People with sickle-cell anemia are homozygous for a DNA mutation that affects hemoglobin, the oxygen-carrying molecule of the blood. Hemoglobin consists of two copies each of the α-globin and β-globin polypeptides. The mutation, which is in the β-globin gene, alters one amino acid in the polypeptide. As a consequence, the function of hemoglobin is significantly impaired in individuals homozygous for the mutation (who have sickle-cell anemia), and mildly impaired in individuals heterozygous for the mutation (who have sickle-cell trait).

Three restriction sites for *Mst*II are associated with the normal β-globin gene, two within the coding sequence of the gene and one upstream of the gene. The sickle-cell mutation eliminates the middle site of the three **(Figure 18.8)**. Cutting the β-globin gene with *Mst*II produces two DNA fragments from the normal gene and one fragment from the mutated gene. Restriction enzyme-generated DNA fragments of different lengths from the same region of the genome, such as in this example, are known as **restriction fragment length polymorphisms** (**RFLPs,** pronounced "riff-lips").

RFLPs can be analyzed using a hybridization-based method called **Southern blot analysis** (named after its inventor, researcher Edwin Southern) **(Figure 18.9).** In this technique, genomic DNA is digested with a restriction enzyme, and the DNA fragments are separated using agarose gel electrophoresis. The fragments are then transferred—blotted—to a nylon membrane, and a labeled probe is used to identify a DNA sequence of interest from among the many thousands of fragments on the nylon membrane. As shown in Figure 18.9 for the sickle-cell anemia example, the DNA bands visualized after hybridization indicate the allele or alleles present in each individual.

A technique called **northern blot analysis** is very similar to Southern blot analysis but is used to study RNA rather than DNA. (The name is a play on words on Southern's name.) In northern blot analysis, RNA molecules extracted from cells or a tissue are separated by size using electrophoresis and then transferred to a filter paper. Hybridization with a labeled probe reveals the RNAs that were present, both their size and concentration. Northern blot analysis is useful, for example, for determining the size of an mRNA encoded by a gene or for quantifying the expression of a gene in different tissues or at different stages of development.

Restriction enzyme digestion and Southern blot analysis may be used to test for a number of human genetic diseases, including phenylketonuria and Duchenne muscular dystrophy (see Section 13.4). In some cases, restriction enzyme digestion is combined with PCR for a quicker, easier analysis because there is no need for a probe or Southern blotting. With that approach, the gene or region of the gene with the restriction enzyme variation is amplified using PCR, and then the amplified DNA is cut with the diagnostic restriction enzyme. Amplification produces enough DNA so that separation by size using agarose gel electrophoresis produces clearly visible bands, positioned according to fragment length. Researchers can then determine whether the fragment lengths match a normal or abnormal RFLP pattern.

DNA Fingerprinting Is Used to Identify Human Individuals as well as Individuals of Other Species

Just as each human has a unique set of fingerprints, each also has unique combinations and variations of DNA sequences (with the exception of identical twins) known as *DNA fingerprints*. **DNA fingerprinting** is a technique used to distinguish between individuals of the same species using DNA samples. Invented by Sir Alec Jeffreys in 1984, DNA fingerprinting has become a mainstream technique for distinguishing human individuals, notably in forensics and paternity testing. The technique is applicable to all kinds of organisms. We focus on humans in the following discussion.

DNA FINGERPRINTING PRINCIPLES In DNA fingerprinting, molecular techniques, typically PCR, are used to analyze DNA variations at various loci in the genome. In the United States, 13 loci in noncoding regions of the genome are the standards for PCR analysis. Each locus is an example of a **short tandem repeat (STR)** sequence (also called a **microsatellite**), meaning that it has a short 2 to 6 bp sequence

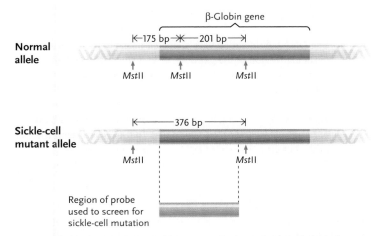

FIGURE 18.8 Restriction site differences between the normal and sickle-cell mutant alleles of the β-globin gene. The figure shows a DNA segment that can be used as a probe to identify these alleles in a subsequent probe-based analysis (see Figure 18.9).

© Cengage Learning 2014

Southern Blot Analysis

Purpose: The Southern blot technique allows researchers to identify DNA fragments of interest after separating DNA fragments on a gel. One application is to compare different samples of genomic DNA cut with a restriction enzyme to detect specific restriction fragment length polymorphisms. Here the technique is used to distinguish between individuals with sickle-cell anemia, individuals with sickle-cell trait, and normal individuals.

Protocol:

1. Isolate genomic DNA and digest with *Mst*II. Here, genomic DNA is from three individuals: **A,** sickle-cell anemia (homozygous for the sickle-cell mutant allele); **B,** normal (homozygous for the normal allele); and **C,** sickle-cell trait (heterozygous for sickle-cell mutant allele).

2. Separate the DNA fragments by agarose gel electrophoresis. The thousands of differently sized DNA fragments produce a smear of DNA down the length of each lane in the gel, which are seen after staining the DNA (see Figure 18.7.)

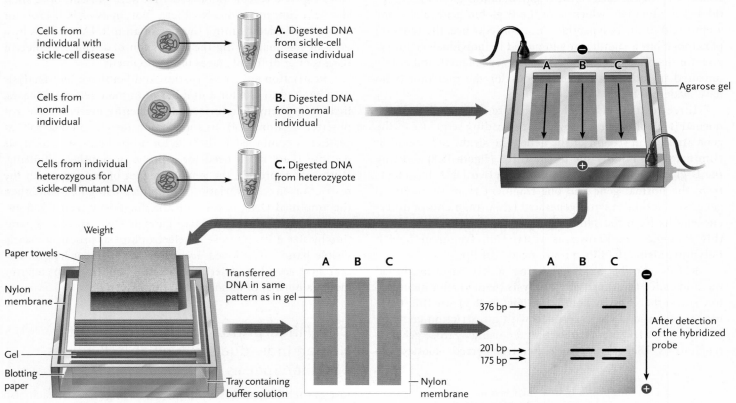

3. Hybridization with a labeled DNA probe to identify DNA fragments of interest cannot be done directly with an agarose gel. Edward Southern devised a method to transfer the DNA fragments from a gel to a nylon membrane. First, treat the gel with a solution to denature the DNA into single strands. Next, place the gel on a piece of blotting paper with ends of the paper in the buffer solution and place the membrane on top of the gel. Capillary action wicks the buffer solution in the tray up the blotting paper, through the gel and membrane, and into the weighted stack of paper towels on top of the gel. The movement of the solution transfers—blots—the single-stranded DNA fragments to the membrane, where they stick. The pattern of DNA fragments is the same as it was in the gel.

4. Use DNA hybridization with a labeled probe to focus on a particular region of the genome. That is, incubate a labeled, single-stranded probe with the membrane and detect hybridization of the probe with DNA fragments on the membrane. For a radioactive probe, place the membrane against X-ray film, which, after development will show a band or bands where the probe hybridized. In this experiment, the probe is a cloned piece of DNA from the area shown in Figure 18.8 (the β-globin gene), that can bind to all three of the *Mst*II fragments of interest.

Interpreting the Results: The hybridization result indicates that the probe has identified a very specific DNA fragment or fragments in the digested genomic DNA. The RFLPs for the β-globin gene can be seen in Figure 18.8. DNA from the sickle-cell anemia individual **(A)** cut with *Mst*II results in a single band of 376 bp detected by the probe, whereas DNA from the normal individual **(B)** results in two bands, of 201 bp and 175 bp. DNA from a sickle-cell trait heterozygote **(C)** results in three bands, of 376 bp (from the sickle-cell mutant allele), and 201 bp and 175 bp (both from the normal allele). This type of analysis in general is useful for distinguishing normal and mutant alleles of genes where the mutation involved alters a restriction site.

of DNA repeated in series. Each locus has a different repeated sequence, and the number of repeats varies among individuals in a population. For example, one STR locus has the sequence 5'-AGAT-3' repeated between 8 and 20 times. As a further source of variation, a given individual is either homozygous or heterozygous for an STR allele; perhaps you are homozygous for an 11-repeat allele, or heterozygous for a 9-repeat allele and a 15-repeat allele. Likely your DNA fingerprint for this locus is different from most of the others in your class. Because each individual has an essentially unique combination of alleles (identical twins are the exception), analysis of multiple STR loci can discriminate between DNA of different individuals and, therefore potentially identify individuals on the basis of DNA analysis. **Figure 18.10** illustrates how PCR is used to obtain a DNA fingerprint for a theoretical STR locus with three alleles of 9, 11, and 15 tandem repeats.

DNA FINGERPRINTING IN FORENSICS DNA fingerprints are used routinely to identify criminals or eliminate innocent persons as suspects in legal proceedings. For example, a DNA fingerprint prepared from a hair found at the scene of a crime, or from a semen sample, might be compared with the DNA fingerprint of a suspect to link the suspect with the crime. Or, a DNA fingerprint of blood found on a suspect's clothing or possessions might be compared with the DNA fingerprint of a vic-

tim. Typically, the evidence is presented in terms of the probability that the particular DNA sample could have come from a random individual. Hence the media report probability values, such as one in several million, or in several billion, that a person other than the accused could have left his or her DNA at the crime scene.

Although courts initially met with legal challenges to the admissibility of DNA fingerprints, experience has shown that they are highly dependable as a line of evidence if DNA samples are collected and prepared with care, and if a sufficient number of loci which have variable numbers of DNA repeat sequences are examined. There is always concern, though, about the possibility of contamination of the sample with DNA from another source during the path from the crime scene to the forensic laboratory for analysis. Moreover, in some cases criminals themselves have planted fake DNA samples at crime scenes to confuse the investigation.

DNA fingerprinting has been used extensively to identify a criminal or rule out a suspect. For example, in a case in England, the DNA fingerprints of more than 5,000 men were made using blood and saliva samples during an investigation of the rape and murder of two teenage girls in 1983 and 1986. The results led to the release of a man wrongly imprisoned for the crimes—the first person exonerated by DNA fingerprinting evidence. But no DNA fingerprints matched the DNA at the

A. Alleles at an STR locus

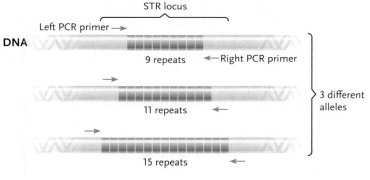

B. DNA fingerprint analysis of the STR locus by PCR

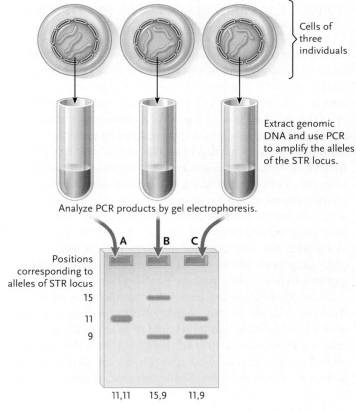

FIGURE 18.10 Using PCR to obtain a DNA fingerprint for an STR locus. (A) Three alleles of the STR locus with 9, 11, and 15 copies of the tandemly repeated sequence. The arrows indicate where left and right PCR primers bind to amplify the STR locus. (B) DNA fingerprint analysis of the STR locus by PCR. The number of bands on the gel and the sizes of the DNA in the bands show the STR alleles that were amplified. One band indicates that the individual was homozygous for an STR allele with a particular number of repeats, while two bands indicates the individual is heterozygous for two STR alleles with different numbers of repeats. Here, individual A is homozygous for an 11-repeat allele (designated 11,11), B is heterozygous for a 15-repeat allele and a 9-repeat allele (15,9), and C is heterozygous for the 11-repeat allele and the 9-repeat allele (11,9).

© Cengage Learning 2014

crime scene. Then a woman overheard a colleague at work saying that he had given his samples in the name of his friend, Colin Pitchfork. Pitchfork was arrested, and his DNA fingerprint subsequently was shown to match the DNA at the crime scene. In 1988 he was convicted of the murders, the first conviction on the basis of DNA evidence. The application of DNA fingerprinting techniques to stored forensic samples has also led to the release of a number of persons wrongly convicted for rape or murder.

DNA FINGERPRINTING IN TESTING PATERNITY AND ESTABLISHING ANCESTRY DNA fingerprints are also widely used as evidence of paternity because parents and their children share common alleles in their DNA fingerprints. That is, each child receives one allele of each locus from one parent and the other allele from the other parent. A comparison of DNA fingerprints for a number of loci can prove almost infallibly whether a child has been fathered or mothered by a given person. DNA fingerprints have also been used for other investigations, such as confirming that remains discovered in a remote region of Russia were actually those of Czar Nicholas II and members of his family, murdered in 1918 during the Russian revolution.

DNA fingerprinting is also used widely in studies of other organisms, including other animals, plants, and bacteria. Examples include testing for pathogenic *E. coli* in food sources such as hamburger meat, investigating cases of wildlife poaching, detecting genetically modified organisms among living organisms or in food, and comparing the DNA of ancient organisms with present-day descendants.

Genetic Engineering Uses DNA Technologies to Alter the Genes of a Cell or Organism

We have seen the many ways scientists use DNA technologies to ask, and answer, questions that were once completely inaccessible. Genetic engineering goes beyond gathering information; it is the use of DNA technologies to modify genes of a cell or organism. The goals of genetic engineering include using prokaryotes, fungi, animals, and plants as factories for the production of proteins needed in medicine and scientific research; correcting hereditary disorders; and improving animals and crop plants of agricultural importance. In many of these areas, genetic engineering has already been highly successful. The successes and potential benefits of genetic engineering, however, are tempered by ethical and social concerns about its use, along with the fear that the methods may produce toxic or damaging foods, or release dangerous and uncontrollable organisms to the environment.

Genetic engineering uses DNA technologies of the kind discussed already in this chapter. DNA—perhaps a modified gene—is introduced into target cells of an organism where that gene is then expressed. In effect, the genotype of the organism is altered by the addition of the gene. Organisms that have undergone a gene transfer are called **transgenic,** meaning that they have been modified to contain genetic information—the **transgene**—from an external source.

The following sections discuss some applications of the genetic engineering of bacteria, animals, and plants.

GENETIC ENGINEERING OF BACTERIA TO PRODUCE PROTEINS
Transgenic bacteria have been made, for example, to synthesize proteins for medical applications, break down toxic wastes such as oil spills, produce industrial chemicals such as alcohols, and process minerals. *E. coli* is the organism of choice for many of these applications of DNA technologies.

Figure 18.11 shows how *E. coli* can be engineered to make a protein from a foreign source. First, the gene for the protein is cloned from the appropriate organism. Then the gene is inserted into an **expression vector,** which is a cloning vector that has extra features, namely the regulatory sequences that allow transcription and translation of a gene. For a bacterial expression vector, this means having a promoter and a transcription terminator that are recognized by the *E. coli* transcriptional machinery, and having the ribosome binding site needed for the bacterial ribosome to recognize the start codon of the transgene (see Section 15.4). The regulatory sequences flank a cluster of restriction sites that are used for cloning so that the inserted gene is placed correctly for transcription and translation when the recombinant plasmid is transformed into *E. coli.*

As was mentioned earlier while discussing DNA libraries, a cDNA copy of a eukaryotic gene is used when we want to use bacteria to express the protein encoded by the gene (see Figure 18.11). This is because a eukaryotic protein-coding gene typically contains introns, which bacteria cannot remove when they transcribe such a gene. However, the eukaryotic mRNA that is copied by reverse transcriptase to synthesize cDNA has no introns; thus, when that cDNA is expressed in bacteria, it can be transcribed and translated to make the encoded eukaryotic protein. The protein is either extracted from the bacterial cells and purified or, if the protein is secreted, it is purified from the culture medium.

Expression vectors are available for a number of organisms. They vary in the regulatory sequences they contain and the selectable marker they carry so that the host organism transformed with the vector carrying a gene of interest can be detected and the host can express that gene.

For example, genetic engineering has been used to produce *E. coli* bacteria that make the human hormone insulin; the commercial product is called Humulin. Insulin is required by persons with some forms of diabetes. Humulin is a perfect copy of the human insulin hormone. Many other proteins, including human growth hormone to treat human growth disorders, tissue plasminogen activator to dissolve blood clots that cause heart attacks, and a vaccine against foot-and-mouth disease of cattle and other cloven-hoofed animals (a highly contagious and sometimes fatal viral disease), have been developed for commercial production in bacteria using similar methods.

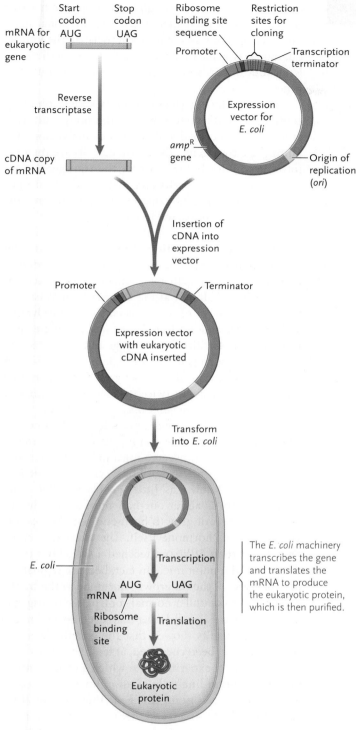

FIGURE 18.11 Using an expression vector to synthesize a eukaryotic protein in *E. coli*.
© Cengage Learning 2014

The *E. coli* machinery transcribes the gene and translates the mRNA to produce the eukaryotic protein, which is then purified.

A concern is that genetically engineered bacteria may be released accidentally into the environment where possible adverse effects of the organisms are currently unknown. Scientists minimize the danger of accidental release by growing the bacteria in laboratories that follow appropriate biosafety protocols. In addition, the bacterial strains typically used are genetic mutants that cannot survive outside the growth media used in the laboratory.

GENETIC ENGINEERING OF ANIMALS Many animals, including fruit flies, fish, mice, pigs, sheep, goats, and cows, have been altered successfully by genetic engineering. Typically the organism has been altered to express a transgene, which means using an expression vector with the appropriate molecular features for gene expression in that organism. There are many purposes for these alterations, including basic research, correcting genetic disorders in humans and other mammals, and producing pharmaceutically important proteins.

Genetic Engineering Methods for Animals. Several methods are used to introduce a gene of interest into animal cells. The gene may be introduced into **germ-line cells,** which develop into sperm or eggs and thus enable the introduced gene to be passed from generation to generation. Or the gene may be introduced into **somatic** ("body") **cells,** differentiated cells that are not part of lines producing sperm or eggs, in which case the gene is not transmitted from generation to generation.

Germ-line cells of embryos are often used as targets for introducing genes, particularly in mammals other than humans **(Figure 18.12).** The treated cells are then cultured in quantity and reintroduced into early embryos. If the technique is successful, some of the introduced cells become founders of cell lines that develop into eggs or sperm with the desired genetic information integrated into their DNA. Individuals produced by crosses using the engineered eggs and sperm then contain the introduced sequences in all of their cells. Several genes have been introduced into the germ lines of mice by this approach, resulting in permanent, heritable changes in the engineered individuals.

A related technique involves introducing desired genes into **stem cells,** which are cells capable of undergoing many divisions in an unspecialized, undifferentiated state, but which also have the ability to differentiate into specialized cell types. In mammals, *embryonic stem cells* are found in a mass of cells inside an early-stage embryo (the blastocyst) (see Figure 18.12, step 3, and Figure 50.12) and can differentiate into all of the tissue types of the embryo, whereas *adult stem cells* function to replace specialized cells in various tissues and organs. In mice and other non-human mammals, transgenes are introduced into embryonic stem cells, which are then injected into early-stage embryos as in Figure 18.12 (step 3). The stem cells then differentiate into a variety of tissues along with cells of the embryo itself, including sperm and egg cells. Males and females are then bred, leading to offspring that are either homozygotes, containing two copies of the introduced gene, or heterozygotes, containing one introduced gene and one gene that was native to the embryo receiving the engineered stem cells.

Introduction of genes into stem cells has been done mostly in mice. One of the highly useful results is the production of a

FIGURE 18.12

Research Method

Introduction of Genes into Mouse Embryos Using Embryonic Germ-Line Cells

Purpose: To make a transgenic animal that can transmit the transgene to offspring. The embryonic germ-line cells that receive the transgene develop into the reproductive cells of the animal.

Protocol:

Germ-line cells derived from mouse embryo

1. Introduce desired gene into germ-line cells from an embryo by injection or electroporation.

Transgene in expression vector

Cell with transgene

2. Clone cell that has the incorporated transgene to produce a pure culture of transgenic cells.

Pure population of transgenic cells

3. Inject transgenic cells into early-stage embryos (called blastocysts).

4. Implant embryos into surrogate (foster) mothers.

5. Allow embryos to grow to maturity and be born.

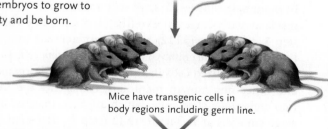

Mice have transgenic cells in body regions including germ line.

6. Interbreed the progeny mice.

Genetically engineered offspring—all cells transgenic

Outcome: The result of the breeding is some offspring in which all cells are transgenic—a genetically engineered animal has been produced.

© Cengage Learning 2014

knockout mouse, a homozygous recessive that has two copies of a gene each altered to a nonfunctional state. The effect of the missing gene's activity on the knockout mouse is a clue to the normal function of the gene. Knockout mice also are used to model some human genetic diseases.

For introducing genes into somatic cells, typically somatic cells are removed from the body, cultured, and then transformed with an expression vector containing the transgene. The modified cells are then reintroduced into the body where the transgene functions. Because germ cells and their products are not involved, the transgene remains in the engineered individual and is not passed to offspring.

Gene Therapy: Correcting Genetic Disorders. The path to **gene therapy**—correcting genetic disorders—in humans began with experiments using mice. In 1982, Richard Palmiter at the University of Washington, Ralph Brinster of the University of Pennsylvania, and their colleagues injected a growth hormone gene from rats into fertilized mouse eggs and implanted the eggs into a surrogate mother. Some normal-sized mouse pups were produced that grew faster than normal and became about twice the size of their normal litter mates: they were *giant mice* **(Figure 18.13).**

Palmiter and Brinster next attempted to cure a genetic disorder by gene therapy. In this experiment, they constructed transgenic mice in which a normal copy of the rat growth hormone gene was added to mutant mice known as little, genotype *lit/lit.* This mouse mutant is about one-half normal size due to decreased levels of growth hormone gene expression and of growth hormone itself. The transgenic mice grew faster than *lit/lit* and reached a size about three times that of mutant mice. In fact, because regulation of expression of the rat growth hormone in the transgenic mice is not under normal control, the transgenic mice grow to be slightly larger than even normal mice. Overall the results showed that the genetic defect in those mice had been corrected, at least partially.

This sort of experiment, in which a gene is introduced into germ-line cells of an animal to correct a genetic disorder, is **germ-line gene therapy.** For ethical reasons, germ-line gene therapy is not permitted with humans. Instead, humans are treated with **somatic gene therapy,** in which genes are introduced into somatic cells (as described in the previous section).

The first successful use of somatic gene therapy with a human subject who had a genetic disorder was carried out in the 1990s by W. French Anderson and his colleagues at the National Institutes of Health (NIH). The subject was a young girl with *adenosine deaminase deficiency (ADA).* Without the adenosine

FIGURE 18.13 A genetically engineered giant mouse (right) produced by the introduction of a rat growth hormone gene into the animal. A mouse of normal size is on the left.

R. Brinster, R. E. Hammer, School of Veterinary Medicine, University of Pennsylvania

deaminase enzyme, white blood cells cannot mature (see Chapter 45). In the absence of normally functioning white blood cells, the body's immune response is so deficient that most children with ADA die of infections before reaching puberty. The researchers were successful in introducing a functional ADA gene into mature white blood cells isolated from the patient. Those cells were reintroduced into the girl, and expression of the ADA gene provided a temporary cure for her ADA deficiency. The cure was not permanent because mature white blood cells, produced by differentiation of stem cells in the bone marrow, are nondividing cells with a finite life time. Therefore, the somatic gene therapy procedure has to be repeated every few months. The subject of this example still receives periodic gene therapy to maintain the necessary levels of the ADA enzyme in her blood. In addition, she receives direct doses of the normal enzyme.

Despite enormous efforts, human somatic gene therapy has not been the panacea people expected. Relatively little progress has been made since the first gene therapy clinical trial for ADA deficiency, and, in fact, there have been major setbacks. In 1999, for example, a teenage patient in a somatic gene therapy trial died as a result of a severe immune response to the viral vector used to introduce a normal gene to correct his genetic deficiency. Furthermore, some children in gene therapy trials using retrovirus vectors to introduce genes into blood stem cells have developed a leukemia-like condition. Research and clinical trials continue as scientists try to circumvent the difficulties.

Turning Domestic Animals into Protein Factories. Another successful application of genetic engineering turns animals into pharmaceutical factories for the production of proteins required to treat human diseases or other medical conditions. Most of these *pharming* projects, as they are called, engineer the animals to produce the desired proteins in milk, making the isolation and purification of the proteins easy, as well as harmless to the animals. ("Pharming" is a word made up from "farming" and "pharmaceuticals.")

One of the first successful applications of this approach was carried out with transgenic sheep that produce a protein required for normal blood clotting in humans. The protein, called a *clotting factor,* is deficient in persons with one form of hemophilia. (The genetics of hemophilia is described in Section 13.2.) These people require frequent injections of the factor to avoid bleeding to death from even minor injuries. Using DNA cloning techniques, researchers joined the gene encoding the normal form of the clotting factor to the promoter sequences of the β-lactoglobulin gene, which encodes a protein secreted in milk, and introduced it into fertilized eggs. Those cells were implanted into a surrogate mother, and the transgenic sheep born were allowed to mature. The β-lactoglobulin promoter controlling the clotting factor gene became activated in mammary gland cells of females. The clotting factor produced was secreted into the milk from which it could be collected and purified.

Other similar projects are being done to produce particular proteins in transgenic mammals. These include a protein to treat cystic fibrosis, collagen to correct scars and wrinkles, human milk proteins to be added to infant formulas, and normal hemoglobin for use as an additive to blood transfusions.

Producing Animal Clones. Making transgenic mammals is expensive and inefficient. And, because only one copy of the transgene typically becomes incorporated into the treated cell, not all progeny of a transgenic animal inherit that gene. Scientists reasoned that an alternative to breeding a valuable transgenic mammal to produce progeny with the transgene would be to clone the mammal. Each clone would be identical to the original, including the expression of the transgene. That this is possible was shown when two Scottish scientists, Ian Wilmut of the Roslin Institute and Keith Campbell of PPL Therapeutics, Roslin, Scotland, announced in 1997 that they were successful in cloning a sheep using a single somatic cell taken from an adult sheep **(Figure 18.14)**—Dolly, the first cloned mammal.

After the successful cloning experiment that produced Dolly, many other mammals have been cloned, including mice, goats, pigs, monkeys, rabbits, dogs, a male calf appropriately named Gene, and a domestic calico cat called *CC* (for *Copy Cat*).

Cloning farm animals has been so successful that several commercial enterprises now provide cloned copies of champion animals. One example is a clone of an American Holstein cow, Zita, who was the U.S. national champion milk producer for many years.

The cloning of domestic animals has its drawbacks. Many cloning attempts fail, leading to the death of the transplanted embryos. Cloned animals often suffer from conditions such as birth defects and poor lung development. Genes may be lost during the cloning process or may be expressed abnormally in the cloned animal. For example, molecular studies have shown

FIGURE 18.14 | **Experimental Research**

The First Cloning of a Mammal

Question: Does the nucleus of an adult mammal contain all the genetic information to specify a new organism? In other words, can mammals be cloned starting with adult cells?

Experiment: Ian Wilmut, Keith Campbell, and their colleagues fused a mammary gland cell from an adult sheep with an unfertilized egg cell from which the nucleus had been removed and tested whether that fused cell could produce a lamb.

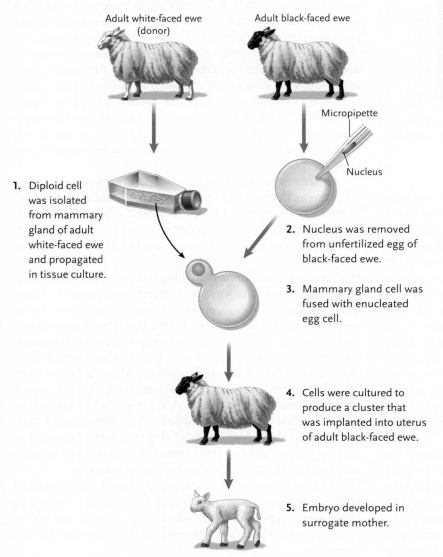

Adult white-faced ewe (donor)

Adult black-faced ewe

Micropipette

Nucleus

1. Diploid cell was isolated from mammary gland of adult white-faced ewe and propagated in tissue culture.

2. Nucleus was removed from unfertilized egg of black-faced ewe.

3. Mammary gland cell was fused with enucleated egg cell.

4. Cells were cultured to produce a cluster that was implanted into uterus of adult black-faced ewe.

5. Embryo developed in surrogate mother.

Result: Dolly was born and grew normally. She was white-faced—a clone of the donor ewe. DNA fingerprinting using STR loci showed her DNA matched that of the donor ewe and neither the ewe who donated the egg, nor the ewe who was the surrogate mother.

Conclusion: An adult nucleus of a mammal contains all the genetic material necessary to direct the development of a normal new organism, a clone of the original. Dolly was the first cloned mammal. The success rate for Wilmut and Campbell's experiment was very low—Dolly represented less than 0.4% of the fused cells they made—but its significance was huge.

THINK LIKE A SCIENTIST Why was it important that the donor ewe and surrogate mother be sheep of different kinds?

Source: I. Wilmut et al. 1997. Viable offspring derived from fetal and adult mammalian cells. *Nature* 385:810–813.

© Cengage Learning 2014

that the expression of perhaps hundreds of genes in the genomes of clones is regulated abnormally.

GENETIC ENGINEERING OF PLANTS Genetic engineering of plants has led to increased resistance to pests and disease; greater tolerance to heat, drought, and salinity; larger crop yields; faster growth; and resistance to herbicides. Another aim is to produce seeds with higher levels of amino acids. The essential amino acid lysine, for example, is present only in limited quantities in cereal grains such as wheat, rice, oats, barley, and corn; the seeds of legumes such as beans, peas, lentils, soybeans, and peanuts are deficient in the essential amino acids methionine or cysteine. Increasing the amounts of the deficient amino acids in plant seeds by genetic engineering would greatly improve the diet of domestic animals and human populations that rely on seeds as a primary food source. Efforts are also under way to increase the vitamin and mineral content of crop plants.

Other possibilities for plant genetic engineering include plant pharming to produce pharmaceutical products. Plants are ideal for this purpose, because they are primary producers at the bottom rung of the food chain and can be grown in huge numbers with maximum conservation of the sun's energy that is captured in photosynthesis.

Some plants, such as *Arabidopsis thaliana* (thale cress), tobacco, potato, cabbage, and carrot, have special advantages for genetic engineering because individual cells can be removed from an adult, altered by the introduction of a desired gene, and then grown in cultures into a multicellular mass of cloned cells called a *callus*. Subsequently, roots, stems, and leaves develop in the callus, forming a young plant that can then be cultivated by standard methods. Each cell in the plant contains the introduced gene. The gametes produced by the transgenic plants can then be used in crosses to produce offspring, some of which will have the transgene, as in the similar animal experiments.

Methods Used to Insert Genes into Plants.
Genes are inserted into plant cells by several techniques. A commonly used method takes advantage of properties of a bacterium,

FIGURE 18.15 | **Research Method**

Using the Ti Plasmid of *Agrobacterium tumefaciens* to Produce Transgenic Plants

Purpose: To make transgenic plants. This technique is one way to introduce a transgene into a plant for genetic engineering purposes.

Protocol:

1. Isolate the Ti plasmid from *Agrobacterium tumefaciens*. The plasmid contains a segment called T DNA (T = transforming), which induces tumors in plants.

2. Digest the Ti plasmid with a restriction enzyme that cuts within the T DNA. Mix with a gene of interest on a DNA fragment that was produced by digesting with the same enzyme. Use DNA ligase to join the two DNA molecules together to produce a recombinant plasmid.

3. Transform the recombinant Ti plasmid into a disarmed *A. tumefaciens* that cannot induce tumors, and use the transformed bacterium to infect cells in plant fragments in a test tube. In infected cells, the T DNA with the inserted gene of interest excises from the Ti plasmid and integrates into the plant cell genome.

4. Culture the transgenic plant fragments to regenerate whole plants.

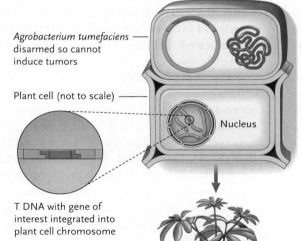

Outcome: The plant has been genetically engineered to contain a new gene. The transgenic plant will express a new trait based on that gene, perhaps resistance to an herbicide or production of an insect toxin, according to the goal of the experiment.

© Cengage Learning 2014

Agrobacterium tumefaciens, which causes crown gall disease. Crown gall disease is characterized by bulbous, irregular growths—tumors, essentially—that develop at wound sites on the trunks and limbs of deciduous trees. *A. tumefaciens* contains a large, circular plasmid called the **Ti (tumor-inducing) plasmid (Figure 18.15).** The interaction between the bacterium and a plant cell it infects stimulates the excision of a segment of the Ti plasmid called *T DNA* (for transforming DNA), which then integrates into the plant cell's genome. Genes on the T DNA are then expressed; the products stimulate the transformed cell to grow and divide, producing a tumor. The tumors provide essential nutrients for the bacterium. The Ti plasmid is used as a vector for making transgenic plants, in much the same way as cloning vectors introduce genes into bacteria.

Successful Plant Genetic Engineering Projects. The most widespread application of genetic engineering of plants involves the production of transgenic crops. Thousands of such crops have been developed and field tested, and many have been approved for commercial use. If you examine the processed, plant-based foods sold at a national supermarket chain, you will likely find that at least two-thirds contain transgenic plants.

In many cases, plants are modified to make them resistant to insect pests, viruses, or herbicides. Crops that have been modified for insect resistance include corn, cotton, and potatoes. The most common approach to making plants resistant to insects is to introduce the gene from the bacterium *Bacillus thuringiensis* that encodes the *Bt* toxin, a natural pesticide. This toxin has been used in powder form to kill insects in agriculture for many years, and now transgenic plants making their own *Bt* toxin are resistant to specific groups of insects that feed on them. Millions of acres of crop plants planted in the United States, amounting to about 70% of the nation's agricultural acreage, are now *Bt*-engineered varieties.

Virus infections cause enormous crop losses worldwide. Virus-resistant transgenic crops would be highly valuable in agriculture. There is some promise in this area. By some unknown process, transgenic plants expressing certain viral proteins become resistant to infections by whole viruses that contain those same proteins. Two virus-resistant, genetically modified crops made so far are papaya and squash.

Several crops have also been engineered to become resistant to herbicides. For example, *glyphosate* (commonly known by its brand name,

Roundup) is a potent herbicide that is widely used in weed control. The herbicide works by inhibiting a particular enzyme in the chloroplast. Unfortunately, it also kills crops. But transgenic crops have been made in which a bacterial form of the chloroplast enzyme was added to the plants. The bacteria-derived enzyme is not affected by Roundup, and farmers who use these herbicide-resistant crops can spray fields of crops to kill weeds without killing the crops. Now most of the corn, soybean, and cotton plants grown in the United States and many other countries are glyphosate-resistant ("Roundup-ready") varieties that were produced by genetic engineering.

Crop plants are also being engineered to alter their nutritional qualities. For example, a strain of rice plants has been produced with seeds rich in β-carotene, a precursor of vitamin A **(Figure 18.16)**. The new rice, which is given a yellow or golden color by its carotene content, may provide improved nutrition for the billions of people that depend on rice as a diet staple. In particular, the rice may help improve the nutrition of children younger than five years of age in southeast Asia, 70% of whom suffer from impaired vision because of vitamin A deficiency. *Insights from the Molecular Revolution* describes an experiment in which rice plants were genetically engineered to develop resistance to a damaging bacterial blight.

Plant pharming is also an active area in both university research labs and biotechnology companies. Plant pharming involves the engineering of transgenic plants to produce medically valuable products. The approach is one described earlier: the gene for the product is cloned into an expression vector, in this case one with a promoter active in plants, and the recombinant DNA molecule is introduced into plants. Products under development include vaccines for various bacterial and viral diseases, protease inhibitors to treat or prevent virus infections, collagen to treat scars and wrinkles, and the protein aprotinin, which is used to reduce bleeding and clotting during heart surgery.

In contrast to modifying animals by genetic engineering techniques, genetically altered plants have been widely developed and appear to have become mainstays of agriculture. But, as the next section discusses, both animal and plant genetic engineering have not proceeded without concerns.

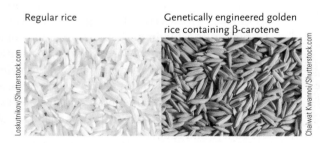

Regular rice

Genetically engineered golden rice containing β-carotene

FIGURE 18.16 Rice genetically engineered to contain β-carotene.

DNA Technologies and Genetic Engineering Are a Subject of Public Concern

When recombinant DNA technology was developed in the early 1970s, researchers were concerned that a bacterium carrying a recombinant DNA molecule might escape into the environment, transfer the recombinant molecule to other bacteria, and produce new, potentially harmful, strains. To address this, and other concerns, the U.S. scientists who developed the technology drew up comprehensive safety guidelines for recombinant DNA research in the United States. Since that time, countless thousands of experiments involving recombinant DNA molecules have shown that recombinant DNA manipulations can be done safely. Over time, therefore, the recombinant DNA guidelines have become more relaxed. Nonetheless, stringent regulations still exist for certain areas of recombinant DNA research that pose significant risk, such as cloning genes from highly pathogenic bacteria or viruses, or gene therapy experiments.

Guidelines for genetic engineering also extend to research in several areas that have been the subject of public concern and debate. While the public is concerned little about genetically engineered microorganisms, for example those cleaning up oil spills and hazardous chemicals, it is concerned about possible problems with **genetically modified organisms (GMOs)** used as food. A GMO is a transgenic organism; the majority of GMOs are crop plants. Issues of concern include the safety of GMO-containing food, and the possible adverse effects of the GMOs on the environment, such as by interbreeding with natural species or by harming beneficial insect species. One example of the latter concern was whether *Bt*-expressing corn would have adverse effects on monarch butterflies that fed on the pollen of the plants. The most recent of a series of independent studies investigating this possibility has indicated that the risk to the butterflies from *Bt* toxin is extremely low.

More broadly, different countries have reacted to GMOs in different ways. In the United States, transgenic crops are planted widely. Before commercialization, such GMOs are evaluated for potential risk by appropriate government regulatory agencies, including the NIH, Food and Drug Administration (FDA), Department of Agriculture, and Environmental Protection Agency (EPA). Usually, the opposition to GMOs has come from particular activist and consumer groups.

Political opposition to GMOs has been greater in Europe, dampening the use of transgenic crop plants in the fields and GMOs in food. In 1999, the European Union (EU) imposed a 6-year moratorium on all GMOs. More recently, the EU has decided that using genetic engineering in agriculture and food production is permissible provided the GMO or food containing it is safe for humans, animals, and the environment.

Globally, an international agreement, the **Cartagena Protocol on Biosafety,** "promotes biosafety by establishing

Rice Blight: Engineering rice for resistance to the disease

Rice is common in the diet of the entire human population; for one-third of humanity, more than 2 billion people, it is the primary nutrient source. Worldwide, 560 million tons of rice are produced annually.

This major human staple is threatened by a rice blight caused by the bacterium *Xanthomonas oryzae*. Rice plants infected by the bacterium turn yellow and wilt **(Figure)**; in many Asian and African rice fields, as much as half of the crop is lost to the bacterial blight. Some wild forms of rice, which are not usable as crop plants, have a natural resistance to the *Xanthomonas* blight, but no cultivated variety is resistant.

Research Goal

To develop a blight-resistant rice strain using a genetic engineering approach.

Experiments

Pamela Ronald and her colleagues at the University of California, Davis used genetic engineering techniques to produce a blight-resistant rice strain. They were aided by the results of genetic experiments in which a gene, *Xa21,* was identified that confers resistance to *Xanthomonas* in a wild rice.

Ronald and her coworkers cloned the *Xa21* gene following these steps:

1. Genetic crosses showed that *Xa21* is located near several known genes, including one that was very close.

2. The DNA of the known gene was used as a probe to find nearby sequences in a genomic library of the resistant wild rice genome. The probe identified 16 DNA chromosome fragments in the library.

3. To determine which fragments included the *Xa21* gene, the researchers introduced each fragment individually into rice crop plants that were susceptible to *Xanthomonas* infection. Out of 1,500 plants that incorporated the fragments, 50 plants, all containing the same 9,600-bp fragment, became resistant to the blight. These plants were the first rice crop plants to be engineered successfully to resist *Xanthomonas* infections.

4. The researchers analyzed the 9,600-base-pair fragment further to identify and isolate the resistance gene from the fragment.

Conclusion

The sequence of the *Xa21* gene revealed that it encodes a plasma membrane receptor protein which in some way triggers an internal cellular response on exposure to *Xanthomonas.* The response alters cell structure or biochemistry to inhibit growth of the bacterium.

Ronald and her coworkers then introduced the *Xa21* gene successfully into three varieties of rice that are widely grown as crops in Asia and Africa. Those varieties became resistant to the bacterial blight, thereby showing that expression of the *Xa21* gene is sufficient to confer resistance.

THINK LIKE A SCIENTIST The *Xa21* gene encodes a plasma membrane receptor. How could that knowledge inform future research on controlling blight?

Source: W.-Y. Song et al. 1995. A receptor kinase-like protein encoded by the rice disease resistance gene, *Xa21. Science* 270:1804–1806.

FIGURE Normal rice plants (left) and rice plants infected by the blight bacterium (right).

practical rules and procedures for the safe transfer [between countries], handling and use of GMOs." Separate procedures have been set up for GMOs that are to be introduced into the environment and those that are to be used as food or feed or for processing. By July 2012, 163 countries had ratified the Protocol; the United States was not one of them.

In sum, the use of DNA technologies in biotechnology has the potential for tremendous benefits to humankind. Such experimentation is not without risk, and so for each experiment, researchers must assess that risk and make a judgment about whether to proceed and, if so, how to do so safely. Furthermore, agreed-upon guidelines and protocols should ensure a level of biosafety that is acceptable to researchers, consumers, politicians, and governments.

In this chapter you have learned about how individual genes can be isolated and manipulated using various DNA technologies. But a gene is just a part of a genome. Researchers also want to know about the set of genes in a complete genome, and how genes and their gene products work together in networks to control life. They also want to know more generally about the organization of the genome with respect to both genes and nongene sequences. Genomes and proteomes (the complete sets of proteins expressed by a genomes) are the subjects of the next chapter.

UNANSWERED QUESTIONS

What are scientists aiming for with "targeted gene therapy"?

The molecular basis of inherited disease resides in your genes. Thus, it is not surprising that gene therapy has evoked such excitement in the field of molecular medicine. Gene therapy is defined as the use of nucleic acid sequences (DNA or RNA) to treat inherited or acquired disease. Gene therapy for monogenic diseases (diseases caused by a single defective gene) has begun to demonstrate great promise. For example, gene therapy for a form of congenital blindness (called Leber congenital amaurosis) has restored vision to blind children. This shot in the arm for the field has renewed enthusiasm for gene therapy of other monogenic diseases such as cystic fibrosis (CF)—a devastating disease caused by a single gene defect in a chloride channel called the cystic fibrosis transmembrane conductance regulator (CFTR) that leads to chronic bacterial infections in the lung (see Chapter 6 *Why It Matters*, and Section 13.4). How can scientists approach gene therapy of CF? What cells in the lung must scientists target to treat this disease? Can gene therapy cure CF? These and other questions about gene therapy for CF are common to a myriad of other genetic diseases and patients' lives are depending on rapid answers. Our work at the University of Iowa focuses on obtaining answers to these questions needed to treat CF using gene therapy.

What are the cellular targets for gene therapy?

The cellular targets for gene therapy of any genetic disease are informed by a basic understanding of how the defective gene product (or protein) produces disease at the cellular and organ level. Using CF as an example, we know the defective gene encodes a chloride channel (CFTR; see Figure 6.1) that is expressed in a variety of cell types in the lung and airways. However, researchers still do not know which of these cell types must be targeted by gene therapy to treat the disease effectively. Defective innate immunity in CF may involve cells lining the surface of the airways and/or mucous-producing cells underneath the airways. Do both of these cellular compartments need to be targeted by gene therapy to reverse CF lung disease? The answer to this question resides in research that bridges gene therapy and the mechanisms of disease progression.

Tissues within the body are made from a complex assortment of unique cell types. These cells perform unique differentiated functions and organize into distinct groups of specialized cell types that act together to perform unique functions in a given organ. Typically, each group has unique microenvironments that house stem cells (called stem cell niches), which replace differentiated cell types within the trophic unit following normal cell turnover or injury. Often, these stem cells lie dormant until a signal from the microenvironment tells them to divide. In the lung, four unique stem cell niches have been identified, each being potentially important targets for gene therapy of CF and other lung diseases. To achieve lasting disease correction, scientists must target the adult stem cells that replace terminally differentiated cell types throughout the patient's lifetime. Therapeutic genes that incorporate within the genome of tissue-specific stem cells will potentially persist indefinitely.

How do scientists deliver genes to treat disease?

Conceptually, gene therapy consists of two strategies: 1) *gene addition*, which incorporates an additional gene into target cells; and 2) *gene correction*, which corrects the gene mutation that exists in the target cells. Although more challenging, the second approach is considered the "Holy Grail" of gene therapy because it corrects the cause of disease—the DNA mutation itself. Viral gene delivery is generally the most efficient means of delivering genetic payloads to cells and thus is the approach used in most current clinical gene therapy trials. In this method, scientists remove most (or all) viral genes from the virus and replace them with the therapeutic gene of interest. Although viruses are generally efficient at gene addition, they can cause the host to mount an immune response. Thus, important areas of research are directed at generating recombinant viruses that can escape the immune response and/or improve efficiency of gene delivery through an increased understanding of viral infection. Certain viruses also have the capability of correcting a gene mutation through homologous recombination, while others insert randomly into the genome, both mechanisms allow therapeutic DNA to persist if stem cells are targeted.

THINK LIKE A SCIENTIST Future gene therapy research must bridge the fields of stem cell biology, disease pathophysiology, virology, immunology, and homologous recombination. Tissue-specific stem cells are important targets of gene therapy, but they are often found in protective microenvironments with unique biologic properties.

Why do you think biological systems have evolved to sequester tissue-specific stem cells into constrained locations?

John F. Engelhardt is the Director of the Center for Gene Therapy, Professor and Head of the Department of Anatomy and Cell Biology at the University of Iowa Roy J. and Lucille A. Carver College of Medicine. Research in the Engelhardt laboratory focuses on the molecular basis of inherited and environmentally induced diseases, and on the development of gene therapies for these disorders.

Courtesy of John Engelhardt

Tom Lynch is a graduate student at the University of Iowa in Dr. Engelhardt's Laboratory. His graduate research focuses on the identification and isolation of tissue-specific stem cells in the airway epithelium. His research is particularly focused on signals that control stem cell dormancy and proliferation in the airway. To learn more about their research, go to: http://elab.genetics.uiowa.edu.

Courtesy of Thomas Lynch

REVIEW KEY CONCEPTS

To access the course materials and companion resources for this text, please visit www.cengagebrain.com.

18.1 DNA Cloning

- Producing multiple copies of genes by cloning is a common first step for studying the structure and function of genes, or for manipulating genes. Cloning involves cutting genomic DNA and a cloning vector with the same restriction enzyme, joining the fragments to produce recombinant plasmids, and introducing those plasmids into a living cell such as a bacterium, where replication of the plasmid takes place (Figures 18.1–18.3).

- A clone containing a gene of interest may be identified among a population of clones by using DNA hybridization with a labeled nucleic acid probe (Figure 18.5).

- A genomic library is a collection of clones that contains a copy of every DNA sequence in the genome. A cDNA (complementary DNA) library is the entire collection of cloned cDNAs made from the mRNAs isolated from a cell. A cDNA library contains only sequences from the genes that are active in the cell when the mRNAs are isolated.

- PCR amplifies a specific target sequence in DNA, such as a gene, which is defined by a pair of primers. PCR increases DNA quantities by successive cycles of denaturing the template DNA, annealing the primers, and extending the primers in a DNA synthesis reaction catalyzed by DNA polymerase. With each cycle of PCR, the amount of DNA doubles (Figure 18.6).

Animation: How to make cDNA

Animation: Base-pairing of DNA fragments

Animation: Polymerase chain reaction (PCR)

Animation: Use of a radioactive probe

Animation: Formation of recombinant DNA

Animation: Restriction enzymes

Animation: Automated DNA sequencing

18.2 Applications of DNA Technologies

- Recombinant DNA and PCR techniques are used in DNA molecular testing for human genetic disease mutations. One approach exploits restriction site differences between normal and mutant alleles of a gene that create restriction fragment length polymorphisms (RFLPs) which are detectable by DNA hybridization with a labeled nucleic acid probe (Figures 18.8 and 18.9).

- Human DNA fingerprints are produced from a number of loci in the genome characterized by short, tandemly repeated sequences that vary in number in all individuals (except identical twins). To produce a DNA fingerprint, the PCR is used to amplify the region of genomic DNA for each locus, and the lengths of the PCR products indicate the alleles an individual has for the repeated sequences at each locus. DNA fingerprints are widely used to establish paternity, ancestry, or criminal guilt (Figure 18.10).

- Genetic engineering is the introduction of new genes or genetic information to alter the genetic makeup of humans, other animals, plants, and microorganisms such as bacteria and yeast. Genetic engineering primarily aims to correct hereditary defects, improve domestic animals and crop plants, and provide proteins for medicine, research, and other applications (Figures 18.11–18.13 and 18.15).

- Genetic engineering has enormous potential for research and applications in medicine, agriculture, and industry. Potential risks include unintended damage to living organisms or to the environment.

Animation: How Dolly was created

Animation: DNA fingerprinting

Animation: Transferring genes into plants

UNDERSTAND & APPLY

Test Your Knowledge

1. A complementary DNA library (cDNA) and a genomic library are similar in that both:
   a. use bacteria to make eukaryotic proteins.
   b. provide information on whether genes are active.
   c. contain all of the DNA of an organism cut into pieces.
   d. clone mRNA.
   e. depend on cloning in a living cell to produce multiple copies of the DNA of interest.

2. Why do the cDNA libraries produced from two different cell types in the human body often contain different cDNAs?
   a. Because different expression vectors must be used to insert cDNAs into different cell types.
   b. Because different cell types contain different numbers of chromosomes.
   c. Because different cell types contain different genomic DNA sequences.
   d. Because different genes are transcribed in different cell types.
   e. Because different cell types contain different restriction enzymes.

3. The point at which a restriction enzyme cuts DNA is determined by
   a. the sequence of nucleotides.
   b. the length of the DNA molecule.
   c. whether it is closer to the 5′ end or 3′ end of the DNA molecule.
   d. the number of copies of the DNA molecule in a bacterial cell.
   e. the location of a start codon in a gene.

4. Restriction endonucleases, ligases, plasmids, *E. coli,* electrophoretic gels, and a bacterial gene resistant to an antibiotic are all required for:
   a. dideoxyribonucleotide analysis.
   b. PCR.
   c. DNA cloning.
   d. DNA fingerprinting.
   e. DNA sequencing.

5. After a polymerase chain reaction (PCR), agarose gel electrophoresis is often used to:
   a. amplify the DNA.
   b. convert cDNA into genomic DNA.
   c. convert cDNA into messenger RNA.
   d. verify that the desired DNA sequence has been amplified.
   e. synthesize primer DNA molecules.

6. Restriction fragment length polymorphisms (RFLPs):
   a. are produced by reaction with restriction endonucleases and are detected by Southern blot analysis.
   b. are of the same length for mutant and normal β-globin alleles.
   c. determine the sequence of bases in a DNA fragment.
   d. have in their middle short fragments of DNA that are palindromic.
   e. are used as vectors.

7. DNA fingerprinting, which is often used in forensics, paternity testing, and for establishing ancestry:
   a. compares one stretch of the same DNA between two or more people.
   b. measures different lengths of DNA from many repeating noncoding regions.
   c. requires the largest DNA lengths to run the greatest distance on a gel.
   d. requires amplification after the gels are run.
   e. can easily differentiate DNA between identical twins.

8. Which of the following is needed both in using bacteria to produce proteins and in genetic engineering of human cells?
   a. DNA fingerprinting based on microsatellite sequences
   b. insertion of a transgene into an expression vector
   c. restriction fragment length polymorphism (RFLP)
   d. screening of a cDNA library by DNA hybridization
   e. antibiotic resistance

9. Dolly, a sheep, was an example of reproductive (germ-line) cloning. Required to perform this process was:
   a. implantation of uterine cells from one strain into the mammary gland of another.
   b. fusion of the mammary cell from one strain with an enucleated egg of another strain.
   c. fusion of an egg from one strain with the egg of a different strain.
   d. fusion of an embryonic diploid cell with an adult haploid cell.
   e. fusion of two nucleated mammary cells from two different strains.

10. Which of the following is *not* true of somatic cell gene therapy?
    a. White blood cells can be used.
    b. Somatic cells are cultured, and the desired DNA is introduced into them.
    c. Cells with the introduced DNA are returned to the body.
    d. The technique is still very experimental.
    e. The inserted genes are passed to the offspring.

Discuss the Concepts

1. What should juries know to be able to interpret DNA evidence? Why might juries sometimes ignore DNA evidence?

2. A forensic scientist obtained a small DNA sample from a crime scene. In order to examine the sample, he increased its quantity by cycling the sample through the polymerase chain reaction. He estimated that there were 50,000 copies of the DNA in his original sample. Derive a simple formula and calculate the number of copies he will have after 15 cycles of the PCR.

Design an Experiment

Suppose a biotechnology company has developed a GMO, a transgenic plant that expresses *Bt* toxin. The company sells its seeds to a farmer under the condition that the farmer may plant the seed, but not collect seed from the plants that grow and use it to produce crops in the subsequent season. The seeds are expensive, and the farmer buys seeds from the company only once. How could the company show experimentally that the farmer has violated the agreement and is using seeds collected from the first crop to grow the next crop?

Interpret the Data

You learned in the chapter that an STR locus is a locus where alleles differ in the number of copies of a short, tandemly repeated DNA sequence. PCR is used to determine the number of alleles present, as shown by the size of the DNA fragment amplified. In the **Figure** to the right are the results of PCR analysis for STR alleles at a locus where the repeat unit length is 9 bp, and alleles are known that have 5 to 11 copies of the repeat. Given the STR alleles present in the adults, state whether each of the four juveniles could or could not be an offspring of those two adults. Explain your answers.

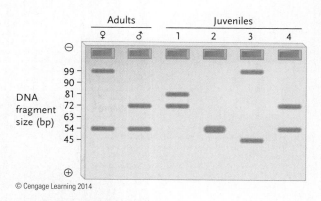

© Cengage Learning 2014

Apply Evolutionary Thinking

In PCR, researchers use a heat-stable form of DNA polymerase from microorganisms that are able to grow in extremely high temperatures. Given what you learned in Chapter 3 about protein folding, and in Chapter 4 about the effects of temperature on enzymes, would you predict that the amino acids of heat-stable DNA polymerase enzymes would have evolved so they can form stronger chemical attractions with each other, or weaker chemical attractions? Explain your answer.

19

Andre Nantel/Shutterstock.com

Results of DNA microarray analysis. DNA microarrays can be used, at a genomic level, to study which protein-coding genes are being expressed and the relative levels of expression of those genes.

Genomes and Proteomes

Why it matters . . . The 1000 Genomes Project (www.1000genomes.org), initiated in 2008, is an international collaborative research effort to find and catalog genetic variation in human genomes. As its name suggests, the Project intends to sequence the genomes of over 1,000 unidentified individuals from around the world. The sequencing data will be valuable in highly focused medical research, as well as in basic research. One interesting research study done as part of the Project analyzed the mutations each of us receives from our parents. As you learned in Chapters 12 and 13, mutations in the germline cells (sperm and egg) of a parent are inherited. Those mutations fall into two groups: mutations inherited by the parent, and new mutations that occur in the germline. In the study, researchers compared the genomes of two parent–offspring trios. They identified 49 and 35 mutations in the two offspring that must have originated as new mutations in their parents' germlines. Let that sink in: we have numerous mutations in our genomes that our parents do not have. Most interestingly, in one family, 92% of the new mutations were from the paternal germline, whereas in the other family 64% of the new mutations were from the maternal germline. These results mean that there is considerable variation in mutation rates within and between families. While this is a relatively limited study (as it only can be at the moment given the limited availability of complete genomes of families to study), nonetheless the results are provocative.

The 1000 Genomes Project research exemplifies the field of **genomics,** the characterization of whole genomes, including their structures (sequences), functions, and evolution. In this chapter, you will learn about genomics and some of the information that has come from genome analysis. You will also learn about **proteomics,** the study of the **proteome,** which is the complete set of proteins that can be produced by a genome. Proteomics involves characterizing the structures and functions of all expressed proteins of an organism, and the interactions among proteins in the cell. The chapter also discusses how comparisons of genome sequences have contributed to our understanding of genome evolution.

19.1 Genomics: An Overview

As defined in *Why It Matters,* genomics is the characterization of whole genomes, including their structures (sequences), functions, and evolution. Having the complete sequence of a genome makes it possible to study the complete set of genes in an organism or a virus, as well as other important sequences of their genomes. Understanding the structures and functions of all the genes in an organism or virus will give us a complete blueprint of the information for the development, structure and function of that organism's life or for that virus's infection cycle. That is an exciting prospect, but we are not there yet.

Before we go further, let us be sure we are clear about genes. As you learned in Chapter 15, some genes encode proteins. Protein-coding genes are the main focus of our discussions. Other so-called noncoding RNA genes are transcribed to produce RNAs that function without being translated into proteins. They include genes for tRNAs, rRNAs, snRNAs, and miRNAs.

Fundamental to genomics is determining the DNA sequences of organismal genomes and the DNA or RNA sequences of viral genomes. We focus here on the DNA genomes of organisms, though most of the techniques described in this chapter have also been applied to the study of viral genomes. Modern DNA sequencing techniques are advances on methods used to analyze the sequences of cloned DNA sequences and DNA sequences amplified by PCR (see Section 18.1). Having the complete sequence of a genome enables researchers to study the organization of genes in the genome as a whole, and to determine how genes function together in networks to control life.

Of natural interest to us is the human genome. The complete sequencing of the approximately 3 billion base-pair human genome—the Human Genome Project (HGP)—began in 1990. The task was completed in 2003 by an international consortium of researchers and by a private company, Celera Genomics (headed by J. Craig Venter). As part of the official HGP, the genomes of several important model organisms commonly used in genetic studies were sequenced for purposes of comparison: *E. coli* (representing prokaryotes); the yeast *Saccharomyces cerevisiae* (representing single-celled eukaryotes); *Drosophila melanogaster* and *Caenorhabditis elegans* (the fruit fly and a nematode worm, representing multicellular invertebrate animals); and *Mus musculus* (the mouse, representing nonhuman mammals). The sequences of the genomes of many organisms and viruses not part of the official Human Genome Project, including plants, have since been completed or are in progress.

A vast amount of DNA sequence data has been generated by genome sequencing projects. For those sequences to be useful, they need to be available centrally for access by all researchers. DNA sequences from genome sequencing projects are deposited into databases that are publicly available via the Internet. For example, GenBank® is an "annotated collection of all publicly available DNA sequences" at the National Institutes of Health (NIH). The Internet link is http://www.ncbi.nlm.nih.gov/genbank/. Currently there are over 150 billion bases in the DNA sequence records at GenBank®. Computational tools at NCBI GenBank® enable researchers, and others, such as students like yourself, to perform various analyses with the sequence data.

Many other genomics databases are accessible using the Internet, with sequence data organized in different ways (perform an Internet search for "DNA sequence database"). For example, there are organism-specific sequence databases as well as databases that include summaries of particular genomics studies. The databases are available for individual researchers to use and also for collaborative efforts involving researchers all over the world. One of the main benefits of collaborative research of this kind is that researchers with different specialities can tackle a particular question or questions at a genomic or multigenomic level.

Genomics consists of three main areas of study:

1. *Genome sequence determination and annotation,* which means obtaining the sequences of complete genomes, and analyzing them to locate putative protein-coding and noncoding RNA genes and other functionally important sequences in the genome.

2. *Determining the functions of genes* (a somewhat outdated term for this is *functional genomics*), which means using genome sequence data as a basis to study and understand the functions of genes and other parts of the genome. With respect to genes, this includes developing an understanding of how their expression is regulated. For protein-coding genes it also includes determining what proteins they encode, and how those proteins function in the organism's metabolic processes.

3. *Studying how genomes have evolved,* which means comparing genome sequence data to develop an understanding of how genes, particularly protein-coding genes, originated and genes and genomes changed over evolutionary time. Studies of genome sequences or large parts of genome sequences for a number of organisms represent an area of genomics known as **comparative genomics.**

Advances in each of these areas of study are accelerating as techniques are developed and improved for automating experimental procedures and more sophisticated computer algorithms for data analysis are generated. Methods that facilitate the handling of many samples simultaneously, whether those samples are DNA molecules for sequencing or genes for analysis, are called **high-throughput techniques.** The next three sections of the chapter discuss each of the three areas of genomics in turn.

STUDY BREAK 19.1

What additional biological questions can be answered if one has the complete sequence of an organism's genome as compared with the sequences of individual genes?

Whole-Genome Shotgun Sequencing

Purpose: Obtain the complete sequence of the genome of an organism.

Protocol:

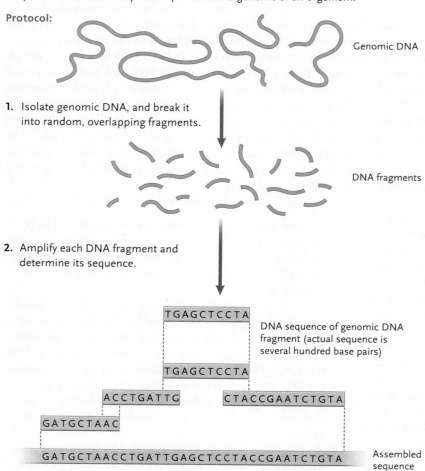

Genomic DNA

1. Isolate genomic DNA, and break it into random, overlapping fragments.

DNA fragments

2. Amplify each DNA fragment and determine its sequence.

TGAGCTCCTA

DNA sequence of genomic DNA fragment (actual sequence is several hundred base pairs)

TGAGCTCCTA

ACCTGATTG   CTACCGAATCTGTA

GATGCTAAC

GATGCTAACCTGATTGAGCTCCTACCGAATCTGTA   Assembled sequence

3. Enter the DNA sequences of the fragments into a computer, and use the computer to assemble overlapping sequences into the continuous sequence of each chromosome of the organism. This technique is analogous to taking 10 copies of a book that has been torn randomly into smaller sets of a few pages each and, by matching overlapping pages of the leaflets, assembling a complete copy of the book with the pages in the correct order.

Interpreting the Results: The method generates the complete sequence of the genome of an organism.

© Cengage Learning 2014

19.2 Genome Sequence Determination and Annotation

Genome sequence determination and annotation means obtaining the sequence of bases in a genome using DNA sequencing techniques, and then analyzing the sequence data using computer-based approaches to identify genes and other se-

quences of interest, which include gene regulatory sequences, origins of replication, repetitive sequences, and transposable elements.

Genome Analysis Begins with DNA Sequencing

DNA sequencing was developed in the late 1970s by Allan M. Maxam, a graduate student, and his mentor, Walter Gilbert of Harvard University. A few years later, Frederick Sanger of Cambridge University designed a method that became the one commonly used in research. Gilbert and Sanger were awarded a Nobel Prize in 1980. DNA sequencing technology has evolved since its development, and particularly rapidly in the past few years.

WHOLE-GENOME SHOTGUN SEQUENCING Before we discuss methods of DNA sequencing, let us consider the strategy generally used to determine the sequence of a genome, **whole-genome shotgun sequencing (Figure 19.1).** In this method, genomic DNA is isolated and purified, and that DNA is broken into thousands to millions of random, overlapping fragments. Each fragment is amplified to produce many copies, and then the sequence of the fragment is determined. The entire genome sequence is then assembled using computer algorithms that search for the sequence overlaps between fragments and stitch together the sequence reads to produce longer contiguous sequences.

DNA SEQUENCING METHODS All DNA sequencing methods have in common the following steps: (1) DNA purification; (2) DNA fragmentation; (3) amplification of fragments; (4) sequencing each fragment; and (5) assembly of fragment sequences into genome sequences. The methods differ in how the amplification is done, the lengths of the fragments, how many fragments are sequenced simultaneously, and how the sequencing reactions themselves are done.

For decades the method devised by Frederick Sanger was by far the most common DNA sequencing technique used. The Sanger method is a DNA synthesis-based method for DNA sequencing. It is based on the properties of nucleotides known as *dideoxyribonucleotides,* which have a —H on the 3′ carbon of the deoxyribose sugar instead of the —OH found in normal deoxyribonucleotides; therefore, the method, explained in **Figure 19.2,** is also called *dideoxy sequencing.*

In recent years the dideoxy sequencing method has been replaced largely, but not completely, by faster, cheaper, and

FIGURE 19.2 **Research Method**

Dideoxy (Sanger) Method for DNA Sequencing

Purpose: Obtain the sequence of a piece of DNA, such as in gene sequencing or genome sequencing. The method is shown here with a typical automated sequencing system.

Protocol:

1. A dideoxy sequencing reaction contains: (1) the fragment of DNA to be sequenced (denatured to single strands); (2) a DNA primer that will bind to the 3′ end of the sequence to be determined; (3) a mixture of the four deoxyribonucleotide precursors for DNA synthesis; (4) a mixture of the four dideoxyribonucleotide (dd) precursors, at about 1/100th of the concentration of the deoxyribonucleotides, each labeled with a different fluorescent molecule; and (5) DNA polymerase to catalyze the DNA synthesis reaction.

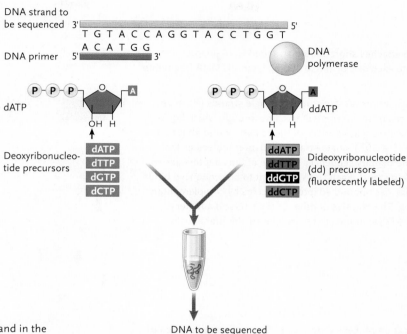

2. DNA polymerase synthesizes the new DNA strand in the 5′→3′ direction starting at the 3′ end of the primer. New synthesis continues until a dideoxyribonucleotide is incorporated randomly into the DNA. The dideoxyribonucleotide acts as a *terminator* for DNA synthesis because it has no 3′-OH group for the addition of the next base (see Section 14.3). For a large population of template DNA strands, the dideoxy sequencing reaction produces a series of new strands, with lengths from one on up. At the 3′ end of each new strand is the fluorescently labeled dideoxyribonucleotide that terminated the synthesis.

3. The labeled strands are separated by electrophoresis using a polyacrylamide gel prepared in a capillary tube. The principle of separation is the same as for agarose gel electrophoresis (see Figure 18.7), but this gel can discriminate between DNA strands that differ in length by one nucleotide. As the bands of DNA fragments move near the bottom of the tube, a laser beam shining through the gel excites the fluorescent labels on each DNA fragment. The fluorescence is registered by a detector, with the wavelength of the fluorescence indicating whether ddA, ddT, ddG or ddC is at the end of the fragment in each case.

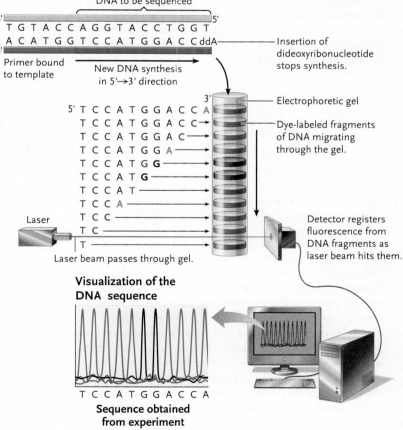

Interpreting the Results: The data from the laser system are sent to a computer that interprets which of the four possible fluorescent labels is at the end of each DNA strand. The results show, on a computer screen or in printouts, colors for the labels as the DNA bands passed the detector. The sequence of the newly synthesized DNA, which is complementary to the template strand, is read from left (5′) to right (3′). (The sequence shown here begins after the primer.)

FIGURE 19.3 | **Research Method**

Illumina/Solexa Method for DNA Sequencing

Purpose: Automated, massively parallel sequencing of up to a billion DNA fragments.

Protocol:

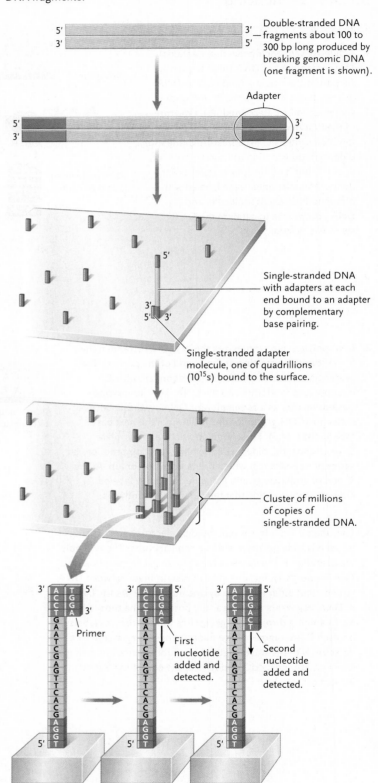

Double-stranded DNA fragments about 100 to 300 bp long produced by breaking genomic DNA (one fragment is shown).

Adapter

Single-stranded DNA with adapters at each end bound to an adapter by complementary base pairing.

Single-stranded adapter molecule, one of quadrillions (10^{15}s) bound to the surface.

Cluster of millions of copies of single-stranded DNA.

Primer

First nucleotide added and detected.

Second nucleotide added and detected.

1. DNA ligase attaches short double-stranded DNA adapter molecules to each end of 100 to 300-bp genomic DNA fragments.

2. The DNA fragments are denatured to single strands which are added to a cell in an automated machine through which liquid can flow. Over the glass surface of the cell are bound about one quadrillion (1×10^{15}) single-stranded adapter molecules that have complementary sequences to those of the adapters added to the DNA in step 1. Each DNA fragment to be sequenced binds to one of the glass-bound adapter molecules by complementary base pairing. The massive number of glass-bound adapters allows many DNA strands to be sequenced simultaneously.

3. An amplification process generates millions of copies of each of the DNA fragments that bound initially to one of the glass-bound adapters, clustered around the place on the cell where that DNA bound. Up to one billion different clusters can be produced in the sequencing cell simultaneously.

4. The DNA fragments are now ready for synthesis-based DNA sequencing. One fragment is shown for the sequencing steps.

5. DNA primers are added to the cell. A primer anneals to each DNA strand (it is the complementary strand to the adapter sequence at its end) and DNA synthesis is done in a cyclic manner one nucleotide at a time using four different fluorescently labeled DNA nucleotide precursors. Each time a labeled nucleotide is added, synthesis stops and the machine uses laser technology to measure the fluorescence so as to identify the base added. That base is the same for all strands in a cluster. By repeated cycles of addition of a nucleotide and laser detection, the sequence of the strand is obtained. Up to about 100 bases of each fragment can be sequenced.

Interpreting the Results: The DNA sequence obtained is complementary to the initial single-stranded DNA strand that paired with the glass-bound adapter. The DNA sequence data from all of the clusters of DNA fragments are analyzed by computer to determine overlaps between fragments and, by the principles described in Figure 19.2, the complete sequence of a genome is assembled.

more automated techniques. In general, these newer high-throughput techniques have decreased sequencing costs by reducing the preparatory steps, automating more of the process, and sequencing up to a billion different DNA fragments in parallel.

Figure 19.3 outlines a next-generation DNA sequencing technique that is widely used in genome sequencing projects, the DNA synthesis-based Illumina/Solexa method. This is an example of *massively parallel DNA sequencing,* because up to one billion different DNA fragments can be sequenced simultaneously.

As a result of the lower cost of sequencing, and the automated massively parallel DNA sequencing methods that are used, the genomes of many species beyond those targeted in the Human Genome Project have been determined. **Table 19.1** summarizes the number of whole genomes sequenced from organisms of the different domains of life. The cost of DNA sequencing is now low enough that sequencing a complete human genome costs only a few thousand dollars. This makes it feasible to sequence thousands of individual human genomes, to learn more about what makes us human and what is responsible for human variation. We are perhaps not too far from the scenario where your own genome sequence may be used to help physicians personalize medical treatments to your specific genetic makeup.

TABLE 19.1	Number of Organismal Genomes Sequenced[a]
Organism	Number of Genomes
Prokaryotes	
Archaea	105
Bacteria	1,978
Total prokaryotes:	**2,083**
Eukaryotes	
Animals	
Mammals	44
Fishes	16
Insects	40
Other animals	46
Total animals:	**146**
Plants	38
Fungi	124
Protists	47
Total eukaryotes:	**355**
Total all organisms:	**2,438**

[a]As of January 2012
© Cengage Learning 2014

Genome Sequences Are Annotated to Identify Genes and Other Sequences of Importance

A raw genome sequence is simply a string of A, T, G, and C letters; it tells us practically nothing about the organism from which it derives, other than the total length of its genome. Therefore, once the complete sequence of a genome has been determined, the next step is *annotation,* the identification of functionally important features in the genome. These include:

- Protein-coding genes.
- Noncoding RNA genes. As mentioned earlier, "noncoding" means that the RNA transcript of the gene is not translated. Rather, the transcript is the functional product of the gene. Noncoding genes include genes for tRNAs, rRNAs, and snRNAs (see Chapter 15), and genes for microRNAs (miRNAs; see Chapter 16).
- Regulatory sequences associated with genes (see Chapter 16).
- Origins of replication (see Chapter 14).
- Transposable elements and sequences related to them (see Chapter 15).
- Pseudogenes. A **pseudogene** is very similar to a functional gene at the DNA sequence level, but one or more inactivating mutations have changed the gene so that it can no longer produce a functional gene product. Most pseudogenes are derived from protein-coding genes and they are recognized by their sequence similarities to functional genes.
- Short repetitive sequences. These are sequences that are repeated a few too many times in the genome. The short tandem repeat (STR) sequences discussed in Section 18.2 are examples of short repetitive sequences.

Annotation is performed by researchers in the field of **bioinformatics,** which is the application of mathematics and computer science to extract information from biological data, including those related to genome structure and function. More broadly, the term bioinformatics may be used to include the experimental work that generates such data. For example, bioinformatics scientists predict the structure and function of gene products and postulate evolutionary relationships of sequences, which are issues for functional and comparative genomics, respectively.

Some examples of genome annotation are illustrated in the following subsections. Protein-coding genes are the focus because they are of particular interest in genome analysis.

IDENTIFYING OPEN READING FRAMES BY COMPUTER SEARCH OF GENOME SEQUENCES Proteins are specified in mRNA molecules by a series of codons starting with the initiation codon AUG and ending with one of the three termination codons, UAG, UAA, or UGA. The span of codons from start to stop codon is called an **open reading frame (ORF),** and ORFs that are longer than 100 codons almost always indicate the presence of a protein-coding gene. Computer algorithms are used to identify possible protein-coding genes in a genome sequence by

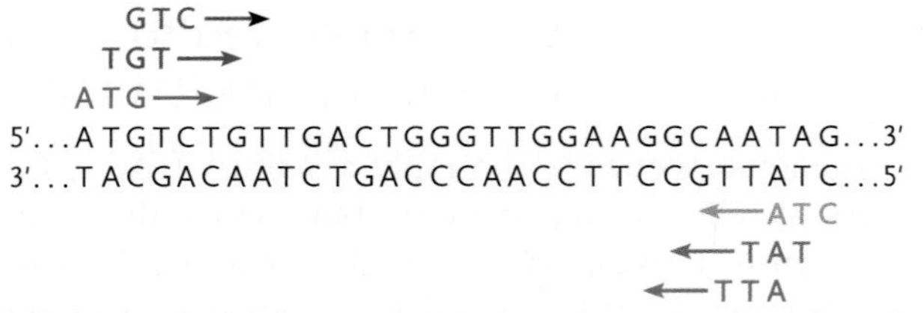

```
GTC ⟶
TGT ⟶
ATG ⟶
5'...ATGTCTGTTGACTGGGTTGGAAGGCAATAG...3'
3'...TACGACAATCTGACCCAACCTTCCGTTATC...5'
                              ⟵ATC
                            ⟵TAT
                          ⟵TTA
```

FIGURE 19.4 The six reading frames of double-stranded DNA. In this particular sequence, one of them is an open reading frame (ORF).
© Cengage Learning 2014

searching for ORFs. In a DNA sequence, this means searching for ATG, separated from a stop codon (TAG, TAA, or TGA) by a multiple of three nucleotides. The search is complicated because which of the two DNA strands is the template strand for transcription is gene-specific. Theoretically, then, each DNA sequence has six reading frames for the three-letter genetic code, three on one strand and three on the other strand, and an ORF can be in any one of those frames. This is illustrated in **Figure 19.4** for a theoretical 30-nucleotide segment of DNA. Note that each single-stranded DNA sequence generated by DNA sequencing can be used to infer the sequence of the complementary DNA strand. If an ORF is present in a particular DNA sequence, it will be in one of those frames.

We can start looking for an ORF going in the 5'-to-3' direction in the top strand of Figure 19.4, starting at the leftmost A nucleotide. In this case, reading in groups of three nucleotides will lead you to the TAG at the right end, which is the stop codon. We have found an ORF in the sequence, and it is coded by the entire length of top strand. However, if instead we start looking for an open reading frame in the top strand starting at either the second nucleotide (the T) or the third nucleotide (the G), we do not find a start codon so we have not found an ORF. For the bottom strand, none of the three frames has a start codon. Computer algorithms can search easily for ORFs in all six reading frames of a DNA sequence.

Searching for protein-coding ORFs is straightforward in prokaryotic genomes because few genes have introns. Eukaryotic protein-coding genes typically have introns and therefore more sophisticated algorithms must be used to identify such genes. For example, the algorithms may search for particular characteristics of protein-coding genes, such as junctions between exons and introns, sequences that are characteristic of eukaryotic promoters, and overrepresentation of certain three-base codons relative to others.

Computer identification typically is the first step in identifying protein-coding genes. However, other evidence is needed before biologists are confident that the sequence they have annotated is a functioning protein-coding gene. Two approaches to obtain such evidence are described in the next two subsections.

IDENTIFYING PROTEIN-CODING GENES BY SEQUENCE SIMILARITY SEARCHES One way of testing whether candidate protein-coding genes found by searching for ORFs are function-ing protein-coding genes is by comparing their sequences with known, identified and verified genes in databases. This is a *sequence similarity search.* Such searches can be done using an Internet browser to access the computer programs and the databases. For example, to use the BLAST (Basic Local Alignment Search Tool) program at the National Center for Biotechnology Information (http://blast.ncbi.nlm.nih.gov), a researcher pastes the putative ORF DNA sequence, or the amino acid sequence of the protein it would encode, into a browser window and sets the program to begin searching. The BLAST program searches the databases of known sequences and returns the best matches, if any. The matches are listed in order, from the closest match to the least likely match. Finding a known gene's sequence that matches the putative ORF sequence closely would be good evidence that the ORF is in fact a protein-coding gene and that it encodes a protein functionally related to that of the matching gene sequence. The principle here is that genes of living organisms tend to be similar to each other because they have evolved from ancestral genes in ancestral organisms. Genes that have highly conserved sequences because they have evolved from a gene in a common ancestor are called **homologous genes.** For example, if a gene in the mouse has been characterized experimentally and its sequence is known, and that gene is evolutionarily conserved in mammals, there should be a match for that gene in the human genome sequence.

With sequence similarity searches, ORFs can be sorted into ones with known and unknown functions. For the latter, experiments are required to show whether they are real protein-coding genes and, if so, what their functions are. For example, analysis of the human genome sequence initially identified more than a thousand putative protein-coding genes with no sequence similarities to known genes. Most of these function-unknown genes have now been shown to be pseudogenes. Such uncertainty makes it difficult to determine the exact number of protein-coding genes in a genome just from its sequence.

IDENTIFYING PROTEIN-CODING GENES FROM SEQUENCES OF GENE TRANSCRIPTS The gold standard for identifying protein-coding genes in a genome sequence is the demonstration that the sequences are transcribed in cells to make mRNAs. One approach to do this makes use of the sequences of transcripts represented in cDNA libraries. Recall from Section 18.1 that a cDNA library is made starting with mRNA molecules isolated from a cell. If the mRNA molecules are isolated under different conditions and from different cell types in a multicellular organism, they will represent the activity of many of the organism's protein-coding genes. However, protein-coding genes that are rarely transcribed or that produce very few mRNA molecules are likely to be missed by this approach.

The single-stranded mRNA molecules are converted to double-stranded DNA molecules using reverse transcriptase, and those DNA molecules are cloned to produce the cDNA library. Some part of each cloned mRNA (as cDNA) is sequenced using a sequencing primer that pairs with the DNA just adjacent to the inserted cDNA fragment in the clone. Using computer algorithms, each cDNA sequence is compared with the genome sequence of the organism to map the location of the sequence and, therefore, the protein-coding gene from which the original transcript was derived.

This approach is also useful for cataloging transcripts in humans and other eukaryotes to identify which genes are alternatively spliced. Recall from Section 16.3 that alternative splicing of pre-mRNA transcripts of protein-coding genes produces different mRNAs by using different splice sites that may remove different combinations of exons, or modify their lengths. The resulting mRNAs are translated to produce a family of related proteins having various combinations of amino acid sequences derived from the remaining exons.

Genome Landscapes Vary Markedly in Size, Gene Number, and Gene Density

With many genomes now sequenced, researchers can compare them to learn about genome sizes, the number of protein-coding genes, and the density of those genes (how widely spaced they are). A vast amount of new information is available about genome landscapes. Here we will present some broad brush strokes and then provide a more detailed description of the *E. coli* genome and the human genome, as examples of prokaryote and eukaryote genomes, respectively.

GENOME SIZES OF VIRUSES, BACTERIA, ARCHAEA, AND EUKARYA **Figure 19.5** shows the ranges of genome sizes for viruses, bacteria, archaea, and different groups of eukaryotes; and **Table 19.2** gives examples of genome sizes and the number of protein-coding genes for some bacteria, archaea, and eukaryotes. We can arrive at some general conclusions about the data. Members of both the Domain Bacteria and Domain Archaea have genomes that vary widely in size. In addition, their genes are densely packed in their genomes, with little noncoding space between them. Thus, the larger genomes of organisms in these

two domains tend to reflect increased gene number. For example, in the Bacteria, *Mycoplasma genitalium* has a genome size of 0.58 Mb with 475 protein-coding genes (and 43 noncoding RNA genes), and *E. coli* has a genome size of 4.6 Mb with 4,146 protein-coding genes (and 176 noncoding RNA genes). In the Archaea, *Thermoplasma acidophilum* has a genome size of 1.56 Mb with 1,484 protein-coding genes, and *Methanosarcina thermophila* has a genome size of 5.75 Mb with 4,540 protein-

TABLE 19.2	Genome Sizes and Estimated Number of Protein-Coding Genes for Selected Members of Domains Bacteria, Archaea, and Eukarya	
Domain and Organism	Genome Size (Mb)	Protein-Coding Genes
Bacteria		
Mycoplasma genitalium	0.58	475
Escherichia coli	4.6	4,146
Archaea		
Thermoplasma acidophilum	1.56	1,484
Methanosarcina acetivorans	5.75	4,540
Eukarya		
Protists		
Tetrahymena thermophila (a ciliated protist)	146	>20,000
Fungi		
Saccharomyces cerevisiae (a budding yeast)	12.1	~6,000
Neurospora crassa (orange bread mold)	40	~10,100
Plants		
Arabidopsis thaliana (thale cress)	120	~26,000
Oryza sativa (rice)	411	~56,000
Invertebrates		
Caenorhabditis elegans (a nematode worm)	100	~20,000
Drosophila melanogaster (fruit fly)	165	~13,700
Vertebrates		
Takifugu rubripes (pufferfish)	393	~27,000
Mus musculus (mouse)	2,600	~22,000
Homo sapiens (human)	3,200	~20,500

© Cengage Learning 2014

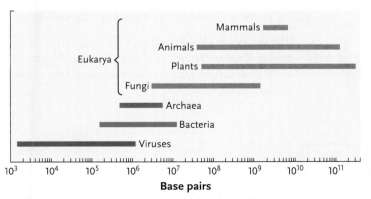

FIGURE 19.5 Ranges of genome sizes for viruses, bacteria, archaea, and eukaryotes. Note that the Domain Archaea has not been studied as extensively as the Domain Bacteria, so there may be representatives with substantially larger or smaller genomes than the range given.

© Cengage Learning 2014

coding genes. More broadly, bacteria and archaea have a fairly consistent density of about 900 genes for every Mb of DNA.

Members of the domain Eukarya vary markedly in form and complexity, and their genomes also show great differences in size. For example, budding yeast, *Saccharomyces cerevisiae,* has a 12.1-Mb genome that is about 0.4% the size of the 3,200-Mb human genome, yet humans, with ~20,500 protein-coding genes, have only a little more than three times the number found in yeast, which has ~6,000 protein-coding genes.

There are no clear rules relating organism complexity and genome size. For instance, the fruit fly, *Drosophila melanogaster,* and the locust, *Schistocerca gregaria,* have similar physiological complexity, but the genome of the locust is 9,300 Mb, which is 52 times the size of the 165-Mb fruit fly genome (and almost three times the size of the human genome). Even within a genus there is not necessarily a consistency. For example, there is a 50-fold variation in the genome size of *Allium* species, which contains onions, leeks, shallots, and garlic. Even among vertebrates, there is great variation in genome size. The pufferfish, *Takifugu rubripes,* for example, has a 393-Mb genome, whereas the genomes of the mouse and humans are about seven times larger. And yet, the pufferfish, with ~27,000 protein-coding genes, has more protein-coding genes than either the mouse

(~22,000 genes) or the human (~20,500). Clearly the genes are spaced more closely in the pufferfish genome than they are in either the mouse or human genomes. But the human genome is by no means the largest among eukaryotes; the genomes of some amphibians and some ferns are about 200 times larger. In general, though, all genes are packed less densely in eukaryotes than they are in prokaryotes, with protein-coding gene density ranging from 500 per Mb (yeast) to fewer than 10 per Mb (humans and other mammals). However, there is no uniformity in the packing, as the pufferfish–mammal comparison shows.

It is important to think critically about the data presented for gene numbers in a genome. As you have learned, the determination of the protein-coding gene number involves both computer and experimental analysis. The outcomes of those analyses are therefore estimates of the number of genes present. Only when an entire genome has been studied experimentally to characterize every gene it contains can we be certain of an organism's exact gene number. For example, you have just learned that the estimated number of protein-coding genes is ~22,000 in the mouse genome and ~20,500 in the human genome. The numbers are not precise. Rather, at this point they likely reflect different extents of progress in annotating the two genomes, and not that 2,000 more protein-coding genes are required necessarily to be a mouse compared to being a human.

A. **Map of the circular *E. coli* K12 genome showing the genes transcribed clockwise (blue) and the genes transcribed counterclockwise (orange) and the location of the origin of replication.**

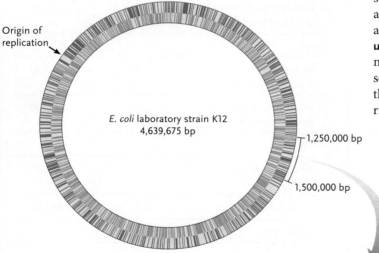

PROFILE OF THE *E. coli* GENOME *E. coli* is one of the most intensively studied model organisms, and the genome of laboratory strain K12 is one of the best annotated. In many ways, *E. coli* has a typical bacterial genome, with the vast majority of its genes on a single circular chromosome with one origin of replication **(Figure 19.6A)**, and the remainder of its genes on one or more plasmids, each of which is much smaller than the circular chromosome. With about 4.6 Mb and about 4,146 protein-coding genes, the *E. coli* K12 genome is in the middle range, sizewise, of bacterial genomes (see Figure 19.5). The noncoding genes are those for rRNAs and tRNAs. There are a small number of transposable elements and repetitive sequences.

Figure 19.6B shows a close up of a 10-kb segment of the *E. coli* genome containing a number of protein-coding genes to illustrate the following characteristics:

- The genes are close together, with little space in between. Promoters for the genes are located immediately upstream of each transcription unit (not shown in the figure).

B. **Detail of a 10-kb region of the *E. coli* K12 genome, from about 3:30 on the genome "clock."**

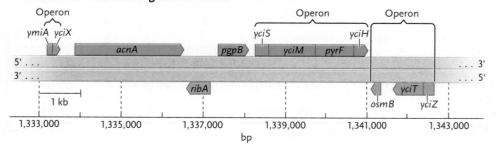

FIGURE 19.6 The genome of *E. coli,* laboratory strain K12.

© Cengage Learning 2014

- Some of the genes are transcribed in the left-to-right direction (using the bottom strand as the template), whereas the others are transcribed in the right-to-left direction (using the top strand as the template). (The two template strands are transcribed in different directions because the two DNA molecules in a double helix are antiparallel; see Chapter 14.)
- Some genes are single transcription units, whereas others are organized into operons (see Chapter 16). In the genome as a whole, about one-half of protein-coding genes are organized into operons.
- The genes vary in length, reflecting the lengths of their encoded proteins.

Table 19.3 summarizes some of what has been learned about the *E. coli* K12 genome to date with respect to its physical aspects, genes, and gene products.

Other bacterial genomes may be larger or smaller than the *E. coli* K12 genome, but their genome landscapes are similar to that of *E. coli* in several ways. For example, typically there is one origin of replication, 85% to 92% of the DNA codes for proteins, there is a mixture of operons and single-gene transcription units, some genes are transcribed using one DNA strand as the template whereas others are transcribed using the other strand, and there are relatively few transposable elements or repetitive sequences.

PROFILE OF THE HUMAN GENOME At about 3.2 billion base pairs, the human genome is about 700 times longer than the *E. coli* genome. Each human individual has 23 pairs of chromo-

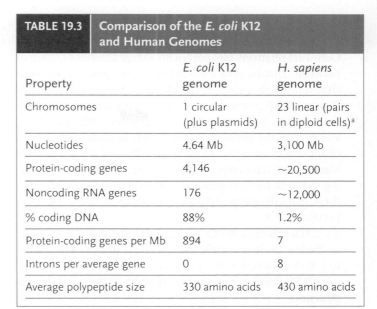

TABLE 19.3 Comparison of the *E. coli* K12 and Human Genomes

Property	*E. coli* K12 genome	*H. sapiens* genome
Chromosomes	1 circular (plus plasmids)	23 linear (pairs in diploid cells)[a]
Nucleotides	4.64 Mb	3,100 Mb
Protein-coding genes	4,146	~20,500
Noncoding RNA genes	176	~12,000
% coding DNA	88%	1.2%
Protein-coding genes per Mb	894	7
Introns per average gene	0	8
Average polypeptide size	330 amino acids	430 amino acids

[a]There are 24 different human chromosomes: 22 autosomes, and the X and Y chromosomes. Each individual has 23 pairs of chromosomes.
© Cengage Learning 2014

somes. Men have 24 different chromosomes, the 22 autosomes and the X and Y chromosomes, whereas women have 23 different chromosomes, the 22 autosomes and the X chromosome. **Figure 19.7A** displays the complete set of human chromosomes, depicting the banding patterns that help researchers identify regions of chromosomes. (Figure 10.7 shows the banding patterns of stained human chromosomes.) **Figure 19.7B** shows chromosome 6 in more detail and then a close-up of a 100-kb segment of the long arm of that chromosome to show the protein-coding genes it contains. Compare this figure with Figure 19.6B, which

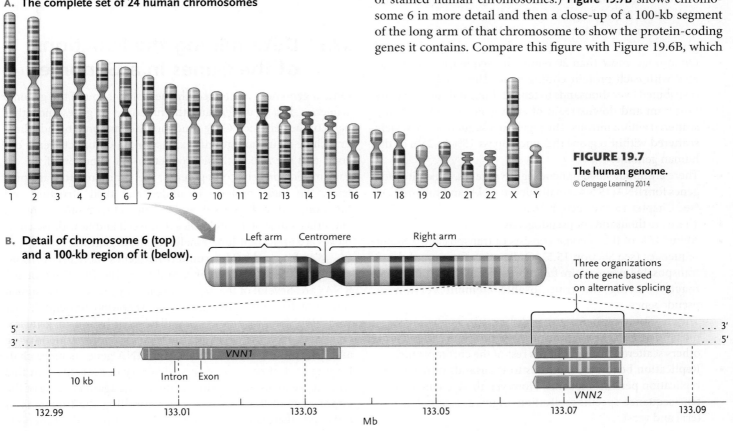

A. The complete set of 24 human chromosomes

1 2 3 4 5 6 7 8 9 10 11 12 13 14 15 16 17 18 19 20 21 22 X Y

FIGURE 19.7
The human genome.
© Cengage Learning 2014

B. Detail of chromosome 6 (top) and a 100-kb region of it (below).

Left arm Centromere Right arm

Three organizations of the gene based on alternative splicing

5' ...   ... 3'
3' ...   ... 5'

VNN1

10 kb   Intron Exon

VNN2

132.99   133.01   133.03   133.05   133.07   133.09
Mb

shows a 10-kb segment of the *E. coli* chromosome and note the following:

- Genes are relatively far apart, with a large amount of space in between. That is, even though the human genome segment shown is ten times longer than the *E. coli* segment shown in Figure 19.6.B, it contains only two protein-coding genes. Each of these genes consists of transcription units that are far longer than the genes in *E. coli*, largely because they consist of about 95% introns and 5% exons. The right-hand gene in Figure 19.7B illustrates at the DNA level the alternative splicing variants for that gene; note the different exons for the three gene drawings. (Alternative splicing is described in Section 16.3.)

- As in the genomes of other organisms, some genes are transcribed in the left-to-right direction and others in the right-to-left direction. For the particular segment shown in Figure 19.7B, both genes are transcribed in the right-to-left direction.

- All the genes are single transcription units. Eukaryotic genes are rarely organized in operons.

Table 19.3 summarizes some of what researchers have learned about the human genome to date with respect to its physical aspects, genes, and gene products. Some of the key features of the human genome are:

- There are about 20,500 protein-coding genes. On average, there are 9 to 10 exons per gene, with some human genes consisting of a single exon and others having over 100 exons. Introns make up about 95% of the average transcription unit. Since about 2% of the genome consists of protein-coding sequences, introns represent about 20% to 25% of the human genome.

- On average, more than 20 regulatory sequences are associated with each protein-coding gene. Those sequences are distributed over thousands to tens of thousands of base pairs upstream and downstream of each gene, as well as being scattered within introns. The regulatory sequences are widely scattered within regions that encompass 15% to 25% of the human genome.

- There are about 9,000 noncoding RNA genes, which include genes for rRNA, tRNA, small nuclear RNA, and microRNAs (see Chapter 15 and Section 16.3).

- There are thousands of pseudogenes.

- About 45% of the genome consists of transposable element sequences (see Section 15.5). Only a tiny fraction of those transposable elements are functionally active. The others are inactive, being the transposable element version of pseudogenes.

- The genome contains a variety of short, repeated sequences, including those at the centromeres and telomeres, as well as others scattered throughout the rest of the chromosome.

- Replication begins at hundreds to thousands of origins of replication per chromosome. However, there are no consistent, sequence-specific replication origins as is found in bacteria and yeast.

Up to this point, we have focused on the features of the enormous portion of the human genome encoded within the linear chromosomes of the cell's nucleus. It is also important to remember that in a eukaryote, each mitochondrion also contains a circular mitochondrial genome, or mtDNA. The human mtDNA is 16.6 kb, much smaller than even a prokaryotic genome. Its 37 genes (13 of them are protein-coding) perform essential functions related to cellular respiration. In addition, photosynthetic eukaryotes, including plants and algae, have a separate circular genome up to several hundred thousand bases in each of their chloroplasts: the cpDNA. Not surprisingly, this genome contains genes involved in photosynthesis.

Other mammalian genomes are very much like the human genome. But for other eukaryotes generally, particular features can vary considerably. For example, eukaryote genomes range from having a very low percentage of protein-coding DNA as in mammals to almost as high a percentage as is seen for prokaryotes.

STUDY BREAK 19.2

1. What is the principle behind whole-genome shotgun sequencing of genomes?
2. What are the key sequences identified by genome annotation?
3. How are possible protein-coding genes identified in a genome sequence of a bacterium? Of a mammal?
4. What general differences are there in the genome landscapes of prokaryotes and eukaryotes?

19.3 Determining the Functions of the Genes in a Genome

Once a genome is annotated, the next step is to use the genome sequence data to understand the functions of genes and other parts of the genome. "Gene function" is considered broadly here to include regulation of gene expression, the products genes encode, and the role of those products in the function of the organism. For protein-coding genes, the gene products are proteins. We study proteins to understand their structure and function, and to discover how they interact with other proteins and other nonprotein molecules in the cell and how those complexes are important functionally for the organism.

We also need to determine what genes there are for noncoding RNAs in the genome, and what the functions of the RNAs are. Several of the noncoding RNA genes can be assigned functions based on evolutionary conservation principles; that is, by looking for sequence similarity with known gene sequences. An example is the rRNA genes. However, identifying and determining the functions of miRNA genes is more challenging in part because of their diversity of sequences. In this section, we focus on protein-coding genes, again because of the importance of their products in controlling the functions of cells and, therefore, of organisms. Determining the functions of protein-coding genes typically relies on computer analysis and on laboratory experiments. The following presents examples of those approaches.

Gene Function May be Assigned by a Sequence Similarity Search of Sequence Databases

You learned earlier that a DNA sequence can be identified as a likely protein-coding gene by using a sequence similarity search of sequences in databases. This approach can also be used to assign the function of a gene. A high degree of similarity between the sequence of a candidate gene of unknown function and the sequence of a gene of known function likely indicates that both sequences evolved from a gene in a common ancestor and that their sequences in the present day have been conserved significantly because they code for proteins that have similar functions. As explained earlier, genes with highly conserved sequences as a result of divergence from a common ancestral gene are homologous genes.

Using sequence similarity searches to determine if a candidate gene and a known gene are homologous is by far the most common method for assigning the functions of genes. Experimental investigation of the functions of genes is considerably more expensive and time-consuming than DNA sequence comparisons; therefore, it is not feasible to repeat experiments in every species whose genome is sequenced. As a result, experimental data are available only for a small fraction of organisms. And because the functions of homologous protein-coding genes are so well conserved during the evolution of organisms, information about the function of a gene in one well-studied species very often applies to the homologous genes in another.

In some cases, the outcome of a sequence similarity search will indicate that the entire candidate gene's sequence is homologous to a known gene's sequence. In other cases, only part of the candidate gene sequence may match closely a sequence in a known gene. Typically this result indicates that the candidate gene encodes a protein with a domain that is related evolutionarily to a domain-encoding region of the known gene. (Protein domains are discussed in Section 3.4 and in the Genome Evolution section later in this chapter.)

Gene Function May be Assigned Using Evidence from Protein Structure

Since the structure of a protein determines its function and the structures and functions of many proteins are known, similarity between the structure of a protein determined for a newly identified gene and a characterized protein will indicate a likely function of the protein product of that gene. (How protein structure may be determined is described later in this section.) However, isolating and purifying a protein and determining its structure experimentally is expensive; therefore, that approach is used for only a small fraction of identified genes.

Gene Function May be Determined Using Experiments That Alter the Expression of a Gene

If a researcher can determine how the phenotype of a cell or organism is affected when the expression of a gene is turned off, or reduced significantly, functional properties of the encoded protein may be inferred. In a simple example, if cells grow larger, the gene may be involved in regulating cell size.

Two main kinds of manipulations are used to turn off or reduce significantly the expression of a gene in genome-scale experiments—gene knockout and gene knockdown:

1. *Gene knockout.* In this approach, researchers replace a normal gene on its chromosome with a defective gene that cannot express a functional protein. Usually, the replacement lacks the ORF that encodes the gene's protein product. In effect, this is a deletion mutation that has been engineered genetically. A deletion mutation is a *null mutation* because there is zero expression of the gene's protein product. For a haploid organism, there is only one copy of each gene to knock out, whereas in diploid organisms both copies of each gene must be knocked out. On a genomic scale, experimental manipulations can be done to knockout each gene systematically one by one. The phenotypic consequences of zero expression of each gene can then be ascertained. Major projects have been done, or are being done, to knock out systematically the function of each gene in the genomes of several organisms, including yeast, the fruit fly, the nematode worm, and the mouse (knockout mice were introduced in Section 18.2).

2. *Gene knockdown.* Knocking down a gene's expression typically is done using RNA interference (RNAi). As discussed in Section 16.3, RNAi reduces the expression of a gene at the translation level. In RNAi, a small regulatory RNA (like a natural miRNA) is transcribed from an expression plasmid introduced into the cell. The sequence of that regulatory RNA forms complementary base pairs with the mRNA of a gene of interest. The base pairing triggers the RNAi molecular mechanisms (see Figure 16.14), which knocks down the expression of the gene by causing degradation of that gene's mRNA or by blocking its translation. For example, RNAi has been used to knock down gene expression of each of the approximately 20,000 genes of the nematode worm one by one. The advantage of RNAi in comparison to gene knockouts is that the decrease in function of a gene can be temporary.

Characterizing genes by studying the effects on phenotype of knockouts or knockdowns can be very expensive and time-consuming. In a genome-wide study, thousands of knockout or knockdown strains have to be engineered genetically, and then each one has to be screened for a battery of possible phenotypic changes. The most ambitious studies of this kind have examined hundreds of phenotypes for each gene, which is only a fraction of the phenotypes that could be characterized. To make this approach really productive in the future will require further development of high-throughput methods that automate the measurement of phenotypic changes.

The study of gene function by identifying changes in phenotypes is called **phenomics.** Knowing the phenotypic effects of knocking out or knocking down a gene usually only gives general information about the function of a gene. But that information is useful for follow-up experiments on the specific protein involved.

Transcriptomics Determines at the Genome Level When and Where Genes Are Transcribed

Some genes are transcribed in all cell types, whereas others are transcribed only when and where they are needed (see Chapter 16). Determining when and where genes are transcribed can shed light on their function. For instance, a researcher might be interested in determining at a genomic scale the gene expression patterns in different cell types, at different stages of embryonic development, at different points of the cell division cycle, or in response to mutation or changes in the environment. A medical example, for instance, would be identifying gene expression differences between normal cells and cells that have become cancerous. The experimental analysis itself may be qualitative—analyzing whether or not genes are expressed—or quantitative—analyzing how the level of expression of genes varies.

The complete set of transcripts in a cell is called the **transcriptome,** and the study of the transcriptome is called **transcriptomics.** Transcriptomics includes cataloging transcripts, and quantifying the changes in expression levels of each transcript during development, in different cell types, under different physiological conditions, and with other variations.

Analysis of transcriptomes is done using high-throughput hybridization or, increasingly, by sequence-based approaches. A hybridization-based approach uses **DNA microarrays,** also called **DNA chips.** The surface of a DNA microarray is divided into a microscopic grid of about 60,000 spaces. On each space of the grid, a computerized system deposits a microscopic spot containing about 10,000,000 copies of a DNA probe that is about 20 nucleotides long.

Studies of gene activity using DNA microarrays involve comparing gene expression under a defined experimental condition with expression under a reference (control) condition. As a theoretical example, **Figure 19.8** shows how a DNA microarray can be used to compare gene expression patterns in normal cells and cancer cells in humans. mRNAs are isolated from each cell type and cDNAs are made from them, incorporating different fluorescent labels: green for one cDNA, red for the other. The two cDNAs are mixed and added to the DNA microarray, where they hybridize with whichever spots on the microarray contain complementary DNA probes. A laser excites the fluorescent labels and the resulting green and red fluorescence is detected and quantified, enabling a researcher to see which genes are expressed in the cells. This technique is semiquantitative because it is also able to quantify differences in gene expression between the two cell types approximately (see *Interpreting the Results* in the figure).

In cancer research, for example, microarray analysis has been used to diagnose a particular cancer type, thereby potentially informing physicians about appropriate treatments. That is, many cancers can be identified based on **signature genes** which are key genes whose expression changes in a way that correlates with a normal cell becoming a cancer cell. *Diffuse large B-cell lymphoma* (DLBCL) is the most common type of non-Hodgkin's lymphoma, a type of cancer of the lymphatic system. Some patients respond to chemotherapy treatment, resulting in prolonged survival, whereas the remainder do not respond to such treatment and show no greater survival than untreated patients. DNA microarray-based transcriptome analysis of DLBCLs in a group of patients showed that the gene expression profile in chemotherapy-responsive tumors is different from that in chemotherapy-nonresponsive tumors. This result indicates that there are two subtypes of DLBCL. Identifying by DNA microarray analysis which type of DLBCL a patient has can, therefore, inform physicians as to the most effective treatment regimen. If the DLBCL is a responder, then chemotherapy is the best approach. If the DLBCL is a nonresponder based on the gene expression profile, then more aggressive treatments can be done.

Other examples of DNA microarray analysis include screening individuals for mutations associated with genetic diseases (such as breast cancer), and studying changing gene expression profiles during *Drosophila* development, when a gingivitis-causing bacterium grows under different conditions (see *Insights from the Molecular Revolution* in Chapter 26), during flower development in plants (see *Focus on Basic Research* in Chapter 37), and when sea urchin larvae are living under different pH conditions (see *Insights from the Molecular Revolution* in Chapter 54).

A newer, sequence-based approach to analyze transcriptomes is **RNA-seq** (whole-transcriptome sequencing). This technique uses high-throughput sequencing of cDNAs to identify and quantify RNA transcripts in a sample. For transcriptomic analysis of protein-coding genes using RNA-seq, mRNAs are isolated and converted to cDNAs (see Figure 18.4). About 30 to 400 nucleotides of each cDNA are sequenced using high-throughput techniques, and the results are aligned with the genome sequence of the organism under study. A single RNA-seq study can identify over 100,000 sequence reads, each one of which indicates the presence of a specific mRNA in the cells being studied. While a relatively new technique, RNA-seq rapidly is becoming the replacement for DNA microarrays in transcriptomics, because of its decreasing cost and its greater precision for quantifying transcripts.

Proteomics Is the Characterization of All Expressed Proteins

Genome research also includes analysis of the proteins encoded by a genome, for proteins are largely responsible for cell function and, therefore, for most of the functions of an organism. The term **proteome** has been coined to refer to the complete set of proteins that can be expressed by an organism's genome. A *cellular proteome* is a subset of those proteins—the collection of

FIGURE 19.8 | **Research Method**

DNA Microarray Analysis of Gene Expression Levels

Purpose: DNA microarrays can be used in various experiments, including comparing the levels of gene expression in two different tissues, as illustrated here. The power of the technique is that the entire set of genes in a genome can be analyzed simultaneously.

Protocol:

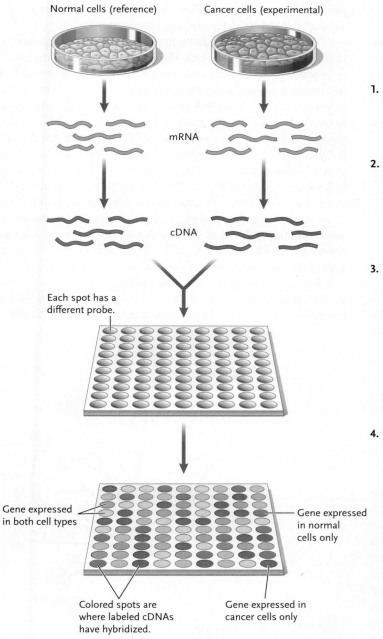

Normal cells (reference)

Cancer cells (experimental)

mRNA

cDNA

Each spot has a different probe.

Gene expressed in both cell types

Gene expressed in normal cells only

Colored spots are where labeled cDNAs have hybridized.

Gene expressed in cancer cells only

1. Isolate mRNAs from a control cell type (here, normal human cells) and an experimental cell type (here, human cancer cells).

2. Prepare cDNA libraries from each mRNA sample. For the normal cell (control) library use nucleotides with a green fluorescent label, and for the cancer cell (experimental) library use nucleotides with a red fluorescent label.

3. Denature the cDNAs to single strands, mix them, and pump them across the surface of a DNA microarray containing a set of single-stranded probes representing every protein-coding gene in the human genome. The probes are spotted on the surface, with each spot containing a probe for a different gene. Allow the labeled cDNAs to hybridize with the gene probes on the surface of the chip, and then wash excess cDNAs off.

4. Locate and quantify the fluorescence of the labels on the hybridized cDNAs with a laser detection system.

Actual DNA microarray result

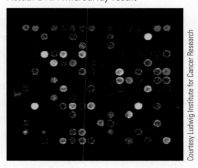

Courtesy Ludwig Institute for Cancer Research

Interpreting the Results: The colored spots on the microarray indicate where the labeled cDNAs have bound to the gene probes attached to the chip and, therefore, which genes were active in normal and/or cancer cells. Moreover, we can quantify the gene expression in the two cell types by the color detected. A purely green spot indicates the gene was active in the normal cell, but not in the cancer cell. A purely red spot indicates the gene was active in the cancer cell, but not in the normal cell. A yellow spot indicates the gene was equally active in the two cell types, and other colors tell us the relative levels of gene expression in the two cell types. For this particular experiment, we would be able to see how many genes have altered expression in the cancer cells, and exactly how their expression was changed.

proteins found in a particular cell type under a particular set of environmental conditions.

The study of the proteome is the field of **proteomics.** The number of possible proteins encoded by the genome is larger than the number of protein-coding genes in the genome, at least in eukaryotes. In eukaryotes, alternative splicing of gene transcripts and variation in protein processing means that expression of a gene may yield more than one protein product. The number of different proteins an organism can produce typically far exceeds the number of protein-coding genes.

Proteomics has three major goals: (1) to determine the structures and functions of all proteins; (2) to determine the location of each protein within or outside the cell; and (3) to identify physical interactions among proteins.

DETERMINING PROTEIN STRUCTURE Protein structure may be determined as follows:

Clone the coding sequence of the gene into an expression vector (see Section 18.2)
↓
Transform the cloned gene into a host to express the protein
↓
Purify the protein
↓
Determine the structure of the protein using X-ray crystallography or nuclear magnetic resonance (NMR)

Protein structure may also be predicted nonexperimentally using computer algorithms based on the known chemistry of amino acids and how they interact.

DETERMINING THE LOCATIONS OF PROTEINS IN CELLS The location of a protein in a cell is important because it is key to its function. The cellular location of a protein can be studied by tagging the protein in some way, and then visualizing the location of the tag microscopically. Different tags are used for visualization using light microscopy or electron microscopy.

IDENTIFYING INTERACTIONS AMONG PROTEINS Many proteins function by interacting with other proteins. In some cases, proteins (actually polypeptides) interact to form the quaternary structure—and therefore the functional form—of a protein (see Section 3.4). Many multi-polypeptide proteins exist, and you have encountered several in this book, for example: (1) hemoglobin is a four-polypeptide protein consisting of two α-globin polypeptides and two β-globin polypeptides and four associated heme groups (see Figure 3.20); (2) RuBP carboxylase/oxygenase (rubisco), the first enzyme of the light-independent reactions of photosynthesis (see Section 9.3), consists of eight copies of a large polypeptide and eight copies of a small polypeptide (see figure in *Insights from the Molecular Revolution,* Chapter 9); and (3) the Lac repressor protein that controls the expression of the *lac* operon in *E. coli* (see Section 16.1) consists of four copies of the same polypeptide.

In other interactions among proteins, the interaction is not permanent, but instead serves to affect the function of one or other of the partners in the interaction. For example, in Chapter 7 you were introduced to protein kinases—enzymes that transfer a phosphate group from ATP to one or more sites on particular target proteins as part of a signal transduction pathway. The phosphorylation of the target proteins occurs as a result of the interaction between the enzymatic protein and each target protein. Once the target protein is phosphorylated, the two proteins no longer interact. Understanding the interactions among proteins is important, then, to help us understand how proteins work individually and together to determine the phenotype of a cell.

Thousands of interactions have been identified experimentally for a variety of organisms. The interaction data are assembled to produce protein-interaction networks, the analysis of which is informing us about the details and complexities of the functions of proteins in cells. **Figure 19.9** shows part of a protein-interaction network centered on the human protein, β-catenin (cadherin-associated protein; the central CTNNB1 sphere in the figure). β-Catenin is involved in the formation of adherens junctions (a type of cell junction; see Section 5.5) in epithelial cells where it links α-catenin (CTNNA1 in the figure) with E-cadherin (CDH1 in the figure; cadherins are discussed in Chapter 50). It also plays a key role in a signaling pathway that is important, for example, in regulating how cell fate is decided during development. Interactions with, for example, APC, LEF1, TCF4, and TCF7L2 occur as part of that signaling pathway's operation.

Just as for genomic DNA sequences, information about various properties of each protein are placed in databases, for example Entrez (http://www.ncbi.nlm.nih.gov) and UniProt

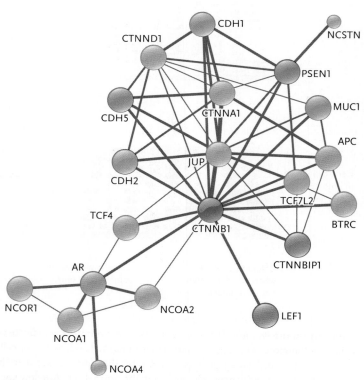

FIGURE 19.9 The protein-interaction network for human β-catenin (CTNNB1). Thicker lines show stronger associations between proteins.
© Cengage Learning 2014

A. Protein classes

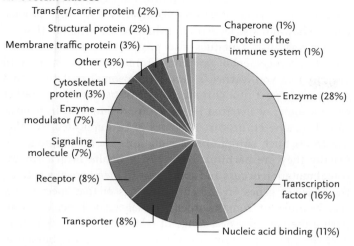

Transfer/carrier protein (2%)
Structural protein (2%)
Membrane traffic protein (3%)
Other (3%)
Cytoskeletal protein (3%)
Enzyme modulator (7%)
Signaling molecule (7%)
Receptor (8%)
Transporter (8%)
Chaperone (1%)
Protein of the immune system (1%)
Enzyme (28%)
Transcription factor (16%)
Nucleic acid binding (11%)

B. Biological classes

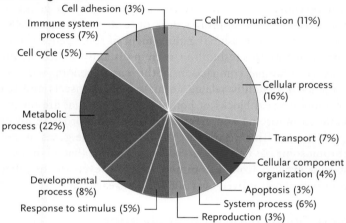

Cell adhesion (3%)
Immune system process (7%)
Cell cycle (5%)
Metabolic process (22%)
Developmental process (8%)
Response to stimulus (5%)
Reproduction (3%)
System process (6%)
Apoptosis (3%)
Cellular component organization (4%)
Transport (7%)
Cellular process (16%)
Cell communication (11%)

FIGURE 19.10 Functions of human protein-coding genes organized with respect to protein classes (A) and biological processes (B).
© Cengage Learning 2014

(http://www.uniprot.org), to create a dossier of that protein that is available to researchers worldwide.

CHARACTERIZING PROTEIN FUNCTION Through various approaches, researchers learn about the functions of proteins. **Figure 19.10A** shows an example of what we have learned about the functions of human protein-coding genes with respect to protein classes, and **Figure 19.10B** shows the functions of human protein-coding genes with respect to the biological processes involving those proteins.

STUDY BREAK 19.3

1. What are the ways by which the function of a gene identified in a genome sequence may be assigned?
2. How would you determine how a steroid hormone affects gene expression in human tissue culture cells?
3. What is the proteome, and what are the major goals of proteomics?

19.4 Genome Evolution

DNA genomes with protein-coding genes are thought to have evolved over 3.5 billion years ago, by the time of the earliest fossil microorganisms that have been discovered (see Chapter 25). Those early cells probably had at most a few hundred protein-coding genes. New genes evolved as life evolved and became more complex, so that most present-day organisms have thousands or tens of thousands of protein-coding genes. In this section you will learn how genes and genomes have evolved, and how genome sequences inform us about the evolutionary history of life.

Comparative Genomics Reveals the Evolutionary History of Genes and Genomes

Understanding how genes evolved and how genomes evolved are major goals of the field of *comparative genomics.* Because the genes in present-day genomes evolved from ancestral genes that were in the genomes of organisms living millions-to-billions of years ago, we can trace the evolutionary history of genes by comparing the genomes of different groups of present-day organisms. From such comparisons, we can estimate when new genes first appeared in ancient organisms, describe how they changed over time, and gain insights into what molecular processes cause new genes to evolve in the first place.

Comparative genomics has shown that some genes are found in the genomes of almost all present-day organisms. Examples are genes involved in core biological processes like transcription and protein synthesis, including genes for some subunits of RNA polymerase, genes for many of the proteins that make up part of the structure of a ribosome, and most of the aminoacyl–tRNA synthetase enzymes that attach amino acids to transfer RNA molecules. The proteins coded for by these genes not only perform the same function in every organism, but they are also related evolutionarily. This conclusion strongly suggests that the single-celled common ancestor of all living organisms had those genes in its genome, and that those genes have been passed down through the generations for billions of years.

Most genes do not appear in the genomes of all organisms but have a more restricted distribution. There are eukaryote-specific genes, bacteria-specific genes, archaea-specific genes, animal-specific genes, plant-specific genes, primate-specific genes, human-specific genes, and so on. For example, mitosis and meiosis genes are eukaryote-specific genes, genes for flowers are plant-specific, and some genes related to brain function are primate-specific.

Analyzing the evolutionary history of genes provides valuable information about how life evolved on the molecular level. For example, by comparing the functions of almost 4,000 evolutionarily related groups of genes in 100 genomes of bacteria, archaea, and eukaryotes, researchers have identified a period about three billion years ago when many new genes evolved. By analyzing the functions of these new genes, the researchers concluded that many of the new genes evolved as adaptations to changes in the amount of oxygen in Earth's atmosphere, following the development of the oxygen-producing photosynthetic reactions (see Chapter 9).

Comparative genomics has also been applied to understanding human evolution. (Human evolution is discussed in Section 32.13.) The human genome was the first mammalian genome sequenced, and researchers now have the sequences of hundreds of human genomes to study and compare to discover what makes us human and what is responsible for human variation. We also have over 40 other mammalian genomes to compare with the human genome. These include genomes of primate species that are closely related to humans, such as the common chimpanzee and the mountain gorilla, as well as less closely related mammals, such as the cow and the duck-billed platypus. Comparing the human genome with the genomes of other primates reveals which features are common to all of these primates and which are unique to humans. The human and chimpanzee genomes are strikingly similar, with 96% DNA sequence identity across the entire genome. The annotation of the chimpanzee genome is not yet complete, but it is likely that these two species share virtually all of their genes, so the genomic changes that occurred in human evolution probably involved only subtle mutations in the protein-coding sequences of genes, and mutations to regulatory sequences that determine how and when each gene is expressed (see Chapter 16). By contrast, comparisons of primate genomes with those of other mammals have identified new genes that evolved only in primates. Further studies of the functions of these genes may shed light on how primates evolved the characteristics that distinguish them from other mammals (see Chapter 32). And, interestingly, the primate-specific genes in the human genome contain the highest fraction of disease-related genes—19.4%—of any group of genes.

DNA microarrays and RNA-seq (see Section 19.3) have been used to compare which genes are transcribed in which parts of the brain in humans, chimpanzees, and rhesus monkeys. Certain groups of genes are expressed in the brain only during embryonic and early postnatal development. In both chimpanzees and rhesus monkeys, many genes involved in the formation of new synapses (communicating junctions between neurons: see Chapter 39) are transcribed only in the first year after birth, while in humans expression of these genes continues up to age five. These differences are most prominent in the prefrontal cortex, which is an area of the brain involved in complex decision-making (see Chapter 40). These comparative findings provide clues to how our species evolved enhanced learning abilities.

Comparative genomics also provides information about how the arrangement of genes on chromosomes has evolved. In Section 13.3 you learned about chromosomal mutations that occur when part of a chromosome is translocated to another chromosome, or inverted in place. Nondisjunction in meiosis can also cause entire chromosomes to be duplicated (see Figure 13.13). Such chromosomal mutations are uncommon, and usually they are harmful. But when a nonharmful chromosomal mutation occurs and spreads to all members of a species, the order of genes on the chromosomes of that species may then be different from the order in closely related species. Comparing the genomes of a range of related species reveals that, over the course of hundreds of millions of years of biological evolution, pieces of chromosomes have changed places repeatedly by translocation and inversion, rearranging the genes on chromosomes like shuffling a deck of cards (see Figure 22.18). Even so, the chromosomal arrangement of some genes is preserved after all this reshuffling, even in distantly related organisms. For example, comparisons of the human genome with the genomes of distantly related animals such as insects and sea anemones reveal blocks of homologous genes that are on the same chromosome and arranged in the same order in all of these species. This means that the order of these genes on the chromosome has been preserved from the time that all of these species evolved from a common ancestor, even though that common ancestor lived over 500 million years ago.

New Genes Evolve by Duplication and Exon Shuffling

The evolution of a new gene is a rare event—much less common than a mutation in an existing gene. But, nonetheless, over millions of years of biological evolution many new genes have been produced. For example, a comparison of genome sequences among four species in the *Drosophila* genus of fruit flies identified over 200 genes that had evolved in just the past 13 million years (a comparatively short time in evolutionary history).

Throughout the history of life on Earth, the evolution of new biological functions has almost always involved the evolution of new genes. For example, photosynthesis became possible only with the evolution of genes coding for proteins that could harness the energy in photons to synthesize ATP and electron carrier molecules. And comparative analysis of mammalian genomes has revealed genes involved in milk production and other biological functions found only in mammals. Evolutionary biologists have described a number of molecular mechanisms to explain where these new genes come from. The most common molecular mechanism to explain the origin of new genes is *gene duplication,* which produces *multigene families* after a series of duplications of genes that all derive from the same ancestral gene. New *types* of genes are produced by a process called *exon shuffling,* which combines parts of two or more genes.

GENE DUPLICATION **Gene duplication** is any process that produces two identical copies of a gene in an organism's genome. Genes can be duplicated either by *unequal crossing-over* of homologous chromosomes during meiosis, or as a result of the replication of transposable elements (see Section 15.5).

Unequal crossing-over is the rare phenomenon in meiosis in which, instead of crossing-over occurring at the exact same point on each homolog of a homologous pair of chromosomes **(Figure 19.11A)**, crossing-over occurs at different points **(Figure 19.11B)**. The result of unequal crossing-over is that one of the recombinant chromosomes is missing one or more genes, while the other has duplicate copies of those genes. Unequal crossing-over produces **tandem duplication** of genes with the duplicate copies clustered together in the same region of the same chromosome.

Gene duplication may occur when a transposable element copies itself and splices the DNA copies elsewhere in a genome. (Transposable element movement is discussed in Section 15.5.) Rarely, transposable elements copy adjacent DNA in addition to their own, producing duplicate copies of any genes in that DNA. This produces **dispersed duplication** of genes, meaning that the copies of the gene are found in different places in the genome—often on two different chromosomes.

At first, the duplicate copies of a gene have the same protein-coding sequences and encode identical proteins. The two genes are functionally redundant, meaning that one could be eliminated from the genome with no loss of biological functionality. Often, one of the redundant copies is mutated into a pseudogene, or lost by deletion. But if both genes remain functional, they will evolve slowly in different ways, as different mutations occur in each gene. Mutations in regulatory sequences may change how each duplicate gene is regulated, or mutations in protein-coding sequences may change the functional properties of the proteins produced by each gene. Over many generations, this evolutionary process can produce two homologous genes with similar but distinct functions.

For example, nitric oxide synthase enzymes catalyze a reaction that produces nitric oxide (NO), a molecule that cells use to communicate with each other (see Section 44.5 for an example). In the human genome, there are three genes for different nitric oxide synthase enzymes. One gene is expressed only in neurons, another only in the endothelial cells lining blood vessels (see Chapter 44), and the third in the liver and in a type of white blood cell called a macrophage (see Chapter 45). Homologs of all three of these genes are found in other mammalian genomes as well as in the genomes of birds and reptiles, so the gene duplications by which these genes evolved must have happened over 200 million years ago. Evolution of tissue-specific expression patterns for each gene must have involved mutations to regulatory sequences that control transcription. The evolution of these genes also involved mutations to the protein-coding sequences, causing the proteins to have subtly different structures and functions. For example, the neuronal and endothelial nitric oxide synthase enzymes are regulated by Ca^{2+} ions, while the third enzyme is not. The three genes are found on three different chromosomes in the human genome, which suggests that they evolved by dispersed duplication.

PRODUCTION OF MULTIGENE FAMILIES Gene duplication is often followed by further duplication of one or both of the original duplicates. When this happens repeatedly over millions of years, a family of homologous genes called a **multigene family** evolves. The nitric oxide synthase enzymes described above are a small gene family, with three members in the human genome. Other multigene families contain tens to hundreds of genes. The members of a multigene family all evolve from one ancestral gene, and therefore have similar DNA sequences and produce proteins with similar structures and functions. But because different mutations occur in each member of a multigene family, the genes gradually evolve subtly different characteristics.

Let us consider the *OPT/YSL* multigene family of the plant, *Arabidopsis thaliana,* as an example. The proteins coded for by these genes are oligopeptide transporters that shuttle short peptide molecules across cell membranes. This multigene family is

A. Normal crossing-over

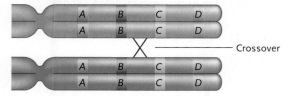

Crossover

Crossing-over occurs between homologous chromatids during prophase I of meiosis (see Figure 11.5). Normally crossing-over occurs at the exact same point on each homolog and results in recombinant chromosomes after meiosis that have the same number of genes in each homolog.

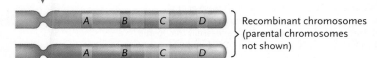

Recombinant chromosomes (parental chromosomes not shown)

B. Unequal crossing-over

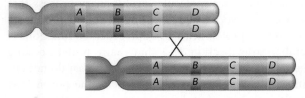

Unequal crossing-over results in recombinant chromosomes after meiosis that have a different number of genes. One (top) has duplicate genes, here *B* and *C*, whereas the other (bottom) has lost genes, here *B* and *C*.

FIGURE 19.11 Duplication of genes by unequal crossing-over.
© Cengage Learning 2014

FIGURE 19.12 Evolution of the plant
OPT/YSL multigene family. **(A)** Family tree
showing evolutionary relationships among *OPT*
and *YSL* genes. **(B)** Distribution of *OPT* genes
on chromosomes in the
Arabidopsis thaliana genome.
© Cengage Learning 2014

A. Family tree showing evolutionary relationships among *OPT* and *YSL* genes

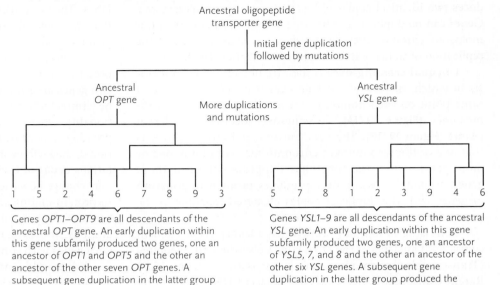

Genes *OPT1–OPT9* are all descendants of the ancestral *OPT* gene. An early duplication within this gene subfamily produced two genes, one an ancestor of *OPT1* and *OPT5* and the other an ancestor of the other seven *OPT* genes. A subsequent gene duplication in the latter group produced the ancestor of *OPT2, 4,* and *6–9,* and the ancestor of *OPT3.*

Genes *YSL1–9* are all descendants of the ancestral *YSL* gene. An early duplication within this gene subfamily produced two genes, one an ancestor of *YSL5, 7,* and *8* and the other an ancestor of the other six *YSL* genes. A subsequent gene duplication in the latter group produced the ancestor of *YSL1–3* and *9,* and the ancestor of *YSL4* and *6.*

B. Distribution of *OPT* genes on chromosomes in the *Arabidopsis thaliana* genome

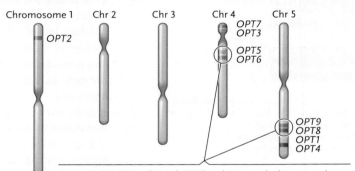

Gene pairs *OPT5* and *6* and *OPT8* and *9* are each close enough that they could have evolved by a recent tandem duplication. But only *OPT8* and *9* are near relatives in the *OPT* gene family tree (part **A** of figure). Therefore, we can hypothesize that *OPT8* and *9* likely resulted from a fairly recent tandem duplication, whereas *OPT6* is a dispersed duplicate of another *OPT* gene.

found in other plant genomes as well as in the genomes of fungi and other eukaryotes. By comparing DNA sequences of *OPT/YSL* genes in the genomes of plants and other organisms, researchers have concluded that, hundreds of millions of years ago, and before the evolution of the plant kingdom, a gene duplication produced the ancestral *OPT* gene and the ancestral *YSL* gene. Mutations in each gene caused them to encode proteins specialized for transporting different oligopeptides. A series of more recent gene duplications occurred since the evolution of plants, but well before the evolution of *Arabidopsis thaliana*. Each duplicate gene in this family accumulated different mutations, producing the functionally diverse set of *OPT* genes and *YSL* genes now found in *Arabidopsis thaliana* and other plants.

Figure 19.12A illustrates the family relationships among the *OPT/YSL* genes using a phylogenetic tree, much as a family tree illustrates relationships in a human family (see Chapter 24 for more information on how phylogenetic trees are constructed). **Figure 19.12B** shows the distribution of the *OPT* genes on *Arabidopsis thaliana* chromosomes and outlines the possible evolutionary history of some of those genes.

The oligopeptide transporter gene family in *Arabidopsis thaliana* is larger than the nitric oxide synthase gene family in the human genome, but other multigene families are even larger. For example, some families of transcription factor proteins and membrane-bound receptor proteins include hundreds of genes. Some of the larger multigene families have members in all kingdoms of eukaryotes, or even in all three domains of living organisms. Such families each evolved from an ancestral gene that first appeared billions of years ago.

EXON SHUFFLING The new genes produced by gene duplication evolve distinct functions, but they retain the same general function as other members of the multigene family into which they have been "born," so to speak. Our examples have illustrated that. By contrast, **exon shuffling**—the duplication and rearrangement of exons—is a molecular evolutionary process that combines exons of two or more existing genes, to produce a gene that encodes a protein with an unprecedented function.

Remember from Section 15.3 that many protein-coding genes in eukaryotes contain introns, sequences that do not encode amino acids. The introns are present in the pre-mRNA transcripts of such genes but, by RNA processing, the introns are removed while the exons—the sequences encoding amino acids in the pre-mRNAs—are spliced together to make the mature mRNAs. In many genes, the junctions between exons fall at points within the protein-coding sequence between major functional regions in the protein. These functional regions correspond to the domains into which many proteins are divided (see Section 3.4).

Exon shuffling can occur in the following way. When a piece of DNA is cut out of a chromosome and reinserted elsewhere in the genome (through the activity of a transposable element, for example), the ends of the piece of DNA that moves may occur within the introns of a gene, causing one or more whole exons to be inserted somewhere else in a chromosome. If those exons are inserted into an intron in another gene, the amino acid sequence encoded by those exons may be added to the amino acid sequence of the encoded protein. Such a transfer of DNA can produce a new gene, coding for a protein that has one or more domains added to the other domains that it already had.

An exon shuffling event occurred very early in the evolution of animals (at least 700 million years ago) that produced a new gene coding for a protein that plays a key role in signaling between cells in animal tissues. Evidence for this exon shuffling event comes from comparing the genome sequence of the choanoflagellate *Monosiga brevicollis* with the sequences of a number of animal genomes, including *Homo sapiens*. Choanoflagellates (see Sections 27.2 and 31.1) are single-celled or colonial protists that are thought to be related evolutionarily to animals. The evolution of multicellularity in the first animals is thought to have involved molecular mechanisms that enabled choanoflagellate-like cells to attach to and communicate with one another, so they could then specialize in performing different functions.

Figure 19.13 shows the exon shuffling event. It involves the Notch family of proteins, which are multidomain, membrane-spanning proteins. The human Notch1 protein, encoded by the *NOTCH1* gene, contains a transmembrane (TM) region, 36 copies of an EGF domain, three copies of an NL domain, and six copies of an ankyrin domain. The TM region anchors the Notch1 protein in the plasma membrane. The three domains are key to the protein's function. The part of the protein that is outside the cell membrane includes the EGF and NL domains, which enables a Notch protein in one cell to bind to other proteins in adjacent cells in a tissue. The ankyrin domains within the cell enable it to attach to the microfilaments that make up part of the cytoskeleton (see Section 5.3).

Figure 19.13 also shows three different genes in the *Monosiga brevicollis* genome, one with a sequence encoding EGF domains, a second with a sequence encoding NL domains, and a third with a sequence encoding ankyrin domains. However, this organism lacks a gene homologous to the gene for the Notch protein. Researchers have hypothesized that, through exon shuffling early in animal evolution, the sequences coding for the EGF, NL, and ankyrin domains in the three genes were combined in one gene, producing the ancestor of the *NOTCH* gene family. At some point, a duplication of the sequence coding for one NL domain occurred, since the *Monosiga brevicollis* gene has sequences coding for only two copies of that domain while animal genes for Notch proteins have sequences coding for three copies.

The genes produced by gene duplication typically encode functional proteins because they are duplicates of an existing functional gene. Most instances of exon shuffling theoretically should produce nonfunctional proteins, because an existing functional protein is interrupted by one or more domains from an unrelated protein. But in a small number of cases, proteins produced by exon shuffling combine the functions of two or more proteins in a new and useful way, like the assembly of the gene for the Notch signaling protein (see Figure 19.13). Interest-

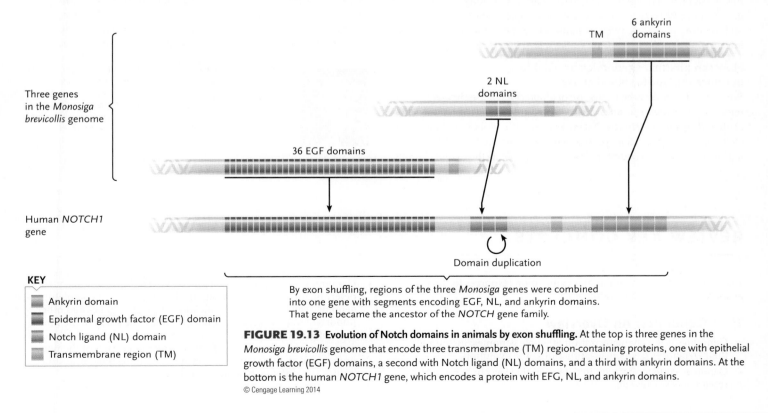

Three genes in the *Monosiga brevicollis* genome

6 ankyrin domains

TM

2 NL domains

36 EGF domains

Human *NOTCH1* gene

Domain duplication

By exon shuffling, regions of the three *Monosiga* genes were combined into one gene with segments encoding EGF, NL, and ankyrin domains. That gene became the ancestor of the *NOTCH* gene family.

KEY

- Ankyrin domain
- Epidermal growth factor (EGF) domain
- Notch ligand (NL) domain
- Transmembrane region (TM)

FIGURE 19.13 Evolution of Notch domains in animals by exon shuffling. At the top is three genes in the *Monosiga brevicollis* genome that encode three transmembrane (TM) region-containing proteins, one with epithelial growth factor (EGF) domains, a second with Notch ligand (NL) domains, and a third with ankyrin domains. At the bottom is the human *NOTCH1* gene, which encodes a protein with EFG, NL, and ankyrin domains.

© Cengage Learning 2014

ingly exon shuffling is thought to account for perhaps 1/3 of newly evolved genes. The genes that are produced by exon shuffling have more novel functions than those produced by gene duplication. In cases like the Notch protein, they become the ancestors of multigene families that evolve through subsequent gene duplications.

The domain structures of proteins in humans and other organisms provide evidence of how common exon shuffling has been in the evolution of proteins. The most widely used do-mains are found in thousands of different proteins, in dozens of different combinations with other domains.

STUDY BREAK 19.4

1. What molecular mechanisms cause tandem duplication of genes and dispersed duplication of genes?
2. Why do new genes produced by exon shuffling have more novel functions than new genes produced by gene duplication?

UNANSWERED QUESTIONS

It is well known that individuals respond differently to medications. For example, a blood pressure lowering medication may effectively lower blood pressure in one patient, have no effect on blood pressure in another patient, and cause intolerable adverse effects in a third patient. Factors such as age, weight, or liver and kidney function may influence response to medication. However, it remains difficult to predict how effective or safe a drug will be for a particular patient based on these factors alone. In clinical medicine, a trial-and-error approach is often used for drug therapy—several drugs or drug doses may be tried before finding the drug and dose that produces optimal response with acceptable tolerability.

Genes encoding proteins involved in the metabolism and transport of a drug through the body and for proteins at the drug target site may greatly influence drug response. Genes involved in disease progression or phenotype may also influence how well a drug works. How can molecular approaches be used to personalize drug therapy? The answer to that question may well come from the field of pharmacogenomics. Pharmacogenomics is the study of how genes affect individual responses to drug therapy. The goal is to choose the most appropriate drug, drug dose, and treatment duration for a particular patient based on genetic information.

How can pharmacogenomics be used to optimize therapy for blood clots?

Prescribed to prevent the formation of blood clots in the circulatory system, the drug warfarin ranks among the top 50 most commonly prescribed drugs in the United States. Because it inhibits the blood from clotting, warfarin also increases the chance of bleeding. The risk for bleeding increases when the warfarin dose is too high, while the risk for clotting

increases when the dose is too low. The dose of warfarin needed to prevent clotting without significantly increasing the chance for bleeding is highly variable among patients, with doses ranging by as much as 20-fold. Age, body size, other diseases the person may have, and medications the person may be taking, all influence the dose of warfarin a person needs. However, considering age, body size, and clinical factors alone is not enough to choose the right warfarin dose for a given patient. Genes encoding for proteins involved in warfarin metabolism (pharmacokinetics) and mechanism of action (pharmacodynamics) are now known to influence warfarin dose requirements. Our group is working with other research groups from around the world to study variations in a person's genome that affect the dose of warfarin that he or she needs. The goal is to develop an algorithm, or equation, for dosing warfarin that includes both genetic and clinical factors and accurately predicts warfarin dose requirements for persons of various racial and ethnic backgrounds.

THINK LIKE A SCIENTIST What are the ethical implications of pharmacogenomics if the genes in question are associated with disease risk as well as drug response? What are the ethical implications of pharmacogenomics if a patient possesses a genotype predictive of poor drug response and there are no alternative treatments available?

Larisa H. Cavallari

Larisa H. Cavallari is an Associate Professor in the Department of Pharmacy Practice at the University of Illinois at Chicago. She is involved in both clinical and basic science research focusing on genetic contributions to cardiovascular drug response. To learn more about Dr. Cavallari's laboratory and research, go to http://pmpr.pharm.uic.edu/pharmacogenomics/index.htm.

REVIEW KEY CONCEPTS

To access the course materials and companion resources for this text, please visit www.cengagebrain.com.

19.1 Genomics: An Overview

- Genomics is the characterization of whole genomes, including their structures (sequences), functions, and evolution.
- Genome sequence data are in databases that may be accessed by researchers worldwide.

- Genomics consists of three main areas of study: genome sequence determination and annotation, the determination of complete genome sequences and identification of putative genes and other important sequences; functional genomics, the study of the functions of genes and other parts of the genome; and comparative genomics, the comparison of entire genomes of parts of them to understand evolutionary relationships and basic biological similarities and differences among species.

19.2 Genome Sequence Determination and Annotation

- The whole-genome shotgun method of sequencing a genome involves breaking up the entire genome into random, overlapping fragments, cloning each fragment, determining the sequence of the fragment in each clone, and using computer algorithms to assemble overlapping sequences into the sequence of the complete genome (Figure 19.1).

- DNA sequencing methods involve DNA purification; DNA fragmentation; amplification of fragments; sequencing each fragment; and assembly of fragment sequences into longer sequences, such as those of a genome (Figures 19.2 and 19.3).

- Once the complete sequence of a genome has been determined, it is annotated to identify key sequences, including protein-coding genes, noncoding RNA genes, regulatory sequences associated with genes, origins of replication, transposable elements, pseudogenes, and short repetitive sequences. Annotation of a genome is the task of researchers in bioinformatics.

- Identifying protein-coding genes in a genome sequence can be done by using a computer search for open reading frames, by searching databases for sequence similarity to genes of known function, and (the gold standard) by studying gene transcripts.

- Complete genome sequences have been obtained for many viruses, a large number of prokaryotes, and many eukaryotes, including the human. For organismal genomes, those of bacteria and archaea are generally smaller than those of eukaryotes and their genes are densely packed in their genomes with little noncoding space in between them. Prokaryotic genes are organized either into single transcription units or into operons. Genomes of eukaryotes vary greatly in size, but there is no correlation between genome size and type of organism. Gene density also varies, but in general it is significantly less than is seen for prokaryotic genes. Eukaryotic genes are organized into single transcription units (Figures 19.5–19.7; Tables 19.2 and 19.3).

19.3 Determining the Functions of the Genes in a Genome

- The function of a protein-coding gene may be assigned by a sequence similarity search of sequence databases, by using evidence from protein structure, or by using knockout or knockdown experiments that alter the expression of a gene.

- Transcriptomics is the study at the genome level of when and where genes are transcribed. One experimental method for studying the transcription of all or many of the genes in a genome simultane-

ously is the DNA microarray (DNA chip); this technique can generate qualitative information about gene transcription, such as the similarities and differences in gene expression in two cell types or in two developmental stages, as well as quantitative information about the relative levels of gene transcription (Figure 19.8).

- Proteomics is the characterization of the complete set of proteins in an organism or in a particular cell type. Protein numbers, protein structures, protein functions, protein locations, and protein interactions are all topics of proteomics (Figures 19.9 and 19.10).

19.4 Genome Evolution

- Comparative genomics traces the evolution of genomes by analyzing similarities and differences in DNA sequences in the genomes of present-day organisms.

- Comparative analysis reveals how homologous protein-coding genes have evolved in groups of organisms, how regulation of gene expression has evolved through mutations in the regulatory sequences of genes, and how chromosome structure has evolved as parts of one chromosome have broken off and been attached to other chromosomes.

- New genes evolve by tandem duplication when chromosomes cross over unequally in meiosis, producing a chromosome containing two copies of the DNA coding for one or more genes (Figure 19.11). New genes evolve by dispersed duplication when transposable elements copy DNA coding for one or more genes, and insert it at another location in the genome.

- Duplicate copies of genes evolve distinct functions as different mutations occur in the two copies. Mutations in protein-coding sequences produce proteins with slightly different structures, while mutations in regulatory sequences cause the genes to be expressed in different cell types and in response to different stimuli.

- Repeated cycles of duplication followed by mutation produce multigene families, which are collections of homologous genes that code for similar but functionally distinct proteins. The largest multigene families comprise hundreds of different genes (Figure 19.12).

- Exon shuffling produces functionally novel proteins by combining parts of two or more different genes. When exons that code for one or more domains of a protein are copied from one gene and inserted into the protein-coding sequence of another gene, those domains are added to the structure of the protein coded for by that gene. Adding new domains to a protein gives the protein new molecular functionalities (Figure 19.13).

UNDERSTAND & APPLY

Test Your Knowledge

1. Why is the Solexa/Illumina DNA sequencing method faster and less expensive than the Sanger method?
   a. It sequences longer fragments of DNA.
   b. It sequences more DNA fragments at the same time.
   c. It does not require the use of fluorescent markers.
   d. It does not require amplification of DNA fragments before sequencing.
   e. It does not require the use of computer algorithms to find places where sequence fragments overlap.

2. How do pseudogenes differ from genes?
   a. They are not transcribed.
   b. They contain longer open reading frames (ORFs).
   c. They do not have introns.
   d. They use a different genetic code.
   e. Their protein-coding sequence contains more than one start codon.

3. What is the main reason that searching for open reading frames (ORFs) is more useful for annotating prokaryote protein-coding genes than it is for annotating eukaryote protein-coding genes?
   a. Eukaryote protein-coding genes contain introns.
   b. The density of protein-coding genes is much higher in eukaryote genomes.
   c. In most prokaryotes, all of the protein-coding genes are located on a single circular chromosome.
   d. Prokaryotes use a different genetic code than eukaryotes.
   e. Prokaryotic protein-coding genes are much longer than eukaryotic protein-coding genes.

4. Which of the following is true about genome size?
   a. Bacteria have genomes that vary widely in size.
   b. The human genome is the largest among eukaryotes.
   c. Organisms with large genomes are always more complex than organisms with small genomes.
   d. As genome size increases in a lineage, the number of genes also always increases.
   e. The smallest known cellular genome is found in a species of Archaea.

5. Which of the following statements about the *E. coli* genome is false?
   a. Most of the genes are located on one circular chromosome.
   b. It has a much higher gene density than the human genome.
   c. It contains fewer genes than the human genome.
   d. All of the genes are transcribed from the same template strand of the DNA double helix.
   e. About half of the genes in the *E. coli* genome are grouped with other genes in operons.

6. Which of the following does *not* characterize the human genome?
   a. Introns occupy 20% to 25% of the genome.
   b. The protein-coding sequences occupy about 2% of the genome.
   c. About 45% of the genome consists of transposable element sequences.
   d. The genome sequence is comprised of approximately 30 million base pairs.
   e. Human cells have about 20,500 different protein-coding genes.

7. About 95% of the average human transcription unit consists of:
   a. short repeat sequences.
   b. protein-coding sequences.
   c. regulatory sequences.
   d. introns.
   e. origins of replication.

8. When the DNA sequences of two protein-coding genes are similar, but only for part of the protein-coding sequence, that suggests:
   a. the two proteins have one or more domains in common.
   b. the two proteins were produced by duplication of an ancestral gene.
   c. the two proteins perform the same function.
   d. one of the two genes is actually a pseudogene.
   e. both genes are pseudogenes.

9. When two protein-coding genes have very similar nucleotide sequences and are located right next to each other on a chromosome, we can hypothesize that:
   a. one of them is a duplicate of the other, copied by a retrotransposon.
   b. they are nonhomologous.
   c. one of them is a pseudogene.
   d. they were produced by unequal crossing-over.
   e. they are transcribed in the same cell types.

10. The proteins coded for by genes in a multigene family begin to evolve distinct functions when:
    a. gene duplication occurs.
    b. exon shuffling occurs.
    c. the genes are expressed by transcription and translation.
    d. different mutations occur in each protein-coding sequence.
    e. the two proteins evolve so they have the same three-dimensional structure.

Discuss the Concepts

1. Why are high-throughput techniques used so much in genomics research? Give examples from Chapter 19 of different uses of high-throughput techniques.

2. Why does the Sanger DNA sequencing method work best when the concentration of dideoxyribonucleotides is much less than the concentration of deoxyribonucleotides? If you wanted to adjust the reaction mixture to produce a greater number of very long complementary sequence fragments, how would you change the relative concentration of dideoxyribonucleotides, and why?

3. Which of the methods for annotating protein-coding genes would you expect to do the best job of distinguishing functioning genes from pseudogenes, and why?

4. The genome of the yeast *Saccharomyces cerevisiae* is only about 0.4% the size of the human genome, yet it contains about 30% as many genes as are in the human genome. Given that, which of the features of the human genome would you expect to find many fewer of in the yeast genome?

5. How does sequencing the genomes of a greater number of animal species help in annotating and determining the functions of human protein-coding genes?

Design an Experiment

You are studying the molecular mechanisms of sex determination in fruit flies *(Drosophila melanogaster)*. How can you determine which genes are expressed at higher levels in male *Drosophila* embryos and which genes are expressed at higher levels in female *Drosophila* embryos? Once you have identified sex-specific genes, what further experiments could you do to test what effects they have on the development of *Drosophila* sex organs?

Interpret the Data

Below is a sequence of 540 bases from a genome. What information would you use to find the beginnings and ends of open reading frames? How many open reading frames can you find in this sequence? Which open reading frame is likely to represent a protein-coding sequence, and why? Which are probably not functioning protein-coding sequences, and why? Note: for simplicity's sake, analyze only this one strand of the DNA double helix, reading from left to right, so you will only be analyzing three of the six reading frames shown in Figure 19.4.

```
5'-AGTTTTATTTAAAAGAGTAGATTAAGAAAAGTAGTATTAGAATTTTATTGATTT
   ATGCAATTAGAGTACCTCAATCTTATTTCTCAAGCTAAAGTTATTGCAGAAAAA
   CAATTTAAAGCTAACCCTTTTTCTTTTGAAACAATTAGAAAAGAAGTAGTTAAA
   CATTTCAAGATTTCAAAACAAGATGAACCAAGCTTAATTGGTCGTTTTTATCAA
   GATTTTCTTGAGGATCCTAACTTTGTCTATTTAGGTGATAGAAAAAGAAAACTT
   CGTGATTTTAGGAAGTTTGATAAATGGAACAAGATATCACAATCTATATTTGTT
   ACAAAGGAGATTTTTGAAGAAGGTTATGAAGATCTTTCCAATAAAAAAGTAGAA
   CCTGAGGAAGGAGTTGGTGATTTCATTATGGGAAATGACGGTGCTGACACTGAA
   ACTGGCAGTGAAATAGTACAAGGTTTAATTAATAATTCATTCAGTGAGGAAAAT
   CAATAGTAGATACGCTTGTTAACTTTAAATTGACGCTTCAAAAAGCAAAGCTAG-3'
```

Apply Evolutionary Thinking

You are studying a multigene family in the mouse genome, and would like to determine which genes evolved as a result of unequal crossing-over of chromosomes and which may have been copied through the action of transposable elements. What information would you use to help answer that question?

Chapter 1

Study Break

STUDY BREAK 1.1
1. The major levels in the hierarchy of life and some of their emergent properties are: cells—life; organisms—learning; populations—birth and death rates; communities, ecosystems, biosphere—diversity and stability.
2. Organisms use energy collected from the external environment for growth (including the production of new molecules and cells), maintenance and repair of body parts, and reproduction.
3. A life cycle is the series of structurally and functionally distinct developmental stages through which organisms pass.

STUDY BREAK 1.2
1. In artificial selection, humans selectively breed individuals with desirable heritable characteristics to enhance those traits in the next generation. In natural selection, genetically-based characteristics that increase survival and reproduction become more common in the next generation.
2. Random changes in DNA—mutations—may change the structure of proteins that contribute to the physical appearance and internal functions of an organism.
3. Being camouflaged may make an animal less likely to be noticed by a predator.

STUDY BREAK 1.3
1. In the cells of prokaryotic organisms, DNA is not separated from other parts of the cell. In the cells of eukaryotic organisms, DNA is enclosed within a nucleus.
2. Humans are classified in Domain Eukarya and Kingdom Animalia.

STUDY BREAK 1.4
1. A scientific hypothesis must be falsifiable. In other words, we must be able to imagine what sort of data would demonstrate that the hypothesis is incorrect.
2. The copper lizard models told the researchers how frequently lizards would perch in the sun just by chance and what the temperatures of nonthermoregulating lizards would be.
3. Model organisms are usually easy to maintain and study, and they have been so well studied that researchers already know a lot about their biology.
4. When scientists describe a set of ideas as a "theory," they recognize that the ideas have already withstood many scientific tests.

Think Like a Scientist

FIGURE 1.11
Animals are more closely related to fungi than they are to plants. According to the phylogenetic tree, animals and fungi shared a common ancestor more recently than did animals and plants.

FIGURE 1.15
To test the effects of two experimental variables simultaneously, you would need to design an experiment that had experimental and control groups for both variables (that is, fertilizer and water). One group might receive both fertilizer and water; a second group would receive fertilizer but no water; a third group would receive water but no fertilizer; and a fourth group would receive neither fertilizer nor water. You could then compare the results of the four treatments to see the effects of both experimental variables, alone and in combination, on your plants.

FIGURE 1.16
The temperature is not very variable among the shaded sites in this environment. Most of the models were in shaded sites, and the temperatures of the shaded sites are clustered around 19° to 20°C.

Test Your Knowledge
1. c 2. b 3. c 4. d 5. b 6. d 7. c 8. a 9. e 10. d

Interpret the Data
Anolis gundlachi does not regulate its body temperature. The graphs show that lizards perch in patches of sun at about the same rate as randomly positioned models and that lizards and models exhibit similar distributions of body temperatures. These data suggest that this lizard species perches at random with respect to environmental factors that might influence its body temperature.

Chapter 2

Study Break

STUDY BREAK 2.1
An element is a pure substance that consists of one type of atom. An atom is the smallest unit of an element that retains its chemical and physical properties. A molecule is a collection of atoms chemically combined in fixed numbers and ratios. Molecules can consist of the same atoms, as is seen for the two oxygen atoms in the molecule oxygen, or of different atoms, as in the combination of two hydrogen atoms and one oxygen atom in a molecule of water. Molecules with component atoms that are different, such as water, are compounds.

STUDY BREAK 2.2
1. Protons and neutrons are found in the nucleus of an atom. Electrons are found in orbitals located in energy levels (shells) that surround the nucleus.
2. Carbon-11 has six protons and five neutrons. Oxygen-15 has eight protons and seven neutrons.
3. The number of valence electrons—the electrons in the outermost shell of an atom—determines its chemical reactivity. If the outermost shell is not completely filled with electrons, the atom tends to be chemically reactive, whereas if that shell is completely filled, the atom is nonreactive.

STUDY BREAK 2.3
1. An ionic bond forms between atoms when those atoms gain or lose electrons completely. An example is the ionic bond in NaCl.
2. A covalent bond forms when atoms share a pair of valence electrons rather than gaining or losing them completely.
3. Electronegativity is a measure of an atom's attraction for the electrons it shares in a chemical bond with another atom. When electrons are shared equally, the atoms remain uncharged and the result is a nonpolar covalent bond. When electrons are shared unequally, one atom carries a partial negative charge and the other atom carries a partial positive charge. The molecule then has polarity, and the bond is a polar covalent bond.
4. In a chemical reaction, atoms or molecules interact to form new chemical bonds or break old ones. Atoms are added to or removed from molecules, or linkages of atoms in molecules are rearranged as a result of bond formation.

STUDY BREAK 2.4
1. Hydrogen bonds between neighboring water molecules produce a water lattice. The constant breakage and reformation of hydrogen bonds in the lattice allows water to flow easily. The polarity of water also contributes to the properties of water; that is, in liquid water, the lattice resists invasion by other molecules unless the invading molecules also contain polar regions that can form competing attractions with water molecules. In that case, the water lattice opens, forming a cavity in which the polar or charged molecule can move. However, nonpolar molecules are unable to affect the water lattice. Hydrogen bonds also give water its unusual ability to resist changes in temperature by absorbing or releasing heat energy, its unusually high boiling point, and its unusually high internal cohesion and surface tension.
2. A solute is a dissolved substance. A solvent is a substance capable of dissolving another substance. A solution is a solute dissolved in a solvent. For example, salt (NaCl) is a solute that can dissolve in the solvent water.

STUDY BREAK 2.5
1. Acids are hydrogen ion (proton, H^+) donors; bases are proton acceptors. An acid dissociates in water to produce a hydrogen ion and an anion. Most bases dissociate in water to give hydroxide ions, which then accept protons to produce water.
2. Buffers act to control the pH of a solution. In living organisms, buffers keep the pH of body and cell fluids within a narrow range, enabling normal cell and body functions to occur. Outside of the normal pH range, the functions of proteins can be affected, thereby adversely affecting the functions of the organism.

Think Like a Scientist

UQ
Properties of natural soils can determine the flow pattern and how much carbon source can be delivered to bacteria, therefore controlling their capability of remediating uranium. In addition, the abundance and spatial distribution of certain minerals can affect the magnitude of energy sources for the bacteria and therefore the efficiency of uranium bioremediation.

Test Your Knowledge
1. d 2. e 3. a 4. b 5. a 6. d 7. d 8. c 9. e 10. b

Interpret the Data
1. About 50 minutes
2. (a) Using a pH of 5.9, H^+ concentration is 0.0000012589 (1.2589×10^{-6}) M and OH^- concentration is 0.00000000789433 (7.9433×10^{-9}) M. (b) Using a pH of 2.5, H^+ concentration is 0.00316228 (3.16228×10^{-3}) M and OH^- concentration is 0.0000000000031623 (3.1623×10^{-12}) M.
3. Using a pH of 2.5 for peak reflux and pH of 4.0 for clinical reflux, the H^+ concentration is 0.003 M greater, and OH^- concentration is 0.0000000000968377 $(9.68377 \times 10^{-11})$ M less than a pH of 4.0.

Chapter 3

Study Break

STUDY BREAK 3.1
1. Organic molecules are molecules based on carbon. Hydrocarbons are a type of organic molecule that consists of carbon linked only to hydrogen atoms.
2. The maximum number of bonds that a carbon atom can form is four.
3. Carboxyl groups donate a hydrogen ion in water and therefore act as acids. Amino groups accept a hydrogen ion in water and therefore act as bases. Phosphate groups donate hydrogen ions in water and therefore act as acids.
4. In a dehydration synthesis reaction, components of water (—H and —OH) are removed. In hydrolysis, the components of a water molecule are added to functional groups as molecules are broken down into smaller subunits.

STUDY BREAK 3.2
A monosaccharide is the structural unit of carbohydrate molecules. Monosaccharides are simple sugars such as trioses, pentoses, and hexoses. Glucose, galactose, and fructose are hexoses. A disaccharide is a molecule assembled from two monosaccharides linked by a dehydration synthesis reaction. Lactose, sucrose, and maltose are disaccharides. A polysac-

charide is a polymer of monosaccharide subunits. The subunits are identical or different, depending on the particular polysaccharide. Glycogen, starch, cellulose, and chitin are polysaccharides.

STUDY BREAK 3.3
The three most common lipids found in living organisms are neutral lipids, phospholipids, and steroids. Most neutral lipids consist of a three-carbon backbone chain formed from glycerol, with each carbon linked to a fatty acid side chain. In the most common phospholipids, glycerol is the backbone, with two of its binding sites linked to fatty acids. The third binding site is linked to a polar phosphate group. Steroids have structures based on a framework of four carbon rings. Differences in side groups attached to the rings distinguish the different types of steroids.

STUDY BREAK 3.4
1. Differences in the side groups (R in the figures) give the amino acids their individual properties.
2. A peptide bond is the bond between the C of the carboxyl group of one amino acid and the N of the amino group of the adjacent amino acid (see Figure 3.19). The bond is formed in a dehydration synthesis reaction between an amino group of one amino acid and a carboxyl group of another amino acid.
3. Domains are distinct structural subdivisions in the final folded forms of proteins. They are the result of the amino acid sequence of the protein (the primary structure of the protein) and the secondary, tertiary, and quaternary (if more than one polypeptide is involved) structures of the protein.

STUDY BREAK 3.5
1. Nucleic acids are formed from nucleotide monomers. A nucleotide consists of a nitrogenous base, a five-carbon sugar, and a phosphate.
2. In DNA, the five-carbon sugar is deoxyribose; in RNA it is ribose. DNA has the pyrimidine nitrogenous base T (thymine), and RNA has U (uracil).

Think Like a Scientist

FIGURE 3.24
In the presence of a high concentration of urea, the formation of the four disulfide bridges of the native enzyme apparently only rarely occurs. As a result, only a low level of enzyme activity is measured in the solution. The other enzymes appear to have formed three-dimensional structures, but likely they contain nonnative disulfide linkages and therefore result in inactive enzymes. Overall, the result strengthens Anfinsen's conclusions because it means that particular disulfide linkages are responsible for the stable, and therefore active, form of the enzyme. If the enzyme instead forms with incorrect pairs of disulfide linkages, the structure is sufficiently different that enzyme activity is lost.

IMR
1. The enzyme that assembles DNA molecules in the bacterium *Escherichia coli* (see Figure 3.26B). The enzyme has two domains, one to assemble the DNA molecules (the polymerase domain) and the other to correct mistakes during DNA assembly (the exonuclease domain).
2. The method of analysis described in *Insights from the Molecular Revolution* was the comparison of domain structure and organization. Fundamentally, the domains consist of particular sequences of amino acids. The function of a domain depends on the sequence of amino acids plus the higher-order structures it forms within the protein. As a domain diverges through evolutionary time, core functional elements must stay the same or highly similar in order that the function of the domain be retained. At the same time, other parts of the domain can experience changes in amino acid sequence. Therefore, through bioinformatics-based computer analysis of proteins sharing a particular function, researchers can determine how the domain has changed over time and, thereby, construct a phylogenetic tree based on the changes.

UQ
In type 2 diabetes, SREBPs in the liver are activated, and this leads to overproduction of fatty acids and triglycerides. Inhibitors of SREBP processing might reduce plasma triglycerides and prevent their toxic effects on blood vessels.

Test Your Knowledge
1. a 2. d 3. c 4. e 5. c 6. d 7. b 8. d 9. b 10. a

Interpret the Data
1. LDL was highest in the saturated fats diet group, and HDL was lowest in the *trans* fatty acids diet group.
2. The *trans* fatty acid group had the highest LDL-to-HDL ratio.
3. From best to worst diet: *cis* fatty acids, saturated fats, *trans* fatty acids.

Chapter 4

Study Break

STUDY BREAK 4.1
1. Kinetic energy is the energy of motion, whereas potential energy is stored energy.
2. An isolated system exchanges neither matter nor energy with its environment. A closed system can exchange energy, but not matter, with its environment. An open system can exchange both energy and matter with its environment.

STUDY BREAK 4.2
1. The change in energy content of a system, and its change in entropy. Reactions tend to be spontaneous when the products have less potential energy than the reactants, and when the products are less ordered than the reactants.
2. The greater the negative value of ΔG, the further a reaction will proceed toward completion and therefore the greater the concentration of product molecules versus reactant molecules.
3. An exergonic reaction releases free energy—ΔG is negative because the products contain less free energy than the reactants. An endergonic reaction requires free energy from the surroundings to run—ΔG is positive because the products contain more free energy than the reactants.

   In a catabolic reaction, energy is released during the breakdown of a complex reactant molecule to a simpler product molecule, whereas in an anabolic reaction, energy is used to create a product molecule that is more complex than the reactant molecule.

   Individual reactions may be exergonic or endergonic. When a series of reactions (each of which may have a positive or a negative ΔG value) forms a metabolic pathway, the pathway is catabolic if the overall sum of individual reaction ΔG values is negative, and it is anabolic if the overall sum of individual reaction ΔG values is positive.

STUDY BREAK 4.3
1. ATP contains the five-carbon sugar ribose, with the nitrogenous base adenine linked to one of the carbons and a chain of three phosphate bonds to another carbon. The phosphate groups are closely associated with each other and their negative charges strongly repel each other, making the bonding arrangements unstable and storing potential energy. Hydrolysis of ATP to remove one or two of the three phosphates is a spontaneous reaction that relieves the repulsion and releases large amounts of free energy.
2. Many individual reactions found in living cells are not spontaneous because they have a positive ΔG. By joining such a reaction to another reaction with a large negative ΔG, the reaction can be completed. The combined reaction is called a coupled reaction. ATP participates in coupled reactions in the enzyme-driven process of energy coupling; that is, ATP comes into close contact with a reactant molecule in an endergonic reaction. When ATP is hydrolyzed, the terminal phosphate group is transferred to the reactant molecule, which makes that molecule less stable so that the reaction continues spontaneously.

STUDY BREAK 4.4
1. Enzymes accelerate reactions by reducing the activation energy of a reaction, the initial input of energy required to start a reaction. Enzymes lower activation energy by rearranging the atoms and bonds of the reacting molecules into the transition state, an activated state that is highly unstable. With relatively little change in energy, the transition state can move forward toward products or backward toward reactants.
2. No.

STUDY BREAK 4.5
1. There is a maximum rate at which an enzyme can combine with substrates and release products. Beyond that point, increasing substrate concentration does not increase the reaction rate any further.
2. In competitive inhibition, the inhibitor competes with the normal substrate molecule for binding to the active site of the enzyme, whereas in noncompetitive inhibition, the inhibitor does not compete directly with the substrate for binding to the active site.
3. As the temperature increases, the increasing kinetic motions of the enzyme's amino acid chains eventually disrupt the enzyme's three-dimensional structure, causing it to unfold and denature. At that point, there is no enzyme activity.

STUDY BREAK 4.6
A ribozyme is an RNA molecule that accelerates the rate of a biological reaction. A ribozyme qualifies as an enzyme because it remains unchanged after the reaction is complete; that is, it is a true catalyst.

Think Like a Scientist

IMR
Both the 5′ and 3′ constant regions are necessary for the active ribozyme structure to form.

UQ
The problem is that each nucleotide in the ribozyme would need to serve as template for polymerization. This means that the ribozyme would have to open up its own structure and inactivate its own catalytic site. Therefore, the existence of such a molecule seems very unlikely. However, it may be possible to design such a molecule if it has two catalytic centers, or if the crucial parts of the ribozyme do not need to template because they exist twice in the molecule.

Test Your Knowledge
1. c 2. d 3. b 4. c 5. e 6. a 7. a 8. b 9. d 10. e

Interpret the Data
1. ΔG is minus 5.5 minus 3.5 = -9 kcal/mol for the mutated polypeptide and minus 2.0 minus 4.0 = -6.0 kcal/mol for the parent polypeptide.
2. Binding of both of the polypeptides is exergonic and spontaneous, and the parent polypeptide would yield more free energy on binding.
3. No, because the ΔG is less negative, it would not be as energetically favorable.

Chapter 5

Study Break

STUDY BREAK 5.1
The plasma membrane is a bilayer of lipid and suspended protein molecules that bounds the cytoplasm of a cell. The lipid bilayer is hydrophobic; therefore, it is a barrier to the passage of water-soluble substances. The membrane has protein channels through which selected water-soluble substances are able to pass.

STUDY BREAK 5.2
The DNA of a prokaryotic cell is located in the nucleoid region, a central area of the cell that has no membrane around it to separate it from the cytoplasm. In most prokaryotes, the DNA is a folded mass. In its unfolded state, it is a circular DNA molecule.

STUDY BREAK 5.3

1. Most of the DNA of a eukaryotic cell is found within a nucleus located roughly in the center of the cell. The nucleus is bounded by the nuclear envelope, a membrane that separates its contents from the cytoplasm. The DNA is complexed with proteins and organized into several linear chromosomes.

2. The nucleolus is an area within the nucleus. It forms around the genes for rRNA in the chromosomes and is the location in which the information in those genes is copied into rRNA. The rRNA combines with proteins in the nucleolus to form the large and small ribosomal subunits. A large and a small ribosomal subunit function together in the cytoplasm to synthesize proteins.

3. The endomembrane system is a collection of organelles, membranous channels, and vesicles that form a major traffic network for the synthesis, distribution, and storage of proteins and other molecules. Its structure includes the endoplasmic reticulum (ER) and the Golgi complex. The rough ER has ribosomes on its outer surface. Proteins synthesized by those ribosomes enter the ER lumen, where they fold into their final shape and may be modified chemically. Then they are delivered to other regions of the cell within vesicles that pinch off from the rough ER. Most of the proteins made on rough ER go to the Golgi complex.

   The smooth ER consists of membranes that lack ribosomes. Functions of the smooth ER include synthesis of lipids that become parts of cell membranes.

   The Golgi complex is a stack of flattened, membranous sacs. The *cis* face of this organelle receives vesicles containing proteins released from the rough ER and continues their chemical modifications. The proteins are then sorted into vesicles that pinch off from the *trans* face of the Golgi complex, the side that faces the plasma membrane. Some of the released vesicles remain in the cytoplasm as storage vesicles of various types, whereas others, called secretory vesicles, release their contents to the outside of the cell.

4. A mitochondrion is an organelle enclosed by two membranes. The outer mitochondrial membrane is smooth and covers the outside of the organelle. The inner mitochondrial membrane is highly folded into cristae. Within the inner membrane is the mitochondrial matrix, which contains DNA and ribosomes. Most of the energy required for eukaryotic cellular activities is generated by reactions in the cristae and the matrix. Those reactions break down sugars, fats, and other fuel molecules into water and carbon dioxide, releasing energy mostly in the form of ATP.

5. The cytoskeleton is an internal cytoplasmic network of filaments and tubules composed mainly of actin and tubulin proteins. The function of the cytoskeleton is to maintain the shape of the cell, reinforce the plasma membrane, and organize internal structures. Changes in the cytoskeleton are responsible for movements of cell organelles, movements of parts of the cell, or movements of the whole cell.

STUDY BREAK 5.4

1. A chloroplast has two membranes: an outer boundary membrane and an inner boundary membrane. The latter, similar to the cristae of the mitochondrion, is highly folded. The two membranes enclose an inner compartment known as the stroma. Within the stroma is a membrane system composed of flattened, closed sacs called thylakoids. Chloroplasts are the sites of photosynthesis in plant cells. The thylakoid membranes contain chlorophyll and other molecules that absorb light energy and convert it into chemical energy. Enzymes in the stroma use the chemical energy to make carbohydrates and other complex organic molecules from water, carbon dioxide, and other simple inorganic precursors.

2. The tonoplast, the membrane surrounding the central vacuole, contains transport proteins that move substances into and out of the vacuole. Central vacuoles also store organic and inorganic salts, organic acids, sugars, storage proteins, pigments, and, in some cells, waste products. Chemical defense molecules are found in the central vacuoles of some plants.

STUDY BREAK 5.5

1. Anchoring junctions are spots or belts that run entirely around cells, effectively sticking adjacent cells together. Microfilaments (in adherens junctions) or intermediate filaments (in desmosomes) anchor the junction in the underlying cytoplasm. Tight junctions involve fusion of a network of junction proteins in the outer halves of the plasma membranes of adjacent cells, forming a tight seal that can keep even ions from moving between the cells. Gap junctions open direct channels between adjacent cells through which ions and small molecules can pass directly. The gap junctions are formed by aligned hollow protein cylinders in the plasma membranes of two cells.

2. The extracellular matrix (ECM) is a complex of proteins and polysaccharides secreted by the cells that it surrounds. Depending on the nature of the network of proteoglycans (carbohydrate-rich glycoproteins) in the ECM, the consistency ranges from soft and jellylike to hard and elastic.

Think Like a Scientist

FIGURE 5.11

Presumably a nuclear envelope-associated component recognizes the nuclear localization signal of proteins destined to enter the nucleus. A logical location for such a component is in association with the nuclear pore complex. Clearly, a very restricted region of a protein can be sufficient to direct nuclear localization. In the same vein, the transport system must be able to be highly selective in recognizing proteins with such a small nuclear localization signal.

FIGURE 5.17

The chemical is inhibiting a step in the pathway for protein secretion from the cell. Possible steps affected are the transport of the protein in vesicles from the ER to the Golgi complex, the production of secretory vesicles by budding from the Golgi membranes, or the release of the protein at the cell surface by exocytosis.

IMR

All the *Methanococcus* genes related to genes in other organisms would have had sequences similar just to those in bacteria rather than some similar to bacterial genes and others similar to eukaryotic genes.

UQ

Perturbing the activity of individual genes and proteins allows cell biologists to determine whether each plays a necessary role in a given cellular process, but does not indicate what mechanistic role the protein performs or how it cooperates with other proteins to perform its functions. On the other hand, reconstituting a process in a test tube allows cell biologists to build a functioning system from individual components and understand how each contributes to the system, but does not address the functional importance of each component in the context of the cell. Both approaches can be technically challenging, but reconstitution is often more challenging because it requires both identification of all of the important players and the isolation of the players in a functional form.

Test Your Knowledge

1. b 2. e 3. c 4. a 5. b 6. c 7. b 8. b 9. b 10. d

Interpret the Data

Most of the cathepsin activity occurs in the fraction containing the lysosomal marker enzymes (fraction 5), thus localizing the worm cathepsin activity to the lysosome.

Chapter 6

Study Break

STUDY BREAK 6.1

1. The fluid mosaic model proposes that the membrane consists of a fluid phospholipid bilayer in which proteins are embedded and float freely.

2. Integral proteins include transport proteins, receptor proteins, recognition proteins, and cell adhesion proteins. Peripheral proteins include microtubules, microfilaments, intermediate filaments, and proteins that link the cytoskeleton together.

STUDY BREAK 6.2

1. In passive transport, molecules and ions move across the membrane from the side with the higher concentration to the side with the lower concentration; that is, the difference in concentration provides the energy for passive transport. In active transport, molecules and ions move across the membrane from the side with the lower concentration to the side with the higher concentration—that is, against the concentration gradient. The energy for active transport comes from the hydrolysis of ATP.

2. The transport of substances through membranes based solely on molecular size and lipid solubility is simple diffusion. The diffusion of polar and charged molecules across membranes with the help of transport proteins is facilitated diffusion.

STUDY BREAK 6.3

1. Osmosis is the passive transport of water across a membrane. This movement follows concentration gradients. For osmosis to occur, there must be a selectively permeable membrane—that is, a membrane that will allow water molecules, but not molecules of the solute, to pass. As long as the solute is at different concentrations on the two sides of the membrane, water movement will occur; that is, it is not necessary for pure water to be present on one side of the membrane.

2. If animal cells are in a hypertonic solution, water molecules will move by osmosis from within the cells to the surrounding solution. If the outward movement of water exceeds the capacity of the cells to replace the lost water, the cells will shrink.

STUDY BREAK 6.4

1. Active transport is the movement of substances across membranes against their concentration gradients by pumps; the energy for active transport comes from ATP hydrolysis. In primary active transport, ATP hydrolysis directly drives the process; that is, the same protein that transports a substance also hydrolyzes ATP. In secondary active transport, ATP hydrolysis indirectly drives the process; that is, the transport proteins themselves do not hydrolyze ATP. Rather, the transporters use a favorable concentration gradient of ions, generated by primary active transport (when ATP hydrolysis is used), as their energy source for active transport of a different ion or molecule.

2. A membrane potential is a voltage difference across a membrane. Ion transport by membrane pumps contributes to this voltage difference. The sodium–potassium pump in the plasma membrane pushes three sodium ions out of the cell and two potassium ions into the cell with each turn of the pump. This leads to an accumulation of positive charges outside the membrane, causing the inside of the cell to become negatively charged with respect to the outside of the cell. In addition, an unequal distribution of ions across the membrane is created by passive transport. The electrical potential difference (voltage) across the plasma membrane is the membrane potential.

STUDY BREAK 6.5

1. In exocytosis, secretory vesicles in the cytoplasm contact and fuse with the plasma membrane, releasing their contents to the outside of the cell.

2. Endocytosis is a mechanism by which substances are brought into the cell from the exterior. The substances become trapped in pitlike depressions that bulge inward from the plasma membrane. The depression pinches off as an endocytic vesicle. In bulk-phase endocytosis, no binding by surface receptors is involved. Extracellular water is taken in together with any other molecules that are in solution in the water. This is the simplest form of endocytosis. In receptor-mediated endocytosis, molecules to be taken in become bound to the outer cell surface by receptor proteins. The receptor proteins are specific in that they recognize and bind only certain molecules from the solution that surrounds the cell. The molecules recognized are mostly proteins or other molecules carried by proteins. Once the receptors have bound their target molecules, the receptors collect into a coated pit, a depression in the plasma membrane. The pits, with the contained

target molecules, pinch off from the plasma membrane to form endocytic vesicles.

Think Like a Scientist

FIGURE 6.6

The protein synthesis inhibitors did not affect the pattern of the membrane proteins. Therefore, new protein synthesis is not necessary for generating the intermixed pattern of membrane proteins. The temperature experiment shows that intermixing depends on the temperature. This makes sense if the intermixing is a property of the membrane because, as you learned earlier in the chapter, the phospholipid bilayer can freeze into a semisolid, gel-like state at low temperatures. Logically, membrane proteins would not be able to move or they would only be able to move a little under those conditions. Together, these experiments support the fluid bilayer model and reject the new protein synthesis model.

FIGURE 6.11

The hydrolysis of ATP to ADP plus phosphate is key to the ability of the channel protein to transport ions. If the rate of hydrolysis of ATP was reduced drastically, the primary active transport pump would only be able to move the ion it transports at a much lower rate.

IMR

The result showed that unlabeled LDL competed with labeled LDL for binding to the cells. The interpretation was that there are specific binding sites for LDL on the cell surface and that their number is limited. The result supported the researchers' overall conclusion because it indicated that specific binding was occurring in the experiments rather than nonspecific binding; that is, there is a specific receptor for LDL.

UQ

Restricting the passage of protons aids in conservation of the membrane's electrochemical potential.

Test Your Knowledge

1. d 2. b 3. a 4. c 5. e 6. d 7. e 8. b 9. a 10. c

Interpret the Data

1. Cell set B is displaying resistance because these cells fail to accumulate paclitaxel. They are using active transport to move the anticancer drug back out across the cell membrane.

2. The drug treatment imatinib has a slight but significant effect on increasing the accumulation of the labeled paclitaxel in set A; however, a much more dramatic effect is seen using either of these drugs on cell set B. This suggests that in cell set B, the active transport mechanism is inhibited by the presence of these additional drugs.

Chapter 7

Study Break

STUDY BREAK 7.1

Specificity of a cellular response depends on the signal–receptor interaction. Specificity starts with the signal molecule; that is, the specific signal molecule is the messenger that elicits a specific cellular response. For example, the hormone epinephrine causes glucose to be released into the bloodstream. Specificity also depends on the target cells; that is, only target cells respond to the signal molecule because they exclusively have receptors for the signal molecule.

STUDY BREAK 7.2

1. Protein kinases are enzymes that add phosphate groups to other proteins. The result of phosphorylation is that the protein will be either stimulated or inhibited in its activity. The cellular responses of signal transduction pathways are produced through the actions of protein kinases.

2. Amplification is the phenomenon of an increase in the magnitude of each step of a signal transduction pathway. Amplification typically occurs because the proteins conducting each step of the pathway are enzymes; that is, each enzyme, when it becomes activated, activates large numbers of molecules entering the next step of the pathway.

STUDY BREAK 7.3

1. For a receptor tyrosine kinase to become activated, the signal molecule first binds to the receptor, which then assembles into a dimer. The receptor adds phosphate groups to tyrosines on the cytoplasmic side of itself, which activates the receptor.

2. A fully activated receptor tyrosine kinase has phosphorylated tyrosines on each of its two monomers. A signaling protein that recognizes a phosphorylated tyrosine as well as surrounding parts of the polypeptide binds to the receptor. Depending on the signaling protein, the binding itself may activate the signaling protein, or it is activated by tyrosine phosphorylation catalyzed by the receptor. In its activated form, the signaling protein initiates a transduction pathway leading to a cellular response. A given receptor can initiate different responses because different combinations of signaling proteins can bind to the receptor.

STUDY BREAK 7.4

1. The first messenger in a G-protein–coupled receptor–controlled pathway is the extracellular signal molecule. When it binds to the G-protein–coupled receptor, it activates a site on the cytoplasmic side of the receptor; the activated receptor, in turn, activates the G protein next to it.

2. The effector is activated by the G protein. The effector is a plasma-associated enzyme that generates a nonprotein signal molecule called the second messenger. The second messenger leads to the activation of protein kinases leading to the cellular responses triggered by the signal molecule.

3. A main way the pathway is turned off is by the conversion of cAMP to 5′-AMP by phosphodiesterase. As long as the receptor is bound by the signal molecule, cAMP is being generated by the activated effector. The continued synthesis of cAMP balances the degradation of cAMP by phosphodiesterase, ensuring that the pathway continues to run. However, if the signal molecule no longer is bound to the receptor, the effector again becomes inactive, cAMP therefore is not generated, and existing cAMP is rapidly degraded by phosphodiesterase. As a result, the protein kinase cascade is shut down and no cellular responses occur.

STUDY BREAK 7.5

1. The steroid receptor is within the cell, whereas the receptor tyrosine kinase and G-protein–coupled receptors are in the membrane or associated with the membrane. Also, the activated steroid receptor directly activates genes, whereas the other two receptors, when active, are just the first steps in pathways that may or may not activate genes.

2. A steroid hormone brings about a specific cellular response because whether a cell responds to a steroid hormone depends on whether it has the internal receptor for the hormone. Then within the cells with the receptor, the specific genes that are controlled are those with regulatory sequences that are recognized by the activated receptor.

STUDY BREAK 7.6

Signal transduction pathways, cellular response systems triggered by cell adhesion molecules, and communication pathways that involve gap junctions between adjacent cells might be integrated in a cross-talk network.

Think Like a Scientist

FIGURE 7.2

The experiment showed that the second messenger was not a protein, or any other large molecule that would be denatured by boiling.

FIGURE 7.7

The transduction pathways controlled by the receptor tyrosine kinase would be active all the time, producing cellular responses in an uncontrolled manner—that is, in the absence of the signal molecule. Mutations such as this are known to be associated with some cancers.

FIGURE 7.9

There are two main possibilities. One is that the receptor would no longer recognize the first messenger signal molecule. In this case, the cellular response(s) controlled by the

signal molecule would not occur. The other is that the receptor would be active even in the absence of the first messenger signal molecule. In this case, cellular response(s) would occur in an uncontrolled way. As you learned in the text, there are a very large number of G-protein–coupled receptors and they control a great many cellular processes. Thus, malfunctions of these receptors can have serious consequences. In fact, mutations in genes for G-protein–coupled receptors have been shown to be responsible for more than 30 human diseases.

IMR

You would test U0126 first in an animal model and, if the results were promising, then in humans in clinical trials. In fact, U0126 has been shown to reduce influenza virus propagation in lungs of infected mice.

UQ

1. Perhaps the male's behavior induces the release of dopamine or another relevant neurotransmitter onto specific neurons in the brain of the female, which contain progestin receptors and are involved in regulation of sexual behavior. The dopamine, in turn, may activate those receptors leading to neuronal changes resulting in the expression of sexual behavior.

2. To answer this question, you have to think of other situations in which the environment might cause the release of a neurotransmitter that then, via cross talk, activates a steroid receptor resulting in changes in behavior. Perhaps stimulation from pups activates neuronal steroid hormone receptors, resulting in changes in maternal behavior in a lactating mother. There are many other examples in which stimulation from the environment or another animal might, via neurotransmitter release, influence the function of a particular steroid hormone receptor (by activating it).

Test Your Knowledge

1. b 2. a 3. d 4. c 5. c 6. c 7. e 8. b 9. c 10. e

Interpret the Data

1. In normal cells the least amount of CFTR was found in vesicles, whereas the greatest amount was found in endoplasmic reticulum. In cells with the CF mutation the least amount of CFTR was found in Golgi bodies, whereas the greatest amount was found in endoplasmic reticulum.

2. The CFTR protein is found in almost the same amount in the endoplasmic reticulum of normal and CF cells.

3. The CFTR protein is held up in the endoplasmic reticulum.

Chapter 8

Study Break

STUDY BREAK 8.1

1. Oxidation is the removal of electrons from a substance; reduction is the addition of electrons to a substance.

2. Cellular respiration refers to the reactions in which oxygen is used as final electron acceptor; it includes the reactions that transfer electrons from organic molecules to oxygen and the reactions that make ATP. Oxidative phosphorylation is the process by which ATP is synthesized using the energy released by electrons as they are transferred to oxygen.

STUDY BREAK 8.2

1. The initial steps of glycolysis, which require 2 ATP, convert glucose to a phosphorylated derivative. The later steps, which release 4 ATP, remove electrons from the glucose derivatives and generate two molecules of pyruvate.

2. The redox reaction in glycolysis is the glyceraldehyde-3-phosphate (G3P) to 1,3-bisphosphoglycerate reaction (see Figure 8.7, step 6).

3. ATP is synthesized in glycolysis by substrate-level phosphorylation occurring in the 1,3-bisphosphoglycerate to 3-phosphoglycerate reaction (see Figure 8.7, step 7), and in the phosphoenolpyruvate (PEP) to pyruvate reaction (see Figure 8.7, step 10).

4. If excess ATP is present in the cytosol, it binds to and inhibits the activity of phosphofructokinase. As a result, the

concentration of fructose-1,6-bisphosphate, the product of the phosphofructokinase reaction, decreases, and the subsequent reactions of glycolysis are slowed or stopped. This is reversed when the ATP level in the cytosol decreases. In the end, this control mechanism helps prevent the needless oxidation of fuel molecules when the cell has an adequate supply of ATP.

STUDY BREAK 8.3

The three-carbon pyruvate molecules are transported from the cytosol into the mitochondria, where they are converted into two-carbon acetyl units through pyruvate oxidation. The citric acid cycle oxidizes the acetyl units completely to carbon dioxide with the transfer of electrons to NAD^+ or FAD.

STUDY BREAK 8.4

1. Each complex contains a unique combination of nonprotein carriers that pick up and release electrons.
2. The proton pumps push protons (H^+) from the mitochondrial matrix to the intermembrane compartment, increasing the proton concentration there. The resulting proton gradient produces an electrical gradient across the inner mitochondrial membrane with the matrix negatively charged with respect to the intermembrane compartment. The charge and proton concentration differences together provide energy for ATP synthesis in what is called proton-motive force.

STUDY BREAK 8.5

Fermentation occurs when oxygen is absent or limited. The electrons carried by the NADH produced in glycolysis are transferred to an organic molecule instead of the electron transfer system. In lactate fermentation, the end product of glycolysis, pyruvate, is converted to lactate. The lactate stores electrons temporarily, transferring them to the mitochondrial electron transfer system when the oxygen content of cells returns to normal. In alcoholic fermentation, pyruvate is converted into ethyl alcohol.

Think Like a Scientist

FIGURE 8.13

The H^+ concentration will increase in the intermembrane compartment. The increase will lead to an increase in ATP production.

FIGURE 8.15

In the light, ATP synthesis would occur and then, in the dark, ATP synthesis would stop. This was in fact observed by Racker and Stoeckenius.

IMR

The observation indicates that the two UCPs are members of the same family of proteins and that presumably they have evolved by diversification from a common ancestral gene.

UQ

One possible hypothesis: there is a direct cause and effect relationship between altered regulation of mitochondrial expression and AD. If a significant number of mice whose genes were not genetically altered were to develop AD, this hypothesis would have to be rejected.

Test Your Knowledge

1. a 2. b 3. a 4. e 5. c 6. b 7. e 8. a 9. d 10. a

Interpret the Data

1.

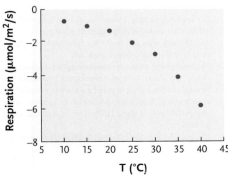

2. Two.
3. A curve represents the data best. A Q10 of 2 indicates a curve because the respiration rate is doubling for every 10°C increase in temperature. This type of relationship is an exponential increase.
4. Temperatures are predicted to increase, which would increase respiration according to these data. The data are insufficient to predict the actual increase in respiration since the plants would likely change their enzyme amounts or types (acclimate) in response to the rising temperature. Further, temperature is not the only factor that will likely change with rising CO_2. Other factors include precipitation, cloud cover, nutrient availability, etc.

Chapter 9

Study Break

STUDY BREAK 9.1

1. The two stages of photosynthesis are: (1) the light-dependent reactions, in which the energy of sunlight is absorbed and converted into chemical energy in the form of ATP and NADPH; and (2) the light-independent (dark) reactions, in which electrons carried by NADPH are used as a source of energy to convert carbon dioxide from inorganic to organic form.
2. In plants, photosynthesis takes place in the chloroplast. The light-dependent reactions are carried out on the thylakoid membranes and stromal lamellae. The light-independent reactions are carried out in the stroma.

STUDY BREAK 9.2

1. The chlorophyll *a* molecules in the antenna complexes are normal molecules of the pigment, consisting of a carbon ring structure with a magnesium atom bound at the center and an attached hydrophobic side chain. These chlorophyll *a* pigments absorb light. The chlorophyll *a* molecules in the reaction centers have modified light absorption properties that result from interactions with particular proteins of the photosystems. The two special chlorophyll *a* molecules of photosystem II are P680; those of photosystem I are P700. These pigment molecules capture light energy from the antenna complex pigments in the form of an excited electron that is passed to a primary acceptor molecule. That electron is passed to the electron transfer system.
2. The making of NADPH begins when electrons derived from water splitting are pushed to higher energies by light absorption in photosystem II. The high-energy electrons pass to a primary acceptor in photosystem II and then down an electron transfer system to P700 in photosystem I, losing energy along the way. Light energy absorbed by photosystem I again excites the electrons, which pass to different electron carriers, ending with ferredoxin. The ferredoxin transfers high-energy electrons to $NADP^+$, which is reduced to NADPH by $NADP^+$ reductase.
3. In the linear electron flow pathway, electrons run through the entire set of photosystems and electron carriers, producing both NADPH and ATP. In the cyclic electron flow pathway, electrons flow cyclically around photosystem I; photosystem II is not involved. The cycle of electrons is through the cytochrome complex and plastocyanin to photosystem I, to ferredoxin, but then back to the cytochrome complex rather than on to $NADP^+$ reductase. Only ATP is produced by this pathway.

STUDY BREAK 9.3

1. Rubisco catalyzes a reaction combining carbon dioxide with RuBP to form two molecules of 3-phosphoglycerate (3PGA). Rubisco, an enzyme unique to photosynthetic organisms, is the key enzyme for producing the world's food because it is responsible for carbon dioxide fixation, a process that ultimately provides organic molecules for most of the world's organisms. Rubisco is the key regulatory site of the Calvin cycle for the following reason: During the daytime, sunlight powers the light-dependent reactions, and the NADPH and ATP produced by those reactions stimulate rubisco, which, in turn, keeps the Calvin cycle running. In darkness, however, NADPH and

ATP levels are low, and as a result, rubisco's activity is inhibited and the Calvin cycle slows down or stops.
2. For each carbon atom that is released from the Calvin cycle in a carbohydrate molecule, one carbon dioxide molecule must enter the cycle. Therefore, to produce a molecule containing 12 carbon atoms, 12 molecules of carbon dioxide must enter the cycle.

STUDY BREAK 9.4

1. Photorespiration uses oxygen and releases CO_2. It occurs when oxygen concentrations are high relative to CO_2 concentrations. In that condition, rubisco acts as a Rubisco rather than a carboxylase, catalyzing the combination of RuBP with O_2 rather than CO_2. The toxic products formed by this reaction cannot be used in photosynthesis and are eliminated from the plant as CO_2. Photorespiration uses energy to salvage the carbons from phosphoglycolate, which greatly reduces the efficiency of energy use in photosynthesis. This can be seen in the reduced growth of plants grown under photorespiration conditions.
2. In the C4 pathway, carbon fixation involves the reaction of CO_2 with phosphoenolpyruvate (PEP) to produce a four-carbon molecule, oxaloacetate. The oxaloacetate is reduced to malate by electrons transferred from NADPH, and malate then is oxidized to pyruvate in a reaction releasing CO_2, which is used in the rubisco-catalyzed first step of the Calvin cycle. In C4 plants, carbon fixation and the Calvin cycle occur in different cell types: carbon fixation in mesophyll cells, and the Calvin cycle in bundle sheath cells. This alternative method of carbon fixation minimizes photorespiration.
3. In C4 plants, carbon fixation and the Calvin cycle occur in different cell types, mesophyll cells and bundle sheath cells, respectively. In CAM plants, carbon fixation and the Calvin cycle occur at different times, at night and during the day, respectively.

STUDY BREAK 9.5

The reactions of photosynthesis and cellular respiration are essentially the reverse of one another, with CO_2 and H_2O the reactants of photosynthesis and the products of cellular respiration. Phosphorylation reactions involving electron transfer systems are part of each process, namely photophosphorylation in photosynthesis and oxidative phosphorylation in cellular respiration. G3P is an intermediate in both pathways: in photosynthesis it is a product of the Calvin cycle, and in cellular respiration it is generated in glycolysis in the conversion of glucose to pyruvate. In photosynthesis, G3P is used for the synthesis of sugars and other fuel molecules, and in cellular respiration, it is part of the catabolism of sugars to simpler organic molecules.

Think Like a Scientist

FIGURE 9.10

Two.

IMR

There are two types of subunit, so there are two genes that encode the complete enzyme. The enzyme consists of eight copies each of the small subunit, encoded by one of the genes, and of the large subunit, encoded by the other gene.

UQ

Imagine two types of desert plants. Some, like the wild watermelon plants of the Kalahari, grow very rapidly, with very robust photosynthesis, as soon as rain falls. They produce as many seeds as they can, quickly before severe drought sets in. This strategy requires rapid photosynthesis, even at the risk of photodamage. Others, such as desert scrub plants like sage, persist throughout the drought. They may grow slowly, with highly protected photosynthesis, even during times of rain. If they grew too fast, their large leaf surface areas would result in high water loss during drought. In some invasive plant species, aggressive photosynthesis and high growth rates have a selective advantage, allowing them to outcompete their rivals for resources like sunlight or growth space. In these cases, high rates of photosynthesis may be an advantage even as they risk photodamage. Other plants may invest for the longer term, building long-lived resilient structures (leaves, wood, etc.). The large investment may render risky photosynthetic strategies less viable, instead preferring more "conservative" down-regulatory strategies. Still other plants may have to contend with lack of key nutrients, in which high photosynthetic rates or the need for rapid repair of the photosynthetic apparatus would be too "expensive" in terms of resources. Finally, humans have selected for traits in crop plants that are not related to photosynthetic yield—for example, tasty fruits, disease resistance, or short growing season. These traits often take precedence over photosynthetic efficiency.

Test Your Knowledge

1. d 2. b 3. a 4. c 5. e 6. a 7. b 8. d 9. d 10. b

Interpret the Data

1. When PPFD is zero, no photosynthesis is occurring, but respiration continues to occur. Based on the definition of net photosynthesis here and respiration from Chapter 8 *Interpret the Data*, respiration alone would have a negative value of CO_2 flux.

    The temperature for a respiration rate of 0.2 μmol/m^2/s is 25°C.

2.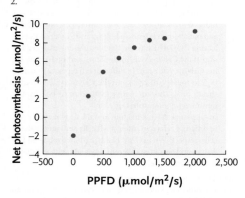

3. A curve represents the data best. Net photosynthesis is saturating with increasing PPFD, indicating that some other factor (such as the amount of photosynthetic enzymes) is limiting photosynthesis.

4. Temperatures are predicted to increase, which would increase respiration according to the data and thus lower net photosynthesis. These data are insufficient to predict the actual impact of CO_2 concentrations on net photosynthesis because the plants would likely change their enzyme amounts or types (acclimate) in response to the rising temperature. Further, temperature is not the only factor that will likely change with rising CO_2. Other factors include precipitation, cloud cover, nutrient availability, etc.

Chapter 10

Study Break

STUDY BREAK 10.1

1. (1) An elaborate master program of molecular checks and balances ensures an orderly and timely progression through the cell cycle. (2) The process of DNA synthesis replicates each DNA chromosome into two copies with almost perfect fidelity. (3) A structural and mechanical web of interwoven "cables" and "motors" of the mitotic cytoskeleton separates the DNA copies precisely into the daughter cells.
2. A linear DNA molecule complexed with proteins
3. In mitosis, DNA replication is followed by the equal separation of the replicated DNA molecules and their delivery to daughter cells. The process ensures that the two cell products of a division have the same DNA content and the same genetic information as the parent cell entering division has.

STUDY BREAK 10.2

1. In order, the stages of mitosis are prophase, prometaphase, metaphase, anaphase, and telophase.
2. Each eukaryotic chromosome has a specialized region known as a centromere. The centromere is where a com-

plex of several proteins, called a kinetochore, forms. During mitosis, some spindle microtubules attach to each kinetochore. These connections determine the outcome of mitosis because they attach the sister chromatids of each chromosome to microtubules leading to the opposite spindle poles; during anaphase, the spindle separates sister chromatids and pulls them to opposite spindle poles. In brief, the centromeres are key to chromosome segregation during mitosis. Although not mentioned in the chapter, this is also apparent when problems occur in which a chromosome fragment without a centromere breaks off from a chromosome. The fragment without a centromere cannot connect to the spindle and, hence, is not segregated properly.

3. Joined sister chromatid pairs attach to kinetochore microtubules during prometaphase and begin their migration to the metaphase plate. The spindle microtubules are also necessary for segregating the sister chromatids to opposite poles of the cell during anaphase. If colchicine is present, no spindle will form and no kinetochore microtubules are present to attach to the sister chromatids. Therefore, the cell will be stuck in mitosis with the condensed pairs of sister chromatids in an unorganized array.

STUDY BREAK 10.3

1. In animal cells, all of which have centrosomes, the spindle forms through division of the cell center. As the dividing centrosome separates into two parts, the microtubules of the spindle form between them. Plant cells lack centrosomes. In plant cells, the spindle microtubules simply assemble around the nucleus. In either case, the microtubules assemble in a parallel array that creates two poles in the dividing cell.
2. Chromosomes (sister chromatids) move apart during anaphase. During the anaphase movements, the kinetochores move along the kinetochore microtubules, which become shorter as anaphase progresses. The nonkinetochore microtubules slide over each other, decreasing the degree of overlap and pushing the poles farther apart. The total distance traveled by the chromosomes is the sum of the two movements.

STUDY BREAK 10.4

1. A Cdk can become active only once it has complexed with a cyclin protein. Each cyclin is present only during a particular segment of the cell cycle, controlled by when it is synthesized and degraded. Thus, the period in the cell cycle when a particular Cdk is active depends on when its activating cyclin is present.
2. When the kinase of the Cdk is activated upon binding to a cyclin, it phosphorylates target proteins in the cell, regulating their activities. Those proteins play roles in initiating or regulating key events of the cell cycle, namely DNA replication, mitosis, and cytokinesis. The progression through the cell cycle is regulated, then, by a succession of cyclin–Cdk complexes, each of which has specific regulatory effects.
3. An oncogene is an altered gene in an organism that contributes to the development of cancer—that is, uncontrolled cell division. Some of the genes that become oncogenes encode components of the cyclin–Cdk system that regulates cell division, whereas others encode proteins that regulate gene activity, form cell-surface receptors, or make up elements of the systems controlled by the receptors.
4. Metastasis is when cells break loose from a tumor, spread through the body, and grow into new tumors in other body regions.

STUDY BREAK 10.5

1. Prokaryotic cell division begins with replication of the bacterial chromosome, starting with duplication of the origin of replication. Once the origin of replication is duplicated, the two origins actively migrate to the two ends of the cells, a process that separates the two replicating chromosomes in the cell. Division of the cytoplasm then occurs by means of a partition of cell wall material that grows inward until the cell is separated into two parts. The cytoplasmic division divides the replicated DNA mole-

cules and cytoplasmic structures between the daughter cells.

2. Present in eukaryotic cell division, but absent from prokaryotic cell division, are the following: the process of mitosis; any form of microtubules for chromosome segregation; a spindle apparatus; cyclin/CDK control proteins.

Think Like a Scientist

FIGURE 10.4
The colchicine would block spindle formation. As a result the cells would not be able to complete mitosis. In fact, colchicine is used to get cells to arrest in metaphase when the chromosomes are at their most condensed. At that stage, the chromosomes are the easiest to visualize by microscopy (see Figure 10.7).

FIGURE 10.12
The result supporting the kinetochore microtubule movement hypothesis would have been the movement of the bleached region toward the pole.

FIGURE 10.14
The result indicates that there are factors in the G_2 nucleus that accelerate the progress of the S nucleus towards mitosis.

IMR
The result indicates that expression of K cyclin (which occurred when the inducer was added) is sufficient to induce S phase entry in quiescent (G_0) cells.

UQ
Although *E. coli* is easy to culture, there is no known prokaryotic equivalent of mitosis; mammalian cells are hard to culture and cannot be kept in long-term culture (see *Focus on Basic Research*). *Focus on Model Organisms* provides abundant rationale for *Saccharomyces* as a model organism for this type of research. Ubiquitin is given this name because it is indeed ubiquitous—present in almost the same form in essentially all eukaryotes (see Chapter 16).

Test Your Knowledge

1. c 2. a 3. e 4. b 5. d 6. d 7. a 8. b 9. b 10. c

Interpret the Data

1. Radium exposure does not have an immediate effect on mitosis; that is, for about two hours after initiating radium exposure, there is not a significant drop in the number of cells in the population undergoing mitosis. (The researchers interpreted this to mean that not only were cells completing mitosis, but other cells commenced mitosis.) However, after about two hours of radium exposure, mitosis in the cell population exhibits a sudden drop to zero. In other words, radium blocks mitosis after a long enough exposure. The corollary is that there was no evidence for radium stimulating mitosis.
2. The effect of radium exposure was not permanent. Mitosis resumed about two hours after radium was removed. (Importantly, this result also shows that the decrease in mitosis as a result of radium exposure was not because of death of the cells.)

Chapter 11

Study Break

STUDY BREAK 11.1

1. Mitosis produces daughter cells that are genetically identical to the parent cell. Either a haploid or a diploid cell can undergo mitosis. Meiosis starts with a diploid cell. There is one round of DNA replication but two rounds of cell division, with the result that four haploid cells are produced from the parent diploid cell.
2. Recombination is the physical exchange of segments between the chromatids of homologous chromosomes. Recombination occurs in prophase I, when homologous chromosomes have each duplicated to produce sister chromatids and are aligned fully in an organization called a tetrad.
3. Meiosis II is the meiotic division that is similar to a mitotic division.

STUDY BREAK 11.2

1. There are three ways in which sexual reproduction generates genetic variability. First, recombination, which involves the physical exchange of segments between homologous chromatids in prophase I of meiosis, generates new combinations of alleles. Second, the random separation of homologous chromosomes during meiosis generates genetic variability; that is, in metaphase I, for each homologous pair of chromosomes, one chromosome makes spindle connections leading to one pole and the other chromosome connects to the opposite pole. This process operates independently for each homologous pair of chromosomes; thus, for each meiosis random combinations of maternal and paternal chromosomes move to the poles during anaphase I. Third, the random joining of male and female gametes produces additional genetic variability.

2. The proportion of gametes that will have chromosomes that originate from the animal's female parent is: $(1/2)^6 = 1/64$.

STUDY BREAK 11.3

In animals, the diploid phase dominates the life cycle; mitotic divisions occur only in this phase. Meiosis in the diploid phase gives rise to products that develop directly into egg and sperm cells without undergoing mitosis.

In most plants, the life cycle alternates between haploid and diploid generations, both of which grow by mitotic divisions. Fertilization produces the diploid sporophyte generation; after growth by mitotic divisions, cells of the sporophyte undergo meiosis and produce haploid spores. The spores germinate and grow by mitotic divisions into the gametophyte generation. After growth of the gametophyte, cells develop directly into egg or sperm nuclei, which fuse in fertilization to produce the diploid sporophyte generation again.

Think Like a Scientist

IMR

The result suggests that CYP26B1 plays an essential role in germ cell development perhaps among all vertebrates, not just mammals.

UQ

The outcome depends on a variety of factors such as the extent of the rearrangement, its location in the genome and the particular organism or species being analyzed. However, very frequently, a phenomenon referred to as "synaptic adjustment" is observed. This allows the duplications and/or inversions to be accommodated (in many cases these just "loop out" from the fully paired homologs) and the homologs still succeed in synapsing. Occasionally, these rearrangements can also lead to nonhomologous synapsis and asynapsis.

Test Your Knowledge

1. b 2. a 3. d 4. c 5. b 6. b 7. b 8. d 9. a 10. b

Interpret the Data

1. At 20°C, one mother produced 21 male offspring; assuming 100 offspring, this would be 21%. This was the highest. The lowest was zero males produced by at least one mother at 20°C and 16-hour photoperiod. 15°C to 20°C seems to result in more male production than either higher or lower temperatures. No mothers produced male offspring at 25°C; some mothers produced no male offspring at 11°C. Less than 10% males were produced at 25°C or 11°C or at 20°C and 16-hour photoperiod.

2. At the moderate temperatures, slightly more males are produced with 12-hour days rather than 16-hour days. However, there is a wide range. If time has an effect, the effect is much less than the effect temperature has on the production of males.

NOTE: The authors concluded that the difference seen for photoperiod was not statistically significant, meaning that the difference seen could be merely because of chance.

Chapter 12

Study Break

STUDY BREAK 12.1

1. The numerical results approximate a $9:3:3:1$ ratio. Therefore, both parents must be heterozygous for both genes. If we designate the alleles for one of the pairs of traits as A and a, and the alleles for the other pair of traits as B and b, each parent has the genotype $Aa\,Bb$.

2. A ratio of $1:1:1:1$ is the typical outcome of a testcross involving a parent who is heterozygous for the alleles of two genes. Using the same allele symbols as in (1), the genotypes of the parents are $Aa\,Bb$ and $aa\,bb$; this is a testcross.

STUDY BREAK 12.2

1. The color pattern involved is an incompletely dominant trait.

2. The fur colors here involve multiple alleles of a single gene. The allele symbols are C for wild type, c^{ch} for chinchilla, c^h for Himalayan, and c for albino, with dominance in the order $C \rightarrow c^{ch} \rightarrow c^h \rightarrow c$: that is, the C allele is completely dominant to the c^{ch} allele, the c^{ch} allele is completely dominant to the c^h allele, and so on. Therefore, we have these genotypes and phenotypes:

C with c^{ch}, c^h, or c = agouti (CC, Cc^{ch}, Cc^h, Cc)
c^{ch} with c^{ch}, c^h, or c = chinchilla ($c^{ch}c^{ch}$, $c^{ch}c^h$, $c^{ch}c$)
c^h with c^h or c = Himalayan (c^hc^h, c^hc)
c with c = albino (cc)

Think Like a Scientist

FIGURE 12.5

Purple-flowered F_2 plants are either PP or Pp in genotype. Only if Pp is selfed will you get purple-flowered and white-flowered progeny (it would be same as the $F_1 \times F_1$ self shown in the figure). There is a ratio of $1\,PP:2\,Pp$ plants as the Punnett square in the figure shows. Therefore, there is a 2 in 3 (two-thirds) chance that you will see both purple-flowered and white-flowered plants in the progeny.

FIGURE 12.8

This problem relates to the testcross shown in the figure. The ratio of the two types of progeny is approximately $1:1$, a result that matches the Experiment 1 results in the figure. Therefore, the white-fat parent must have been heterozygous. The yellow-fat parent has to be homozygous recessive because of the information given about the fat-color phenotype. In short, the parents were Ww (white fat) $\times ww$ (yellow fat), which gives an expected $1:1$ ratio of Ww (white fat) : ww (yellow fat).

FIGURE 12.9

The cross would be $Rr\,Yy \times Rr\,yy$. The progeny are determined by constructing a Punnett square and filling in the squares. For the $Rr\,Yy$ parent, there are four types of gametes just as in the figure; that is, $1/4\,R\,Y$, $1/4\,R\,y$, $1/4\,r\,Y$, and $1/4\,r\,y$. For the $Rr\,yy$ parent, there are two types of gametes; that is, $1/2\,R\,y$ and $1/2\,r\,y$. This gives you 8 squares in this particular Punnett square with these progeny: $1/8\,RR\,Yy$, $1/8\,RR\,yy$, $1/8\,Rr\,Yy$, $1/8\,Rr\,yy$, $1/8\,Rr\,Yy$, $1/8\,Rr\,yy$, $1/8\,rr\,Yy$, and $1/8\,rr\,yy$. Collecting the genotypes with the same phenotypes together we get $3/8$ round, yellow ($1/8\,RR\,Yy + 1/8\,Rr\,Yy + 1/8\,Rr\,Yy$), $3/8$ round, green ($1/8\,RR\,yy + 1/8\,Rr\,yy + 1/8\,Rr\,yy$), $1/8$ wrinkled, yellow ($1/8\,rr\,Yy$), and $1/8$ wrinkled, green ($1/8\,rr\,yy$).

IMR

In Chapter 4, you learned that the substrate(s) for an enzyme interact with a small region of an enzyme called the *active site*. It is at the active site where enzyme catalysis occurs. The active site of an enzyme is formed by a particular array of amino acids in the polypeptide chain. Those amino acids have evolved to be optimal for enzyme activity. A mutation that alters an amino acid in or near the active site can affect the structure of the active site and, therefore, adversely affect the enzyme's catalytic activity. The extent of the deleterious effect depends on the particular amino acid substitution and the chemical changes brought about in the protein by the substitution.

UQ

If you sequence lots of species you will identify lots of regions of strong evolutionary constraint. If you look at many individuals within a species, you can look at the rate of fast-evolving alleles that might have responded to population-specific pressures or drift and start asking questions pertaining to the genome of the human. Ultimately, your choice will depend on the question you are trying to answer. If you are trying to understand developmental processes, cross-species comparisons are extremely useful, but if you are looking at immunity, maybe the within-species comparison is better.

Test Your Knowledge

1. (a) The CC parent produces all C gametes, and the Cc parent produces $1/2\,C$ and $1/2\,c$ gametes. All offspring would have colored seeds—half homozygous CC and half heterozygous Cc. (b) Both parents produce $1/2\,C$ and $1/2\,c$ gametes. Of the offspring, three-fourths would have colored seeds ($1/4\,CC + 1/2\,Cc$) and one-fourth would have colorless seeds ($1/4\,cc$). (c) The Cc parent produces $1/2\,C$ gametes and $1/2\,c$ gametes, and the cc parent produces all c gametes. Half of the offspring are colored ($1/2\,Cc$) and half are colorless ($1/2\,cc$).

2. The genotypes of the parents are $Lele$ and $lele$.

3. The taster parents could have a nontaster child, but nontaster parents are not expected to have a child who can taste PTC. The chance that they might have a taster child is $3/4$. The chance of a nontaster child being born to the taster couple is $1/4$. Because each combination of gametes is an independent event, the chance of the couple having a second child, or any child, who cannot taste PTC is expected to be $1/4$.

4. (a) All $A\,B$. (b) $1/2\,A\,b + 1/2\,a\,b$. (c) $1/2\,A\,B + 1/2\,a\,B$. (d) $1/4\,A\,B + 1/4\,A\,b + 1/4\,a\,B + 1/4\,a\,b$.

5. (a) All $Aa\,BB$. (b) $1/4\,AA\,BB + 1/4\,Aa\,Bb + 1/4\,Aa\,BB + 1/4\,Aa\,Bb$. (c) $1/4\,Aa\,Bb + 1/4\,Aa\,Bb + 1/4\,aa\,Bb + 1/4\,aa\,bb$. (d) $1/4\,Aa\,Bb + 1/8\,AA\,Bb + 1/8\,Aa\,BB + 1/8\,Aa\,bb + 1/8\,aa\,Bb + 1/16\,AA\,BB + 1/16\,AA\,bb + 1/16\,aa\,BB + 1/16\,aa\,bb$.

6. (a) All $A\,B\,C$. (b) $1/2\,A\,B\,c + 1/2\,a\,B\,c$. (c) $1/4\,A\,B\,C + 1/4\,A\,B\,c + 1/4\,a\,B\,C + 1/4\,a\,B\,c$. (d) $1/8\,A\,B\,C + 1/8\,A\,B\,c + 1/8\,A\,b\,C + 1/8\,A\,b\,c + 1/8\,a\,B\,C + 1/8\,a\,B\,c + 1/8\,a\,b\,C + 1/8\,a\,b\,c$.

7. Because the man can produce only 1 type of allele for each of the 10 genes, he can produce only 1 type of sperm cell with respect to these genes. The woman can produce 2 types of alleles for each of her 2 heterozygous genes, so she can produce $2 \times 2 = 4$ different types of eggs with respect to the 10 genes. In general, as the number of heterozygous genes increases, the number of possible types of gametes increases as 2^n, where n = the number of heterozygous genes.

8. Use a standard testcross; that is, cross the guinea pig with rough, black fur with a double recessive individual, $rr\,bb$ (smooth, white fur). If your animal is homozygous $RR\,BB$, you would expect all the offspring to have rough, black fur.

9. One gene probably controls pod color. One allele, for green pods, is dominant; the other allele, for yellow pods, is recessive.

10. The cross $RR \times Rr$ will produce $1/2\,RR$ and $1/2\,Rr$ offspring. The cross $Rr \times Rr$ will produce $1/4\,RR$, $1/2\,Rr$, and $1/4\,rr$ as combinations of alleles. However, the $1/4\,rr$ combination is lethal, so it does not appear among the offspring. Therefore, the offspring will be born with only two types, RR and Rr, with twice as many Rr as RR in a $1:2$ ratio (or $1/3\,RR + 2/3\,Rr$).

11. The parental cross is $GG\,LeLe\,RR \times gg\,lele\,rr$. All offspring of this cross are expected to be tall plants with green pods and round seeds, or $Gg\,Lele\,Rr$. When crossed, this heterozygous F_1 generation is expected to produce eight different phenotypes among the offspring: green–tall–round, green–dwarf–round, yellow–tall–round, green–tall–wrinkled, yellow–dwarf–round, green–dwarf–wrinkled, yellow–tall–wrinkled, yellow–dwarf–wrinkled, in a $27:9:9:9:3:3:3:1$ ratio.

12. The genotypes are: bird 1, *Ff Pp*; bird 2, *FF PP*; bird 3, *Ff PP*; bird 4, *Ff Pp*.

13. Yes, it can be determined that the child is not hers, because the father must be AB to have both an A and B child with a type O wife; none of the woman's children could have type O blood with an AB father.

14. The cross is expected to produce white, tabby, and black kittens in a 12 : 3 : 1 ratio.

15. The mother is homozygous recessive for both genes, and the father must be heterozygous for both genes. The child is homozygous recessive for both genes. The chance of having a child with normal hands is 1/2, and that of having a child with woolly hair is 1/2. Using the product rule of probability, the probability of having a child with normal hands and woolly hair is 1/2 × 1/2 = 1/4.

Interpret the Data

Yes, the data support the hypothesis that the tolerance trait is dominant. Let us assign *T* as the symbol for the dominant allele for being tolerant, and *t* as the symbol for the recessive allele for being intolerant.

Cross 1: The crosses of true-breeding tolerant plants with true-breeding intolerant plants are *TT* × *tt*. The F₁ offspring are *Tt*. If the tolerance trait is dominant, the F₁ × F₁ crosses are genotypically *Tt* × *Tt*, which produces an F₂ generation with a genotypic ratio of 1 *TT*:2 *Tt*:1 *tt*, which gives a phenotypic ratio of 3 alive (tolerant):1 dead (intolerant), which approximately matches the results.

Cross 2: As described for Cross 1, the F₁ plants have the *Tt* genotype. An F₁ × intolerant parent cross is *Tt* × *tt*. The offspring from this cross are expected to have a genotypic ratio of 1 *Tt*:1 *tt*, which gives a phenotypic ratio of 1 alive:1 dead, which approximately matches the results.

Cross 3: An F₁ × tolerant parent cross is *Tt* × *TT*. The offspring from this cross are expected to have a genotypic ratio of 1 *TT*:1 *Tt*; so all of the progeny are expected to be alive, which matches the results.

Chapter 13

Study Break

STUDY BREAK 13.1

The cross to use is the testcross. Here, the testcross would be *Aa Bb* × *aa bb*. A testcross is used so that you can follow the meiotic events in the dihybrid parent (including the consequences of crossing over between linked genes) because all of the gametes from the testcross parent carry recessive alleles for the genes in the cross. A testcross shows linkage when the ratio of 1 : 1 : 1 : 1 for the four possible phenotypes is not seen; that is, the 1 : 1 : 1 : 1 ratio result occurs when two genes assort independently. However, if two genes are linked, there will be excess of the two parental classes of progeny compared with the two recombinant classes.

STUDY BREAK 13.2

The differences between sex-linked inheritance and autosomal inheritance are seen clearly when reciprocal crosses are made and followed through to the F₂ generation. If sex-linked inheritance is involved, a cross of miniature-winged female × normal-winged male flies will give an F₁ generation of all normal-winged female and all miniature-winged male flies. (This result would not be found with autosomal inheritance. Instead, all F₁ flies—both males and females—would have normal wings.) Selfing the F₁ flies will give an F₂ generation with 1:1 normal-winged:miniature-winged flies in both sexes. (For autosomal inheritance, you would see a 3:1 ratio of normal-winged:miniature-winged flies in both sexes.)

In the reciprocal cross of true-breeding normal-winged female × miniature-winged male, the F₁ flies will all have normal wings if sex-linked inheritance is involved. (This result is the same as for autosomal inheritance.) Selfing the F₁ flies will give an F₂ generation in which all females have normal wings and the males will be 1/2 normal-winged and 1/2 miniature-winged. (For autosomal inheritance, you would see a 3:1 ratio of normal-winged:miniature-winged flies in both sexes.)

In summary, reciprocal crosses show different segregation patterns of phenotypes for sex-linked inheritance and

autosomal inheritance. The two modes of inheritance are easiest to distinguish in a cross of a mutant female × wild-type male because then, in the F₁ generation, all males show the mutant phenotype when sex-linked inheritance is involved.

STUDY BREAK 13.3

(a) Duplication of a chromosome segment occurs when a segment breaks from one chromosome and is inserted into its homolog.

(b) A Down syndrome individual results when nondisjunction of chromosome 21 during meiosis occurs (usually in females), producing gametes with two copies of chromosome 21 and one copy of every other chromosome. When such a gamete fuses with a normal gamete, the result is a zygote with three copies of chromosome 21 and two copies of the other chromosomes. This individual will have Down syndrome. Aneuploidy is the term for the condition of extra or missing chromosomes.

(c) A translocation occurs when a broken segment of a chromosome becomes attached to a different, nonhomologous chromosome.

(d) Polyploidy means that there are more sets of chromosomes than the typical diploid set. Polyploidy may result if the spindle fails to function properly in mitosis of cell lines leading to gametes. Cells affected in this way will have twice the normal number of sets of chromosomes. When meiosis subsequently occurs, the gametes produced will have two sets of chromosomes. Fusion of these gametes with, for instance, a gamete with one set of chromosomes will produce a zygote with three sets of chromosomes—a triploid cell.

STUDY BREAK 13.4

1. Autosomal recessive inheritance: For a child to exhibit an autosomal recessive trait, he or she must inherit one recessive allele from each parent. For autosomal recessive inheritance to explain Simpson syndrome in the family, the father must be homozygous for the Simpson syndrome allele, *ss*, and the mother must be heterozygous, *Ss*. The expectation would be that 1/2 of the children would be *Ss* and 1/2 would be *ss*, regardless of sex, and that is what is found. Therefore, on the assumption that the mother is heterozygous, the syndrome could be an autosomal recessive trait. Sex-linked recessive inheritance: One characteristic of sex-linked recessive inheritance is that affected females pass on the trait to all their sons. Here, we start with an affected male, and he would have to be X^sY if it is a sex-linked recessive trait. To explain the children, we would have to assume that the mother is heterozygous, X^SX^s. The cross of X^SX^s × X^sY is expected to give 1/2 females with the syndrome and 1/2 males with the syndrome, which is what is described. Therefore, the syndrome could be a sex-linked recessive trait.

2. Autosomal recessive inheritance: In pedigrees of autosomal recessive traits, the appearance of progeny with a trait when both parents do not have the trait is one common feature. If we assume that both parents are heterozygous, *Ww*, then the children can be explained; that is, you can get both wiggly-eared children (*ww*, expected frequency 1/4) and nonwiggly-eared children (*WW* or *Ww*, combined expected frequency 3/4), regardless of sex. Therefore, wiggly ears could be an autosomal recessive trait based on this family. Sex-linked recessive inheritance: Because a male individual has only one X chromosome, it is not possible for two nonwiggler parents to produce a wiggler daughter. To get a wiggler daughter, the male would have to be X^WY (wiggler) and the female would have to be X^WX^w (nonwiggler), which is not the case here. Therefore, we cannot conclude that ear wiggling is a sex-linked recessive trait based on this family.

STUDY BREAK 13.5

A mutant trait that shows cytoplasmic inheritance is caused by an alteration in the DNA of an organelle, either the mitochondrion or the chloroplast. A key property of cytoplasmic inheritance is that a trait is transmitted by a parent to all offspring, regardless of sex. The most common form of this is maternal inheritance, in which the progeny inherit the trait

from their mother, paralleling the inheritance of mitochondria and mitochondrial DNA from the female parent and not from the male parent. This pattern of inheritance would not be seen for genes on chromosomes in the nucleus (as explained in Chapter 11).

Think Like a Scientist

FIGURE 13.2

The two largest classes in the progeny of the testcross would be red, vestigial and purple, normal—that is, the parental phenotypes. In this case the parental phenotypes are not a wild-type flies and flies with both mutations, but flies each with just one of the mutant phenotypes and the other phenotype wild-type.

The two smallest classes in the progeny of the testcross would be wild-type flies and flies that are both purple and vestigial.

FIGURE 13.8

The woman with normal vision must have at least one *c⁺* allele. One of her X chromosomes is inherited from her father, who is color blind and therefore *cY* in genotype. Therefore, the woman is heterozygous *c⁺c*. The man has normal vision. He has one X and one Y chromosome so the only possibility for his genotype is *c⁺Y*; the fact that his father was color blind is not a factor in this case. For the cross *c⁺c* × *c⁺Y*, the expected progeny are 1/4 *c⁺c⁺* (normal vision female):1/4 *c⁺c* (normal vision female):1/4 *c⁺Y* (normal vision male):1/4 *cY* (color blind male). In sum, all female offspring will have normal vision and 1/2 of the males will have normal vision and 1/2 will be color blind.

IMR

You learned about amino acids and protein structure in Section 3.4. Figure 3.17 shows the 20 amino acids. Glycine is a nonpolar amino acid, whereas arginine is positively charged polar amino acid. Since the varied properties and functions of proteins depend on the types and locations of the different amino acid side groups in their structures, it would be expected that a change from a nonpolar amino acid to a charged polar amino acid in a functionally important part of a protein would adversely affect the protein's function. Both glycine and alanine are nonpolar amino acids; the two amino acids are therefore similar, chemically speaking. You would speculate that substituting alanine for glycine would have a minimal effect on protein function.

UQ

No, the genes in the del(5q) region still have some activity. In these patients, only one copy of the gene is lacking due to a chromosomal deletion. The other copy, or allele, is present and appears to have its normal function. Since these patients now only have one copy of the genes in this deleted region, there is a lowered gene dosage, referred to as "haploinsufficiency." A lowered gene dosage (and resultant production of lower protein levels) can have profound effects on cell growth. Scientists currently believe that the lowered gene expression of *multiple* genes in the del(5q) region work together to play a role in disease development.

Test Your Knowledge

1. All sons will be color blind, but none of the daughters will be. However, all daughters will be heterozygous carriers of the trait.

2. The chance that her son will be color blind is 1/2, regardless of whether she marries a normal or color-blind male.

3. All these questions can be answered from the pedigree. Polydactyly is caused by a dominant allele, and the trait is not sex-linked. The genotypes of each person are:

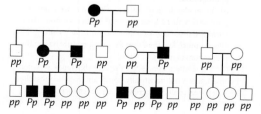

4. The sequence of the genes is ADBC.
5. Let the allele for wild-type gray body color = b^+, and the allele for black body = b. Let the allele for wild-type red eye color = p^+, and the allele for purple eyes = p. Then the parents are:

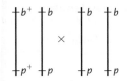

The F_1 flies with black bodies and red eyes are:

and the flies with gray bodies and purple eyes are:

6. The genes are linked by their presence on the same chromosome (an autosome), but they are not sex-linked. Because the F_1 females must have produced 600 gametes to give these 600 progeny, and because $42 + 30$ of these were recombinant, the percentage of recombinant gametes is 72/600, or 12%, which implies that 12 map units separate the two genes.
7. The initial cross is $X^w X^w$ white female $\times$ $X^{w^+} Y$ red male. The F_1 females produced are $X^{w^+} X^w$, with red eyes, and the F1 males are $X^w Y$, with white eyes.
    The cross of an F_1 female with a parental male, therefore, is $X^{w^+} X^w$ (red) $\times$ $X^{w^+} Y$ (red). The offspring are $X^{w^+} X^{w^+}$ and $X^{w^+} X^w$ females, both of which have red eyes, and $X^{w^+} Y$ (red) and $X^w Y$ (white) males in equal proportions. Thus the phenotypic ratio for females is 2 red : 0 white, and that for males is 1 red : 1 white.
    The cross of an F_1 male with a parental female is $X^w Y$ (white) $\times X^w X^w$ (white). All progeny of both sexes will be white-eyed.
8. You might suspect that a recessive allele is sex-linked and is carried on one of the two X chromosomes of the female parent in the cross. When present on the single X of the male (or if present on both X chromosomes of a female), the gene is lethal.

Interpret the Data

1. Data are expressed as µg of DNA to control for the amount of sample analyzed from each individual.
2. Smoking promotes the formation of DNA adducts. The more the individual smoked, the more adducts were observed.
3. Stopping smoking lowered the level of adduct formation.
4. The highest point in the data from the nonsmoker most probably reflects an exposure to high levels of passive or second-hand smoke in the nearby environment.

Chapter 14

Study Break

STUDY BREAK 14.1

[35]S-labeled phages in this scenario will have labeled protein coats and unlabeled DNA. When these phages infect bacteria, radioactivity enters the cell and is found in the progeny phages. In addition, radioactivity is found in the phage material removed by the blender. ^{32}P-labeled phages in this scenario are like the phages in Hershey and Chase's experiment; they have labeled DNA but unlabeled protein. When these phages infect bacteria, radioactivity enters the cell and is found in the progeny phages. No radioactivity is found in the phage material removed by the blender.

STUDY BREAK 14.2

1. Adenine and guanine are purines. Thymine and cytosine are pyrimidines.
2. Complementary base pairs are held together by hydrogen bonds. Each base is attached to the deoxyribose sugar by a covalent bond.
3. Watson and Crick described the right-handed double helix that consists of two sugar–phosphate backbones on the outside and complementary base pairs between the two backbones. A complementary base pair is a purine paired with a pyrimidine, more specifically, an A with a T, and a G with a C. The two strands of DNA are antiparallel. The key dimensions of the molecule are: diameter = 2 nm; 1 base pair = 0.34 nm; 1 turn of the helix = 10 base pairs = 3.4 nm.
4. The question focuses on the complementary base-pairing rules: A = T and G = C. If A = 20%, then T = 20%, giving 40% of the DNA as A–T base pairs. Therefore, 60% of the base pairs in this DNA molecule are G–C, and the percentage of C is 30%.

STUDY BREAK 14.3

1. Complementary base pairing ensures that the new DNA double helix is a faithful copy of the parental DNA double helix. For whatever base is exposed on the template strand, the DNA polymerase inserts the nucleotide with the complementary base.
2. DNA polymerases cannot initiate a DNA strand; they can add DNA nucleotides only to the 3′ end of an existing strand. The primer serves to provide a short stretch of nucleic acid that can be extended by DNA polymerase. The primer consists of RNA, rather than DNA, and is made by primase.
3. DNA polymerase III is the main DNA polymerase for replication in *E. coli*. This enzyme extends each primer that is synthesized on the lagging strand template, and synthesizes the leading strand. DNA polymerase I is used in lagging strand DNA synthesis. This enzyme replaces DNA polymerase when a new DNA fragment reaches the 5′ end of the Okazaki fragment that was made previously. With its 5′→3′ exonuclease activity, DNA polymerase I removes the RNA primer of that Okazaki fragment, and with its 5′→3′ polymerizing activity it replaces that primer with DNA nucleotides. In this way the Okazaki fragment is converted from an RNA–DNA hybrid into a DNA fragment.
4. Telomeres are buffers against the progressive loss of the ends of chromosomes by repeated rounds of replication. Only when the hundreds to thousands of copies of the telomere repeats have been lost are genes exposed. When those genes are lost by continued chromosome shortening and/or when chromosomes break down in the absence of telomeres, the cell is severely damaged.

STUDY BREAK 14.4

Proofreading prevents errors from being introduced into the DNA sequence. The DNA sequence in an organism's genome specifies everything about that organism—most notably, its function and reproduction. If significant errors occur during replication, gene sequences could be changed and the function of the organism could be adversely affected. Particularly if there is a high rate of errors, as there would be in the absence of proofreading, these errors would have potentially lethal consequences.

As part of the proofreading process, DNA polymerase reverses and removes the mispaired nucleotide. The enzyme then continues to move forward, inserting the correct nucleotide.

DNA repair mechanisms then look for and correct any errors that were not detected by proofreading. For example, repair enzymes remove a section of the newly synthesized DNA with the mismatch, and DNA polymerase synthesizes a replacement section with the correct base pairing.

STUDY BREAK 14.5

1. The nucleosome consists of two molecules each of histones H2A, H2B, H3, and H4 assembled into a nucleosome core particle wrapped with almost two turns of DNA. The diameter of the nucleosome is 10 nm.
2. Histone H1 is responsible for the next level of chromosome packing above the nucleosome. H1 binds to the exit/entry point of DNA on the nucleosome and to the linker DNA and brings about a coiling of the chromatin into the 30-nm chromatin fiber. The coiled structure is called the solenoid.

Think Like a Scientist

FIGURE 14.2

Yes, theoretically this could have been done; that is, the transforming principle is the genetic material. Given that *R* and *S* cells are genetically different, the transforming principle released from heat-killed *R* cells could convert living *S* cells to *R* cells. However, only if all *S* cells in a mouse were converted to *R* cells would the mouse live. This is unlikely so, practically speaking, this proposed experimental design is not a good one.

FIGURE 14.3

Carbon is found in both protein and DNA. Therefore, the radioisotope would be found in phage coats on phage-infected cells, within phage-infected cells, and in progeny phages. The success of the Hershey and Chase experiment relied on using radioisotopes that were specific for DNA and protein which is why the isotopes of phosphorus and sulfur were used. The carbon radioisotope would have not been useful in the experiment to answer the question asked.

FIGURE 14.9

First consider the data in the Figure. After one generation in ^{14}N, all the DNA is [15]N–[14]N. After two generations, 1/2 the DNA is [15]N–[14]N and 1/2 is [14]N–[14]N. Therefore, by understanding how the semiconservative replication mechanism works, after four replications you would expect 1/8 of the DNA to be [15]N–[14]N, and 7/8 to be [14]N–[14]N. After five replications you would expect 1/16 of the DNA to be [15]N–[14]N and 15/16 to be [14]N–[14]N.

FIGURE 14.15

Without normal DNA ligase activity, Okazaki fragments could not be joined and synthesis of a continuous lagging strand could not be completed. In actuality, a mutant DNA ligase of this kind would be lethal to the cell. Some temperature-sensitive mutants have been isolated for bacteria, though, in which DNA ligase activity is normal at normal growth temperatures but nonfunctional or only partially functional at high temperatures. At the high temperatures, such mutants accumulate Okazaki fragments because they cannot be joined into longer chains.

IMR

Males have one X chromosome whereas females have two. For a female to express an X-linked recessive trait, both of her X chromosomes must carry the mutant gene. However, for a male to express such a trait, he just needs one copy of the mutant chromosome. On probability grounds, therefore, it is more likely for a male to have an X-linked recessive trait because he only has to inherit one mutant chromosome in contrast to the situation for a female who must inherit two mutant copies.

UQ

Telomerase alone is not the sole determinant of telomere length. The complex of proteins at the telomere regulates access of telomerase to the telomere. Thus, even if the amount of telomerase in the cell is unchanged, certain conditions could alter the function of one or more proteins at the telomere to increase the ability of telomerase to elongate the chromosome ends.

Test Your Knowledge

1. b 2. d 3. a 4. a 5. d 6. c 7. b 8. a 9. c 10. b

Interpret the Data

1. Schedule B, CDDP (8h) followed by 5-FU, was the most effective treatment because it resulted in the lowest percentage of cell proliferation.
2. B was the only schedule in which cells were first treated with CDDP alone.

Chapter 15

Study Break

STUDY BREAK 15.1

1. Although most enzymes are proteins, not all proteins are enzymes. And, some proteins consist of more than one polypeptide subunit. Each different polypeptide is encoded by a different gene, hence the one gene–one polypeptide hypothesis.
2. There are four different letters in the code (A, U, G, C), so a five-letter code would have 4^5 possible combinations = 1,024 codons.

STUDY BREAK 15.2

1. 5′-GUUUAACCGAAUAAUGGCCUAC-3′
2. The promoter determines where transcription of a gene will begin. In prokaryotes, RNA polymerase binds to the nucleotide sequence of the promoter and orients in the correct way to transcribe the associated gene. In eukaryotes, transcription factors bind to the promoter and then recruit RNA polymerase, which then orients properly for transcription from the transcription start point.

STUDY BREAK 15.3

1. Both pre-mRNAs and mRNAs have a 5′ cap, exons, and a 3′ poly(A) tail. Only pre-mRNAs have introns, which are removed from pre-mRNAs to produce mRNAs.
2. Particular snRNPs bind to the ends of an intron using their contained RNAs to recognize the boundary sequences of the intron. Other snRNPs then bind, causing the intron to loop out, and completing the active spliceosome. Cleavage at each intron-exon junction, looping back of the intron on itself, and joining the two exons together completes the splicing event; the intron and snRNPs are then released.

STUDY BREAK 15.4

1. In eukaryotes, a complex of the small ribosomal subunit, initiator tRNA, initiation factors, and GTP binds to the 5′ cap of the mRNA and scans along the mRNA until it reaches the first AUG codon, which is the start codon. The anticodon of the initiator tRNA binds to the start codon, the large ribosomal subunit binds, and the initiation factors are released when GTP is hydrolyzed.

    In prokaryotes, a complex of the small ribosomal subunit, initiator tRNA, initiator factors, and GTP binds to the region of the mRNA where the AUG start codon is located, directed by a specific RNA sequence upstream of the start codon. The other steps are the same as those in eukaryotes.
2. The P site is where the tRNA with the growing polypeptide is located. Downstream of the P site is the A site. An incoming aminoacyl–tRNA enters the A site and its anticodon base pairs with the codon of the mRNA in that site. When the polypeptide is transferred to the amino acid on the tRNA in the A site, the ribosome translocates one codon along the mRNA. As translocation takes place, the empty tRNA that was in the P site is moved to the E site. It remains there, blocking a new aminoacyl–tRNA entering the A site until translocation is finished. Then the empty tRNA is released from the ribosome.
3. Proteins found in the cytosol are made on free ribosomes.

    Proteins are sorted to the endomembrane system by cotranslational import. The proteins begin their synthesis on free ribosomes. These proteins have signal sequences at their N-terminal ends that direct them and the ribosome to dock with a receptor on the rough ER membrane. Continued translation inserts the growing polypeptide into the lumen of the ER, and the signal sequence is removed by signal peptidase. The proteins are then tagged

to target them for sorting to their final destinations. Some proteins remain in the ER, whereas others are transported via the Golgi to vesicles for sorting to lysosomes, secreting them from the cell, or depositing them in the plasma membrane.

Proteins are sorted to mitochondria, chloroplasts, microbodies, and the nucleus by posttranslational import. Proteins destined for the mitochondria, chloroplasts, and microbodies have short N-terminal transit sequences that target them to the organelle. A transit sequence interacts with a specific transport complexes on the organelle, and the polypeptide is then taken into the organelle. The transit sequence is then removed. Proteins destined for the nucleus have a nuclear localization signal which is bound by a cytosolic protein. The complex then interacts with the nuclear pore complex and the polypeptide then is transported into the nucleus through the pore.

STUDY BREAK 15.5

1. A missense mutation involves a change from a sense codon to another sense codon that specifies a different amino acid. A silent mutation involves a change from one sense codon to another sense codon, but where both codons specify the same amino acid.
2. Genetic recombination occurs by crossing over between two homologous sequences. TE transposition occurs by integration into a new location with which the TE has no sequence homology.
3. Both: (1) have inverted repeats at their ends; (2) contain a transposase gene; and (3) integrate into target sites and cause a duplication of the target site.
4. A transposon is a mobile genetic element that moves from one location to another in the genome as a DNA molecule.

    A retrotransposon is a mobile genetic element that moves from one location to another in the genome using an RNA intermediate; that is, the integrated DNA element is transcribed to produce an RNA copy. The RNA copy is reverse-transcribed into DNA, which then integrates at a new location in that genome.

Think Like a Scientist

FIGURE 15.2

Skin color depends on the synthesis of a pigment. It is reasonable to assume that synthesis of a pigment involves a biosynthesis pathway. A hypothesis to explain the results is that mutations in genes that encode enzymes that control different steps in that pathway can result in albinism. Therefore, the first couple consists of two individuals each of whom are homozygous for the albinism mutation in one of those genes. That pairing can produce offspring who have the same genotypes as the parents with respect to albinism genes, and, thus, they have albinism.

The individuals of the second couple, however, must be homozygous mutant for different genes involved in albinism. However, their other albinism gene is homozygous normal. Therefore, each offspring of this couple inherits a normal allele and a mutant allele of one albinism gene and a normal allele and a mutant allele of the other albinism gene. Because albinism is a recessive trait, the presence of the two normal genes of the genes results in the children having normal skin color.

FIGURE 15.6

No. Although transcription occurs in highly similar ways in eukaryotes and prokaryotes, the initiation processes in particular are specific; that is, different promoter sequences are used, and eukaryotic RNA polymerases typically recognize only eukaryotic promoter sequences, and prokaryotic RNA polymerases typically recognize only prokaryotic promoter sequences. Therefore, *E. coli* RNA polymerase is unlikely to be able to recognize the promoter of the eukaryotic protein-coding gene in the figure.

IMR

The result provides support for the hypothesis that the conserved bases likely play an important functional role in peptide bond formation and, therefore, supports the conclusions of the study.

UQ

1. The first ribosomes were made only of RNA.
2. Not to make proteins, but to make simple protein fragments (peptides) that bind to RNA to help RNA carry out its biological functions before the evolution of proteins.

Test Your Knowledge

1. b 2. a 3. e 4. d 5. b 6. d 7. b 8. a 9. e 10. d

Interpret the Data

The codons specifying the amino acids are shown in the following figure:

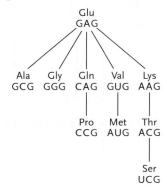

Chapter 16

Study Break

STUDY BREAK 16.1

1. The Lac repressor is active when it is made. In normal cells, in the absence of lactose, the Lac repressor binds to the operator, blocking transcription. In the mutant, the Lac repressor is not made, so transcription can never be blocked because no repressor is available to bind to the operator. As a consequence, the lactose metabolizing enzymes will be made both in the presence and absence of lactose in the medium. The mutation involved is an example of a regulatory mutant. Mutations such as this were valuable to Jacob and Monod in developing the operon model for gene regulation.
2. The Trp repressor is inactive when it is made. In normal cells, in the presence of tryptophan, the Trp repressor is activated, binding to the operator and blocking transcription. In the mutant, the Trp repressor is not made, so the operon cannot be turned off when tryptophan is present. This means that the tryptophan biosynthesis enzymes will be produced both in the presence and absence of tryptophan.

STUDY BREAK 16.2

1. General transcription factors bind to the promoter and recruit RNA polymerase II, orienting the enzyme so that it will begin transcription at the beginning of the gene. Activators bind to regulatory sequences associated with genes and increase the rate of transcription. Activators that bind to regulatory sequences in the proximal promoter region interact directly with the general transcription factors at the promoter to exert their action. Activators that bind to regulatory sequences in the enhancer stimulate transcription indirectly. These latter activators bind to a coactivator that also binds to the complex of proteins at the promoter, and transcription then occurs at the maximal possible rate.
2. Histones are general negative regulators of gene expression. When DNA is complexed with histones in normal chromatin, gene promoters typically are not very accessible to the transcription machinery. By acetylating histones, the chromatin is remodeled, making the promoter now accessible to the transcription machinery. Acetylation of histones occurs in response to the binding of an activator to a regulatory sequence associated with the gene.

STUDY BREAK 16.3

1. A microRNA (miRNA), in a complex with particular proteins, binds to an mRNA by complementary base pairing. Either the proteins cut the mRNA in the region of pairing, thereby destroying that molecule, or the double-stranded RNA region blocks translation.

2. Removal of the poly(A) tail would result in the mRNA not being translated.

STUDY BREAK 16.4

1. Determination is the process by which the developmental fate of a cell is set. Differentiation is the establishment of a cell-specific developmental program in cells. Determination and differentiation are under molecular control. Regulatory genes encode regulatory proteins that bind to promoters of the genes they control, switching the genes on or off depending on the interaction.

2. The segmentation genes subdivide the embryo progressively into regions, thereby determining the segments of the embryo and the adult. In essence, they organize the embryo into segments. The homeotic genes specify the identity of each segment with respect to the body part it will become.

STUDY BREAK 16.5

1. A tumor suppressor gene encodes a product that has an inhibitory role in the cell division cycle. If a tumor suppressor gene is mutated so that its product is nonfunctional or its function is significantly diminished, then the gene's inhibitory control of cell division is lost or reduced. As a result, the cell may progress toward division.

2. A proto-oncogene encodes a product that has a stimulatory role in cell division. Mutations that lead to increased levels of that product have converted a proto-oncogene to an oncogene. The increased amount of product is stimulatory to cell division.

3. Some miRNAs have been shown to regulate the expression of mRNA transcripts of particular tumor suppressor genes. Overexpression of those miRNAs can abnormally inhibit the activity of those tumor suppressor genes, thereby removing or decreasing inhibitory signals for cell proliferation. Other miRNAs have been shown to regulate the expression of mRNA transcripts of particular proto-oncogenes. Inactivation of those miRNAs means that the proto-oncogenes are expressed at higher than normal levels, which can stimulate cell proliferation.

Think Like a Scientist

FIGURE 16.3

If the Lac repressor cannot bind to the operator, then the *lac* operon structural genes will be transcribed even in the absence of lactose because there is no block to the binding of RNA polymerase to the promoter. By contrast, in a normal cell, the Lac repressor binds to the operator and blocks RNA polymerase from transcribing the *lac* structural genes as long as lactose is absent from the medium.

FIGURE 16.5

If the Trp repressor cannot bind tryptophan, then the repressor could not be activated when tryptophan is present in the medium. Without an active repressor in this condition, RNA polymerase cannot be blocked from binding to the promoter. Therefore, the Trp structural genes will be transcribed. By contrast, in a normal cell, the presence of tryptophan in the medium activates the Trp repressor which then binds to the operator and blocks transcription of the *trp* structural genes by RNA polymerase.

IMR

First, the CREB monomer must have a binding site for itself so that it can form a dimer. Second, the dimer must have a binding site for CRE (cAMP response element) sequences in the enhancers for regulatory genes that it controls. Third, CREB must have a binding site for the Tax protein.

UQ

Dr. Kay's lab uses mice to test RNAi therapies against hepatitis C virus. You could test various RNAi agents via an intravenous infusion to find a dose where the viral load (or level) is reduced substantially without causing any dangerous toxicities that have been determined by standard laboratory testing of blood samples. Kay's lab aims to reduce viral load by at least 100-fold.

Test Your Knowledge

1. d 2. a 3. b 4. c 5. e 6. b 7. a 8. d 9. a 10. b

Interpret the Data

1. The woman would have an 8.6% chance of dying of cancer.

2. Her risk of dying of cancer is 18.0% if she carries a mutated *BRCA1* gene.

3. This study suggests a *BRCA1* mutation is more dangerous than a *BRCA2* mutation (in effect, *BRCA1* mutations are associated with a higher percentage of deaths). Sixteen out of 89 women with the *BRCA1* mutation died whereas only 1 out of 35 women with the *BRCA2* mutation died.

4. You would need to know the number of deaths of women who had a preventive mastectomy and the number of deaths of women who had a preventive oophorectomy.

Chapter 17

Study Break

STUDY BREAK 17.1

1. An F$^+$ cell contains the F factor plasmid in addition to the chromosomal DNA. The F factor enables an F$^+$ cell to conjugate with an F$^-$ cell, which lacks the F factor. By a special replication mechanism, a copy of the F factor is transferred to the recipient F$^-$ cell in an F$^+$ × F$^-$ conjugation, so the recipient is converted to an F$^+$ cell.

   An Hfr cell has the F factor integrated into the chromosomal DNA. When an Hfr cell conjugates with an F$^-$ cell, the F factor begins its replicative transfer into the recipient as in an F$^+$ × F$^-$ conjugation and, by that transfer mechanism brings in chromosomal genes from the Hfr donor. Those chromosomal genes can recombine with the genes in the recipient. Because replication of the F factor begins in the middle of the plasmid, the entire F factor cannot be transferred to the recipient unless the entire chromosome is transferred to the recipient, which occurs only rarely. Hence, the recipient remains F$^-$ in this conjugation experiment.

2. In gene segregation in sexually reproducing organisms, genetic material moves from one generation to the next, basically by descent. In horizontal gene transfer, the movement of genetic material between organisms is other than by descent. For instance, in transformation, a DNA molecule in the environment is taken up by a recipient cell—the DNA has moved horizontally from a donor cell (from which the DNA was released) to the recipient cell. Genetic recombination in the recipient can alter the phenotype of that cell.

STUDY BREAK 17.2

1. A virulent phage always enters the lytic cycle when it infects a bacterial cell. The end result is the assembly of many progeny phages and their release into the surroundings when the cell breaks open.

   A temperate phage enters either the lytic cycle or the lysogenic cycle when it infects a cell. The lytic cycle is the same as that for a virulent phage. The lysogenic cycle involves integration of the phage's chromosome into the bacterial chromosome. In the integrated state the phage—now called the prophage—is inactive and replicates only when the bacterial chromosome replicates. In response to an adverse environmental signal to the cell, the phage chromosome can excise itself from the bacterial chromosome and enter the lytic cycle.

2. Animal cells: Viruses without an envelope bind by their recognition proteins to receptor proteins in the host cell's plasma membrane and are then taken into the cell by receptor-mediated endocytosis. For some enveloped viruses, the genome-containing capsid enters the cell when the envelope fuses with the host cell's plasma membrane. For other enveloped viruses, the complete virus, with the envelope, enters the cell by endocytosis.

   Plant cells: All plant viruses lack envelopes. They enter cells either through mechanical injuries to leaves and stems or by the action of biting and feeding insects.

3. Non-retrovirus RNA viruses have RNA genomes that are replicated in an RNA-to-RNA manner. Retroviruses have RNA genomes that are replicated via a DNA intermediate: that is, the RNA genome is copied to double-stranded DNA by reverse transcriptase and the DNA molecule integrates into the host cell's nuclear chromosomes, from which location new RNA viral genomes are transcribed.

STUDY BREAK 17.3

Although all are infectious agents, viruses have protein coats, whereas viroids and prions do not. A virus consists of a nucleic acid genome surrounded by a protein coat and, in some cases, an envelope. Viroids are naked, single-stranded RNA circular molecules. Prions are proteins.

Think Like a Scientist

FIGURE 17.1

The control plates were used to check that the mutant alleles in the two strains did not mutate back to normal alleles; that is, mutation back to the nonmutant state would allow the cells to grow on the minimal medium plates. By having more than one mutant allele in each strain (two in strain 1 and three in strain 2), Lederberg and Tatum were making it extremely unlikely to see colonies on the minimal plate; that is, for strain 1, two mutant alleles would have to mutate to normal alleles, and for strain 2, three mutant alleles would have to mutate to normal alleles. If we made the simple assumption that the rate of mutation from mutant to normal is the same for all genes, then the probability of having two genes mutate from mutant to normal is the square of the probability for one to mutate. Moreover, the probability of having three genes mutate from mutant to normal is the cube of the probability for one to mutate. Therefore, having just one mutant allele in each strain would mean that the probability of generating a nonmutant cell by mutation would be much higher than for the two- or three-mutant allele strains.

FIGURE 17.4

The transfer of chromosomal genes by conjugation is dependent on the F factor. Genes on the F factor direct the one-way transfer of a copy of the F factor to a recipient in F$^+$ × F$^-$ conjugations. When the F factor is integrated into the chromosome to produce an Hfr, again the F factor genes direct a one-way transfer of DNA from donor to recipient, in this case involving the F factor attached to the bacterial chromosome. There is no mechanism, therefore, for transfer of chromosome genes in the opposite direction—that is, from recipient to donor.

FIGURE 17.8

As you learned in the text, the prophage state is maintained by a phage-encoded repressor molecule that prevents expression of phage genes for the lytic cycle. When the prophage is transferred into a recipient cell, that cell will not contain repressor proteins and, because it takes time for new repressor proteins to be synthesized, the result will be that the λ lytic cycle genes will be expressed and the recipient will be killed by phage release. If the recipient already contained a λ prophage, then repressor proteins would already be present and the transferred prophage would remain repressed with respect to the lytic cycle genes.

IMR

That the DNase converted the radioactive macromolecules from acid-insoluble to acid-soluble, whereas RNase did not, affirms that the macromolecules synthesized must be DNA and not RNA.

UQ

Researchers can further study the interactions between normal prion protein and prion aggregates and the effect of these interactions on neurons. One possible function for the normal prion protein is to act as a signaling molecule to regulate apoptosis. Interactions with prion particles may induce aberrant signaling through the normal prion protein. Since the normal prion protein also binds to many other molecules (for example, extracellular matrix components), one could also investigate the effects of the conversion process on the interactions of normal prion protein with other molecules.

Test Your Knowledge

1. e 2. a 3. b 4. c 5. d 6. a 7. c 8. c 9. a 10. b

Interpret the Data

The answer is shown in the figure. The segments of donor chromosomes transferred by each Hfr strain are shown, with the arrowheads showing the beginning.

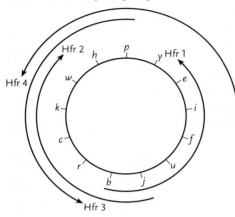

Chapter 18

Study Break

STUDY BREAK 18.1

1. Each restriction enzyme recognizes a specific sequence in DNA, typically in the range of 4 to 6 bp, and cuts both strands of the DNA within the sequence. Restriction enzymes differ with respect to the DNA sequences—the restriction sites—they recognize and cut. The enzymes most useful for cloning produce sticky ends.
2. Replication origin so that the plasmid will replicate in *E. coli*, an antibiotic resistance gene to allow selection of bacteria containing the plasmid, a cluster of restriction sites (a multiple cloning site) to provide choices for inserting fragments of DNA, and the *lacZ*$^+$ gene to use in blue–white screening, in which colonies containing recombinant plasmids (white) are distinguished from colonies with vectors lacking inserted DNA (blue).
3. A cDNA library is a collection of clones in which DNAs that are complementary to the mRNAs in the cell have been inserted into a cloning vector. The array of particular clones in a cDNA library is directly related to the genes expressed in the cells from which the mRNAs are isolated. Therefore, not all genes are represented in any given cDNA library. By contrast, a genomic library is a collection of clones containing all the sequences of an organism inserted into a cloning vector. A genomic library contains all of the genes of an organism, as well as noncoding sequences found between and within genes.
4. PCR is a method to amplify a specific segment of DNA. The amplification process depends on DNA replication and, therefore, requires primers. A limitation of PCR, then, is that DNA sequence information must be available for the segment of DNA to be amplified. Otherwise, the primers cannot be synthesized. The ingredients of a PCR are the template DNA containing the sequence to be amplified, the two primers, a buffer, and a DNA polymerase that is tolerant to the high temperature used to denature the DNA repeatedly during the cycles of the reaction.

STUDY BREAK 18.2

1. Each human (or other organism) has unique combinations and variations of DNA sequences. DNA fingerprinting exploits these combinations and variations to distinguish between different individuals (with the exception of identical twins). Using DNA technologies to analyze the particular regions of the genome showing sequence variation, it is possible to compare two DNA samples to

see if they are from the same person or not. DNA fingerprinting is used in several ways, including forensics, paternity testing, and basic research.
2. A transgenic organism is one into which a gene or genes from an external source have been introduced as a means to modify the organism genetically.
3. Germ-line cells develop into reproductive cells, so modifying this cell type genetically will lead to the genetic modification being passed to offspring. In somatic cell gene therapy, germ-line cells and their products are not involved, so the inserted genes remain with the individual and are not passed to offspring.

Think Like a Scientist

FIGURE 18.14

To demonstrate that the experiment was a success, it was necessary to show that a clone matched that of the donor genetically and, importantly, did not match that of the surrogate mother. Only by starting with genetically distinguishable kinds of sheep could that be demonstrated.

IMR

As discussed in Chapter 7, plasma membrane receptors, when activated by a signal molecule, transduce the signal into the cell and through a signal transduction pathway to bring about a cellular response. By knowing that signal transduction pathway is involved, researchers could set out to elucidate the pathway thereby providing information that could be useful for the design of disease control strategies.

UQ

Tissue-specific stem cell niches serve to both protect stem cells against environmental insults and control stem cell responses. Cellular and neuronal signals that control stem cell dormancy and proliferation are often uniquely adapted to these microenvironments. This ensures that tissue-specific stem cells receive signals in a tightly controlled manner.

Test Your Knowledge

1. e 2. d 3. a 4. a 5. d 6. b 7. e 8. b 9. a 10. d

Interpret the Data

An offspring must have one allele of each gene inherited from the male parent, and the other allele of each gene inherited from the female parent. The same goes for STR locus alleles.

Juvenile 1: Not the offspring of the two adults. The juvenile has one allele in common with the male parent (8 repeats = 72 bp), but its other allele came from neither adult (9 repeats = 81 bp). In other words, it is heterozygous for 8- and 9-repeat alleles.

Juvenile 2: Could be the offspring of the two adults. There was only one DNA fragment size amplified by PCR. The interpretation is that this juvenile is homozygous for a 6-repeat (54-bp) allele, which is why there is more DNA amplified (a thicker band). Each adult has a 6-repeat allele that could have passed to the juvenile.

Juvenile 3: Not the offspring of the two adults. This juvenile is heterozygous for 5-repeat (45-bp) and 11-repeat (99-bp) alleles. The 11-repeat allele could have been inherited from the female adult, but neither adult has the 5-repeat allele.

Juvenile 4: Could be the offspring. This juvenile is heterozygous for 6-repeat (54-bp) and 8-repeat (72-bp) alleles. It could have inherited the 6-repeat allele from either parent, but could have inherited the 8-repeat allele only from the male parent. Therefore, the 6-repeat allele would have had to come from the female parent.

Chapter 19

Study Break

STUDY BREAK 19.1

Having the complete sequence of a genome enables a researcher to ask questions about the organization of genes and of other sequences of importance in the genome. The most important question is how the functions of all of the genes in a genome direct that organism's life. It also allows researchers to compare gene content and gene order between organisms, which provides insight into how genomes evolve.

STUDY BREAK 19.2

1. In whole-genome shotgun genome sequencing, the entire genome is broken into thousands to millions of random, overlapping fragments, each fragment is cloned and sequenced, and then the genome sequence is assembled by computer on the basis of sequence overlaps between fragments.
2. Protein-coding genes, noncoding RNA genes, regulatory sequences associated with genes, origins of replication, transposable elements and related sequences, pseudogenes, and short repetitive sequences.
3. To find protein-coding genes in a bacterium, computer algorithms search a genome sequence for open reading frames (ORFs), which are defined as the DNA equivalent of a start codon (ATG) in frame (a multiple of three away) from a stop codon (TAG, TAA, or TGA). Such a segment of DNA could potentially produce an mRNA that could be translated into a protein. Other analyses would be required to show if any given ORF identified in this way is an actual protein-coding gene.

   To find protein-coding genes in a mammal, one also needs to find ORFs. However, the presence of introns in eukaryotic protein-coding genes complicates matters. Therefore, more sophisticated computer algorithms are needed—in this case ones that attempt to locate exon-intron boundaries while they search for ORFs.
4. Prokaryotic genomes vary widely in size. Nonetheless, their genes are densely packed in their genomes (a fairly consistent density of 900 genes per Mb), with little non-coding space between them. Genomes of eukaryotes also show a great range of sizes; overall, their genomes are much larger than those of prokaryotes. Gene density also varies considerably in eukaryotes, ranging from 500 per Mb for the single-celled yeast to fewer than 10 per Mb in mammals. There is no uniformity to the packing of genes, however.

   Prokaryotic genes are arranged either singly or in operons, depending on the genes. All eukaryotic genes are arranged singly; no operons are found in eukaryotes. The number of regulatory sequences typically is much higher for eukaryotic protein-coding genes than for prokaryotic protein-coding genes. There are no introns in prokaryotic protein-coding genes, whereas most eukaryotic protein-coding genes contain introns. Introns can make up significant proportions of eukaryotic genomes—for example, about 20% in humans. In contrast to prokaryotes, eukaryotes contain many pseudogenes, many transposable element sequences, and a variety of short repeated sequences. A prokaryotic chromosome has one origin of replication, whereas a eukaryotic chromosome may have hundreds to thousands of origins of replication.

STUDY BREAK 19.3

1. The function of a gene identified in a genome sequence may be assigned by a sequence similarity search of sequence databases to find a match with a gene of known function, by comparing the structure of the protein encoded by the gene with the structures of proteins with known function, and by experiments that alter the expression of the gene to see how the phenotype of a cell or organism is affected.
2. A genome-wide analysis of gene expression here could involve DNA microarray assays. The mRNAs could be isolated from untreated tissue culture cells and from cells treated with a steroid hormone. Convert each mRNA preparation to cDNAs using reverse transcriptase, using nucleotide precursors with different fluorescent labels for the two batches; for example, green for the untreated (reference) sample and red for the treated (experimental) sample. Mix the two cDNAs and pump them through a DNA chip prepared to have spots with DNA representing every gene in the genome. Allow hybridization to occur, and analyze the hybridization by laser detection. The colors of the spots indicate which genes are affected by the steroid hormone. Purely red spots represent genes that are active only in hormone-treated cells. Purely green spots represent genes that are

active only in untreated cells. Spots that are a mixture of green and red represent genes that are active in both types of cells. Based on controls, it would be possible to see if any of the genes have higher or lower expression levels in the treated cells.

3. The proteome is complete set of proteins that can be expressed by an organism's genome. Proteomics has the three major goals of determining the number and structures and functions of proteins of the proteome, determining the location of each protein within or outside the cell, and determining the physical interactions between proteins.

STUDY BREAK 19.4

1. Tandem duplication happens when homologous pairs of chromosomes do not line up properly in meiosis. As a result of unequal crossing over, one gamete inherits a chromosome with two copies of the stretch of DNA between the two points of crossing over, and therefore duplicate copies of any genes on that stretch of DNA. Dispersed duplication happens when transposable elements copy the DNA for a gene and insert the new copy somewhere else in the genome.

2. Gene duplication produces two identical copies of a gene that only evolve distinct functions as different mutations occur in each copy. Exon shuffling, by contrast, combines parts of two or more genes in an evolutionarily novel way. The protein coded for by the new gene contains a combination of domains that may never before have existed in one protein. Because a protein's domains usually each perform a different molecular function, the newly evolved protein may have a combination of functionalities unlike any previously existing protein.

Think Like a Scientist

UQ

Knowledge that an individual is genetically predisposed to developing a disease or is at greater risk for the aftereffects of disease based on genotype may cause emotional distress, and fear of discrimination by employers or insurance companies. Similar concerns may exist if genetic information suggests that disease treatment options for an individual are limited. The Genetic Information Nondiscrimination Act (GINA) was recently enacted to help address these concerns.

Test Your Knowledge

1. b 2. a 3. a 4. a 5. d 6. d 7. d 8. a 9. d 10. d

Interpret the Data

To find an open reading frame, you first find a start codon (see Figure 15.5), and then count three nucleotides at a time until you find a stop codon in the same reading frame (see Figure 19.4). This sequence contains 6 start codons and therefore 6 open reading frames. The first open reading frame begins with the start codon at the very beginning of the second line and has 145 codons between the start and stop codon, which makes it large enough to code for a functional protein. The other open reading frames have only 5, 15, 8, 28, and 11 codons between the start and stop codon, so they are very unlikely to code for functional proteins.

Appendix B: Classification System

The classification system presented here is based on a combination of organismal and molecular characters and is a composite of several systems developed by micro-biologists, botanists, and zoologists. This classification reflects current trends toward a phylogenetic approach to taxonomy, one that incorporates the ever more detailed information about the relationships of monophyletic lineages provided by new molecular sequence data. In keeping with these trends, we have omitted reference to the traditional taxonomic categories, such as "class" and "order." Instead, we present the major monophyletic lineages in each of the three domains, and we indicate their relationships within a nested hierarchy that parallels that of traditional Linnaean classification.

Although researchers generally agree on the identity of the major monophyletic lineages, the biologists who study different groups have not established universal criteria for identifying the somewhat arbitrary taxonomic categories included in the traditional Linnaean hierarchy. As a result, a "class" or an "order" of flowering plants may not be the equivalent of a "class" or an "order" of animals. In fact, as described in *Unanswered Questions* at the end of Chapter 24, systematic biologists are shifting toward a more phylogenetic approach to taxonomy and classification, such as the one represented here.

Bear in mind that we include this appendix to introduce the diversity of life and illustrate many of the evolutionary relationships that link monophyletic groups. Like all phylogenetic hypotheses, this classification is open to revision as new information becomes available. Moreover, the classification is incomplete because it includes only those lineages that are described in Unit Four.

Prokaryotes and Eukaryotes

Organisms fall into two groups, prokaryotes and eukaryotes, based on the organization of their cells. Prokaryotes consist of the Domains Bacteria and Archaea and are characterized by a central region, the nucleoid, which has no boundary membrane separating it from the cytoplasm, and by membranes typically limited to the plasma membrane. Most prokaryotes are single-celled, although some are found in simple associations. All other organisms are eukaryotes, which make up the Domain Eukarya. Eukaryotes are characterized by cells with a central, membrane-bound nucleus, and an extensive membrane system. Some eukaryotes are single-celled, while others are multicellular.

Domain Bacteria

The largest and most diverse group of prokaryotes. Includes photoautotrophs, chemoautotrophs, and heterotrophs.

PROTEOBACTERIA Gram-negative bacteria that include chemoautotraphs, chemoheterotrophs, and photoautotrophs. Some are aerobic and some are anaerobic. Five subgroups of proteobacteria are recognized.

SPIROCHETES Helically spiraled bacteria that move by twisting in a corkscrew pattern

CHLAMYDIAS Gram-negative intracellular parasites of animals, with cell walls that lack peptidoglycans

GRAM-POSITIVE BACTERIA Chemoheterotrophic bacteria with thick cell walls

CYANOBACTERIA Photoautotrophic Gram-negative bacteria that carry out photosynthesis using the same chlorophyll as in plants and release oxygen as a by-product

Domain Archaea

Prokaryotes that are evolutionarily between eukaryotic cells and the bacteria. Most are chemoautotrophs. None is photosynthetic. Originally discovered in extreme habitats, they are now known to be widely dispersed. Compared with bacteria, the Archaea have a distinctive cell wall structure and unique membrane lipids, ribosomes, and RNA sequences. Some are symbiotic with animals, but none is known to be pathogenic.

EURYARCHAEOTA Includes methanogens, extreme halophiles, and some extreme thermophiles

NANOARCHAEOTA Extreme thermophile (only one member known). Lives as a symbiont of a thermophilic member of the Crenarchaeota.

CRENARCHAEOTA Includes most of the archaean extreme thermophiles, as well as psychrophiles; mesophilic species comprise a large part of plankton in cool, marine waters

THAUMARCHAEOTA All living species (four) are mesophilic chemoautotrophs. Other members are represented by DNA sequences in environmental samples.

KORARCHAEOTA Only living member is a thermophile. Other members are represented by DNA sequences in environmental samples.

Domain Eukarya

PROTISTS A collection of single-celled and multicelled lineages, which are almost certainly not a monophyletic group.

Excavata (excavates) Single-celled animal parasites that have greatly reduced mitochondria, or organelles derived from mitochondria, and move by means of flagella; most have a scooped out (excavated) feeding apparatus on the ventral surface of the cell

Metamonada (metamonads)—consist of the Diplomonadida and the Parabasala

Diplomonadida (diplomonads)—cells have two nuclei; move by multiple freely beating flagella

Parabasala (parasabala)—move by freely beating flagella and an undulating membrane formed by a flagellum buried in a fold of cytoplasm

Euglenozoa (euglenozoans)—mostly single-celled, highly motile cells that swim using flagella and have disc-shaped inner mitochondrial membranes

Euglenids—free-living with anterior flagella; most are photosynthetic autotrophs

Kinetoplastids—nonphotosynthetic heterotrophs that live as animal parasites. Their single mitochondrion contains a large DNA–protein deposit called a kinetoplast.

Chromalveolata (chromalveolates) Heterogeneous group with a range of forms and life styles

Alveolata (alveolates)—characterized by small membrane-bound vesicles called alveoli in a layer under the plasma membrane

Ciliophora (ciliates)—motile, primarily single-celled highly complex heterotrophs; swim by means of cilia

Dinoflagellata (dinoflagellates)—nonmotile single-celled marine heterotrophs or autotrophs; shell formed from cellulose plates

Apicomplexa (apicomplexans)—nonmotile parasites of animals with apical complex for attachment and invasion of host cells

Stramenopila (stramenopiles)—characterized by two different flagella, one with hollow tripartite projections, and the other plain

Oomycota (oomycetes)—water molds, white rusts, and mildews; all are funguslike heterotrophs lacking chloroplasts

Bacillariophyta (diatoms)—single-celled autotrophs that carry out photosynthesis by pathways similar to those of plants; covered by a glassy silica shell; most are free-living, whereas some are symbionts

Chrysophyta (golden algae)—colonial; each cell of the colony has a pair of flagella and is covered by a glassy shell consisting of plates or scales; most are autotrophs that carry out photosynthesis by pathways similar to those of plants

Phaeophyta (brown algae)—photosynthetic autotrophs

Rhizaria (rhizarians) Amoebas with stiff, filamentous pseudopodia; some with outer shells

Radiolaria (radiolarians)—marine heterotrophs; glassy internal skeleton with projecting axopods—raylike strands of cytoplasm supported internally by long bundles of microtubules

Foraminifera (forams)—marine heterotrophs; shells, most of which are chambered, spiral structures, consist of organic matter reinforced by calcium carbonate

Chlorarachniophyta (chlorarachniophytes)—green, photosynthetic amoebas

Archaeplastida (archaeplastids) Red algae, green algae, and land plants *(viridaeplantae),* photosynthesizers with a common evolutionary origin

Rhodophyta (red algae)—marine seaweeds, typically multicellular with plantlike bodies; most are reddish in color

Chlorophyta (green algae)—green single-celled, colonial, and multicellular autotrophs that carry out photosynthesis using the same pigments as plants; likely ancestor of land plants

Amoebozoa (amoebozoans) Includes most of the amoebas as well as the slime molds. All use pseudopods for locomotion and for feeding.

Amoebas—single-celled; use lobose (non-stiffened) pseudopods for locomotion and feeding

Cellular slime molds—heterotrophs; primarily individual cells; move by amoeboid motion, or as a multicellular mass

Plasmodial slime molds—heterotrophs; live as plasmodium, a large composite mass with nuclei in a common cytoplasm, that moves and feeds like a giant amoeba

Opisthokonta (opisthokonts) A single posterior flagellum at some stage in the life cycle; consist of the fungi, animals, and two protist groups, the choanoflagellates and the nucleariids.

Choanoflagellata (choanoflagellates)—motile protists with a single flagellum surrounded by collar of closely packed microvilli; likely ancestor of animals and fungi

Nucleariidae (nucleariids)—heterotrophic, mainly spherical amoebae with radiating, fine pseudopods that are not supported by microtubules; more closely related to fungi than to animals

PLANTAE Multicellular autotrophs, mostly terrestrial, and most of which gain energy via photosynthesis; life cycle characterized by alternation of a gametophyte (gamete-producing) generation and sporophyte (spore-producing) generation

Bryophytes (seedless plants)—nonvascular plants and seedless vascular plants

Hepatophyta (liverworts)—leafy or simple flattened thallus with rhizoids; no true leaves, stems, roots, or stomata (porelike openings for gas exchange); spores in capsules[1]

Anthocerophyta (hornworts)—simple flattened thallus, hornlike sporangia[1]

Bryophyta (mosses)—feathery or cushiony thallus; some with hydroids; spores in capsules[1]

Lycophyta (club mosses)—simple leaves, cuticle, stomata, true roots; most species have sporangia on sporophylls; fertilization by swimming sperm[2]

Pterophyta (ferns, whisk ferns, horsetails)—*Ferns:* Finely divided leaves; sporangia in sori. *Whisk ferns:* Branching stem from rhizomes; sporangia on stem scales. *Horsetails:* hollow stem, scale-like leaves, sporangia in strobili.[2]

Spermatophyta (seed plants)—vascular plants in which embryos develop within seeds

Gymnosperms—seeds born on stems, on leaves, or under scales

Cycadophyta (cycads)—shrubby or treelike with palmlike leaves; male and female strobili on separate plants

Ginkgophyta (ginkgoes)—lineage with a single living species *(Ginkgo biloba);* tree with deciduous, fan-shaped leaves; male, female reproductive structures on separate plants

Gnetophyta (gnetophytes)—shrubs or woody vinelike plants; male and female strobili on separate plants

Coniferophyta (conifers)—predominant extant gymnosperm group; mostly evergreen trees and shrubs with needlelike or scalelike leaves; male and female cones usually on the same plant

Anthophyta (angiosperms/flowering plants)—reproductive structures in flowers

Monocotyledones (monocots)—grasses, palms, lilies, orchids and their relatives; a single cotyledon (seed leaf); pollen grains have one groove

Eudicotyledones (eudicots)—roses, melons, beans, potatoes, most fruit trees, others; two cotyledons; pollen grains have three grooves

Other major angiosperm lineages: magnoliids (magnolias and relatives); star anise (Family Illicium); water lilies (Family Nymphaeaceae); *Amborella* (Family Amborellaceae)

FUNGI Heterotrophic, mostly multicellular organisms with cell wall containing chitin and cell nuclei occurring in threadlike hyphae; life cycle typically includes both asexual and sexual phases, with sexual structures used as the basis for phylum-level classification. Single-celled species are known as yeasts.

Chytridiomycota (chytrids)—mostly aquatic; asexual reproduction by way of motile zoospores; sexual reproduction via gametes produced in gametangia; hyphae mostly aseptate

Zygomycota (zygomycetes)—terrestrial; asexual reproduction via nonmotile haploid spores formed in sporangia; sexual spores (zygospores) form in zygosporangia; aseptate hyphae

Glomeromycota (glomeromycetes)—terrestrial; asexual reproduction via spores at the tips of hyphae; form mycorrhizal associations with plant roots

Ascomycota (ascomycetes/sac fungi)—terrestrial and aquatic; sexual spores form in asci; asexual reproduction occurs via conidia (nonmotile spores); septate hyphae

Basidiomycota (basidiomycetes)—terrestrial; reproduction usually via sexual basidiospores produced by basidia; septate hyphae

Basidiomycetes: mushroom-forming fungi and relatives

Teliomycetes: rusts

Ustomycetes: smuts

Conidial fungi—not a true phylum but a convenience grouping of species for which no sexual phase is known

Microsporidia—single-celled sporelike parasites of animals, other groups; phylogeny uncertain

---

[1]Nonvascular plants (bryophytes)—A polyphyletic group of plants with no specialized structures for transporting water and nutrients; swimming sperm require liquid water for sexual reproduction
[2]Seedless vascular plants—A polyphyletic group of plants in which embryos are not housed inside seeds

ANIMALIA Multicellular heterotrophs; nearly all with tissues, organs, and organ systems; motile during at least part of the life cycle; sexual reproduction in most; embryos develop through a series of stages; many with larval and adult stages in life cycle

Parazoa Animals lacking tissues and body symmetry

Porifera (sponges)—multicellular; extract oxygen and particulate food from water drawn into a central cavity

Eumetazoa Animals with tissues and either radial or bilateral symmetry

Radiata—acoelomate animals with radial symmetry and two tissue layers

Cnidaria (cnidarians)—two tissue layers; single opening into gastrovascular cavity; nerve net; nematocysts for defense and predation; some sessile, some motile; most are predatory, some with photosynthetic endosymbionts; freshwater and marine

Hydrozoa: hydrozoans

Scyphozoa: jellyfishes

Cubozoa: box jellyfishes

Anthozoa: sea anemones, corals

Ctenophora (comb jellies)—two (possibly three) tissue layers; feeding tentacles capture particulate food; beating cilia provide weak locomotion; marine

Bilateria—animals with bilateral symmetry and three tissue layers

Protostomia—acoelomate, pseudocoelomate, or schizocoelomate; many with spiral, indeterminate cleavage; blastopore forms mouth; nervous system on ventral side

Lophotrochozoa—many with either a lophophore for feeding and gas exchange or a trochophore larva

Ectoprocta (bryozoans)—coelomate; colonial; secrete hard covering over soft tissues; lophophore; sessile; particulate feeders; marine

Brachiopoda (lamp shells)—coelomate; dorsal and ventral shells; lophophore; sessile; particulate feeders; marine

Phoronida (phoronid worms)—coelomate; secrete tubes around soft tissues; lophophore; sessile; particulate feeders; marine

Platyhelminthes (flatworms)—acoelomate; dorsoventrally flattened; complex reproductive, excretory, and nervous systems; gastrovascular cavity in many; free-living or parasitic, often with multiple hosts; terrestrial, freshwater, and marine

Turbellaria: free-living flatworms

Trematoda: flukes

Monogenoidea: flukes

Cestoda: tapeworms

Rotifera (wheel animals)—pseudocoelomate; microscopic; complete digestive system; well-developed reproductive, excretory, and nervous systems; particulate feeders; major components of marine and freshwater plankton

Nemertea (ribbon worms)—schizocoelomate; proboscis housed within rhynchocoel; complete digestive tract; circulatory system; predatory; mostly marine

Mollusca (mollusks)—schizocoelomate; many with trochophore larva; many with shell secreted by mantle; body divided into head–foot, visceral mass, and mantle; well-developed organ systems; variable locomotion; herbivorous or predatory; terrestrial, freshwater, and marine

Polyplacophora: chitons

Gastropoda: snails, sea slugs, land slugs

Bivalvia: clams, mussels, scallops, oysters

Cephalopoda: squids, octopuses, cuttlefish, nautiluses

Annelida (segmented worms)—schizocoelomate; many with trochophore larva; segmented body and organ systems; well-developed organ systems; many use hydrostatic skeleton for locomotion; some predatory, some particulate feeders, some detritivores; terrestrial, freshwater, and marine

Polychaeta: marine worms

Oligochaeta: freshwater and terrestrial worms

Hirudinea: leeches

Ecdysozoa—cuticle or exoskeleton is shed periodically

Nematoda (roundworms)—pseudocoelomate; body covered with tough cuticle that is shed periodically; well-developed organ systems; thrashing locomotion; many are parasitic on plants or animals; mostly terrestrial

Onychophora (velvet worms)—schizocoelomate; segmented body covered with cuticle; locomotion by many unjointed legs; complex organ systems; predatory; terrestrial

Arthropoda (arthropods)—schizocoelomate; jointed exoskeleton made of chitin; segmented body, some with fusion of segments in head, thorax, or abdomen; complex organ systems; variable modes of locomotion, including flight; specialization of numerous appendages; herbivorous, predatory, or parasitic; terrestrial, freshwater, and marine

Trilobita: trilobites (extinct)

Chelicerata: horseshoe crabs, spiders, scorpions, ticks, mites

Crustacea: shrimps, crayfishes, lobsters, crabs, barnacles, copepods, isopods

Myriapoda: centipedes, millipedes

Hexapoda: springtails and insects

DEUTEROSTOMIA—enterocoelomate; many with radial, determinate cleavage; blastopore forms anus; nervous system on dorsal side in many

Echinodermata (echinoderms)—secondary radial symmetry, often organized around five radii; hard internal skeleton; unique water vascular system with tube feet; complete digestive system; simple nervous system; no circulatory or respiratory system; generally slow locomotion using tube feet; predatory, herbivorous, particulate feeders, detritivores; exclusively marine

Asteroidea: sea stars

Ophiuroidea: brittle stars

Echinoidea: sea urchins, sand dollars

Holothuroidea: sea cucumbers

Crinoidea: feather stars, sea lilies

Concentricycloidea: sea daisies

Hemichordata (acorn worms)—pharynx perforated with branchial slits; proboscis; complex organ systems; tube-dwelling in soft sediments; particulate or deposit feeders; exclusively marine

Chordata (chordates)—notochord; segmental body wall and tail muscles; dorsal hollow nerve chord; perforated pharynx; complex organ systems; variable modes of locomotion; extremely varied diets; terrestrial, freshwater, and marine

Cephalochordata: lancelets

Urochordata: tunicates, sea squirts

Vertebrata: vertebrates

Myxinoidea: hagfishes

Petromyzontoidea: lampreys

Placodermi: placoderms (extinct)

Chondrichthyes: sharks, skates, and rays

Acanthodii: acanthodians

Actinopterygii: ray-finned fishes

Sarcopterygii: fleshy-finned fishes

Amphibia: salamanders, frogs, caecilians

Synapsida: mammals

Anapsida: turtles

Diapsida: sphenodontids, lizards, snakes, crocodilians, birds

3′ end The end of a polynucleotide chain at which a hydroxyl group is bonded to the 3′ carbon of a deoxyribose sugar.

3′ untranslated region (3′ UTR) The part of an mRNA between the stop codon and the 3′ end of the molecule; this region does not code for amino acids.

5′ cap In eukaryotes, a guanine-containing nucleotide attached in a reverse orientation to the 5′ end of pre-mRNA and retained in the mRNA produced from it. The 5′ cap on an mRNA is the site where ribosomes attach to initiate translation.

5′ end The end of a polynucleotide chain at which a phosphate group is bound to the 5′ carbon of a deoxyribose sugar.

5′ untranslated region (5′ UTR) The part of an mRNA between the 5′ end of the molecule and the start codon; this region does not code for amino acids.

10-nm chromatin fiber The most fundamental level of chromatin packing of a eukaryotic chromosome in which DNA winds for almost two turns around an eight-protein nucleosome core particle to form a nucleosome and linker DNA extends between adjacent nucleosomes. The result is a beads-on-a-string type of structure with a 10-nm diameter.

30-nm chromatin fiber Level of chromatin packing of a eukaryotic chromosome in which histone H1 binds to the 10-nm chromatin fiber causing it to package into a coiled structure about 30 nm in diameter and with about six nucleosomes per turn. Also referred to as a *solenoid*.

A site The site where the incoming aminoacyl–tRNA carrying the next amino acid to be added to the polypeptide chain binds to the mRNA.

absorption spectrum Curve representing the amount of light absorbed at each wavelength.

acid Proton donor that releases H⁺ (and anions) when dissolved in water.

acidity The concentration of H⁺ in a water solution, as compared with the concentration of OH⁻.

acid precipitation Rainfall with low pH, primarily created when gaseous sulfur dioxide (SO_2) dissolves in water vapor in the atmosphere, forming sulfuric acid.

action spectrum Graph produced by plotting the effectiveness of light at each wavelength in driving photosynthesis.

activation energy The initial input of energy required to start a reaction.

activator A regulatory protein that controls the expression of one or more genes.

active site The region of an enzyme to which substrate(s) bind and where catalysis occurs.

active transport The mechanism by which ions and molecules move against the concentration gradient across a membrane, from the side with the lower concentration to the side with the higher concentration.

adaptation Characteristic that helps an organism survive longer or reproduce more under a particular set of environmental conditions.

adenine A purine that base-pairs with either thymine in DNA or uracil in RNA.

adherens junction Animal cell junction in which intermediate filaments are the anchoring cytoskeletal component.

adhesion The adherence of molecules to the walls of conducting tubes, as in plants.

aerobic respiration The form of cellular respiration found in eukaryotes and many prokaryotes in which oxygen is a reactant in the ATP-producing process.

agarose gel electrophoresis Technique by which DNA, RNA, or protein molecules are separated in an agarose gel subjected to an electric field.

alcohol A molecule of the form R—OH in which R is a chain of one or more carbon atoms, each of which is linked to hydrogen atoms.

alcoholic fermentation Reaction in which pyruvate is converted into ethyl alcohol and CO_2 in a two-step series that also converts NADH into NAD^+.

aldehyde Molecule in which the carbonyl group is linked to a carbon atom at the end of a carbon chain, along with a hydrogen atom.

allele One of two or more versions of a gene.

allosteric activator Molecule that converts an enzyme with an allosteric site, a regulatory site outside the active site, from the inactive form to the active form.

allosteric inhibitor Molecule that converts an enzyme with an allosteric site, a regulatory site outside the active site, from the active form to the inactive form.

allosteric regulation Specialized control mechanism for enzymes with an allosteric site, a regulatory site outside the active site, that may either slow or accelerate activity depending on the enzyme.

allosteric site A regulatory site outside the active site.

alpha (α) helix A type of secondary structure of a polypeptide in which the amino acid chain is twisted into a regular, right-hand spiral.

alternative hypothesis An explanation of an observed phenomenon that is different from the explanation being tested.

alternative splicing Mechanism by which a pre-mRNA in a eukaryotic cell is processed by reactions that join exons in different combinations to produce different mRNAs from a single gene.

amino acid A molecule that contains both an amino and a carboxyl group.

amino group Group that acts as an organic base, consisting of a nitrogen atom bonded on one side to two hydrogen atoms and on the other side to a carbon chain.

aminoacylation The process of adding an amino acid to a tRNA. Also referred to as *charging*.

aminoacyl–tRNA A tRNA linked to its "correct" amino acid, which is the finished product of aminoacylation.

aminoacyl–tRNA synthetase An enzyme that catalyzes aminoacylation.

amniocentesis Technique of prenatal diagnosis in which cells are obtained from the amniotic fluid for DNA testing, for biochemical analysis, or to test for the presence of chromosomal mutations.

amplification An increase in the magnitude of each step as a signal transduction pathway proceeds.

amyloplast Colorless plastid that stores starch in plants.

anabolic pathway A metabolic pathway in which energy is used to build complicated molecules from simpler ones; also called a *biosynthetic pathway*. An individual reaction in an anabolic pathway is an anabolic reaction, also called a *biosynthetic reaction*.

anabolic reaction Metabolic reaction that requires energy to assemble simple substances into more complex molecules.

anaerobic respiration The form of cellular respiration found in some prokaryotes in which a molecule other than oxygen is used in the ATP-producing process.

anaphase The phase of mitosis during which the spindle separates sister chromatids and pulls them to opposite spindle poles.

anaphase-promoting complex (APC) An enzyme complex activated by M phase–promoting factor that controls the separation of sister chromatids and the onset of daughter chromosome separation in anaphase of mitosis.

anchoring junction Cell junction that forms belts that run entirely around cells, "welding" adjacent cells together.

aneuploid An individual with extra or missing chromosomes.

anion A negatively charged ion.

antenna complex (light-harvesting complex) In photosystems, the sites at which light is absorbed and converted into chemical energy during photosynthesis, an aggregate of many chlorophyll pigments and a number of carotenoid pigments that serves as the primary site of absorbing light energy in the form of photons.

anticodon The three-nucleotide segment in a tRNA that pairs with a codon in an mRNA.

antiparallel Strands of double-stranded DNA that run in opposite directions with the 3′ end of one strand opposite the 5′ end of the other strand.

antiport A secondary active transport mechanism in which a molecule moves through a membrane channel into a cell and powers the active transport of a second molecule out of the cell. Also referred to as *exchange diffusion.*

applied research Research conducted with the goal of solving specific practical problems.

aquaporin A specialized protein channel that facilitates diffusion of water through cell membranes.

Archaea One of two domains of prokaryotes; archaeans have some unique molecular and biochemical traits, but they also share some traits with Bacteria and other traits with Eukarya.

archaeal chromosome DNA molecule in archaeans in which hereditary information is encoded.

artificial selection Selective breeding of organisms to ensure that certain desirable traits appear at higher frequency in successive generations.

asexual reproduction Any mode of reproduction in which a single individual gives rise to offspring without fusion of gametes; that is, without genetic input from another individual. *See also* **vegetative reproduction.**

aster Radiating array produced as microtubules extending from the centrosomes of cells grow in length and extent.

atom The smallest unit that retains the chemical and physical properties of an element.

atomic nucleus The nucleus of an atom, containing protons and neutrons.

atomic number The number of protons in the nucleus of an atom.

atomic weight The weight of an element in grams, equal to the mass number.

ATP (adenosine triphosphate) The primary agent that couples exergonic and endergonic reactions.

ATP/ADP cycle The continual hydrolysis and resynthesis of ATP in living cells.

ATP synthase A membrane-spanning protein complex that couples the energetically favorable transport of protons across a membrane to the synthesis of ATP.

autosomal dominant inheritance Pattern in which the allele that causes a trait is dominant, and only homozygous recessives are unaffected.

autosomal recessive inheritance Pattern in which individuals with a trait are homozygous for a recessive allele.

autosome Chromosome other than a sex chromosome.

autotroph An organism that produces its own food using CO_2 and other simple inorganic compounds from its environment and energy from the sun or from oxidation of inorganic substances.

auxotroph A mutant strain that requires for its growth a nutrient supplement that is not needed by the wild-type strain.

Avogadro's number The number 6.022×10^{23}, derived by dividing the atomic weight of any element by the weight of an atom of that element.

Bacteria One of the two domains of prokaryotes; collectively, bacteria are the most metabolically diverse organisms.

bacterial chromosome DNA molecule in bacteria in which hereditary information is encoded.

bacterial flagellum *See* **flagellum.**

bacteriophage A virus that infects bacteria. Also referred to as a *phage.*

Barr body The inactive, condensed X chromosome seen in the nucleus of female placental mammals.

basal body Structure that anchors cilia and flagella to the surface of a cell.

base Proton acceptor that reduces the H^+ concentration of a solution.

base-pair mismatch An error in the assembly of a new nucleotide chain in which bases other than the correct ones pair together.

basic research Research conducted to search for explanations about natural phenomena in order to satisfy curiosity and to advance collective knowledge of living systems.

beta (β) sheet A type of primary structure in a polypeptide in which the amino acid chain zigzags in a flat plane to form a beta strand, and beta strands then align side by side in the same or opposite direction.

bioinformatics The application of mathematics and computer science to extract information from biological data, including those related to genome structure and function.

biological evolution The process by which some individuals in a population experience changes in their DNA and pass those modified instructions to their offspring.

biological research The collective effort of individuals who have worked to understand how living systems function.

bioremediation Applications of chemical and biological knowledge to decontaminate polluted environments.

biosphere All regions of Earth's crust, waters, and atmosphere that sustain life.

biosynthetic reaction An individual reaction in an anabolic pathway (biosynthetic pathway).

biotechnology The manipulation of living organisms to produce useful products.

blending theory of inheritance Theory suggesting that hereditary traits blend evenly in offspring through mixing of the blood of the two parents.

buffer Substance that compensates for pH changes by absorbing or releasing H^+.

bulk-phase endocytosis Mechanism by which extracellular water is taken into a cell together with any molecules that happen to be in solution in the water. Also referred to as *pinocytosis.*

C₃ pathway *See* **light-independent reaction;** also referred to as the *Calvin cycle.*

C₄ pathway In C₄ plants the pathway to fix CO_2 into oxaloacetate in mesophyll cells and then produce CO_2 for the Calvin cycle in bundle sheath cells.

Ca²⁺ pump (calcium pump) Pump that pushes Ca^{2+} from the cytoplasm to the cell exterior, and also from the cytosol into the vesicles of the endoplasmic reticulum.

calorie (cal) The amount of heat required to raise 1 g of water by 1°C, known as a "small" calorie; when capitalized, a unit equal to 1,000 small calories.

Calvin cycle *See* **light-independent reaction.**

CAM pathway In CAM plants the pathway to fix CO_2 into oxaloacetate and then produce CO_2 for the Calvin cycle, both occurring in mesophyll cells, but separated by time of day. CAM stands for "crassulacean acid metabolism."

cancer When a tumor becomes malignant and its cells invade and disrupt surrounding tissues.

CAP (catabolite activator protein) Key regulatory molecule involved in positive gene regulation of the *lac* operon.

CAP site Region in the promoter of the *lac* operon and in the promoters of a large number of other operons that control the catabolism of many sugars to which activated catabolite activator protein (CAP) binds, thereby enabling RNA polymerase to bind and transcribe the operon's structural genes.

capsid The protective layer of protein that surrounds the nucleic acid core of a virus in free form; also known as a *coat.*

capsule An external layer of sticky or slimy polysaccharides coating the cell wall in many prokaryotes.

carbonyl group The reactive part of aldehydes and ketones, consisting of an oxygen atom linked to a carbon atom by a double bond.

carboxyl group The characteristic functional group of organic acids, formed by the combination of carbonyl and hydroxyl groups.

carotenoid Molecule of yellow-orange pigment by which light is absorbed in photosynthesis.

carrier A heterozygote—an individual who carries a recessive mutant allele and could pass it on to offspring, but does not display its symptoms.

carrier protein Transport protein that binds a specific single solute and transports it across the lipid bilayer.

Cartagena Protocol on Biosafety An international agreement that promotes biosafety as it relates to the handling and use of genetically modified organisms.

catabolic pathway A metabolic pathway in which energy is released by the breakdown of complex molecules to simpler compounds. An individual reaction in a catabolic pathway is a catabolic reaction.

catabolic reaction Cellular reaction that breaks down complex molecules such as sugar to make their energy available for cellular work.

catalysis The process of accelerating a chemical reaction with a catalyst.

catalyst Substance with the ability to accelerate a spontaneous reaction without being changed by the reaction.

cation A positively charged ion.

cDNA library The entire collection of cloned cDNAs made from the mRNAs isolated from a cell.

cell Smallest unit with the capacity to live and reproduce.

cell adhesion molecule A cell surface protein responsible for selectively binding cells together.

cell adhesion protein Protein that binds cells together by recognizing and binding receptors or chemical groups on other cells or on the extracellular matrix.

cell center *See* **centrosome.**

cell culture Living cells growing in a growth medium in a laboratory vessel.

cell cycle The sequence of events during which a cell experiences a period of growth followed by nuclear division and cytokinesis.

cell fractionation Technique that divides cells into fractions containing a single cell component.

cell junction Junction that seals the spaces between cells and provides direct communication between cells.

cell lineage Cell derivation from the undifferentiated tissues of the embryo.

cell plate In cytokinesis in plants, a new cell wall that forms between the daughter nuclei and grows laterally until it divides the cytoplasm.

cell signaling The system of communication between cells through signaling pathways.

cell theory Three generalizations yielded by microscopic observations: all organisms are composed of one or more cells; the cell is the smallest unit that has the properties of life; and cells arise only from the growth and division of preexisting cells.

cell wall A rigid external layer of material surrounding the plasma membrane of cells in plants, fungi, bacteria, and some protists, providing cell protection and support.

cellular respiration The process by which energy-rich molecules are broken down to produce energy in the form of ATP.

cellulose One of the primary constituents of plant cell walls, formed by chains of carbohydrate subunits.

centimorgan *See* **map unit.**

central dogma The name given by Francis Crick to the flow of information from DNA to RNA to protein.

central vacuole A large, water-filled organelle in plant cells that maintains the turgor of the cell and controls movement of molecules between the cytosol and sap.

centriole A cylindrical structure consisting of nine triplets of microtubules in the centrosomes of most animal cells.

centromere A specialized chromosomal region that connects sister chromatids and attaches them to the mitotic spindle.

centrosome (cell center) The main microtubule organizing center of a cell, which organizes the microtubule cytoskeleton during interphase and positions many of the cytoplasmic organelles.

channel protein Transport protein that forms a hydrophilic channel in a cell membrane through which water, ions, or other molecules can pass, depending on the protein.

chaperone protein (chaperonin) "Guide" protein that binds temporarily with newly synthesized proteins, directing their conformation toward the correct tertiary structure and inhibiting incorrect arrangements as the new proteins fold.

character A specific heritable attribute or property of an organism.

character differences Alternative forms of characters. Also called a *trait.*

charging *See* **aminoacylation.**

checkpoint Internal control of the cell cycle that prevents a critical phase from beginning until the previous phase is complete.

chemical bond Link formed when atoms of reactive elements combine into molecules.

chemical equation A chemical reaction written in balanced form.

chemical reaction A reaction that occurs when atoms or molecules interact to form new chemical bonds or break old ones.

chemiosmotic hypothesis Model proposing that mitochondrial electron transfer produces an H^+ gradient and that the gradient powers ATP synthesis by ATP synthase.

chiasmata *See* **crossover.**

chlorophyll Molecule of green pigment that absorbs photons of light in photosynthesis.

chloroplast The site of photosynthesis in plant cells.

cholesterol The predominant sterol of animal cell membranes.

chorionic villus sampling Technique of prenatal diagnosis in which cells are obtained from portions of the placenta that develop from tissues of the embryo for DNA testing, for biochemical analysis, or to test for the presence of chromosomal mutations.

chromatin Any assemblage of eukaryotic nuclear DNA molecules and their associated proteins.

chromoplast Plastid containing red and yellow pigments.

chromosomal mutation A variation from the normal condition in chromosome structure or chromosome number.

chromosomal protein A histone and nonhistone protein associated with DNA in a eukaryotic nuclear chromosome.

chromosome In eukaryotic cells, a linear structure composed of a single DNA molecule complexed with protein. Each eukaryotic species has a characteristic number of chromosomes in the nucleus. Most prokaryotes have a single, usually circular chromosome with few or no associated proteins.

chromosome segregation The equal distribution of daughter chromosomes to each of the two cells that result from cell division.

chromosome theory of inheritance The principle that genes and their alleles are carried on the chromosomes.

cisternae Membranous channels and vesicles that make up the endoplasmic reticulum.

citric acid cycle Series of reactions in which acetyl groups are oxidized completely to carbon dioxide and some ATP molecules are synthesized. Also referred to as *Krebs cycle* and *tricarboxylic acid cycle.*

class A Linnaean taxonomic category that ranks below a phylum and above an order.

clathrin The network of proteins that coat and reinforce the cytoplasmic surface of cell membranes.

clone An individual genetically identical to an original cell from which it descended.

cloning vector DNA molecule into which a DNA fragment can be inserted to form a recombinant DNA molecule for the purpose of cloning.

CO_2 fixation Process in which electrons are used as a source of energy to convert inorganic CO_2 to an organic form.

coactivator (mediator) In eukaryotes, a large multiprotein complex that bridges between activators at an enhancer and proteins at the promoter and promoter proximal region to stimulate transcription.

coat *See* **capsid.**

coated pit A depression in the plasma membrane that contains receptors for macromolecules to be taken up by endocytosis.

codominance Condition in which alleles have approximately equal effects in individuals, making the alleles equally detectable in heterozygotes.

codon Each three-letter word (triplet) of the genetic code.

coenzymes Organic cofactors that include complex chemical groups of various kinds.

cofactor An inorganic or organic nonprotein group that is necessary for catalysis to take place.

cohesion The high resistance of water molecules to separation.

combinatorial gene regulation The combining of a few regulatory proteins in particular ways so that the transcription of a wide array of genes can be controlled and a large number of cell types can be specified.

commaless The sequential nature of the words of the nucleic acid code, with no indicators such as commas or spaces to mark the end of one codon and the beginning of the next.

community Populations of all species that occupy the same area.

comparative genomics Comparison of the sequences of entire genomes (or extensive portions of them) to understand evolutionary relationships and the basic biological similarities and differences among species.

competitive inhibition Inhibition of an enzyme reaction by an inhibitor molecule that resembles the normal substrate closely enough so that it fits into the active site of the enzyme.

complementary base pairing Feature of DNA in which the specific purine–pyrimidine base pairs A–T (adenine–thymine) and G–C (guanine–cytosine) occur to bridge the two sugar–phosphate backbones.

complementary DNA (cDNA) A DNA molecule that is complementary to an mRNA molecule, synthesized by reverse transcriptase.

complete medium A growth medium containing a full complement of nutrient substances that a wild-type microorganism can make for itself.

complex virus A bacteriophage with a DNA genome that has a tail attached at one side of a polyhedral head.

compound A molecule whose component atoms are different.

concentration The number of molecules or ions of a substance in a unit volume of space.

concentration gradient A difference in concentration of molecules or ions between two areas.

condensation reaction Reaction during which the components of a water molecule are removed, usually as part of the assembly of a larger molecule from smaller subunits. Also referred to as *dehydration synthesis reaction.*

conformation The overall three-dimensional shape of a protein.

conformational change Alteration in the three-dimensional shape of a protein.

conjugation In bacteria, the process by which DNA of the donor cell moves through the cytoplasmic bridge into the recipient cell. With some types of donor cells, this can lead to genetic recombination in the recipient cell. In ciliate protists, a process of sexual reproduction in which individuals of the same species temporarily couple and exchange genetic material.

consumer An organism that consumes other organisms in a community or ecosystem.

contact inhibition The inhibition of movement or proliferation of normal cells that results from cell–cell contact.

control Treatment that tells what would be seen in the absence of the experimental manipulation.

corepressor In the regulation of gene expression in bacteria, a regulatory molecule that combines with a repressor to activate it and shut off an operon.

cotranslational import A mechanism by which a polypeptide being sorted via the endomembrane system in a eukaryotic cell begins its import into the endoplasmic reticulum simultaneously with translation of the mRNA encoding the polypeptide.

cotransport *See* **symport.**

covalent bond Bond formed by electron sharing between atoms.

crassulacean acid metabolism (CAM) A biochemical variation of photosynthesis that was discovered in a member of the plant family Crassulaceae. Carbon dioxide is taken up and stored during the night to allow the stomata to remain closed during the daytime, decreasing water loss.

criss-cross inheritance The transmission pattern characteristic of an X-linked allele from a parent of one sex to a "child" of the opposite sex to a "grandchild" of the first sex.

crista Fold that expands the surface area of the inner mitochondrial membrane.

cross-fertilization Fertilization of one plant by a different plant.

crossing-over The recombination process in meiosis, in which chromatids exchange segments.

crossover Site of recombination during meiosis. Also referred to as *chiasmata.*

cross-pollination *See* **cross-fertilization.**

cross-talk Interaction by which cell signaling pathways communicate with one another to integrate their responses to cellular signals.

C-terminal end The end of an amino acid chain with a —COO⁻ group.

cyclic AMP (cAMP) In particular signal transduction pathways, a second messenger that activates protein kinases, which elicit the cellular response by adding phosphate groups to specific target proteins. cAMP functions in one of two major G-protein–coupled receptor–response pathways.

cyclic electron flow An electron transport pathway associated with photosystem I in photosynthesis that produces ATP without the synthesis of NADPH.

cyclin In eukaryotes, protein that regulates the activity of CDK (cyclin-dependent kinase) and controls progression through the cell cycle.

cyclin-dependent kinase (CDK) A protein kinase that controls the cell cycle in eukaryotes.

cytochrome Protein with a heme prosthetic group that contains an iron atom.

cytokinesis Division of the cytoplasm into two daughter cells following nuclear division in mitosis or meiosis.

cytoplasm All the parts of the cell that surround the central nucleus (eukaryotes) or nucleoid region (prokaryotes).

cytoplasmic determinants The mRNA and proteins stored in the egg cytoplasm that direct the early stages of animal development in the period before genes of the zygote become active.

cytoplasmic inheritance Pattern in which inheritance follows that of genes in the cytoplasmic organelles, mitochondria, or chloroplasts.

cytosine A pyrimidine that base-pairs with guanine in nucleic acids.

cytoskeleton The interconnected system of protein fibers and tubes that extends throughout the cytoplasm of a eukaryotic cell.

cytosol Aqueous solution in the cytoplasm containing ions and various organic molecules.

dalton A standard unit of mass, about 1.66×10^{-24} grams.

decomposer A small organism, such as a bacterium or fungus, that feeds on the remains of dead organisms, breaking down complex biological molecules or structures into simpler raw materials.

degeneracy (redundancy) The feature of the genetic code in which, with two exceptions, more than one codon represents each amino acid.

dehydration synthesis reaction *See* **condensation reaction.**

deletion Chromosomal alteration that occurs if a broken segment is lost from a chromosome.

denaturation A loss of both the structure and function of a protein due to extreme conditions that unfold it from its normal conformation.

deoxyribonucleic acid (DNA) The large, double-stranded, helical molecule that contains the genetic material of all living organisms.

deoxyribonucleotide Nucleotide containing deoxyribose as the sugar; deoxyribonucleotides are components of DNA.

deoxyribose A 5-carbon sugar to which a nitrogenous base and a phosphate group link covalently in a nucleotide of DNA.

desmosome Anchoring junction for which microfilaments anchor the junction in the underlying cytoplasm.

determination Mechanism in which the developmental fate of a cell is set.

development A series of programmed changes encoded in DNA, through which a fertilized egg divides into many cells that ultimately are transformed into an adult, which is itself capable of reproduction.

diacylglycerol (DAG) In particular signal transduction pathways, a second messenger that activates protein kinases, which elicit the cellular response by adding phosphate groups to specific target proteins. DAG is involved in one of two major G-protein–coupled receptor–response pathways.

differentiation Process by which cells that have been committed to a particular developmental fate by the determination process now develop into specialized cell types with distinct structures and functions.

diffusion The net movement of ions or molecules from a region of higher concentration to a region of lower concentration.

dihybrid A zygote produced from a cross that involves two characters.

dihybrid cross A cross between two individuals that are heterozygous for two pairs of alleles.

discontinuous replication Replication in which a DNA strand is formed in short lengths that are synthesized in the direction opposite of DNA unwinding.

dissociation The separation of water to produce hydrogen ions and hydroxide ions.

disulfide linkage Linkage that occurs when two sulfhydryl groups interact during a linking reaction.

DNA *See* **deoxyribonucleic acid.**

DNA chip *See* **DNA microarray.**

DNA fingerprinting Technique in which DNA samples are used to distinguish between individuals of the same species.

DNA helicase An enzyme that catalyzes the unwinding of DNA template strands.

DNA hybridization Technique in which a gene or sequence of interest is identified in a set of clones when it base pairs with a single-stranded DNA or RNA molecule called a nucleic acid probe.

DNA methylation Process in which a methyl group is added enzymatically to cytosine bases in the DNA.

DNA microarray A solid surface divided into a microscopic grid of thousands of spaces each containing thousands of copies of a DNA probe. DNA chips are used commonly for analysis of gene activity and for detecting differences between cell types. Also referred to as a *DNA chip*.

DNA polymerase I In *E. coli*, the replication enzyme that replaces the RNA primer at the start of a new DNA segment with DNA.

DNA polymerase III The principal replication polymerase in *E. coli* that synthesizes the majority of the new DNA.

DNA repair mechanism Mechanism to correct base-pair mismatches that escape proofreading.

DNA technologies Techniques to isolate, purify, analyze, and manipulate DNA sequences.

domain In protein structure, a distinct, large structural subdivision produced in many proteins by the folding of the amino acid chain. In systematics, the highest taxonomic category; a group of cellular organisms with characteristics that set it apart as a major branch of the evolutionary tree.

dominance The masking effect of one allele over another.

dominant The allele expressed when paired with a recessive allele.

dosage compensation mechanism Mechanism in placental mammals by which the effects of most genes carried on the X chromosome in females are equalized in females (who have two X chromosomes) and males (who have one X chromosome).

double helix Two nucleotide chains wrapped around each other in a spiral.

double-helix model Model of DNA consisting of two polynucleotide strands twisted around each other.

duplication Chromosomal alteration that occurs if a segment is broken from one chromosome and inserted into its homolog.

E site The site where an exiting tRNA binds before its release from the ribosome in translation.

ecosystem Group of biological communities interacting with their shared physical environment.

effector In signal transduction, a plasma membrane–associated enzyme, activated by a G protein, that generates one or more second messengers. In homeostatic feedback, the system that returns the condition to the set point if it has strayed away.

electrochemical gradient A difference in chemical concentration and electric potential across a membrane.

electromagnetic spectrum The range of wavelengths or frequencies of electromagnetic radiation extending from gamma rays to the longest radio waves and including visible light.

electron Negatively charged particle outside the nucleus of an atom.

electron microscope Microscope that uses electrons to illuminate the specimen.

electron transfer system Stage of cellular respiration in which high-energy electrons produced from glycolysis, pyruvate oxidation, and the citric acid cycle are delivered to oxygen by a sequence of electron carriers.

electronegativity The measure of an atom's attraction for the electrons it shares in a chemical bond with another atom.

element A pure substance that cannot be broken down into simpler substances by ordinary chemical or physical techniques.

elongation In transcription, the step in which RNA polymerase (RNA polymerase II in eukaryotes) moves along the gene extending the RNA chain, with the DNA continuing to unwind ahead of the enzyme. In translation, the step in which the assembled translation complex reads the string of codons in the mRNA one at a time while joining the specified amino acids into the polypeptide.

elongation factor (EF) A protein that aids in an elongation step of translation.

embryogenesis Stages of development from a fertilized egg to an embryo.

emergent property Characteristic that depends on the level of organization of matter, but does not exist at lower levels of organization.

enantiomers Isomers that are mirror images of each other. Also referred to as *optical isomers*.

endergonic reaction Reaction that can proceed only if free energy is supplied.

endocytic vesicle Vesicle that carries proteins and other molecules from the plasma membrane to destinations within the cell.

endomembrane system In eukaryotes, a collection of interrelated internal membranous sacs that divide a cell into functional and structural compartments.

endoplasmic reticulum (ER) In eukaryotes, an extensive interconnected network of cisternae that is responsible for the synthesis, transport, and initial modification of proteins and lipids.

endothermic Referring to a reaction that absorbs energy, that is, a reaction in which the products have more potential energy than the reactants.

end-product inhibition *See* **feedback inhibition.**

energy coupling The process in living cells by which the hydrolysis of ATP is coupled to an endergonic reaction so that energy is not wasted as heat.

energy levels Regions of space within an atom where electrons are found. Also referred to as *shells*.

enhancer In eukaryotes, a region at a significant distance from the beginning of some protein-coding genes that contain regulatory sequences to which regulatory proteins bind to stimulate or inhibit the rate of transcription initiation over and above that seen for regulatory events at the promoter proximal elements.

enthalpy The potential energy in a system.

entropy Disorder, in thermodynamics.

enveloped virus A virus that has a surface membrane derived from its host cell.

enzyme Protein that accelerates the rate of a cellular reaction.

enzyme specificity The ability of an enzyme to catalyze the reaction of only a single type of molecule or group of closely related molecules.

epistasis Interaction of genes, with one or more alleles of a gene at one locus inhibiting or masking the effects of one or more alleles of a gene at a different locus.

ER (endoplasmic reticulum) lumen The enclosed space surrounded by a cisterna.

Eukarya The domain that includes all eukaryotes, organisms that contain a membrane-bound nucleus within each of their cells; all protists, plants, fungi, and animals.

eukaryote Organism in which the DNA is enclosed in a nucleus.

eukaryotic chromosome A DNA molecule, with its associated proteins, in the nucleus of a eukaryotic cell.

euploid An individual with a normal set of chromosomes.

exchange diffusion *See* **antiport.**

exergonic reaction Reaction that has a negative ΔG because it releases free energy.

exocytosis In eukaryotes, the process by which a secretory vesicle fuses with the plasma membrane and releases the vesicle contents to the exterior.

exon An amino acid-coding sequence present in pre-mRNA that is retained in a spliced mRNA that is translated to produce a polypeptide.

exon shuffling Molecular evolutionary process that combines exons of two or more existing genes to produce a gene that encodes a protein with an unprecedented function.

exothermic Referring to a reaction that releases energy, that is, a reaction in which the products have less potential energy than the reactants.

experimental data Information that describes the result of a careful manipulation of the system under study.

experimental variable The variable in a scientific study that is manipulated by the experimenter.

expression vector Cloning vector that, in addition to normal features, contains regulatory sequences for transcription and translation of a gene.

extracellular matrix (ECM) A molecular system that supports and protects cells and provides mechanical linkages.

F factor Plasmid in a donor bacterial cell that confers on that cell the ability to conjugate with a recipient bacterial cell.

F pilus Structure on the cell surface that allows an F^+ donor bacterial cell (a cell containing an F factor) to attach to an F^- recipient bacterial cell (a cell lacking an F factor). Also referred to as a *sex pilus*.

F^- cell Recipient cell in conjugation between bacteria; it lacks an F factor.

F^+ cell Donor cell in conjugation between bacteria; it contains an F factor.

F_1 generation The first generation of offspring from a genetic cross.

F₂ generation The second generation of offspring from a genetic cross produced by interbreeding F₁ individuals.

facilitated diffusion Mechanism by which polar and charged molecules diffuse across membranes with the help of transport proteins.

facultative anaerobe An organism that can live in the presence or absence of oxygen, using oxygen when it is present and living by fermentation under anaerobic conditions.

familial (hereditary) cancer Cancer that runs in a family.

family A Linnaean taxonomic category that ranks below an order and above a genus.

fat Neutral lipid that is semisolid at biological temperatures.

fatty acid One of two components of a neutral lipid, containing a single hydrocarbon chain with a carboxyl group linked at one end.

feedback inhibition In enzyme reactions, regulation in which the product of a reaction acts as a regulator of the reaction. Also referred to as *end-product inhibition.*

fermentation Process in which electrons carried by NADH are transferred to an organic acceptor molecule rather than to the electron transfer system.

fertilization The fusion of the nuclei of an egg and sperm cell, which initiates development of a new individual.

first law of thermodynamics The principle that energy can be transferred and transformed but it cannot be created or destroyed.

first messenger The extracellular signal molecule in signal transduction pathways controlled by G-protein–coupled receptors.

flagellum (plural, *flagella*) A long, threadlike cellular appendage responsible for movement; found in both prokaryotes and eukaryotes, but with different structures and modes of locomotion.

fluid mosaic model Model proposing that the membrane consists of a fluid phospholipid bilayer in which proteins are embedded and float freely.

formula The name of a molecule written in chemical shorthand.

frameshift mutation Mutation in a protein-coding gene that causes the reading frame of an mRNA transcribed from the gene to be altered, resulting in the production of a different, and nonfunctional, amino acid sequence in the polypeptide.

free energy The energy in a system that is available to do work.

freeze-fracture technique Technique in which experimenters freeze a block of cells rapidly, then fracture the block to split the lipid bilayer and expose the hydrophobic membrane interior.

functional genomics The study of the functions of genes and of other parts of the genome.

functional groups The atoms in reactive groups.

furrow In cytokinesis, a groove that girdles the cell and gradually deepens until it cuts the cytoplasm into two parts.

G₀ phase The phase of the cell cycle in eukaryotes in which many cell types stop dividing.

G₁ phase The initial growth stage of the cell cycle in eukaryotes, during which the cell makes proteins and other types of cellular molecules, but not nuclear DNA.

G₂ phase The phase of the cell cycle in eukaryotes during which the cell continues to synthesize proteins and to grow, completing interphase.

gamete A haploid cell, an egg or sperm. Haploid cells fuse during sexual reproduction to form a diploid zygote.

gametogenesis The formation of male and female gametes.

gametophyte In organisms in which alternation of generations occurs, notably plants and green algae, the multicellular haploid generation that produces gametes. *Compare* **sporophyte.**

gap gene In *Drosophila* embryonic development, one of the first activated set of segmentation genes that progressively subdivide the embryo into regions, determining the segments of the embryo and the adult. Gap genes subdivide the embryo along the anterior–posterior axis into broad regions that later develop into several distinct segments.

gap junction Junction that opens direct channels allowing ions and small molecules to pass directly from one cell to another.

gated channel Ion transporter in a membrane that switches between open, closed, or intermediate states.

gene A unit containing the code for a protein molecule or one of its parts, or for functioning RNA molecules such as tRNA and rRNA.

gene cloning DNA cloning when it involves a gene.

gene duplication Any process that produces two identical copies of a gene in an organism's genome. It is the simplest and most common mechanism for the evolution of new genes.

gene therapy Correction of genetic disorders using genetic engineering techniques.

general transcription factor (basal transcription factor) In eukaryotes, a protein that binds to the promoter of a gene in the area of the TATA box and recruits and orients RNA polymerase II to initiate transcription at the correct place.

generalized transduction Transfer of bacterial genes between bacteria using virulent phages that have incorporated random DNA fragments of the bacterial genome.

genetic code The nucleotide information that specifies the amino acid sequence of a polypeptide.

genetic counseling Counseling that allows prospective parents to assess the possibility that they might have a child affected by a genetic disorder.

genetic engineering The use of DNA technologies to alter genes for practical purposes.

genetic recombinants Nonparental combinations of alleles. In eukaryotes they result from crossing-over in meiosis.

genetic recombination The process by which the combinations of alleles for different genes in two parental individuals become shuffled into new combinations in offspring individuals.

genetic screening Biochemical or molecular tests for identifying inherited disorders after a child is born.

genetically modified organism (GMO) A transgenic organism.

genomic imprinting Pattern of inheritance in which the expression of a nuclear gene is based on whether an individual organism inherits the gene from the male or female parent.

genomic library A set of clones that contains a copy of every DNA sequence in a genome.

genotype The genetic constitution of an organism in terms of its genes and alleles.

genus A Linnaean taxonomic category ranking below a family and above a species.

germ-line cells Cells that develop into sperm or eggs.

germ-line gene therapy Experiment in which a gene is introduced into germ-line cells of an animal to correct a genetic disorder.

glycocalyx A carbohydrate coat covering the cell surface.

glycogen Energy-providing carbohydrates stored in animal cells.

glycolipid A lipid molecule with carbohydrate groups attached.

glycolysis Stage of cellular respiration in which sugars such as glucose are partially oxidized and broken down into smaller molecules.

glycoprotein A protein with carbohydrate groups attached.

glycosidic bond Bond formed by the linkage of two α-glucose molecules with oxygen as a bridge between a carbon of the first glucose unit and a carbon of the second glucose unit.

Golgi complex In eukaryotes, the organelle responsible for the final modification, sorting, and distribution of proteins and lipids.

gonad A specialized gamete-producing organ in which the germ cells collect. Gonads are the primary source of sex hormones in vertebrates: ovaries in the female and testes in the male.

G-protein–coupled receptor In signal transduction, a surface receptor that responds to a signal by activating a G protein.

granum Structure in the chloroplasts of higher plants formed by thylakoids stacked one on top of another.

growth factor A molecule (typically a peptide hormone) that stimulates cell division of a target cell.

growth-inhibiting factor A molecule that inhibits cell division of a target cell.

guanine A purine that base-pairs with cytosine in nucleic acids.

H⁺ pump *See* **proton pump.**

haploid An organism or cell with only one copy of each type of chromosome in its nuclei.

head The anteriormost part of an organism's body, containing the brain, sensory structures, and feeding apparatus. For a bacteriophage, the usually polyhedral part of the virus containing the genetic material.

heat of vaporization The heat required to give water molecules enough energy of motion to break loose from liquid water and form a gas.

helical virus A virus in which the protein subunits of the coat assemble in a rodlike spiral around the genome.

heterogametic sex The sex that produces two types of gametes with respect to the sex chromosomes.

heterotroph An organism that acquires energy and nutrients by eating other organisms or their remains.

heterozygote An individual with two different alleles of a gene.

heterozygous State of possessing two different alleles of a gene.

Hfr cell A special donor cell that can transfer genes on a bacterial chromosome to a recipient bacterium.

high-throughput technique A method that facilitates the handling of many samples simultaneously, such as DNA molecules for sequencing or genes for analysis.

histone A small, positively charged (basic) protein that is complexed with DNA in the chromosomes of eukaryotes.

histone code A regulatory mechanism for altering chromatin structure and, therefore, gene activity, based on signals in histone tails represented by chemical modification patterns.

homeobox A region of a homeotic gene that corresponds to an amino acid section of the homeodomain.

homeodomain An encoded transcription factor of each protein that binds to a region in the promoters of the genes whose transcription it regulates.

homeostasis A steady internal condition maintained by responses that compensate for changes in the external environment.

homeotic gene Any of the family of genes that determines overall body plan (the structure of body parts) during embryonic development.

homogametic sex The sex that produces only one type of gamete with respect to the sex chromosomes.

homologous chromosomes The two chromosomes of each pair in a diploid cell—one of the pair derives from the maternal parent and the other derives from the paternal parent. Homologous chromosomes have the same genes, in the same order, in their DNA.

homozygote An individual with two copies of the same allele.

homozygous State of possessing two copies of the same allele.

horizontal gene transfer Movement of genetic material between organisms other than by descent.

hormone A signaling molecule secreted by a cell that can alter the activities of any cell with receptors for it; in animals, typically a molecule produced by one tissue and transported via the bloodstream to another specific tissue to alter its physiological activity.

Hox gene A type of homeotic gene that controls an organism's overall body plan.

human immunodeficiency virus (HIV) A retrovirus that causes acquired immunodeficiency syndrome (AIDS).

hydration layer A surface coat of water molecules that covers other polar and charged molecules and ions.

hydrocarbon Molecule consisting of carbon linked only to hydrogen atoms.

hydrogen bond Noncovalent bond formed by unequal electron sharing between hydrogen atoms and oxygen, nitrogen, or sulfur atoms.

hydrolysis Reaction in which the components of a water molecule are added to functional groups as molecules are broken into smaller subunits.

hydrophilic In chemistry and biology, referring to polar molecules that associate readily with water.

hydrophobic In chemistry and biology, referring to nonpolar substances that are excluded by water and other polar molecules.

hydroxyl group Group consisting of an oxygen atom linked to a hydrogen atom on one side and to a carbon chain on the other side.

hypertonic Solution containing dissolved substances at higher concentrations than the cells it surrounds.

hypothesis A tentative explanation for an observation, phenomenon, or scientific problem that can be tested by further investigation.

hypotonic Solution containing dissolved substances at lower concentrations than the cells it surrounds.

ice lattice A rigid, crystalline structure formed when a water molecule in ice forms four hydrogen bonds with neighboring molecules.

incomplete dominance Condition in which the effects of recessive alleles can be detected to some extent in heterozygotes.

independent assortment Mendel's principle that the alleles of the genes that govern two characters assort independently during formation of gametes. Mechanistically this is the case because any combination of chromosomes may be segregated to the spindle poles during meiosis I.

inducer Concerning regulation of gene expression in bacteria, a molecule that turns on the transcription of the genes in an operon.

inducible operon Operon whose expression is increased by an inducer molecule.

induction A mechanism in which one group of cells (the inducer cells) causes or influences another nearby group of cells (the responder cells) to follow a particular developmental pathway.

inheritance The transmission of DNA (that is, genetic information) from one generation to the next.

initiation In transcription, the step in which the molecular machinery that carries out transcription assembles at the promoter and begins synthesizing an RNA copy of the gene. In translation, the step in which the translation components assemble on the start codon of the mRNA.

initiation factor (IF) A protein that aids an initiation step of translation.

initiator codon *See* **start codon.**

inner boundary membrane Membrane lying just inside the outer boundary membrane of a chloroplast, enclosing the stroma.

inner mitochondrial membrane Membrane surrounding the mitochondrial matrix.

inorganic molecule Molecule without carbon atoms in its structure.

inositol triphosphate (IP$_3$) In particular signal transduction pathways, a second messenger that activates transport proteins in the endoplasmic reticulum to release Ca^{2+} into the cytoplasm. IP$_3$ is involved in one of two major G-protein–coupled receptor–response pathways.

insertion sequence A transposable element that contains only genes for its transposition.

integral protein Protein embedded in a phospholipid bilayer.

interkinesis A brief interphase separating the two meiotic divisions.

intermediate filament A cytoskeletal filament about 10 nm in diameter that provides mechanical strength to cells in tissues.

interphase The first stage of the mitotic cell cycle, during which the cell grows and replicates its DNA before undergoing mitosis and cytokinesis.

intron A non-protein-coding sequence that interrupts the protein-coding sequence in a eukaryotic gene. Introns are removed by splicing in the processing of pre-mRNA to mRNA.

inversion Chromosomal alteration that occurs if a broken segment reattaches to the same chromosome from which it was lost, but in reversed orientation, so that the order of genes in the segment is reversed with respect to the other genes of the chromosome.

ion A positively or negatively charged atom.

ionic bond Bond that results from electrical attractions between atoms that have lost or gained electrons.

isomers Two or more molecules with the same chemical formula but different molecular structures.

isotonic Equal concentration of water inside and outside cells.

isotope A distinct form of the atoms of an element, with the same number of protons but different number of neutrons.

karyotype A characteristic of a species consisting of the shapes and sizes of all the chromosomes at metaphase.

ketone Molecule in which the carbonyl group is linked to a carbon atom in the interior of a carbon chain.

kilocalorie (kcal) The scientific unit equivalent to a Calorie and equal to 1,000 small calories.

kinetic energy The energy of motion.

kinetochore A specialized structure consisting of proteins attached to a centromere that mediates the attachment and movement of chromosomes along the mitotic spindle.

kingdom A Linnaean taxonomic category that ranks below a domain and above a phylum.

Kingdom Animalia The taxonomic kingdom that includes all living and extinct animals.

Kingdom Fungi The taxonomic kingdom that includes all living or extinct fungi.

Kingdom Plantae The taxonomic kingdom encompassing all living or extinct plants.

Krebs cycle *See* **citric acid cycle.**

***K*-selected species** Long-lived species that thrive in more stable environments.

lactate fermentation Reaction in which pyruvate is converted into lactate.

lagging strand The new DNA strand synthesized discontinuously during replication in the direction opposite to that of DNA unwinding.

lagging strand template The DNA template strand for the lagging strand.

latent state The time during which a pathogen such as a virus remains in an infected organism in an inactive form and cannot be isolated and identified.

leading strand The new DNA strand synthesized during replication in the direction of DNA unwinding.

leading strand template The DNA template strand for the leading strand.

life cycle The sequential stages through which individuals develop, grow, maintain themselves, and reproduce.

light microscope Microscope that uses light to illuminate the specimen.

light-dependent reaction The first stage of photosynthesis, in which the energy of sunlight is absorbed and converted into chemical energy in the form of ATP and NADPH.

light-independent reaction The second stage of photosynthesis, in which electrons are used as a source of energy to convert inorganic CO_2 to an organic form. Also referred to as the *Calvin cycle.*

linkage The phenomenon of genes being located on the same chromosome.

linkage map Map of a chromosome showing the relative locations of genes based on recombination frequencies.

linked genes Genes on the same chromosome.

linker A short segment of DNA extending between one nucleosome and the next in a eukaryotic chromosome.

locus The particular site on a chromosome at which a gene is located.

lysogenic cycle Cycle in which the DNA of the bacteriophage is integrated into the DNA of the host bacterial cell and may remain for many generations.

lysosome Membrane-bound vesicle containing hydrolytic enzymes for the digestion of many complex molecules.

lytic cycle The series of events from infection of one bacterial cell by a phage through the release of progeny phages from lysed cells.

M phase–promoting factor (MPF) A complex of M cyclin and cyclin-dependent kinase 1 (Cdk1). The complex initiates mitosis and orchestrates some of its key events.

magnification The ratio of an object as viewed to its real size.

map unit The unit of a linkage map, equivalent to a recombination frequency of 1%. Also referred to as a *centimorgan.*

mass The amount of matter in an object.

mass number The total number of protons and neutrons in the atomic nucleus.

maternal chromosome The chromosome derived from the female parent of an organism.

maternal-effect gene One of a class of genes that regulate the expression of other genes expressed by the mother during oogenesis and that control the polarity of the egg and, therefore, of the embryo.

maternal inheritance A type of uniparental inheritance in which all progeny (both males and females) have the phenotype of the female parent.

matter Anything that occupies space and has mass.

meiosis The division of diploid cells to haploid progeny, consisting of two sequential rounds of nuclear and cellular division.

meiosis I The first division of the meiotic cell cycle in which homologous chromosomes pair and undergo an exchange of chromosome segments, and then the homologous chromosomes separate, resulting in two cells, each with the haploid number of chromosomes and with each chromosome still consisting of two chromatids.

meiosis II The second division of the meiotic cell cycle in which the sister chromatids in each of the two cells produced by meiosis I separate and segregate into different cells, resulting in four cells each with the haploid number of chromosomes.

membrane potential An electrical voltage that measures the potential inside a cell membrane relative to the fluid just outside; it is negative under resting conditions and becomes positive during an action potential.

messenger RNA (mRNA) An RNA molecule that serves as a template for protein synthesis.

metabolism The biochemical reactions that allow a cell or organism to extract energy from its surroundings and use that energy to maintain itself, grow, and reproduce.

metagenomics Analysis of DNA sequences of the genomes in entire communities of microbes that have been isolated from the environment.

metaphase The phase of mitosis during which the spindle reaches its final form and the spindle microtubules move the chromosomes into alignment at the spindle midpoint.

microbody Small, membrane-bound organelle that carries out vital reactions linking metabolic pathways.

microfilament A cytoskeletal filament composed of actin.

microRNA (miRNA) One of the major types of small regulatory RNAs in eukaryotes involved in RNA interference (RNAi).

microsatellite *See* **short tandem repeat (STR).**

microscope Instrument of microscopy with different magnifications and resolutions of specimens.

microscopy Technique for producing visible images of objects that are too small to be seen by the human eye.

microtubule A cytoskeletal component formed by the polymerization of tubulin into rigid, hollow rods about 25 nm in diameter.

microtubule organizing center (MTOC) An anchoring point near the center of a eukaryotic cell from which most microtubules extend outward.

middle lamella Layer of gel-like polysaccharides that holds together walls of adjacent plant cells.

minimal medium A growth medium containing the minimal ingredients that enable a nonmutant organism, such as *E. coli*, to grow.

miRNA-induced silencing complex (miRISC) Protein complex containing an miRNA that binds to sequences in the 3′ UTRs of target mRNAs, resulting in either inhibition of translation of the mRNAs or their degradation.

mismatch repair Repair system that removes mismatched bases from newly synthesized DNA strands.

missense mutation A base-pair substitution mutation in a protein-coding gene that results in a different amino acid in the encoded polypeptide than the normal one.

mitochondrial electron transfer system Series of electron carriers that alternately pick up and release electrons, ultimately transferring them to their final acceptor, oxygen.

mitochondrial matrix The innermost compartment of the mitochondrion.

mitochondrion Membrane-bound organelle responsible for synthesis of most of the ATP in eukaryotic cells.

mitosis Nuclear division that produces daughter nuclei that are exact genetic copies of the parental nucleus.

model organism An organism with characteristics that make it a particularly useful subject of research because it is likely to produce results widely applicable to other organisms.

molarity *(M)* The number of moles of a substance dissolved in 1 L of solution.

mole (mol) Amount of substance that contains as many atoms or molecules as there are atoms in exactly 12 g of carbon-12, which is 6.022×10^{23}.

molecular phylogenetics Approach of using DNA or amino acid sequence comparisons to determine evolutionary relationships among organisms.

molecular weight The weight of a molecule in grams, equal to the total mass number of its atoms.

molecule A unit composed of atoms combined chemically in fixed numbers and ratios.

monohybrid An F_1 heterozygote produced from a genetic cross that involves a single character.

monohybrid cross A genetic cross between two individuals that are each heterozygous for the same pair of alleles.

monoploid An individual with one set of chromosomes instead of the usual two.

monosaccharides The smallest carbohydrates, containing three to seven carbon atoms.

monounsaturated Fatty acids with one double bond.

motif A highly specialized region in a protein produced by the three-dimensional arrangement of amino acid chains within and between domains.

mRNA splicing Process that removes introns from pre-mRNAs and joins exons together.

multicellular organism Individual consisting of interdependent cells.

multigene family A family of homologous genes in a genome. The members of a multigene family have all evolved from one ancestral gene, and therefore have similar DNA sequences and produce proteins with similar structures and functions.

multiple alleles More than two different alleles of a gene.

mutation A spontaneous and heritable change in DNA.

Na⁺/K⁺-ATPase *See* **Na⁺/K⁺ pump.**

Na⁺/K⁺ pump Pump that pushes 3 Na⁺ out of the cell and 2 K⁺ into the cell in the same pumping cycle. Also referred to as the *sodium–potassium pump* or as *Na⁺/K⁺-ATPase.*

natural selection The evolutionary process by which alleles that increase the likelihood of survival and the reproductive output of the individuals that carry them become more common in subsequent generations.

neutral lipid Energy-storing molecule consisting of a glycerol backbone and three fatty acid chains.

neutron Uncharged particle in the nucleus of an atom.

nicotinamide adenine dinucleotide (NAD⁺) A coenzyme that serves as an electron carrier.

nitrogenous base A nitrogen-containing molecule with the properties of a base.

noncoding RNA gene A gene encoding an RNA that is not translated; that is, a gene other than a protein-coding gene.

noncompetitive inhibition Inhibition of an enzyme reaction by an inhibitor molecule that binds to the enzyme at a site other than the active site and, therefore, does not compete directly with the substrate for binding to the active site.

noncyclic electron flow Pathway in photosynthesis in which electrons travel in a one-way direction from H_2O to $NADP^+$.

nondisjunction The failure of homologous pairs to separate during the first meiotic division or of chromatids to separate during the second meiotic division.

nonhistone protein All the proteins associated with DNA in a eukaryotic chromosome that are not histones.

nonpolar association Association that occurs when nonpolar molecules clump together.

nonpolar covalent bond Bond in which electrons are shared equally.

nonsense codon *See* **stop codon.**

nonsense mutation A base-pair substitution mutation in a gene in which the base-pair change results in a change from a sense codon to a nonsense codon in the mRNA. The polypeptide translated from the mRNA is shorter than the normal polypeptide because of the mutation.

northern blot analysis Method of analysis in which RNA molecules extracted from cells or a tissue are separated by size using electrophoresis, transferred to a filter paper, and then probed to reveal the presence of particular RNAs.

N-terminal end The end of a polypeptide chain with an —NH₃⁺ group.

nuclear envelope In eukaryotes, membranes separating the nucleus from the cytoplasm.

nuclear localization signal A short amino acid sequence in a protein that directs the protein to the nucleus.

nuclear pore complex A large, octagonally symmetrical, cylindrical structure that functions to exchange molecules between the nucleus and cytoplasm and prevents the transport of material not meant to cross the nuclear membrane. A nuclear pore—a channel through the complex—is the path for the exchange of molecules.

nucleoid The central region of a prokaryotic cell with no boundary membrane separating it from the cytoplasm, where DNA replication and RNA transcription occur.

nucleolus The nuclear site of rRNA transcription, processing, and ribosome assembly in eukaryotes.

nucleoplasm The liquid or semiliquid substance within the nucleus.

nucleoside Chemical structure containing only a nitrogenous base and a five-carbon sugar.

nucleosome The basic structural unit of chromatin in eukaryotes, consisting of DNA wrapped around a histone core.

nucleosome core particle An eight-protein particle formed by the combination of two molecules each of H2A, H2B, H3, and H4, around which DNA winds for almost two turns.

nucleosome remodeling complex Multiprotein complex that uses the energy of ATP hydrolysis to slide a nucleosome along the DNA of a eukaryotic chromosome to expose the promoter of a gene, or to restructure the nucleosome without moving it to allow transcription factors to bind.

nucleotide The monomer of nucleic acids consisting of a five-carbon sugar, a nitrogenous base, and a phosphate.

nucleus The central region of eukaryotic cells, separated by membranes from the surrounding cytoplasm, where DNA replication and messenger RNA transcription occur. In the nervous system, a nucleus is a concentration of nerve cells within the central nervous system that have related functions.

null hypothesis A statement of what would be seen if the hypothesis being tested were wrong.

observational data Basic information on biological structures or the details of biological processes.

oil Neutral lipid that is liquid at biological temperatures.

Okazaki fragments The short lengths of lagging strand DNA produced by discontinuous replication.

oncogene A gene capable of inducing one or more characteristics of cancer cells.

one gene–one enzyme hypothesis Hypothesis showing the direct relationship between genes and enzymes.

one gene–one polypeptide hypothesis Restatement of the one gene–one enzyme hypothesis, taking into account that some proteins consist of more than one polypeptide and not all proteins are enzymes.

open reading frame (ORF) Segment of a protein-coding gene or an mRNA transcribed from such a gene that involves a start codon separated by a multiple of three nucleotides from one of the stop codons. The ORF is a potential polypeptide-coding sequence.

operator A DNA regulatory sequence that controls transcription of an operon.

operon A cluster of prokaryotic genes organized into a single transcription unit and their associated regulatory sequences.

optical isomers *See* **enantiomers.**

orbital The region of space where the electron "lives" most of the time.

order A Linnaean taxonomic category of organisms that ranks above a family and below a class.

organelles The nucleus and other specialized internal structures and compartments of eukaryotic cells.

organic acid (carboxylic acid) Acid for which the characteristic functional group is a carboxyl group (—COOH).

organic molecule Molecule based on carbon.

origin of replication A specific region at which DNA replication commences. Bacterial chromosomes have single origins of replication (*ori*) whereas eukaryotic chromosomes have multiple origins.

osmosis The passive transport of water across a selectively permeable membrane in response to solute concentration gradients, a pressure gradient, or both.

osmotic pressure A state of dynamic equilibrium in which the pressure of the solution on one side of a selectively permeable membrane exactly balances the tendency of water molecules to diffuse passively from the other side of the membrane due to a concentration gradient.

outer membrane In Gram-negative bacteria, an additional boundary membrane that covers the peptidoglycan layer of the cell wall.

outer mitochondrial membrane The smooth membrane covering the outside of a mitochondrion.

ova *See* **ovum.**

ovary In animals, the female gonad, which produces female gametes and reproductive hormones. In flowering plants, the enlarged base of a carpel in which one or more ovules develop into seeds.

ovum (plural, *ova*) A female sex cell, or egg.

oxidation The partial or full loss of electrons from a substance.

oxidative phosphorylation Synthesis of ATP in which ATP synthase uses an H^+ gradient built by the electron transfer system as the energy source to make the ATP.

oxidized Substance—the electron donor—from which the electrons are lost during oxidation.

P generation The parental individuals used in an initial genetic cross.

P site The site in the ribosome where the tRNA carrying the growing polypeptide chain is bound during translation.

pairing Process in meiosis in which homologous chromosomes come together and pair. Also referred to as *synapsis*.

pair-rule gene In *Drosophila* embryonic development, one of the second activated set of segmentation genes that progressively subdivide the embryo into regions, determining the segments of the embryo and the adult. Pair-rule genes control the division of the embryo into units of two segments each.

parental Phenotypes identical to the original parental individuals.

partial diploid A condition in which part of the genome of a haploid organism is diploid. Recipients in bacterial conjugation between an Hfr and an F^- cell become partial diploids for part of the Hfr bacterial chromosome.

passive transport The transport of substances across cell membranes without expenditure of energy, as in diffusion.

paternal chromosome The chromosome derived from the male parent of an organism.

pattern formation The arrangement of organs and body structures in their proper three-dimensional relationships.

pedigree Chart that shows all parents and offspring for as many generations as possible, the sex of individuals in the different generations, and the presence or absence of a trait of interest.

peptide bond A link formed by a dehydration synthesis reaction between the —NH_2 group of one amino acid and the —COOH group of a second.

peptidyl transferase An enzyme that catalyzes the reaction in which an amino acid is cleaved from the

tRNA in the P site of the ribosome and forms a peptide bond with the amino acid on the tRNA in the A site of the ribosome.

peptidyl–tRNA A tRNA linked to a growing polypeptide chain containing two or more amino acids.

peripheral protein Protein held to membrane surfaces by noncovalent bonds formed with the polar parts of integral membrane proteins or membrane lipids.

peroxisome Microbody that produces hydrogen peroxide as a by-product.

pH scale The numerical scale used by scientists to measure acidity.

phage *See* **bacteriophage.**

phagocytosis Process in which some types of cells engulf bacteria or other cellular debris to break them down.

phenotype The observable or measurable (biochemical, molecular) characteristics of an organism that are produced by an interaction between the genotype and the environment.

phosphate group Group consisting of a central phosphorus atom held in four linkages: two that bind —OH groups to the central phosphorus atom, a third that binds an oxygen atom to the central phosphorus atom, and a fourth that links the phosphate group to an oxygen atom.

phosphodiester bond The linkage of nucleotides in polynucleotide chains by a bridging phosphate group between the 5′ carbon of one sugar and the 3′ carbon of the next sugar in line.

phospholipid A phosphate-containing lipid.

phosphorylation The addition of a phosphate group to a molecule.

photoautotroph Photosynthetic organism that uses light as its energy source and carbon dioxide as its carbon source.

photophosphorylation The synthesis of ATP coupled to the transfer of electrons energized by photons of light.

photorespiration A process that metabolizes a by-product of photosynthesis.

photosynthesis The conversion of light energy to chemical energy in the form of sugar and other organic molecules.

photosystem A large complex into which the light-absorbing pigments for photosynthesis are organized with proteins and other molecules.

photosystem I In photosynthesis, a protein complex in the thylakoid membrane that uses energy absorbed from sunlight to synthesize NADPH.

photosystem II In photosynthesis, a protein complex in the thylakoid membrane that uses energy absorbed from sunlight to synthesize ATP.

phylogenetic tree A branching diagram depicting the evolutionary relationships of groups of organisms.

phylum (plural, *phyla*) A major Linnaean division of a kingdom, ranking above a class.

phytosterol A sterol that occurs in plant cell membranes.

pilus (plural, *pili*) A hair or hairlike appendage on the surface of a prokaryote.

pinocytosis *See* **bulk-phase endocytosis.**

plasma membrane The outer limit of the cytoplasm responsible for the regulation of substances moving into and out of cells.

plasmid A DNA molecule in the cytoplasm of certain prokaryotes, which often contains genes with functions that supplement those in the nucleoid and which can replicate independently of the nucleoid DNA and be passed along during cell division.

plasmodesma (plural, *plasmodesmata*) A minute channel that perforates a cell wall and contains extensions of the cytoplasm that directly connect adjacent plant cells.

plasmolysis Condition due to outward osmotic movement of water, in which plant cells shrink so much that they retract from their walls.

plastids A family of plant organelles that includes chloroplasts, amyloplasts, and chromoplasts.

pleiotropy Condition in which single genes affect more than one character of an organism.

ploidy The number of chromosome sets of a cell or species.

polar association Association that occurs when polar molecules attract and align themselves with other polar molecules and with charged ions and molecules.

polar body A nonfunctional cell produced in oogenesis.

polar covalent bond Bond in which electrons are shared unequally.

poly(A) polymerase The enzyme in eukaryotes that adds a chain of adenine nucleotides—the poly(A) tail—to the cleaved 3′ end of a pre-mRNA.

poly(A) tail The string of A nucleotides added posttranscriptionally to the 3′ end of a cleaved pre-mRNA molecule and retained in the mRNA produced from it that enables the mRNA to be translated efficiently and protects it from attack by RNA-digesting enzymes in the cytoplasm.

polyadenylation signal Sequence near the 3′ end of a eukaryotic gene which, in the pre-mRNA transcript of the gene, specifies where the transcript should be cleaved. Once cleaved, a poly(A) tail is added to the 3′ end of the RNA.

polygenic inheritance Inheritance in which several to many different genes contribute to the same character.

polyhedral virus A virus in which the coat proteins form triangular units that fit together like the parts of a geodesic sphere.

polymerase chain reaction (PCR) Process that amplifies a specific DNA sequence from a DNA mixture to an extremely large number of copies.

polypeptide The chain of amino acids formed by sequential peptide bonds.

polyploid An individual with one or more extra copies of the entire haploid complement of chromosomes.

polysaccharide Chain with more than 10 linked monosaccharide subunits.

polysome The entire structure of an mRNA molecule and the multiple associated ribosomes that are translating it simultaneously.

polyunsaturated Fatty acid with more than one double bond.

population All the individuals of a single species that live together in the same place and time.

posttranslational import A mechanism by which proteins are sorted to their final cellular locations in a eukaryotic cell after they have been made on free ribosomes in the cytosol.

potential energy Stored energy.

prediction A statement about what the researcher expects to happen to one variable if another variable changes.

pre-mRNA (precursor-mRNA) The primary transcript of a eukaryotic protein-coding gene, which is processed to form messenger RNA.

prenatal diagnosis Techniques in which cells derived from a developing embryo or its surrounding tissues or fluids are tested for the presence of mutant alleles or chromosomal alterations.

primary active transport Transport in which the same protein that transports a substance also hydrolyzes ATP to power the transport directly.

primary cell wall The initial cell wall laid down by a plant cell.

primary producer An autotroph, usually a photosynthetic organism, a member of the first trophic level.

primary structure The sequence of amino acids in a protein.

primase An enzyme that assembles the primer for a new DNA strand during DNA replication.

primer A short nucleotide chain made of RNA that is laid down as the first series of nucleotides in a new DNA strand, or made of DNA for use in the polymerase chain reaction (PCR).

Principle of Independent Assortment Mendel's principle that the alleles of the genes that govern two characters assort independently during formation of gametes.

Principle of Segregation Mendel's principle that the pairs of alleles that control a character segregate as gametes are formed, with half the gametes carrying one allele, and the other half carrying the other allele.

prion An infectious agent that contains only protein and does not include a nucleic acid molecule.

probability The possibility that an outcome will occur if it is a matter of chance.

product An atom or molecule leaving a chemical reaction.

product rule Mathematical rule in which the final probability is found by multiplying individual probabilities.

prokaryote Organism in which the DNA is suspended in the cell interior without separation from other cellular components by a discrete membrane.

prokaryotic chromosome The genetic material of prokaryotes, in most cases a single, circular DNA molecule.

prometaphase A transition period between prophase and metaphase during which the microtubules of the mitotic spindle attach to the kinetochores and the chromosomes shuffle until they align in the center of the cell.

promoter The site to which RNA polymerase binds (prokaryotes) or to which general transcription factors bind and recruit RNA polymerase (eukaryotes) for initiating transcription of a gene.

promoter proximal element Regulatory sequence within the promoter proximal region, a region upstream of the promoter of a eukaryotic protein-coding gene. Regulatory proteins bind to promoter proximal elements and stimulate or inhibit the rate of transcription initiation.

promoter proximal region Upstream of a eukaryotic gene, a region containing regulatory sequences—promoter proximal elements—for transcription called promoter proximal elements.

proofreading mechanism Mechanism during DNA replication in which DNA polymerase backs up and removes a mispaired nucleotide from a newly synthesized DNA strand and then adds the correct nucleotide to the growing chain.

prophage A viral genome inserted in the host cell DNA.

prophase The beginning phase of mitosis during which the duplicated chromosomes within the nucleus condense from a greatly extended state into compact, rodlike structures.

proteasome Large cytoplasmic protein complex in eukaryotic cells that degrades ubiquitinylated proteins.

protein Molecules that carry out most of the activities of life, including the synthesis of all other biological molecules. A protein consists of one or more polypeptides depending on the protein.

protein chip *See* **protein microarray.**

protein-coding gene A gene encoding a protein.

protein kinase Enzyme that transfers a phosphate group from ATP to one or more sites on particular proteins.

protein microarray Similar in concept to a DNA microarray, a solid surface with a microscopic grid with thousands of spaces containing probes for analyzing the proteome, the complete set of proteins encoded by the genome of an organism. Also referred to as a *protein chip.*

protein phosphatase Enzyme that removes phosphate groups from target proteins.

proteome The complete set of proteins that can be expressed by the genome of an organism.

proteomics The study of the proteome.

protists A diverse and polyphyletic group of single-celled and multicellular eukaryotic species.

proton Positively charged particle in the nucleus of an atom.

proton pump Pump that moves hydrogen ions across membranes and pushes hydrogen ions across

the plasma membrane from the cytoplasm to the cell exterior. Also referred to as H^+ *pump*.

proton-motive force Stored energy that contributes to ATP synthesis, as well as to the cotransport of substances to and from mitochondria.

proto-oncogene A gene that encodes various kinds of proteins that stimulate cell division. Mutated proto-oncogenes—oncogenes—contribute to the development of cancer.

protoplast fusion A plant breeding process in which protoplasts are fused into a single cell.

provirus DNA copy of a retrovirus RNA genome that becomes inserted into host DNA.

pseudogene A gene that is very similar to a functional gene at the DNA sequence level but that has one or more inactivating mutations that prevent it from producing a functional gene product.

Punnett square Method for determining the genotypes and phenotypes of offspring and their expected proportions by combining gametes and their probabilities of occurrence in a matrix table.

purine A type of nitrogenous base with two carbon–nitrogen rings.

pyrimidine A type of nitrogenous base with one carbon–nitrogen ring.

pyruvate oxidation (pyruvic acid oxidation) Stage of cellular respiration in which the three-carbon molecule pyruvate is converted into a two-carbon acetyl group that is completely oxidized to carbon dioxide.

quantitative trait A character that displays a continuous distribution of the phenotype involved, typically resulting from several to many contributing genes.

quantitative trait loci (QTLs) The individual genes that contribute to a quantitative trait.

quaternary structure The arrangement of polypeptide chains in a protein that contains more than one chain.

R plasmid A bacterial plasmid containing genes that provide resistance to unfavorable conditions.

radioactivity The giving off of particles of matter and energy by decaying nuclei.

radioisotope An unstable, radioactive isotope.

random coil An arrangement of the amino acid chain providing flexible regions that allow sections of the chain to bend.

reactants The atoms or molecules entering a chemical reaction.

reaction center Part of photosystems I and II in chloroplasts of plants. In the light-dependent reactions of photosynthesis, the reaction center receives light energy absorbed by the antenna complex in the same photosystem.

reading frame The series of codons for a polypeptide encoded by the mRNA.

reception In signal transduction, the binding of a signal molecule with a specific receptor in a target cell. In neural signaling, the first of four components in which a stimulus is detected by specialized sensory receptors.

receptor protein Protein that recognizes and binds molecules from other cells that act as chemical signals.

receptor tyrosine kinase In signal transduction, a surface receptor with built-in protein kinase activity.

receptor-mediated endocytosis The selective uptake of macromolecules that bind to cell surface receptors concentrated in clathrin-coated pits.

recessive An allele that is masked by a dominant allele.

reciprocal cross A genetic cross in which the two parents are switched with respect to which trait is associated with each sex.

recognition protein Protein in the plasma membrane that identifies a cell as part of the same individual or as foreign.

recombinant Phenotype with a different combination of traits from those of the original parents.

recombinant chromosomes Chromosomes that contain nonparental combinations of alleles. In eukaryotes they are generated by crossing-over in meiosis.

recombinant DNA DNA from two or more different sources joined together.

recombination *See* **genetic recombination.**

recombination frequency In the construction of linkage maps of diploid eukaryotic organisms, the percentage of testcross progeny that are recombinants.

redox reaction Coupled oxidation–reduction reaction in which electrons are removed from a donor molecule and simultaneously added to an acceptor molecule.

reduced Substance—the electron acceptor—that gains electrons during reduction.

reduction The partial or full gain of electrons to a substance.

regulatory gene Gene that encodes a protein that regulates the expression of a structural gene or genes.

regulatory protein DNA-binding protein that binds to a regulatory sequence and affects the expression of an associated gene or genes.

regulatory sequence DNA sequence involved in the regulation of a gene or genes to which a regulatory protein binds to control the transcription of the gene or genes.

release factor A protein that recognizes stop codons in the A site of a ribosome translating an mRNA and terminates translation. Also referred to as the *termination factor*.

renaturation The reformation of a denatured protein into its folded, functional state.

replica plating Technique for identifying and counting genetic recombinants in conjugation, transformation, or transduction experiments in which the colony pattern on a plate containing solid growth medium is pressed onto sterile velveteen and transferred to other plates containing different combinations of nutrients.

replicates Multiple subjects that receive either the same experimental treatment or the same control treatment.

replication bubble The two Y-shaped replication forks joined together at the tops of the Ys after DNA is unwound at an origin of replication.

replication fork The region of DNA synthesis where the parental strands separate and two new daughter strands elongate.

repressible operon Operon whose expression is prevented by a repressor molecule.

repressor A regulatory protein that prevents the operon genes from being expressed.

reproduction The process in which parents produce offspring.

resolution The minimum distance two points in a specimen can be separated and still be seen as two points.

response In signal transduction, the last stage in which the transduced signal causes the cell to change according to the signal and to the receptors on the cell. In neural signaling, the fourth and last component involving the action resulting from the integration of neural messages.

restriction endonuclease (restriction enzyme) An enzyme that cuts DNA at a specific sequence.

restriction fragment A DNA fragment produced by cutting a long DNA molecule with a restriction enzyme.

restriction fragment length polymorphisms When comparing different individuals, restriction enzyme–generated DNA fragments of different lengths from the same region of the genome.

retrotransposon A transposable element that transposes via an intermediate RNA copy of the transposable element.

retrovirus A virus with an RNA genome that replicates via a DNA intermediate.

reverse transcriptase An enzyme that uses RNA as a template to make a DNA copy of the retrotransposon. Reverse transcriptase is used to make DNA copies of RNA in test tube reactions.

reversible The term indicating that a reaction may go from left to right or from right to left, depending on conditions.

ribonucleic acid (RNA) A polymer assembled from repeating nucleotide monomers in which the five-carbon sugar is ribose. Cellular RNAs include mRNA (which is translated to produce a polypeptide), tRNA (which brings an amino acid to the ribosome for assembly into a polypeptide during translation), and rRNA (which is a structural component of ribosomes). The genetic material of some viruses is RNA.

ribonucleotide Nucleotide containing ribose as the sugar; ribonucleotides are components of RNA.

ribose A five-carbon sugar to which the nitrogenous bases in nucleotides link covalently.

ribosomal RNA (rRNA) The RNA component of ribosomes.

ribosome A ribonucleoprotein particle that carries out protein synthesis by translating mRNA into chains of amino acids.

ribosome binding site In translation initiation in prokaryotes, a sequence just upstream of the start codon that directs the small ribosomal subunit to

bind and orient correctly for the complete ribosome to assemble and start translating in the correct spot.

ribozyme An RNA-based catalyst that is part of the biochemical machinery of all cells.

RNA *See* **ribonucleic acid.**

RNA interference (RNAi) The phenomenon of silencing a gene posttranscriptionally by a small, single-stranded RNA that is complementary to part of an mRNA.

RNA polymerase An enzyme that catalyzes the assembly of ribonucleotides into an RNA strand.

rough ER Endoplasmic reticulum with many ribosomes studding its outer surface.

RuBP carboxylase/oxygenase (rubisco) An enzyme that catalyzes the key reaction of the Calvin cycle, carbon fixation, in which CO_2 combines with RuBP (ribulose 1,5-bisphosphate) to form 3-phosphoglycerate.

S phase The phase of the eukaryotic cell cycle during which DNA replication occurs.

saturated enzymes Enzymes for which increases in substrate concentration have no effect on the reaction rate.

saturated fatty acid Fatty acid with only single bonds linking the carbon atoms.

scientific method An investigative approach in which scientists make observations about the natural world, develop working explanations about what they observe, and then test those explanations by collecting more information.

scientific name A two-part name identifying the genus to which a species belongs and designating a particular species within that genus.

scientific theory A broadly applicable idea or hypothesis that has been confirmed by every conceivable test.

second law of thermodynamics Principle that for any process in which a system changes from an initial to a final state, the total disorder of the system and its surroundings always increases.

second messenger In particular signal transduction pathways, an internal, nonprotein signal molecule that directly or indirectly activates protein kinases, which elicit the cellular response.

secondary active transport Transport indirectly driven by ATP hydrolysis.

secondary cell wall A layer added to the cell wall of plants that is more rigid and may become many times thicker than the primary cell wall.

secondary structure Regions of alpha helix, beta strand, or random coil in a polypeptide chain.

secretory vesicle Vesicle that transports proteins to the plasma membrane.

segment polarity gene In *Drosophila* embryonic development, one of the third activated set of segmentation genes that progressively subdivide the embryo into regions, determining the segments of the embryo and the adult. Segment polarity genes set the boundaries and anterior–posterior axis of each segment in the embryo, thereby determining the regions that become segments of larvae and adults.

segmentation genes Genes that work sequentially, progressively subdividing the embryo into regions, determining the segments of the embryo and the adult.

segregation The separation of the pairs of alleles that control a character as gametes are formed.

selectively permeable Membranes that selectively allow, impede, or block the passage of atoms and molecules.

self-fertilization (self-pollination) Fertilization in which sperm nuclei in pollen fertilize egg cells of the same plant.

semiconservative replication The process of DNA replication in which the two parental strands separate and each serves as a template for the synthesis of new progeny double-stranded DNA molecules.

sense codon A codon that specifies an amino acid.

sex chromosomes Chromosomes that are different in male and female individuals of the same species.

sex pilus *See* **F pilus.**

sex-linked gene Gene located on a sex chromosome.

sexual reproduction The mode of reproduction in which male and female parents produce offspring through the union of egg and sperm generated by meiosis.

shells *See* **energy levels.**

short tandem repeat (STR) Short 2- to 6-bp DNA sequence repeated in series.

signal peptide A short segment of amino acids to which the signal recognition particle binds, temporarily blocking further translation. A signal peptide is found on polypeptides that are sorted to the endoplasmic reticulum. Also referred to as *signal sequence.*

signal sequence *See* **signal peptide.**

silencing Phenomenon in which methylation of cytosines in eukaryotic promoters inhibits transcription and turns the genes off.

silent mutation A base-pair substitution mutation in a protein-coding gene that does not alter the amino acid specified by the gene.

simple diffusion Mechanism by which certain small substances diffuse through the lipid part of a biological membrane.

single-stranded binding protein (SSB) Protein that coats single-stranded segments of DNA, stabilizing the DNA for the replication process.

sink Any region of a plant where organic substances are being unloaded from the phloem and used or stored.

siRNA-induced silencing complex (siRISC) Protein complex containing an siRNA that binds to a sequence in a target RNA resulting in cleavage of that RNA.

sister chromatid One of two exact copies of a chromosome duplicated during replication.

sliding DNA clamp A protein that encircles the DNA and binds to the DNA polymerase to tether the enzyme to the template, thereby making replication more efficient.

slime layer A coat typically composed of polysaccharides that is loosely associated with bacterial cells.

small interfering RNA (siRNA) One of the major types of small, single-stranded regulatory RNAs in eukaryotes involved in RNA interference (RNAi).

small ribonucleoprotein particle A complex of RNA and proteins.

smooth ER Endoplasmic reticulum with no ribosomes attached to its membrane surfaces. Smooth ER has various functions, including synthesis of lipids that become part of cell membranes.

sodium–potassium pump *See* **Na⁺/K⁺ pump.**

solenoid *See* **30-nm chromatin fiber.**

solute The molecules of a substance dissolved in water.

solution Substance formed when molecules and ions separate and are suspended individually, surrounded by water molecules.

solvent The water in a solution in which the hydration layer prevents polar molecules or ions from reassociating.

somatic cell Any of the cells of an organism's body other than reproductive cells.

somatic gene therapy Gene therapy in which genes are introduced into somatic cells.

Southern blot analysis Technique in which labeled probes are used to detect specific DNA fragments that have been separated by gel electrophoresis.

specialized transduction Transfer of bacterial genes between bacteria using temperate phages that have incorporated fragments of the bacterial genome as they make the transition from the lysogenic cycle to the lytic cycle.

species A group of populations in which the individuals are so closely related in structure, biochemistry, and behavior that they can successfully interbreed.

spermatozoan Also called *sperm;* a haploid cell that develops into a mature sperm cell when meiosis is complete.

spindle The structure that separates sister chromatids and moves them to opposite spindle poles.

spindle pole One of the pair of centrosomes in a cell undergoing mitosis from which bundles of microtubules radiate to form the part of the spindle from that pole.

spliceosome A complex formed between the pre-mRNA and small ribonucleoprotein particles, in which mRNA splicing takes place.

spontaneous reaction Chemical or physical reaction that occurs without outside help.

sporadic (nonhereditary) cancer Cancer that is not inherited.

spore A haploid reproductive structure, usually a single cell, that can develop into a new individual without fusing with another cell; found in plants, fungi, and certain protists.

sporophyte An individual of the diploid generation produced through fertilization in organisms that undergo alternation of generations; it produces haploid spores.

SRP (signal recognition particle) receptor A protein on the membrane of the endoplasmic reticulum that binds the signal recognition particle.

starch A storage polysaccharide in plants consisting of branched or unbranched chains of glucose subunits.

start codon The first codon read in an mRNA in translation—AUG. Also referred to as the *initiator codon.*

stem cells Cells capable of undergoing many divisions in an unspecialized, undifferentiated state and that also have the ability to differentiate into specialized cell types.

stereoisomers Molecules that are mirror images of one another are an example of stereoisomers.

steroid A type of lipid derived from cholesterol.

steroid hormone receptor Internal receptor that turns on specific genes when it is activated by binding a signal molecule.

steroid hormone response element The DNA sequence to which the hormone-receptor complex binds.

sterol Steroid with a single polar —OH group linked to one end of the ring framework and a complex, nonpolar hydrocarbon chain at the other end.

sticky end End of a DNA fragment generated by digestion with a restriction enzyme, with a single-stranded structure that can form hydrogen bonds with a complementary sticky end on any other DNA molecule cut with the same enzyme.

stop codon A codon that does not specify amino acids. The three nonsense codons are UAG, UAA, and UGA. Also referred to as the *nonsense codon* and *termination codon.*

strict aerobe Cell with an absolute requirement for oxygen to survive, unable to live solely by fermentations.

strict anaerobe Organism in which fermentation is the only source of ATP.

structural gene Gene that encodes a protein that has a function other than gene regulation.

structural isomers Two molecules with the same chemical formula but atoms that are arranged in different ways.

substrate The particular reacting molecule or molecular group that an enzyme catalyzes.

substrate-level phosphorylation An enzyme-catalyzed reaction that transfers a phosphate group from a substrate to ADP.

sugar–phosphate backbone Structure in a polynucleotide chain that is formed when deoxyribose sugars (in DNA) or ribose sugars (in RNA) are linked by phosphate groups in an alternating sugar–phosphate–sugar–phosphate pattern.

sulfhydryl group Group that works as a molecular fastener, consisting of a sulfur atom linked on one side to a hydrogen atom and on the other side to a carbon chain.

sum rule Mathematical rule in which final probability is found by summing individual probabilities.

surface tension The force that places surface water molecules under tension, making them more resistant to separation than the underlying water molecules.

symport The transport of two molecules in the same direction across a membrane. Also referred to as *cotransport*.

synapsis *See* **pairing.**

synaptonemal complex A protein framework that tightly holds together homologous chromosomes as they pair.

systems biology An area of biology that studies the organism as a whole to unravel the integrated and interacting network of genes, proteins, and biochemical reactions responsible for life.

tandem duplication Duplicate genes clustered together in the same region of the same chromosome. Unequal crossing-over produces such an arrangement.

target site The location in a genome to which a transposable element moves when it transposes.

telomerase An enzyme that adds telomere repeats to chromosome ends.

telomeres Repeats of simple-sequence DNA that maintain the ends of linear chromosomes.

telophase The final phase of mitosis, during which the spindle disassembles, the chromosomes decondense, and the nuclei reform.

temperate bacteriophage Bacteriophage that may enter an inactive phase (lysogenic cycle) in which the host cell replicates and passes on the bacteriophage DNA for generations before the phage becomes active and kills the host (lytic cycle).

template A nucleotide chain used in DNA replication for the assembly of a complementary chain.

template strand The DNA strand that is copied into an RNA molecule during gene transcription.

termination In transcription, the step in which transcription ends and the RNA transcript and RNA polymerase (RNA polymerase II in the case of eukaryotes) are released from the DNA template. In translation, the step in which the translation complex disassembles after the last amino acid of the polypeptide specified by the mRNA has been added to the polypeptide.

termination codon *See* **stop codon.**

termination factor *See* **release factor.**

terminator Specific DNA sequence for a gene that signals the end of transcription of a gene. Terminators are common for prokaryotic genes.

tertiary structure The overall three-dimensional folding of a polypeptide chain.

testcross A genetic cross between an individual with the dominant phenotype and a homozygous recessive individual.

testis (plural, *testes*) The male gonad. In male vertebrates, they secrete androgens and steroid hormones that stimulate and control the development and maintenance of male reproductive systems.

tetrad Homologous pair of eukaryotic chromosomes during the first meiotic division consisting of four chromatids.

T-even bacteriophage Virulent bacteriophages, T2, T4, and T6, that have been valuable for genetic studies of bacteriophage structure and function.

thermodynamics The study of the energy flow during chemical and physical reactions.

thymine A pyrimidine that base-pairs with adenine.

Ti (tumor-inducing) plasmid A plasmid used to make transgenic plants.

tight junction Region of tight connection between membranes of adjacent cells.

tonicity The effect a solution has on a cell when the solution surrounds it.

tonoplast The membrane that surrounds the central vacuole in a plant cell.

topoisomerase An enzyme that relieves the overtwisting and strain of DNA ahead of the replication fork.

totipotent Having the capacity to produce cells that can develop into or generate a new organism or body part.

trace element An element that occurs in organisms in very small quantities (less than 0.01%); in nutrition, a mineral required by organisms only in small amounts.

tracer Isotope used to label molecules so that they can be tracked as they pass through biochemical reactions.

trait One of the forms of a genetic character.

transcription The mechanism by which the information encoded in DNA is made into a complementary RNA copy.

transcription factor (TF) In eukaryotes, the proteins required for RNA polymerase to initiate transcription or that regulate that process. One class of transcription factors recognizes and binds to the promoter in the area of the TATA box and then recruit RNA polymerase.

transcription initiation complex Combination of general transcription factors with RNA polymerase II.

transcription unit A region of DNA that transcribes a single primary transcript.

transcriptional regulation The fundamental level of control of gene expression that determines which genes are transcribed into mRNA.

transduction In cell signaling, the process of changing a signal into the form necessary to cause the cellular response. In prokaryotes, the process in which DNA is transferred from donor to recipient bacterial cells by an infecting bacteriophage.

transfer cell Any of the specialized cells that form when large amounts of solutes must be loaded or unloaded into the phloem; they facilitate the short-distance transport of organic solutes from the apoplast into the symplast.

transfer RNA (tRNA) The RNA that brings amino acids to the ribosome for addition to the polypeptide chain.

transformation The conversion of the hereditary type of a cell by the uptake of DNA released by the breakdown of another cell.

transgene A gene introduced into an organism by genetic manipulation to alter its genotype.

transgenic An organism that has been modified to contain genetic information from an external source.

transit sequence Short amino acid sequence on a protein that serves to direct the protein to the appropriate organelle (other than the nucleus or ER) in a eukaryotic cell.

transition state An intermediate arrangement of atoms and bonds that both the reactants and the products of a reaction can assume.

translation The use of the information encoded in mRNA to assemble amino acids into a polypeptide.

translocation In genetics, a chromosomal alteration that occurs if a broken segment is attached to a different, nonhomologous chromosome. In vascular plants, the long-distance transport of substances by xylem and phloem.

transport The controlled movement of ions and molecules from one side of a membrane to the other.

transport epithelium A layer of cells with specialized transport proteins in their plasma membranes.

transport protein A protein embedded in the cell membrane that forms a channel allowing selected polar molecules and ions to pass across the membrane.

transposable element (TE) A sequence of DNA that can move from one place to another within the genome of a cell.

transposase An enzyme that catalyzes some of the reactions inserting or removing the transposable element from the DNA.

transposition The movement of a transposable element from one site to another in a genome.

transposon A bacterial transposable element with an inverted repeat sequence at each end enclosing a central region with one or more genes.

tricarboxylic acid cycle *See* **citric acid cycle.**

triglyceride A nonpolar compound produced when a fatty acid binds by a dehydration synthesis reaction at each of glycerol's three —OH-bearing sites.

true-breeding Individual that passes traits without change from one generation to the next.

tumor Tissue mass that results with differentiated cells of complex multicellular organisms deviate from their normal genetic program and being to grow and divide.

tumor-suppressor gene A gene that encodes proteins that inhibit cell division.

turgor pressure The internal hydrostatic pressure within plant cells.

unequal crossing-over The rare phenomenon in meiosis in which, instead of crossing-over occurring at the exact same point on each homolog of a homologous pair of chromosomes, crossing-over occurs at different points.

unicellular organism Individual consisting of a single cell.

uniparental inheritance A pattern of inheritance in which all progeny (both males and females) have the phenotype of only one of the parents.

universal A feature of the nucleic acid code, with the same codons specifying the same amino acids in all living organisms.

unsaturated fatty acid Fatty acid with one or more double bonds linking the carbons.

valence electron An electron in the outermost energy level of an atom.

Van der Waals forces Weak molecular attractions over short distances.

variable An environmental factor that may differ among places or an organismal characteristic that may differ among individuals.

vegetative reproduction Asexual reproduction in plants by which new individuals arise (or are created) without seeds or spores; examples include fragmentation from the parent plant or the use of cuttings by gardeners.

vesicle A small, membrane-bound compartment that transfers substances between parts of the endomembrane system.

virion A complete virus particle.

viroid A plant pathogen that consists of strands or circles of RNA, smaller than any viral DNA or RNA molecule, that have no protein coat.

virulent bacteriophage Bacteriophage that kills its host bacterial cells during each cycle of infection.

virus An infectious agent that contains either DNA or RNA surrounded by a protein coat.

water lattice An arrangement formed when a water molecule in liquid water establishes an average of 3.4 hydrogen bonds with its neighbors.

wax A substance insoluble in water that is formed when fatty acids combine with long-chain alcohols or hydrocarbon structures.

weight A measure of the pull of gravity on an object.

wobble hypothesis Hypothesis stating that the complete set of 61 sense codons can be read by fewer than 61 distinct tRNAs because of particular pairing properties of the bases in the anticodons.

X chromosome Sex chromosome that occurs paired in female cells and single in male cells.

X-linked gene A gene on the X chromosome.

X-linked inheritance The pattern of inheritance of an X-linked gene.

X-linked recessive inheritance Pattern in which displayed traits are due to inheritance of recessive alleles carried on the X chromosome.

X-ray diffraction Method for deducing the position of atoms in a molecule.

Y chromosome Sex chromosome that is paired with an X chromosome in male cells.

zygote A fertilized egg.

Experimental data, 13
Experimental variable, 15
Expression vector, 404–406, 405i
Extracellular fluid (ECF), 142
  pinocytosis of, 134
Extracellular matrix (ECM), 113, 115, 115i

F+ cell, 372, 373i
F– cell, 372, 373i, 374
F factor, 371, 372, 373i, 374
F pilus, 372, 373i
F1 generation, 243
F2 generation, 243
Facilitated diffusion, 126t, 127, 128i, 129
Facultative anaerobe, 177
FAD. See Flavin adenine dinucleotide
Fallén, C. F., 264
Faloona, F., 398
Familial (hereditary) cancers, 361
Familial hypercholesteremia, 136
Family, 9
Fat, 50, 51, 52
Fats, 171, 171i
Fatty acid
  defined, 50
  saturated, 50–52, 50i
  trans, 52
  unsaturated, 50–52, 50i
Feedback inhibition, 82, 83i
Fermentation, 176–179
  alcoholic, 177, 177i
  by anaerobes, 177
  defined, 176
  lactate, 176–177, 177i
Ferredoxin, 190i, 191, 192i
Fertilization, 225, 226i, 233, 234–235, 234i
Fe/S protein, 172i, 173i
FGFRs. See Fibroblast growth factor
  receptors
Fibroblast growth factor receptors
  (FGFRs), 278
Fibronectins, 115, 115i
Final acceptor for electrons, 163, 190i
Fire, Andrew, 352
First law of thermodynamics, 72–73
First messenger, 142, 149, 150i
Fischer, Edmond, 147
Fishing spider (Dolomedes species), 12i
5′ cap, 318, 319i
5′ end, 290, 290i, 317
5′ untranslated region (5′ UTR), 318, 319i
Flagella
  beating patterns, 110i
  eukaryotic, 109–110, 110i
  prokaryotic, 92i, 95i, 96
Flavin adenine dinucleotide (FAD),
  168–169, 168i, 170i–173i, 171, 173,
  175–176, 175i
Flavin mononucleotide (FMN), 172i, 173i
Fluid mosaic model, 122i, 123–124
Fluorescence, 186, 186i
Fluorescence microscopy, 93i
FMN. See Flavin mononucleotide
FMR-1 gene, 301
Folding, protein, 328
Forensics, DNA fingerprinting in,
  391–392, 403–404
Formula, 24
Four-o' clock (Mirabilis) plant, 280, 280i
Fragile X syndrome, 301
Fragment reaction, 327
Frameshift mutation, 331i, 332
Franklin, Rosalind, 65, 286, 291, 291i, 292
Free energy
  changes, 75–76
  in coupled reactions, 77–78, 77i
  defined, 74
  exergonic and endergonic reactions, 75,
    76i
  release with ATP hydrolysis, 77–78, 77i
  in reversible reactions, 75
Free water, 129
Freeze-fracture technique, 124–125, 125i
Fructose, 46i
Frye, L. David, 124, 124i

Fumarase, 170i
Fumarate, 169, 170i
Functional groups, 43–44, 44t–45t, 46
  amino, 44t
  carbonyl, 43, 44t
  carboxyl, 43, 44t
  defined, 43
  hydroxyl, 43, 44t
  phosphate, 43, 45t
  sulfhydryl, 43, 45t
Fungal cells, 92i
Fungi
  genomes sequenced, 421t
  Kingdom Fungi, 11, 12i
  meiosis, time and place of, 234–235,
    234i
Furrow, 211, 212i
Fused Toes (FTO) gene, 258

G0 phase, 207, 207i, 219
G1 phase, 207, 207i, 208i, 209i, 214,
  216–220, 216i, 217i, 222
G1/S checkpoint, 216–217, 217i
G2 phase, 207, 207i, 208i, 214, 215, 217,
  217i, 218
G2/M checkpoint, 217, 217i
Gametes, 225, 226, 231, 233–235, 234i
Gametogenesis, 226
Gametophyte, 234–235, 234i
Gap genes, 358i, 359
Gap junction, 114, 114i, 142, 157
Garden pea (Pisum sativum), 240, 241i
Garrod, Archibald, 311
Gated channel, 127, 128i
Gecko, 31, 32i
GenBank, 417
Gene(s)
  alleles, 243, 249–251, 250i, 251i
  cloning, 392–393, 392i, 394, 395i
  defined, 7
  homologous, 422
  linked, 262–266, 263i, 265i, 266i
  locus, 251, 251i
  as Mendel's factors, 243, 249
  non-protein coding, 318
  organization, 346, 346i
  protein-coding, 311, 313
  relationship to enzymes, 311, 312i–313i
  sex-linked, 266–271, 268i–271i
  trapped, 334
Gene activation, steroid hormones and,
  154, 155–156, 156i
Gene cloning, 392, 392i, 393, 394, 395i. See
  also DNA cloning
Gene duplication, 432, 433
Gene expression
  nonhistone proteins, role of, 306
  regulation of, 359–365
    chromatin remodeling, 351, 351i
    in eukaryotes, 344–351, 346i–351i
    posttranscriptional regulation,
      351–354, 353i
    posttranslational regulation, 354,
      354i
    in prokaryotes, 340–344, 341i–345i
    transcriptional regulation, 340–351
    translational regulation, 353–354
Gene function determination, 426–431
  by gene knockout/gene knockdown,
    427–428
  phenomics, 428
  from protein structure, 427
  proteomics, 428, 430–431
  by sequence similarity search, 427
  transcriptomics, 428
Gene knockdown, 427–428
Gene knockout, 427–428
Gene of interest, 392–393, 392i, 394, 395i,
  397, 397i
Gene regulation
  combinatorial, 348–349, 349i
  negative, 344
  positive, 343, 343i, 344
Gene therapy, 406–407, 412
Gene transfer, 370–377

General transcription factors, 347, 347i,
  349i
Generalized transduction, 380
Genetic code, 313–315, 315i, 317
Genetic counseling, 279
Genetic disorders
  cancer as, 361–362
  correcting with gene therapy, 406–407
  molecular testing for, 401
Genetic engineering, 392, 404–412
  of animals, 405–408, 406i–408i
  of bacteria to produce proteins,
    404–405, 405i
  of plants, 408–412, 409i, 410i
  public concern over, 410–412
Genetic recombinants, 232, 232i
Genetic recombination, 232, 232i, 264,
  282. See also Recombination
  in Escherichia coli, 371–377, 371i, 374
Genetic screening, 279
Genetic variability
  generation by independent assortment
    of chromosomes, 232–233, 232i
  generation by random joining of
    gametes in fertilization, 233
  generation by recombination, 231–232,
    232i
Genetically modified organisms (GMOs),
  410
Genetics. See also Inheritance
  bacterial, 369–377
  of cancer, 360–364
  human, 276–279, 276i, 277i, 279i
  of human disease, 258
  viral, 377–387
Genome
  Escherichia coli, 423i, 424–425, 424i,
    425t
  human, 423i, 424, 425–426, 425i,
    425t
  landscapes, 423–426
  number sequenced, 421t
  sequence determination and annota-
    tion, 417, 418–426
  size, 423–424, 423i, 423t
Genome annotation
  identifying open reading frames by
    computer search of genome
    sequences, 421–422
  identifying protein-coding genes from
    sequences of gene transcripts,
    422–423
Genome evolution, 375–377
Genomic imprinting, 280–281, 281i, 351
Genomic library, 394
Genomics
  comparative, 417, 431–436
  defined, 416
  evolution of genome, 417, 431–436
  gene function determination,
    426–431
  genome sequence determination and
    annotation, 417, 418–426
  overview, 417
Genotype, 243
Genus, 9
Germ-line cells, genetic engineering of,
  405–406, 406i
Germ-line gene therapy, 406
Gibberellin, 252
Gibbs, Josiah Willard, 74
Gilbert, Walter, 418
Gilman, Alfred G., 149
Gland cells, 142
Glucagon, 152
Glucocorticoid, 350
Glucose
  phosphorylation of, 80, 80i
  from photosynthesis, 184, 194
  ring formation, 47, 48i
Glucose-1-phosphate, 75, 75i
Glucose-6-phosphate, 75, 75i, 80, 80i
Glutamine, 77, 77i
Glyceraldehyde, 47i

Glyceraldehyde-3-phosphate (G3P), 45t,
  166, 166i, 167i, 193–194, 195i, 197,
  198, 199
Glycerol, 50, 51, 51i, 53
Glyceryl palmitate, 51i
Glycocalyx, 95, 122
Glycogen, 47, 49, 49i, 170
Glycogen phosphorylase, 142, 143, 144i
Glycogen synthase, 152
Glycolate, 197
Glycolipid, 55, 122, 122i
Glycolysis, 164i, 165–166, 168, 175–176,
  175i
  defined, 164, 166
  energy-requiring and energy-releasing
    steps, 166
  reactions, 166i, 167i
  regulation of, 166, 168
Glycoprotein
  description, 63
  extracellular matrix, 115
  membrane, 122, 122i
  surface receptors, 145, 149
Glycosidic bond, 48, 48i
Glyoxysome (glycosome), 107
Glyphosate, 409–410
GMOs. See Genetically modified organisms
Goiter, 23–24
Goldstein, Joseph L., 67, 136
Golgi, Camillo, 103
Golgi complex, 98i, 99i, 103–104, 103i
Gonads, 226
Gorbsky, G.S., 215i
G-protein-coupled receptor, 149–154,
  149i–152i
G3P. See Glyceraldehyde-3-phosphate
Grana, 111, 112i, 184, 185i
Green, Michael R., 362
Green algae, photosynthesis in, 182–183,
  182i
Green sulfur bacteria, 192
Greider, Carol W., 303
Griffith, Frederick, 287, 288i, 374
Ground state, electron, 186–187, 186i
Growth factor, 149, 219, 360, 361i
Growth hormone gene, 406, 407i
Growth-inhibiting factor, 360, 361i
GTP. See Guanosine triphosphate
Guanine, 63, 64i, 66i, 290, 290i
Guanosine triphosphate (GTP), 64,
  149–150, 150i–152i, 153, 154i, 157i
Gymnophilus, 12i

Haber, Edgar, 60, 61i
Haplinsufficiency, 282
Haploid, 206, 226, 234–235, 235i
Hartwell, Leland, 214
HDL. See High-density lipoprotein
Head, virus, 378, 378i, 380
Heart disease, cholesterol and, 52
Heat
  specific, 33
  units of measurement, 33–34
  of vaporization, 34
Heat energy, 73, 73i
  uncoupling proteins and, 176, 178
Hedden, Peter, 252
HeLa cells, 214, 216i
Helianthus annuus, 92i
Helical viruses, 377, 378i
Helicase, 296, 297i, 298i, 300t
Helicobacter pylori, 12i
Helix-turn-helix, 348, 348i
Hemoglobin
  gene duplication, 272
  genes, 339, 350
  sickle-cell, 239–240, 254, 257, 257i, 276,
    332, 332i, 401, 402i
  structure of, 57, 58i, 61
Hemophilia, 270, 270i
Hereditary enamel hypoplasia, 279, 279i
Herhsko, Avram, 354
Herpesvirus, 220, 378i, 382, 383
Herrick, James, 239–240
Hershey, Alfred D., 288–289, 289i